Species, Species Concepts, and Primate Evolution

ADVANCES IN PRIMATOLOGY

MOLECULAR ANTHROPOLOGY: Genes and Proteins in the Evolutionary Ascent of the Primates
Edited by Morris Goodman and Richard E. Tashian

SENSORY SYSTEMS OF PRIMATES
Edited by Charles R. Noback

NURSERY CARE OF NONHUMAN PRIMATES
Edited by Gerald C. Ruppenthal

COMPARATIVE BIOLOGY AND EVOLUTIONARY RELATIONSHIPS OF TREE SHREWS
Edited by W. Patrick Luckett

EVOLUTIONARY BIOLOGY OF THE NEW WORLD MONKEYS AND CONTINENTAL DRIFT
Edited by Russell L. Ciochon and A. Brunetto Chiarelli

NEW INTERPRETATIONS OF APE AND HUMAN ANCESTRY
Edited by Russell L. Ciochon and Robert S. Corruccini

SIZE AND SCALING IN PRIMATE BIOLOGY
Edited by William L. Jungers

SPECIES, SPECIES CONCEPTS, AND PRIMATE EVOLUTION
Edited by William H. Kimbel and Lawrence B. Martin

Species, Species Concepts, and Primate Evolution

Edited by

WILLIAM H. KIMBEL
Institute of Human Origins
Berkeley, California

and

LAWRENCE B. MARTIN
State University of New York
Stony Brook, New York

PLENUM PRESS • NEW YORK AND LONDON

Library of Congress Cataloging-in-Publication Data

Species, species concepts, and primate evolution / edited by William H. Kimbel and Lawrence B. Martin.
p. cm. -- (Advances in primatology)
Includes bibliographical references (p.) and index.
ISBN 0-306-44297-3
1. Primates--Speciation. 2. Species. 3. Primates--Variation. 4. Primates--Evolution. I. Kimbel, William H. II. Martin, Lawrence, 1955- . III. Series.
QL737.P9S69 1993
599.8'0415--dc20 93-6920
CIP

Cover design by Wendy Martin

ISBN 0-306-44297-3

A Division of Plenum Publishing Corporation
233 Spring Street, New York, N.Y. 10013

Printed in the United States of America

To the memory of my father,
Philip Kimbel, M.D.

—W. H. K.

To my dear wife, Wendy,
with gratitude for her sacrifices and support

—L. B. M.

Contributors

Gene H. Albrecht
Department of Anatomy and Cell Biology
University of Southern California
Los Angeles, California 90033

Peter Andrews
Department of Palaeontology
The Natural History Museum
London SW7 5BD
England

S. Boinski
Laboratory of Comparative Ethology
National Institutes of Health—Animal Center
Poolesville, Maryland 20837
and
Department of Anthropology
University of Florida
Gainesville, Florida 32611

Thomas M. Bown
Branch of Paleontology and Stratigraphy
U.S. Geological Survey
Denver, Colorado 80225

Dana A. Cope
Department of Sociology and Anthropology
College of Charleston
Charleston, South Carolina 29424

Robert K. Costello
Doctoral Program in Anthropology
The Graduate Center of the City University of New York
New York, New York 10021

C. Dickinson
Department of Anthropology
University of Chicago
Chicago, Illinois 60637

Niles Eldredge
Department of Invertebrates
American Museum of Natural History
New York, New York 10024

Colin P. Groves
Department of Prehistory and Anthropology
Australian National University
Canberra, ACT 2601
Australia

Terry Harrison
Department of Anthropology
New York University
New York, New York 10003

Clifford J. Jolly
Department of Anthropology
New York University
New York, New York 10003

Jay Kelley
Department of Oral Biology
College of Dentistry
The University of Illinois at Chicago
Chicago, Illinois 60612

William H. Kimbel
Institute of Human Origins
Berkeley, California 94709

Leonard Krishtalka
Section of Vertebrate Paleontology
Carnegie Museum of Natural History
Pittsburgh, Pennsylvania 15213

Steven R. Leigh
Department of Anthropology
Northwestern University
Evanston, Illinois 60628
Present address:
Department of Anatomical Sciences
State University of New York
at Stony Brook
Stony Brook, New York 11794

Lawrence B. Martin
Departments of Anthropology and Anatomical Sciences
State University of New York at Stony Brook
Stony Brook, New York 11794

J. C. Masters
Population Genetics Laboratory
Museum of Comparative Zoology
Harvard University
Cambridge, Massachusetts 02138
Present address:
University of the Witwatersrand
Wits 2050, South Africa

Joseph M. A. Miller
Department of Anatomy and Cell Biology
University of Southern California
Los Angeles, California 90033

G. S. Mugaisi
National Museums of Kenya
Nairobi
Kenya

J. Michael Plavcan
Department of Biological Sciences
University of Cincinnati
Cincinnati, Ohio 45221

Yoel Rak
Department of Anatomy
Sackler Faculty of Medicine
Tel Aviv University
Tel Aviv
Israel
and
Institute of Human Origins
Berkeley, California 94709

Kenneth D. Rose
Department of Cell Biology and Anatomy
The Johns Hopkins University School of Medicine
Baltimore, Maryland 21205

A. L. Rosenberger
U.S. National Zoological Park
Smithsonian Institution
Washington, D.C. 20008

Brian T. Shea
Departments of Cell, Molecular and Structural Biology/Anthropology
Northwestern University
Chicago, Illinois 60611

Frederick S. Szalay
Department of Anthropology
Hunter College
New York, New York 10021

Ian Tattersall
Department of Anthropology
American Museum of Natural History
New York, New York 10024

M. F. Teaford
Department of Cell Biology and Anatomy
The Johns Hopkins University School of Medicine
Baltimore, Maryland 21205

A. Walker
Department of Cell Biology and Anatomy
The Johns Hopkins University School of Medicine
Baltimore, Maryland 21205

Bernard Wood
Hominid Paleontology Research Group
Department of Human Anatomy and Cell Biology
University of Liverpool
Liverpool L69 3BX
England

Preface

A world of categories devoid of spirit waits for life to return.

Saul Bellow, *Humboldt's Gift*

The stock-in-trade of communicating hypotheses about the historical path of evolution is a graphical representation called a *phylogenetic tree*. In most such graphics, pairs of branches diverge from other branches, successively marching across abstract time toward the present. To each branch is tied a tag with a name, a binominal symbol that functions as does the name given to an individual human being. On phylogenetic trees the names symbolize species. What exactly do these names signify? What kind of information is communicated when we claim to have knowledge of the following types?

"*Tetonius mathewii* was ancestral to *Pseudotetonius ambiguus*."

"The sample of fossils attributed to *Homo habilis* is too variable to contain only one species."

"Interbreeding populations of savanna baboons all belong to *Papio anubis*."

"*Hylobates lar* and *H. pileatus* interbreed in zones of geographic overlap."

While there is nearly universal agreement that the notion of *the species* is fundamental to our understanding of how evolution works, there is a very wide range of opinion on the conceptual content and meaning of such particular statements regarding species. This is because, oddly enough, evolutionary biologists are quite far from agreement on what *a species* is, how it attains this status, and what role it plays in evolution over the long term.

Investigators of the order Primates have devoted considerable research effort to developing knowledge both of the extant primate fauna and of the paleontological record of primate evolution. In recent years a large (and still growing) molecular data set bearing on primate phylogenetic relationships down to the species level has been amassed. In short, primate biologists can boast of one of the better mammalian data bases for basic research in systematics, population biology, and paleobiology. However, a look back over the literature on species and speciation will clearly show that, with a few notable exceptions, such conceptual issues have not often been of central concern in primate evolutionary research.

It is only during the past few years that a subtle turn toward introspection has begun to affect researchers in primate evolutionary biology. There is evidence of a greater concern for the necessity of grounding our explanations in well-corroborated phylogenetic hypotheses, a growing appreciation for the importance of the historical–narrative aspects of our explanations, and a better acquaintance with the conceptual underpinnings of the species. It is out of this ferment that *Species, Species Concepts, and Primate Evolution* has developed.

In compiling this book, we asked both theoreticians and specialists in one or another extant or extinct primate group to relate their research explicitly to the conceptual foundation of the species, as they see it. Our goal was to demonstrate that many of the debates in primate evolutionary biology today are not resolvable simply by adding new data to the stockpile. Instead, the essence of any number of these debates can be understood by analyzing the marked differences in the underlying interpretation of the species concept.

The chapters are sorted into four thematic sections and a summary, although the reader will encounter a good deal of thematic overlap between sections, in keeping with our stated goal. Part 1 focuses on ontological issues by examining the role of species in systematics and evolutionary theory. The second part is devoted to studies of species and speciation among living primates, with special emphasis on the detection and interpretation of taxonomic diversity. Part 3 looks at the interface of evolutionary theory and species recognition from the perspective of the primate fossil record. The chapters of Part 4 address the ever-controversial subject of species and hominid systematics. Finally, in the summary chapter, we attempt to distill the diversity of conceptual perspectives and to extract the essence of their relevance for primate evolutionary studies.

Each chapter of this book was reviewed by two other chapter authors as well as one external referee. We thank all reviewers for taking the time to provide us with careful and thoughtful comments on the manuscripts, even while many of them were busy preparing their own contributions. Manuscripts were also circulated to nonreviewing chapter authors with the expectation that the sharing of ideas and data would promote discussion and debate. We think this was largely successful.

Some of the arguments appearing in this volume were aired initially in a symposium entitled "Species, Species Concepts, and Primate Evolution" convened by us at the 1991 meetings of the American Association of Physical Anthropologists in Milwaukee. We thank Dr. Lorna Moore, then AAPA Program Chair, for helping to ensure the success of the symposium.

We thank our colleagues Drs. Ross MacPhee and John Fleagle, Series Editors of *Advances in Primatology,* and Mary Born, Senior Editor at Plenum, for their early enthusiasm for this project, which resulted in its publication in this series. We wish to thank Dr. Eric Meikle (Institute of Human Origins) for generous assistance with critical editorial matters and for compiling the index. We are grateful to Larissa Smith (IHO) and Helen Giles (Stony Brook) for help in the administrative arena. We are especially grateful to Wendy Martin for her expert execution of the cover art.

William H. Kimbel
Lawrence B. Martin

Berkeley and New York

Contents

Part 1. *Species in Evolutionary Theory* 1

1
What, If Anything, Is a Species? 3
Niles Eldredge

Introduction 3 Some Basic Ideas 4 The Biological Species Concept 6 Paterson's Recognition Concept 8 Species, Reproductive Continuity, and Geologic Time 9 Differentiated Taxa, Reproductive Cohesion, and Species 14 References 18

2
Species Concepts: The Tested, the Untestable, and the Redundant 21
Frederick S. Szalay

Introduction 22 The Biological Species Concept Is Part of Evolutionary Theory 25 Taxonomic Species Notions and Some Consequences 30 Species, Speciation, and Accountability in Constructing Species Taxa 36 Conclusions 38 References 39

3
Primates and Paradigms: Problems with the Identification of Genetic Species 43
J. C. Masters

Introduction 43 Why *Genetic* Species? 44 How Do We Identify Genetic Species? 45 Is There a Common Factor in the Origin and Maintenance of Genetic Species? 46 What Are Isolating Mechanisms and How Do They Arise? 47 What Is the Recognition Concept of Species? 48 How Effective Has the Recognition Concept Been in Identifying Primate Species? 50 Are the Recognition Concept and the Phylogenetic Species Concept Mutually Exclusive? 57 How Useful Is the Recognition Concept to Paleontology? 59 References 61

Part 2. *Speciation and Variation among the Living Primates* 65

4
Species, Subspecies, and Baboon Systematics 67
Clifford J. Jolly

Introduction 67 Some Concepts and Definitions 68 The Phenostructure and Zygostructure of Extant Baboons 72 Translating Population Structure into Classification: Alternative Schemes 83 A Paleontologist's Eye View of Baboon Systematics 100 Summary and Conclusions 103 References 105

5
Speciation in Living Hominoid Primates 109
Colin P. Groves

Introduction 109 Speciation in the Hylobatidae 110 Speciation in the Hominidae 118 Conclusions 119 References 120

6
Geographic Variation in Primates: A Review with Implications for Interpreting Fossils 123
Gene H. Albrecht and Joseph M. A. Miller

Introduction 123 Geographic Variation in Living Primates 128 Summary and Synthesis 147 Implications for Interpreting Fossils 152 References 157

7
Speciation and Morphological Differentiation in the Genus Lemur 163
Ian Tattersall

Introduction 163 The Species and Subspecies of the Genus *Lemur* 164 Morphological Variation among *Lemur* Species 167 Discussion 173 Summary 175 References 176

8
Squirrel Monkey (Genus Saimiri) Taxonomy: A Multidisciplinary Study of the Biology of Species 177
Robert K. Costello, C. Dickinson, A. L. Rosenberger, S. Boinski, and Frederick S. Szalay

Introduction 177 Review of the Current Status of *Saimiri* 178 Methods and Materials 192 New Evidence Bearing on Species of *Saimiri:* Data and Discussion 196 Conclusions 205 Summary 205 References 206

9

Measures of Dental Variation as Indicators of Multiple Taxa in Samples of Sympatric Cercopithecus Species 211

Dana A. Cope

Introduction 211 Materials and Methods 213 Results 216 Discussion 232 Conclusions 233 References 236

10

Catarrhine Dental Variability and Species Recognition in the Fossil Record 239

J. Michael Plavcan

Introduction 239 Materials and Methods 241 Results 247 Discussion 253 Summary 260 References 261

11

Multivariate Craniometric Variation in Chimpanzees: Implications for Species Identification 265

Brian T. Shea, Steven R. Leigh, and Colin P. Groves

Introduction 265 Selected Examples 267 Within- and Between-Group Cranial Variation in Chimpanzees 269 Materials and Methods 270 Results 272 Discussion 277 Why Intraspecific Variation Must Be Investigated 280 Is Intraspecific Variation Relevant to Speciation? 284 Special Issues with the Fossil Record 286 Current Debates in Paleoanthropology 287 A Paleontological Example 289 Conclusions 290 References 292

Part 3. *Species and Species Recognition in the Primate Fossil Record* 297

12

Species Concepts and Species Recognition in Eocene Primates 299

Kenneth D. Rose and Thomas Bown

Introduction 299 Species Concepts 302 Species in Paleontology 309 Nomenclature of Evolutionary Intermediates 321 Discussion and Conclusions 323 References 326

13

Anagenetic Angst: Species Boundaries in Eocene Primates 331

Leonard Krishtalka

Introduction 331 Paleontological Species and Evolutionary Patterns 332 Early Eocene Primates 334 Anagenetic Angst 335 Other Anagenetic Issues 337 Other Solutions 339 Modest Proposals 341 References 342

14
Cladistic Concepts and the Species Problem in Hominoid Evolution 345
Terry Harrison

Introduction 345 The Family-Group Level and Above 348 The Genus Group and Species Levels 357 Species Recognition 360 Conclusions 364 References 367

15
Species Discrimination in Proconsul from Rusinga and Mfangano Islands, Kenya 373
M. F. Teaford, A. Walker, and G. S. Mugaisi

Introduction 373 Materials and Methods 377 Results 380 Discussion 380 Conclusions 390 References 390

16
Species Recognition in Middle Miocene Hominoids 393
Lawrence B. Martin and Peter Andrews

Introduction 393 Materials and Methods 398 The Paşalar Example 413 Conclusions 424 References 425

17
Taxonomic Implications of Sexual Dimorphism in Lufengpithecus 429
Jay Kelley

Introduction 429 Species Number at Lufeng 431 Sexual Dimorphism in *Lufengpithecus* 440 Implications of *Lufengpithecus* Sexual Dimorphism for Miocene Hominoid Taxonomy 446 Uniformitarianism, Operational Limits, and Falsifiability 450 References 453

Part 4. *Species and Species Recognition in the Hominid Fossil Record* 459

18
The Importance of Species Taxa in Paleoanthropology and an Argument for the Phylogenetic Concept of the Species Category 461
William H. Kimbel and Yoel Rak

Introduction 461 The Nature of Species Taxa 463 Defining the Species Category: The Phylogenetic Concept 465 Discovering Phylogenetic Species in Paleontology 469 Sexual Dimorphism in *Australopithecus boisei* 472 The Two Species of *Homo habilis* 473 The Problem of Variation in *Australopithecus africanus* 475 Clades, Grades, and *Homo erectus* 478 Conclusions 480 References 482

19
Early Homo: How Many Species? 485
Bernard Wood

Introduction 485 Early *Homo:* The Problem 486 Taxonomic Hypotheses 488 Hypothesis Testing 492 Degree of Variation 494 Pattern of Variation 494 Temporal Trends in Character States 500 One, or More, Early *Homo* Taxa? 507 Multiple Taxa—Geographic and Temporal Distributions 508 Taxonomic Implications 511 Identification of Early *Homo* Taxa from Koobi Fora 514 Postcranial Evidence 516 Summary 517 References 518

20
Morphological Variation in Homo neanderthalensis and Homo sapiens in the Levant: A Biogeographic Model 523
Yoel Rak

Introduction 523 Morphological and Temporal Considerations 524 A Biogeographic Model 531 Conclusions 533 References 535

Part 5. *Summary* 537

21
Species and Speciation: Conceptual Issues and Their Relevance for Primate Evolutionary Biology 539
William H. Kimbel and Lawrence B. Martin

Introduction 539 Conceptual Foundations of the Species Notion 541 Reproductive Cohesion or Reproductive Isolation? 542 Species and Lineage: the Temporal Dimension 545 Origins and Extinctions: Recognizing Species in the Fossil Record 547 Methodological Issues 548 Conclusions 551 References 552

Index 555

Species in Evolutionary Theory

1

What, If Anything, Is a Species?

1

NILES ELDREDGE

Introduction

Each generation, evolutionary biologists seem doomed to grapple with the "species question" epitomized in my title. The task is often seen as Sisyphean, with no final consensus on what really constitutes a "species" ever emerging. In contrast, I believe that important strides in understanding what species may or may not be have been taken in the last 50 years, including much important work underway at the present. In particular, we have hope, if not for consensus, at least for a detailed understanding of why consensus on the nature of species has persistently eluded evolutionary biologists.

This chapter focuses primarily on ontological issues: what we think species *are*. Are they in some sense real? Or are species simply the (more-or-less) artificial constructs of the human mind, devised and recognized for purposes of convenience, or to serve some further purpose—be that purpose classificatory, biostratigraphic, ecological, or even phylogenetic? I will be especially concerned with characterizing various viewpoints that take species to be "real," exploring in what sense they can be said to be real, and, especially, what their natures have been said to be, again, from various viewpoints. I shall be far less occupied with epistemological points: how we recognize species. To my mind, as individual evolutionary biologists, whether as paleontologists, systematists of the living biota, or geneticists, we each need a coherent set of ideas about what a species is before we can adopt a set of rules that might enable us to recognize a species when we see one. I am especially concerned to avoid the approach epitomized,

---

NILES ELDREDGE • Department of Invertebrates, American Museum of Natural History, New York, New York 10024.

Species, Species Concepts, and Primate Evolution, edited by William H. Kimbel and Lawrence B. Martin. Plenum Press, New York, 1993.

for example, by Sokal and Crovello (1970) in their negatively critical assessment of the biological species concept (BSC) on grounds that were virtually entirely epistemological. Sokal and Crovello argued that the BSC is flawed because it is difficult to apply in rigorous, repeatable fashion to the problem of species recognition in nature.

In contrast, I believe it is our fundamental task as scientists to characterize the natures of classes, or categories, of entities that we have reason to believe exist in nature. Because quarks and other subatomic particles present formidable epistemological problems in the apprehension in any one particular instance is no reason to abandon a conceptualization of them as a category of entities that exist and exhibit various properties in nature. Physicists, after all, are prone to positing the existence of such entities as a necessity arising out of the current state of theory, long before empirical evidence confirming (or denying) their existence can be generated in a cyclotron. We *are*, I hasten to add, in the business of describing nature as accurately and completely as possible; we must abandon, or at least revise, our ontologies in the light of our experiences with nature. Ultimately, ontology stands or falls on such experience, and epistemology is the link between our purely conceptual views and our (more-or-less) direct experiences of nature. I am merely taking the position here that it is unwise to abandon a particular view of the nature of species just because the requisite criteria for recognizing particular instances may be lacking on a case-by-case basis.

The most important conclusion of this chapter is that the confusion over what species really *are* does not merely reflect a Tower of Babel-ish competition of different theoretical perspectives, with little hope of final resolution. The species problem at base reflects what I take to be the ontological fact that biotic nature is organized, simultaneously, into at least two (possibly more) contrasting sorts of entities at or about the "species level." In other words, biologists of varying perspective indeed "see" different sorts of species in biotic nature (and, epistemologically, even in the same data set) but do so, at least in part, not out of slavish adherence to a particular theoretical perspective but because biotic nature is indeed "packaged" into discrete or quasi-discrete entities of different sorts. It is nature itself that is providing the source of disagreement. That is the message, as I take it, from the past 50 years or so of wrangling over the species question in evolutionary biology. If so, this represents some progress: We need no longer look for consensus on any one particular view of what a species *is*. There are several sorts of entities that have, quite legitimately, been called species. The following analysis seeks to explore the various ways nature seems to be organized at what we have been loosely calling the species level. In instances when I have a preference for one view over another, I will so state and explain why, but any reader seeking a final, definitive statement about what a species really is will seek in vain. Nature, or so it seems to me, does not work that way.

Some Basic Ideas

Species is a category term, the lowest rung of the original and still traditional Linnaean hierarchical ladder. Originally so conceived, a species is a taxon no different in kind, in principle, from a taxon at any other given rank. It was Simpson (1963) who appears to have been the first to make explicit the distinction between taxa and categories: A taxon is an instance, an example, one of many possible members of a particular category. Categories, in contrast, are classes, membership in which depends solely on any taxon fitting certain criteria.

To a first approximation, the general criterion for category membership, even up to the present day, is phenotypic resemblance among organisms: Phyla are collections of classes, classes of orders, orders of families, families of genera, genera of species, and species of similar organisms. It all boils down, at first glance, to the association of similar organisms into groups, the smallest of which are called *species*.

From the outset, however, there has been one aspect of the system that seems, however slightly, to set species apart from other taxa: Biologists long before Darwin had taken to classifying very dissimilar organisms into the same species if it was discovered that they were linked reproductively; that is, if they were highly dimorphic members of a reproductive plexus (males and females, say, or other forms of polymorphism), they would be considered the same species. Conversely, zoologists have long separated species, no matter how similar their constituent organisms appeared, if it was known that there were no reproductive connections among them.

It is this tension between reproductive connectedness and aspects of morphological resemblance that lies at the heart of all debates about what species are. Modern positions all concede that among sexually reproducing organisms species include only those organisms that can, in theory (and see below), engage in reproductive activities among themselves. The issue then is, are *all* such organisms members of the same species or are there reasons to view species composition more restrictively? In a nutshell, all departures (save Paterson's recognition concept—see below) from the biological species concept are based on ontological claims of structure more narrowly distributed than reproductive communality: Nearly all serious rivals to the biological species concept, in other words, see species as differentiated taxa *within* reproductively defined "species" entities. As we shall see, there are several sets of such alternative species conceptualizations.*

There is a related issue complexly interwoven with the reproductive vs. morphological differentiation problem: discreteness. Though in each generation of evolutionary biologists there seems to be a minority opinion that sees a totally continuous spectrum of morphological variation within the biota, pre-Darwinian biologists as a rule saw discreteness in the biota. *Species tot sunt quot diversas formas ab initio produxit Infinitum Ens,* that is, there are as many species as the Creator (Infinite Being) originally fashioned diverse forms, according to Linnaeus. William Whewell, the prominent British philosopher, was simply fol-

*Because all modern concepts of species see such entities as fundamentally historical, genealogical units, I note in passing an important aspect of what species seem not to be: Despite the fact that Mayr (1982) explicitly added the proviso that species have ecological niches, I have reviewed elsewhere (Eldredge, 1985a; 1989; Eldredge and Salthe, 1985) the arguments that species are parts of the genealogical hierarchy but are not parts of economic systems (such as ecosystems or communities). Though I shall not discuss this issue further, I note here that species, as well as higher level taxa, were often taken to play concerted economic roles (occupying niches or adaptive peaks) in the ontology of the modern synthesis (see especially Eldredge, 1985a, for review).

lowing suit when he wrote, "Species have a real existence in Nature, and a transition from one to another does not exist" (Whewell, 1837). Species, to the early naturalists, were tolerably discrete one from another, though there can be little doubt that the discreteness was emphasized in resistance to the very idea of evolution, as Whewell's pronouncement actually makes quite clear.

Darwin (1859), of course, changed all that radically. In convincing his contemporaries that life had descended "with modification" from a single common ancestor, Darwin was essentially arguing that the species that look so discrete at the present were not so discrete in the past, as they diverged from a common ancestor. Moreover, species as we see them today are merely status reports in the flux of morphological change that is bound to continue to accrue as geological time progresses. The temporal discreteness between species was shaken utterly in the Darwinian view. Even the contemporary discreteness between species, if sampled near the point of divergence, would seem to crumble.

Yet Darwin had another view of species in mind when he spoke of species as "permanent varieties" (Darwin, 1871). In that context, he saw variation within species as potential for future evolution, but with an uncertain fate. It is species that lend "permanence" to morphological change, meaning that it is discreteness, explicitly reproductive discontinuity, that enables morphological change to be incorporated into the phylogenetic stream.

Though Darwin, in effect, took both sides of the discreteness issue, there is little doubt this his vision replaced a world view of static, discrete entities with a notion of biotic flux. Species were little more than minimally differentiated collections of organisms in the eyes of many immediately post-Darwinian biologists, a position certainly adopted by paleontologists (and retained to the present, see below), but also adopted by a number of zoological colleagues as well. There was, however, a small number of late 19th- and early 20th-century biologists (e.g., M. Wagner, J. T. Gulick, G. J. Romanes, D. S. Jordan) who stressed the importance of reproductive discontinuity between closely related species.

Thus the issue of spatial and temporal discreteness has always been intimately commingled with the problem of tokogeny (reticulation of lineages through sexual reproduction) and its relation to various aspects of morphological similarity and differentiation in any consideration of the nature of species. As can readily be imagined, virtually all permutations and combinations are not only possible, but have been realized in debates on this issue: Species may be wholly reproductive entities, or morphologically defined subsets within such reproductive entities; they may not be entities in any real sense at all; and if they are, they may be discrete spatially, temporally, or both, or even neither! But there is, generally speaking, logic behind all positions, and even some order in this apparent chaos.

The Biological Species Concept

The biological species concept was born out of an explicit attempt to integrate the element of discontinuity with the Darwinian evolutionary view, which had hitherto placed such emphasis on continuity. It was Dobzhansky (1937) who first pointed out that Darwinian theory primarily addresses the origin, maintenance, and further modification of organismic traits: It is quintessentially a theory of (phenotypic) *diversity.* Species arise as a simple outgrowth of accumulation of (genetically based) phenotypic change, a view that embodies an ontological conceptualization that emphasizes species as groups of similar organisms, over and above any concomitant notion of reproductive community.

Natural selection, as Dobzhansky stressed, produces a continuum, an unbroken spectrum of morphological variation. Yet nature comes packaged into more-or-less discrete entities that have been called "species." Thus Dobzhansky (1937, Chapter 1) specifically invoked a second theme in evolutionary biology: *discontinuity.* Diversity and discontinuity became the twin themes in evolutionary biology, strongly reinforced by Mayr's (1942) adoption of precisely the same terminology.

Dobzhansky (1937) and Mayr (1942) both thought that discontinuity must reflect a primary outcome of the evolutionary process. Especially in Dobzhansky's writings, it is evident that discontinuity must serve some role or purpose. Specifically, Dobzhansky saw discontinuity as a means whereby species could focus on particular "adaptive peaks" (see Eldredge, 1985a, Chapter 2 for extensive discussion). As Paterson (e.g., 1978, 1980, 1985) has taken pains to point out, Dobzhansky saw the entire process of speciation through the action and further evolution of "isolating mechanisms" as a means to drive two embryonic species completely apart—the better to focus the adaptations of organisms within each of the two to specific niches, or "adaptive peaks." Whatever its genesis, to both Dobzhansky and Mayr discontinuity reflects more than mere extinction of "intermediates" in a continuum of morphological differentiation generated by natural selection working on a groundmass of available genetic variation.

It is because natural selection would, in general, be expected to generate a smooth continuum of adaptive diversity that Dobzhansky (1935, 1937), followed by Mayr (1942), adopted a *reproductive* conceptualization of species. That is to say, their definitions emphasized the tokogenetic relations within, and lack thereof beyond, the boundaries of a "species" as the prime component of the very definition of "species." This is a subtle reversal of the older Darwinian notion of species as clusters of similar organisms that shared the ability to interbreed: The BSC sees species first as reproductive communities, members of which may well all be more similar to one another than to members of other such reproductive communities—though any given species may be highly di- or polymorphic, or virtually identical with another species [as in sibling species, to which Mayr (1942) actually devoted considerable attention].

In my opinion, Dobzhansky and Mayr opted for a "reproductive community" notion of species precisely because both saw the origin of discontinuity as quintessentially the superposition of "reproductive isolation" on what would otherwise be a smooth adaptive continuum of phenotypic diversity. Reproductive discontinuity is the necessary first step towards more general phenotypic discreteness. It was only logical, then, to see species as discrete reproductive communities. The species definitions of Dobzhansky (1935, 1937) and Mayr (1942), as epitomized in Mayr's (1942, p. 120) famous short version, were nearly identical: "Species are groups of actually or potentially interbreeding natural populations, reproductively isolated from other such groups."

Though dictionaries still cite phenotypic similarity as the prime component of the definition of species, it was Dobzhansky and Mayr who, at least up until now, changed the order of importance to (1) reproductive communality, followed by (2) phenotypic similarity, as ingredients of a species ontology. Indeed, neither mentioned degree of phenotypic similarity or discreteness in their definitions *per se.*

All discussions of species in the past 50 years spring from these works, even those that adopt strongly divergent views and appear not even to address the BSC *per se,* as in much of the recent debate within (phylogenetic) systematics circles. Ever since reproductive communality was firmly in place as the *sine qua non* of the conceptualization of species, the issue of whether or not to recognize entities *within* a reproductive community as appropriate kinds of entities to be themselves called "species" has been a major component of discussions of the "species question."

Paterson's Recognition Concept

Paterson (see 1985 and earlier references) has been a persistent critic of the BSC; yet many biologists have failed to grasp the distinction between his and the Dobzhansky–Mayr view of the nature of species, and for good reason: On the most fundamental level, Paterson's view of species (which he terms the *recognition concept,* or simply *RC*), like the BSC, sees species as inclusive reproductive communities.

But there are important refinements to the "reproductive community" notion of the BSC to be found in Paterson's ideas. Paterson defines species as "the most inclusive group that shares a fertilization system." He terms the fertilization system the *specific mate-recognition system,* (SMRS), which encompasses all interactive aspects of male–female reproductive physiology. Species, to Paterson, are explicitly natural, *self-defining* entities in nature. They are real entities, and moreover, generally quite discrete. There is a critically important epistemological ramification of Paterson's position: Because species are real, self-defining entities (because it is actual organismic reproductive behavior that coheres a species, incidentally setting its limits), biologists can objectively recognize species through study of the SMRS, aspects of which may even be fossilizable (cf. Vrba, 1980).

It is more in the realm of speciation theory, rather than species definitions *per se,* that Paterson departs from earlier work, especially Dobzhansky's ideas on speciation. Paterson's SMRS corresponds to a degree with the Dobzhansky (and Mayr) notion of isolating mechanisms. But Dobzhansky saw "isolating mechanisms" as evolving to divide and keep two species apart. Paterson, in contrast, sees the SMRS as a means of maintaining successful reproduction. If an ancestral SMRS diverges (e.g., in allopatry), such that reproduction does not take place in neosympatry, such a phenomenon, in Paterson's view, is merely a side effect of selection's maintaining successful mating within isolated populations.

In my view, Paterson's notion of the SMRS sharpens and extends the original BSC concept of species as reproductive communities. It may well also clarify basic epistemological issues of species recognition, as well as questions on the speciation process, topics beyond the immediate concern of this review. At base, however, it does not significantly alter the fundamental ontological stance of the BSC. In turning now to conceptualizations of species in time and space, we shall encounter additional examples that seem to conform in all essentials to the "reproductive community" notion. But we shall also encounter marked deviations, all of which reflect a retention of, or a return to, a pre-BSC view that morphological similarity should take precedence over reproductive communality in our ontological conceptualization of the very nature of species.

Species, Reproductive Continuity, and Geologic Time

Returning to Darwin's characterization of species as "permanent varieties," it has been commonplace in evolutionary biology in general to see the significance of the development of "reproductive isolation" as the beginning of independent evolutionary history: Once a species buds off from its ancestor, or once a single species divides into two or more reproductively independent, descendant species, each is consequently subject to its own evolutionary history. It is further widely assumed that a species may accumulate a great deal of either geographic or chronologic (or both) genetically based phenotypic differentiation once it has appeared: Species have origins, histories, and ultimately terminations, during which time they may (or may not) (1) give rise to descendant species and (2) accumulate significant phenotypic evolutionary change. As we shall see below, this rather straightforward and simple ontological extrapolation of the BSC into space, and particularly time, is by no means universally held, even by the very architects of the BSC. It does conform, however, to the conceptualizations of a number of biologists and paleontologists, including myself (cf. Eldredge, 1989).

Ghiselin (1974), followed closely by Hull (1976, 1978, 1980), proposed that species be considered as "individuals" rather than as "classes." In a sense, this suggestion simply reiterates Simpson's (1963) distinction between taxon and category: The species category is a class; any entity conforming to the criteria of "species" (whatever such criteria may be) belongs in the category "species." But any *particular* species is an entity, a named thing, an "individual." Gold, for example, is a class: Any atom with atomic number 79 belongs to the category "gold," wherever and whenever it occurs in the material universe. Relevant here, too, is the fact that the natural alchemy of the material universe sees gold forming independently over and over in space-time.

Not so with species, which are spatiotemporally localized entities. Ghiselin and Hull were especially concerned to characterize species as "individuals" precisely because species are prone to evolutionary modification throughout their existence (M. Grene, personal communication). According to Ghiselin and Hull, classes have immanent and unchanging properties: If an atom of gold is modified by particle gain or loss, it is no longer an atom of gold. But a species may change and still be the same species—the way an individual *Homo sapiens* organism, for example, changes ontogenetically through the developmental process, from conception to grave. It is *because* species change through time that they are individuals: Species cannot be classes, membership in which depends on holding certain attributes (e.g., "cranial capacity approximately 1400 cc" as a criterion for membership in the species *Homo sapiens*) because such criteria themselves are subject to modification.

Some philosophers deny that there is a hard and fast dichotomy between "classes" and "individuals" as developed in Ghiselin and Hull's work. But, in the main, their distinctions are germane to discussions of species ontology, and their characterization of species represents, at the most abstract level, the purest visualization of the BSC-RC "reproductive community" ontological conception of the nature of species as spatiotemporally localized bounded entities.

Along these lines, both Simpson (1961, p. 153) and Wiley (1978) have made important contributions to an "evolutionary" notion of species in time. Simpson, when asked when he first saw the importance of the BSC (on a questionnaire developed by Mayr; see Mayr, 1980, p. 462) pointed out that the BSC has no temporal dimension, an unsatisfactory situation (in his view) for a supposedly evolutionary concept. To offset this egregious lack of a temporal dimension to a supposedly evolutionary concept, Simpson (1951; but see especially 1961, p. 153) proposed his "evolutionary species" definition: "*An evolutionary species is a lineage (an ancestral-descendant sequence of populations) evolving separately from others and with its own unitary evolutionary role and tendencies.*" Note that Simpson definitely had a notion of tokogenetic reticulation within his lineages—the prime factor lending internal coherence to an evolutionary species, and at the same time lending it its own evolutionary independence, as no sharing of genetic information *between* such lineages was imagined to occur as the overwhelming rule.*

Simpson's formal characterization of species takes the notion of reproductive communality and extends it through time. Through the phrase, "separate roles and tendencies," Simpson acknowledges not only evolutionary independence among species, but the definite possibility, as well, that species may acquire a great deal of change in the course of their history. Note, though, that there is little said explicitly on the issue of discreteness: One might infer from Simpson's evolutionary species definition that new species appear when reproductive communality is disrupted (i.e., just as in the BSC-RC ontology). One might further surmise that extinction terminating such a species lineage would lend even more discreteness to the upper end of a lineage.

But it should be remembered that Simpson, virtually everywhere other than in his actual definition of the evolutionary species concept, adopted an ontological stance no different from that of Mayr (1942), Bock (1979, 1986), Gingerich (1974, 1976, 1979), and scores of other biologists, paleontologists, and paleoanthropologists: He held that there are two kinds of extinction. True extinction is unequivocal termination of a lineage, "extinction without issue" in

*Simpson, of course, was a vertebrate paleontologist, aware of, but not overly concerned by, the sorts of hybridization problems that routinely trouble botanists. Also, I reiterate here that this entire paper discusses only sexually reproductive systems, though I note that the vast majority of asexual, clone-forming organisms either have sexual stages or are otherwise capable of exchanging genetic information with other organisms—whether or not of the same "species."

Simpson's terms. But there is also what Simpson referred to as "pseudo-extinction": A taxon becomes extinct by evolving into a descendant, a concept reviewed in detail in the following section.

Wiley (1978, 1981) was far more explicit than Simpson on issues of discreteness. Like Simpson, Wiley extended the reproductive communality gist of the BSC, seeing species as reproductive lineages through time that persist as entities, regardless of the amount of phenotypic change accrued. Species originate through phylogenetic branching, which is to say, through the splitting of reproductive communities into two or more lineages. Species terminate by true extinction. Some arguments in the cladistic literature of the 1970s espoused a special form of "pseudoextinction," whereby species were said to become extinct when split into two (or potentially more) "daughter" species—an ontological position adopted, in the main, to obviate epistemological problems in cladistics. It is interesting that Simpson himself (1961, p. 168, Fig. 14A) anticipated precisely the same formulation as a subset of ways of characterizing species identity in the context of cladogenesis. Indeed, Simpson wrote, "In principle, the best solution is available when the lineages can be divided into three species, one ancestral and two descendant, separated at the point of branching" (Simpson, 1961, p. 168). Wiley, in contrast, conceives of species as originating through speciation in the BSC-RC sense and persisting beyond cladogenetic events (i.e., as a daughter buds off from a parental species) until eventually claimed by true extinction.

Finally, I note that the ontological position on the nature of species that I have myself adopted falls squarely in the present category. It is implicit in the notion of "punctuated equlibria" (Eldredge, 1971; Eldredge and Gould, 1972) and is made explicit especially in later works (Eldredge, 1985a, 1985b, 1989 and references therein) that species are spatiotemporal bounded entities, "individuals" in the Ghiselin–Hull sense. They are held together through reproductive interactions (explicitly, through shared possession of a fertilization system, or SMRS, following Paterson). Species arise through changes in such fertilization systems, entailing a spectrum ranging from virtually no to a great deal of "economic" adaptive change. Species persist as tokogenetic lineages until true extinction claims them. They remain individuals even upon giving rise, cladogenetically, to descendant reproductive communities (though epistemological problems of recognizing which is the ancestor and which the descendant may arise if the ancestral species is divided roughly equally in half). Species are thus taxa that engage in "more-making": They may give rise to descendant individual taxa of like kind (unlike taxa at other hierarchical levels; see Eldredge, 1989, for details of this position).

Again, in my view, species are discrete temporally through the process of speciation (origination) and extinction (true termination). I confess that the prevalence of stasis (relative lack of change in most species once they appear in the fossil record) and the relative rapidity of speciation (where most morphological change in evolution appears to me to be concentrated) accentuates the discreteness of individual species in time and space. Be that as it may be, note that the underlying concept agrees with Wiley's in all essentials: Even if species were to routinely accrue vast amounts of evolutionary change through time (i.e., in the phenotypic properties of its component organisms), species would still be held to be reproductive communities, with beginnings, histories, and eventually, true terminations.

Such a view, once again, strikes me as the logical and relatively straightforward extrapolation of BSC reproductive community notions of what species are into a temporal framework. It may therefore come as something of a surprise that, beginning with the very architects of the BSC within the "modern synthesis," and carried straight on through to the present day, no such projection of the BSC has been formulated or adopted by the vast majority of biologists who have articulated views on this subject, or who leave us to infer their ontological stance through their characterizations of evolutionary history.

Mayr (1942, especially p. 153), as Simpson appreciated, never left the pre-BSC Darwinian picture of what species look like in evolutionary time. On one page, Mayr (1942, p. 152) characterizes synchronic (and especially sympatric) species as akin to *Paramecium* "individuals," that is, quite discrete. Even when species are caught in the act of dividing, such that the status of differentiated allopatric populations may, in any instance, be difficult to assess, the situation still mirrors *Paramecium* in a petri dish: Some paramecia may be caught in the act of dividing, too, but there had been a single individual just prior to the division, and there will soon be two discrete paramecia where once there had been but one. Splitting—whether of paramecia or species—can be epistemologically messy when it comes to assessing individuality, but the overall ontology is clear. In this discussion, we see the roots of the Ghiselin–Hull position on species as "individuals."

Mayr's notion of synchronic individuality of species falls directly and completely out of his conceptualization of species strictly as reproductive communities. Thus it comes as something of a complete contrast to read his views, on the very next page, of the nature of species in evolutionary time:

> The species of each period are the descendants of the species of the previous period and the ancestors of those of the next period. The change is slight and gradual and should, at least theoretically, not permit the delimitation of definite species. (Mayr, 1942, p. 153)

Here, the concept of a species as a reproductive plexus moving through time—the notion of species ever since Darwin, and the fundamental quality of the BSC in the atemporal sense (and in the temporal formulations reviewed immediately above) is indeed present. But gone is any semblance of discreteness *between* species as they succeed one another in wholesale transformation through geologic time. Indeed, the conceptualization of what a species *is* differs radically from the picture Mayr paints in the synchronic case, as Mayr himself (1942, p. 154) is aware.

Although couched in an epistemological context, that is, how do we divide species up in an evolving continuum, Mayr is positing *subdivisions* of such lineages that he is perfectly content *also* to call "species." Species here are arbitrarily recognized subdivisions of the lineage. They are spatiotemporally localized collections of individuals that resemble one another more closely than they do other organisms of other segments—other "species"—of the evolving lineage. As Gingerich (1979) (whose ontological views on species in geologic time coincide precisely with those of Mayr discussed here) has put it, recognized subdivisions of such continua should be about as morphologically differentiated as, say, contemporaneous closely related species, such as three species of canids living in contemporary Michigan, as cited by Gingerich (1979).

Species conceived of as (arbitrarily delineated) subdivisions of an evolving lineage are the "chronospecies" of many authors. There are several aspects crucial to this ontological conception of nature. First, many authors, in fact, realize that species so conceived do not represent the same sort of notion as the BSC, though it must be said that much of the controversy about species, speciation, and their relation to transformational evolutionary change of phenotypic properties stems from different ontological concepts of species held as underlying assumptions; [see the debate on punctuated equilibria (e.g., Eldredge and Gould, 1972; Gingerich, 1974, 1976; Gould and Eldredge, 1977) as but one of a number of examples of such discussion arguing, in effect, from different premises on the very nature of species].

Secondly, note that discontinuity was addressed by Dobzhansky, and subsequently by Mayr, because nature seemed to them to be organized into synchronically discrete reproductive "packages"—their "species." Reproductive isolation, though it may not precede phenotypic differentiation, nonetheless is essential to render discrete the continuum of phenotypic differentiation produced by natural selection. Mayr (as a personification of this prevalent stance) saw species *in time* as lacking such discreteness: After all, we are talking about reproductive *continuity* within lineages through time.

Why then revert to an arbitrary subdivision of such lineages into quite different sorts of entities, calling them by the very same term: *species?* It is not clear, but in general Mayr (who did not discuss paleontological data at length in his 1942 book) seems to have recognized that paleontologists commonly recognize different species in successive stratigraphic horizons or units. Such species are commonly separated by stratigraphic gaps that, in effect, chop a lineage into separate segments, the very task that the systematist would theoretically have to perform, using arbitrary criteria, were the fossil record complete. Indeed, it is this line of reasoning that Gingerich (in the references cited and elsewhere) follows in his discussions of relatively complete stratigraphic data, which he feels force him to divide fossil mammal lineages arbitrarily into "species."

Such taxa are in no way to be construed as "classes" in the Ghiselin–Hull sense (neither are they discrete individuals!): Chronospecies are held to be subdivisions of lineages, which are themselves assumed to be internally reticulate (through reproduction) and independent of other such lineages. Significantly, when direction of morphological change reverses in one of Gingerich's lineages (e.g., in the *Hyopsodus* lineage, Gingerich, 1976, Fig. 5), he does not synonymize species, but rather keeps them separate, as they are temporally disjunct segments of the lineage, regardless of their morphological identity (i.e., in terms of the parameters Gingerich is measuring). Atoms of gold may be formed independently, in different spatiotemporal contexts, but, save in one example known to me (discussed below), chronospecies are generally not seen as classes, with membership depending strictly on attainment of one or more phenotypic properties, which might be acquired at different times and in different places.

Yet arbitrary divisions of an evolving lineage—divisions based on the distributions of phenotypic properties—do have "class-like" attributes: Species are defined as clusters of anatomically similar organisms. Within an array of arbitrarily recognized species, membership depends on two criteria: place in time and degree of morphological similarity, dispersed around a mean.

There is nothing "wrong," of course, with calling such arbitrary divisions of evolving lineages *species,* so long, that is, as their ontological status (and difference from the BSC) is duly recognized. Biostratigraphers, for example, continue to maintain the utility of such subdivisions, though some paleontologists (myself certainly included) think the position more theoretical than real, as there is a dearth of examples of such evolving species-lineage continua actually documented in the fossil record.

It is in the ranks of paleoanthropology that the notion of chronospecies converges most dramatically with the ontological notion that species are simply classes of similar organisms. Here, we must infer ontology from reconstructions of phylogenetic history. A number of paleoanthropologists (especially Coon, 1962 and, more recently, Wolpoff, 1986, and Brace, 1991) see mid-Pleistocene phylogeny in the genus *Homo* as entailing the independent evolution of *Homo sapiens,* via archaic stages, from an ancestral *Homo erectus,* separately in different regions at distinct times. *Homo erectus,* the archaic intermediates, and *Homo sapiens,* so conceived, are strictly grade-group categories. Note here, though, that there is an added, geographic element that further departs from the BSC notion of within-species cohesion supplied by reproductive interaction: Subspecific differentiation and separateness is emphasized greatly and held to be maintained throughout the 1.5 million year duration of the *Homo* transformational lineage.

Seeing *Homo sapiens* evolving polyphyletically three separate times, in three separate places, yet somehow maintaining a mutually shared fertilization system, plays hob with most ontological conceptions of species in biology over the last 50 years. Alternative scenarios, such as the emerging notion that *Homo sapiens* arose once, in Africa, and subsequently spread throughout the world, are currently gaining wide acceptance. Whatever the historical truth of the phylogenetic details, I merely note here that the underlying ontology of the latter scenario is far more in accord with BSC-RC species ontology, however much the dispute over (monophyletic!) chronospecies may still be with us.

Differentiated Taxa, Reproductive Cohesion, and Species

Phylogenetic systematics, or cladistics, is, at base, a methodology, an epistemological construct for the discovery of phylogenetic history (or the hierarchical array of similarity interlinking all members of the biota, which most biologists agree amounts to the same thing). As such, cladistics would ordinarily lie outside the realm of this review of species ontology. But, from its inception, phylogenetic systematics has adopted various ontological positions on the nature of taxa generally, and on species in particular, that have significance to the overall question of what species are.

Early discussions of cladistics (cf. Nelson, 1970) emphasized the difficulty of recognizing ancestors—which immediately focused attention on species-level taxa. Many early formulations (including one by the paleontologists Schaeffer *et al.*, 1972) recommended treating taxa of all levels as in principle the same type of entity; specifically, no taxon was to be taken to be ancestral to any other. Though some followers of the New Systematics had continued to recognized ancestry and descent among taxa of any conceivable rank, in practice, the issue involved species-level taxa because it is only at that level that a process—speciation, the derivation of descendant from ancestral species—was thought to actually occur in nature. Higher taxa, by near-universal agreement, themselves arise as species.

The issue soon crystallized on the question of monophyly; evolution produces skeins of monophyletic taxa, defined as all species descended from a single ancestral species. Monophyletic taxa include the stem and all descendant species. How, then, can the concept of monophyly be applied to individual species? Though not a universal sentiment among cladists, there was nonetheless considerable emphasis placed on moving away from the BSC and seeing species-level taxa as merely the smallest group of diagnosably distinct taxa, defined and recognized in essentially the same fashion as any other taxon. It is noteworthy, however, that even in Nelson and Platnick (1981, p. 12), species are still acknowledged to include known males and females (i.e., regardless of patterns of shared similarity), and, tellingly, species themselves, unlike any higher level taxon, may not, in fact, be diagnosable in terms of autapomorphies (i.e., evolutionary specializations shared by all members of the taxon, and not shared with organisms from any other taxon—apparent tacit recognition of ancestry and descent among species-level taxa.)

The ontologically most interesting point raised by cladists is that species, when defined "biologically" (i.e., in the BSC-RC reproductive community sense), may well be paraphyletic. This is the central issue in the recent flurry of papers and rebuttals within phylogenetic systematics circles (e.g., Cracraft, 1987, 1989; Eldredge, 1989; Mishler and Donoghue, 1982; Mishler and Brandon, 1987; Nixon and Wheeler, 1990; Wheeler and Nixon, 1990; de Queiroz and Donoghue, 1988, 1990). In other words, looking at fertilization systems as characters to be analyzed for phylogenetic content, a shared SMRS may be a symplesiomorphy. Lieberman (in press), however, makes the intriguing point that the SMRS may well be a taxon-level (specifically, species-level) "character," and thus outside the realm of conventional character analysis. Be that as it may, from a standard cladistics point of view, the situation is best considered in conjunction with a three-taxon statement cladogram (Fig. 1). Taxa A and B share an SMRS; yet, on the basis of other (nonreproductive) characters, it can be shown that taxa B and C actually share a more recent ancestry than either does with taxon A. What, then, are the species? Here we leave the realm of epistemology and ask what a species *is*. To portray accurately the course of evolutionary history, many phylogeneticists these days would recognize three species: Species A and B are not conspecific, because they are diagnosably different, and only B shares certain synapomorphies with C; yet B and C cannot be conspecific because they do not share a fertilization system. The BSC-RC formulation would recognize only two species: A + B (shared fertilization systems) and C (separate fertilization system).

Both the possibility of paraphyly if taxa are recognized solely on the basis of an SMRS, and the potential recognition of morphologically defined subunits ("chronospecies") within a tokogenetic lineage represent instances whereby non-

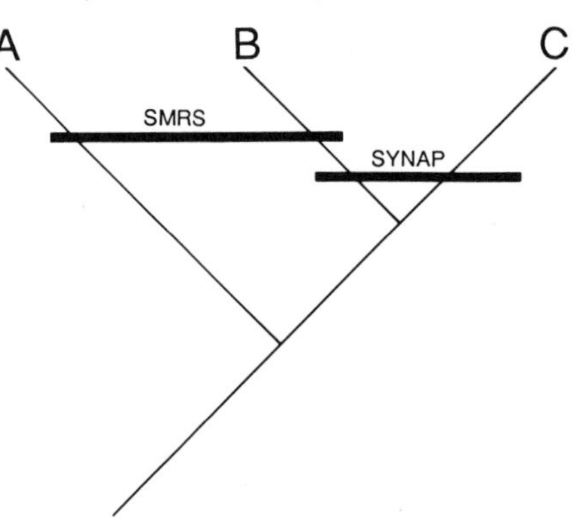

Fig. 1. Phylogenetic relationships among three taxa. B and C share synapomorphous resemblance, implying that they share a more recent ancestry than either does to A, yet A and B share an SMRS. Which are the species? A, B, and C? Or A+B and C?

reproductive phenotypic differentiation is finer than reproductive differentiation. There is, however, at least one profound difference between the two categories. Chronospecies are held to be nondiscrete temporally, whereas discreteness (as well as demonstration of symplesiomorphy) is held to underlie the recognition of two species that share a fertilization system (as in Fig. 1).

What the two types of cases do share, however, is the recognition that nonreproductive ("economic" see Eldredge, 1989, for use of this term) features may differentiate at a finer level than reproductive features.* There is, after all, no necessary correlation of rates of evolution of economic and reproductive attributes (see Eldredge, 1989, especially Chapter 4, for extensive discussion). Much the same consideration, in my view, underlies the notorious resistance of the great majority of botanists to acceptance of the BSC: Many perfectly good botanical species—well-differentiated taxa defined and recognized on the basis of economic attributes—nonetheless are infamous for their readiness to hybridize with other "good" species.

Phylogenetic systematists are concerned with historical entities, including the discreteness and also the permanence (or historical independence) of entities called "species." The issue becomes, at what level of differentiation *within* an SMRS-defined "species" do we have entities that can be said to be, not only (morphologically) discrete, but also phylogeneticaly discrete; that is, not subject to absorption or dissolution by virtue of a shared SMRS with another such entity? I have said nothing, up to now, on how geneticists (i.e., other than Paterson) view species. The genetic structure of species is in itself a fairly labyrinthine topic, and I will select only one aspect that bears directly on the discussions so far in this paper—the views of Sewall Wright. Wright, though a theoretician and mathematical analyst of experimental data, was nonetheless the geneticist who, more than any other biologist, has given us our received view of the structure of species in nature. Wright (e.g., 1931, 1932) saw species as typically broken up into many (though minimally but one) local populations that he, at first, called "colonies." These quasi-isolated local populations are now variously

*Mayr (personal communication), has remarked that, in evolutionary lineages, we may suppose that the ability to interbreed may be expected to become modified such that early members would not be anticipated to be able to reproduce successfully with later members; reproductive differentiation, in other words, may well keep pace with economic temporal differentiation. Nonetheless, reproductive *continuity* is assumed, and the emphasis is on economic differentiation in the face of a reproductive continuum.

called "demes" or "avatars," depending on whether their context is seen to be reproductive (demes) or as parts of local ecosystems (avatars); see, e.g., Eldredge, 1985a, 1989, for use of these terms in the explicit context of genealogical and ecological hierarchies.

The point is that no species is uniformly distributed. Each colony undergoes a quasi-independent evolutionary history, where genetic events by chance alone (as well as through selection) will lead to differentiation. Yet such differentiation is *within* a shared fertilization system. It is the fate of local populations variously to become extinct or to be absorbed into adjoining populations. Uniqueness is counterbalanced by gene flow. Demes are not like whole species, as tokogeny will prevent separate demes from diverging very far. Indeed, Wright was right, in my view, when he realized, late in life, that his shifting balance theory, based on the multidemic structure of species, should have predicted stasis for most species: Demic differentiation is the opposite of wholesale, concerted change, and the predictable loss of demic differentiation acts against long-term accrual of adaptive differentiation.

Sexual reproduction at the local, within-demic level produces tokogenetic reticulation. Because demes within a general area do not as a rule exhibit great interdemic differentiation, at this level, too, by near-universal agreement, there is no reason to suppose that small collections of demes constitute "species." We are looking for the point where the noisy signal of reticulating tokogeny is replaced by clear, hierarchically structured, phylogenetic "signal" (a point especially clearly addressed by de Queiroz and Donoghue in papers cited above). But there is a gray area between the local deme level and the regionally differentiated collections of demes—corresponding, as a rule, with the entities that, for example, Mayr (1942), would call "subspecies." It is at this level that we encounter arguments that such morphologically differentiated, more-or-less phenotypically discrete taxa, which nonetheless share an SMRS with other such allopatric taxa, are to be called true "species." Many older "species" of North American passerine birds have been synonymized in recent decades (e.g., several species of morphologically well-differentiated juncos, now all lumped into the dark-eyed junco, *Junco hyemalis*). Such actions reflect adherence to the BSC, and additionally reflect field evidence that along parapatric boundaries of differentiated taxa, hybridization occurs, and that the SMRS (meaning songs, primarily, in this instance) appear to be the "same," or at least nearly so. That contemporary ornithologists are, bit by bit, recognizing once again such allopatric, or parapatric, taxa as distinct species likewise reflects a shift in ontological position rather than the simple dictates of taste or a changing epistemology.

There are real issues here, ones that will not be resolved by polemic. Nature is organized into reproductive communities, shared fertilization systems. It also *may* be organized into differentiated subunits—where the differentiation involves predominantly (though not exclusively) economic, somatic characters—which may or may not reflect adaptive ecological differentiation. One argument in favor of retention of the BSC-RC reproductive view of species is that it pertains to all cases within sexually reproducing organisms, which, as I have already remarked (see footnote on p. 10), constitute the overwhelming number of cases in the biota.

Yet there is much to be said for recognizing differentiated subentities within BSC-RC entities. The case for chronospecies recognition is less solid (in my view), as the underlying ontological conceptualization explicitly rejects discreteness. Yet when the data, through accident of preservation or discovery, present us with prepackaged, hence discrete, chronologically serial "species," there is the pragmatic (epistemologic) consideration that *not* to treat such diagnosably different entities as discrete species would be potentially to commit the same error; that is, it would tend to minimize morphological information, thus potentially "lose" historical information, just because such serial chronologic taxa *may* be only sample points within a continuous species-level lineage. After all, such chronospecies may, with further data, prove to have been connected through true cladogenetic events.

Where, then, do we draw the line and say, "Here is where the limits of species are?" Few dispute that cladogenetic events produce the beginning of species (though some would say that the ancestral species ceases to exist, and two or more daughter species take its place—the minority view, to be sure). No one disputes that extinction of a BSC-RC lineage is the end of a species. But when can we be sure that tokogeny stops and phylogenetic—true historical—independence begins? If such phylogenetic, historical independence is our benchmark criterion of what a species is, adhering to a BSC-RC ontological conception, extended to embrace spatiotemporal distributions as discussed above, is the safe and conservative choice. No one can tell the future: The whole point of Darwin's remark that species are "permanent varieties" was that subspecies, no matter how widely distributed and well differentiated, are not necessarily to be expected to survive as independent entities to be injected into the "phylogenetic mainstream." (Indeed, even many new species, meaning "reproductively isolated" taxa, in my view, suffer extinction soon after arising; see Eldredge, 1989, p. 150 for discussion.) That is why my own preference tends towards the conservative retention of (a fully spatiotemporal version of) the BSC-RC. On the other hand, I am persuaded, as in the case with the dark-eyed junco cited previously, that well-differentiated taxa within BSC-RC species may well display phylogenetic structure and thus merit recognition as separate species, rather than as divisions of a single "biological" species. I am only sure of one thing: It is nature, not my own vacillation, nor the collective inability of biologists over the ages to agree, that lies at the roots of our difficulties in deciding what species actually *are*.

Acknowledgments

I thank Marjorie Grene and Bruce Lieberman, as well as reviewers Douglas Futuyma, Judith Masters, and Frederick Szalay, for genuinely constructive discussion and commentary on the first draft of this chapter.

References

Bock, W. J., 1979. The synthetic explanation of macroevolutionary change—a reductionist approach, in: J. H. Schwartz and H. B. Rollins (eds.), *Models and Methodologies in Evolutionary Theory. Bull. Carnegie Mus. Nat. Hist.* **13**:20–69.

Bock, W. J. 1986. Species concepts, speciation and macroevolution, in: K. Iwatsuki, P. H. Raven, and W. J. Bock (eds.), *Modern Aspects of Species*, pp. 31–57. University of Tokyo Press, Tokyo.

Brace, C. L. 1991. *The Stages of Human Evolution.* Prentice Hall, Englewood Cliffs, NJ.

Coon, C. S. 1962. *The Origin of Races.* Knopf, New York.

Cracraft, J. 1987. Species concepts and the ontology of evolution. *Biol. Phil.* **2:**329–346.

Cracraft, J. 1989. Speciation and its ontology, in: D. Otte and J. A. Endler (eds.), *Speciation and its Consequences,* pp. 28–59. Sinauer Associates, Sunderland, MA.

Darwin, C. 1859. *On the Origin of Species.* John Murray, London.

Darwin, C. 1871. *The Descent of Man, and Selection in Relation to Sex.* John Murray, London.

de Queiroz, K. and Donoghue, M. J. 1988. Phylogenetic systematics and the species problem. *Cladistics* **4:**317–338.

de Queiroz, K. and Donoghue, M. J. 1990. Phylogenetic systematics and species revisited. *Cladistics* **6:**83–90.

Dobzhansky, T. 1935. A critique of the species concept in biology. *Phil. Sci.* **2:**344–355.

Dobzhansky, T. 1937. *Genetics and the Origin of Species.* Reprint ed., 1982. Columbia University Press, New York.

Eldredge, N. 1971. The allopatric model and phylogeny in Paleozoic invertebrates. *Evolution* **25:**156–167.

Eldredge, N. 1985a. *Unfinished Synthesis. Biological Hierarchies and Modern Evolutionary Thought.* Oxford University Press, New York.

Eldredge, N. 1985b. *Time Frames.* Simon and Schuster, New York. Reprint ed., 1989, Princeton University Press, Princeton, NJ.

Eldredge, N. 1989. *Macroevolutionary Dynamics.* McGraw-Hill, New York.

Eldredge, N. and S. J. Gould. 1972. Punctuated equilibria: an alternative to phyletic gradualism, in: T. J. M. Schopf (ed.), *Models in Paleobiology,* pp. 82–115. Freeman, Cooper, San Francisco.

Eldredge, N. and Salthe, S. N. 1984. Hierarchy and evolution. *Oxford Surv. Evol. Biol.* **1:**182–206.

Ghiselin, M. T. 1974. A radical solution to the species problem. *Syst. Zool.* **23:**536–544.

Gingerich, P. D. 1974. Stratigraphic record of Early Eocene *Hyopsodus* and the geometry of mammalian phylogeny. *Nature* **248:**107–109.

Gingerich, P. D. 1976. Paleontology and phylogeny: patterns of evolution at the species level in Early Tertiary mammals. *Am. J. Sci.* **276:**1-28.

Gingerich, P. D. 1979. The stratophenetic approach to phylogeny reconstruction in vertebrate paleontology, in: J. Cracraft and N. Eldredge (eds.), *Phylogenetic Analysis and Paleontology,* pp. 41–77. Columbia University Press, New York.

Gould, S. J. and Eldredge, N. 1977. Punctuated equilibria: The tempo and mode of evolution reconsidered. *Paleobiology* **3:**115–151.

Hull, D. L. 1976. Are species really individuals? *Syst. Zool.* **25:**174–191.

Hull, D. L. 1978. A matter of individuality. *Phil. Sci.* **45:**335–360.

Hull, D. L. 1980. Individuality and selection. *Ann. Rev. Ecol. Syst.* **11:**311–332.

Lieberman, B. S. 1992. An extension of the SMRS into a phylogenetic context. *Evolutionary Theory* (in press).

Mayr, E. 1942. *Systematics and the Origin of Species.* Reprint ed., 1982. Columbia University Press, New York.

Mayr, E. 1980. Biographical essays. G. G. Simpson, in: E. Mayr and W. B. Provine (eds.), *The Evolutionary Synthesis,* pp. 452–463. Harvard University Press, Cambridge.

Mayr, E. 1982. *The Growth of Biological Thought.* Harvard University Press, Cambridge.

Mishler, B. D. and Brandon, R. N. 1987. Individuality, pluralism, and the phylogenetic species concept. *Biol. Phil.* **2:**397–414.

Mishler, B. D. and Donoghue, M. J. 1982. Species concepts: a case for pluralism. *System. Zool.* **31:**491–503.

Nelson, G. J. 1970. Outline of a theory of comparative biology. *Syst. Zool.* **19:**373–384.

Nelson, G. J. and Platnick, N. I. 1981. *Systematics and Biogeography. Cladistics and Vicariance.* Columbia University Press, New York.

Nixon, K. C. and Wheeler, Q. D. 1990. An amplification of the phylogenetic species concept. *Cladistics* **6:**211–223.

Paterson, H. E. H. 1978. More evidence against speciation by reinforcement. *S. Afr. J. Sci.* **74:**369–371.

Paterson, H. E. H. 1980. A comment on "mate recognition systems." *Evolution* **34:**330–331.

Paterson, H. E. H. 1985. The recognition concept of species, in: E. S. Vrba (ed.), *Species and Speciation. Transvaal Mus. Monogr.* **4:**21–29.

Schaeffer, B. M., Hecht, K. and Eldredge, N. 1972. Phylogeny and paleontology. *Evol. Biol.* **6:**31–46.

Simpson, G. G. 1951. The species concept. *Evolution* **5:**285–298.

Simpson, G. G. 1961. *Principles of Animal Taxonomy.* Columbia University Press, New York.

Simpson, G. G. 1963. The meaning of taxonomic statements, in: S. L. Washburn (ed.), *Classification and Human Evolution,* pp. 1–31. Aldine, Chicago.

Sokal, R. R. and Crovello, T. J. 1970. The biological species concept: a critical evaluation. *Am. Naturalist* **104:**127–153.

Vrba, E. S. 1980. Evolution, species and fossils: How does life evolve? *S. Afr. J. Sci.* **76:**61–84.

Wheeler, Q. D. and Nixon, K. C. 1990. Another way of *looking at* the species problem: a reply to de Queiroz and Donoghue. *Cladistics* **6:**77–81.

Whewell, W. 1837. *History of the Inductive Sciences.* Parker, London.

Wiley, E. O. 1978. The evolutionary species concept reconsidered. *System. Zool.* **27:**17–26.

Wiley, E. O. 1981. *Phylogenetics: The Theory and Practice of Phylogenetic Systematics.* John Wiley, New York.

Wolpoff, M. H. 1986. Describing anatomically modern *Homo sapiens.* A distinction without a definable difference. *Anthropos* (Brno) **23:**41–53.

Wright, S. 1931. Evolution in Mendelian populations. *Genetics* **16:**97–159.

Wright, S. 1932. The roles of mutation, inbreeding, crossbreeding, and selection in evolution. *Proc. Sixth Int. Congr. Genetics* **1:**356–366.

Species Concepts — 2

The Tested, the Untestable, and the Redundant

FREDERICK S. SZALAY

> Concepts of method designate systematic courses of action devised by men for the purpose of achieving certain goals . . . [these] . . . represent part of man's conceptual equipment. Epistemology is a science devoted to the discovery of the proper methods of acquiring and validating knowledge.
>
> Ayn Rand (1966, p. 36)

> If a taxon is defined (as a morphospecies) in such a way that it coincides with the phenon, the taxonomist may facilitate the task of sorting specimens, but this activity will result in "species" that are biologically, and hence scientifically, meaningless. The objective of a scientifically sound concept of the species category is to facilitate the assembling of phena into biologically meaningful taxa on the species level.
>
> Mayr and Ashlock (1991, p. 24)

> In the time dimension, therefore, species can only be defined arbitrarily or with reference to compartment limits set by contemporary organisms, or organisms from some other, arbitrarily selected transect of time-character space.
>
> Andersson (1990, p. 378)

FREDERICK S. SZALAY • Department of Anthropology, Hunter College, New York, New York 10021.

Species, Species Concepts, and Primate Evolution, edited by William H. Kimbel and Lawrence B. Martin. Plenum Press, New York, 1993.

Introduction

The Random House College Dictionary of 1968 gives as the first of several listed definitions of the word *theory:* "a coherent group of general propositions used as principles of explanation for a class of phenomena." It is in this sense that I will use the concept of theory in this chapter. I would like to add that all sorts of connections between carefully tested concepts, which serve as units of thought and which have objectively established perceptual bases in reality, are usually built into a framework that is often called a theory. These theories, and others within them, are all ultimately tested against sundry aspects of objective reality. The methodology itself employed in *taxonomic practice,* which is an invented human activity and not a reality of nature, should be based on, follow, or fall out of the tested (not merely testable) theories of evolutionary biology.

The terms that I use in the discussions below have equivalence to the evolutionary process that manifests itself in a biogeographic context in several obviously inseparable, yet conceptually discrete, phenomena (see especially Stuessy, 1980, 1987). The data of taxonomy reflect these phenomena; taxonomic hypotheses (particularly phylogenetic ones) that deal with the products of evolution should be subjected to tests from all aspects of the evolutionary process. The evolutionary (phylogenetic) process and its consequence, which is phylogeny (Fig. 1), is divisible only for heuristic purposes, although it may be said to manifest itself in

1. phyletic (anagenetic or patristic) relationships, which are tested by vertical comparisons;
2. cladistic (branching or splitting) relationships, which are tested by horizontal comparisons;
3. phenetic distance, described (and measured) either vertically or horizontally (this expresses the combined results of both phyletics and cladistics);
4. temporal (chronistic) distance, measured in terms of either relative or absolute time scales; and a
5. biogeographic context that is inexorably tied to earth history.

Evolutionary change in sexually reproducing organisms occurs in species and their populations, and the biological species concept (BSC) is a product of a tradition of population-based thinking that culminated in the discovery and enunciation of natural selection (see especially, Mayr, 1982). The BSC was clarified and established during the development of the Modern Synthesis (e.g., Mayr, 1942). This theoretical concept is a cornerstone of evolutionary theory and *subsequently* for taxonomic methodology. Its significance is evolutionary, it has been tested beyond reasonable doubt, but without the BSC having solved all operational problems of microtaxonomy! It is the application of the BSC that is a matter of fundamental concern for the practice of taxonomy. It is the theoretical nature of the BSC, and its pivotal connecting role between evolutionary biology and various taxonomic axioms, that has been repeatedly misunderstood or misconstrued, with the consequent unnecessary confusion about species concepts.

Needless to say, competing ideas relating to species should be examined

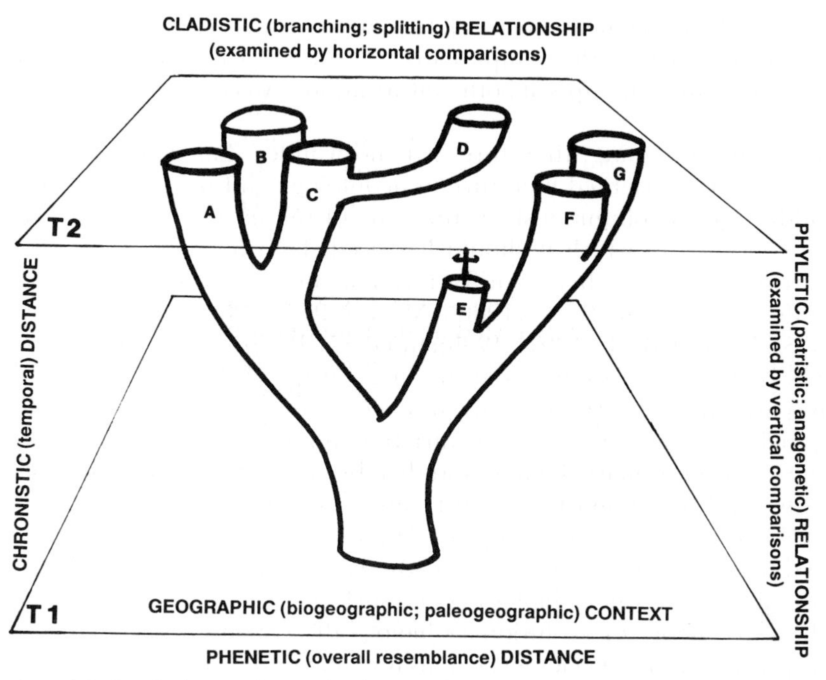

Fig. 1. A model of evolutionary change that is not taxic (i.e., not based on taxonomic preconceptions) illustrates the nature of evolutionary relationships of species and lineages and underscores that a phylogeny (which is the result of specific phyletic and cladistic events) contains a variety of information (modified after Stuessy, 1987). The *tested* BSC, a theoretical concept as discussed in the text, is not defined or conceived in relation to a lineage—it is defined in relation to other species in sympatry. A to G represent seven lineages, and not species, which meet between time planes T1 and T2 and continue together with others back into time. The evolutionary relationships between and within these lineages manifest themselves both in species differences tested in sympatry at any one level of time and in chronistic and phenetic distances. At T2, six distinct species (E is extinct before that time) are clearly tested, if they become sympatric, as valid biological species [whether or not the speciation process is completed or not, as discussed by Bock (1966)]. Such species can be taxonomically delineated with confidence. Given a good fossil record for the lineages depicted in this model, however, many other taxonomic species may be recognized without implying that these units represent distinct lineages. Such valid multidimensional taxonomic species (e.g., such as a well-documented series of chronospecies of one lineage) are not the "same" species; they are instants of their respective lineages. The number of speciations obviously does not necessarily correspond to the valid taxonomic species, which reflect the evolutionary differentiation of such a monophyletic group as that depicted here. Taxonomic species in the fossil record must be additionally interpreted through phylogenetic analysis that employs both phyletic and cladistic approaches whether they represent one or more known or new lineages.

both ontologically as well as epistemologically before these serve as the foundational theory for the methods of taxonomic activities. There is necessity not only for the establishment of the existence of certain realities of the world (in this case, the ontology of a concept of species), but also the means by which one can have access to this objective reality. If a certain notion of species is real, how do we then test this reality? If species are, or should be viewed as, "monophyletic" [in the particular Hennigian sense that they cease to exist following a speciation event, or, in the sense of Eldredge and Novacek (1985), whose monophyletic concept is wedded to true extinction], then what means are there to ascertain or negate the notion that the samples we call species are in fact such entities? Science and testing (attempts at both "falsification" *and* corroboration) are inseparable.

Given the foregoing, therefore, it is no wonder that there are few other areas of organism-oriented literature in neontology and paleontology that have been addressed as voluminously as the issue of the *species category* and the theoretical concepts on which it should be based. I do not intend to review this literature in great depth here, but for recent accounts and additional references see Andersson (1990), Bock (1986), Mayr (1942, 1963), Mayr and Ashlock (1991), Szalay and Bock (1991), Willman (1985, 1989), and Vrba (1985).

It is really not necessary to argue in great detail that the practice of taxonomy, particularly microtaxonomy, has been profoundly altered as a result of conceptual advances in the study of variation in populations, and the increasing awareness of evolutionary dynamics studied by naturalist-population geneticists, ecologists, and taxonomist-naturalists and paleontologists. The well-known tomes of Dobzhansky, Mayr, Simpson, Huxley, and others that ushered in the still solid framework of an ongoing neo-Darwinian synthesis amply testify to that fact (see also evolutionary biology texts, such as that of Futuyma, 1986). The growth in the theory of evolutionary biology directly led to the change of taxonomic practice at the species level. This microtaxonomy is an activity of clustering organisms based on a variety of theoretical concepts and methods derived from these. As alluded to above, *in science methods should fall out of existing and highly tested (corroborated) theory.* In systematic biology such an overarching theoretical framework (at least the broad but well-corroborated outlines of it) was the start of the neo-Darwinian synthesis. It explained descent and diversification by harmonizing the advances in genetics, ecology, evolutionary morphology and physiology, paleontology, and population biology.

With all this said, I now briefly return to the critical idea of testability as this concerns the related but obviously nonindentical concepts of species versus lineage. As spelled out below I use the concept of speciation in its biological (and only valid) sense: division of one species into two or more species. I reject at the outset (see more detailed justification below) the often referred to "vertical" or "morphological" speciation (which is really the proper *taxonomizing* of samples) because of the conceptual imprecision such terms perpetuate in theoretical evolutionary biology. Horizontal comparisons of either traits or groups of organisms (see Bock, 1977) in sympatry yield the tested theoretical biological base for the BSC. Vertical comparisons (and the tests these supply) of samples, on the other hand, are at the heart of the theoretical lineage concept (see below), but not of a species notion, such as the evolutionary species concept (EVSC), which in fact refers to a lineage segment. As I justify below, the Hennigian, cladistic, or phylogenetic species concept (PSC) cannot be tested because *there are no known empirical means to ascertain that branching was responsible for or accompanied the evolution of sample differences* unless the samples are assuredly sympatric and synchronic. All morphological differences are simply not synonymous with specific, speciation-related "results of branching," not even in different populations of the same species. In spite of these clear theoretical evolutionary objections, the PSC is based on the notion of cladistic differentiation (see below). Furthermore, the

PSC, sometimes confusingly labeled the *monophyletic species*, has been used in two different senses. One sense implies a species' "birth" at its branching origin and its "death" (pseudoextinction) after its splitting. The other sense is genuinely holophyletic (e.g., *sensu* Eldredge and Novacek, 1985) in that it ties punctuation and true extinction to the notion of the monophyletic species. In contrast to the BSC, as it is developed below, the EVSC and PSC are taxonomic species notions rather than theoretical concepts that can be tested (i.e., they lack epistemology; Bock, 1986; Szalay and Bock, 1991).

Both the recognition concept (RC) and the ecological species concept (ECSC), which I also briefly discuss below, are integral components of a relational, biological species concept. Although the ideas emphasized in these are important aids in delineating allopatric samples, they are redundant species concepts.

It is obviously of some general importance whether the taxonomic ordering of the record of life should be based on tested theory derived from evolutionary biology or on particular practices of taxonomy and the theory derived from the latter. These latter are primarily operational efforts for grouping organisms. Many such procedures are dependent on a definition of "species," based on (1) preevolutionary (essentialistic) traditions of morphology, (2) assumed dichotomized splitting events of species taxa (as originally suggested by Hennig), or (3) the monophyletic species notions (see above). Such definitions of species, as discussed below briefly, are based, on uncorroborated and untestable assumptions about (1) the meaning of morphological distance; (2) the nature of morphological change before, during, and after (allopatric) speciation; and (3) the choreography of lineages.

In light of the above, my intention in this chapter is to (1) argue for the theoretical primacy of *well-tested* evolutionary theory as it relates to the biology of populations for the derivation of the microtaxonomic (species-related) method; (2) argue that the ECSC and RC are redundant because their tests are conceptually inseparable from that of the BSC; (3) show through a critique of selected contributions from the literature that the mere delineation of a theoretical model for a particular taxonomic species notion is not sufficient and that it is unacceptable in the absence of an epistemology to examine critical components of such a proposed ontology; and (4) touch on some connections between theory and taxonomic practice.

The Biological Species Concept Is Part of Evolutionary Theory

Our quest to understand the evolutionary history of "becoming" is shaped, fueled, and delimited by the various concepts and techniques of science, and in the case of sexually reproducing organisms specifically by the various concept-bound methods of evolutionary biology and the earth sciences. This search for, and ordering of, knowledge about organisms into a cohesive objective data base is clearly both partly obtained and interpreted by the most tested available theories of evolutionary biology. Thus a major goal of taxonomy is to establish biologically significant patterns (species patterns and lineage patterns) of organisms in order to have, as a result, the most enduring building blocks for the construction of the edifice of the history of life. The attainment of such patterns without the constraints supplied by investment of understanding of biological and geological mechanisms and processes is very doubtful (Szalay and Bock, 1991).

In the science of taxonomy a name must stand for an objective aspect of the organic world in space and time, and the Linnaean system is widely accepted to serve that purpose, in spite of the fact that it was invented in preevolutionary days. A species *nomen,* like *nomina* on other categorical levels, should reflect unambiguous and tested realities; otherwise, its significance is questionable. While a species nomen is a limiting designation, evolutionary history at any one point in time is the result of *unbroken reproductive continuity.* It follows therefore, that the *temporal* and *spatial aspects* of a concept embodied in a nomen is always subject to an arbitrary (not artificial) taxonomic delineation, as the *vertical (temporal) continuity of life dictates* (see below). It does not matter how cleverly or ideally we supply definitions for theoretical concepts of taxa, we must deal with the objective world of organisms in space and time. Ontology without epistemology is of dubious value in science. These statements do not deny that a proper taxonomic delineation is connected to such foundational aspects of evolutionary theory as anagenetic and cladogenetic descent, as well as the nature of the information available.

The apparent terminological paradox [introduced by Bock (1979), reiterated in Szalay and Bock (1991), and discussed below] between the *evolving* species and the concept of lineage (which is the time path of species) defines the ontology of these two distinct concepts and sets the epistemological boundaries for their examination. This paradox is a necessary combination of the limits of language and the demands of testability. With this perspective on the relevant theoretical concepts, given a binomial taxonomic name, what range of meanings can it confer? The diversity of organisms, and their inability and/or unwillingness to interbreed, is an unshakable testimony to the *reality of species* for sexually reproducing organisms. Sympatrically occurring species are absolutely real in any instant of time. The BSC is based on this fact and on the fact of biological evolution, which manifests itself in the species category of sexually reproducing organisms. The (multidimensional) taxon, as it is delineated, is the taxonomist's best attempt to make objective at least part of this evolutionary reality. This attempt is especially difficult for allopatrically and allochronically distributed populations.

The elucidation of the relationship of the biological species concept to taxonomy was one of Mayr's (1942, 1963) major contributions to evolutionary biology. As noted in a recent text (Mayr and Ashlock, 1991, p. 26): "This species concept is called biological not because it deals with biological taxa but because the definition is biological. It utilizes criteria that are meaningless in the inanimate world." This concept continues to be misunderstood by some because of confusion about the nature of the connection of theory to practice. It is also often rejected because of its inconvenient contradiction of particular perceptions of *taxonomic analysis* from which a taxic (i.e., discontinuous) evolutionary theory is generated. The latter phenomenon is the source of an ideological power struggle, over the alleged conceptual independence of cladistic taxonomy from evolutionary biology (see comments in Szalay, 1991). The BSC is a tested foundation of evolutionary theory (Szalay and Bock, 1991, p. 10), and it is the base of valid methods in the practice of taxonomy that deal with sexually reproducing organisms.

Mayr's (1982) sweeping survey and analysis of the growth of these concepts serve as the necessary background for an appreciation of the nature of change in taxonomic practice after the establishment of evolutionism in the 19th century. It should be noted that evolutionary taxonomy is called *evolutionary* not because it deals with the products of evolution, which is the only professed connection of cladistic taxonomy with evolutionary mechanisms. It is called that because all known evolutionary mechanisms related to objective characters and populations, as well as their transformations, are potentially part of the testing procedures to which taxonomic hypotheses must be subjected.

I also add here that I reject the revisionist suggestion by de Queiroz (1988) that taxonomy did not become evolutionary until the rise of cladism, a view echoed by others recently even in the popular media. Based on this claim, de Queiroz attempted to rename the cladistic theoretical approach as *evolutionary taxonomy*. Evolutionary taxonomy, particularly at the species level, at its inception was based on the foundations of the Modern Synthesis, and in spite of its own evolution since then, often has been called in print "outmoded" and "simplistic" (to cite only the milder invectives). The charge of "simplistic" is, of course, by taxonomists who simply employ outgroup-biased parsimony algorithms, instead of testing taxonomic properties (as outlined in Bock, 1981), to arrive at taxograms (cladogram + axiomatically delineated taxa).

Mayr and Ashlock (1991, p. 27) state (their emphasis is retained) that it is ". . . quite clear that *the word* species *in biology is a relational term:* A is a species in relation to B and C because it is reproductively isolated from them." The BSC is a robust theoretical concept (Bock, 1986; Szalay and Bock, 1991) that is not only testable but far more often than not is tested in sympatric and synchronic situations (hence Mayr's original 1963 term for the species category as the *nondimensional species*). As developed by Mayr (1963), the taxonomic applicability of the concept results in the taxonomic species, the species taxon, a multidimensional unit (the term I prefer to the concept of species *individuality*) delineated by taxonomists. The species taxon is sometimes easy to delineate (e.g., sympatric coexistence of chimps and gorillas, or any other closely related species), but becomes increasingly difficult with an increase in distance and time from the nondimensional situation (see Costello *et al.*, on squirrel monkeys, this volume; Jolly on baboons, this volume). I want to emphasize the need for understanding the *theoretical concept* of biological species as it is *tested relationally*. I am less concerned with the definition of what a biological species is (for a recent account, see Bock, 1986).

Why the RC and ECSC Are Part of the BSC

There are two species concepts that are not, in my view, separable from the BSC. I will now briefly comment on the recognition concept and the ecological concept. Ideas developed around these add greatly to the realistic construction of multidimensional species taxa, but as independent concepts they can only be tested under the same criteria that validate the BSC.

The recognition concept (RC) and its juxtaposition to the isolation concept of the BSC has been a contentious issue recently. Paterson (1985) has given a full account of the recognition concept, or as it has become known, the species concept based on the specific mate-recognition system (SMRS). Others have dealt with this issue in detail (Bock, 1986; Coyne *et al.,* 1988; Mayr, 1988; Raubenheimer and Crowe, 1987; Szalay and Bock, 1991), and it appears that the RC is fully accommodated within the BSC, so such a new foundational concept is redundant. The RC does not solve the epistemological problem of allochronic and allopatric samples—the only test remains sympatry, and the specific taxonomic judgments necessary for the recognition of a multidimensional taxon remain the same. Paterson's attempt to tie the RC to the "reality" of punctuation merely detracts from his arguments that the recognition concept is a notion different from the BSC. Scoble (1985), while a supporter of the recognition concept, is somewhat less enthusiastic about its panacea-like qualities.

Andersson (1990) states that the BSC cannot deal with uniparentally reproducing organisms (for which it was obviously not intended when proposed) and that it ". . . fails to identify the driving force of speciation" (p. 375). Citing Van Valen (1976) as the first clear formulator of the ecological species concept, the arguments put forward by Andersson for the ECSC reveal that the concept is dependent on the assumption that ecological niches exist independent of the organisms themselves. The role of ecology in speciation is much older than a formally defined ECSC. Simpson's (1951, 1961) formulation of the evolutionary species concept contains a large dose of ecological characterization, and as clearly enunciated by Mayr (1982) and Bock (1986), species, under the BSC, can also be ecological units. In my view, the reality of ecological niches or adaptive zones independent of the organisms themselves—while certain heuristics of the words live on—has been effectively put to rest by Bock and von Wahlert (1965). It is quite obvious that close ecological scrutiny of organisms supplies additional and powerful arguments for the delineation of multidimensional species taxa.

The BSC and Terminology in Evolutionary Theory and Taxonomy

Let me now briefly review some aspects of the BSC that reflect the relations of concepts to evolutionary theory, and the consequences for the delineation of unambiguous terminology in taxonomy. Based on the foundations of the BSC and the arguments for the necessity of tested evolutionary theory-based concepts for the taxonomic method, Bock (1979), and more recently Szalay and Bock (1991), delineated the *theoretical nature* of species and lineage, their different ontologies, properties, and their significance in evolutionary theory (see also Bock, 1986; Mayr and Ashlock, 1991; and numerous references therein). First of all, it must be realized that ordering samples of organisms into species taxa and theorizing about "species" or "lineage" are activities in which the methods of delineation should follow the theoretical foundations. The taxa themselves are the hypotheses about the specific limits of the taxon. The taxa are relationally delineated locally and horizontally, particularly well when populations of extensive geographical areas are well connected, whereas they can be delineated only arbitrarily vertically. That is what taxonomic species must be if they are to be based on tested evolutionary theory.

Discrete boundaries between temporally continuous populations of lineages do not exist—no "species boundaries" were crossed in the existing lineages of life from their extinct or existing tips back to the origin of life. This type of statement usually rattles the psyche of some taxonomists, yet it should not, as it is an inescapable consequence deduced from the theory of Darwinian descent. This obviously does not mean that taxonomic species delineated anywhere in this continuum based on objective traits lack ontological validity. A slice of any sexually reproducing lineage at any one point in time is a biological species, but the different slices of this lineage cannot be considered the same or different species in a *theoretical sense* or in any epistemologically meaningful way (Bock, 1979). Taxonomically, however, they may well be considered the same or different taxa depending whether the usual gaps necessary for taxonomic species boundaries are present. Andersson's (1990) quote (epigraph, p. 21) expresses the operational aspect of the *taxonomic species* in time.

Szalay and Bock (1991, p. 16) sum up what they considered to be the significant aspects of the theoretical concepts of species and lineage for evolutionary biology, and subsequently for the practice of evolutionary (=phylogenetic) taxonomy:

> . . . species are the units of sexually reproducing organisms which consist of individuals which interact with the several causes of evolutionary change. Species are the units which are involved in the processes of evolution, namely phyletic evolution and speciation. Phyletic lineages are the time paths (the record) resulting from the outcomes of these processes in species taxa. Phyletic lineages are history and as such are not involved in the ongoing process of evolutionary change; they do not have a role in the process itself. Species, not lineages, evolve . . .

The reference to several causes of evolutionary change refers to natural selection, ecological factors, competition, and other numerous components, in addition to the genetic, epigenetic (developmental), and other ("internal") mechanisms and processes.

Taxonomic Species Notions and Some Consequences

Szalay and Bock (1991) have recently referred to a number of propositions relating to the nature of species as *taxonomic species notions,* in order to emphasize that biological taxonomic practice should follow the conceptual developments of *tested* evolutionary theory, and not the reverse. Of the several taxonomic species notions discussed here, the EVSC, first introduced by Simpson (1951, 1961) and nearly completely endorsed by Wiley (1978, 1981), is in fact a description of a part of a *lineage* (see discussion in Szalay and Bock, 1991), more specifically a (multidimensional) *taxonomic species* with a geologically appreciable time dimension. The other taxonomic species notion, the PSC, was specifically tied by Hennig (1966) to the delineation of species taxa in accordance with the history of furcation of lineages.

Hennig's taxic view of evolutionary history theoretically "unitizes" all branchings taxonomically. But because of its taxic view of evolutionary change, perception of all phenon differences (the data or evidence) has come to be attributed to dichotomous branching events. Consequently today there is general concern with the cladistic aspect of phylogeny but not with phyletics. Taxonomy and cladogenesis (and not phylogeny; see Introduction) have become synonymous in cladistic theory and method as a consequence of cladogram-related testing of either a "phylogeny" or the taxa themselves. Taxograms (taxa *cum* cladogeny expressed as cladogram) and the methods based on them, in a strict theoretical sense, are circular.

As noted, species taxa based on the PSC cease to exist when they speciate (Hennig, 1966; Ax, 1987). Not only does this "pseudoextinction" of evolving lineages make absolutely no sense, but also the epistemology to examine such an ontology simply does not exist. Such a taxonomic species notion is untestable in the real world of sample discontinuity. Yet the cladistic idea of species is widely followed by influential texts, such as Eldredge and Cracraft (1980) with the subsequent dilemma of the "monophyletic species."

The quandary of having all taxa holophyletic ("monophyletic" in the term-switched sense of cladistics), the species included, has caused scholastic havoc among cladistic theoreticians (students who deduce theory from the cladistic component of the evolutionary process) because the ancestor of *any taxon* cannot be holophyletic, a widely accepted notion. The concept of "monophyly" in regard to species, however, was rectified through the wedding of punctuation with the PSC, and when Eldredge and Novacek (1985) discuss "monphyletic species" they mean truly holophyletic entities. In either sense, the advocates of the "monophyletic species" notions have, I believe, taken an empirically unexaminable concept and on these have based both taxic edifices and macroevolutionary speculations (see Szalay, 1991). The multidimensional taxonomic species notion, and not an independently tested phylogenetic base, then, is the real foundation of the practice of clustering groups of similar specimens (phena) into "monophyletic species" (or even subsets of a species in order to logically follow Hennig's dictum at all costs; see, for example, discussions in de Queiroz and Donoghue, 1988, 1990a,b; Wheeler and Nixon, 1990).

In light of the preceding discussion, the current search, for example, by paleoanthropologists for taxonomic species representing different lineages in the existing record of hominids should, I believe, also deal with the conceptual and practical issues as means for framing their empirical research. Without careful modeling based on specifically targeted human intra- and interpopulational *morphological* variation, such an undertaking will reveal no evolutionary reality. Assumptions that geographically widespread, allopatric, gracile hominids of any one time plane had the same survival strategies, unlike the diversity seen in humans and even *Pan troglodytes,* are not warranted, I believe. Different assumptions will seriously affect taxonomic judgments. Taxonomically recognizable units (species) of the past are obviously not all representatives of new lineages. Similarly, cladistic techniques are beginning to produce "species" in the genus *Homo* that are entirely the consequences of cladistic assumptions about what a phenon is.

The Hennig principle (which states that synapomorphies are positive tests for the most recent cladistic relationship of taxa at a designated categorical level) is a powerful concept, and it is properly derived from the theory of descent. Synapomorphies must be tested, however, in order to add meaning to the practice (Szalay and Bock, 1991). Unfortunately there are no adequate process-related provisions for such testing in the post-Hennigian cladistic (taxic) conception of evolutionary history. It is important to remember that distilled Hennigian concepts offered by cladistic theory have emerged from not only the Darwinian heritage of concern with descent, but also from the tradition of typologically essentialistic, atemporal, *natural* groups of pre-Darwinian taxonomy which had no theory of causal change. Such theory is taxon- rather than trait- and process-based in its epistemology, and this renders testing phylogenies against independently established character transformations problematic. The theory derived from the taxic framework is also sharply curtailed by the discontinuities of living and fossil samples, a must for any practical taxonomy. A taxically based theoretical framework for the testing of evolutionary history, one that is not based on the independent testing against character transformations, is, in my view, simply the wrong one for the understanding of evolutionary change, and therefore so are the methods derived from it. Unlike functional–adaptive analysis (F-AA) (see Bock, 1981), the taxic approach does not provide appropriate conceptual foundations for analyzing evolutionary transformation. [Note that in the significant character-coding related analysis of Mickevich and Weller (1990, p. 161), they state, "Cladogram construction requires transmodal characters. An initial set of character state trees is needed to construct the first cladogram." They define transmodal characters on p. 168, as "Character state trees whose transformations are produced by the application of a model of character evolution." This statement represents, to me, another clear recognition that without independent analysis of character evolution, there are no evolutionarily meaningful phylogenetic patterns of taxa to be obtained; see Szalay and Bock (1991), for discussion.]

The taxic view of evolutionary history, rooted in the PSC, offers only taxon-based concepts, embodied in the slogan that depicts "decreasing science" from "cladogram, to tree, to scenario." The practice based on such a taxic conceptualization of evolution (e.g., Eldredge and Cracraft, 1980; Cracraft, 1983; Mishler and Brandon, 1987) distorts the nature and meaning of diversity described and analyzed by its aid. As has been stated many times for a variety of problems in science, the method chosen can become impervious to or obscure the important questions it is supposed to answer.

The pseudoextinction variety of the "monophyletic species" depends on two consecutive phyletic splittings or buddings (speciations), which are the (untestable) points in time (who can say when this occurred?) at which the "individuality" of a "species" begins and then suffers demise. Yet a case may be made that this ontology (a postulated reality) of successive lineage-producing isolation events, basically that of branching evolution, should be the foundation of a "species" notion. The theoretical foundation of a "species," then, would be based on the segment of a lineage between two speciation "events." In fact, if we had a *complete record* of speciations of sexually reproducing organisms and related events that result in phyletic evolution at the periphery without speciation (*the complete story of descent*), then this theoretical base for interpreting organic diversity would be an excellent one. This would be so regardless of how samples of continuous or separate lineages might appear (phenotypically), interact, or how they have become altered between or without any speciations. But again, with such knowledge who would need to worry about the problems of phylogeny and the other goals of taxonomy?

So what really is a speciation at the time it is happening? And if we do not have the blueprint for descent, then how do we test for the necessary speciations that define the postulated "spatiotemporally bounded entities" of "monophyletic species"? This is where the issue of epistemology about the postulated "monophyletic species" should come into play. But does it ever? The notion of "monophyletic species," the very base of Hennigian cladistic theory, and its macroevolutionary extension is, I believe, untestable. It asserts (by definition and without evidence) that recognized taxonomic species are the result of (1) either "birth" by speciation and "death" by speciation (and consequently should have the specified duration defined by these events) or (2) true holophyly of a lineage without splitting after its origin.

In either case, the definition of and access to examine (epistemology) such an "individual" is theoretically phylogeny based; it is said that in practice it can be delineated by morphological (taxonomic) criteria. Plainly, the ontology of the "monophyletic species" (either variety) is based on the cladistic component of phylogeny, but that ontology can be ascertained only through characters that are the products of phylogeny, *either* phyletic *or* cladistic.

It must be realized that if such a theoretical structure about "species" is to have objectivity or logical consistency, then severe strictures must be postulated that would set artificial limits on the nature of the evolutionary process. Characters, according to such a logically restricted view of evolution (which would, by definition, circumscribe the nature of a "monophyletic species-individual"), should not evolve continually (particularly at different rates), because if they did, then there would be no means by which "species birth–death" events could be ascertained. In cladistic practice and in several macroevolutionary theories, the use of the PSC does axiomatically mean that morphological change (the origin of diversity reflected in the fossil record) should be associated only with speciation events and not with phyletic evolution as well. From such patently false strictures of a taxonomic theory that is founded on taxonomic species notions, it follows that morphologically distinct, stratigraphically separated samples are not considered as parts of an unsplit lineage, but rather are routinely regarded as monophyletic (i.e., holophyletic) species.

The consequences of the ontologically perfectly acceptable notion of "units," which are the equal "segments between branchings," or of the holophyletic species, are important to ponder. From such ontologies it follows that clusters of life must become the very tests on which this theory is based. This, however, is circular, as the never complete data set (and its preponderant discontinuity) is the supposed expression of the split-related discreteness (punctuation) that these data are purported to test. Such an approach yields unacceptable theory because the notion of testing is subtly excised from the taxonomic practice itself. For example, Archibald (1990, p. 13A), while paradoxically endorsing the PSC in referring to speciation as pseudoextinction, has sounded concerns, stating that "Treating all discrepencies [in the fossil record] as true extinction may have a profound effect on interpreting [faunal] turnover, such as during mass extinctions." The unrecognized concern is really with the eradication of the habit of vertical comparisons in search for patristic relationships; exclusively cladistic analysis is followed even by many mammalian paleontologists, who often have access to a good fossil record.

It should not be overlooked that the replacement for the so-called "simplistic" neo-Darwinian synthesis is a theory of evolution, which for its interpretation of phylogenetic patterns depends on a taxic perspective of species, that is circular and untestable. The philosophically derived notion of "species as individuals" (whether holophyletic or not) is, of course, quite in harmony with the notion of either of the two varieties of the "monophyletic species." I will now examine a number of selected papers that consider the *theoretical issue of species* from a multidimensional taxic perspective, that is, based on a taxonomic notion of the species.

The ontology (but not the epistemology) of both the "species-individual" and the "monophyletic species" has been approached cautiously by Eldredge (1985), but more boldly by Eldredge and Novacek (1985). Eldredge's (1985; see also this volume) examination of the ontology of species concludes (p. 17) that "The significance of seeing species as 'individuals' in an evolutionary context, i.e., the role that species play in evolution, lies in their function as reservoirs of genetic information pertaining to organismic adaptations." He also states that (p. 17) ". . . construal of species as 'individuals' formally requires abandonment of the relatively simple structure of the modern synthesis in favour of a more expressly hierarchically structured evolutionary theory." For Eldredge (p. 17), "It is no great conceptual leap to *postulate* (italics supplied) that species might be construed as spatiotemporally bounded historical entities, i.e., as individuals in their own right, with births, histories, and deaths, thus analogous in these specific respects with organisms—the prototypical 'individuals.'"

In what I consider a potentially confusing contribution, Willman (1989) has proposed an extension of the BSC, as he understood it. Willman's contention that "The existence of a reproductive gap is a biological phenomenon and not related to space; therefore spatial distribution is not inherent in the biological species concept" (p. 97) demonstrates that he overlooked the relational and tested aspect of the theoretical definition of the species category as defined by Mayr. There is no testability of the relational concept without sympatry. Furthermore, in the same vein, his belief that (p. 98) "If the populations do not interbreed because of geographical separation they belong to one and the same species as long as reproductive isolation—the biological species criterion—has not occurred" is problematical. It is inexplicable, unless one assumes that "one knows" (based on the morphological, i.e., the taxonomic, concept) what has or has not occurred. Willman, in arguing for the superiority (of his particular understanding) of the BSC over an evolutionary concept (such as the taxonomy-dependent EVSC proposed by Simpson), has misconstrued the BSC. This unwarranted implication prompted him to redescribe the BSC in historical terms as follows (p. 106): ". . . biospecies originate by speciation . . . and they are always terminated by the same process if they do not go extinct. . . ."

Willman, then, has returned to the taxonomically based view of species by Hennig, overlooking the biological realities of testability and the persistence of *species taxa* regardless of the peripatric buds that may or may not have branched off them and did or did not merge back. Willman's praise for the BSC is apparently not for its foundational role in evolutionary theory of sexually reproducing organisms, but for his own conception of its connectability to the phylogenetic species notion. While he supports frequent paraphyly of species (and he should not if he consistently followed his definition; Willman, 1985), he does not come to terms with the necessity for the foundation of the species concept in evolutionary theory.

There have been a number of other poorly constructed criticisms of the BSC. In one volume, for example, Vrba (1985), who in several of her papers has clearly accepted the foundational role of the phylogenetic species notion, has also presented a somewhat confusing view concerning the ontology of species. She states in the introductory chapter of a volume entitled *Species and Speciation* (p. ix) that

> The concept that species are discrete 'limbs' in the sexual part of life's phylogeny makes sense. After all, the notion that all life forms on earth are related and situated on a single tree of life is a hypothesis at the highest level of generality. It is appropriate that the boundaries of species in time and space should have a one to one relationship with discrete parts of the tree of life.

It appears that Vrba has confused the notion of "limb" with segments of branching (splitting or budding), which may or may not characterize such limbs, a minor *lapsus*. She has, however, demonstrated the use of the taxonomic perspective that simply assumes that data represent theory, rather than against data we test all theories. Vrba has tried carefully (p. xvii) to evade the limits imposed by epistemology concerning "species individuals" based on the PSC. She notes the inappropriateness of the notion of "progress" (a term that stands for a number of distinct concepts) for biological evolution. The idea that teleological explanations are inappropriate in evolutionary theory and causal analysis is widely accepted among many evolutionists, myself included. Vrba, however, implies that a notion of "progress" is at the core of the view ". . . that life's trends are smoothly continuous and progressive, and that arbitrary subdivision along the branches may be labelled as successive 'species.' Thus, most trends in this view can be explained by simple extrapolation from progressive selection within species to better and better adapted organisms towards the pinnacles of trends." She apparently criticizes the multidimensional species taxon based on the BSC and implies that it is based on some ill-conceived notion of old-fashioned gradualism—the best that I am able to understand. This is a transluscent straw man (in place of what the tested theoretical structure of the related, but distinct, concepts of species and lineage are in *evolutionary biology*), and it cannot be a substitute for tested theoretical contributions (yet to appear) that would give *evolutionary legitimacy* either to the evolutionary or the cladistic species notion(s). Nor can the following statement serve that purpose:

> Rather, the current concepts of species and speciation suggest lineages whose "chunks of the genealogical nexus" . . . are precariously connected by the results of disruptive

> episodes. The changes accompanying speciation are chaotic, each deviation from the established fertilization system is a disaster, from the "point of view" of ancestral organisms. (p. xvii)

There is an often puzzling, obtuse, and steadily growing literature related to species-level taxonomy. It may be characterized as contributions to a "cladistic species theory," for the lack of a better designation. Meaning is difficult to decipher in these papers because of the chaotic use of the ever-shifting taxonomic jargon and the fairly regular redefinition of terms to stand for different concepts. The various concepts themselves are often difficult to judge as to whether or not they correspond to some scientific reality. The contours of a language derived from the concepts of cladistic theory to the exlusion of the other components of evolution (see Fig. 1) simply fail to match the realities of the biological world. This steady growth of scholasticism, as it relates to the theoretical dichotomous unfolding of life, is at best interesting, but, as I see it, irrelevant to evolutionary biology. Amid such efforts in cladistic taxonomic theory, the agonizingly equivocating attempts to "evolutionize" (but also to retain the taxically based precepts) usually result in bewildering attempts to perpetuate the conflation of evolutionary theory and a specific theoretical view of taxonomy (e.g., de Queiroz, 1988; de Queiroz and Donoghue, 1988, 1990a,b; Wheeler and Nixon, 1990). Some of these contributors, while using the same language often with different meaning, predictably come to logically opposing views. The unfulfillable demands of the classificatory stricture of holophyly, spawned in a universe viewed as empirically atemporal, and its exegesis to the various species notions, appears to be a burden on students of cladistic taxonomic species theory. For example, de Queiroz and Donoghue (1988, p. 335), after their reexamination of the species problem, conclude that

> . . . each of the species concepts we have considered designates units that can be used as terminal taxa and each one also has consequences . . . Finally, if all species are to be monophyletic, then some organisms are not parts of species, although in contrast with species concepts based on interbreeding, these organisms are not asexuals but members of ancestral populations. In considering these consequences, a given reader may see some of the insurmountable difficulties and others as simple facts of life. However, which consequences are viewed as problems and which ones as facts will depend on one's point of view. This is the species problem.

Is it? Or rather, is it a mere choice that substitutes for tested theory? Suffice it to say that theoretically oriented cladists continue to seek solutions to the "species problem" that are divorced from evolutionary biology, closely heeding the plea of Eldredge and Cracraft (1980) for the separation of "phylogenetic patterns and the evolutionary process."

To conclude this section I would like to note that there are few areas of confusion as unnecessary as those created by purely scholastically induced issues in evolutionary theory and taxonomy. These include the biologically irrelevant philosophical issues of cladistically delineated "individuality," and the metaphor of "birth and death" of "species" used in several recent macroevolutionary studies (e.g., Vrba and Eldredge, 1984; Eldredge and Novacek, 1985). The heuristics themselves rest on a cladistically conceived taxic view of the history of life; no room is permitted for phyletic considerations and the obvious lineage continuity required by the dicta of evolutionary theory. Van Valen (1988) has recently presented strong arguments as to why the philosophically delineated concepts of "individuals" and "classes" are inappropriate imports into evolutionary theory. He proposed the applicability of the concept of "fuzzy sets" (derived by him from set theory) for what I understand to be recognizable taxonomic species or morphospecies, and for any other taxon. This is, of course, distinct from the concept of the "monophyletic species-individual" (of either variety). In fact, the taxonomic units of nature (multidimensional species taxa, which are the results of inspection, biological judgment based on quantitatively supported evidence, and taxonomic grouping of data), as outlined by Simpson, Mayr, and others, and persuasively reaffirmed by Van Valen, do provide us with a comparative measure of taxonomic and adaptive diversity of organisms. These empirically determined units, based on the guidance provided by living models based on the BSC (see below), are free of the burden of an implied cladistic significance of phena taxonomically delineated as species. Thus such species taxa, the fuzzy sets, stand for *evolutionary differentiation,* whether phyletic or branching related. This view of species as the crucibles of evolutionary change is identical to the *multidimensional species* taxon established by Mayr (1942, 1963; for recent discussions, see Bock, 1986; Mayr and Ashlock, 1991; Szalay and Bock, 1991).

Species, Speciation, and Accountability in Constructing Species Taxa

I did not discuss speciation in this chapter, yet the concept is of some importance. Speciation is a theoretical concept from evolutionary biology, rather than from the taxonomic sphere. It is important to realize (and I am again pleading for the clear conceptual, theoretical, meaning of terms) that there is no "morphospeciation," or "speciation through phyletic evolution without branching," in spite of the fact that this is often implied. This kind of conceptualization (and language) should not be adopted, regardless of its venerable history beginning with Darwin and followed by many taxonomists, particularly paleontologists. The phenomenon of speciation is simply the eventual result of either the reproductive or genetic isolation of different living populations. Successive populations (while isolated in time) cannot meet such a definition of testing. The speciation process reaches its full conclusion when the respective lineages are genetically isolated from one another (see particularly Bock, 1986). While speciation is obviously a real phenomenon, it is the testable outcomes, the species themselves, which are inseparable from the BSC.

It should be emphasized that speciation, in terms of *evolutionary* realities, as opposed to its taxonomically expressed results, is often through peripheral isolates (the metaphor of *budding* has been used, in contrast to the more general concept of "splitting"), i.e., when peripatric populations become increasingly differentiated from the remaining bulk of the other populations. At some point the once cohesive populations no longer constitute the single gene pool of sexually reproducing organisms (see Mayr and Ashlock, 1991). The new bud(s) of species(s) may give rise to a radiation of lineages. It is also quite feasible that these may merge back into the mother lineage, in which case no new species (hence lineage) is formed. While such peripheral isolation occurs, the *original source* may (or may not) continue without comparable or even appreciable evolution; it may (or may not) remain the same unit that can be taxonomically delineated. If no change occurred in the main populations, then this mother lineage would continue to be unmistakably recognized as a single *taxonomic* species (see especially Mayr and Ashlock, 1991). If appreciable change is coupled with a discontinuous fossil record, then several taxonomic species may be (should be) recognized without an overburden of cladistic beliefs. Microtaxonomy should be kept independent of phylogenetic analysis. As argued above, basing taxonomic method on taxonomic species notions tied to the assumption of holophyly (coupled either with real extinction or pseudoextinction) is an enterprise that lacks testable theoretical foundations, unless workers *assume* punctuation at that mysterious "geological moment" called speciation (see discussion in Szalay and Bock, 1991; see also Kellog, 1988, on the "weak links in the chain of arguments from punctuation to hierarchy").

Taxonomists, particularly paleontologists, rather than considering species taxa to be the untestable "monophyletic species" (of either variety) should carefully delineate taxonomic species *based on a combination of all characters in relation to other relevant species known.* To do that one must use theoretical sophistication and extensive field information, based on both intra- and interpopulational variation in carefully chosen and researched appropriate living models of biological species. A morphological understanding in the fullest sense is also essential (see Bock, 1981), using judgments about adaptation and ecology, in other words, in the most eclectic and astute biological manner possible. This microtaxonomy should not, under any circumstance, be conflated with branching-based cladistic analysis, even if some or all of the same objective information is used to test phylogenetic (phyletic and cladistic) hypotheses. Furthermore, genetics, biogeography, ontogenetic and other polymorphisms, and adaptations are part of the context of examination and enter most phases of species recognition and delineation. I maintain that the allegedly process-free scientific delineation of (biological) species and phylogenetic patterns (*sensu* Eldredge and Cracraft, 1980) has yet to be demonstrated (Szalay and Bock, 1991).

There are no magic algorithms to yield universal answers for the unequivocal recognition of species, and there never will be, as long as lineages will differ from one another and continue to be allopatrically and allochronically distributed. There is no escaping theory- and process-rooted judgment in taxonomy, i.e., tests that are *probability,* and not *parsimony,* based. I believe, as do many others, that in paleontology it is proper taxonomic practice to consider a series of stratigraphically superimposed samples that are distinct and not excessively overlapping, and change through time, as discrete taxonomic species (chronospecies); such evidence is usually convincing of the notion that these represent an uninterrupted lineage (as argued above, however, one can never know about peripheral isolates and their history of connectivity). *This is in no conceivable way contradictory to the nondimensional BSC, as some have thought.* While the latter is the theoretical entity with its potential to be tested and examined, the former is its spatiotemporal specific extension based on judgments related to the probability of variation in extant models of biological species. The reality of the *multidimensional taxonomic species* in the fossil record is tied to the complex phenetics resulting from the temporal interplay of phyletics and cladistics. As noted in the introduction, cladistics can supply only part of the evolutionary history of evolving species. While there are always difficulties in expressing a lineage taxonomically, that is a problem of practical delineation.

I suggest that recognition and construction of species taxa and lineages should be based on a number of activities. These procedures depend on whether the organisms are living or fossil, or both. For fossils, the application of particular species models should come from geographically, phylogenetically, and adaptively (ecologically) appropriate neospecies. Often knowledge about such living models is nonexistent (but see several, noncladistic, microtaxonomic endeavors in this volume). Quite simply, we should create (i.e., actively pursue the gathering of) such knowledge (see Costello *et al.*, this volume). The undertakings to delineate the parameters of biological species taxonomically, and only *subsequently* their phylogenetic (evolutionary) relationships, should involve the following:

1. Establishment of species identity, which can be tested beginning at a specific local level (either reproductive or genetic isolation).
2. Conduct of research on the nature of form–function variations (both morphological and behavioral) in tested biological species as these relate to adaptation.
3. Application of the obtained patterns of variation in well-established, unambiguous species taxa (extended and interconnected populations; and biogeographical or synspecies as well, in contrast to those in individual allopatric or parapatric taxonomic groupings of the same genus) to spatiotemporally isolated living and fossil samples; and, only after the issues invariably raised in 1–3 have become settled, carry out
4. Analysis of vertical and branching relationships of species (after their identity has been independently established) by *noncircular phylogenetic research* based on character analysis (Bock, 1977, 1981; Hecht and Edwards, 1977; Neff, 1986; Szalay and Bock, 1991).

Points 1–3 are obviously not sequential activities, but their logical subordination in analysis is critical in the defense of any conclusion.

Conclusions

Finally, I would like to reiterate, as noted by Szalay and Bock (1991, p. 10), that, "Clearly, the species concept, which is a theoretical idea, does not belong to systematics or taxonomy—that is to the basic theory of systematics whatever that may be—and never has. Rather, the species concept is an integral part of basic evolutionary theory which is a corollary of our firm position that all of taxonomy, including classification and phylogeny, is grounded on evolutionary theory." Platonic essentialism, typology, and the original morphological (taxonomic) species concept were virtual synonyms of systematics *cum* taxonomy prior to

Darwin. The BSC, however, has grown out of the union between Darwinian descent and population thinking, which culminated during the laying of the foundations of the modern synthesis of the several subfields of evolutionary biology. Any *multidimensional species taxon* that paleontologists delineate may represent separate and divergent lineages, or segments of a single lineage—the time paths of evolving species. While such taxa should be firmly based on the approaches suggested above, their connections and phylogenetic interpretations in relation to other species taxa are subject to the testing procedures used in historical–narrative explanations (see Bock, 1981; Szalay and Bock, 1991). As repeatedly noted, phylogenetic analysis should not be conceived as an axiomatic corollary of assumptions about "monophyletic species."

Species-level taxonomy is critical, and species taxa make up the structural framework on which we base our understanding of the evolutionary history of organisms. Nevertheless, this microtaxonomy cannot be equated with phylogenetic analysis, which properly consists of sundry hypothesis formulation and testing procedures related to characters, taxonomic properties, and the taxa themselves.

ACKNOWLEDGMENTS

I want to express my gratitude for being asked by the organizers and editors, Bill Kimbel and Lawrence Martin, to cochair the symposium from which this volume originated. I also thank Walter J. Bock and David Dean for their thorough and critical reading the manuscript. I am grateful to Bill Kimbel for his helpful suggestions in editing the manuscript. PSC-CUNY grant no. 661222 has helped defray costs of the manuscript preparation.

References

Andersson, L. 1990. The driving force: species concepts and ecology. *Taxon* **39:**375–382.

Archibald, J. D. 1990. Metaspecies and modes of speciation in the analysis of faunal turnover of Judithian–Clarkforkian mammals. *J. Vert. Paleo.* **9:**13A.

Ax, P. 1987. *The Phylogenetic System. The Systematization of Organisms on the Basis of their Phylogenesis.* John Wiley & Sons, Chichester.

Bock, W. J. 1977. Adaptation and the comparative method, in: *Major Patterns in Vertebrate Evolution,* M. K. Hecht, P. Goody, and B. M. Hecht (eds.), *NATO Adv. Inst. Ser. A* **14:**57–82.

Bock, W. J. 1979. A synthetic explanation of macroevolutionary change—a reductionistic approach. *Bull. Carn. Mus.* **13:**20–69.

Bock, W. J. 1981. Functional-adaptive analysis in evolutionary classification. *Am. Zool.* **21:**5–20.

Bock, W. J. 1986. Species concepts, speciation, and macroevolution, in: *Modern Aspects of Species,* K. Iwatsuki, P. H. Raven, and W. J. Bock (eds.), pp. 31–57. Tokyo Univ. of Tokyo Press.

Bock, W. J. and von Wahlert, G. 1965. Adaptation and the form-function complex. *Evolution* **19:**269–299.

Coyne, J. A., Orr, H. A., and Futuyma, D. J. 1988. Do we need a new species concept? *Syst. Zool.* **37:**190–200.

Cracraft, J. 1983. Species concepts and speciation analysis. *Curr. Ornith.* **1:**159–187.

de Queiroz, K. 1988. Systematics and the Darwinian revolution. *Phil. Sci.* **55:**238–259.

de Queiroz, K. and Donoghue, M. J. 1988. Phylogenetic systematics and the species problem. *Cladistics* **4:**317–338.

de Queiroz, K. and Donoghue, M. J. 1990a. Phylogenetic systematics or Nelson's version of cladistics. *Cladistics* **6:**61–75.

de Queiroz, K. and Donoghue, M. J. 1990b. Phylogenetic systematics and species revisited. *Cladistics* **6:**83–90.

Futuyma, D. J. 1986. *Evolutionary Biology,* 2nd ed. Sinauer Associates, Inc., Sunderland, MA.

Eldredge, N. 1985. The ontology of species, in: E. S. Vrba (ed.), *Species and Speciation,* pp. 17–20. Transvaal Museum Monograph No. 4. Transvaal Museum, Pretoria.

Eldredge, N. and Cracraft, J. 1980. *Phylogenetic Patterns and the Evolutionary Process. Method and Theory in Comparative Biology.* Columbia University Press, New York.

Eldredge, N. and Novacek, M. J. 1985. Systematics and paleobiology. *Paleobiology* **11:**65–74.

Ghiselin, M. T. 1991. Classical and molecular phylogenetics. *Boll. Zool.* **58:**289–294.

Hecht, M. K. and Edwards, J. L. 1977. The methodology of phylogenetic inference above the species level, in: M. K. Hecht, P. Goody, and B. M. Hecht (eds.), *Major Patterns in Vertebrate Evolution, NATO Adv. Inst. Ser. A,* **14:**3–51.

Hennig, W. 1966. Phylogenetic systematics. University of Illinois Press, Urbana.

Kellog, D. 1988. "And then a miracle occurs"—weak links in the chain of arguments from punctuation to hierarchy. Biol. Phil. **3:**3–28.

Mayr, E. 1942. *Systematics and the Origin of Species.* Columbia University Press, New York.

Mayr, E. 1963. *Animal Species and Evolution.* Harvard University Press, Cambridge.

Mayr, E. 1982. *The Growth of Biological Thought. Diversity, Evolution, and Inheritance.* Harvard University Press, Cambridge.

Mayr, E. 1988. The why and how of species. Biol. Phil. **3:**431–441.

Mayr, E. and Ashlock, P. D. 1991. *Principles of Systematic Zoology,* 2nd ed. McGraw-Hill, New York.

Mickevich, M. F. and Weller, S. J. 1990. Evolutionary character analysis: tracing character change on a cladogram. Cladistics **6:**137–170.

Mishler, B. D. and Brandon, R. N. 1987. Individuality, pluralism, and the phylogenetic species concept. Biol. Phil. **2:**397–414.

Neff, N. A. 1986. A rational basis for a priori character weighting. Syst. Zool. **35:**110–123.

Paterson, H. E. H. 1985. The recognition concept of species, in: E. S. Vrba (ed.), *Species and Speciation,* pp. 21–29. Transvaal Museum Monograph No. 4. Transvaal Museum, Pretoria.

Rand, A. 1966. *Introduction to Objectivist Epistemology.* The Objectivist, Inc., New York.

Raubenheimer, D. and Crowe, T. M. 1987. The recognition species concept: is it really an alternative? South Afr. J. Sci. **83:**530–534.

Scoble, M. J. 1985. The species in systematics, in: E. S. Vrba (ed.), *Species and Speciation,* pp. 31–34. Transvaal Museum Monograph No. 4. Transvaal Museum, Pretoria.

Simpson, G. G. 1951. The species concept. Evolution **5:**285–298.

Simpson, G. G. 1961. *Principles of Animal Taxonomy.* Columbia University Press, New York.

Stuessy, T. F. 1980. Cladistics and plant systematics: problems and prospects. Introduction. Syst. Bot. **5:**109–111.

Stuessy, T. F. 1987. Explicit approaches for evolutionary classification. Syst. Bot. **12:**251–262.

Szalay, F. S. 1991. The unresolved world between taxonomy and population biology: what is, and what is not, macroevolution? J. Hum. Evol. **20:**271–280.

Szalay, F. S. and Bock, W. J. 1991. Evolutionary theory and systematics: relationships between process and patterns. Z. Zool. Syst. Evolut.-Forsch. **29:**1–39.

Van Valen, L. M. 1976. Ecological species, multispecies, and oaks. Taxon **25:**233–239.

Van Valen, L. M. 1988. Species, sets, and the derivative nature of philosophy. Bio. Phil. **3:**49–66.

Vrba, E. S. 1985. Introductory comments on species and speciation, in: E. S. Vrba (ed.), *Species and Speciation,* pp. ix–xviii. Transvaal Museum Monograph No. 4. Transvaal Museum, Pretoria.

Vrba, E. S. and Eldredge, N. 1984. Individuals, hierarchies and processes: toward a more complete evolutionary theory. Paleobiology **10:**146–171.

Wheeler, Q. D. and Nixon, K. C. 1990. Another way of looking at the species problem: a reply to de Queiroz and Donoghue. Cladistics **6:**77–81.

Wiley, E. O. 1978. The evolutionary species concept reconsidered. Syst. Zool. **27:**17–26.
Wiley, E. O. 1981. *Phylogenetics. The Theory and Practice of Phylogenetic Systematics.* J. Wiley & Sons, New York.
Willman, R. 1985. *Die Art in Raum und Zeit.* Parey, Berlin and Hamburg.
Willman, R. 1989. Evolutionary or biological species? Abh. Naturwiss. Ver. Hamburg **28:**95–110.

Primates and Paradigms 3

Problems with the Identification of Genetic Species

J. C. MASTERS

Introduction

Sir Charles Lyell opened his first notebook on "the species question" in 1855 (Wilson, 1970). A hundred years later, Mayr (1957) edited a volume entitled *The Species Problem.* Now, 34 years on, Eldredge has opened this volume with a title that emphasizes, once again, that crucial step in understanding natural organization: "What, if Anything, Is a Species?" In recognition of this probing tradition, I have constructed this chapter around a series of questions and possible answers relating to the application of species concepts, particularly to primates. I should point out that while Eldredge has tended to deal with the ontological side of the question—what we think species really *are*—this contribution is directed more toward epistemological issues—how do we identify the kinds of species that we believe to exist?

These two levels of inquiry bear an obvious relationship, although there have been recommendations that the recognition of pattern in nature be kept separate from theories of process in order to avoid circular reasoning (see the next section). The stance taken in this chapter is that pattern and process exist in a dialectical relationship to one another, as described by Levins and Lewontin

---

J. C. MASTERS • Population Genetics Laboratory, Museum of Comparative Zoology, Harvard University, Cambridge, Massachusetts 02138. *Present address:* University of the Witwatersrand, Wits 2050, South Africa.

Species, Species Concepts, and Primate Evolution, edited by William H. Kimbel and Lawrence B. Martin. Plenum Press, New York, 1993.

(1985), where each level of understanding informs and modifies the other. I believe, with Szalay (this volume), that this relationship is basic to the hypothetico-deductive method, and I trust this chapter will illustrate its heuristic value.

To begin from a relatively uncontroversial starting point, species may be viewed as clusters in phenotypic and genotypic space. Taxonomic species concepts emphasize the phenotypic aspect, while biological or genetic species concepts emphasize the genotypic; specifically, genetic concepts identify species in terms of gene exchange (or, in the case of allopatric populations, the potential for gene exchange). This chapter concentrates on genetic species, although phenotypic concepts will also be considered. I shall not discuss ecological or evolutionary species concepts, since they have been adequately dealt with elsewhere (e.g., Eldredge, 1985; Kimbel, 1991).

Why Genetic Species?

The best answer to this question lies in the ontological nature of species. Evolutionary biologists are crucially interested in the origin of diversity, and (apparently) irreversible change; genetic species are the measurable units of this change. In the words of Mayr (1963, p. 11), "The origin of new species, signifying the origin of essentially irreversible discontinuities with entirely new potentialities, is the most important single event in evolution." Futuyma (1987, p. 467) has provided a cogent and succinct argument for why this should be so:

> . . . the evidence from geographic variation tells us that character evolution does not require speciation. But in the absence of speciation, much of the geographical variation we observe is ephemeral, leaving little imprint on evolution in the long term. . . . over even moderately short spans of evolutionary time (tens or hundreds of thousands of years), the habitats to which populations are adapted shift, often over large distances, in consequence of climatic change. . . . the consequence is occasional (perhaps at intervals of thousands of years) but massive gene flow (i.e., hybridization), on a scale far greater than the trickle that often characterizes populations at equilibrium. . .

More eloquently still, Futuyma (1987, p. 470) stated, "If we envision an adaptive landscape . . . of character values, successive speciation events are the pitons affixed to the slopes of an adaptive peak."

Genetic species thus derive their significance from their role in the process of evolutionary diversification. To proponents of pattern (or transformed) cladistics, however, their "process dependence" renders them undesirable (Nelson and Platnick, 1981; Cracraft, 1987; Patterson, 1988; Nixon and Wheeler, 1990). These authors, among others, recommend a species concept based on the distribution of character states, that is, a strictly phenotypic species definition, which has only the loosest relation to our ontological view of species:

> . . . a process is that which is the cause of a pattern. No more, no less. Pattern analysis is, in its own right, both primary and independent of theories of process, and is a necessary prerequisite to any analysis of process. (Nelson and Platnick, 1981, p. 35)

They support the phylogenetic species concept (PSC), which will be discussed further in a later section.

The chief argument of the pattern cladists against genetic concepts and other "process-dependent" theories, as I mentioned earlier, is one of circular reasoning: "If the causal explanation is to be convincing and efficient, the pattern is better not perceived in terms of the explanatory process" (Patterson, 1988, p. 72); and "When the perception of pattern is process-dependent, that pattern may not reveal alternative processes" (Nixon and Wheeler, 1990, p. 212). The beauty of cladograms constructed without reference to process theories, Patterson tells us, is that they make no assumptions about evolutionary relationships. They do, however, make a prediction, that "further samples of information will show the same pattern" (p. 79).

I question the utility of such an approach to explaining natural organization. The essence of the hypothetico-deductive method is falsifiability, which requires strong predictions that may relatively easily be refuted. While I stand with Patterson in decrying nonfalsifiable pattern-process arguments (see the next section), one of Patterson's pattern cladograms carries a very limited range of further research problems. A particular pattern can indeed be falsified by a new set of characters that fail to support the original character distribution, but then what? Which set of characters do we trust? Presumably, we go out and collect some more until we have a range of distributions, one of which shows a greater degree of character consensus than the others. This will generate a phenetic classification, upon which the evolutionary biologists must structure their theories of process.

Alternatively, one can take a theory of process that makes falsifiable predictions and apply it to a range of patterns that can be falsified, allowing pattern and process theories to shape and respond to each other. This is my preferred method.

How Do We Identify Genetic Species?

It is not unreasonable to expect that, as a general rule, genetic discontinuities should be reflected in the divergent morphologies of species. (I use the term *morphology* here in its more restrictive sense, to apply to those gross features most commonly available to taxonomists: pelage, skull and skeletal features, and aspects of anatomy and physiology.) Unfortunately, there are many examples that violate this expectation. Many species show high levels of variability, sometimes to the point of easily distinguishable geographic races (Mayr, 1963). This is true of several species of nocturnal strepsirhines (Hill, 1953). On the other hand, this suborder also includes many sibling or morphocryptic species that have only recently begun to be investigated (Olson, 1979; Masters, 1985, 1988; Zimmermann *et al.*, 1988; see also Nash *et al.*, 1988, and papers therein).

Obviously, gross morphology alone is not always a dependable indicator of genetic species. Does this imply that data concerning gene frequencies would be more reliable?

In the early days of molecular biology, several authors supported the notion that five hypothetical stages in a gradualistic model of speciation [local populations, subspecies, semispecies, sibling (morphocryptic) species, morphologically different species] could be identified according to levels of genetic differentiation (Selander and Johnson, 1973; Avise, 1974, 1976; Ayala, 1975). In a classic example of circular reasoning, they arranged populations displaying such levels of divergence into ordered series and used these to support their speciation model. This hypothesis is not testable, since there is no result that cannot be shoe-horned into it.

It has also been strongly challenged. Mayr (quoted in Avise, 1974) argued that "an approach that merely counts the number of gene differences is meaningless, if not misleading," and "it is the total system of developmental interactions, the totality of feedbacks and canalizations, that makes a species." The inclusion of sibling species as a stage in divergence has been undermined by several authors. Lambert and Paterson (1982) reviewed the literature on morphological resemblance and genetic distance, and found no consistent relationship between the two measures for genetic species. King and Wilson's (1975) study of the genetic distance between humans and chimps is a graphic illustration of this decoupling. More importantly, Lambert and Paterson concluded that "no set amount of genetic divergence can be found to accompany speciation events" (1982, p. 296), which calls the entire speciation model into question.

As a consequence of their observations, Wilson and his colleagues (King and Wilson, 1975; Wilson, 1976) have suggested that changes in structural genes (like those monitored in enzyme electrophoresis) are of little importance to organismal evolution. They have suggested instead that "[e]volution at the organismal level may depend primarily on regulatory mutations, which alter patterns of gene expression. Mutations affecting the arrangement of genes on chromosomes may be a common source of these altered patterns of gene expression" (Wilson, 1976, p. 233). While it is certainly true that many closely related species differ karyotypically (White, 1973, 1978), Charlesworth *et al.* (1982) have argued that there is no necessary correlation between morphological evolution and/or speciation, and chromosomal rearrangement. Similarly, Sites and Moritz (1987, p. 154) have described claims that chromosomal rearrangements have a direct effect on the phenotype, as "poorly substantiated." The existence of homosequential species in some groups [e.g., the speciose Hawaiian *Drosophila* (Carson, 1970)] and chromosomal polymorphisms in others [e.g., the prosimian primates (Ying and Butler, 1971; de Boer, 1973; Stanyon *et al.*, 1987)] suggests caution in the interpretation of karyotypic data.

It is therefore apparent that data gleaned from studies of morphology, genetic distances, and karyology are all less absolute than we would wish them to be as indicators of genetic species. If speciation is indeed a genetic event, why is it so difficult to characterize? Or, to put it another way,

Is There a Common Factor in the Origin and Maintenance of Genetic Species?

According to the traditional neo-Darwinian interpretation of genetic species, there is indeed such a common factor: reproductive isolation. Dobzhansky's original definition reveals what he regarded as the necessary and sufficient characters for species identification: "Species are . . . groups of populations the gene exchange between which is limited or prevented in nature by one, or a combination of several, reproductive isolating mechanisms" (Dobzhansky, 1951, p. 262). This opinion was echoed by Mayr (1957, p. 6): "[Species] are more succinctly defined by isolation from non-conspecific populations than by the relation of conspecific individuals to each other. The crucial species criterion is thus not the fertility of individuals, but rather the reproductive isolation of populations." Furthermore, "[isolating mechanisms] are perhaps the most important set of attributes a species has, because they are, by definition, the species criteria" (Mayr, 1963, p. 89). Thus, our research program is obvious: In order to identify genetic species, we have only to identify the factors or "mechanisms" responsible for reproductive isolation.

As will become apparent, the identification of isolating mechanisms is not clear cut and constitutes a major problem in the application of the Dobzhansky–Mayr Biological Species Concept.

What Are Isolating Mechanisms and How Do They Arise?

Reproductive isolating mechanisms comprise a mixed bag of characters, including morphological, behavioral, ecological, physiological, and genetic features, that present barriers to gene exchange. In the light of the current argument, they have two important characteristics: (1) They have a species-wide distribution, since they are the species-defining criteria*; (2) according to the classical view, they originate as "a by-product of genetic differences accumulated between (allopatric) populations" (Charlesworth *et al.*, 1982, p. 482).

These two features do not fit easily into a single model of the origin of species. On the one hand, in order for a species to be recognizable as such, we require it to display a suite of characters that is maintained with reasonable consistency throughout its range. On the other hand, we expect these characters to have been acquired according to some haphazard, ill-defined genetic process, termed *divergence in allopatry*. In my view, it is not surprising that, half a century on from the birth of the New Synthesis, we still lack a formal theory of the population genetics of speciation (see Lewontin, 1974; Dover, 1982).

If speciation is viewed as occurring relatively rapidly in small populations (Mayr, 1954; Carson, 1968, 1970, 1971, 1982; Templeton, 1980), then the fixation of species-wide characters is not a problem. On the other hand, such models are no more explicit about the kinds of changes occurring at speciation than are the classical divergence models. In Dobzhansky's original formulation of his species concept, isolating mechanisms had coherence because they were the products of positive selection for traits that would limit gene exchange (Masters *et al.*, 1987a; Masters and Spencer, 1989). For example,

> . . . an accumulation of genic differences between the parental forms does not necessarily produce either hybrid incompatibility or sterility. The genetic factors responsible

*It has been argued that it is reproductive *isolation* and not the mechanisms themselves that are important; hence different mechanisms may function in various parts of the species' range. However, this is contrary to Mayr's meaning (above), and renders the concept of isolating mechanisms—and the species they define—even more nebulous and less testable than it is at the best of times.

> for the production of these isolating mechanisms appear to constitute rather a class by themselves. (Dobzhansky, 1935, p. 352)

Since then, however, the ability of natural selection to produce isolating mechanisms has been challenged (Paterson, 1978, 1981, 1985; Lambert and Paterson, 1984; Lambert *et al.*, 1984; Spencer *et al.*, 1986), effectively removing their one unifying attribute. The characterization of isolating mechanisms is now much more problematic. Must we accord any trait that contributes to the reduction of gene exchange the status of an isolating mechanism? What if reproductive isolation is incomplete, and hybrids show an intermediate degree of fertility; how much isolation is necessary for species status? Genetic barriers to fertility will be acted upon negatively by natural selection and so will have an ephemeral existence; should these be regarded as indicators of "irreversible change"? How closely related do species have to be for them to require isolating mechanisms?

Such difficulties and inconsistencies have led several workers, including myself, to support an alternative genetic species concept, i.e., Paterson's recognition concept (RC).

What Is the Recognition Concept of Species?

Paterson (1985, p. 25) has proposed the following definition of genetic species: "We can . . . regard as a species that most inclusive population of individual biparental organisms which share a common fertilization system." The fertilization system consists of those characters under direct selection to facilitate syngamy (i.e., gametic fusion). In mobile organisms, an important component of this system is the means by which the animals attract and recognize each other—the specific-mate recognition system (SMRS). The important point to note, which is ignored by proponents of the idea that the RC is accommodated within or inseparable from the isolation concept (IC) (e.g., Szalay, and Rose and Bown, this volume), is the role envisaged for natural selection. According to the IC, the main selective pressures are negative ones, operating upon those individuals that mismate. In the case of RC, the main selective force operates positively to achieve mating and syngamy. This distinction may appear too subtle to be important to anyone but a population geneticist, but its consequences are significant.

Elements of the SMRS will be closely fitted to the environment in which speciation occurred and to the organisms' way of life within that environment. For example, forest animals will tend to use loud vocalizations for distance communication and either quieter calls or highly contrasting visual signals (or a combination of both) for close recognition. The frequency structure and temporal patterning of the calls will reflect the attenuation properties of the habitat (Waser and Brown, 1986) and the range that the signal is intended to travel. Where visual signals are involved, e.g., in animals like ourselves and other diurnal anthropoids, aspects of external morphology will be useful indicators to taxonomists of species identity. A striking example of this is to be found in the African guenons, where facial patterns play an important role in intraspecific communication (Kingdon, 1988) and in guenon systematics. However, in animals that employ mainly vocal or olfactory signals, like the nocturnal prosimians, a change in the system of specific-mate recognition will not necessarily imply changes in morphology, and morphocryptic species will occur (Masters, 1985, p. 15).

Just how useful is this prediction for the identification of fossil species? The answer to this must be, "useful in some instances, in others not at all." Although some visual signals will be encoded in skeletal structures, for example, the large canines that form an important part of intraspecific communication in savanna baboons, others will only be apparent in the pelage and soft tissues. Thus, species identifications employing elements of anterior dentition are likely to be more reliable in groups such as baboons than in organisms that do not use such teeth in display. Turner and Chamberlain (1989) have compared systems of fossil taxonomy that do and do not employ SMRS-related characters and found those that do to be less contested.

An important aspect of the recognition concept for systematists is that it predicts that the fertilization system will be subjected to strong stabilizing selection, since any aberrant signalers or receivers will simply not mate. As a consequence, the essential components of the SMRS will show low levels of variation over time and space. This means (*contra* Szalay, this volume) that it should be possible to assign specific status to allopatric populations, provided that enough is known about their reproductive biology. Furthermore, since the function of mate recognition is to achieve syngamy rather than to avoid mismating, the presence or absence of sympatric heterospecifics will be irrelevant to the structure of the SMRS. This is another area in which the RC differs markedly from the IC. In order for true isolating mechanisms to be effective, they must generally undergo geographic variation, and Mayr (1963, p. 492) has stated that ". . . any character ever described as distinguishing species is also known to be subject to geographic variation." Additionally, "[w]here no other closely related species occur, all courtship signals can 'afford' to be general, nonspecific, and variable" (Mayr, 1963, p. 109).

There is some evidence to support the RC expectation of stasis in the primate loud calls that have been analyzed. Waser (1982) demonstrated both temporal and spatial stability in the "whoopgobbles" of African mangabey species. Populations of grey-cheeked mangabeys (*Cercocebus albigena*) separated by more than 2000 km have indistinguishable calls. Similarly, loud calls of crested mangabeys (*C. galeritus*) differ only in minor details, although the populations occur on opposite sides of the continent and have been separated at least since the early Pleistocene. The stability of the *C. galeritus* calls was independent of the presence or absence of sympatric mangabey species (Waser, 1982). Marshall and Marshall (1976) studied the territorial songs of gibbons and found them to be consistent throughout the species' ranges, regardless of whether they were sympatric with or geographically separated from other taxa. My own studies of loud calls in greater galagos (Masters, 1985, 1991) indicate stability in call structures in conspecific, allopatric populations.

Not all aspects of an organism's reproductive behavior, or indeed all aspects of a particular signal, will be of equal significance to the function of mate recognition. Hence, not all aspects will be expected to show the same limited variability. Noncrucial elements may be highly variable, resulting in regional dialects or varieties, as well as the encoding of individual variation. In my study of greater galago loud calls (Masters, 1991), species recognition appeared to be encoded in the most energetic frequency bands as well as the duration of the individual units of which the calls were comprised, and possibly the interunit intervals. Individual recognition appeared to be contained in the relative intensities of higher level harmonics. These predictions can be tested using artificially constructed calls and playback experiments.

An important result for demonstrating the difference in approach between the RC and the IC is the fact that the call element that showed the lowest variability in the two species, and hence the greatest significance for mate recognition, was the most energetic frequency band. This element was also extremely similar between the species, with a mean of 1.39 kHz (n = 27) for *G. crassicaudatus* and 1.32 kHz (n = 10) for *G. garnettii*. Investigators searching for premating isolating mechanisms emphasize those aspects of the communication system that are most dissimilar between species, as indeed they should if these traits have evolved to prevent hybridization. If, on the other hand, traits have evolved to secure syngamy, the crucial aspect is not that they should be divergent but that they should be received as transmitted. Thus, authors such as Ryan and Wilczynski (1988), who claim to have demonstrated regional variation in mate recognition systems, may have investigated subsidiary elements that have little significance for mate recognition. It is notable that these authors, in particular, only investigated the dominant frequencies of advertisement calls; all other aspects were held constant.

The RC has encountered a range of responses from the evolutionary community, many of which have been critical (e.g., Butlin, 1987; Donoghue, 1987; Raubenheimer and Crowe, 1987; Templeton, 1987; Coyne *et al.*, 1988; Mayr, 1988; Verrell, 1988; Chandler and Gromko, 1989). Among primatologists and anthropologists, however, its reception has been more cordial, and it has been applied to primate species problems from prosimians through hominids, with varying degrees of success.

How Effective Has the Recognition Concept Been in Identifying Primate Species?

Identification of signals appropriate to mate recognition in primates is not quite as straightforward as it may sound. For a start, recognition of conspecific mates does not appear to be a short-term process taking place directly before the breeding season (Masters, 1985, p. 84). Rather it appears to be part of a broader, more complex social system, lasting throughout the breeding cycle of the population. Eventual acceptance or rejection of a mate may well have more to do with familiarity, and thus individual recognition, than simple conspecific mate recognition [e.g., see Dixson's (1978) mating experiments with familiar and unfamiliar males]. This is true even among prosimians, as Charles-Dominique (1977, p. 218) has described:

> As a rule, one tends to associate courtship with the act of copulation, since the two components of behaviour are closely linked in numerous groups of animals. But among the lorisids courtship serves to establish social bonds between individual males and females which will not in fact mate with one another until much later. . .

Tattersall (1989, p. 122) has similarly indicated that, in the captive breeding of various lemurs, the wider social context may influence the breeding compatibility of couples during the mating season significantly.

Thus, mate-recognition signals must be interpreted in a far broader sense in primates than is usually implied. They include those communicatory signals responsible for both the establishment and maintenance of the social structure within which sexual recognition and mating take place (social bonding), as well as the more intimate and less obvious signals exchanged just prior to copulation. In fact, mate recognition *per se* has already been accomplished once this advanced stage is reached. The close interdependence of social and sexual signals has been discussed further for prosimians by Schilling (1979) and by Brockleman and Gittins (quoted in Groves, this volume) for gibbons.

Prosimians

The first application of the RC to a primate species problem was that of Masters (1985, 1986, 1988, 1991), who investigated the genetic status of populations within the greater galagos, now generally accepted as comprising the species *Galago crassicaudatus* and *G. garnettii*. Results from a morphometric study of craniodental anatomy (Masters and Lubinsky, 1988), comparative karyology (Masters, 1986; Masters *et al.*, 1987b), and the electrophoresis of erythrocytic allozymes (Masters and Dunn, 1988) all pointed to a genetic discontinuity, which was then tested against potential specific-mate recognition signals. Although chemical signals provided tantalizing evidence of their involvement in both mate and individual recognition, their volatile and short-lived nature created logistic impediments to further analysis. The loud calls of the two taxa proved far more amenable and demonstrated the heuristic value of the recognition approach to species identification (see above). Call structures also carried implications for the types of habitat frequented by the two species.

In the light of the success of the above study, several other workers have applied similar techniques to species problems within the Galaginae (see below). Crompton (1989) became an RC convert because of its nonrelational nature, which enabled him to reinterpret his data on posture and locomotion in prosimians in terms of adaptive processes *within* populations, rather than in terms of competition or niche partitioning *between* them. He noted that, among galagos, species-specific locomotor patterns appear to have become fixed along with systems of specific-mate recognition and may hence serve as useful species indicators.

Using a combination of the approaches of Masters (1985) and Crompton (1989), Courtenay and Bearder (1989) investigated the identity of the "small forest bushbabies in northern Malawi." They compared morphological characters (pelage coloration and body measurements) of museum specimens with locomotor patterns and the structure of advertisement calls of animals in the wild, and identified the animals as *G. zanzibaricus.* There are, however, some major problems with this study. First, the morphological sample of the population in question comprised a single animal collected some decades ago (exact date unknown). Second, it is not clear how the authors came to be sure that this specimen was from the same taxon as the one they were observing in the field, or indeed if the field animals were all members of a single taxon, since the study sites were distributed along the length of Malawi. A more rigorous study is obviously called for, preferably one involving genetic techniques.

A similar study was undertaken by Harcourt and Bearder (1989), comparing *G. moholi* in South Africa with *G. zanzibaricus* in Kenya. This study involved quantitative analyses of morphological characters (head-body length, weight) and frequency of multiple births, as well as more anecdotal comparisons of pelage coloration, locomotor patterns, call structures, social organization, and karyology. Once again, a detailed analysis including genetic data as well as some control for habitat differences would assist in applying this information to the more general systematic problems besetting the lesser galagos.

Zimmermann *et al.* (1988) conducted a quantitative analysis of the call repertoires in Senegalese and South African lesser galagos, which offered strong support for the recognition of these two taxa as distinct genetic species. As was true of other studies of species-specific vocalizations in primates, the loud calls proved most useful for species discrimination. Zimmermann and her colleagues argued that the differences in fundamental frequencies and harmonic composition observed between the calls of the two taxa reflect dissimilarities in laryngeal and supralaryngeal structure, as well as in the corresponding functional areas of the brain. An alteration in the signals employed in intraspecific communication, hence, has repercussions throughout the structure of the organism.

In addition, Zimmermann (1990) conducted a comparative bioacoustic analysis of eight galagine taxa, using both univariate techniques and a stepwise discriminant analysis. She repeated the analysis of Masters (1985, 1991) on *G. crassicaudatus* and *G. garnettii,* and confirmed the previous result. In an extension of her earlier work she compared the loud calls of *G. moholi, G. senegalensis,* and *G. zanzibaricus,* again confirming their recognition as distinct species. The three other species investigated were *G. alleni, G. matschiei,* and *G. demidoff.* The discriminant analysis revealed eight more-or-less discrete clusters, leading Zimmermann to the conclusion that a high degree of species specificity existed in the acoustic variables of loud calls that she had examined. She viewed this as strong support for the usefulness of loud calls in the identification of galago species, particularly in the presence of morphological similarity.

The success of the RC in identifying galago species suggests that it may be equally rewarding if applied to the prosimians of Madagascar. This, however, has still to be done. Although Tattersall (1989) has expressed his support for using the RC to identify genetic species in primates, he has not used it in his own systematic research. His most recent review of lemur systematics (Tattersall, this volume) relies strongly on the isolation concept, and on postmating isolation in particular. Data on the mating activities of captive animals confined in small cages and isolated from important environmental and social contexts, as well as information on hybrid viability or fertility, can be highly misleading as to species identity (Paterson, 1988). My personal belief is that a study of specific-mate recognition signals would reveal several morphocryptic species among the nocturnal and crepuscular lemurids, particularly within the highly variable species, *Lemur fulvus*. In the genus *Lepilemur*, several morphologically similar taxa that had previously been distinguished only on cytogenetic grounds have been shown to display differences in their long-distance advertisement calls (Petter and Charles-Dominique, 1979).

Anthropoids

The RC has yet to be applied to the Old and New World monkeys, although there is every reason to suspect that such an exercise would be fruitful. For example, Marler (1978) reported that the long-range spacing calls of adult male *Cercopithecus* monkeys are the most species specific of their vocal repertoire and indicated that this may be a general principle in primate vocal communication. The work of Waser (1982, cited above) on *Cercocebus* supports such an idea. An investigation of the loud calls emitted by male black-and-white colobus monkeys, considered together with data on cranial dimensions, coat pattern, and geographic distribution, led Oates and Trocco (1983) to affirm the presence of five genetic species within the group. Fujita (1987) has demonstrated the significance of visual clues to conspecific recognition in macaques.

Groves (this volume) suggests that facial and pelage patterns, as well as vocalizations, are likely to be components of the SMRSs of gibbons. This interpretation of gibbon calls is strongly supported by the work of Marshall and Marshall (1976, see above) and Mitani (1987), who confirmed through playback experiments the significance of loud calls for species recognition in *H. muelleri* and *H. agilis albibarbis*.

The only author to have applied the RC to humans, or indeed to hominids in general (used here to include fossil and extant great apes and hominins), is Turner (1985, 1986; Turner and Chamberlain, 1989). His thesis is as follows: Despite claims to the contrary (Masters *et al.*, 1984), ". . . it is abundantly clear that humans today recognize mates from anywhere within the modern range of variability, and that the diversity in external physical appearance is not reflected in disjunctions of the modern SMRS" (Turner, 1986, p. 422).

This led him to the conclusion that the visual component of our SMRS is unlikely to be based on gross external features. He deduced further evidence for the deemphasis of vision in hominin mate recognition from the absence of major skeletal differences between *Homo* species. On the other hand, since chimps and hominins are morphologically quite distinct, this suggested that "an SMRS with a visual component largely based on gross external features may be a primitive condition for the hominoids. Our epigamic features are in this respect a primitive retention" (1986, p. 422). Similarly, a relatively large morphological gap between the australopithecines and *Homo habilis* "implies the greater importance at that time of the gross visual component of the SMRS" (1986, p. 427).

Turner posited the development of language as a potential reason for the deemphasis of visual signaling in hominins. As the underlying cause of this behavioral shift, he suggested the adoption of clothing during the intensely cold glacial conditions that marked the Middle Pleistocene: "Once hidden by clothing, if only partly, the human figure offers considerably fewer over-cues and signals to potential mates" (1986, p. 427). This implies a northern origin for most of our species-defining traits, and thus presumably for *Homo sapiens*.

Although I support Turner's general approach and applaud his attempt to apply the RC to a complex problem, I must take issue with aspects of his interpretation. First, the fact that characters used by some as indicators of "race" are not part of our SMRS, does not imply that we deemphasize gross morphology in mate recognition. Who can deny the significance of body form and posture in sexual attraction? Second, the acquisition of a new communication channel—language—does not necessitate the reduction of a previous modality. If there is one thing that marks our system of intraspecific communication, it is the number of channels that are employed at any one time, and the amount of redundancy encoded in our signals. A person who is blind or deaf, or even both, may be severely disadvantaged in social situations, but is certainly not incapable of emitting and receiving signals. When it comes to communication, we are supreme generalists. (Note, however, that this does not imply that our ancestors must also have been such generalists.)

Third, I do not believe that our highly species-specific epigamic characters are simply a hangover from earlier, language-free days. They are intimately associated with our social structure. Finally, his hypothesis is at odds with what we know of hominid habitats both now and in the past. Since chimps live in densely vegetated areas, and many workers believe australopithecines did too (Ciochon and Corruccini, 1976; Vrba, 1979; Stern and Susman, 1983; Susman *et al.*, 1984; Cadman and Rayner, 1989), a system of communication that relied on visual signaling would not have been very appropriate, except for relatively close individual identification. Distance identification is more likely to have been achieved by loud whoops and cries. As hominins moved out of the forests, they may have been expected to rely more—not less—on visual signals.

Words of Caution

Gratifying as the primatological reception of the RC has been, it is clear that not all of the authors who employ it are fully conversant with Paterson's theory or with the aspects that distinguish it from other species concepts. For instance, there is a strong tendency to conflate SMRSs with premating isolating mechanisms. Nash *et al.* (1988, pp. 503–4), in attempting to apply the RC to prosimian species, have suggested "In particular, for nocturnal taxa, commonly used *visible* morphological features may suggest that the forms are the same, but behavioral features suggest differences in 'mate recognition systems' (Paterson, 1985), for example, vocalizations, which would keep the gene pools separate." Foley (1991, p. 414) has described the RC as follows: "More recently Paterson . . . has proposed that as species form *through reproductive isolation* . . . then only characteristics relating to mate selection and fertilization . . . are relevant to defining species morphologically" (my emphasis). Groves (this volume) treats the SMRS throughout as a set of mechanisms to protect a gene pool from introgression.

There is also a suggestion among some would-be RC supporters that the demonstration of differences in loud calls obviates the need for further genetic validation of species; for example, "[the present study] shows how it is possible to use behavioral data, particularly vocalizations, to identify a species without having to collect further specimens or to disturb the animals in the wild" (Courtenay and Bearder, 1989, p. 32). I do not believe that many adherents of the RC would be happy with such an approach. While behavioral differences, particularly relating to intraspecific communication, will certainly indicate the possibility of separate species and the direction for future research, they need to be backed up by other kinds of relevant information—where possible, genetic information.

Although Crompton (1989, p. 105) correctly identified environmental stress and population fragmentation as the precipitating factors in speciation according to the RC, his subsequent genetic model is not a valid deduction from the theory. He states, "In such small, stressed, populations events such as chromosome inversions can rapidly become fixed. The communication and other systems by which mates are recognized become disrupted, *and massive changes to the genome are required to stabilize them in new configurations*" (my emphasis). Neither Paterson nor any of his followers has, to my knowledge, suggested that a change in the system of mate recognition requires major readjustments to the genome. In the light of our knowledge of species with homosequential chromosomal banding patterns (discussed earlier), this would be unwise in the extreme. In fact it is not even clear that Crompton's chosen mechanism for change—heterochrony—need be linked to major chromosomal reorganization.

Crompton also applauded the ability of the RC "to explain the process of allopatric speciation" (p. 105). Would that this were the case. What the RC *has* done, is to provide a set of heuristic predictions with which the mysterious process of speciation may be probed. After the fall of the theory of speciation by reinforcement, the IC has had little or nothing to offer in this area. [Note: Rose and Bown (this volume) have dismissed the RC because of its dependence on a particular process of speciation. Exactly what process this refers to—other than the requirement for allopatry—I am not sure. I am also surprised that this should be a criticism.]

By far the most unlikely interpretation of the RC that has come to my notice is that of Aich *et al.* (1990). On the subject of the phylogenetic affinities of *Theropithecus*, these authors suggest, "According to Paterson [1978; 1985] *Papio* is closely related to *Theropithecus*, because there are hybrids between *T. gelada* and *P. anubis* in the wild" (p. 130). I believe I am safe in saying that Paterson has never commented on the relationships of baboon genera in his life, but their statement contains something more. I interpret it as suggesting that since the members of the two taxa share sufficient elements of their SMRSs to bring about hybridization, they must be closely related. I dispute the logic of this argument. As I pointed out at the beginning of this section, the SMRS in primates is a complex system of signals and responses enmeshed in a social context of familiarity, rank orders, and even, in some cases, friendship (Smuts, 1985). Burton and Chan (1987) have described cross-species infant care between long-tail macaque infants (*Macaca fascicularis*) and stump-tail macaque "baby sitters" (*Macaca thibetana*) in a polyspecific group on the Kowloon Peninsula, Hong Kong. They suggest that such behavior might facilitate interbreeding between macaque species "through the enlargement and amalgamation of the communication systems of the two groups" (p. 754). Though the factors that facilitate hybridization in primates are certainly interesting and deserving of study, there is no reason to believe they will be important to phylogenetic reconstruction. Indeed, such interbreeding may well be dependent on the presence of unchanged plesiomorphic aspects of the system of specific-mate recognition (Rosen, 1979).

Kimbel (1991) rightly pointed out that the issue of gene exchange between genetically, morphologically, ecologically, or behaviorally divergent populations is a crucial one for genetic species concepts, and one over which proponents of the IC and RC are in substantial agreement. The particular example to which he drew attention is the hybridization between anubis and hamadryas baboons in the Awash National Park, Ethiopia (Phillips-Conroy and Jolly, 1986; Jolly, this volume). "According to the Recognition Concept," Kimbel wrote, "such taxa must be considered conspecific because they share a common fertilization system, despite divergence in SMRS-related morphology and behavior" (p. 363).

My initial response on reading this was to agree with Kimbel: Where there is large-scale gene exchange, there is usually one genetic species. On reflection, though, I must disagree. If the SMRS of a primate species is indeed a complex of sexual signals and responses embedded in a matrix of social organization, as was argued earlier, then *P. anubis* and *P. hamadryas* must be viewed as distinct species. These populations do not "share a common fertilization system," as Kimbel intimates, since the SMRS is an integral part of that system. As Phillips-Conroy and Jolly (1986, p. 347) have described, ". . . the behavioral and anatomical specializations of anubis and hamadryas baboons comprise two distinct functional complexes adapted, respectively, to wetter and more arid habitats." While anubis inhabits woodlands and riverine forests, and its social organization comprises large, matrilineally based troops in which males and females associate in temporary consortships, hamadryas occurs in semidesert scrub and rocky ravines, and forms permanent, male-centered, one-male groups with strong male–female bonding.

Contrary to some RC critics (e.g., Coyne *et al.*, 1988), the occurrence of hybridization does not fly in the face of RC predictions (see Masters and Spencer, 1989). This particular example is a problem for the IC, however, since, despite the obvious existence of two different adaptive complexes, no isolating mechanisms have developed, in line with Paterson's (1978) prediction. According to Phillips-Conroy and Jolly (1986), the cline has in fact flattened rather than steepened since it was first investigated, because of the successful integration of hybrid males.

How, then, would an RC advocate interpret the situation? There are two important considerations here. First, mating between anubis and hamadryas individuals is not at random. In the words of Phillips-Conroy and Jolly (1986, p. 345), "Although the first hybrids in the Awash were obviously the product of matings between the parental types, we suspect that within a generation or two most hybrids were the product of matings in which at least one, and often both, parents were themselves hybrids. Certainly this seems to be the case today." Second, there is a marked asymmetry in the interbreeding between these taxa, which is reminiscent of asymmetries in hybridization between *Drosophila* species (Kaneshiro, 1976; Watanabe and Kawanishi, 1979; Lambert, 1984). Most hybrids are born to hamadryas fathers and anubis or hybrid females; reciprocal matings are rare. This implies that the hamadryas SMRS contains a substantial proportion of the components of the anubis SMRS, along with some added extras.

Phillips-Conroy and Jolly suggest that *P. hamadryas* is descended from an anubis-like ancestor, which would accord well with the hypothesis of Watanabe and Kawanishi (above). However, Kimbel (1991) cites data to suggest otherwise, and it may simply be that the shared components are plesiomorphic, brought into prominence because the area of available habitat is limited, and because many baboons show a propensity for extensive intertroop migration.

Kimbel has suggested that the PSC (as described earlier) is a more appropriate tool for decoding this particular problem, because it would not underestimate the amount of evolutionary diversification that has occurred. According to my analysis, the RC also indicates the presence of two species and thus yields a conclusion more consistent with that of the phenotypic PSC (Kimbel, 1991) than with that of the genetic IC. My penultimate question, hence, is as follows.

Are the Recognition Concept and the Phylogenetic Species Concept Mutually Exclusive?

To evolutionary biologists interested in mechanism, species are the crucial units of interest because speciation is a fundamentally different event from the processes occurring at other systematic levels. To those interested in systematic pattern, however, a species is a taxon like any other and should be treated as such. Ehrlich (1961) argued this position from a phenetic viewpoint; more recently, Nelson (1989) has argued it for pattern cladistics. Hence, although these two schools have had their differences, they are in agreement on one issue at least: genetic species concepts are not useful to systematists. I shall contest this below.

The phenetic species concept was based on morphological distance (Sokal and Crovello, 1970) but is not in common use. The dominant phenotypic concept of recent years is one amenable to cladistic analysis, the phylogenetic species concept (PSC). There are several versions of the PSC [see Nixon and Wheeler (1990) for a summary], and a consensus definition has yet to stabilize. The most recent example derives from Nixon and Wheeler (1990, p. 218), who state that a species is "the smallest aggregation of populations (sexual) or lineages (asexual) diagnosable by a unique combination of character states in comparable individuals (semaphoronts) [i.e., individuals of comparable sex and ontogenetic stage]." The crucial element in all PSC definitions is their allusion to the smallest diagnosable systematic units, although some authors (e.g., Donoghue, 1985; Mishler and Brandon, 1987) have included a requirement for monophyly.

The quest for a strictly pattern-oriented species concept has obviously been a contentious topic for some time. In 1896, Jordan argued cogently that in fact there was no such thing as a truly morphological species concept:

> The systematist puts *male* and *female* together in one species, and hence makes at once the concession that his term "species" is not a purely morphological one, but that the higher criterion of the term is of a physiological kind. (Jordan, 1896, p. 436)

Eldredge (this volume) has made a similar point. Obviously, then, some element of process will be allowed to intrude upon the pattern definition. The question is, how much? Gaffney (1979) has argued that only two process theories of the highest generality are necessary in order to test phylogenetic hypotheses: first, that taxa have evolved from a common origin, and second, that new taxa are often characterized by evolutionary novelties (synapomorphies). However, I believe this underestimates the difficulties in recognizing pattern in nature. When investigating a particular phylogenetic hypothesis, many other assumptions will be required and, if not stated explicitly, will be smuggled in implicitly. For example, several lower level "process" theories are implicit in the recognition of homology. I believe our interests are best served by making all assumptions explicit.

Not all characters will be equally important in the maintenance of species cohesion, or equally useful in the identification of patterns of irreversible change. As I have indicated above, a theory of process provides important information as to the reliability of a particular character. Failure to distinguish inter- from intraspecific variation confounds the search for generalities that may provide clues to an understanding of evolutionary mechanism. Theories of both pattern and process stand to benefit from a dialectical approach.

Genetic species concepts have been criticized by advocates of phenotypic concepts on two main grounds. The first of these entails their depiction by Mayr (1942, p. 120) as "actually or potentially interbreeding natural populations." Sokal and Crovello (1970, p. 149) of the phenetics school described the IC as "neither operational nor heuristic nor of any practical value" on the grounds that "potential interbreeding" was impossible to test. More recently, several cladists have slated the IC because the "potential to interbreed" constitutes a plesiomorphic character and will yield inaccurate estimations of evolutionary units (Bremer and Wanntorp, 1979; Rosen, 1979; Donoghue, 1985; Cracraft, 1987; Mishler and Brandon, 1987; De Queiroz and Donoghue, 1988; McKitrick and Zink, 1988). Several of the latter dismissed the RC as suffering from the same defect.

I believe that both of these groups have abandoned genetic species concepts prematurely, because of an overly restrictive interpretation of "interbreeding potential." Sokal and Crovello's criterion for species membership was interfertility, and they stated at great length the impracticability of measuring the presence or absence of this feature between all potential breeding partners. This is not a valid criterion. For example, 1 in 10 couples in the United States is not capable of producing children; according to Sokal and Crovello, this would be because, under the IC, they were heterospecific.

The cladistic criticism of "interbreeding potential" as pleisomorphic appears similarly to depend on the absence of postmating isolating mechanisms, although I have not seen this stated explicitly anywhere. If my reading of their argument is correct, then this criticism falls away when genetic species are approached from the recognition perspective (Masters and Spencer, 1989). To an RC advocate, "potential to interbreed" means possession of a common system of fertilization mechanisms. In Paterson's (1988, p. 69) words, "[t]hese characters are diverse and include such characters in the mating partners as the design features of the gametes, those determining synchrony in the achievement of reproductive condition, the coadapted signals and receivers of mating partners, and their coadapted organs of gamete delivery and reception." No doubt, because of the diversity of elements in this system, some aspects will be plesiomorphic; on the other hand, because speciation entails a change in the system, some aspects must also be apomorphic. A person interested in species biology or speciation will focus on the system as a whole; a systematist intent on treating a species like any other rank in the systematic hierarchy may focus only on the apomorphic elements in the system. Under these circumstances, the RC and PSC will identify the same species, and the answer to the question, "Are the RC and the PSC mutually exclusive?" will be "No."

The second criticism leveled at genetic species concepts is their inapplicability to uniparental or asexual organisms. A morphological species concept could be applied universally. In defense of the PSC, Mishler and Brandon (1987, p. 407) make the following point:

> One of us . . . happens to work on a genus of mosses . . . in which frequently sexual, rarely sexual, and entirely asexual lineages occur. The interesting thing is that the asexual lineages form species that seem comparable in all important ways with species recognized in the mostly asexual lineages and even in the sexual lineages. It just happens in this case that potential interbreeding or lack thereof seems of little or no importance in the origin and maintenance of diversity. The application of the PSC here is able to reflect an underlying unity that the biological species concept could not.

There is a logical problem with this statement. Because these authors have observed a variety of systems displaying an apparently similar degree of stability, they have assumed that the cause of the stability in the various systems must be the same, and "reflect[s] an underlying unity." In reality, the genetics of the systems are in no way comparable. In obligate sexual organisms, genomes are recombined and reshuffled each generation; in asexual taxa, mutation is the sole source of genetic change. The fact that sexual species retain a similar degree of stability to asexual lineages is indeed remarkable, but the mechanisms behind this similarity are unlikely to have much in common.

The PSC may well have the advantage of wider application over the RC. However, "species" thus defined may have a tokogenetic relationship to one another rather than a phylogenetic one (see Eldredge, this volume). Where systematists have access to information concerning mating relationships and organismal biology, I recommend that they employ a combination of the PSC and RC, as outlined above. Hypotheses regarding specific identity must be corroborated by biology, or else remain hypotheses. For paleontologists, the situation is somewhat different. My final question, therefore, is as follows.

How Useful Is the Recognition Concept to Paleontology?

Fossil species must be discriminated on morphological criteria. As I mentioned earlier, in some cases these characters will include elements of the SMRS, and the morphologically defined taxa will coincide with genetic species (e.g., Vrba, 1980). In other instances, however, mate-recognition signals may not have been encoded morphologically, or the appropriate pieces of morphology may not have been preserved. In long-extinct groups there may not even be extant close relatives to which to refer in attempting to identify SMRS components. In such situations, the RC will be of no help whatsoever.

I do not believe that the strength of the RC lies in its ability to justify the taxa particular paleontologists wish to recognize, despite Turner and Chamberlain's (1989) attempts to use it in this way to validate the lumping of African and Asian *Homo erectus*. There is nothing in the RC, for example, that would argue against Tattersall's (1986) recognition of several Middle-to-Late Pleistocene species of *Homo*. If anything, the obvious morphological differences displayed by these taxa should cause RC advocates to support his proposal, since vision is so important in hominin signaling. However, it would be too easy for the RC to become yet another adaptational "just-so story" in interpreting the communicatory behavior of extinct species, for such an exercise to be undertaken without caution and reference to scientific method.

The strength of the RC lies in its ability to identify and to make predictions about the processes underlying patterns in the fossil record, which may then be subjected to further investigation. The RC predicts stasis over time and space in characters associated with specific-mate recognition, but not necessarily in traits unconnected with this function. It hence provides a plausible explanation for why some lineages appear to evolve gradually (see Rose and Bown, and Krishtalka, this volume), while others support punctuated equilibria. It depends on the part of the organism that forms the basis of taxonomic allocation.

The RC views environmental change as the driving force behind speciation. This means sister species will be expected to show ecological differences in addition to the observed divergence in their mate-recognition systems. It also implies that sister species are unlikely to be close ecological competitors. The fit between the SMRS and the habitat at speciation suggests an interesting way of using paleoecological data to infer biological information on fossil taxa. A reconstruction of the kinds of habitat changes that occurred could provide clues to the kinds of signaling systems that were in use and to the reliability of morphological data in identifying genetic species.

Since paleontologists are restricted in their investigations to morphological characters, a phenotypic concept like the PSC is best suited to their pursuits. Although it is true that genetic divergence will be underestimated in the case of morphocryptic species and overestimated with respect to polytypic species by the use of this concept, such errors are unavoidable in paleontology. Workers developing theories of mechanism from PSC-defined species should be conscious of the potential pitfalls and should exercise appropriate caution.

I conclude by agreeing with Szalay (this volume) that the RC is not a panacea. However, it *is* an effective cure for a specific biological problem.

Acknowledgments

I thank Niles Eldredge, Clifford Jolly, Lawrence Martin, Richard Rayner, and Frederick Szalay for their constructive comments on an earlier version of this paper. Thanks also to Lawrence Martin and William Kimbel for inviting me to take part in their symposium, and to Ruth Hubbard, Richard Lewontin, James Maki, and Richard Rayner, without whom that participation would not have been possible. Hugh Paterson continues to be a source of guidance and intellectual stimulation, and his encouragement is gratefully acknowledged.

References

Aich, H., Moos-Heilen, R., and Zimmermann, E. 1990. Vocalizations of adult gelada baboons (*Theropithecus gelada*): acoustic structure and behavioural context. *Folia Primatol.* **55:**109–132.
Avise, J. C. 1974. Systematic value of electrophoretic data. *Syst. Zool.* **23:**465–481.
Avise, J. C. 1976. Genetic differentiation during speciation, in: F. J. Ayala (ed.), *Molecular Evolution,* pp. 106–122. Sinauer Associates, Inc., Sunderland, MA.
Ayala, F. J. 1975. Genetic differentiation during the speciation process. *Evol. Biol.* **8:**1–78.
Bremer, K., and Wanntorp, H.-E. 1979. Geographic populations or biological species in phylogeny reconstruction? *Syst. Zool.* **28:**220–224.
Burton, F. D., and Chan, L. K. W. 1987. Notes on the care of long-tail macaque (*Macaca fascicularis*) infants by stump-tail macaques (*Macaca thibetana*). *Can. J. Zool.* **65:**752–755.
Butlin, R. K. 1987. Species, speciation, and reinforcement. *Am. Natur.* **130:**461–464.
Cadman, A., and Rayner, R. J. 1989. Climatic change and the appearance of *Australopithecus africanus* in the Makapansgat sediments. *J. Hum. Evol.* **18:**107–113.
Carson, H. L. 1968. The population flush and its genetic consequences, in: R. C. Lewontin (ed.), *Population Biology and Evolution,* pp. 123–137. Syracuse University Press, Syracuse, New York.
Carson, H. L. 1970. Chromosome tracers of the origin of species. *Science* **168:**1414–1418.
Carson, H. L. 1971. Speciation and the founder principle. *Stadler Genet. Symp.* **3:**51–70.
Carson, H. L. 1982. Speciation as a major reorganization of polygenic balances, in: C. Barigozzi (ed.), *Mechanisms of Speciation,* pp. 411–433. Alan R. Liss, New York.
Chandler, C. R., and Gromko, M. H. 1989. On the relationship between species concepts and speciation processes. *Syst. Zool.* **38:**116–125.
Charles-Dominique, P. 1977. Ecology and behaviour of nocturnal primates. Duckworth, London.
Charlesworth, B., Lande, R., and Slatkin, M. 1982. A neo-Darwinian commentary on macroevolution. *Evolution* **36:**474–498.
Ciochon, R. L., and Corruccini, R. S. 1976. Shoulder joint of Sterkfontein *Australopithecus. S. Afr. J. Sci.* **72:**80–82.
Courtenay, D. O., and Bearder, S. K. 1989. The taxonomic status and distribution of bushbabies in Malawi with emphasis on the significance of vocalizations. *Int. J. Primatol.* **10:**17–34.
Coyne, J. A., Orr, H. A., and Futuyma, D. J. 1988. Do we need a new species concept? *Syst. Zool.* **37:**190–200.
Cracraft, J. 1987. Species concepts and the ontology of evolution. *Biol. Philos.* **2:**329–346.
Crompton, R. H. 1989. Mechanisms for speciation in *Galago* and *Tarsius. Hum. Evol.* **4:**105–116.
De Boer, L. E. M. 1973. Cytotaxonomy of the Lorisoidea (Primates: Prosimii). I. Chromosome studies and karyological relationships in the Galagidae. *Genetica* 44:155–193.
De Queiroz, K., and Donoghue, M. 1988. Phylogenetic systematics and the species problem. *Cladistics* **4:**317-338.
Dixson, A. F. 1978. Effects of ovariectomy and estradiol replacement therapy upon the sexual and aggressive behavior of the greater galago *Galago crassicaudatus crassicaudatus. Horm. Behav.* **10:**61–70.
Dobzhansky, T. 1935. A critique of the species concept in biology. *Philos. Sci.* **2:**344–355.
Dobzhansky, T. 1951. *Genetics and the Origin of Species,* 3rd ed. Columbia University Press, New York.
Donoghue, M. J. 1985. A critique of the Biological Species Concept and recommendations for a phylogenetic alternative. *Bryologist* **88:**172–181.
Donoghue, M. J. 1987. South African perspectives on species: an evaluation of the recognition concept. *Cladistics* **2:**285–294.
Dover, G. 1982. Molecular drive: a cohesive model of species evolution. *Nature* **299:**111–117.
Ehrlich, P. R. 1961. Has the biological species concept outlived its usefulness? *Syst. Zool.* **10:**167–176.
Eldredge, N. 1985. The ontology of species, in: E. S. Vrba (ed.), *Species and Speciation,* pp. 17–20. Transvaal Museum Monograph No. 4. Transvaal Museum, Pretoria.
Foley, R. A. 1991. How many species of hominid should there be? *J. Hum. Evol.* **20:**413–427.
Fujita, K. 1987. Species recognition by five macaque monkeys. *Primates* **28:**353–366.
Futuyma, D. J. 1987. On the role of species in anagenesis. *Am. Natur.* **130:**465–473.
Gaffney, E. S. 1979. An introduction to the logic of phylogeny reconstruction, in: J. Cracraft and N. Eldredge (eds.), *Phylogenetic Analysis and Paleontology,* pp. 79–111. Columbia University Press, New York.

Harcourt, C. S., and Bearder, S. K. 1989. A comparison of *Galago moholi* in South Africa with *Galago zanzibaricus* in Kenya. *Int. J. Primatol.* **10:**35–45.

Hill, W. C. O. 1953. *Primates: Comparative Anatomy and Taxonomy, Vol. I. Strepsirhini.* University of Edinburgh Press, Edinburgh.

Jordan, K. 1896. On mechanical selection and other problems. *Novit. Zool.* **3:**426–525.

Kaneshiro, K. Y. 1976. Ethological isolation and phylogeny in the *Planitibia* subgroup of Hawaiian *Drosophila. Evolution* **30:**740–745.

Kimbel, W. H. 1991. Species, species concepts and hominid evolution. *J. Hum. Evol.* **20:**355–371.

King, M. -C., and Wilson, A. C. 1975. Evolution at two levels in humans and chimpanzees. *Science* **188:**107–116.

Kingdon, J. 1988. What are face patterns and do they contribute to reproductive isolation in guenons?, in: A. Gautier-Hion, F. Bourliere, J. -P. Gautier, and J. Kingdon (eds.), *A Primate Radiation: Evolutionary Biology of the African Guenons,* pp. 227–245. Cambridge University Press, Cambridge.

Lambert, D. M. 1984. Specific-Mate Recognition Systems, phylogenies and asymmetrical evolution. *J. Theor. Biol.* **109:**147–156.

Lambert, D. M., and Paterson, H. E. 1982. Morphological resemblance and its relationship to genetic distance measures. *Evol. Theory* **5:**291–300.

Lambert, D. M., and Paterson, H. E. H. 1984. On 'Bridging the gap between race and species': the Isolation Concept and an alternative. *Proc. Linn. Soc. N. S. W.* **107:**501–514.

Lambert, D. M., Centner, M. R., and Paterson, H. E. H. 1984. Simulation of the conditions necessary for the evolution of species by reinforcement. *S. Afr. J. Sci.* **80:**308–311.

Levins, R., and Lewontin, R. 1985. *The Dialectical Biologist.* Harvard University Press, Cambridge, MA.

Lewontin, R. C. 1974. *The Genetic Basis of Evolutionary Change.* Columbia University Press, New York.

Marler, P. 1978. Vocal ethology of primates: implications for psychophysics and psychophysiology, in: D. J. Chivers and J. Herbert (eds.), *Recent Advances in Primatology,* Vol. I, pp. 795–801. Academic Press, London.

Marshall, J. T., and Marshall, E. R. 1976. Gibbons and their territorial songs. *Science* **193:**235–237.

Masters, J. C. 1985. Species within the Taxon *Galago crassicaudatus* E. Geoffroy. Unpublished Ph.D. thesis, University of the Witwatersrand, Johannesburg.

Masters, J. 1986. Geographic distributions of karyotypes and morphotypes within the greater galagines. *Folia Primatol.* **46:**127–141.

Masters, J. 1988. Speciation in the greater galagos (Prosimii: Galaginae): review and synthesis. *Biol. J. Linn. Soc.* **34:**149–174.

Masters, J. C. 1991. Loud calls of *Galago crassicaudatus* and *G. garnettii* and their relation to habitat structure. *Primates* **32:**153–167.

Masters, J. C., and Dunn, D. S. 1988. Distribution of erythrocytic allozymes in two sibling species of greater galago [*Galago crassicaudatus* E. Geoffroy 1812 and *G. garnettii* (Ogilby 1838)]. *Am. J. Primatol.* **14:**235–245.

Masters, J., and Lubinsky, D. 1988. Morphological clues to genetic species: multivariate analysis of greater galago sibling species. *Am. J. Phys. Anthropol.* **75:**37–52.

Masters, J. C., and Spencer, H. G. 1989. Why we need a new genetic species concept. *Syst. Zool.* **38:**270–279.

Masters, J., Lambert, D., and Paterson, H. 1984. Scientific prejudice, reproductive isolation and *apartheid. Persp. Biol. Med.* **28:**107–116.

Masters, J. C., Rayner, R. J., McKay, I. J., Potts, A. D., Nails, D., Ferguson, J. W., Weissenbacher, B. K., Allsopp, M., and Anderson, M. L. 1987a. The concept of species: Recognition versus Isolation. *S. Afr. J. Sci.* **83:**534–537.

Masters, J. C., Stanyon, R., and Romagno, D. 1987b. Standardized karyotypes for the greater Galagos, *Galago crassicaudatus* E. Geoffroy, 1812 and *G. garnettii* (Ogilby, 1838) (Primates: Prosimii). *Genetica* **75:**123–129.

Mayr, E. 1942. *Systematics and the Origin of Species.* Columbia University Press, New York.

Mayr, E. 1954. Change of genetic environment and evolution, in: J. Huxley, A. C. Hardy, and E. B. Ford (eds.), *Evolution as a Process,* pp. 188–213. Allen and Unwin, London.

Mayr, E. (ed.) 1957. *The Species Problem.* American Association for the Advancement of Science Publication No. 50, Washington, D.C.

Mayr, E. 1963. *Animal Species and Evolution.* Harvard University Press, Cambridge, MA.

Mayr, E. 1988. The why and how of species. *Biol. Philos.* **3:**431–441.
McKitrick, M. C., and Zink, R. M. 1988. Species concepts in ornithology. *Condor* **90:**1–14.
Mishler, B. D., and Brandon, R. N. 1987. Individuality, pluralism, and the phylogenetic species concept. *Biol. Philos.* **2:**397–414.
Mitani, J. C. 1987. Species discrimination of male songs in gibbons. *Am. J. Primatol.* **13:**413–423.
Nash, L. T., Pitts, R. S., and Bearder, S. K. 1988. Proceedings of a symposium entitled "Variability within Galagos," held at the XIth Congress of the International Primatological Society, Göttingen, Federal Republic of Germany, July 1986. *Int. J. Primatol.* **9:**503–505.
Nelson, G. 1989. Cladistics and evolutionary models. *Cladistics* **5:**275–289.
Nelson, G., and Platnick, N. 1981. *Systematics and Biogeography: Cladistics and Vicariance.* Columbia University Press, New York.
Nixon, K. C., and Wheeler, Q. D. 1990. An amplification of the phylogenetic species concept. *Cladistics* **6:**211-223.
Oates, J. F., and Trocco, T. F. 1983. Taxonomy and phylogeny of black-and-white colobus monkeys: inferences from an analysis of loud call variation. *Folia Primatol.* **40:**83–113.
Olson, T. R. 1979. Studies on Aspects of the Morphology of the Genus *Otolemur.* Unpublished PhD thesis, University of London, London.
Paterson, H. E. H. 1978. More evidence against speciation by reinforcement. *S. Afr. J. Sci.* **74:**369–371.
Paterson, H. E. H. 1981. The continuing search for the unknown and unknowable: a critique of contemporary ideas on speciation. *S. Afr. J. Sci.* **77:**113–119.
Paterson, H. E. H. 1985. The recognition concept of species, in: E. S. Vrba (ed.), *Species and Speciation,* pp. 21–29. Transvaal Museum Monograph No. 4. Transvaal Museum, Pretoria.
Paterson, H. 1988. On defining species in terms of sterility: problems and alternatives. *Pacific Sci.* **42:**65–71.
Patterson, C. 1988. The impact of evolutionary theories on systematics, in: D. L. Hawksworth (ed.), *Prospects in Systematics,* pp. 59–91. Clarendon Press, Oxford.
Petter, J. -J., and Charles-Dominique, P. 1979. Vocal communication in prosimians, in: G. A. Doyle and R. D. Martin (eds.). *The Study of Prosimian Behavior,* pp. 247–305. Academic Press, New York.
Phillips-Conroy, J. E., and Jolly, C. J. 1986. Changes in the structure of the baboon hybrid zone in the Awash National Park, Ethiopia. *Am. J. Phys. Anthropol.* **71:**337–350.
Raubenheimer, D., and Crowe, T. M. 1987. The Recognition Species Concept: is it really an alternative? *S. Afr. J. Sci.* **83:**530–534.
Rosen, D. E. 1979. Fishes from the uplands and intermontane basins of Guatemala: revisionary studies and comparative geography. *Bull. Am. Mus. Nat. Hist.* **162:**267–376.
Ryan, M. J., and Wilczynski, W. 1988. Coevolution of sender and receiver: effect on local mate preference in cricket frogs. *Science* **240:**1786-1788.
Schilling, A. 1979. Olfactory communication in prosimians, in: G. A. Doyle, and R. D. Martin (eds.), *The Study of Prosimian Behavior,* pp. 461–542. Academic Press, New York.
Selander, R. K., and Johnson, W. E. 1973. Genetic variation among vertebrate species. *Ann. Rev. Ecol. Syst.* **4:**75-91.
Sites, J. W., Jr., and Moritz, C. 1987. Chromosomal evolution and speciation revisited. *Syst. Zool.* **36:**153–174.
Smuts, B. B. 1985. *Sex and Friendship in Baboons.* Aldine Press, New York.
Sokal, R. R., and Crovello, T. J. 1970. The biological species concept: a critical evaluation. *Am. Natur.* **104:**127–153.
Spencer, H. G., McArdle, B. H., and Lambert, D. M. 1986. A theoretical investigation of speciation by reinforcement. *Am. Natur.* **128:**241–262.
Stanyon, R., Masters, J. C., and Romagno, D. 1987. The chromosomes of *Nycticebus coucang* (Boddaert, 1785) (Primates: Prosimii). *Genetica* **75:**145–152.
Stern, J. T., and Susman, R. L. 1983. The locomotor anatomy of *Australopithecus afarensis. Am. J. Phys. Anthropol.* **60:**279–317.
Susman, R. L., Stern, J. T., and Jungers, W. L. 1984. Arboreality and bipedality in the Hadar hominids. *Folia Primatol.* **43:**113–156.
Tattersall, I. 1986. Species recognition in human paleontology. *J. Hum. Evol.* **15:**165–175.
Tattersall, I. 1989. The roles of ecological and behavioral observation in species recognition among primates. *Hum. Evol.* **4:**117–124.
Templeton, A. R. 1980. The theory of speciation *via* the founder principle. *Genetics* **94:**1011–1038.

Templeton, A. R. 1987. Species and speciation. *Evolution* **41:**233–235.

Turner, A. 1985. The Recognition Concept of species in palaeontology, with special consideration of some issues in hominid evolution, in: E. S. Vrba (ed.), *Species and Speciation,* pp. 153–158. Transvaal Museum Monograph No. 4. Transvaal Museum, Pretoria.

Turner, A. 1986. Species, speciation and human evolution. *Hum. Evol.* **1:**419–430.

Turner, A., and Chamberlain, A. 1989. Speciation, morphological change and the status of *Homo erectus. J. Hum. Evol.* **18:**115–130.

Verrell, P. A. 1988. Stabilizing selection, sexual selection and speciation: a view of Specific-mate Recognition Systems. *Syst. Zool.* **37:**209–215.

Vrba, E. S. 1979. A new study of the scapula of *Australopithecus africanus* from Sterkfontein. *Am. J. Phys. Anthropol.* **51:**117–130.

Vrba, E. S. 1980. Evolution, species and fossils: how does life evolve? *S. Afr. J. Sci.* **76:**61–84.

Watanabe, T. K., and Kawanishi, M. 1979. Mating preference and the direction of evolution in *Drosophila. Science* **205:**906–907.

Waser, P. M. 1982. The evolution of male loud calls among mangabeys and baboons, in: C. T. Snowdon, C. H. Brown, and M. R. Petersen (eds.), *Primate Communication,* pp. 117–143. Cambridge University Press, Cambridge.

Waser, P. M., and Brown, C. H. 1986. Habitat acoustics and primate communication. *Am. J. Primatol.* **10:**135–154.

White, M. J. D. 1973. *Animal Cytology and Evolution,* 3rd ed. Cambridge University Press, Cambridge.

White, M. J. D. 1978. *Modes of Speciation.* W. H. Freeman, San Francisco.

Wilson, A. C. 1976. Gene regulation in evolution, in: F. J. Ayala (ed.), *Molecular Evolution,* pp. 225–234. Sinauer Associates, Inc., Sunderland, MA.

Wilson, L. G. (ed.). 1970. *Sir Charles Lyell's Scientific Journals on the Species Question.* Yale University Press, New Haven, CT.

Ying, K. L., and Butler, H. 1971. Chromosomal polymorphism in the lesser bush babies (*Galago senegalensis*). *Can. J. Genet. Cytol.* **13:**793–800.

Zimmermann, E. 1990. Differentiation of vocalizations in bushbabies (Galaginae, Prosimiae, Primates) and the significance for assessing phylogenetic relationships. *Z. Zool. Syst. Evolut. -Forsch.* **28:**217–239.

Zimmermann, E., Bearder, S. K., Doyle, G. A., and Anderson, A. B. 1988. Variations in vocal patterns of Senegal and South African lesser bushbabies and their implications for taxonomic relationships. *Folia Primatol.* **51:**87–105.

2

Speciation and Variation among the Living Primates

Species, Subspecies, and Baboon Systematics

4

CLIFFORD J. JOLLY

Introduction

The baboons of the genus *Papio* [excluding the gelada (*Theropithecus*) and (*pace* Delson, 1975), the mandrills, and drills (*Mandrillus*)] comprise a cluster of parapatric populations spread across most of the Ethiopian faunal zone. The present paper uses baboon diversity to explore some aspects of species definition and diagnosis, without attempting a comprehensive revision of the group or an exhaustive exploration of the species concept. The baboons are well suited to this purpose, because the various phenotypically distinct "forms" (which I call *subspecies*) have some but not all of the attributes commonly used to define one or another variant of the species concept. Another advantage is their quasi-continuous distribution, mostly undivided by extrinsic barriers that would avoid the problem of delineating natural units yet including some populations that are geographically isolated but not phenetically distinct, which illustrate the problems of definition raised by extrinsic isolation.

Finally, relevant aspects of baboon biology are quite well known. The geographical distribution and intrapopulational variation of skin-and-skull characters are extensively documented in museum collections (most of which were examined by the author, or by Dr. Colin Groves, in a yet unpublished collaboration) and systematic publications. Some of the remaining gaps can be filled from sources such as illustrated natural histories, travel guides, and wildlife documentaries enabling us to draw most of the geographical boundaries between different baboon morphs quite precisely. Furthermore, the socioecology of baboons has probably been studied in more depth and contexts than that of any other nonhuman primate, allowing the application of species concepts that involve behavioral criteria.

CLIFFORD J. JOLLY • Department of Anthropology, New York University, New York, New York 10003

Species, Species Concepts, and Primate Evolution, edited by William H. Kimbel and Lawrence B. Martin. Plenum Press, New York, 1993.

Some Concepts and Definitions

Definitions, in general, are subject to both ontological and epistemological evaluation. A definition that is *ontologically* strong provides a pithy and logically consistent expression, usually referring to a more diffuse but fundamental underlying concept. *Epistemological* strength implies that the definition is readily applicable to the diagnosis of real cases. For example, the pre-Darwinian definition of species as units corresponding to templates in the mind of the Creator is ontologically unassailable, being unchanging, significant, and unambiguous (the Almighty, presumably, knowing what she was about), but it is epistemologically weak, being operational only by those systematists with a hot line to divinity. Conversely, a definition of a species as a group of organisms with the same chromosome number would be epistemologically easy, but ontologically meaningless as a reflection of evolutionary significance. The referent concept underlying the notion of classification in general, and species in particular, is the belief that there is an inherent structure of natural diversity. This structure consists of lumps, discontinuities, and clusters in two distinct but closely related "landscapes" (which I propose to call *zygostructure* and *phenostructure*), representing, respectively, patterns of interbreeding of organisms and patterns of distribution of their properties. Species are commonly defined by referring to both structures: Organisms belonging to different species are both reproductively isolated and different from each other; those that are conspecific comprise a reproductive community and share a unique cluster of attributes.

Since Darwin's day, it has been accepted that this structure was produced by organic evolution and should be explained in evolutionary terms, so that species should be "significant units in evolution" (Hull, 1970). There is an essential relationship between the two major aspects of population structure and the cumulative nature of evolution; reproductive and hence genetic isolation allow morphologically distinct populations to persist long enough to become the basis for further morphological change (Futuyma, 1987; Masters, this volume). Most practicing systematists, I think, carry a mental image of "good" species as lines on a phylogenetic diagram—separate, isolated, unambiguous, boldly charting their course through time and across adaptive space, and splitting, again unambiguously, to yield new species. That was how Darwin drew species, and we want our species to be entities of that kind, even if we are able to perceive them only in cross section at one time level. An ontologically effective definition of the category "species" should encapsulate this idea by distinguishing populations with evolutionary "significance" of this sort from those that lack it.

The classical biological species concept (BSC) is perhaps the most successful attempt to embody the essential aspects of "evolutionary importance." Though epitomized by Mayr's (1942) definition of species as "groups of actually or potentially interbreeding natural populations, which are reproductively isolated from other such groups," the BSC carries implications that go well beyond this core definition. Systematists working in the "synthetic" tradition—including Mayr himself—use a working concept of a "good biological species" that includes ecological, behavioral, and phenetic dimensions, as well as reproductive isolation. Much of this baggage has evidently been acquired in the quest for ontological perfection, inspired by the belief in the reality and universality of the biological species, as opposed to other kind of species (especially "typological," non-population-based concepts) and other categories. It is thus very much a hallmark of the BSC's role as a cornerstone of the synthetic revision of evolutionary theory, which was centrally concerned with demonstrating the ontological unity of evolutionary biology, whether practiced by neontologists, geneticists, or paleontologists. However, the very ontological richness of the BSC may impede its rigorous application to the practical task of species recognition. The more numerous the diagnostic criteria, the more frequently ambiguous cases will be discovered. In such cases, the imprecision of the BSC allows the practicing systematist to mix phenetic, tokogenetic, and even ecological criteria at will to yield the desired result, an expression of a preconceived "real species."

This chapter will not attempt to tackle comprehensively the question of the objective "reality" of species taxa. On the one hand, one must accept that persistent, distinctive populations of organisms certainly exist and new ones must come into being by natural processes. If they did not, cumulative, mechanistic evolutionary change would be impossible. At any time level (including the present) it is possible to point to many entities that undoubtedly are "good species" by any definition. To this extent, then, we can accept the philosophical position that "good species" are real and "out there," waiting to be discovered, and the "species problem" is largely epistemological: how to design a strategy to detect them. For present purposes, we can assume that there are three major, current species-finding strategies (as explained much more fully by other contributors to this volume; for example, Kimbel and Rak, Eldredge, and Masters): the isolation concept (the core of Mayr's definition of a biological species), the phylogenetic species concept, and the recognition concept.

However, though the division of the living world into species taxa—significant and persistent units of evolution—may be reality, it is not directly observable reality. The primary data consist of the observed and inferred morphology (in the widest sense, including molecular structure) and breeding behavior of individual organisms. Species (like Mendelian populations) represent a more highly abstracted level of reality. A species taxon is real in the same sense as an epidemic (an abstraction from the incidence of disease) or a cloud (a name we give a cluster of water vapor droplets). Such entities, like species, are "individuals," which have a finite life span and can be given an identifying name ("The 1919 flu") (Hull, 1970).

They can also be defined in such a way that independent observers arrive at the same conclusion about their number and limits, but only if there is advance agreement on a boundary value delineating the edges of the class. For example, the statement "an epidemic has started when [N] new cases of disease X are diagnosed daily" defines a class of real phenomena that can be reproducibly identified, but only *if* the critical value of "N" is agreed. Although we can treat species as real, we must recognize that each of the three current species concepts depends upon criteria (reproductive isolation, homologous morphological distinctiveness, and mate recognition, respectively) that in nature are continuously distributed and multifactorially determined. Such data can easily be expressed as a list of entities called species, and all taxonomists should agree on the number and limits of these, as long as they also agree upon the diagnostic cut-off point on the appropriate continuum: how absolutely isolated, how distinctive, how similar in the mate-recognition system, do two populations have to be, to be called different species?

Inevitably, when describing the landscapes of natural diversity solely in terms of what populations are or are not "good species," we lose information. Although we can (in theory) agree to threshold values for the critical variables (alone or in combination), thereby defining species in a reproducible way, we should not be under the illusion that by doing so we have revealed entities more "real" than the data from which they are derived, nor that we have described a "reality" more important than the many data that will inevitably remain unexpressed. This suggests that there is little to be lost in abandoning, or at least deemphasizing, three favorite pursuits of systematists: the debate about what species "really are," the search for a perfect species definition that is both comprehensive and precise, and a research strategy in systematics that sees as its primary goal the discovery and delineation of "good species" in nature. I suggest that energy would be more usefully directed toward description and analysis of the two landscapes of natural population structure at a less abstract, pretaxonomic level. Afterwards, it is possible, though not essential, to apply these data to whatever general species definition is preferred. However, this second step, in which multidimensional patterns of variation and gene flow are translated into a one-dimensional classification, must necessarily involve a collapse and loss of information—a process in which "reality" is caricatured, not "discovered."

Whatever species definition is adopted, it would seem obvious that it should be applied across the board, regardless of its outcome. In other words, even though we accept the reality of the structure of nature that we express as species, we should not insist upon species status for certain particular taxa, no matter how hallowed by usage. Above all, we should resist the temptation to "test" species definitions by matching their results against what we "know" are good species and then tinkering with the definition so that it produces the desired result.

Although definition and diagnosis are clearly distinct levels of analysis, it has been forcefully argued (for example, by Sokal and Crovello, 1970) that classifications are stronger when cases (species taxa) are diagnosed in the same terms as the category (*the* species) is defined. Thus, if a biological species is defined according to the core definition, the only admissible evidence for the limits of a species taxon is the observed reproductive behavior of its members. As emphasized by systematists from Simpson to Mayr to Kimbel, this restriction is unnecessary. After all, the law *defines* murder as an act of unlawful killing, but if Miss Marple *recognizes* a murder from the *post hoc* evidence—a bloodstain on the drawing-room carpet—we do not therefore have to redefine murder as a dirty rug. Similarly, the *definition* of a species as a reproductive isolate (or reproductive community) is not compromised by the fact that actual species taxa must almost always be *recognized* by the distribution of phenotypic characters resulting from the reproductive isolation, and not from direct observation of noninterbreeding or of mate selection.

The main objective of this chapter is to match observable data about baboon population biology against the criteria of various species definitions. This material will be described in terms of the two structures mentioned earlier.

Phenostructure and Zygostructure

These two structures are aspects of a pervasive duality in evolutionary theory, here termed the *pheno-zygo* contrast. *Pheno* terms and concepts concern the observable characteristics (phenotypes) of organisms, at any level from the supraindividual (as in the case of social organization and gene frequencies, both attributes of populations), down to the DNA base pair. (It may seem perverse to call genomic features *phenetic*, in view of the fundamental distinction between genotype and phenotype in developmental biology. If there were another term for "everything observable, including genes," it could be substituted. However, in the present context, the important attribute of the genotype is that it consists of observable attributes, and thus logically belongs to the pheno category.) Information about such characters is phenodata; the patterns of distribution of phenetic variation (phenostructure) can be used to distinguish phenotaxa, of which the kind advocated by Sokal and Crovello (1970), based upon overall resemblance, are one variety. A phenospecies is therefore very close, to a morphological species or morphospecies, but the neologism is preferable because it expresses its basis in data from organizational levels both higher and lower than the gross structure (morphology) of individual organisms.

By contrast, terms and concepts of the zygo cluster are based upon the probability of formation of viable zygotes (this is thus very close to the concept of *tokogeny*, used by Eldredge (this volume), in logical opposition to morphology). Whereas phenodata, phenodefinitions, and phenostructure ultimately reflect the genes carried by organisms at some particular (past or present) time, zygodata, zygodefinitions, and zygostructure of populations are defined by what is happening, or not happening, or happened in the past, to the transmission of genes. Thus zygostructure includes all aspects of population structure in the geneticist's sense; rates of gene flow and migration (but not gene or genotype frequencies themselves, which are an aspect of phenostructure); frequency of cross-mating, intensity of positive and negative assortative mating by phenotype, avoidance of inbreeding, and so on. According to this scheme, any population of organisms (in the general sense, as well as the narrower, genetic one) has both a phenostructure and a zygostructure. The former is the distribution of phenetic characters (in the fullest sense), while the latter is the distribution of probabilities of producing viable zygotes.

Although conceptually distinct, the two structures are obviously intimately related. Patterns of breeding behavior in the past (zygostructure) are responsible for present genotypic (and hence phenotypic) distributions (phenostructure), and present zygostructure predicts further phenostructure. In many actual instances, the two data sets are related to each other like the bloodstain and the murder: Phenotypic intergradation in zones of geographical overlap or parapatry can be used to infer rates of interbreeding in the immediate past, which are attributes of the population's zygostructure. The theory of speciation and population genetics predicts that zygostructure and phenostructure should tend towards concordance; the more zygoisolated two populations are, the more likely they are to become phenetically distinct. Conversely, the more a population lacks internal (zygo)structure, the more homogeneous its distribution of gene frequencies, and hence phenotypes, is expected to become.

However, in the real world, the expected concordance between the two structures will be periodically disrupted by "events" of population history, and herein lies one of the major values of separate description and analysis. For example, at the lowest, intrademic level, "structuredness" can be represented by N_e (effective population number, a zygostructural measure) and F_{st} probability of two homologous genes being identical by descent, a phenostructural concept). At equilibrium F_{st} can be used to estimate N_e and other aspects of zygostructure that are difficult to determine empirically. However, an event (such as the drastic reduction of a large, panmictic population) would result in a population of individuals far more outbred than their (new) N_e would predict and a prognosis of loss of variation (see discussion in Chepko-Sade *et al.*, 1987). Similarly, at a higher level, a narrow hybrid zone between distinct, contiguous populations (a phenostructural feature) is consistent with several alternative zygostructures. It might represent an equilibrium, with assortative mating and hybrid disadvantage balancing gene flow from either side. In this case, of course, present zygostructure would predict an unchanging phenostructure, as long as environmental variables remained constant. On the other hand, it might be due to a recent ("historical") suturing of population ranges, with gene frequencies still far from equilibrium, or a more complex situation involving environmentally driven cycles. Only an on-the-ground investigation of *both* population-structural dimensions, treated independently, could elucidate the situation.

Just as full account of contemporary zygostructure and phenostructure should encompass all the data generally used to diagnose neontological populations and species, so a description of changes in zygostructure and phenostructure would be an account of evolutionary pattern. The traditional phylogenetic tree represents both aspects. In its sequence of branch points, it depicts zygostructural events deemed radical enough to be called "speciation." The "differentiation" axis of a phylogenetic tree depicts the evolution of phenostructure. A phylogenetic tree can, of course, be used to delineate higher taxa, either "traditionally," by a mixture of zygo and phenocriteria, or cladistically, by excluding the phenetic dimension from consideration.

The Phenostructure and Zygostructure of Extant Baboons

Although the baboons are one of the most extensively studied primate groups, there are gaps in our knowledge of both the primary aspects of population structure. Analyzing the two structures separately emphasizes the paucity of direct zygostructural data, often concealed by phenodata-based speculation (Figs. 1 and 2).

Phenostructure: The "Forms" of Baboon and Their Distribution

Accounts of baboon systematics generally distinguish five parapatric "forms," distinguishable on external characters (Table 1). These are the Guinea (or Western), Anubis (or olive), Hamadryas (or sacred), Yellow and Chacma baboons. To these can be added two less widely known, yet equally distinct and apparently stable as phenotypes. One is the Kinda (rhymes with Linda, from the name of its type locality in Zaire) baboon, a small version of the Yellow, which has no widely used English vernacular name. The other is the Gray-footed baboon, which is generally lumped with the Chacma. Other recognizable types, known only from small collections gathered in poorly explored regions, may represent

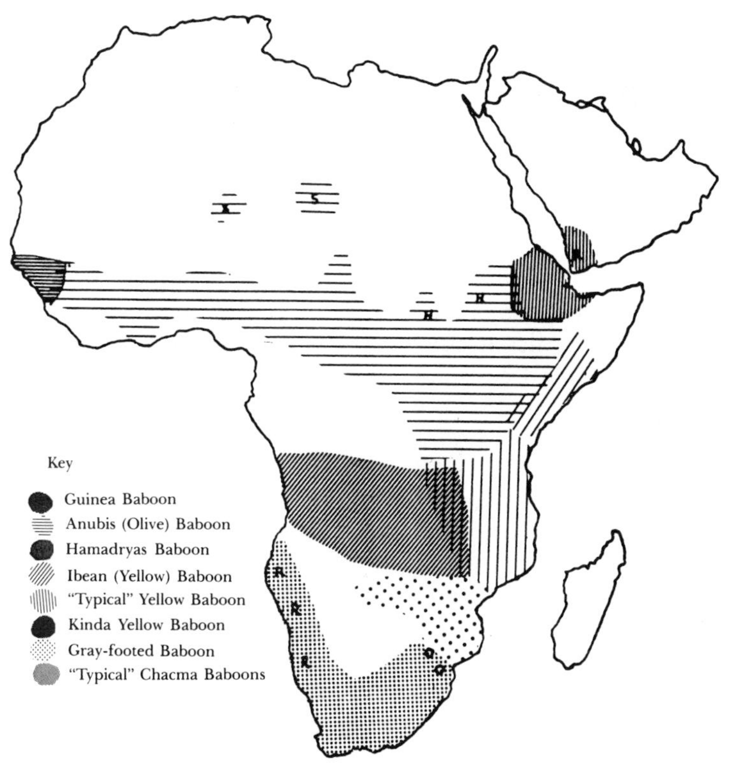

Fig. 1. Recent phenostructure of *Papio* baboons. See key for details. Overlapping textures indicate zones of hybridization or intergradation. S = Saharan (small) forms of olive baboon; H = sites of collection of "Heuglin's baboons"; A = Arabian (perhaps smaller) forms of Hamadryas; O = sites of collection of "Transvaal Chacmas"; R = general distribution of (small) Kalahari Chacmas.

widespread and phenotypically stable populations, but are described here as variants within the forms.

All the recognized forms (except the Kinda baboon, which differs in cranial size) are defined primarily on pelage characters, which are most pronounced in adult males, but can also be seen in juveniles and females. Accounts of these features are to be found in Jolly (1965), Jolly and Brett (1973), and Hill (1967) and are summarized in Table 1. Adult males of the five commonly recognized forms are illustrated in Dorst and Dandelot (1970). These pelage characters are quite stable, varying little within local populations, except the marginal demes that clearly show the effects of ongoing secondary hybridization (e.g., Maples and McKern, 1967; Nagel, 1973; Phillips-Conroy and Jolly, 1986). It is possible that observation of a hybrid zone led DeVore and Washburn (1963) to state that typical representatives of more than one form are to be observed within a single social group.

In these days of genetic markers, it may seem quaint to describe the major population structure of a species on the basis of pelage features. Certainly, these features are of unknown genetic etiology and are not susceptible to direct, quantitative population-genetic analysis, as protein or RFLP markers would be. Nevertheless, it is clear that they are genetically determined, and the diversity of naturally occurring, F_{1+n} hybrids in zones of intergradation (Phillips-Conroy

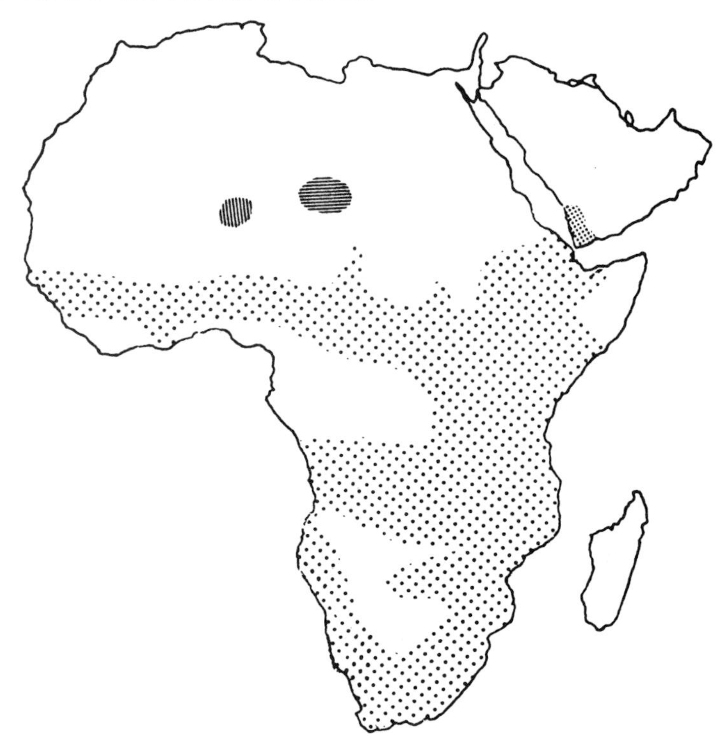

Fig. 2. Recent zygostructure of *Papio* baboons. Large dots: distribution of populations apparently in genetic continuity. Vertical stripes: Air isolate of olive baboons. Horizontal stripes: Tibesti isolate of olive baboons. Small dots: Arabian isolate of Hamadryas baboons.

and Jolly, 1986) indicates that several unlinked, genetic loci are involved. Moreover, on a zoogeographical scale, these features have the very real advantage of being easily visible in conventional museum specimens, some of which were collected in areas where baboons no longer occur or where their distributions may have changed. By contrast, in spite of active programs of research into the molecular genetics of baboon populations, each major form is still represented in these studies by only one or two demes at most. Finally, pelage features have the advantage that they are obviously visible to the baboons themselves; if baboons were to evaluate potential mates on the basis of recognition features (in the sense of Paterson, 1985), then these would presumably be among the features they would use.

In this account, baboon distribution is discussed in the zoogeographic present, as though the distributions observed at the time of description and collection still prevailed. The forms are discussed under their vernacular names, which are taxonomically neutral and also, in this case, often less ambiguous than a formal bi- (or tri-) nominal.

Guinea Baboon

This small, maned, reddish-furred form is confined to the extreme west of the northern savanna belt, from Senegal to western Guinea. Field observers

Table 1. Phenotypic Characters in Subspecies of *Papio hamadryas*

Character	Subspecies									
	Guinea	Hamadryas	Anubis	Heuglin's	Ibean	Typical yellow	Kinda	Gray-footed	Transvaal	Chacma
General color	Reddish-brown	Gray-brown	Olive-brown	Olive-brown	Yellowish-brown	Yellowish-brown	Yellowish-brown	Gray-brown	Gray-brown	Dark brown
Cheek hair color	Like back	Lighter	Like back	Lighter	Lighter	Lighter	Lighter	Like back	Like back	Like back
Belly hair color	Like back	Like back or darker	Like back	Lighter	Lighter	Lighter	Lighter	Lighter	Lighter	Lighter
Hand/foot hair color	Self (like arms)	Self (like arms)	Black or self	Self (like arms)	Self (like arms)	Self (like arms)	Self (like arms)	Self (like arms)	Black	Black
Silvery fringe on hands/feet	Absent	Absent	Absent	Absent	Present	Present	Present	Absent	Absent	Absent
Hair rings	Multiple	Multiple	One or two (agouti)	One or two (agouti)	Black tip	Black tip	Black tip	Black tip, black ring	Black tip, black ring	One light ring
Mane hair relief	Moderate	Strong	Moderate	Moderate	Weak	Absent	Absent	Absent	Absent	Absent
"Fringes" on nape and flanks	Absent	Absent	Absent	Absent	Yellow	Yellow	Yellow	Black	Black	Black
Texture, long back hairs	Waved	Waved	Waved	Waved	Waved	Straight	Straight	Straight	Straight	Straight
Male anal field	Medium, pink	Large, pink	Small, dark	Small, dark	Small, dark	Small, dark	Small, dark	Small, dark	Small, dark	Small, dark
Facial color	Purple-black	Pink-red	Purple-black	Purple-black	Purple-black	Purple-black	Purple-black	Purple-black	Purple-black	Purple-black
Natal coat	?	Black	Black	?	Black	Black	Reddish	Black	?	Black
Tail shape	Arched	Arched	Bent	?	Bent	Bent but variable	Arched	Bent	Bent	Bent
Skull size	Medium	Medium	Medium-large	Large	Large	Medium-large	Small	Large	Large	Large
Facial orientation	"Normal"	"Normal"	"Normal"	"Normal"	"Normal"	"Normal"	"Normal"	Downwardly flexed	Downwardly flexed	Downwardly flexed
Upper facial shape	"Normal"	"Normal"	Often broad	?	"Normal"	"Normal"	"Normal"	Narrow, deep	Narrow, deep	Narrow, deep
Postorbital constriction	"Normal"	"Normal"	"Normal"	"Normal"	?	Less constricted	Less constricted	Somewhat less	"Normal"	"Normal"
Temporal lines	"Normal"	Anteriorly placed	"Normal"	"Normal"	"Normal"	"Normal"	Weak	"Normal"	"Normal"	"Normal"

(J. Bert, personal communication) have noted that animals from the eastern part of the range are distinctly darker and less bright red in color than those from the west. The nature of the boundary between these distinct forms, if such they are, has not yet been investigated, nor has the boundary between the Guinea baboon and its one neighbor, the Anubis.

Olive or Anubis Baboon

Anubis baboons are found throughout the savannas and woodlands lying between the Sahara and the rainforest belt of west and central Africa, from Sierra Leone in the west to the Sudan and Ethiopia in the east, north to about 13°N. Their range extends along mountain ranges into higher latitudes: Aïr in Niger, Ouadaï and Ennedi in Chad, and the mountains of Darfur in the Sudan. An isolated population is found in the Tibesti Massif. Anubis are found throughout western and southern Ethiopia, western Kenya, Uganda, and northwestern Tanzania in a variety of savanna and forest habitats. In eastern Zaire, Uganda, and Kenya, they occur in evergreen forest as well as savannas and woodlands. Due to this wide range, typologically based splitting, and the persistent notion that baboons fell into "eastern" and "western" species, many redundant names appear in the earlier literature. Elliot (1913), for example, lists no less than eight full typological species on the basis of specimens of olive baboons. However, although there is some intrademic variation in coat color and the distribution of black fur on the hands and feet, olive baboons from Sierra Leone, the Ethiopian highlands, the forests of eastern Zaire, or the Serengeti Plain are remarkably similar. Cutting across this general uniformity, there is some geographical variation in cranial dimensions, which presumably reflects differences in overall body size. The largest skulled populations seem to be found in the forests of Uganda and eastern Zaire, while the smallest are seen in the Saharan isolates and in marginal populations whose ranges adjoin those of Yellow baboons and Hamadryas baboons in Tanzania and Ethiopia, respectively.

A little-known form ("Heuglin's baboon") is found in southern Sudan and southwest Ethiopia. The few skins available resemble those of typical Anubis baboons in having waved hair and a mane, but are distinctly light in color, and, unlike typical Anubis, have pale cheeks and undersides.

Hamadryas Baboon

The range of the Hamadryas, the familiar "sacred" baboon of ancient Egypt, adjoins that of the Anubis to the east, including the eastern highlands and semidesert eastern lowlands of Ethiopia and Somalia. Hamadryas are also found in the comparatively well-watered highlands of the southwest Arabian peninsula, including part of Yemen and Saudi Arabia. Written records indicate that baboons have been present in Arabia throughout the islamic era (D. Varisco, personal communication). However, the Hamadryas kept by the ancient Egyptians were probably imported, and there must be doubt whether the Arabian baboons represent a truly indigenous population or the descendants of captives shipwrecked some time within the past 4000 years. Although sometimes described as a smaller race, the Arabian Hamadryas falls within the size range of the African sample (which may, however, itself be heterogeneous in this respect) and is outwardly indistinguishable from populations of the adjoining mainland.

On the mainland, variation in pelage color is evident in museum specimens and has also been noted by field observers (Kummer, 1968). Generally, animals from the western part of the range are darker in pelage, while populations in which the males have more typical, white manes are found in the East, in Somalia and the Ethiopian lowlands.

The range of the Hamadryas baboon meets that of the Anubis in central Ethiopia (see below), and (presumably) that of the Yellow baboon in southern Ethiopia or central Somalia in the valley of the Webe Shebeli River.

Yellow Baboons (Typical, Ibean, and Kinda)

Baboons with typical yellow baboon coloration have a range that extends from the Webe Shebeli valley in southeastern Ethiopia, in an arc running east and south of the equatorial forest belt and the East African highlands, to Benguela Province in central Angola. Some confusion exists in the literature (Ansell, 1960, 1970; Hayes *et al.*, 1990) about the identity of baboons found in Malawi, northern Mozambique, and northwestern Zambia. These animals have been called *Papio ursinus jubilaeus,* and "dwarf Chacma," both of which imply that they are allied with forms found to the south. Judging from the localities, these are also the same animals that Hayes *et al.* (1990) called hybrids between *P. ursinus* and *P. cynocephalus.* However, all skulls and skins examined (except the type of *P. pruinosus,* which is abnormal in coloration and presumably is an aberrant individual) are indistinguishable from those of "typical" Yellow baboons from north-central Tanzania. Hayes *et al.* (1990), presumably following Ansell (1960, 1970), imply that the small yellow baboons from western Zambia (here discussed under the name of Kinda baboons) are typical *P. cynocephalus,* which is true in a morphological but not a nomenclatorial sense.

Geographical variation among "Yellow baboons" takes two forms: in the texture of the pelage and in cranio-dental size (again, presumably a reflection of overall dimensions). "Typical" Yellow baboons from Tanzania to Mozambique are large, and the long hairs of the back are straight and silky in texture, and brownish-yellow in color. In the lowlands of Kenya and in Somalia, baboons with the same general coloration and size have coarser, more wavy, mane hair (more like that of Anubis baboons). The name *ibeanus* (Ibean baboon) is available for this form, which is familiar to primatologists as the populations studied by the Altmanns' group (e.g., Altmann and Altmann, 1970).

To the west of typical Yellow baboons, in Angola and western Zambia, is found a very distinct form, the Kinda baboon. Its pelage is like the typical yellow's in color, but is very soft and silky in texture, and the tail is said to be arched rather than "broken" in form (Ansell, 1970). According to Ansell (1970) the natal coat of the newborn Kinda baboon is reddish-brown, a radical departure from the black infant of other forms. Most striking is the small size of the skull and teeth; the skull of an adult male is about the same size and shape as that of an adult female of the "full-sized" Yellow baboons, and the female is proportionately smaller. Between the range of the Kinda and that of "typical," large yellows, in central and northeastern Zambia, southern Zaire, and southwestern Tanzania, are found populations intermediate in cranial size (Freedman, 1963; Jolly 1965; Ansell, 1970; Lönnberg, 1919; Hill and Carter, 1941) to which no distinct name has been applied. Although many gaps in the distribution need to be filled, it seems likely that the intermediate-sized populations form a steep west-east cline linking Kinda baboon populations with large-skulled populations of "typical" Yellow baboons.

Chacma Baboons (Typical, Gray-Footed, Transvaal, and Kalahari)

The large baboons with black nape-fringes, generally grouped as "Chacmas," are found to the south of the range of Yellow and Kinda baboons, from a line immediately north of the Zambezi River to the Cape of Good Hope. Their distribution appears to be much more continuous in the well-watered east of the range than in the desertic west; in Botswana and Namibia baboons have been collected and reported mostly from river valleys and oases, and large tracts are probably unpopulated. Several distinct geographical forms can be distinguished.

The "typical" Chacma is dark-brown in color, with black fur on the hands and feet, and a comparatively short tail. Animals of this pelage type are found in the more southerly and westerly parts of the species range: most extensively in South Africa, and perhaps adjoining parts of Botswana, but also extending northward into western Namibia and southwest Angola. Within this range, there is variation in size; animals from the arid west are distinctly smaller in size than those from further south and east. The name *ruacana* is available for the small, but otherwise typical, Kalahari Chacma.

Specimens from the Transvaal are distinctly lighter in ground color than typical Chacmas, though resembling them in size and in other pelage characters. More investigation is needed to determine whether they represent a widespread and stable population (for which the name *orientalis* is available), or whether they are drawn from a clinal transition (or a hybrid zone) between the typical Chacma and the Gray-footed baboon.

The latter is found from the Limpopo River north to the valley of the Zambezi, the area including southern Mozambique, Zimbabwe, and the extreme south of Zambia, the Caprivi Strip of Namibia, and the Okavango region of Botswana. Gray-footed baboons are slightly smaller than typical Chacmas (in cranial dimensions) and lighter in color, with hands and feet the same color as the limbs and a longer tail. In all these features, they approach Yellow baboons. The name *griseipes* is available for such animals.

Patterns of Phenostructure

Traditionally, variation in *Papio* has been described in terms of these five "major forms." However, all the recognizable forms comprise a single, continuous series, in which, for the most part, neighboring populations differ from each other in orderly, stepped-cline fashion. Thus, each step in the series Anubis—Heuglin's—Ibean—"typical" Yellow involves the replacement of a character state typical of Anubis by one characteristic of Yellow baboons. The series then continues south as typical Yellows—Gray-footed—Transvaal—typical Chacma, with the features of typical Chacmas replacing those of Yellow baboons, again in a stepwise fashion. From this north-south chain, four shorter side branches project, leading, respectively, to typical Hamadryas, typical Guineas, Kinda baboons and Kalahari Chacmas. The first two involve transitions in pelage features as well as size; the latter two primarily involve size.

Only one case violates the general rule that the most similar populations are neighbors. Hamadryas and Guinea baboons, whose present ranges are separated by over 5000 km of Anubis country, resemble each other and differ from Anubis in being smaller, with a long tail that is arched rather than broken, more heavily maned, and having multiple rings on the mane hairs (in Hamadryas, only in males; in females the hairs are plain). There may also be some resemblances in social behavior (Dunbar and Nathan, 1972) and some details of cranial and dental structure (Jolly, 1965), although both of these need to be confirmed by more extensive materials.

Hybrid Zones

Between the boundaries of forms are found narrow zones in which groups show unusual phenotypic diversity; these are most reasonably interpreted as hybrid swarms resulting from ongoing crossbreeding. The best-documented zone lies between Hamadryas and Anubis in Ethiopia (Nagel, 1973; Sugawara, 1979; Phillips-Conroy *et al.*, 1991), but the one between Ibean (Yellow) baboons and Anubis in Kenya has also been noted and mapped (Maples and McKern, 1967; Kingdon, 1971; Samuels and Altmann, 1986; C. Jolly, personal observation.) Of the 12 boundary zones between contiguous forms, only these two have been directly investigated, and on-the-ground survey of each of the critical border areas would obviously be of great interest. For instance, as Ansell (1970) has observed, the known ranges of two very distinct forms, Gray-footed and Kinda baboons, approach each other closely in south-central Zambia, and no intermediate or hybrid populations have been observed. In west Africa, the brown eastern Guinea populations could well be the result of the introgression of Anubis genes, pointing to a Guinea-Anubis hybrid zone in the same way as the brown western Hamadryas point to the Hamadryas-Anubis hybrid zone that includes Awash National Park.

Phenostructure and Habitat

Since ecological divergence is often cited as a diagnostic of good biological species, I (Jolly, 1965) used published information to document the ecological context for each locality known to be inhabited by a particular form of baboon. As a general rule, there is no clear distinction: Each of the major forms lives in a variety of habitats. However, the boundary between forms tends to lie at the junction between distinct ecovegetational zones, so that regional natural histories often describe neighboring forms of baboon as differing in their choice of habitat. However, these choices are not manifested throughout the forms' ranges. The preferred habitat of a given form at its boundary with one of its neighbors may be different from that shown where it meets another. For example, in West and Central Africa, olive baboons appear to be absent from lowland rainforests (where other large, semiterrestrial forest monkeys occur), but north and east of the Zaire-Lualaba river systems, where *Mandrillus* is absent and the local semiterrestrial mangabey is ecologically specialized, olive baboons are found well within rainforests of low and medium altitude, as well as in both moist and dry woodlands and savannas. However, in Kenya, where their range abuts that of Yellow baboons, and Ethiopia, where it meets that of Hamadryas, olive baboons are considered to be typically animals of highland forest and woodland. Typical Yellow baboons are the "lowland" form in Kenya, but inhabit highlands in Tanzania and Malawi. In Namibia (Shortridge, 1934), the small, desert chacma is a "rock baboon," frequenting outcrops and cliffs, while the Gray-footed baboon is found more frequently in woodlands. Ansell (1960) makes the same ecological distinction in Zambia, but here it is the Kinda baboon (which he calls *P. cynocephalus*) that is the woodland form, while the Gray-footed baboon (Ansell's *P. ursinus*) is the "rock dweller."

The most consistent ecological association is exhibited by the Hamadryas, which is apparently found only in arid habitats of various kinds and shows some minor physical adaptations to resources found in such environments (Jolly, 1970). The distinctive male behaviors that lead ontogenetically to the formation of one-male units (Kummer *et al.,* 1970), and the striking male epigamic pelage features that were probably favored by sexual selection in this social milieu (Jolly, 1963), can also be seen as consequences of foraging in a habitat of low productivity, but they are not economic adaptations in the sense that an animal lacking them would be at a disadvantage *vis-à-vis* its habitat. Indeed, other forms, especially Saharan Anubis and Kalahari Chacmas, live in habitats as arid as those typical of Hamadryas, and (although they forage in small groups) they do not exhibit the characteristic one-male group, fission-fusion social structure (or, of course, the striking male epigamic features) of Hamadryas. Moreover, although the Hamadryas-Anubis boundary in Ethiopia does indeed correspond to a steep gradient between semidesert and highland forest, the behavior of individual, migrant Hamadryas males in the hybrid zone (Phillips-Conroy *et al.,* 1991) gives no indication that they prefer, or survive better in, arid habitats.

Thus, none of the baboon forms exhibits ecological preferences or (apart from the Hamadryas, to a minor degree) ecological adaptations in any usual sense of those terms. I would hypothesize that the tendency for phenetic boundaries to coincide with habitat edges is determined more by the history of the populations involved than by their "adaptations" in any Darwinian sense. The observed pattern would be predicted if populations of different forms, when expanding by group fission into a previously unoccupied habitat, tended preferentially to occupy microhabitats familiar to them. Populations expanding from opposite edges of the unoccupied landscape would then tend, when they met, to establish a boundary corresponding to a local ecotone. Obviously, such a scenario implies a dynamic zoogeographic history that has yet to be reconstructed in detail.

The Contemporary Zygostructure of *Papio*

A general picture of the low-level population structure of *Papio* baboons emerges from field studies that have monitored the demography and behavior of particular local populations (e.g., Packer, 1979; Altmann *et al.*, 1977; Sigg *et al.*, 1982; Abegglen, 1984). Except Hamadryas, *Papio* baboons, like most other cercopithecines, apparently live in groups with a core of female kin. Most males leave their natal group in early adulthood and take up residence in another group, where they breed. There are, however, some apparent exceptions to this general rule of male migration and female philopatry. Most Hamadryas males reportedly remain within their natal group (Kummer, 1990; though this rule, too, evidently has frequent exceptions; see discussion in Phillips-Conroy *et al.*, 1991), while females leave their natal one-male group and sometimes their natal band also (Kummer, 1990). Intertroop migration of females has been seen in Chacma baboons (Anderson, 1987). As Melnick and Pearl (1986) have emphasized, (male) prereproductive emigration in papionins results in local groups that are outbred, although they often differ from one another in gene frequencies due to a kind of founders' effect among breeding males.

Much less is known about the higher level zygostructure of baboons; for example, the large-scale clumping and habitat-dependent distribution of demes. However, from what is known of baboon ecology and habitat-dependent differences in population density, one can infer that a good deal of structure must exist at this level: The distribution of demes over the landscape is likely to be much more clumped, patchy, and discontinuous than most theoretical population-genetic models assume. Similarly, we know a good deal (in a few local populations) about the rates of immigration and emigration of individuals from troops and the age- and sex-related probability of such events, but very little about the fate of emigrants, the probability of their survival, or the course of their travels. Thus we cannot reliably estimate such variables as the mean and variance of the distance separating an individual's birthplace and the birthplace(s) of its offspring. Unfortunately, these parameters define aspects of population structure, such as the size and integrity of local demes, and, thus, the potential of gene flow to homogenize populations of various levels, up to the whole species.

Lacking information about "normal" gene flow among demes, we have no well-defined baseline from which to estimate the evolutionary impact of a geographical barrier (as in the case of the Arabian and Saharan baboons), or of interform gene flow in hybrid zones. However, the latter do allow us to look at the concordance between zygostructure and high-level phenostructure (i.e., the extent to which the geographical edges of phenetically distinct populations are matched by discontinuities in gene flow). Field observations yield both primary zygodata (patterns of mating behavior and breeding success of individuals of the two parental forms and their hybrid descendants) and detailed phenodata with immediate zyostructural relevance (e.g., the occurrence of mixed social groups, including both parental types, and the frequency of hybrids of various degrees of intermediacy.)

Of the known hybrid zones, only that between Anubis and Hamadryas in the Awash National Park, Ethiopia, has been investigated in any detail (see Phillips-Conroy *et al.*, 1991, for a review of studies), and even here much remains to be learned. The results depict a local zygostructure that is complicated by motivational and behavioral differences between the two forms. Nevertheless, it is clear that (contrary to early reports) male Hamadryas baboons not only migrate into Anubis groups, they remain there for periods comparable to the residence times of Anubis males and become fully integrated into the social structure of the host group (Phillips-Conroy *et al.*, 1991). Although they are not able to complete a Hamadryas-type social agenda (they do not, apparently, form long-term, exclusive harem bonds, in spite of continual efforts to do so), this does not prevent them from forming close relationships with females (Nystrom, 1991) and fathering hybrid offspring. We have no reason to doubt that these hybrids are themselves fully fertile and socially integrated.

Similarly, migrant Anubis males attach themselves to Hamadryas groups (Sugawara, 1979; Brett, unpublished observations), although their breeding success has not yet been investigated. Since both Kimbel (1991) and Masters (this volume) have drawn taxonomic conclusions from our account of the distribution of phenotypes in the Awash (Phillips-Conroy and Jolly, 1986), it should be emphasized that we cannot as yet discern any patterning that cannot be explained most simply in terms of the demic structure of the Awash populations. There is no evidence that Awash hybrids are at any social or ecological disadvantage. We do not know how hybrids would fare in habitats more "typical" of the two parental types, although we do know that immigrant Hamadryas males live long and well in Awash Anubis groups (Phillips-Conroy *et al.*, 1991) and that Anubis baboons certainly survive in semidesert environments such as Tibesti.

Second, our statement that most hybrids now born in the Awash have at least one hybrid parent does not imply nonrandom mating. The preponderance, among hybrids, of animals that seem to be F_{1+n} rather than F_1 is simply a consequence of a multigenerational history of hybridization, which has interposed a zone in which groups consist mostly of hybrids, between the ranges of "pure" Hamadryas and Anubis groups. These groups produce *relatively* few F_1 offspring, just because they contain few "pure" parents. However, these hybrid groups exchange migrants (mostly males, as far as we can tell) with Anubis groups upstream and perhaps with Hamadryas groups downstream also. Moreover, phenotypically Hamadryas males enter groups with a majority of phenotypically Anubis members and show no signs of mating selectively by phenotype. Third, the assumption that over the whole hybrid zone most hybrids are now born to Hamadryas fathers and Anubis (or hybrid) mothers is untested, and probably incorrect, except trivially in the case of an Anubis group receiving Hamadryas male immigrants early in the hybridization process. Even there, the "Hamadryas" immigrants are often actually Hamadryas-like hybrids (Phillips-Conroy *et al.*, 1991). As hybrid natal females mature in such groups, they mate with resident males—most of which are Anubis—producing hybrid offspring. Such backcrosses are the obvious source of the large number of hybrids with Anubis characteristics predominating, which are found in these groups (Phillips-Conroy and Jolly, 1986) and to which Kimbel (1991) draws attention. The important point is that there is no substantial barrier to gene flow corresponding geographically to the relatively sharp Anubis-Hamadryas phenotypic transition, and members of each of the two forms have no compunction about recognizing members of the other as appropriate mates [although this apparently does not mean that they share an specific-mate recognition system (Masters, this volume)].

The much more limited information about the Anubis-Ibean hybrid zone (Maples and McKern, 1967; Samuels and Altmann, 1986) suggests that here too male baboons migrate across the "boundary" just as readily as they move between local groups of their own form, and that they breed just as successfully as any other immigrant. In other words, here too there is apparently no break in the present-day zygostructure corresponding to the phenostructural boundary.

The hybrid zones represent one kind of discordance between zygostructure and phenostructure. The geographical isolates (Arabian Hamadryas and Saharan Anubis) represent the opposite case, where zygostructure is highly discontinuous, but there is little or no corresponding phenostructural discontinuity. Since the zygostructural discontinuity corresponds to an extrinsic, geographical barrier, there is no way of knowing whether it also corresponds to an intrinsic "barrier" of any kind, although we presume that it does not.

Finally, we should note that the present-day zygostructure of the *Papio* baboons is not bounded by the limits of the genus *Papio,* as this is presently recognized, since occasional interbreeding in the wild, with production of offspring, apparently occurs between Anubis baboons and geladas (Dunbar and Dunbar, 1974).

Translating Population Structure into Classification: Alternative Schemes

The classifications proposed before extensive fieldwork had elucidated the broad outlines of baboon phenostructure are now only of historic interest and need not be reviewed in detail. For example, Elliot (1913) recognized three subgenera and 16 species within the genus *Papio* as defined here. Most of his species can be fairly characterized as typological and his subgenera as arbitrary. Schwarz (1928), on the other hand, anticipated later schemes by recognizing populations of Yellow and Chacma baboons as a *Rassenkreis*—a chain of subspecies—within a single polytypic biological species. Other works of the period that attempted to express variation within *Papio* in manageable units (e.g., Ellerman *et al.,* 1953) did so by treating variables such as size and coloration in an essentially arbitrary manner (Jolly, 1965).

Baboons and the Biological Species

About 25 years ago, I (Jolly, 1965) suggested a classification of baboons within the framework of the BSC. Each of five "major forms" (Guinea, Olive, Yellow, Chacma, and Hamadryas) was distinguished as a species ("semispecies" would probably have been more correct). The distinctiveness of the Hamadryas (which was assigned its own genus *Comopithecus* in some contemporary classifications, e.g., Simpson, 1945) was recognized by its separation from the others at the superspecies level. Forms (such as the Gray-footed baboon), commonly regarded as geographical races within the "species," were designated as subspecies. This comparatively conservative scheme was adopted by systematists (e.g., Hill, 1967).

With the description of the Awash hybrid zone (Kummer, 1968; Nagel, 1973), it became apparent that a separation of the Hamadryas at the superspecies level (or higher) was inappropriate, and this usage dropped out of favor. Three alternatives evolved from it:

1. The two-species scheme retains the Hamadryas-vs.-the-rest dichotomy of the previous classification but reduces all ranks by one level; the result is two species, one (*P. hamadryas*) for the Hamadryas, the other (*P. cynocephalus*) for all the rest. Major forms, except for the Hamadryas, become subspecies, and geographical variants are no longer named (e.g., contributors to Smuts *et al.*, 1986). Members of the non-Hamadryas "species" are often called "savanna baboons."
2. The five-species scheme eliminates the Hamadryas-vs.-the rest separation, but the five major forms retain their specific (or semispecific) rank (Altmann and Altmann, 1970; Kummer, 1970; Thorington and Groves, 1970).
3. The one-species scheme (Groves in Thorington and Groves, 1970; Szalay and Delson, 1979) combines both elements: The *Hamadryas-Cynocephalus* dichotomy is eliminated and all forms are reduced to subspecies.

That all three schemes are still current among primatologists who more or less explicitly accept the usages of the BSC is testimony to the latter's flexibility (some might say its slipperiness). Because of the Awash hybrid zone, both the five-species and the two-species models must accommodate the existence of extensive marginal interbreeding and hybrid swarms between named species. According to the canonical definition, the existence of such zones does not necessarily disqualify the "forms" as separate biological species (Mayr, 1942) as long as cross-boundary gene flow is insufficient to threaten their essential coherence as units of evolution (Hull, 1970). While this criterion is ontologically elegant, it obviously leaves much to be desired epistemologically, because (like the famous "potentially interbreeding" clause) it requires the systematist to speculate about the evolutionary future. Theoretically, it might be possible to estimate the time needed for the forms to merge together if we know current values of all the relevant variables (rate of flow, disadvantage of hybrids, and so on). But even then, we would not be able to predict whether these values would be maintained until equilibrium was attained, rather than be altered by extrinsic, environmental change. This criterion of the BSC is obviously of little practical help in such cases.

However, the three current classifications are not equally adequate expressions of the BSC and contemporary taxonomic practice. The least satisfactory is the two-species model, widely espoused by socioecologists (e.g., Smuts *et al.*, 1986; Richard, 1985), because it separates at the species level the very forms that interbreed freely in the best-documented hybrid zone. If any two forms belong to the same biological species, on the criterion of demonstrated gene flow, these are Hamadryas and Anubis. The two-species scheme separates these, while combining other, pairs of neighbors (for example, Kinda and Gray-footed) whose intergradation has not yet been proved by on-the-ground investigation. Moreover, the twofold scheme is very likely to be paraphyletic, which some would consider a weakness. The current phenostructure of *Papio* does not permit an unambiguous reconstruction of its zygostructural history (i.e., its cladistics), but nothing suggests that the Hamadryas is the sister taxon of all other baboons. The

Hamadryas' peculiarities are clearly autapomorphic and probably part of a single adaptive complex. Its synapomorphies link it with the Anubis and the Guinea baboons, in the "maned" group, and is likely to have been derived from a form most closely resembling the Guinea baboon. The "savanna baboon species" can only be defined by the *absence* of the derived features of Hamadryas baboons. Finally, the error of this classification is compounded by the misleading vernacular "savanna baboon." Many "savanna baboons" (including all "typical" chacmas) do not live in savannas; indeed, some live in habitats as arid as the Hamadryas'.

Why should the two-species classification have gained wide currency, in spite of its paraphyletic taxa and its weakness as an expression of any coherent, zygo-based species concept, including the classic BSC? Perhaps the answer lies in an excessive reverence for "the species" as a universal currency of evolutionary analysis. The socioecologists are not unaware of hybrid zones, of the core definition of the BSC, or of accepted taxonomic practice, but their agenda is to discern patterns of adaptation linking ecology with behavior and structure. They have been told, by the same tradition that gave rise to the BSC, that the universal, all-purpose operational unit for the comparative study of evolution is the biological species. Moreover, one of the attributes of a good biological species (though not part of its formal definition) is ecological distinctiveness (Mayr, 1963). Thus, by backwards logic, the Hamadryas baboon, ecologically distinct from the other forms, must be a separate biological species, while the various forms of "savanna baboon," to which primatological folk wisdom (probably simplistically) attributes a common ecology and social organization, are lumped in a single species (Fig. 3A).

The socioecologists' taxonomic scheme expresses a reality (though an oversimplified one) about baboon ecodiversity that is significant for their analyses. Its error is its implication that biological species are the only valid units for comparative analyses. The prime fact for evolutionary socioecology is that ancestral Hamadryas baboons evidently acquired derived attributes, physical and behavioral, arguably in adaptation to a semidesert habitat. It is of no consequence in this context that this evolutionary change in the population's phenostructure demonstrably did *not* entail the appearance of genetic isolation. (In fact, the question of how the Hamadryas became specialized without speciating is an interesting, but separate issue.) The Hamadryas does not need to be a "good" species to be significant. To put it another way, had the narrow Hamadryas-Anubis hybrid zone never been discovered, or never existed, the implications of the Hamadryas' adaptive complex for comparative ecology would not have been one whit different, or less significant. The "demotion" of the Hamadryas from species status, which in my view is the best expression of current zygostructure is BSC terms, is simply irrelevant.

The five-species scheme is less objectionable, but its nested arrangement of subspecies within species still misrepresents the continuity within the whole *Rassenkreis*. Some neighboring populations assigned to different species—for example, *P. anubis heuglini* and *P. cynocephalus ibeanus*—are in fact very similar. Calling the Gray-footed baboon *P. ursinus griseipes* (i.e., grouping it with the Chacma) obscures the fact that phenetically it is at least as close to the "typical" Yellow baboon, *P. cynocephalus cynocephalus*. The populations supposedly typical

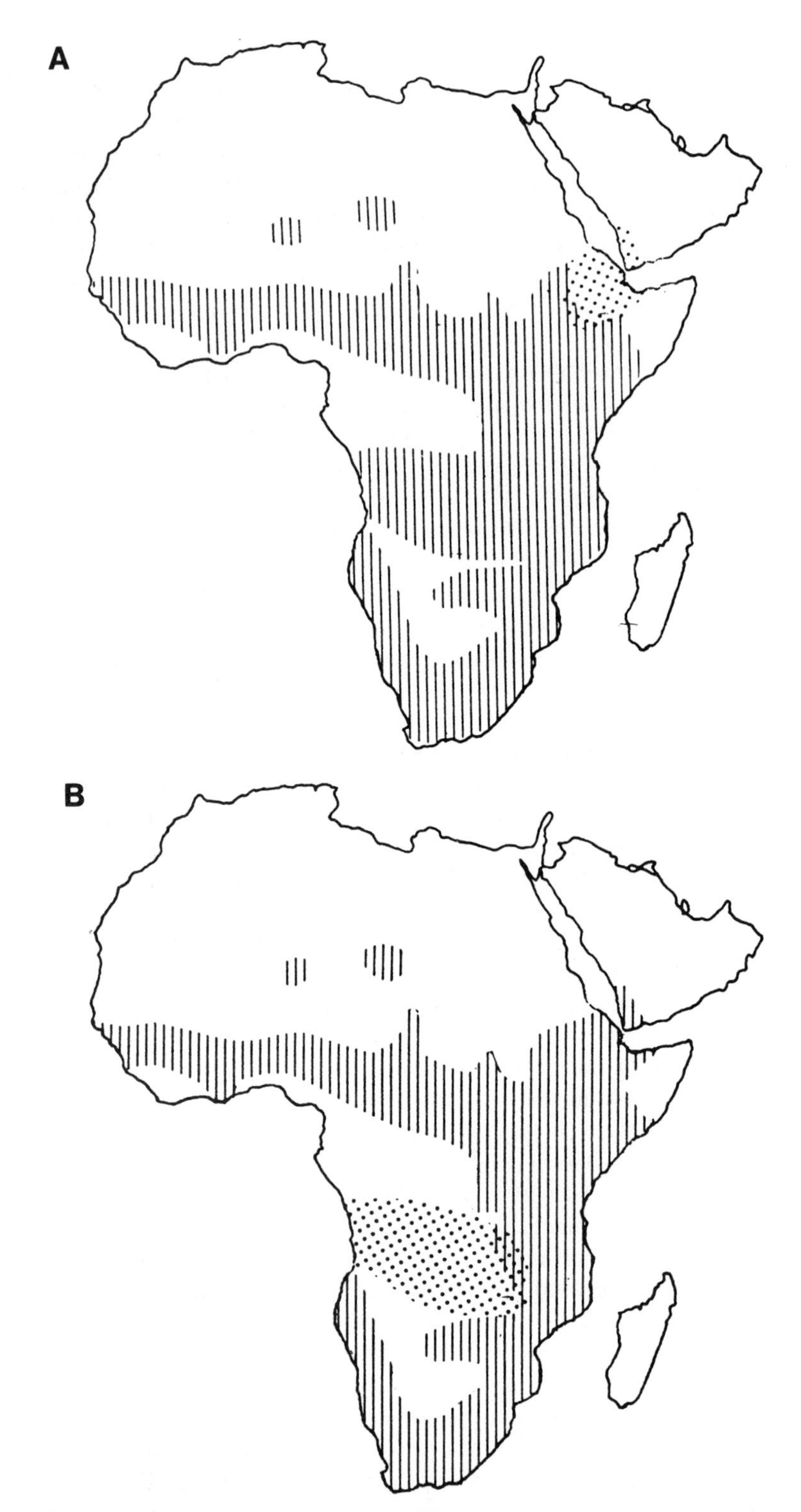

Fig. 3. "Socioecologists' species" and "paleontologists' species" of baboons. Overlapping textures indicate zones of hybridization or intergradation. (A) Distribution of species recognized by the "Hamadryas-vs.-the-rest" classification. Stripes: "Savanna baboon" (*Papio cynocephalus*). Dots: "Desert baboon" (*Papio hamadryas*). (B) Distribution of hypothetical "paleontologists' species," recognized by discontinuity in dental size. Stripes: "Big baboon species." Dots: "Small baboon species."

of major forms are, at most, the end populations of stepped clines. (The "typical" Yellow baboon does not even have this distinction; phenetically, it lies squarely in the middle of the typical Chacma-Kinda baboon series. It is an unhappy accident of nomenclature that this is also the type population for *P. cynocephalus,* the valid name for the non-Hamadryas "species" in the twofold scheme.) The end-point forms are the most extreme, but the populations that link them (especially, the Gray-footed, "typical" Yellow, and Ibean baboons, but probably the Transvaal and Heuglin's baboon also) are just as "real"—just as phenotypically stable within extensive ranges. They are not recombinant types picked arbitrarily from variable populations or hybrid swarms. Moreover, the hierarchical structure of the five-species scheme implies as particular history of diversification—that the five "species" were the products of an initial radiation and that the "subspecies" were subsequently produced by internal diversification. While this may have occurred in some cases, it is not the only possible explanation for the observed phenostructure, and until alternative scenarios can be tested we should avoid enshrining one of them in classification.

No classification could represent all the complexities of relationships among these populations (even if we knew them). However, assigning all recognizable populations equal rank at least carries no misleading cladistic implications, and assigning them all to one species expresses the lack of intrinsic barriers to gene flow, a major element in the BSC. Apart from the Saharan and Arabian isolates, all populations are linked through marginal interbreeding, observed or inferred. It would therefore theoretically be possible for a baboon gene to "flow" from Dakar in Senegal, to Dire Dawa in Ethiopia, to Cape Town in South Africa. It is another question, of course, whether in fact any gene is likely to make that journey and whether biologically realistic levels of gene flow would have any significant genetic or evolutionary impact. This is a question that could, in theory, be answered, but the necessary zygostructural parameters (migration rates and distances, demic distribution) simply cannot be estimated from data presently available. This being so, it is unfortunate that many systematists set great store by such tenuous and hypothetical links of gene flow, seeing them as the hallmark of the biological species' evolutionary coherence.

The fiction that a biological species is necessarily "united by gene flow" is even more obvious in the case of the Saharan and Arabian populations, which are totally surrounded by extrinsic but highly effective barriers to interbreeding, yet are invariably classified as Anubis and Hamadryas, respectively. To assign the isolates to their "correct" place, the BSC must either invoke "potential to interbreed" (which, since interbreeding in captivity is explicitly excluded as evidence, is epistemologically weak) or simply fall back upon phenetic and/or ecological resemblance.

What about the gelada, with which olive baboons apparently hybridize successfully, though rarely, in the wild (Dunbar and Dunbar, 1974)? Should it therefore be assigned to the same biological species? Geladas are invariably assigned to a different species, almost always to a different genus, and by some (e.g., Szalay and Delson, 1979) to a different papionin subtribe, Theropithecina. Certainly, the BSC does not require that geladas and Anubis be considered conspecific, as it permits slight marginal gene flow between species. This reinforces the point that BSC species boundaries are not, in practice, defined by the criterion of contemporary reproductive isolation, when the latter yields results incompatible with preconceived species taxa. Once again, however, the taxonomic decision is much less interesting than the zygostructure-phenostructure discrepancy itself. What is truly remarkable is that an Anubis living in central Ethiopia is more likely to produce an offspring with a gelada than with an Anubis living in Ghana or Tibesti, let alone with a Chacma in the Kalahari.

This particular zygo-pheno discordance may be very old. The gelada belongs to a clade that, from a copious fossil record, is known to have been phenetically distinct from *Papio* for at least 4 million years, during which it underwent a modest but undeniable radiation (Jolly, 1972; Jablonski, 1991). During most of this time, various species of *Theropithecus* were widespread in the lowlands of Africa, well beyond the present range of *T. gelada,* and at some sites were sympatric with *Papio.* This raises an intriguing possibility, that *Theropithecus* exchanged genes with *Papio,* albeit at a low level, while both clades were undergoing adaptive radiation and internal cladogenesis. The implications of this for molecular divergence, and "clocks" derived therefrom (e.g., Cronin and Meikle, 1979), deserve consideration. For present purposes, we can simply observe that this scenario provides a nice demonstration of the extent of possible decoupling between zygo- and pheno-divergence, and the consequent inadequacy of a single taxonomic system as an expression of both forms of population structure and their evolution.

These arguments may appear trivial, in that they debate taxonomic cases that are not seriously disputed. All contemporary systematists would probably agree that Ethiopian Anubis are conspecific with Tibesti Anubis (although they can never mate with them), but not with geladas (with which they sometimes do). But they illustrate the elasticity of the BSC; within reason, the mix and stringency of zygo and phenocriteria can be adjusted until the population delineated coincides with the systematist's underlying preconception of the "real" species taxon. This, in turn argues that if the biological species concept is to be retained, its formal definition should reflect the subjective way it is actually used—perhaps "a cluster of phenetically and/or zygotically related populations, usefully considered a unit of actual or potential evolutionary importance." Obviously, such a "fuzzy" definition conveys only an imprecise idea of the degree of zygoisolation or phenodivergence exhibited by any particular species. That is not necessarily undesirable, however, if it reinforces the point that these phenomena can be, and should be, described in their own right, not as adjuncts to species definition.

Assuming, then, that the baboon forms are good examples of the category, what is a subspecies? First of all, it is clearly *not* a deme, even a special subcategory of deme (*contra* Kimbel and Rak, this volume). Demes are lowest level units of zygostructure—minimal Mendelian populations. Thus, deme structure is zygodefined and time specific. A subspecies (especially early in its history) may include only one deme, but widespread subspecies include many demes, some totally zygoisolated (like Tibesti baboons). Unlike demes, polytypy—the existence of subspecies—is primarily feature of phenostructure, not zygostructure. The relationship of subspecies to zygostructure is historical or cladistic; populations of a subspecies are phenotypically similar because they are *descended* from a single deme (or from a multidemic population with internal genetic continuity) and have inherited a diagnostic set of characters from that ancestor. Subspecies are clusters of phenotypically indistinguishable populations, but these clusters are no more "held together by gene flow," than species are.

Where subspecies are parapatric, the contemporary zygostructure, consisting of demes and deme clusters linked by migration and gene flow, may show no break that corresponds to the major phenostructural disconformity at the edge of the subspecies. This is the case in baboon hybrid zones, where *individuals* whose phenotypes identify them as belonging to two different subspecies can often be found mingling in the same breeding group, and therefore by definition belonging to the same deme. On the other hand, populations belonging to the same (pheno)subspecies may be genetically isolated by a major (extrinsic) zygostructural barrier (such as the Red Sea). Such cases of the disarticulation of zygo- and phenostructures, due to history, should not simply be dismissed as abnormal or transient; they may be more common in nature than is generally assumed and (as the gelada case suggests) may persist over stretches of geological time long enough to become a consideration in the interpretation of fossils.

This description presents a quite different view of subspecies from the one that sees them emerging only as ephemeral clusters of local adaptations within in a widespread, static species. This is not to say, of course, that local adaptations do not arise in this way. For example, Dunbar (personal communication) has shown that gross body size in baboons at a locality is highly predictable from mean annual rainfall (tellingly, this relationship held good regardless of subspecies, although Kinda baboons were not included). But it is hard to see how adaptations of this kind could account for the major features of baboon phenostructure—the weak concordance between habitat and phenotype, the apparently nonadaptive nature of the defining pelage features, and so on. If these inferences are correct, they imply that subspecies can and do originate in "events," that they have a life history as individuals (in the philosophical sense), and they have a dual definition, in which zygostructural history is the major component but contemporary phenostructure is also invoked. This makes them much more like species and, for that matter, like supraspecific taxa, than is often conceded. Fortunately, this is one set of hypotheses in evolutionary theory that may actually be testable. The development of new screening techniques makes it possible to examine large numbers of unlinked genetic markers that have a low probability of functional association. Concordant distribution of such markers with each other and with the external features identifying subspecies membership would be good *prima facie* evidence for common ancestry and an event-type origin.

If this hypothesis of subspecies origins is supported, it will perhaps weaken one source of prejudice against subspecies, especially among theorists of a punctuationist bent: their persistent association with the gradualist idea that subspecies are *incipient species*. The logic (e.g., Tattersall, 1991; Kimbel and Rak, this volume) seems to be as follows: Subspecies are defined as incipient species, but species origins are fast, local "events"; therefore, incipient species should hardly ever be observed; therefore, subspecies should not exist, or, if they do, they are ephemeral entities appearing in the static phase of species that have nothing to do with cladogenesis.

Certainly, the idea that subspecies are incipient species is inherent in the writings of the founders of the Synthesis; for example, Huxley's (1942, p. 167) statement, "... since most evolution is a gradual process, borderline cases must occur," and Mayr's (1963) extended discussion and documentation of populations showing incomplete speciation. But the evidence can be accommodated just as easily within a "punctuated" framework; for example, Mayr's definition of semispecies, as "... natural populations that have acquired some but not all properties of species" contains no implication of evolutionary mode. A gradualist gloss appears only with the next sentence: "It is employed when an *incipient* species has not *yet* fully completed the process of speciation" (Mayr, 1970, p. 287, italics added). Omit the italicized words, and substitute *population* for *species,* and the gradualist implication is gone. The painstaking documentation of the reality of geographical subspecies and semispecies (not least, by Tattersall himself, e.g., this volume) seems like a very large baby to eject with so little bath water.

Subspecies (or semispecies) do not need to be regarded as incipient species, any more than an airliner is an incipient space shuttle. Each takes off (an event) in the same direction (away from the planet), propelled by a force that is similar in kind but different in degree. Each has a lifetime during which it is independent. But one, being launched with sufficient force, achieves an indefinite, independent orbit, while the other eventually has to land. It is the landing, not the takeoff, that defines their qualitative difference. Analogously, it is a subspecies' failure to attain complete independence, not the mode of its origin, that defines its difference from a species. In this view, species, subspecies, and semispecies all can, and probably do, originate in episodes of zygostructural change that are essentially similar. However, whereas a "full" species can disappear only by extinction, a subspecies can disappear either by extinction or by Cheshire-cat-like fading away, losing its distinctiveness as it merges into neighboring populations.

This aspect of subspecies, often described as their "ephemeral" quality, is routinely cited as a reason for regarding them as units of lesser evolutionary importance, justifying investigation of *speciation* in particular, rather than the more general evolutionary phenomenon of *divergence* (see, for example, Masters, this volume). The case is well stated by Futuyma's (1987) analogy of species as pitons in the slope of an adaptive peak, footholds that can be trusted not to disappear before the next adaptive step is taken, unlike the shaky distinctiveness of subspecies. While the point is well taken, it also deserves further scrutiny. Is the difference really that absolute; is a subspecies *always* too ephemeral to be a "unit of evolutionary importance"?

We must start from the a realization that, ultimately, all taxa are "ephemeral"—all species and subspecies alike eventually disappear. The question about any taxon is thus not an absolute, Will it disappear?, but rather a relative, "How likely is it to persist long enough to provide a foothold for another evolutionary step—another event of divergence?" It should be noted that the ephemerality of subspecies has to be measured against that of all species, not just widespread, successful ones. If the punctuationists are correct in their interpretation of macroevolution as driven largely by the production of swarms of hopeful peripheral isolates, rapid extinction must be the fate of the vast majority of protospecies. Each *might* become a new evolutionary step, but its chances must be extremely small. A measure of security against extinction, both by competition and by accident, is attained only by the small minority of species that reach maturity as comparatively widespread, large-sized, and stable populations.

However, as the baboons show, subspecies too can be widespread, large, and stable. The gene flow occurring at their boundaries cannot be considered a major factor in their vulnerability to extinction, which must depend largely upon ecological and demographic factors. For purely ecological reasons, a subspecies such as *P. hamadryas anubis* will probably last longer than many "good"—but local and threatened—species, such as, for example, *Mandrillus sphinx*. Erosion by gene flow will occur only at the edges of its range, where it is in parapatry with others. The larger its range and gross population size, the lower the ratio of edge to middle, and the less the impact of gene flow will be. We lack the zygostructural data to quantify marginal gene flow, but the time needed for complete homogenization of (for instance) Anubis baboons and Ibean (Yellow) baboons via the east African hybrid zone would surely be considerable.

Thus subspecies such as those of baboons have evolutionary importance, in the sense that, although, like all taxa, they are ultimately ephemeral, they are likely to persist long enough to give rise to other, differentiated populations if circumstances of zoogeography are favorable. This would explain, for example, the stepped cline extending across the whole baboon species if each population, in turn, differentiated from its ancestor and then gave rise to a more derived offshoot. This model, again, is susceptible to testing by the use of genetic markers, but the data are not yet in. Among the closely related macaques, however, analysis of mitochondrial DNA polymorphisms indicates that just such a pattern occurred; the sister taxon of Japanese macaques (*Macaca fuscata*) is not simply rhesus macaques (*M. mulatta*), but specifically the populations of rhesus macaques closest to Japan (D. Melnick, personal communication).

Baboons and the Phylogenetic Species Concept

If the subspecies of baboons have so many species-like qualities, why not simply call them species? This would apparently be approved by the growing number of evolutionary theorists (Eldredge, this volume) who define a species as "an irreducible cluster . . . of organisms, diagnosably distinct from other such clusters, and within which there is a parental pattern of ancestry and descent" (Cracraft, 1989, p. 34) or, even more tersely, as "the smallest aggregation of populations . . . diagnosable by a unique combination of character states" (Nixon and Wheeler, 1990, p. 218). This definition fits not only many "good" BSC species (*M. sphinx*, for example), but also entities traditionally called subspecies of a polytypic species. For example, Groves (1986, p. 205) informally describes the subspecies level as "the place at which we stop lumping populations" (i.e., because adding any others would produce phenetic heterogeneity), which seems to coincide exactly with Nixon and Wheeler's definition of a species. With all clearly defined subspecies elevated to species rank, the polytypic species is effectively defined away. The remaining intraspecific variation will consist of deme-level adaptations whose distribution is likely to be geographically nonconcordant, and that will almost certainly prove ephemeral, the kind of patterns of variation that led Wilson and Brown (1953) to question the utility of the subspecies concept as a way of describing infraspecific variation.

This phylogenetic species concept (PSC) is a logical extension of the cladistic approach to the definition of higher taxa (just as, in the "synthetic" tradition of systematics, the BSC generates "compromise" taxa that match the higher taxa of Simpsonian classifications). Phylogenetic species are defined as clades and recognized by their unique cluster of shared, derived features. Whether or not this is a shift from a zygodefinition ("isolation") to a phenetic one (as Masters, this volume, suggests) depends upon one's definition of a clade, a matter over which cladists themselves seem divided. At any rate, the PSC shifts the defining criterion from the result of divergence (complete or near-complete genetic isolation) to its cause (cladogenesis). In terms of the aircraft analogy, the PSC groups the airliner with the space shuttle (because they both take off), rather than with (say) a bicycle (because neither enters orbit).

One salutary effect is that no forced arguments about hypothetical gene flow and potential interbreeding are needed to delineate taxa that include allopatric populations. The PSC unambiguously assigns the Tibesti and Arabian baboons to their "correct" clade, because they exhibit no substantial derived characters not shared with mainland members of their (sub)species.

A second advantage of the PSC is that it provides a more rational, universal unit for comparative socioecology, since each species taxon now corresponds to the origin event of a set of derived conditions. For instance, socioecologists seeking correlates of one-male-group social structure can infer two relevant origin events among the Papionina (the African papionin monkeys.) One occurred in the ancestry of the gelada, the other in the ancestry of the Hamadryas. Appropriate socioecological questions would be: Under what conditions did this evolutionary novelty arise? Are there any common features in the ecological contexts of the two origin events? The Papionina retaining the ancestral, multimale organization (also seen in most Macacina, their sister taxon) include many species assigned to four different genera—*Cercocebus, Mandrillus, Lophocebus,* and *Papio*—but this taxonomic diversity is irrelevant to the origins of the derived condition. Alternatively, one might take a different tack and assume that the *persistence* of multimale social organization through zygostructural events (including, but not confined to, complete zygospeciation) is also meaningful. In this case, one would also take into account the populations that have retained the ancestral condition, presumably judging the significance of this fact by the number and intensity of divergence events in their ancestry.

The application of the PSC to the baboons is generally discussed in terms of the familiar two (Masters, this volume) or five (Kimbel, 1991) species, but in fact it would require us to recognize at least eight, and perhaps more than 11, full species, of all degrees of zygoseparation and morphological distinctness. Aside from the despair of students required to learn the unfamiliar names of the "new" species, would this matter? At one level, probably not very much. Reality, after all, consists of the patterns and history of population structure, not their taxonomic expression. Any species definition has an arbitrary component to it, since degrees of zygo- and phenoseparation form a continuous series (though not necessarily a transformationist-type sequence) of stages. However, there are real disadvantages to the change in taxonomic usage.

First, considerable confusion is likely to result from the fact that the BSC is still the dominant paradigm. Many nonsystematists would undoubtedly miss the point that this revision is a redefinition of the species category and treat it as though it were a major factual discovery or assertion about taxa: that all the entities hitherto called *subspecies* are in fact good species in the sense of the BSC's core criterion of genetic isolation and integrity. Kimbel (1991, p. 366) points out that when, following the BSC, reproductive criteria are used to diagnose conspecificity (he has in mind Hamadryas and "savanna" baboons), this "may obscure evolutionarily meaningful taxa." However, this veil of obscurity will descend only if subspecies taxa are regarded, by definition, as meaningless ephemera. For the majority of nonsystematist biologists (as well as many systematists), reared on the polytypic species, subspecies are real features of natural variation and no more "obscure" than species are. Kimbel's statement might as easily be turned on its head; if populations that are in substantial genetic communication are named as full species, this is likely to "obscure" the evolutionarily meaningful existence of substantial, continuing, gene flow between them.

This is well illustrated by Kimbel's own analysis. He remarks that (1991, p. 364) "immunological distance, protein electrophoresis and DNA hybridization data show that Anubis and Hamadryas baboons are not sister taxa [*P. anubis* and *P. cynocephalus* are more closely related than either is to *P. hamadryas* (Cronin and Meikle, 1979) . . .]." There are a number of problems with drawing this conclusion from the data cited,* but the most important is that the authors give no clue to the identity or provenience of the animals used in their study. [In spite of pleas for such information (Jolly, 1971), this is normal practice in molecular systematics].

We can, however, be reasonably certain that little of the very extensive ranges of the "species" was sampled. In fact, given the source of most baboons imported during the 1960s and 1970s, it is not at all unlikely that the "*P. cynocephalus*" and "*P. anubis*" were, respectively, an Ibean and an olive baboon, captured within a few tens of miles of each other, close to the south Kenyan hybrid zone (Vagtborg, 1967). Far from proving the sister group status of two widespread baboon "species," these data may show no more than the phenetic resemblance of two individuals, part of the same zygostructural nexus, and perhaps no more different (in the few proteins examined) than members of the same deme.

This example is not egregiously wrong (in fact, its data were state of the art for their time, and its conclusion may be correct), but it illustrates the difference between species and subspecies. Sarich and Cronin were quite justified in using a few animals, of unstated provenience, when comparing guenons, geladas, ma-

*The cited work is primarily concerned with the phylogenetic history of *Theropithecus*. Its few data on *Papio* are drawn from earlier publications: Cronin and Sarich (1976, mostly about mangabeys) and Sarich and Cronin (1976, a general review). They consist of two measures of phenetic distance (electrophoretic band matching and immunological distance; I can find no allusion to DNA hybridization) that are by their nature not susceptible to cladistic analysis. So conclusions about "sister taxa" may be thought premature. Only one of the data sets (electrophoretic band matching) permits the three-way comparison. The full extent of the results is, "We cannot differentiate *P. papio* and *P. anubis* (D = 0), but all the other "species" in the genus can be discriminated from one another, with the D values ranging from 0.35 for the *cynocephalus-anubis* comparison to 0.5 for those involving *ursinus* and *hamadryas*" (Sarich and Cronin, 1976, p. 162) [quotes in original]. Since we do not know how many animals were tested to yield this result, we have no intra-"species" calibration to tell us whether a D value of 0.35 is significantly different from one of 0.5. In any event, the pattern scarcely supports the twofold division, as Kimbel implied.

caques, mandrills, mangabeys, and (generic) baboons, which was the main point of these papers, because the genetic distinctness of these taxa is not an issue. But when comparing members of different baboon subspecies, precisely because they are not reproductively isolated from each other, the provenience and representativeness of the individuals used in the analysis immediately become an issue.

This is well illustrated by Williams-Blangero *et al.* (1990), who analyze genetic markers in representative captive populations of each of the five major forms distinguished as subspecies. The analysis succeeds in its primary objective of pointing out the pitfalls awaiting the biomedical researcher who ignores the internal diversity of *P. hamadryas* (*sensu lato*). Of more interest in this context is the pattern of generalized distances among samples, which suggest an Anubis-Yellow-Hamadryas cluster, with Guineas and Chacmas as distant outliers.

As Williams-Blangero *et al.* emphasize, this finding applies only to these particular samples of baboons, all resident at the Southwest Foundation for Biomedical Research. Given that these Anubis and Yellow baboon populations were probably both derived from Kenya (Vagtborg, 1967), the pattern of genetic distances seems to approximate geographical distance quite closely and perhaps could be explained without reference to form membership at all. Since these major forms are neither panmictic nor isolated, neither common population-genetic models nor cladistic analysis can adequately describe their genetic diversity, which is why, I would argue, the subspecies category is best conserved to label such entities.

Another advantage in retaining the subspecies category, and consequently emphasizing the fact of postdifferentiation gene flow, is that this draws attention to the possibility that new, distinct taxa may originate by hybridization as well as by conventional cladogenesis. For example, if current trends continue, we can anticipate the continued spread of Anubis genes into the Ethiopian Hamadryas population and vice versa, with the eventual amalgamation of their gene pools and the disappearance from Ethiopia of both "pure" phenotypes. (Under present climatic conditions, very few Anubis genes can be expected to flow into the Ethiopian pool across the arid areas of northern Kenya.)

If so, the result will be the appearance of a new link in the *Rassenkreis,* a population intermediate phenotypically between Anubis of Kenya and the Sudan, and "pure" Hamadryas, which will then be confined to Arabia. There is a distinct possibility that some of the existing forms may have originated in this way. For example, "typical" Yellow baboons may well have originated from the fusion of Kinda and Gray-footed components (much as Hayes *et al.* (1990) suggest is now occurring in Malawi), with subsequent northward expansion of the stabilized hybrid population. Such scenarios must necessarily remain hypothetical until tested against more detailed genetic analysis. The point is that they should not be overlooked as possibilities and that if the entities involved are called subspecies rather than species, this is less likely to occur.

The fact that subspecies (i.e., PSC "species" with substantial gene flow between them) may share apomorphies resulting from gene flow rather than common ancestry will inevitably complicate their cladistic analysis, as Harrison (this volume) points out. However, this need not rule out cladistic analysis altogether. If a subspecies were to appear by a divergence event qualitatively similar to those that produce species, a "signal" from that event should be present in the derived population. Gradually, this signal would decay, as synapomorphies are overlaid by autapomorphies and as genes flow in from populations that are neighbors, but not necessarily sister taxa. However, a reasonable scenario combining genealogy and gene flow should be retrievable for a time. Characteristics dating from the subspecies' origin event should be nearly universal in their distribution, while those introduced by subsequent gene flow should reveal their origin by a clinal distribution, with lowest frequencies in the "pure" center of the species' range.

Once again, this emphasizes the point that when dealing with entities that are permeable to gene flow and not panmictic, it is essential to investigate their internal structure and variation (an option that is rarely available to the paleontologist). And it may be worth reemphasizing that all this phenostructural complexity due to gene flow after incomplete zygodivergence is a fact of nature; no redefinition of subspecies to species will make it go away (though it may well "obscure" it!) This is a message to be taken to heart, especially by paleontologists, whose assumptions about zygoseparation of distinguishable morphs can rarely, if ever, be checked by direct observation of zygostructure.

Although I am convinced of the value of maintaining the subspecies category, even in the context of the PSC, one potential drawback must be examined—the apparent loss of the easily defined universal currency for comparative studies that is provided by a more literal interpretation of the PSC. The subspecies belonging to a single polytypic species may be more "specieslike" in some ways than the nominal polytypic species enclosing them. In this case, it is they, and not the more inclusive and diffuse taxa (the polytypic species) that should, in many contexts, be entered into comparative studies along with coherent, monotypic species. No problem will result, however, if, as suggested previously, the analysis uses *significant events of phenodivergence* rather than species *per se* as units of comparison. A classification including subspecies (either BSC or PSC based) will only become problematic if concepts such as "intraspecific variation" are applied too mechanically, without regard to the entities and the problem concerned. The answer is attention to experimental design, not tinkering with classification.

Some opposition to the use of subspecies categories comes from advocates of the phylogenetic species concept who are also paleontologists (e.g., Kimbel, 1991). They claim, with some justification, that the subspecies concept has been misused in a diachronic context to overextend the reasonable bounds of species taxa, and thus to avoid confronting the problem of the speciation process. This topic is beyond the scope of the present paper; we may merely observe that even if correct, the argument is vulnerable to the "NRA defense," that a useful instrument should not be banned simply because it might be abused.

Thus, even if the basic premise of the phylogenetic species concept is accepted—and there are good reasons for doing so—I would suggest retaining a taxonomic usage recognizing the evolutionarily important differences between "full" species and subspecies or semispecies. Denied the use of the subspecies concept, we would be left without a term for phenotypically distinct entities ("phylogenetic subspecies?") that are widespread and internally stable enough to have a substantial time depth, and perhaps even to persist long enough to give rise to other, more derived entities, or to appear in the fossil record; yet—given the opportunity—to hybridize freely at their borders, to have SMRSs that are indistinguishable (at least to the animals themselves), and to tend toward genetic coalescence. If we lack a term for them, it is a short step to denying the existence of such entities in general, and consequently imposing an unnecessary limit on explanatory scenarios available to both neontology and (by extension) to paleontology. However, in arguing for the utility of the subspecies concept, we must not forget the seamlessness of the zygoisolation continuum. Unless defined by an *absolute* lack of gene flow (as well as phenetic differentiation), species are quantitatively but not qualitatively different from semispecies and subspecies.

Baboons and the Recognition Concept

The recognition concept—according to which species are diagnosed by possession of a specific-mate recognition system (SMRS) (Paterson, 1985)—and its application to primate systematics has been discussed by Masters (this volume). The following treatment will be based upon Masters' exegesis, which includes baboons as an illustrative case.

As Eldredge (this volume) points out, using the criterion of an animal's own recognition of appropriate mates seems a useful refinement and sharpening of the BSC. Like the PSC, it will recognize a distinction between intrinsic, behaviorally defined cohesion and cohesion, such as that of the Tibesti or Arabian baboons, that is (as far as we know) enforced only by being surrounded by an uncrossable barrier. Speciation, too, is more pointedly defined, as the breakdown of a shared system of mate finding.

Any discussion of the recognition concept must start from the realization that it is not a new species *definition.* It refers explicitly to the concept of a species (genetic species) as a population sharing a gene pool, which is derived directly from Carson (1957; see also Masters and Spencer, 1989) and probably ultimately owes much to the species concepts (though less, apparently, to the speciation theory) of T. Dobzhansky. The invention and application of the recognition concept has undoubtedly led to insights in evolutionary biology. Problems arise, however, in its application to species diagnosis. As far as I can see, there is little in the theory that cannot be included in one of the following general statements:

1. Speciation rarely if ever involves the selective reinforcement of intrinsic isolating mechanisms and often involves the reorganization of mate-recognition systems.
2. Animals tend to find their mates by using their dominant sensory modalities, which may not be the same as the taxonomists'; hence, morphocryptic (sibling) species exist. Understanding of this fact provides the systematist with clues for detecting such morphocryptic species.
3. Signals used by animals to identify conspecifics (for mating or other purposes) tend to be conserved in evolution, even through cladogenetic events. It is therefore a good idea for systematists to look at such features.
4. Members of a species share a specific-mate recognition system. Members of different species do not.

Of these, the first three are insightful, testable propositions that have led to fruitful research in evolutionary biology, but they are not directly relevant to the problem of species definition and diagnosis. The trouble lies with the fourth proposition.

When it comes to the use of the RC as a diagnostic tool for detecting "good" genetic species, I must confess an inability to understand, from Masters' reasoning and primate examples, how the specific part of a particular *specific*-mate recognition system can be distinguished from the more widely shared, "nonspecific" components, without referring to previously diagnosed species taxa. If this is in fact the case, the RC can diagnose species only by tautology. The second problem concerns the edges of the SMRS; how do we tell, in a particular case, which characteristics of a species are part of its SMRS and which are not? It is hard to imagine any aspect of a species' biology that does not affect its mate-finding ability in some way, and Masters appears to encourage a broad definition that includes not only social behavior and epigamic features, but also habitat and (in the case of the baboons) dental proportions. The trouble with this liberal approach is that it effectively reduces the statement that "each species has its own SMRS" to the considerably less original, "species are different."

Paterson's species definition (perhaps more exactly described as a diagnostic criterion of a genetic species) is as follows (1985, p. 25; cited by Masters, this volume): the "most inclusive population of individual biparental organisms which share a common fertilization system." In mobile animals, such as primates, a fertilization system consists largely, though not entirely, of the "means by which the animals attract and recognize each other—the specific-mate recognition system" (Masters, this volume). To operationalize the concept, we must answer two questions: First, and most obviously, what does "attract and recognize" mean? Second, more crucially, what animals are "the animals"?

Most systematists have assumed, I think, that "attract and recognize" means attract and recognize as mates, and that the criterion for attraction and recognition is sexual interest and mating (or their proxy, the production of offspring). However, this may not be an entirely correct reading. For example, the "whoop-gobble" of mangabeys (Masters, this volume) is considered to be a conserved signal that conveys membership in a reproductive community. By implication, a population lacking the whoop-gobble could be assumed to have broken out of the mangabey SMRS, and hence to belong to a different species. But it appears not to matter whether the message conveyed to another mangabey is "here is a potential mate," or "here is a potential sexual rival," or, simply, "here be mangabeys, keep out of our food patch." One would have thought that only the first of these, or, at a stretch, the second, could be relevant to a concept of a mate-recognition system. However, any species-specific communicative behavior, or even any prominently visible structure thought to be used in communication (for example, large male canine teeth in baboons; Masters, this volume), seems to be grist for the RC mill, whether or not it can be shown to have any relevance to mate recognition or attraction *per se*.

As for the second question, some have assumed that the answer should be "any animals." In this case, captive and experimental studies could be used, and the entities diagnosed as species would mostly be more inclusive even than conventional biological species. (In the baboon case, it would probably include macaques and mangabeys, as well as geladas). However, mate recognition in captivity is explicitly excluded by Masters (this volume). Another possibility would be "animals in 'the wild,'" which should yield taxa closely resembling biological species. However, this also is evidently too broad, since the natural interbreeding of Hamadryas and Anubis baboons in the wild is also considered inadmissable, on the grounds of their differences in structure and social organization. The only acceptable answer to the "what animals?" query would seem to be "animals that interbreed in the wild *and* are indistinguishable in structure and behavior," which, by any definition, must mean they are conspecific. Thus, the definition of an SMRS would seem to reduce to something like, "the behaviors and structures shared by members of a species and used for social communication." This would be a perfectly workable and useful concept to apply to the first three propositions listed previously but is obviously unhelpful in species definition or diagnosis. Since the working definition of a species seems to be any population that differs in any aspect of its appearance or social behavior, it is unsurprising that the recognition concept seems to diagnose entities equivalent to phylogenetic species (or sometimes what I would call phylogenetic subspecies). It would be more to the point, if one is to demonstrate the usefulness of a distinct RC concept for species diagnosis, to present a real or hypothetical case in which its indications would differ from those of the PSC and would more closely reflect the referent "reality."

If evidence having nothing directly to do with finding and recognizing mates is included as part of the SMRS, then any interpopulational differences could be interpreted the same way. (Applying the same criteria as Masters uses to infer that Hamadryas and "other" baboons are distinct species, it would be easy to conclude, for example, that different SMRSs are found within modern *Homo sapiens*.) If mate recognition is not to be defined by the recognition of mates, it might perhaps be better to find an alternative name for it.

Finally, we should consider the practice of using phenodata to extend the RC beyond the range of behavioral observation. Since it is practically impossible to determine the stimulus value of all potential components of the SMRS of every population (especially those that are known only as fossils), many characteristics have to be evaluated on commonsense grounds or by analogy with similar features in other species. This is reasonable, but the baboon example shows potential pitfalls in assuming this kind of functional link. Each baboon subspecies is externally quite different, and the animals themselves are certainly capable of perceiving these differences. Moreover, I have argued (1963) that in the Hamadryas case at least, features such as the luxuriant male mane evolved by classic intersexual selection (female choice) in a particular ecological and social setting. Nevertheless, in the hybrid zones there is no evidence that the physical differences affect the responses of potential mates in any way. [The suggestion of assortative mating presented by Masters (this volume) seems to be based upon a misreading of published accounts of the Awash (Hamadryas × Anubis) hybrid zone, as discussed above]. Whether or not one calls them part of the SMRS is irrelevant; what matters is that one cannot assume that even striking, population-specific external features that have evolved by sexual selection indicate membership of a distinct "genetic" species.

Another problem with linking the issue of mate recognition directly to species diagnosis is that it creates a false dichotomy between "intraspecific" and "interspecific" mating. While the former is considered *the* essential diagnostic attribute of species, the latter merits only a brief dismissal: "SMRSs sometimes have similar features that may be misconstrued [by the animals]" (Masters and Spencer (1989, p. 275). This is essentially the traditional BSC (IC) explanation, differing only in that the latter would predict selection in favor of behaviors tending to prevent such "mistakes," while the RC would not.* Both the IC and the RC see interspecies gene flow as an aberration, and a closed, integrated species gene pool as the norm.

However, if we forget species definitions and simply look at the distribution of mate recognition (defined simply as recognition of mates) in higher primates, we find complex continua that stretch well beyond conventional species boundaries. As Masters (this volume) emphasizes, mate recognition in anthropoids seems to be very often a matter of degree and is influenced by a variety of contextual factors, as well as by the age and sex of the decision makers. For example, *Macaca nemestrina* males apparently consider sympatric *M. fascicularis* females to be legitimate mates, but only when *M. nemestrina* females are absent or in short supply (Bernstein, 1966). This particular case was documented in a disturbed area and can therefore be swept under the rug of "human interference," but this is not the case in the forests of western Uganda, where male *Cercopithecus mitis* sometimes hybridize with female *C. ascanius* (Aldridge-Blake, 1968).

In any case, even if human influence is sometimes implicated in the environmental changes that seem to precipitate hybridization between "good" species, this does not remove the phenomenon from the realm of nature. Such instances would only be irrelevant if they were never duplicated under "natural" circumstances. But, given the known dynamism of climate, landform, and vegetation, it is surely more profitable to regard instances of hybridization due to anthropogenic modification of the environment as valuable examples of a natural phenomenon that would happen, though perhaps more rarely, even without human intervention.

Of course, it might be argued that these hybridizations are unimportant because they do not disrupt the essential integrity of the gene pool of the species involved. However, we cannot assume that such cases involve a simple lapse of judgment on the part of the participants. *C. mitis* and *C. ascanius* are morphologically quite distinct, have widely overlapping ranges, and are universally regarded as specifically separate. Nevertheless, cross-mating may in fact be a strategic reproductive decision by an individual that would not otherwise find a mate at all. The behavioral and genetic dynamics of hybridization, and its likely long-term effects, must be investigated empirically on a case-by-case basis. They cannot be inferred from the degree of phenetic distance between the interbreeding populations, much less from the taxonomic rank ("good species" or conspecific subspecies) to which the populations are conventionally assigned. This is particularly obvious in the case of the gelada-Anubis hybridization.

*It is surely stretching a point to imply (Masters, this volume) that the IC would predict that the Awash baboons would evolve isolation mechanisms *by 1973* (the date of the information reported in Phillips-Conroy and Jolly, 1986) and that their failure to do so is a "problem for the IC." In any event, the observed flattening of the intersubspecies phenocline in the Awash between the 1960s and 1973 can be explained by local, climatically driven effects.

A Paleontologist's Eye View of Baboon Systematics

As Harrison (this volume) shows, evolutionary paleontologists usually wear two hats. On the one hand, as paleobiologists, they have concerns that parallel those of the socioecologists; the need to discern and to describe "important" genetic and ecological discontinuities in assemblages representing (roughly) synchronic biotas of the past and to draw general conclusions about environment–biota relationships. On the other, as systematists, they are concerned with the definition and phyletic relationships of evolutionary entities that can be regarded as analogous to the species of neontology, but with the important addition of a documented diachronic record.

Variation in an extant group, like the baboons, provides the kind of analogy used to interpret the biased, fragmentary, and entirely phenetic evidence of fossils. For those who follow the BSC concept, paleospecies are defined as zygospecies, but diagnosed on phenetic data, using rules about zygo–pheno relationships that have been deduced from the extant fauna. The diagnosis of fossil (zygo)species necessarily means inferring paleozygostructure from a very limited set of phenodata—the evidence of those fragmentary hard parts that happen to have been preserved as fossils. Moreover, the interpretation of such data in zygostructural terms is only possible if there are general rules of correspondence between zygodiversity and phenodiversity. These rules must be derived from living forms, whose zygostructure is (supposedly) known from evidence independent of simple phenetic difference. Since this topic is discussed by others in this symposium (Cope, this volume; Plavcan, this volume), it will not be examined in depth here.

On the whole, variation among baboons supports the contention, recently emphasized by Tattersall (e.g., this volume), that inferences drawn from the limited subsample of hard-part phenodata present in fossils are likely to underestimate the zygodiversity of a fossil group, as well as variation in the external "soft part" features that would be visible in the living animal. For example, individual Guinea, Anubis, and Hamadryas baboons, which differ obviously in external appearance, can barely be distinguished in mixed samples of skulls and dentitions, especially if such tests are performed "blind" (Jolly, unpublished).

However, the baboons add a twist to this story. Tattersall (1991) goes on to argue that since hard-part phenodata, analogous to paleophenodata, consistently fail to distinguish closely related, but "good" (zygo)species among the living, it follows that such morphs as *can* be distinguished in a fossil assemblage are very likely to represent good species. The logic of this argument seems to embody a *non sequitur,* but in any case the baboons show that empirically there are limits to this line of reasoning. Suppose that a paleontologist of the future were judging extant baboons purely on hard-part analysis, without prior knowledge of external characteristics or zygostructure. Judged only by dental and cranial size, and size-related shape characters, they would fall clearly into two nonoverlapping groups. One consists only of Kinda baboons and the other of all other baboons, including the full-sized Yellow baboons that must certainly be the Kinda baboon's close relatives. In this case, the paleontologist would probably *overestimate* the number of "biological species" in the sample by a factor of two (Fig. 3B).

Moreover, even if the range of variation exhibited by the total sample of teeth and skulls were not judged too large to be accommodated within a single species, the paleontologist's most likely primary division (big vs. small, Kinda vs. the rest) would not recognize the most probable pattern of relationship (zygostructural history) among the constituent populations. This is not to say that the "large" forms of *Papio* (i.e., all except *P. h. kindae*) are identical or indistinguishable in hard-part structure.

Populations defined on the basis of external characters can often be distinguished on the basis of simple bivariate indices of cranial shape and dental proportions (Jolly, 1965) and would probably be even more distinct if subjected to multivariate methods, as illustrated by Albrecht and Miller (this volume). When they are considered as a analogy of fossil collections, however, the issue is whether "blind" sorting by eye or by multivariate methods would yield clusters conforming to natural populations. This has not yet been rigorously examined. Actual paleontological collections are hardly ever as representative and evenly spread as the sample used here. If the patchiness of fossil collections were to be simulated by subsampling the total geographical spread of extant baboon material, it would be possible to mimic a variety of phenostructural patterns, none of which could be unambiguously translated into the group's present-day zygostructure or its inferred zygostructural history.

In short, whereas *Cercopithecus* (Cope, this volume) and *Lemur* (Tattersall, this volume) evidently have (zygo)speciated extensively with little hard-part diversification, *P. hamadryas* shows just the opposite: a taxon with marked phenodiversity whose zygodiversification stopped short of complete reproductive isolation. This situation may not be unusual among primates, as Albrecht and Miller (this volume) demonstrate. It may perhaps be especially prevalent among species whose ecology permits them a wide and semicontinuous distribution. The gray langurs (*Presbytis entellus* subsp.) are a particularly striking case in point (Albrecht and Miller, this volume). Thus, taken together, it would seem that these nonhuman primate cases fail to support Tattersall's statement (1991, p. 80) that ". . . when clearly identifiable hard-tissue autapomorphies are found consistently to discriminate between two fossil sister populations it is reasonable to conclude that speciation has intervened." They are more consistent with his less sanguine opinion, on the same page, that "There appears to be no necessary relationship whatever between speciation, the establishment of genetic isolation between closely related populations . . . and any specifiable degree of morphological shift."

What this means is that the number of zygospecies ("good biological species") represented by a fossil sample simply cannot be determined from its phenostructure (distribution of morphological variation) alone, even if (*pace* L. Martin) we limit permissible intraspecific variation to that seen among living primates.

Note that this does not mean that we can say nothing at all about the zygostructure of fossil forms. We can certainly infer that contemporary, closely related but consistently separable morphs represent *populations* at some level; the uncertainty pertains merely to the *degree* of zygoseparation—were they subspecies or species? The importance of this question depends, of course, upon how different the two kinds of entities are conceived to be. As implied previously, I believe that in some ways contemporary, species-centered theory tends to overstate both the qualitative nature of this distinction and its theoretical importance. Suppose that the remains of Kinda baboons and Gray-footed baboons found their way into the same deposit. Given the coarse time grain of many fossil catchments, and the rapidity with which intersubspecific boundaries can fluctuate, it is not unlikely that remains of each would be found sharing a final resting place, even though they are parapatric or allopatric in life. When controlled for sexual dimorphism (easier in baboons than hominoids!), they would show no overlap in cranial or dental dimensions. We could infer that they represented populations separated by at least one substantial divergence "event," and that in the time between the event(s) and the fossilization, gene flow had been insufficient to restore homogeneity. We could not tell, from phenostructure alone, whether the "event" had established *complete* zygoseparation, nor whether an SMRS had been ruptured, let alone predict the subsequent course of zygostructural evolution that would determine whether the distinctness of the "forms" would be maintained or fade away. Perhaps these details are of limited paleontological interest, even though they diagnose a species in one or another of its guises. Suppose, however, we insist (as compilers of paleofaunal lists are apt to do) on knowing what the species were.

There are two possible strategies, both of which are represented by contributors to this volume. One, espoused by Tattersall (1991), is simply to define (and to diagnose) fossil species morphologically, according to the phylogenetic species concept. This approach has the advantage that the progress of phylogenetic species so delineated can be charted through time and space, since neither chronological nor spatial distribution is involved in their diagnosis. However, the price that is paid is that closely related species recognized morphologically cannot be assumed to have the attributes of BSC or RC species. The baboons (and langurs) show that the phenostructure–zygostructure equivalence is too weak for that convenient assumption to be valid.

Specifically, we can legitimately define *Homo neanderthalensis* as a phylogenetic species, and we can make the same kind of assumptions about its origins (in an event) and zygoisolation (at every time horizon during its documented existence) that we made about the hypothetical fossil samples of extant baboons. However, we cannot assume that it necessarily was *completely* zygoisolated from other human species at all times or that it had its own distinct SMRS. Most important, we cannot assume that, just because it persisted unchanged over a relatively extensive period of time, it must eventually have disappeared by extinction rather than by being absorbed or genetically swamped. In other words, the possibility that it was a phylogenetic subspecies remains open and can only be investigated by adding distributional data to the analysis, and, even then, only in probabilistic terms.

The latter approach is exemplified by Kimbel and Rak (this volume), who use distribution as an adjunct to the diagnosis of (zygo)species status for two "forms" of early *Homo* that are obviously close relatives. Here, the conclusion that two populations are being sampled is based upon phenostructure: a "fairly consistent pattern of morphological differentiation." The second conclusion, that the two populations were fully isolated (zygo)species (rather than subspecies), is based upon their apparent coexistence in two widely separated regions. This inference is probably valid. It is theoretically possible that two parapatric subspecies could be represented in a single fossil site that happened to be situated on or close to the boundary zone between their ranges. However, the possibility that *two* sites both sampled the boundary must be considered less likely than the alternative explanation, that the two forms were widely sympatric. If they were widely sympatric, then they were probably reproductively isolated, if not absolutely, at least to the degree that *Cercopithecus ascanius* is from *C. mitis, Macaca fascicularis* from *M. nemestrina,* or *Theropithecus gelada* from *Papio hamadryas.*

Summary and Conclusions

To provide data for evaluation of alternative species concepts, variation among the baboons has been described in terms of the taxon-free concepts of zygostructure and phenostructure. The description reveals the relative paucity of zygostructural information, especially of the kind needed to evaluate hypotheses about the evolutionary role of gene flow. Nevertheless, a reasonably clear picture emerges, showing a zygostructure characterized by gene flow that is apparently unrestricted by intrinsic barriers and structured only by social organization, geographical impediments, and distance. By contrast, the phenostructure of the group, at least as represented by external morphology, is much less continuous, with sharp discontinuities defining forms. At a descriptive level, the combination of the two landscapes of variation forms as pattern that is thoroughly familiar from works such as Mayr (1963): a *Rassenkreis* of forms that are morphologically distinct but not reproductively isolated. There seems little or no reason to avoid the traditional and obvious taxonomic expression of this pattern, namely, a single polytypic species including nine or more definable subspecies. Recognizing them as an equivalent number of full (phylogenetic) species would obscure the important attributes of subspecies, conceal the overall closeness of their relationship, and bias interpretations of their past and predictions about their future evolution.

Developments in evolutionary theory over the past two decades have had the salutary effect of directing attention towards the process of speciation, emphasizing the difference between adaptation and the evolution of reproductive isolation, and pointing to the importance of comparatively sudden and even revolutionary "events" of genetic reorganization. A less salutary side effect has been the tendency to dismiss intraspecific variation as inconsequential ephemera, perhaps tainted by association with "gradualist" evolutionary philosophy. The evident stability and success achieved by baboon subspecies, which was probably due to events in their zygostructural history, but is demonstrably unrelated to irreversible genetic isolation, suggests that this view should be reevaluated. Subspecies and semispecies—entities, as Mayr described them, with some but not all of the features of species—do not have to be regarded as "incipient species" to be important in evolution. Moreover, a population does not have to achieve full reproductive isolation (or, except trivially and by definition, a new SMRS) in order to give rise to a new generation of isolates.

If there is a general conclusion to be drawn, it is that many concepts involved in species definition that are usually discussed as though dichotomous and absolute are more realistically treated as continuous and relative. These concepts include *isolation* (genetic and reproductive), *ephemerality* (vulnerability to extinction or loss of distinctiveness to gene flow), *mate recognition*, "*evolutionary importance*" (i.e., probability of founding a substantial clade), and *evolutionary divergence* (including, as a special case, speciation). This would seem to indicate that the species concept itself may be less absolute than it has, of late, tended to be regarded, and less special as a unit of analysis. Species, like other taxa, are most usefully defined by a combination of phenostructure and zygostructural history (genealogy), and there is a case to be made for recognizing subspecies (and perhaps semispecies) as true taxa, with a zygohistorical basis.

As for the baboons themselves, they amply illustrate Mayr's (1963) statement that taxonomic organization inevitably forces an oversimplified structure upon a complex, evolutionarily dynamic situation. Specifically, no arrangement of species and subspecies can express the following complications of the baboon case:

1. The ecological/adaptive distinctiveness of the Hamadryas baboon, but (apparently) no other subspecies
2. The more marked osteodontic distinctiveness of *kindae* which is combined with external characters much like other Yellow baboons
3. The recognition, as mates, of members of a much wider entity that probably includes most papionine monkeys
4. The present zygostructure of the group, including the total isolation (but lack of phenetic distinctiveness) of Saharan and Arabian baboons, and the marginal interbreeding with geladas
5. The phylogenetic relations among subspecies within the group, inferred from the persistent signatures of historically sequential events of differentiation.

This is not, however, a counsel of despair. It simply suggests two conclusions, one particular, one general:

1. That we call the baboons a single species, including (at present) nine recognizable phylogenetic subspecies, but do not have any illusions that we have, thereby, fully described the complexities of the present or past zygostructure of the group, much less about its potential as a single evolutionary entity or a cluster of such entities.

2. That in this and similar complex cases, which are probably far from exceptional, we should be less concerned with the formal systematics—admitting that they will never be perfect, either in general or in particular—and instead concentrate upon filling the gaps in knowledge about the two major axes of populational diversity—phenostructure and zygostructure—which between them encompass most if not all of that underlying structure of the natural world that systematists strive to express.

Acknowledgments

Many of the ideas expressed in this chapter have developed from conversations with colleagues in the Awash Baboon Project, especially Fred Brett and

Jane Phillips-Conroy (its codirectors). This work was supported by grants from the National Science Foundation, NIH, the Harry Frank Guggenheim Foundation, and Earthwatch. The critical reading of an earlier version of this chapter by Judith Masters, Bill Kimbel, Gene Albrecht, Terry Harrison, and Lawrence Martin contributed significantly to the final result. I would like to express my gratitude to Bill and Lawrence for the opportunity to participate in the lively session and the subsequent learning experience, and for their patience in awaiting the manuscript. Someone has to be last.

References

Abegglen, J.-J. 1984. *On Socialization in Hamadryas Baboons.* Associated University Presses, London.

Aldrich-Blake, F. P. G. 1968. A fertile hybrid between two *Cercopithecus* spp. in the Budongo Forest, Uganda. *Folia Primatol.* **9:**15–21.

Altmann, J., Altmann, S. A., Hausfater, G., and McCuskey, S. A. 1977. Life-history of yellow baboons: physical development, reproductive parameters and infant mortality. *Primates* **18:**315–330.

Altmann, S. A., and Altmann, J. 1970. *Baboon Ecology.* University Press, Chicago.

Anderson, C. 1987. Female transfer in baboons. *Am. J. Phys. Anthropol.* **73:**241–250.

Ansell, W. F. H. 1960. *Mammals of Northern Rhodesia.* Govt. Printer, Lusaka.

Ansell, W. F. H. 1978. *Mammals of Zambia.* Govt. Printer, Lusaka.

Bernstein, I. S. 1966. Naturally-occurring primate hybrid. *Science* **154:**1559–1560.

Carson, H. L. 1957. The species as a field for gene recombination, in: E. Mayr (ed.), *The Species Problem,* pp. 23–38. American Academy for the Advancement of Science, Washington, D.C.

Chepko-Sade, B. D., Shields, W. M., Berger, J., Halpin, Z. T., Jones, W. T., Rogers, L. L., Rood, J. L., and Smith, A. T. 1987. The effects of dispersal and social structure on effective population size, in: B. D. Chepko-Sade and Z. T. Halpin (eds.), *Mammalian Dispersal Patterns,* pp. 287–321. University Press, Chicago.

Cracraft, J. 1989. Speciation and its ontology: the empirical consequences of alternative species concepts for understanding patterns and processes of differentiation, in: D. Otte and J. A. Endler (eds.), *Speciation and its Consequences,* pp. 28–59. Sinauer, Sunderland, MA [cited by Kimbel, 1991].

Cronin, J. E., and Meikle, W. E. 1979. The phyletic position of *Theropithecus:* congruence among molecular, morphological and paleontological evidence. *Syst. Zool.* **28:**259–269.

Cronin, J. E., and Sarich, V. S. 1976. Molecular evidence for dual origin of mangabeys among Old World Monkeys. *Nature* **260:**700–702.

Delson, E. 1975. Evolutionary history of the Cercopithecidae, in: F. S. Szalay (ed.), *Approaches to Primate Paleobiology. Contrib. Primatol.* **5:**167–217.

DeVore, I., and Washburn, S. L. 1963. Baboon ecology and human evolution, in: F. C. Howell and F. Bourlière (eds.), *African Ecology and Human Evolution.* Viking Fund, New York.

Dorst, J., and Dandelot, P. 1970. *A Field Guide to the Larger Mammals of Africa.* Collins, London.

Dunbar, R. I. M., and Dunbar, P. 1974. On hybridization between *Theropithecus gelada* and *Papio anubis* in the wild. *J. Hum. Evol.* **3:**187–192.

Dunbar, R. I. M., and Nathan, M. F. 1972. Social organization of the Guinea baboon, *Papio papio. Folia Primatol.* **17:**321–334.

Ellerman, J. R., Morrison-Scott, T. C. S., and Hayman, R. W. 1953. *Southern African Mammals, 1758–1951: A Reclassification.* British Museum, London.

Elliot, D. G. 1913. *A Review of the Primates.* American Museum of Natural History, New York.

Freedman, L. 1963. A biometric study of Papio cynocephalus skulls from Northern Rhodesia and Nyasaland. *J. Mammal.* **44:**24–43.

Futuyma, D. J. 1987. On the role of species in anagenesis. *Am. Nat.* **130:**465–473.

Groves, C. P. 1986. Systematics of the great apes, in: D. R. Swindler and J. Erwin (eds.), *Comparative Primate Biology I: Systematics, Evolution and Anatomy,* pp. 187–217. Alan R. Liss, New York.

Hall, K. R. L. 1965. Behavior and ecology of baboons, patas and vervet monkeys in Uganda, in: H. Vagtborg (ed.), *The Baboon in Medical Research*, pp. 43–61. University of Texas Press, San Antonio.

Hayes, V. J., Freedman, L., and Oxnard, C. E. 1990. The taxonomy of savannah baboons: an odontometric analysis. *Am. J. Primatol.* **22:**171–190.

Hill, J. E., and Carter, T. D. 1941. The mammals of Angola, Africa. *Bull. Am. Mus. Nat. Hist.* **78:**66–67.

Hill, W. C. O. 1967. Taxonomy of the baboon, in: H. Vagtborg (ed.), *The Baboon in Medical Research*, Vol. 2, pp. 4–11. University of Texas Press, San Antonio.

Hull, D. L. 1970. Contemporary systematic philosophies. *Ann. Rev. Ecol. Syst.* **1:**19–53.

Huxley, J. S. 1942. *Evolution, the Modern Synthesis.* Allen and Unwin, London.

Jolly, C. J. 1963. A suggested case of sexual selection in primates. *Man* **222:**177–178.

Jolly, C. J. 1965. The Origins and Specialisations of the Long-Faced Cercopithecoidea. PhD Thesis, University of London.

Jolly, C. J. 1970. The large African monkeys as an adaptive array, in: J. R. Napier and P. H. Napier (eds.), *The Old World Monkeys*, pp. 139–174. Academic Press, New York.

Jolly, C. J. 1971. Protein variation and primate systematics, in: J. Moor-Jankowski and D. E. Goldsmith (eds.), *Medical Primatology.* Karger, Basel.

Jolly, C. J. 1972. The classification and natural history of *Theropithecus* (*Simopithecus*) (Andrews, 1916), Baboons of the African Plio-Pleistocene. *Bull. Brit. Mus. Nat. Hist. (Geol)* **22:**1–123.

Jolly, C. J., and Brett, F. L. 1973. Genetic markers and baboon biology. *J. Med. Primatol.* **2:**85–99.

Kimbel, W. H. 1991. Species, species concepts and human evolution. *J. Hum. Evol.* **20:**355–371.

Kingdon, J. 1971. *East African Mammals: An Atlas of Evolution in Africa.* Academic Press, New York.

Kummer, H. 1968. *Social Organisation of Hamadryas Baboons. A Field Study.* Karger, Basel and University Press, Chicago.

Kummer, H., Götz, W., and Angst, W. 1970. Cross-species modifications of social behavior in baboons, in: J. R. Napier and P. H. Napier (eds.), *The Old World Monkeys*, pp. 351–363. Academic Press, New York.

Kummer, H. 1990. The social system of hamadryas baboons and its presumable evolution, in: M. T. deMello, A. White, and R. W. Byrne (eds.), *Baboons: Behavioral Ecology, Use and Care.* IPS, Brasilia.

Lönnberg, E. 1919. Contributions to the knowledge about the Monkeys of the Belgian Congo. *Rev. Zool. Afr.* **7:**145–149.

Maples, W. R., and McKern, T. W. 1967. A preliminary report on classification of the Kenya baboon, in: H. Vagtborg (ed.), *The Baboon in Medical Research*, Vol. 2, pp. 13–22. University of Texas Press, San Antonio.

Masters, J. C., and Spencer, H. G. 1989. Why we need a new genetic species concept. *Syst. Zool.* **38:**270–279.

Mayr, E. 1942. *Systematics and the Origin of Species.* Columbia University Press, New York.

Mayr, E. 1963. *Animal Species and Evolution.* Harvard University Press, Cambridge.

Mayr, E. 1970. *Populations, Species, and Evolution.* Harvard University Press, Cambridge.

Melnick, D. J., and Pearl, M. C. 1986. Cercopithecines in multimale groups: genetic diversity and population structure, in: B. B. Smuts, R. M. Seyfarth, R. W. Wrangham, and T. T. Struhsaker (eds.), *Primate Societies*, pp. 121–135. University Press, Chicago.

Nagel, U. 1973. A comparison of hamadryas baboons, anubis baboons, and their hybrids at a species border in Ethiopia. *Folia Primatol.* **19:**104–165.

Nixon, K. C., and Wheeler, Q. D. 1990. An amplification of the phylogenetic species concept. *Cladistics* **6:**211–223.

Packer, C. 1979. Inter-troop transfer and inbreeding avoidance in *Papio anubis. Anim. Behav.* **27:**1–36.

Paterson, H. E. H. 1985. The recognition concept of species, in: E. S. Vrba (ed.), *Species and Speciation.* Transvaal Museum Monograph 4. Pretoria.

Phillips-Conroy, J. E., and Jolly, C. J. 1986. Changes in the structure of the baboon hybrid zone in the Awash National Park, Ethiopia. *Am. J. Phys. Anthropol.* **71:**337–350.

Phillips-Conroy, J. E., Jolly, C. J., and Brett, F. L. 1991. The characteristics of hamadryas-like male baboons living in anubis baboon troops in the Awash Hybrid Zone, Ethiopia. *Am. J. Phys. Anthrop.*

Richard, A. 1985. *Primates in Nature.* W. H. Freeman, San Francisco.

Samuels, A., and Altmann, J. 1986. Immigration of a male *Papio anubis* into a group of Papio cynocephalus baboons and evidence for an *anubis-cynocephalus* hybrid zone in Amboseli, Kenya. *Int. J. Primatol.* **7:**131–133.

Sarich, V. M., and Cronin, J. E. 1976, in: M. Goodman and R. E. Tashian (eds.), *Molecular Anthropology*, pp. 141–170. Plenum, New York.

Schwarz, E. 1928. Ein neuer Pavian aus Nord-Rhodesia. *Zeitschr. Säugertierk.* **3**:211–212.

Shortridge, G. C. 1934. *The Mammals of South West Africa.* I. Heineman, London.

Sigg, H., Stolba, A., Abegglen, J. J., and Dasser, V. 1982. Life history of hamadryas baboons: physical development, infant mortality, reproductive parameters and family relationships. *Primates* **23**:473–487.

Simpson, G. G. 1945. *Tempo and Mode in Evolution.* Columbia University Press, New York.

Smuts, B. B., Seyfarth, R. M., Wrangham, R. W., and Struhsaker, T. T. (eds.), 1986. *Primate Societies,* pp. 121–135. University Press, Chicago.

Sokal, R. R., and Crovello, T. J. 1970. The biological species concept: a critical evaluation. *Am. Natur.* **104**:127–153.

Sugawara, K. 1979. Sociological study of a wild group of hybrid baboons between *Papio anubis* and *Papio hamadryas* in the Awash Valley, Ethiopia. *Primates* **20**:21–56.

Szalay, F. S., and Delson, E. 1979. *Evolutionary History of the Primates.* Academic Press, New York.

Tattersall, I. 1991. What was the human revolution? *J. Hum. Evol.* **20**:77–83.

Thorington, R. W., and Groves, C. P. 1970. An annotated classification of the Cercopithecoidea, in: J. R. Napier and P. H. Napier (eds.), *The Old World Monkeys,* pp. 629–647. Academic Press, New York.

Vagtborg, H. 1967. *The Baboon in Medical Research,* Vol. 2. University of Texas Press, San Antonio.

Williams-Blangero, S., Vanderberg, J. L., Blangero, J., Konigsberg, L., and Dyke, B. 1990. Genetic differentiation between baboon subspecies. Relevance for biomedical research. *Am. J. Primatol.* **20**:67–81.

Wilson, E. O., and Brown, W. L. 1953. The subspecies concept and its taxonomic application. *Syst. Zool.* **2**:97–111.

Speciation in Living Hominoid Primates

5

COLIN P. GROVES

Introduction

The most widely discussed review of models of speciation in animals (White, 1978) lists seven possible modes:

1. Gradual divergence of two large populations after their geographic isolation
2. The founder principle
3. De facto differentiation of two races following extinction of the geographic intermediates
4. Local steepening of a cline
5. Selection against hybridization of strongly differentiated local races (area effects)
6. Stasipatric speciation: spread of chromosomal rearrangements aided by high levels of inbreeding
7. Sympatric speciation, in the strict sense, by assortative mating

Discussion of these models by Groves (1989) and examination of the mammalian evidence suggested that strictly allopatric (1–3) and parapatric (4–5) models do not necessarily explain observed patterns better than modes 6 and 7. A common pattern, in which modes 6 and 7 would work very well, is the modified centrifugal speciation pattern of Brown (1957).

It seems worthwhile to examine the applicability of these modes in the Hominoidea. Of the two families to be recognized in a cladistic schema (Groves,

---

COLIN P. GROVES • Department of Archaeology and Anthropology, Australian National University, Canberra, A.C.T. 2601, Australia.

Species, Species Concepts, and Primate Evolution, edited by William H. Kimbel and Lawrence B. Martin. Plenum Press, New York, 1993.

1986, 1989), the Hylobatidae offer more evidence than do the Hominidae, but as we shall see a certain amount of inference is possible with regard to the Hominidae also.

Speciation in the Hylobatidae

The Hylobatidae have been assigned to a single genus, *Hylobates,* since the revision of Groves (1972). Of the four subgenera, the distribution of one (*Symphalangus*) is entirely contained within that of one of the others (nominotypical *Hylobates*); the other three are allopatric to each other.

The initial divisions within the Hylobatidae are deep, perhaps unexpectedly so: PPED (plasma protein electrophoretic distance) units between the subgenera *Nomascus, Symphalangus,* and *Hylobates* average 1.7 ± 0.1, which is comparable to divisions within the Hominidae and implies a radiation beginning 4–5 million years ago (Cronin *et al.,* 1984). Work in progress by S. Easteal and the author will test the amount of differentiation on nuclear and mitochondrial DNA sequences; it may have implications for genus-level taxonomy. By contrast, PPED units within the *H. (H.) lar* group are of the order of 0.3–0.4, implying separation times under half a million years. It is interesting that in Chivers' (1977) model of speciation in the genus, although the initial diversification phase is set much too late according to the PPED criteria (Chivers puts it about 800,000 B.P.), the radiations of the *lar* group are placed, according to sea-level reconstructions, at about the right time period. Indeed, in that the members of the *lar* group are even now rather incompletely separated, a very late diversification is strongly implied.

Two of the four subgenera of *Hylobates* have sufficient diversity at the species/subspecies level to be of interest to the present discussion: *Nomascus* and *Hylobates.* Although species boundaries within these two subgenera appear to be superficially similar to each other, in being parapatric or at most interdigitating, the species-maintaining mechanisms seem on closer examination to be rather different.

Nomascus

Groves (1972) classified all taxa of this subgenus into a single species, *H. (N.) concolor,* but anticipated its possible dismemberment at some future date when relationships became better known. The described taxa, whether species or subspecies, are mapped in Fig.1. Since then, Dao (1983) and Ma and Wang (1986) have both proposed to separate *H. leucogenys* as a full species on the grounds of its marginal sympatry with *H. concolor sensu stricto* in Vietnam and Yunnan, respectively. Ma and Wang (1986) also supported their argument with reference to morphological findings, especially the shape of the baculum, which is a simple rod in *H. concolor,* but more complex in shape (probably more derived) in *H. leucogenys.* Groves and Wang (1990) have proposed, in addition, that a southern

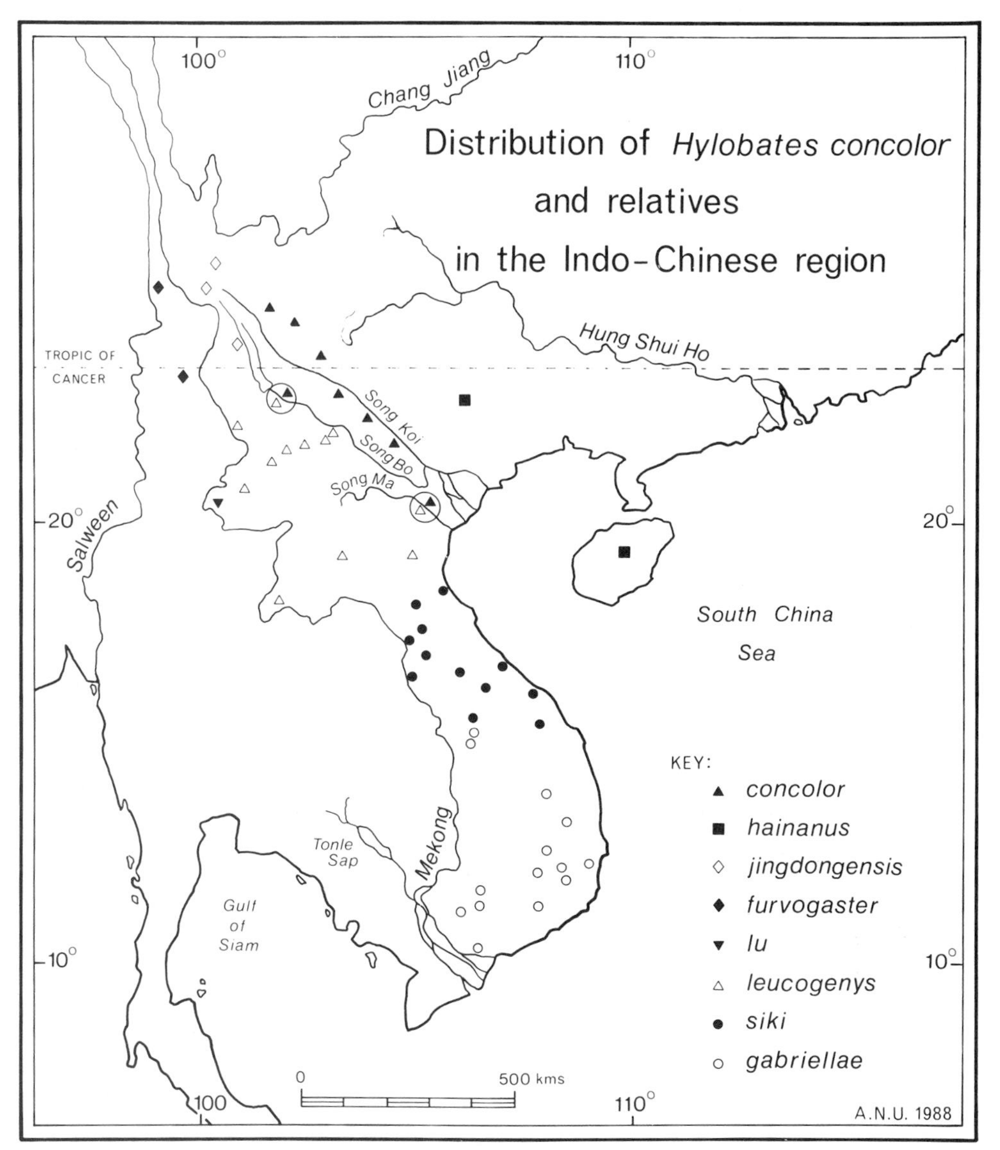

Fig. 1. Distribution of species and subspecies of the subgroup *Nomascus*.

species, *H. gabriellae*, is also distinguishable, appealing again to the morphology of the baculum, which is simple and resembles that of *H. concolor*. Boundary-zone data for *H. gabriellae* and *H. leucogenys* are lacking, but the two subspecies of the latter, *H. g. gabriellae* and *H. g. siki*, are known to interbreed where their ranges meet (Groves, 1972), though at what frequency is unknown.

Couturier and Lernould (1991) have examined karyotypes of four taxa in this taxonomically now rather complex subgenus: listing them north to south, they are *H. concolor ?hainanus*, *H. leucogenys*, *H. g. siki*, and *H. g. gabriellae* (taxonomy after Groves and Wang, 1990). Compared to *H. leucogenys*, whose karyotype has been well studied, *H. g. siki* differs by a reciprocal translocation

(1;22); *H. g. gabriellae* has the same translocation as *siki,* as well as a pericentric inversion of chromosome 7; and *H. leucogenys* × *H. concolor ?hainanus* hybrids are heterozygous for inversion 7 (implying that it is present in homozygous state in the latter) and lack the 1;22 translocation. This oddly mosaic pattern may imply homoplasy; alternatively, there have been considerable distributional changes in the past to explain the sharing of an inversion between the northernmost (*H. concolor*) and southernmost (*H. g. gabriellae*) taxa and its absence in the two more centrally distributed taxa.

The subspecies *H. c. lu* is cut off from its conspecifices by a bloc of *H. leucogenys* territory (Fig. 1); this may provide some support for the distributional reshuffling hypothesis; if, as I suspect, most or all of the character states of *H. leucogenys* are derived, then a centrifugal mechanism would explain it best. Again, the bacular similarity between *H. gabriellae* and *H. concolor* may imply past geographic musical chairs, but this similarity too seems likely to be primitive (given the simpler form of the baculum and its resemblance to that in other subgenera), so the same centrifugal mechanism may suffice. The most parsimonious model would seem to be that:

1. A widespread ancestral species resembled *H. concolor* in being all black in the male and juvenile, and with a black venter in the adult female, and possessing the small baculum and pericentric inversion 7
2. The southern populations of this species (proto-*H. gabriellae*) developed light-colored cheeks in the male and juvenile, and lost the ventral blackening of the adult female
3. The northern part (proto-*siki*/*leucogenys*) of this southern form lost pericentric inversion 7
4. The northern part (proto-*leucogenys*) of this latter form developed a new style of cheek whiskers and a highly derived bacular morphology, and this morphology began to spread at the expense of preexisting ones
5. The remainder (proto-*gabriellae*/*siki*) of the southern populations homogenized for reciprocal translocation 1;22; this and the bacular difference prevented introgression by *leucogenys.* Probably the striking facial differences of the males and ventral pattern differences of the females acted as specific mate-recognition systems protecting *H. concolor* from *leucogenys* introgression, and the bacular difference acted as a *de facto* reinforcement.

In this model speciation does not follow cladistic lines. Groves (1989) has argued that, indeed, there is no reason why it should.

It is interesting, in conclusion, that the translocation, whose possession differentiates *H. gabriellae* from *H. leucogenys,* may reinforce (or enforce) reproductive isolation, while the pericentric inversion does not (since it differentiates the two subspecies of *H. gabriellae,* known to intergrade in the wild). This, in turn, requires that isolation between *H. leucogenys* and *H. concolor,* separated chromosomally only by the inversion, must depend on some other factor; above I have suggested simple color pattern.

Field work—if sufficient forest survives in Vietnam and Laos—may of course reveal that the picture is quite different from that suggested by the museum data. Recall that to Groves (1972) the picture in the *H. lar* complex appeared quite different from the way we now know it to be.

Hylobates

The *H. lar* group is a very tight-knit one, with a mosaic distribution of characters, involving mainly color polymorphism and the pattern of face and extremities; their distribution forms a complex of crosscutting north–south and east–west axes. The distributions of the putative species tend to be defined by rivers (Fig. 2), with different levels of interbreeding where their ranges meet across the headwaters. Although deciding how many species there are in this complex has some purely practical importance, one can only agree with Creel and Preuschoft (1984), who stated that "of much greater importance . . . is the opportunity afforded by the lesser apes to study the *process* of systematic differentiation, including speciation, among primates." They are, in reality, semispecies, of which some may have technically (on reproductive criteria) passed the point at which we could award them species status, and others have not.

At the headwaters of the Takhong River in the Khao Yai National Park, Thailand, two members of the group, *H. lar* and *H. pileatus,* meet and occasionally hybridize. Brockelman and Srikosamatara (1984) and Brockelman and Gittins (1984) have described the picture. More than 210 gibbon groups have been mapped in the area, though only 61 have been observed well, and just 18 of these appear to contain hybrids, although second-generation hybrids are not always recognizable. Among them are four "trios," groups of one male with two females of different genotypes, the junior female never having a dependent infant; at the time it appeared that such groups are unique to the hybrid zone, but subsequent fieldwork (Srikosamatara and Brockelman, 1987) has found such a group in pure *H. pileatus* in Khao Soi Dao, though the high frequency of trios in the hybrid zone remains a matter of significance.

In the Khao Yai overlap zone, pure *H. lar* can be found 4 km into the *pileatus* side of the zone, and pure *H. pileatus* go 5 km into the *lar* side; hybrids go further—9–12 km into the *lar* side and 6 km into the *pileatus* side (though the former is better studied). The change of the morphological index from 90% *lar* to 90% *pileatus* occurs over about 9 km (Brockelman and Gittins, 1984).

Hylobates lar comes into contact with another species of the group, *H. agilis,* at Ulu Mudah, on the shores of an artificial lake in a logged forest in northern peninsular Malaysia (Gittins, 1984; Brockelman and Gittins, 1984). In general, *H. lar* occurs on the south side of the lake and *H. agilis* on the north, but there are two mixed groups, one on either side (interestingly it is the male, in each case, that has strayed onto the wrong side), as well as a group that may be mixed or may be an *agilis* group on the wrong side.

Marshall *et al.* (1984) and Marshall and Sugardjito (1986) briefly describe a hybrid and backcross zone between what they term *H. muelleri* and *H. agilis albibarbis* (two taxa that are difficult to differentiate by pelage characters but easy to distinguish vocally) at the headwaters of the Barito River, Kalimantan; they characterize the hybrid population as "sparse," in contrast to denser populations

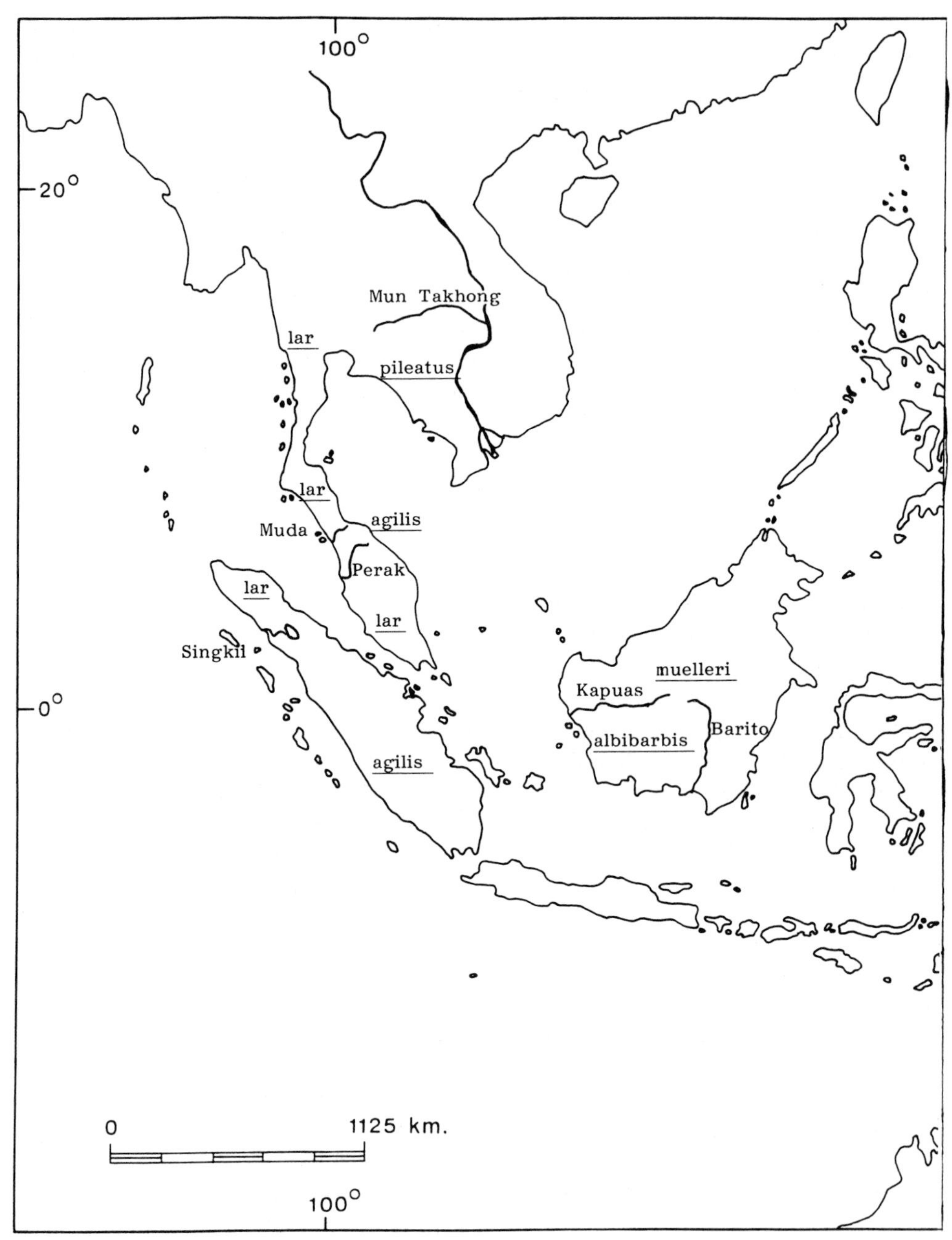

Fig. 2. Distribution of semispecies of the *lar* group, showing river barriers.

of the pure forms on either side. This zone has not yet been fully described, but additional information is given by Bodmer *et al.* (1991). The hybrid zone is between the rivers Busang (to the west) and Murung (to the east), upper feeder streams of the Barito; it covers several thousand square kilometers, with a density of 2.9 groups per square kilometer, which indeed is not high. West of the

Busang, towards the Joloi River, more hybrids occur; these are more *albibarbis*-like in the southern parts and more *muelleri*-like in the north. This doubtless explains Marshall and co-worker's (1984) mention of vocalizations typical of *H. muelleri* heard west of the hybrid zone amid the "dense, pure population of *H. agilis*." A brief personal communication from R. J. Mather (August, 1990) confirms that the hybrid zone in this case is quite unbroken, with no "pure forms" to be found within it.

Of these three hybrid zones, two (the *lar/pileatus* and *lar/agilis* zones) would appear to produce low frequencies of hybrids only, leaving the parental forms in the majority, though the evidence in the second case is extremely poor due to past disturbance and disruption. On the other hand, the *albibarbis/muelleri* hybrid zone goes much further: only hybrids are encountered within it. The question to be asked is, why should this be? What factors keep them apart in the first and (presumably) second cases, but not in the third?

Members of the *H. lar* group differ from one another essentially only in two parameters: pelage and vocalization.

Pelage

Hylobates lar has white hands and feet and a full white face ring; either sex (in the Thai subspecies) may be either black or buffy in general body color. *H. pileatus* also has white feet, but the face ring is reduced essentially to a brow band; there is a whitish crown ring, surrounding a black patch, and the crown hair is more flattened and flares out sideways. In *H. pileatus* the mature male is predominantly black; the female is buffy with a black ventral surface.

In *H. agilis* the polymorphism resembles that of Thai *H. lar*, i.e., nonsexually dichromatic, but in Malay *H. lar* there is a wide spectrum of color variation, centered on a medium-brown tone. *H. agilis* lacks the white hands and feet of *H. lar*, and the face ring is reduced to a brow band with whitish cheek whiskers in the male but not usually in the female (Groves, 1984; Marshall *et al.*, 1984).

Between *H. muelleri* and *H.* (cf. *agilis*) *albibarbis* there is no such clear-cut color differentiation: Both tend to a gray-brown tone (in the Barito region, *H. muelleri* is the grayer of the two), with a darker crown and ventral region.

To arrive at some idea of how pelage color might act in mate recognition, the data of Fooden (1969) were reexamined for evidence of assortative mating. Fooden listed the color phase composition of gibbon pairs observed by him and Carpenter (1940), noting that there seemed to be positive assortative mating, i.e., dark × dark and pale × pale pairings predominated over dark × pale. He performed no statistical tests, however, so to test his claim I calculated chi^2 on the basis of his tables. From Fooden's own data (5 d × d, 10 p × p, 1 d × p), we get chi^2 = 12.12 at 1 *df*, $p < 0.001$; from the data of Carpenter (1940)—11 d × d, 5 p × p, 3 d × p, 2 p × d—we get chi^2 = 4.95 at 1 *df*, $p < 0.05$. In both cases, therefore, the claim of positive assortative mating is substantiated: in *H. lar entelloides* and *H. l. carpenteri*, like tends to mate with like, and this differentiates both these subspecies of *H. lar* from *H. pileatus*, in which (there being sexual dichromatism) the pair are always unlike in color.

Turning to the data of Brockelman and Srikosamatara (1984), we find that matings within mixed groups are as follows (irrespective of species or hybrid status): 6 d × d, 3 p × p, 1 d × p, 2 p × d. This gives chi^2 = 5.53 at 1 *df*, $p < 0.05$. In the overlap zone, therefore, gibbons tend to behave more like *H. lar*, i.e., like mates with like. This is even the case for the two *H. pileatus* × hybrid matings: in both cases the male *H. pileatus* had paired with a black hybrid (TO2 and SY3, respectively, in Brockelman and Gittins' Table 41.2), implying that pelage color may not be the most crucial factor in this species' mate preference, although it does seem to be so for *H. lar*.

The two mixed *H. lar/agilis* groups (Brockelman and Gittins, 1984) were one d × p, one p × p; there is no basis for determining assortative mating tendencies here.

The two protagonists in the Borneo hybrid zone do not differ strongly in pelage characters. It may or may not be significant that here alone hybridization is more than merely occasional, but first the data from vocalizations must be examined.

Vocalizations

Taxa of the *H. lar* group differ most strikingly in the vocalizations of both sexes; not only in the form of the vocalizations themselves, but also in the structure of the male/female duet and in the daily timing of the calls. The male's solo is typically given around dawn in *H. agilis* and *H. muelleri* (in which it is very frequent) and in *H. moloch* (in which it is rarer), but later (8–9 AM) in *H. lar* and *H. pileatus*, in both of which it is rarely given; whereas the duet (great call) is usually somewhat after dawn in the first three species and in *H. lar* (where it follows on from the male solos), but later in the morning in *H. pileatus* (Haimoff *et al.*, 1984). All these differences are potentially significant as Specific Mate-Recognition Systems (see below).

Mitani (1987a,b) studied the significance of vocalizations and of the difference between *H. muelleri* and *H.* (cf. *agilis*) *albibarbis* by playback experiments. When a group call or a solitary female call was played back, a nearby pair would approach and duet; but when a male solo was played, only mated males approached (Mitani, 1987a). This seems to imply a greater tolerance by females of males' vocalizations than vice versa. When male songs of *H. muelleri* were played to these same *albibarbis* groups (Mitani, 1987b), males did not invariably approach and females usually gave an alarm call. He concludes that the male song is strictly a spacing call, so that in this sense they are less inclined to recognize *H. muelleri* as competitors; but females are alarmed by *H. muelleri* males rather than regarding them as potential mates.

These findings now have to be placed in the context of the finding that the two taxa concerned interbreed and form a complete hybrid swarm. It may be that the differences in frequency of the various response behaviors are inadequate to prevent interbreeding, given the slightness of the visual (pelage) differences and/or the identity in timing and frequency of the male solos between the two.

H. agilis and *H. lar* differ in both these respects much more substantially than do the two Bornean taxa; while *H. lar* and *H. pileatus* differ in visual appearance, but neither makes a high frequency of male solos. To test which factor is more important, it would be necessary to determine whether the hybridization rates in the two cases are similar (which would suggest that the visual factor is crucial) or lower in *H. agilis* (implying that timing and frequency of male solos are crucial). The very limited evidence suggests the latter, but as the contact zones between *H. lar* and *H. agilis* seem to be completely disrupted now, the evidence may never be fully satisfactory. It must be noted, finally, that duet and female solitary playback experiments also need to be conducted, and all such experiments need to be repeated on different pairs of species.

Perspective

It is certainly the case that taxonomic differentiation within species groups of *Hylobates* runs through an exceptionally wide gamut, from 70-km wide intergrade zones (as between *H. lar entelloides* and *H. l. carpenteri;* Groves, 1972) to the very narrow ones, as surveyed above. It is tempting to suggest that the presence or absence of vocal differences may be a crucial factor; on the other hand, the two subspecies of *H. hoolock,* which are not known to differ vocally, also hybridize over a very restricted area (Groves, 1972). Again, it is tempting to see the deciding factor as the presence of rivers as barriers, as rivers intervene between all the pairs of taxa in the *lar* group described above; yet the *Nomascus* taxa are not separated by rivers (it may be, of course, that in this case the differentiation between these taxa is so ancient that the rivers have long since altered their courses).

If vocalizations and perhaps pelage or facial patterns are such good Specific Mate-Recognition Systems (SMRS), how did they originate? Marshall *et al.* (1984) make the point that, starting from Javanese *H. moloch* as the putatively primitive taxon, we can distinguish an eastern and a western lineage of increasing northward differentiation: soaring great calls and white face rings to the west (*lar* and *agilis*), and bubbling calls and fancy hair-dos to the east (*muelleri* and *pileatus*). At the same time there are crosscutting strands: black morphs, white hands and feet, and also three-rooted P_3s (see Groves, 1972) in the north (*pileatus* and the northern races of *lar*), dorsoventral contrasts in the south (*muelleri* and light-phase *agilis*) (see also Groves, 1984). It looks like the result of an original clinal pattern being disturbed by the development of barriers, followed by homogenization on either side of them, an imprinting mechanism being implied.

Geissman (1984) has made interesting observations of two captive *H. lar* × *H. pileatus* hybrids. The male's vocalization was almost identical to that of *H. agilis,* and the female's to that of *H. moloch.* Geissmann reports that he consulted those familiar with the natural hybrid zones (Brockelman and Marshall) and discovered that this was noticeable in some wild hybrids also. The implication of a predominately northward differentiation (see above) seems to be confirmed.

Brockelman and Gittins (1984) discuss the gibbon data, with comments on speciation, which they note is poorly or not at all correlated with ecological boundaries. The specific mate-recognition system concept is but a "starting point in defining a broader concept that includes any social communication character that affects reproductive success . . . [it] is under strong stabilising selection": the demonstrative communication paradigm (Brockelman, 1984: p. 289), consisting of displays and other signals affecting potential mate recognition, resource defense, intrasexual interaction, and social bonding. Such a paradigm, they argue, would have to arise at the point of speciation itself, when the founder population was very small, given that new forms of such conservative characters would have no chance of spreading by natural selection in a large population; only in small isolated populations, with no near neighbors, would there be any chance of new variants becoming widespread and finally fixed.

This is fair enough, but given the observations and analyses of Groves (1972), Geissman (1984), and Marshall *et al.* (1984), that pelage character states and vocalizations are progressively more derived (from south to north), we can view each of them, to a degree, as elaborations on the one before: if a more southerly (= more primitive) state is a sign stimulus, then its more northerly (= more derived) version is the supernormal stimulus and so would spread. The stimuli form constellations (the total demonstrative communication paradigms, henceforth, DCPs), and their new, supernormal versions would spread in concert, their progress braked only by barriers such as seaways or river systems. Such a model opens up the possibilities that the recorded hybrid zones could be moving frontiers, shifting in time but temporarily held up by the river barriers. This would imply that the upper Barito feeder streams are, in effect, a genetic bottlenecking system, acting as a temporary brake on the spread of the *H. muelleri* DCP. The case of the *H. lar/pileatus* boundary is more complex, because *H. pileatus* is a combination of both eastern and western evolutionary streams, not merely a "supernormalization" of *H. lar,* so that the latter retains its own DCP validity and is not subject to invasion by that of *H. pileatus.*

Speciation in the Hominidae

In contrast to the Hylobatidae, the Hominidae offer no discernable evidence of the mechanics of speciation. Instead, we have a group whose phylogeny is probably the most intensively researched of all living organisms, offering a wealth of inferential clues.

As summarized by Groves (1986), the evidence to the mid-1980s suggested a scenario approximately as follows. The earliest split in the hominoid line was between the orangutan lineage (Ponginae) and the others (Homininae). In the Homininae, the gorilla would be the sister group of the human/chimpanzee clade, a still controversial view, supported by some analyses (Diamond, 1988; Ruvolo *et al.,* 1991) but challenged by others (Andrews, 1987). The view that sees evolutionary change as concentrated around speciation events—hence fast-evolving lines are more speciose than slower-evolving ones—was supported, and in particular there seemed to be much greater time depth within any one of the three great ape genera (*Pongo, Gorilla, Pan*) than within *Homo.* More recent findings have, however, indicated that in many respects, such as a chromosome morphology, humans have actually changed less from the common ancestor than have chimpanzees (Stanyon *et al.,* 1986).

The divergence between the two recognized orangutan subspecies is unexpectedly great, certainly way above the 10,000 years since the terminal Pleistocene sea-level rise separated Borneo from Sumatra. The possibility of craniometric differences between intra-Bornean populations as great as those between any of these and the Sumatran population (Courtenay *et al.*, 1988; Groves *et al.*, 1992) is a reason for being cautious about recommending species separation for Bornean and Sumatran subspecies. For the moment, it may be noted that the genetic distance between catchall Bornean and Sumatran subspecies is considerably greater than between subspecies of other mammals, such as big cats (Janczewski *et al.*, 1990) and infers a separation date of some 1.13 ma (based on a 13-ma Ponginae/Homininae separation date).

Similarly, the indicated time depth between eastern and western lowland gorillas, and the implied time depth between eastern lowland and mountain gorillas (Groves, 1986) is startling. There is now a need to examine the genetic differences between population isolates within each subspecies of great apes.

According to the data cited in Groves (1986), the genetic distance between *Pan paniscus* and *P. troglodytes* is somewhat less than that between the two subspecies of *Pongo pygmaeus,* although this is not confirmed by the most recent study (Janczewski *et al.*, 1990). There are also indications of considerable differences among the populations of *P. troglodytes* traditionally given subspecific status, even though these are very poorly defined craniometrically (Groves *et al.*, 1992). Favoring a 4-million-year split between human and chimpanzee lines, Hasegawa and Horai (1991) calculate a separation time of between 1.70 and 2.45 ma between pygmy and common chimpanzees based on three data sets from noncoding portions of mtDNA (the standard errors of the three estimates overlap). These would be somewhat increased were a 5-million-year human/chimpanzee separation accepted instead, a probable maximum according to Hasegawa and Horai (1991).

These data all suggest that, morphologically at least, evolution in the nonhuman Hominidae (the so-called great apes) has been very slow. When their fossil records become better known, they will furnish excellent tests for the view that slowly evolving lineages have low speciation rates. Again, we must qualify this as "morphologically" (i.e., not in molecular factors; see above).

The only speciation mode indicated is geographical. Water barriers have traditionally been implicated in quasi-speciation among the great apes, and there seems no question of interbreeding in the wild between any two taxa (except perhaps the mysterious "subspecies" of *Pan troglodytes*). There is a rather striking contrast to gibbons: over the vast period of time since their separation, reproductive isolation has not evolved between Bornean and Sumatran orangutans (which interbreed confusingly in captivity), whereas to different degrees this has occurred between pairs of the far less well-differentiated taxa of *lar*-group gibbons.

Conclusions

This brief survey of the nonhuman Hominoidea serves to remind us of the diversity of possible pathways for speciation. In *Nomascus,* speciation has been accompanied by chromosome change; in nominotypical *Hylobates,* it has not, and it may be significant that these two different patterns define separate supraspecific groups (see also Groves, 1989, 47–8). Specific mate-recognition systems take several forms, and the hybrid zones of *lar*-group gibbons offer opportunities for elucidating their relative importance. In general, the differences among the quasi-specific taxa of the Hylobatidae are considerably less than among the subspecies of *Gorilla gorilla* or *Pongo pygmaeus*, and far less than, say, among the semispecies of the genus *Papio*. It seems very doubtful whether such taxa could be differentiated if they were found as fossils, reinforcing the conclusions of several contributors to this symposium that species-level diversity in the fossil record is likely to be underestimated. The sorts of characters that differentiate the *lar*-group species are in no way different from those that differentiate subspecies, or even morphs. As discussed above, the most likely mode of origin for species in this group is precisely as derived polymorphic variants that spread as supernormal stimuli and, indeed, may still be "predatory" upon their plesiomorphic parents. On the other hand, the great ape data show us the other end of the spectrum: gibbon (at least *lar*-group) speciation is young and still active; great ape speciation and even subspeciation is ancient and stable.

References

Andrews, P. 1987. Aspects of hominoid phylogeny, in: C. Patterson (ed.), *Molecules and Morphology in Evolution: Conflict or Compromise?*, pp. 21–53. Cambridge University Press, Cambridge, England.

Bodmer, R. E., Mather, R. J. and Chivers, D. J. 1991. Rain forests of central Borneo–threatened by modern development. *Oryx* **25:**21–26.

Brockelman, W. Y. 1984. Social behavior of gibbons: Introduction, in: H. Preuschoft, D. J. Chivers, W. Y. Brockelman and N. Creel (eds.), *The Lesser Apes*, pp. 285–290, Edinburgh University Press, Edinburgh.

Brockelman, W. Y., and Gittins, S. P. 1984. Natural hybridization in the Hylobates lar species group: implications for speciation in gibbons, in: H. Preuschoft, D. J. Chivers, W. Y. Brockelman, and N. Creel (eds.), *The Lesser Apes*, pp. 498–532. Edinburgh University Press, Edinburgh.

Brockelman, W. Y., and Srikosamatara, S. 1984. Maintenance and evolution of social structure in gibbons, in: H. Preuschoft, D. J. Chivers, W. Y. Brockelman, and N. Creel (eds.), *The Lesser Apes*, pp. 298–323. Edinburgh University Press, Edinburgh.

Brown, W. L. 1957. Centrifugal speciation. *Q. Rev. Biol.* **32:**247–277.

Carpenter, C. R. 1940. A field study in Siam of the behavior and social relations of the gibbon (*Hylobates lar*), *Comp. Psychol. Monogr.* **16:**1–212.

Chivers, D. J. 1977. The lesser apes, in: HRH Ranier and G. A. Bourne (eds.), *Primate Conservation*, pp. 539–98. Academic Press, London.

Couturier, J., and Lernould, J. M. 1991. Karyotypic study of four gibbon forms previously considered as subspecies of *Hylobates (Nomascus) concolor* (Primates, Pongidae, Hylobatidae). *Folia Primatol.* **56:**95–104.

Courtney, J., Groves, C., and Andrews, P. 1988. Inter- and intra-island variation? An assessment of the differences between Bornean and Sumatran Orang-utans, in: J. H. Schwartz (ed.), *Orang-utan Biology*, pp. 19–29, Oxford University Press, New York.

Creel, N., and Preuschoft, H. 1984. Pathways of speciation: an introduction in: H. Preuschoft, D. J. Chivers, W. Y. Brockelman, and N. Creel (eds.), *The Lesser Apes*, pp. 427–430. Edinburgh University Press, Edinburgh.

Cronin, J. E., Sarich, V. M., and Ryder, O. 1984. Molecular evolution and speciation in the lesser apes, in: H. Preuschoft, D. J. Chivers, W. Y. Brockelman, and N. Creel (eds.), *The Lesser Apes*, pp. 467–485. Edinburgh University Press, Edinburgh.

Dao, V. T. 1983. On the north Indochinese gibbons (*Hylobates concolor*)(Primates: Hylobatidae) in North Vietnam. *J. Hum. Evol.* **12:**367–72.

Diamond, J. M. 1988. DNA-based phylogenies of the three chimpanzees. *Nature* **332:**6856.
Fooden, J. 1969. Color phase in gibbons. *Evolution* **23:**627–644.
Geissmann, T. 1984. Inheritance of song parameters in the gibbon song, analysed in 2 hybrid gibbons (*Hylobates pileatus* × *H.lar*). *Folia Primatol.* **42:**216–235.
Gittens, S. P. 1984. Territorial advertisement and defence in gibbons, in: H. Prenschaft, D. J. Chivers, W. Y. Brockelman, and N. Creel (eds.), *The Lesser Apes,* pp. 420–424. Edinburgh University Press, Edinburgh.
Groves, C. P. 1972. Systematics and phylogeny of gibbons, in: D. M. Rumbaugh (ed.), *Gibbon and Siamang,* Vol. 1, pp. 1–89. S. Karger, Basel.
Groves, C. P. 1984. A new look at the taxonomy and phylogeny of the gibbons, in: H. Preuschoft, D. J. Chivers, W. Y. Brockelman, and N. Creel (eds.), *The Lesser Apes,* pp. 542–561. Edinburgh University Press, Edinburgh.
Groves, C. P. 1986. Systematics of the Great Apes, in: D. R. Swindler and J. Erwin (eds.), *Comparative Primate Biology. 1. Systematics, Evolution and Anatomy,* pp. 187–217, Alan R. Liss, New York.
Groves, C. P. 1989. *A Theory of Human and Primate Evolution.* Oxford University Press, Oxford.
Groves, C. P., and Wang Y. 1990. The gibbons of the subgenus *Nomascus* (Primates, Mammalia). *Zool. Res.* **11:**147–154.
Groves, C. P., Westwood, C., and Shea, B. T. 1992. Unfinished business: Mahalanobis and a clockwork orang. *J. Hum. Evol.* **22:**327–340.
Haimoff, E. H., Gittins, S. P., Whitten, A. J., and Chivers, D. J. 1984. A phylogeny and classification of gibbons based on morphology and ethology, in: H. Prenschaft, D. J. Chivers, W. Y. Brockelman, and N. Creel (eds.), *The Lesser Apes,* pp. 614–632. Edinburgh University Press, Edinburgh.
Hasegawa, M., and Horai S. 1991. Time of the deepest root for polymorphism in human mitochondrial DNA. *J. Mol. Evol.* **32:**37–42.
Janczewski, D. N., Goldman, D., and O'Brien, S. J. 1990. Molecular genetic divergence of Orang Utan (*Pongo pygmaeus*) subspecies based on isozyme and two-dimensional gel electrophoresis. *J. Hered.* **81:**375–387.
Ma S., and Wang Y. 1986. The taxonomy and distribution of the gibbons in southern China and its adjacent region, with description of three new subspecies. *Zool. Res.* **7:**393–410.
Marshall, J., and Sugardjito, J. 1986. Gibbon systematics, in: J. Erwin and D. R. Swindler (eds.), *Comparative Primate Biology, 1: Systematics, Evolution and Anatomy,* pp. 137–185. Alan R. Liss, New York.
Marshall, J. T., Sugardjito, J., and Markaya, M. 1984. Gibbons of the Lar Group: relationship based on voice, in: H. Preuschoft, D. J. Chivers, W. Y. Brockelman, and N. Creel (eds.), *The Lesser Apes,* pp. 533–541. Edinburgh University Press, Edinburgh.
Mitani, J. C. 1987a. Territoriality and monogamy among agile gibbons (*Hylobates agilis*). *Behav. Ecol. Sociobiol.* **20:**265–269.
Mitani, J. C. 1987b. Species discrimination of male song in gibbons. *Am. J. Primatol.* **13:**413–423.
Ruvolo, M., Disotell, T. R., Allard, M. W., Brown, W. M., and Honeycutt, R. L. 1991. Resolution of the African hominoid trichotomy by use of mitochondrial gene sequence. *Proc. Natl. Acad. Sci. USA* **88:**1570–1574.
Srikosamatara, S., and Brockelman, W. Y. 1987. Polygyny in a group of pileated gibbons via a familial route. *Int. J. Primatol.* **8:**389–393.
Stanyon, R., Chiarelli, B., Gottlieb, K., and Patton, W. H. 1986. The phylogenetic and taxonomic status of *Pan paniscus:* a chromosomal perspective. *Am. J. Phys. Anthrop.* **69:**489–498.
White, M. J. D. 1978. *Modes of Speciation.* W. H. Freeman, San Francisco.

Geographic Variation in Primates

6

A Review with Implications for Interpreting Fossils

GENE H. ALBRECHT AND
JOSEPH M. A. MILLER

Introduction

Knowledge of geographic variation is fundamental to recognizing the kinds and numbers of both living and extinct primate species. This review focuses on geographic variation in the craniodental anatomy of living primates. We limit our attention to multivariate studies that used craniometric and/or odontometric data to investigate differences among demes, subspecies, and closely related species. Our intent is to document the nature of geographic variation in the skulls and teeth of living primates and, then, to discuss the problems and prospects of dealing with such variation among fossils. We begin with some general background observations about geographic variation and speciation in primates.

There is no consensus among biologists about what a species is, nor about how a species arises (e.g., see the diversity of opinions expressed in Otte and Endler, 1989). Primatologists echo these same arguments about species and speciation in their own group (e.g., Tattersall, 1986, Kimbel, 1991; Godfrey and Marks, 1991; and the various contributions to this volume). Species defini-

GENE H. ALBRECHT AND JOSEPH M. A. MILLER • Department of Anatomy and Cell Biology, University of Southern California, Los Angeles, California 90033.

Species, Species Concepts, and Primate Evolution, edited by William H. Kimbel and Lawrence B. Martin. Plenum Press, New York, 1993.

tions abound, including the biological, evolutionary, morphological, and recognition species concepts, among others. Endler (1989) produced a useful categorization of species definitions based on their aims: (1) taxonomic vs. evolutionary species concepts, (2) theoretical vs. operational species concepts, (3) contemporaneous vs. clade (paleontological) species concepts, and (4) reproductive (isolation) vs. cohesive species concepts. We endorse his emphasis on the utility of different concepts for answering different questions about different organisms, and we refrain from advocacy of one concept over another. However, no matter which species concepts are applied, inferences about primate species depend largely on interpreting morphological variation within and among primate groups, thereby taking into account relative differences among individuals, sexes, demes, subspecies, and species (see Shea *et al.,* this volume, on this same point). Our review of craniodental morphology is directed at documenting the place of geographic variation within this comparative framework. Doing so is a necessary prerequisite for better understanding of speciation in primates, whether the intent is (1) practical, as in delimiting primate taxa (i.e., what kinds of primates are there?), or (2) theoretical, as in elucidating the mechanisms active in the process of primate speciation (i.e., how do primates speciate?).

Defined as the "occurrence of differences among spatially segregated populations of a species," geographic variation is a universal phenomenon and represents a cornerstone of evolutionary studies (Mayr, 1963, p. 297; Gould and Johnston, 1972; Endler, 1977; Moore, 1991). Our emphasis will be on geographic variation as an integral part of what we call the hierarchy of morphological variability among organisms (Fig. 1). Interpreting the significance of variation at any level of this hierarchy can only be done in light of the magnitude and pattern of variation at other levels (i.e., somewhat analogous to analysis of variance in statistics). Thus, ideally, studies designed to investigate geographic variation should attempt to control for ontogenetic, individual, sexual, and interspecific variation; likewise, studies at the species level should incorporate geographic and all lower levels of variation. Geographic variation itself subsumes the entire intraspecific continuum, from differences among local breeding groups of individuals (e.g., troops of primates), to differences among interbreeding geographic populations (demes), to differences among interbreeding, taxonomically distinct groups of demes (subspecies). These categories are distinguished from higher levels of variation corresponding to all the familiar categories of biological classification from species on up.

The practical difficulty for the primatologist interested in delimiting species is drawing the line between geographic and interspecific variation. The use of binomial species designations imposes an unrealistic constraint of having to make dichotomous decisions about whether different populations are members of the same species or not at any one time. However, such decisions about taxonomic rank should not obscure our understanding of the complexity of speciation. As Godfrey and Marks (1991) emphasized, speciation involves many processes leading to a continuous range in the degree to which primates are isolated reproductively and/or genetically from one another. At one extreme are freely interbreeding local populations. At the other extreme are sympatric populations that never interbreed. In between are the many primates that are genotypically and/or phenotypically distinct yet hybridize to some extent in the wild.

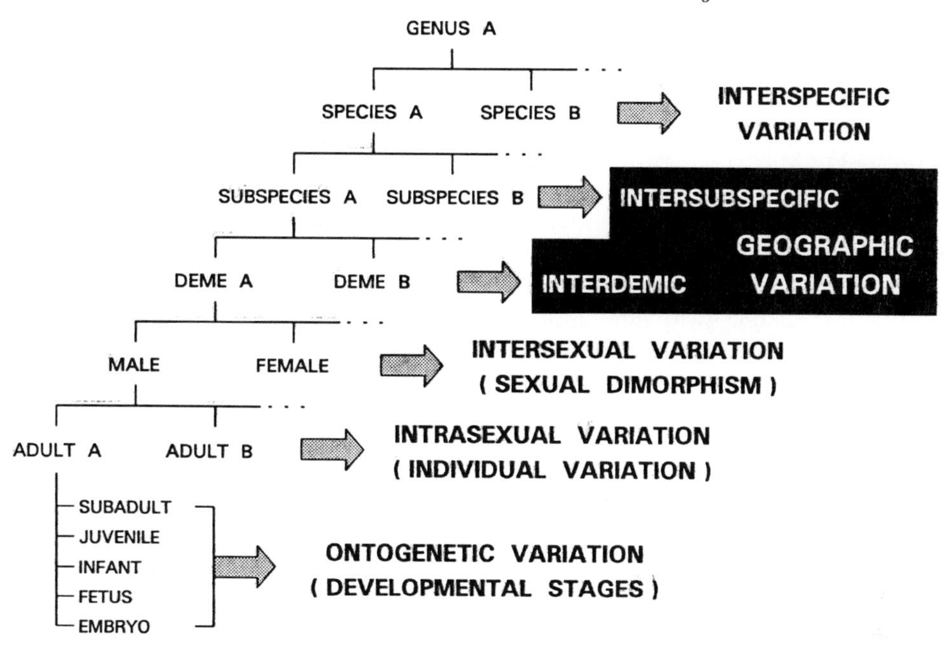

Fig. 1. Hierarchial classification of morphometric variation at the species level and below.

It is not always possible, therefore, to make sharp, unequivocal demarcations as to the species status of many primates. Nor is it desirable to do so if speciation is of scientific interest as a biological process, rather than being considered just an instantaneous event of historical importance marking a node of phyletic evolution. We refer the reader to the persuasive contribution by Jolly (this volume), who uses baboons to illustrate the constraints of formal systematics and narrowly applied species concepts, neither of which can be expected to characterize perfectly the "phenostructural" and "zygostructural" complexity of population diversity in the natural world.

The range of reproductive interactions among primates is illustrated by the macaques (hybridization in captivity is not considered here). Sometimes interbreeding between sympatric species does not occur, even though they may be found in mixed-species troops (e.g., *Macaca radiata* and *M. mulatta* intermix in central India; Fooden, 1988). Sometimes hybridization represents aberrant matings between distinct species in atypical situations (e.g., hybrids of *M. fascicularis* and *M. nemestrina* are known from areas of disturbed habitat in West Malaysia; Bernstein, 1966). Sometimes hybridization is an apparently rare event between parapatric or narrowly sympatric, closely related forms (e.g., *M. fascicularis* and *M. mulatta* hybridize in small, discontinuous areas along their common boundary in Indochina; Fooden, 1964, 1971). Sometimes hybridization is inferred from narrow zones of morphological intergradation between neighboring forms (e.g., subspecies of *M. nemestrina* are strongly differentiated, except where they meet on the Thai-Malay Peninsula; Fooden, 1975, Albrecht, 1980a). Sometimes there is widespread intergradation over broad areas between weakly distinguished forms (e.g., there is a wide transition zone between subspecies of *M. sinica* in Sri Lanka; Fooden, 1979). Sometimes the outcome of interbreeding remains theoretical when macaques, closely related or not, are disjunct (allopatric) in their geographic distributions (e.g., the many insular forms of *M. fascicularis* found throughout Southeast Asia are all reproductively isolated from one another, as is *M. sylvanus* from North Africa from all the Asian macaques). The endemic radiation of macaques on Sulawesi, as outlined in a later discussion of the group, provide further testimony about the range of interbreeding typically found among evolving primates. Godfrey and Marks' (1991) review, Groves' (this volume) report on gibbons, and Jolly's (this volume) account of baboons confirm the commonplace nature of intermediate levels of reproductive isolation in taxonomically distinct primates.

As macaques, baboons, and gibbons illustrate, there can be a diversity of relationships among primates living at any one time, either present or past. Our listing of various states of reproductive isolation in macaques is not meant to imply anything about degrees, stages, or sequences in the process of genetic divergence during speciation. Our point is simply that relationships among primates are complex, and this presents problems for primate taxonomists who are trying to determine whether the animals they study are species or not. This point is applicable no matter what species concepts are invoked and no matter what modes of speciation are thought to be active.

Given unlimited scientific resources, researchers would be addressing questions about speciation in the field by investigating the details of interactions among local populations. Questions of interest would relate to interbreeding, gene exchange, selection, and fine-grained analyses of behavioral and morphological adaptations. Practicality, of course, imposes a different reality with respect to species determinations, that is, taxonomic studies usually rely on assessments of morphological variation in specimens collected from the wild and preserved in museum collections (i.e., the "phenodata" of Jolly, this volume). Morphological variation is then used to make statements about genetic divergence, reproductive isolation, or other issues of interest. However, as also emphasized by Shea *et al.* (this volume), these statements are only inferences, since museum studies, based as they are on static samples drawn from a changing world, can provide only indirect evidence about the dynamic evolutionary processes of speciation. Actual genetic data and field observations on wild populations are welcome supplements, but such information is available only in a minority of cases and, even then, often is limited in scope to a few localities. Moreover, such information is often as problematic as inferences drawn from the morphological analysis of museum specimens.

Geographic variation is common in primates. Our modern classification of primates would not exist without the cumulative effect of hundreds of studies documenting geographic variation. Primate taxonomists rely on identifying and then assessing the significance of geographic differences in external morphology, especially color and pattern of the pelage. The overwhelming importance of external appearance has been emphasized historically (e.g., Pocock, 1925a,b) and is equally evident from the descriptions, tables, and keys of diagnostic characters in recent taxonomic surveys of primates (e.g., Hershkovitz, 1977; Napier, 1976, 1981, 1985; Tattersall, 1982; Jenkins, 1987). Such traits are expected to be of primary importance among animals that are, for the most part, diurnal and visually oriented. Experimental work on the ability of macaques to discriminate among closely related species suggests the importance of visual stimuli for species recognizing their own kind in the field (Fujita, 1987). Skeletal features, as our review shows, also exhibit significant geographic variation. However, they are less often used in classificatory schemes because external morphology provides such clear-cut identification in living animals. It may be that some skeletal features are modified in direct response to their importance as visual cues in primate communication (e.g., anatomy of tail carriage in pigtail macaques described by Fooden, 1975, or facial specializations in Sulawesi macaques described by Fooden, 1969, and Albrecht, 1978). Other characters, including those of genotype, behavior, and internal anatomy (including molecules), may be geographically variable but are generally less amenable to study.

Paleoprimatologists are greatly disadvantaged when dealing with questions of species and speciation among extinct animals. Consider the cumulative effect of compounding the following problems and questions. (1) The scarcity of fossil specimens makes it difficult to appreciate levels of variability. Does the variation among a handful of fossils represent differences among individuals, sexes, geographic demes, subspecies, species, or most probably, some combination of these? (2) The incomplete, biased samplings of fossil localities, which inadequately represent species ranges, make it difficult to interpret the meaning of geographic differences. Do distinct morphs from two fossil sites represent valid species that were sympatric at some geographically intermediate, nonfossiliferous locality, or do the morphs represent demes of a single species that has undiscovered, geographically intermediate populations forming a continuum of morphological variation? (3) The addition of temporal effects makes it difficult to distinguish evolutionary changes from shifting geographic distributions of taxa through time. Do distinct morphs from two sites of different ages represent one chronospecies that evolved and shifted its geographic distribution, or do the two morphs represent stasigenetic, allopatric, geographically variable taxa sampled across different places and times?

The most serious morphological deficit facing the paleoprimatologist interested in delimiting species is knowledge about the external appearance of animals now known only from fossilized skeletal materials. Reconstructions of what fossil animals actually looked like in life are speculative, aside from some soft tissues whose anatomy might be inferred from skeletal features. All that can be said, if our knowledge of living taxa is any guide, is that subspecies and species of extinct primates probably exhibited the same sorts of striking differences in pelage and other external features of visual impact. The paleontologist, however, must rely on variation in skeletal and dental characters for distinguishing fossil species. Some workers question the degree to which skeletal differences are observable among primate subspecies or closely related species (e.g., Tattersall, 1986, 1991; Kimbel, 1991). If differences among such sister taxa are small and, for the most part, unobservable in living primates, then paleoprimatologists are justified in designating each recognizable morph as a different species (excluding sexual differences). Otherwise, if morphological differentiation of hard parts is common and recognizable within living species, then distinguishable morphs in a fossil sample may represent nothing more than geographic variants, as opposed to distinct species. It is clearly important for primate paleontologists to understand the nature of craniodental geographic variation in living primates before attacking the less tractable problem of interpreting skeletal variation in fossil assemblages.

Geographic Variation in Living Primates

We review all studies that used canonical variate analysis (discriminant functions) and/or generalized distances to investigate craniometric and/or odontometric differences among demes, subspecies, and closely related species of living primates. These two methods are appropriate for analyzing differences among taxa because within-group variation is used as a standard of comparison (Albrecht, 1980b, 1993; Reyment *et al.*, 1984). This comparative analysis of within and between-group variation conforms to the methodological prescription recommended so strongly by Shea *et al.* (this volume). Both canonical variates and generalized distances provide quantitative summaries of relationships among individuals and groups based on overall morphological pattern, as defined by the suite of characters chosen for analysis. Multivariate methods are simply modern, quantitative analogs of classic comparative anatomy with metric summaries substituted for qualitative, "by eye" assessments of primate anatomies (Albrecht, 1980b).

We do not consider the many primate studies presenting univariate comparisons of craniodental variables. For the most part, such analyses are not very successful compared to multivariate approaches in characterizing geographic variation (e.g., compare Hershkovitz's, 1977, univariate summaries of craniodental dimensions with the multivariate analyses of *Callithrix jacchus* and *Saguinus fuscicollis* discussed below). Nor do we consider the many primate studies based on principal components, Penrose size and shape distances, and other such multivariate methods for analyzing mean differences among primate taxa (e.g., see Natori and Hanihara, 1988, 1989). Most such studies are directed at determining relationships among rather than within species, and are not directly comparable to multivariate studies incorporating measures of within-group variation.

Canonical variate results of published studies are converted to a standard format in our figures. Each group centroid (mean) is shown surrounded by a circle, which theoretically encloses 90% of the dispersion of specimens belonging to that group. These population confidence limits have a radius of 2.15 standard deviation units. The units of canonical variates, as well as generalized distances, are equal to the standard deviation of the within-group distribution adjusted to unit variance in all directions. Generally, this involves pooled within-group distributions, the use of which assumes homogeneity of the individual within-group dispersions across the samples included in any one study. The circles allow the distinctiveness of demes, subspecies, and species to be quickly assessed.

The generalized distance (D) between two samples is simply the linear distance between group centroids measured in the canonical variate data space (the square root of Mahalanobis' D^2; $\bar{D}$ signifies an average of two or more generalized distances). This means generalized distances are expressed in standard deviation units of the standardized (circular) within-group dispersion. They allow rough comparisons among studies, although this should be done with great caution, since studies are based on (1) variable sets, which differ in number as well as kind of measurements, and (2) within-group samples, which differ in composition as well as cohesiveness. Some studies published only generalized distances without the ordination obtained by canonical variate analysis. In those cases, we used principal coordinate analysis of the generalized distance matrix to derive canonical variates (Gower, 1966a,b). The results are exactly the same as canonical variate analysis using an unweighted among-group covariance matrix (i.e., "unweighted" in that sample sizes do not affect the construction of canonical variate axes; Albrecht, 1980b, 1993).

Our survey demonstrates the significant amounts of craniodental variation found whenever comparisons are made among geographic populations of primates. The empirical evidence also shows that the degree of variation differs from taxon to taxon, but in general, there is correspondence between morphological and taxonomic differentiation. Moreover, the pattern of morphological variation often reflects zoogeographic, ecogeographic, taxonomic, and hypothesized historic relationships among geographic populations.

Prosimians

Sportive Lemurs (Lepilemur)

The seven forms of lemurs usually recognized are distributed in the coastal regions of Madagascar (Fig. 2). These are classified as subspecies by some workers or, as we do here, as species by others, but all agree on the need for further systematic work (Tattersall, 1982; Jenkins, 1987). The external characters distinguishing these nocturnal prosimians include subtle and quite variable color differences in the tail and trunk. The relative proportions of teeth, ears, tails, and overall body sizes may also be important. Drenhaus (1975) performed a multivariate analysis of 17 skull dimensions on 59 specimens of 13 samples. Canonical variate analysis shows the demes arranged in a ring of overlapping distributions, a pattern having striking resemblance to their geographic distribution around the periphery of the island. This suggests a polytypic ring species but is equally compatible with a zoogeographic series of speciation events. Our analysis of generalized distances reveals a hierarchy of relationships among demes: (1) $\bar{D} = 3.40$ (range of 2.75–4.03) among demes of the same species (e.g., between demes 1 and 2), (2) $\bar{D} = 5.78$ (3.27–8.61) between demes of neighboring species (e.g., between demes 2 and 3), and (3) $\bar{D} = 7.33$ (5.02–11.23) between demes of non-neighboring species (e.g., between demes 2 and 6).

The smallest distances between neighboring demes of neighboring species involve those sportive lemurs inhabiting similar ecogeographic regions of Madagascar. These are also the most problematic taxa in terms of their distinctiveness as species: (1) $D = 3.81$ between demes 2 and 3, representing *L. mustelinus* and *L. microdon*, respectively, from the humid, tropical forests of the east, and (2) $D = 3.27$ between demes 8 and 9, representing *L. ruficaudatus* and *L. edwardsi*, respectively, from the dry deciduous forests of the west. These distances are comparable to the aforementioned intraspecific values.

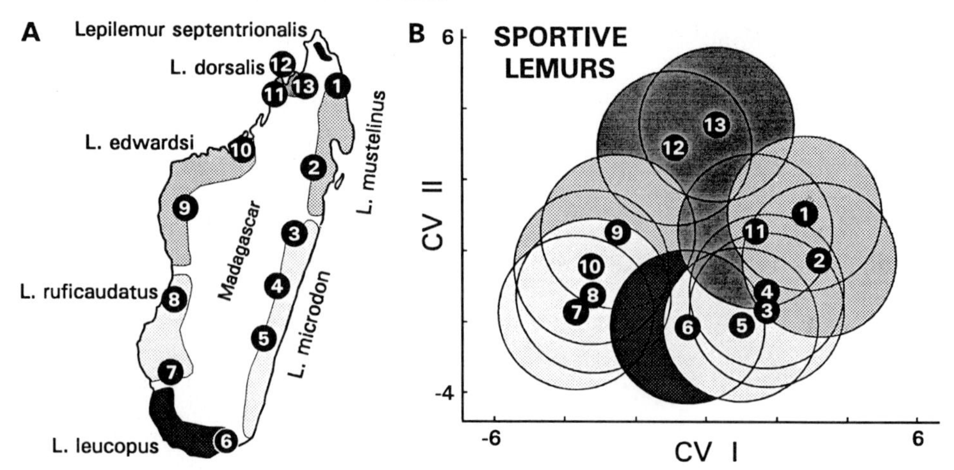

Fig. 2. Geographic variation in sportive lemurs (*Lepilemur*). (A) Distribution of species on Madagascar (some workers consider these to be subspecies; adapted from Drenhaus, 1975, Table 7, Fig. 1, and Albrecht *et al.*, 1990, Fig. 3). Markers indicate approximate location of demes used in multivariate analysis: 1–2 = *L. mustelinus*, 3–5 = *L. microdon*, 6 = *L. leucopus*, 7–8 = *L. ruficaudatus*, 9–10 = *L. edwardsi*, 11–13 = *L. dorsalis* (no sample of *L. septentrionalis*). (B) Canonical variate analysis based on 17 skull dimensions for 13 demes representing all species of sportive lemurs except *L. septentrionalis* (adapted from Drenhaus, 1975, Fig. 17a).

In contrast, neighboring demes of the more distinctive sportive lemurs living in different ecogeographic zones are more distant in their skull morphology: (1) D = 7.09 between *L. microdon* and *L. leucopus*, (2) D = 5.26 between *L. leucopus* and *L. ruficaudatus*, (3) D = 6.08 between *L. edwardsi* and *L. dorsalis*, and (4) D = 8.61 between *L. dorsalis* and *L. mustelinus* if these two are considered neighbors in the absence of *L. septentrionalis*. Such values are similar to interspecific distances between nonneighboring taxa. It would be interesting to complete the morphometric picture by including *L. septentrionalis* from the north (see Jungers and Rumpler, 1976, for univariate comparisons) and extinct forms that probably once lived in the now uninhabitable central regions of Madagascar (small animals such as sportive lemurs may have been overlooked because of past paleontological focus on giant prosimians).

Greater Galagos (Otolemur)

Recent classifications recognize two species and treat them as generically distinct from other galagos (Jenkins, 1987; Nash *et al.*, 1989). Masters and Lubinsky (1988) measured 24 craniofacial dimensions on 222 specimens of *O. crassicaudatus* and *O. garnetti* from southern Africa. Their canonical variate analysis differentiated the two species but, unfortunately for the study of geographic variation, finer groupings of specimens at demic or subspecific levels were not considered. Geographic variation in galagos (*Otolemur, Galago,* and *Galagoides*) warrants more attention, given the thousands of museum specimens available for quantitative studies of morphological differences in a taxon that includes as many as 3 genera, 4 subgenera, 11 species, and 30 subspecies dispersed widely across sub-Saharan Africa.

New World Monkeys

Tufted-Ear Marmosets (Callithrix jacchus)

The five subspecies generally recognized are distinguished by striking differences in facial pattern and development of hair around their ears (Hershkovitz, 1977) (Fig. 3A,B). Tufted-ear marmosets are treated as distinct

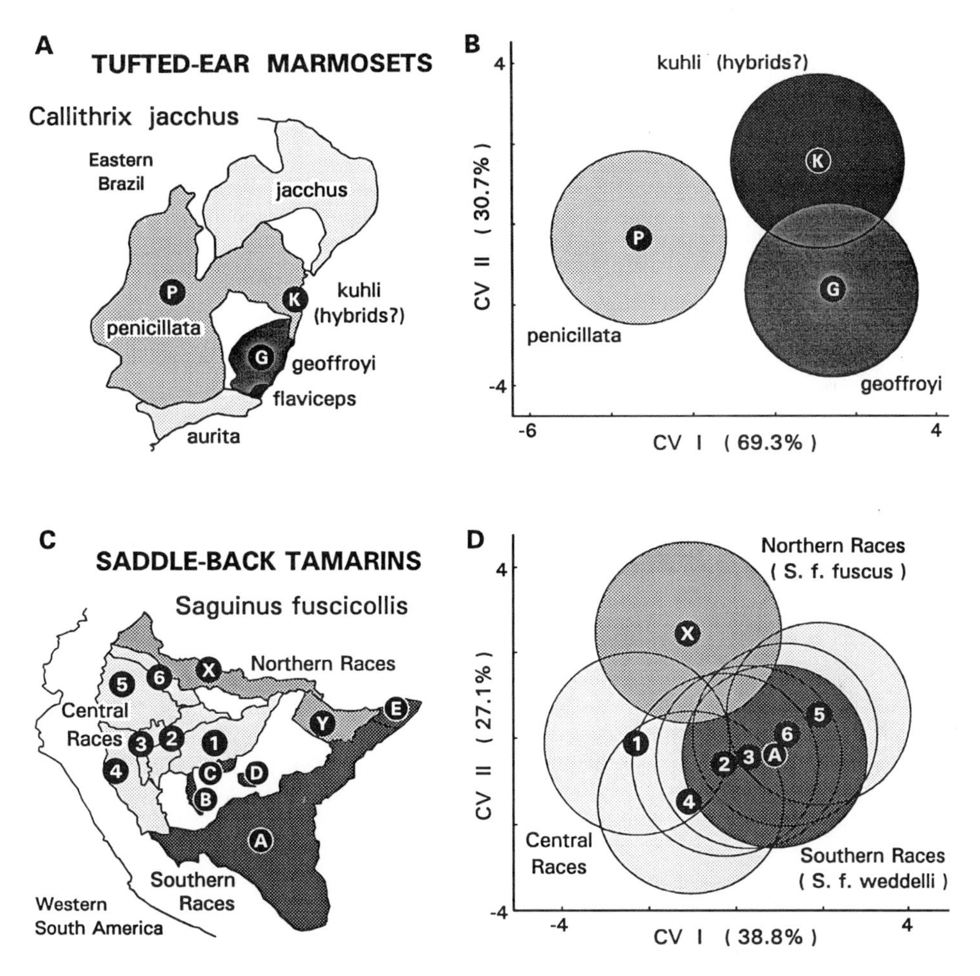

Fig. 3. Geographic variation in tufted-ear marmosets (*Callithrix jacchus*) and saddle-back tamarins (*Saguinus fuscicollis*). (A) Distribution of subspecies of *C. jacchus* in eastern Brazil (some workers consider these to be species; adapted from Hershkovitz, 1977, Fig. IX.5). Markers indicate subspecies (*geoffroyi* and *penicillata*) and a suspected hybrid sample (*kuhli*) used in multivariate analysis. (B) Canonical variate analysis based on 20 dental variables for the two subspecies and a hybrid sample of tufted-ear marmosets (adapted from Natori, 1990, Fig. 1). (C) Distribution of subspecies of *S. fuscicollis* in Amazon basin (adapted from Hershkovitz, 1977, Fig. X.24). Markers indicate subspecies: northern races (X = *fuscus*, Y = *avilapiresi*), central races (1 = *fuscus*, 2 = *nigrifrons*, 3 = *illigeri*, 4 = *leucogenys*, 5 = *lagonotus*, 6 = *tripartitus*), and southern races (A = *weddelli*, B = *acrensis*, C = *melanoleucus*, D = *cruzlimai*, E = *crandalli*). (D) Canonical variate analysis based on principal coordinate analysis of generalized distances, calculated from 11 facial variables, among 8 of 13 subspecies (distances from Cheverud and Moore, 1990, Table IV).

species by some workers who interpret evidence for sympatry and intergradation differently (Hershkovitz, 1975, 1977 vs. Coimbra-Filho and Mittermeier, 1973; Mittermeier and Coimbra-Filho, 1981; Mittermeier *et al.*, 1988). One controversy involves whether marmosets from the coastal regions of the *C. j. penicillata* range represent a distinct taxon (*kuhli*) or morphological intergrades between *C. j. penicillata* and *C. j. geoffroyi*. Natori (1990) analyzed 20 dimensions of the postcanine dentition in 69 specimens representing three populations (the two recognized subspecies and the *kuhli* sample). Canonical variate analysis placed the *kuhli* sample intermediate between *C. j. geoffroyi* and *C. j. penicillata* but separate from both. These odontometric results were interpreted as supporting the taxonomic distinctiveness of *kuhli*. However, the results remain problematic because specimens from throughout the ranges of the subspecies were lumped together without regard for their proximity to the *kuhli* sample. Evaluating the continuity of morphological variation requires a finer-grained analysis of geographic variants based on more specimens from more localities extending across the region in question.

Saddle-Back Tamarins (Saguinus fuscicollis)

Thirteen subspecies are distributed in contiguous areas separated by the major rivers of the upper Amazon basin (Hershkovitz, 1977) (Fig. 3C,D). These forms are sharply defined by suites of subtle to remarkable differences in the coloration and color pattern of the face, trunk, limbs, and tail (Hershkovitz, 1977, Plates III–IV). There is little evidence for intergradation between neighboring populations. Based on progressive changes in pelage and zoogeographic relationships (geographic metachromism), Hershkovitz (1977) proposed evolutionary relationships among the subspecies, which he divided into two northern, six central, and five southern races. Cheverud and Moore (1990) analyzed 11 facial dimensions in 103 saddle-back tamarins of eight subspecies (sexes pooled). Their multivariate analysis revealed significant morphological variation among the subspecies with an intersubspecific $\bar{D} = 3.07$ (1.74–4.39). The morphometric distances were significantly correlated with: (1) evolutionary distances measured by the number of founder events separating subspecies in Hershkovitz's scheme based on geographic metachromism ($r = 0.62$); and (2) geographic distances measured by the number of major rivers separating subspecies ($r = 0.36$). The contiguity of geographic ranges was not important, but this is expected, given the complexity of zoogeographic relationships posited by Hershkovitz. Our principal coordinate analysis of the published generalized distance matrix places the single representative of the northern races as the most outlying subspecies (*S. f. fuscus;* $\bar{D} = 3.61$ compared to the other seven subspecies). The central races are divided into those south (*S. f. fuscicollis, S. f. nigrifrons, S. f. illigeri,* and *S. f. leucogenys*) and north (*S. f. lagonotus* and *S. f. tripartitus*) of the Amazonas. The distances within each of these groups ($\bar{D} = 2.60$ and $D = 2.74$, respectively) are less than the distances between groups ($\bar{D} = 3.36$). The only southern race (*S. f. weddelli*) represented is intermediate in its facial morphology compared to the central races. It would be interesting to obtain measurements on the other three southern races to see if these most chromatically distinct forms are also the most derived in their facial morphology.

Squirrel Monkeys (Saimiri)

There is broad agreement about the identity, nomenclature, taxonomic levels, and relationships of some forms and controversy about others (Hershkovitz, 1984; Thorington, 1985; Costello *et al.*, this volume). Some taxa have distinctive external features (e.g., the "roman arch" of *S. boliviensis*'s supraorbital region and *S. ustus*'s bare ears), while others are separated by karyotypic differences only (*S. sciureus sciureus* and *S. s. macrodon*). Thorington (1985) conducted several multivariate analyses based on 19 skull variables to test hypotheses derived from his study of coat-color variation. In general, the two anatomical systems gave concordant results. One analysis was based on 114 males representing six demes of three subspecies of *S. sciureus* (Fig. 4A,B, which uses the taxonomic designations of Hershkovitz, 1984). Canonical variate analysis emphasizes the closeness of *S. s. sciureus* and *S. s. macrodon,* considered to be one subspecies by Thorington, and confirms the morphological distinctiveness of *S. s. cassiquiarensis,* which is geographically interposed between the two.

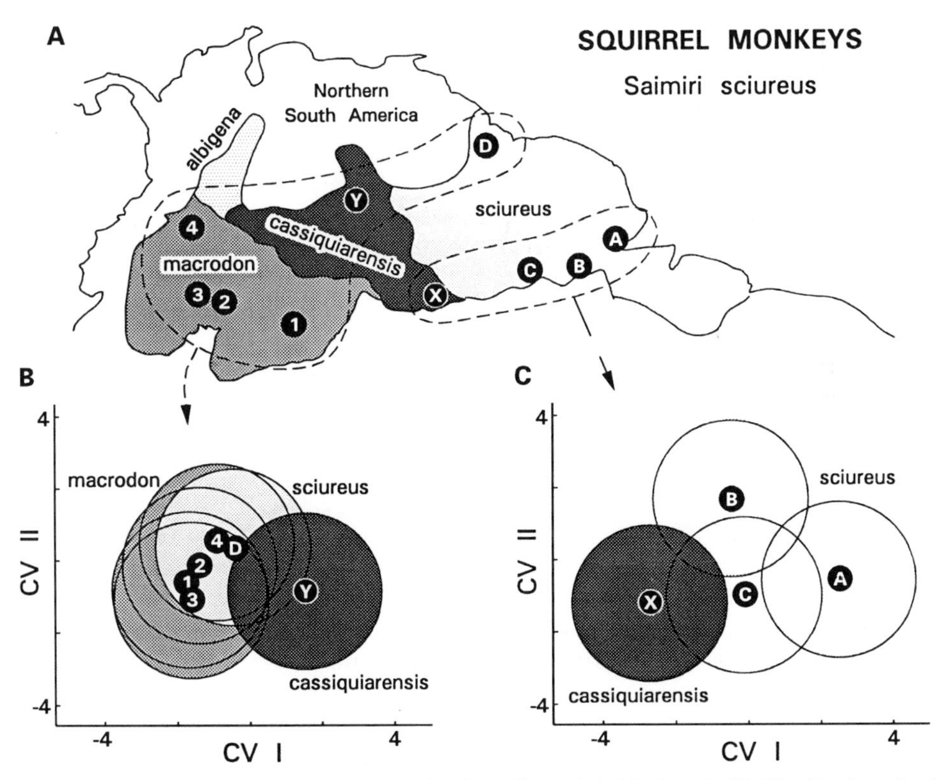

Fig. 4. Geographic variation in common squirrel monkeys (*Saimiri sciureus*). (A) Distribution of subspecies of *S. sciureus* in northern South America (shading adapted from Hershkovitz, 1984, Fig. 2). Markers indicate approximate location of demes used in multivariate analysis: 1–4 = *macrodon,* A–D = *sciureus,* Y, X = *cassiquiarensis.* (B) Canonical variate analysis based on 19 skull variables for 6 demes representing 3 of 4 subspecies of common squirrel monkeys (adapted from Thorington, 1985, Fig. 4). (C) Canonical variate analysis based on 19 skull variables for 4 demes representing 2 of 4 subspecies of common squirrel monkeys (adapted from Thorington, 1985, Fig. 6).

Another analysis was based on 43 males representing four demes of two subspecies of *S. sciureus* distributed along the north bank of the Amazon (Fig. 4A,C). Canonical variate analysis places the demes in approximate geographic order, with the most western and eastern samples flanking the geographically and morphologically intermediate samples. The lability of morphological resemblances among demes is evident when the two analyses are compared. In the first analysis, four demes of *S. s. macrodon* are tightly clustered; in the second analysis, three demes of *S. s. sciureus* are as distinct from one another as the subspecific separation of *S. s. cassiquiarensis* in either analysis.

Spider Monkeys (Ateles)

The taxonomy of this genus requires revision to clarify uncertainty about the ranks and relationships of the recognized forms (Froehlich *et al.*, 1991) (Fig. 5). One classification recognizes eight kinds of spider monkeys in Central America (one species) and seven in South America (three species), these being distinguished by either coloration of the body or distinctive markings of the face and head (Napier, 1976). Froehlich *et al.* (1991) measured 76 craniodental dimensions on 284 spider monkeys representing 25 demes of the seven South American spider monkeys and one deme from Central America (*geoffroyi*). Their final multivariate ordination, based on 50 variables controlled for sex differences, is shown here with the canonical variate axes redrawn to a common scale so as to make circular the within-group dispersions. The role of geographic variation is evident, even though the first two canonical variates account for only 41% of the total among-group variation. Demes of recognized forms are clustered together and separated from demic clusters of other forms (demes of *paniscus, marginatus,* and *chamek* in the center, as well as *geoffroyi, fusciceps,* and *robustus* in the upper left, are separated by subsequent canonical variates). The demes marked by lowercase letters are both geographic and morphologic intermediates, suggesting clinal intergradations between some neighboring forms.

Froehlich *et al.* (1991) provided the following systematic and biogeographic interpretations of their morphometric results. Amazonian spider monkeys form a ring species in the terra firms forests around regions of the central Amazon basin that are seasonally inundated and were flooded over long periods of time in the past (*paniscus* → *marginatus* → *chamek* → *belzebuth*); *paniscus* and *belzebuth,* the morphologically distinct end forms of the ring, are geographic neighbors but are separated by the nutrient-poor, black-water river drainages of the Guiana highlands. The Guianan spider monkey (*A. paniscus*), with its red face and unique karyotype, is treated as having evolved species-level distinctions compared to other members of the ring (*A. belzebuth*). South American spider monkeys from the extreme northwest of the continent probably represent a single species, as suggested by morphological intergradation between forms whose subspecific distinctions can be explained by a refugia model (*fusciceps, robustus,* and *hybridus* as subspecies of *A. hybridus*). Contrary to the suggestion by Froehlich *et al.* (1991), it is premature to treat these three forms as conspecific with Central American spider monkeys, since only a single deme of the latter was analyzed (*A. geoffroyi*).

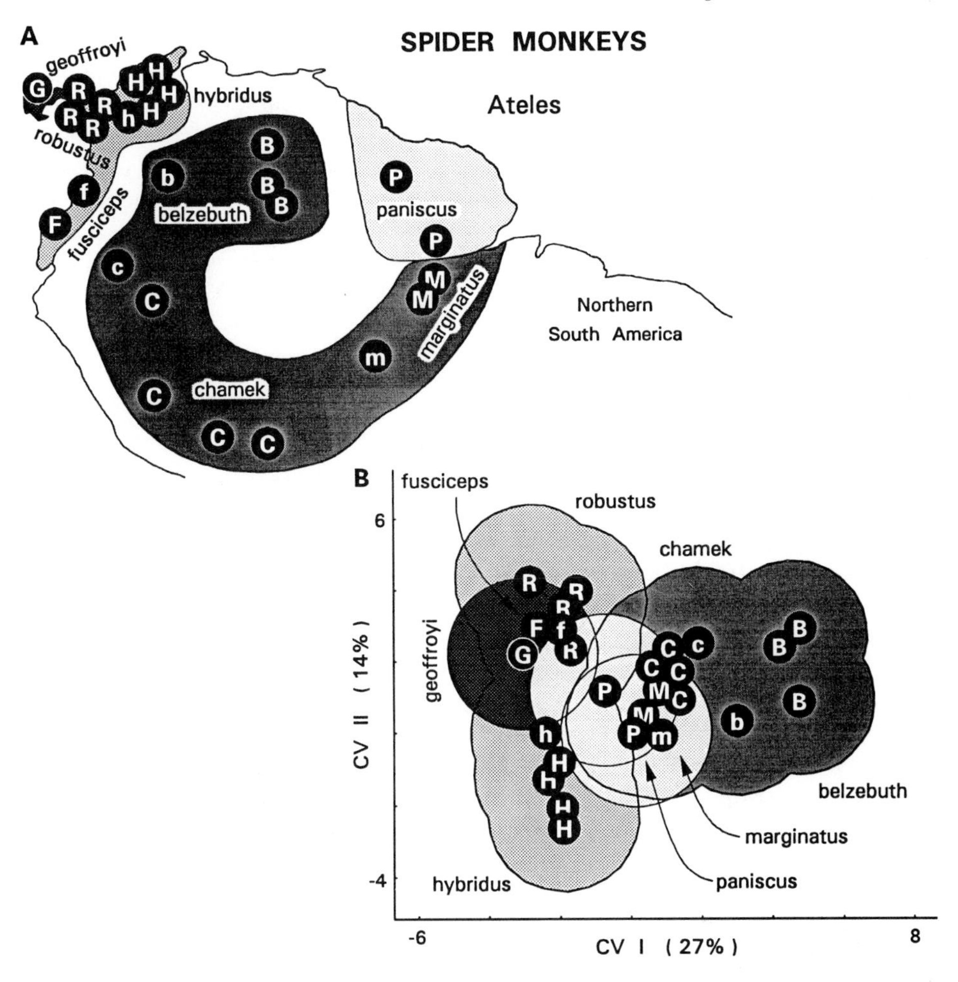

Fig. 5. Geographic variation in South American spider monkeys (*Ateles*). (A) Distribution of spider monkeys in northern South America, with shading indicating species affiliations as determined by multivariate analysis (adapted from Froehlich *et al.*, 1991, Fig. 1). Markers with capital letters indicate approximate location of 26 demes used in multivariate analysis: G = *A. geoffroyi* (the only representative of the Central American spider monkeys), P = *A. paniscus*, *A. hybridus* (H = *hybridus*, F = *fusciceps*, R = *robustus*), and *A. belzebuth* (B = *belzebuth*, C = *chamek*, M = *marginatus*). Markers with lowercase letters indicate closest morphometric affinity of demes that are geographically and morphologically intermediate between neighboring subspecies. (B) Canonical variate analysis based on 50 craniodental variables for 26 demes of spider monkeys (adapted from Froehlich *et al.*, 1991, Fig. 4). The shaded regions include the outer limits of the overlapping 90% confidence circles for individual demes of the different species; complete circles are shown only for demes of *geoffroyi*, *marginatus*, and *paniscus*.

Howler Monkeys (Alouatta)

Even a superficial examination of skulls leaves an impression of great variability within and between the seven species of this genus (*A. pigra* plus those given by Napier, 1976). Watanabe (1982) used 30 dimensions of 85 skulls to differentiate three demes of *A. caraya*, the red howler. His multivariate analysis places the three samples in geographic order with D = 5.41 between northern and southern Colombia (sexes averaged), D = 12.01 between southern Colombia and northern Bolivia (males only), and D = 14.23 between the extremes of northern Colombia and Bolivia (males only). A sex-averaged D = 28.80 between these red howlers and a small Colombian sample of *A. palliata,* the mantled howler, is the largest interspecific difference yet found for closely related primates (indeed, the distance is so extreme as to suggest an error in the analysis). These results are suggestive of the opportunity howlers offer for studying geographic variation.

Old World Monkeys

Black-and-White Colobus Monkeys (Subgenus Colobus)

The subgenus *Colobus* is comprised of four polytypic species distributed widely across Africa (Napier, 1985) (Fig. 6). The taxonomy of both subspecies and species is based on variations in the pattern and hair length of the black and white coat. Hull (1979) used 76 measurements of 592 male specimens to investigate craniodental differences among 33 populations representing all but 2 of 22 subspecies, including multiple demes of some races. The positioning of samples on the first two canonical variates alone confirms the distinctiveness of the four species (results for 460 females representing 31 demes are similarly patterned but with tighter groupings of demes at the species level). Most demes of a subspecies do not differ from one another at statistically significant levels, but most comparisons between conspecific subspecies are statistically significant. In all cases, cranial morphology is distinctive enough to minimize misidentification of specimens (just 28 of 592 skulls are not correctly placed in their own deme, and only 10 of these are misidentified as the wrong subspecies).

As Hull emphasized, craniodental variation in black-and-white colobus monkeys illustrates the range of possible variation from indistinguishable local populations through distinct biological species. The metric relationships among conspecific subspecies often match zoogeographic distributions. This is clearly seen in *C. polykomos,* where *C. p. dollmani* is intermediate between the other two subspecies, and in *C. angolensis,* where the subspecies from central Africa are arranged metrically in approximate geographic order and the two subspecies of eastern Africa are displaced both metrically and geographically from their conspecifics. In *C. guereza,* when generalized distances are used to untangle the cluster seen in the canonical variate plot, subspecies from central African forests have the closest craniodental associations (*C. g. occidentalis* and *C. g. uellensis*). Otherwise, interpretations of morphometric affinities in *C. guereza* are confounded by the complexity of historical and biotic factors behind the evolution and dispersal of colobus monkeys in the now-fragmented forests in and around the Rift Valley of eastern Africa.

Non-Sulawesi Macaques (Macaca)

Macaques are divided here into those from the island of Sulawesi, discussed below, and all others, which include species distributed across southern Asia and a relict species in northwestern Africa. Various combinations of charac-

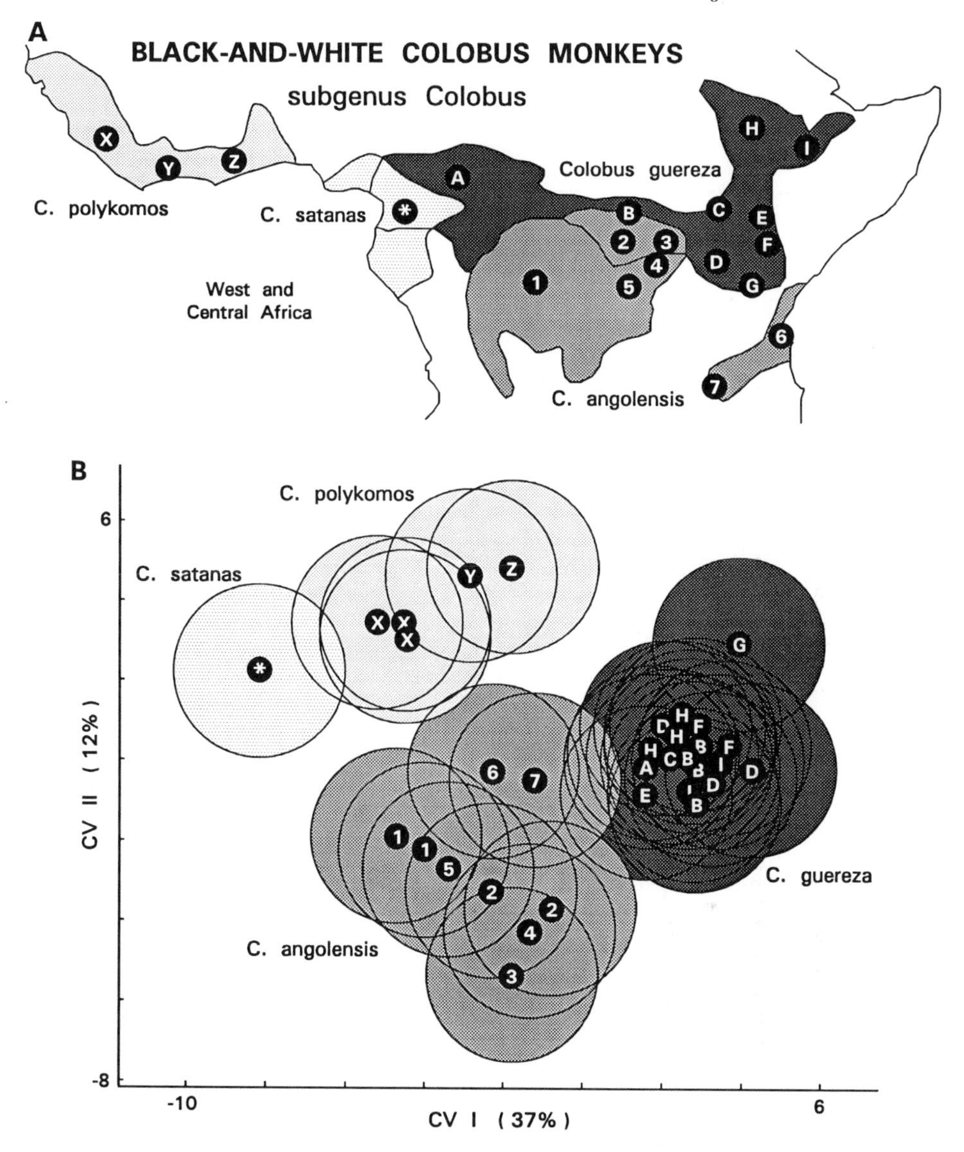

Fig. 6. Geographic variation in black-and-white colobus monkeys (subgenus *Colobus*). (A) Distributions of species in western, central, and eastern Africa (adapted from Hull, 1979, Fig. 1). Markers indicate subspecies represented in multivariate analysis: *Colobus angolensis* (1 = *angolensis*, 2 = *cottoni*, 3 = *ruwenzorii*, 4 = *adolfifrederici*, 5 = *cordieri*, 6 = *palliatus*, 7 = *sharpei*), *C. guereza* (A = *occidentalis*, B = *uellensis*, C = *dodingae*, D = *matschiei*, E = *percivali*, F = *kikuyuensis*, G = *caudatus*, H = *guereza*, I = *gallarum*), *C. polykomos* (X = *polykomos*, Y = *dollmani*, Z = *vellerosus*), and *C. satanas* (* = *anthracinus*). (B) Canonical variate analysis based on 76 craniodental variables for 33 demes of males (multiple demes of some subspecies; adapted from Hull, 1979, Fig. 3).

ters divide the non-Sulawesi macaques into four species groups (based on reproductive anatomy and copulatory behavior), 12 species (based on size, tail morphology, color and pattern of pelage, facial coloration, condition of female perineum, ecology, and behavior), and numerous subspecies (based on size, tail morphology, and color and pattern of pelage) (Fooden, 1969, 1975, 1976, 1979,

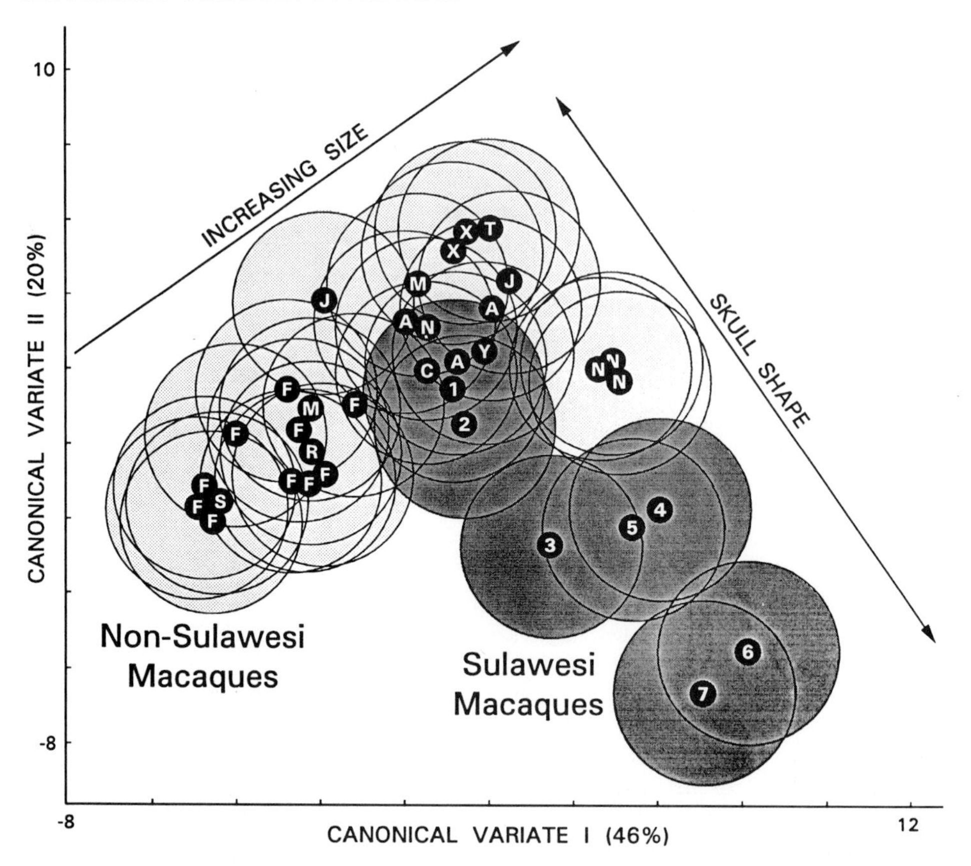

Fig. 7. Geographic variation in craniofacial morphology of macaques (*Macaca*). Canonical variate analysis based on 24 craniofacial variables for 35 samples representing all macaque species except *M. silenus* (adapted from Albrecht, 1978, Fig. 26). Markers indicate demes (multiple demes of some species): *arctoides* species group (X = *M. arctoides*), *fascicularis* species group (F = *M. fascicularis*, M = *M. mulatta*, C = *M. cyclopis*, J = *M. fuscata*), *sinica* species group (S = *M. sinica*, R = *M. radiata*, A = *M. assamensis*, T = *M. thibetana*), *silenus-sylvanus* species group (N = *M. nemestrina*, Y = *M. sylvanus*; 1–7 = Sulawesi macaques as in Fig. 8). The size gradient was identified using the method described by Albrecht (1979); the shape vector was drawn perpendicular to the size vector.

1980, 1981, 1982, 1983, 1986, 1990, 1991; Fooden *et al.*, 1985; Napier, 1981; Fa, 1989).

Albrecht (1976, 1977, 1978) studied craniofacial variation using 24 linear measurements taken on 400 males of 35 geographically defined samples representing all but one macaque species (*M. silenus;* this includes the seven Sulawesi macaques discussed below) (Fig. 7). Sulawesi and non-Sulawesi macaques are about equally variable despite the tremendous disparity between the two groups in geographic, taxonomic, and adaptive diversity (greatest D = 13.57 among Sulawesi males compared to 12.16 among non-Sulawesi males). Sulawesi macaques are characterized by differences in skull shape. In contrast, non-Sulawesi macaques are differentiated primarily along a gradient reflecting latitudinal size effects within and between species (Albrecht, 1978, 1980a, 1983; Albrecht *et al.*, 1990). Members of different species groups broadly overlap but are increasingly differentiated in skull shape as size increases. This is evident from generalized distances, which include the shape information of higher canonical variates: $\bar{D}$ = 3.40 among the smallest macaques (*M. sinica* and three smallest demes of *M. fascicularis*) vs. $\bar{D}$ = 8.35 among the largest members of each species group (*M. arctoides, M. thibetana, M. fuscata,* and the three largest demes of *M. nemestrina*).

Within species groups, intraspecific relationships are closer on average than distances among demes of different species (Table 1). The six species represented by multiple samples have the following intraspecific relationships. (1) The generalized distance is 2.89 between *M. a. arctoides* and *M. a. melanota,* bear macaques from the northern and southern parts of the species range, respectively. (2) Three geographic populations of Sundaic pigtail macaques, *M. nemestrina nemestrina,* have skulls very similar to one another (intrasubspecific $\bar{D}$ = 2.74, with a range of 2.12–3.14) but well differentiated in size and shape from Indochinese pigtails, *M. n. leonina* (intersubspecific $\bar{D}$ = 6.52, with a range of 6.16–7.00). (3) The 10 demes of crab-eating macaques (*M. fascicularis*) exhibit similarities between craniometric and zoogeographic relationships (e.g., D = 1.94 between neighboring Sumatran and Bornean samples, while D = 7.29 between widely separated Luzon and Thai-Burmese samples). (4) An insular isolate of Japanese macaque from Yaku Island, *M. fuscata yakui,* is smaller than those of the nominate race, *M. f. fuscata,* from the main islands of Kyushu and Honshu (D = 6.46). (5) Two demes of rhesus macaques (*M. mulatta*), at the morphological and geographic extremes of a latitudinal size gradient (Albrecht, 1978), show differences in skull morphology that would be bridged, presumably, by geographical intermediates (D = 6.06). (6) Three demes of Assamese macaques (two *M. assamensis assamensis* and one *M. a. pelops*) are approximately equidistant from each other in skull morphology ($\bar{D}$ = 4.04). The potential of more detailed study in understanding geographic variation is evident from pre-

Table 1. Generalized Distances among Demes of Non-Sulawesi Macaques Arranged by Taxonomic Rank of Comparison[a]

		Average generalized distances (D) among			
		Demes of same species		Demes of different species in same group	
Species group	Species (#demes)	$\bar{D}$	Range	$\bar{D}$	Range
arctoides	*M. arctoides* (2)	2.89			
silenus-sylvanus[b]	*M. nemestrina* (4)	4.63	(2.12–7.00)	8.08	(6.86–9.07)
	M. sylvanus (1)				
fascicularis	*M. fascicularis* (10)	4.84	(1.94–7.29)	7.66	(4.42–10.84)
	M. fuscata (2)	6.46			
	M. mulatta (2)	6.06			
	M. cyclopis (1)				
sinica	*M. assamensis* (3)	4.04	(3.52–4.70)	7.62	(4.50–12.16)
	M. sinica (1)				
	M. radiata (1)				
	M. thibetana (1)				

[a]Generalized distances (D) from Albrecht (1978, Table XVI).
[b]No data available for *Macaca silenus*.

liminary craniometric analysis of 32 *M. fascicularis* demes, whose interrelationships imply interactions among dispersal patterns, zoogeographic affiliations, morphological differentiation, latitudinal size clines, and insular size effects (Albrecht, 1983).

Sulawesi Macaques (Macaca)

The seven macaques from different parts of the island are differentiated by tail length, head crests, trunk and limb coloration, and features of the buttocks region (Fooden, 1969) (Fig. 8). Albrecht (1976, 1977, 1978) measured 24 craniofacial dimensions on all known museum specimens of wild-caught animals (79 males, 76 females). As shown above, craniometric differentiation is unique among macaques in being so strongly related so skull shape, especially of the facial skeleton. Multivariate analyses of both males and females support the partitioning of Sulawesi macaques into seven well-defined groups. Sex-averaged generalized distances are $\bar{D} = 7.39$ (4.74–9.36) among neighboring forms and $\bar{D} = 10.09$ (6.12–14.27) among nonneighbors. The closest resemblance is between *nigra* of Sulawesi and an introduced population on Bacan (Batjan) Island some 300 km east of Sulawesi ($D = 4.57$).

The pattern of craniometric relationships is consistent with the zoogeogra-

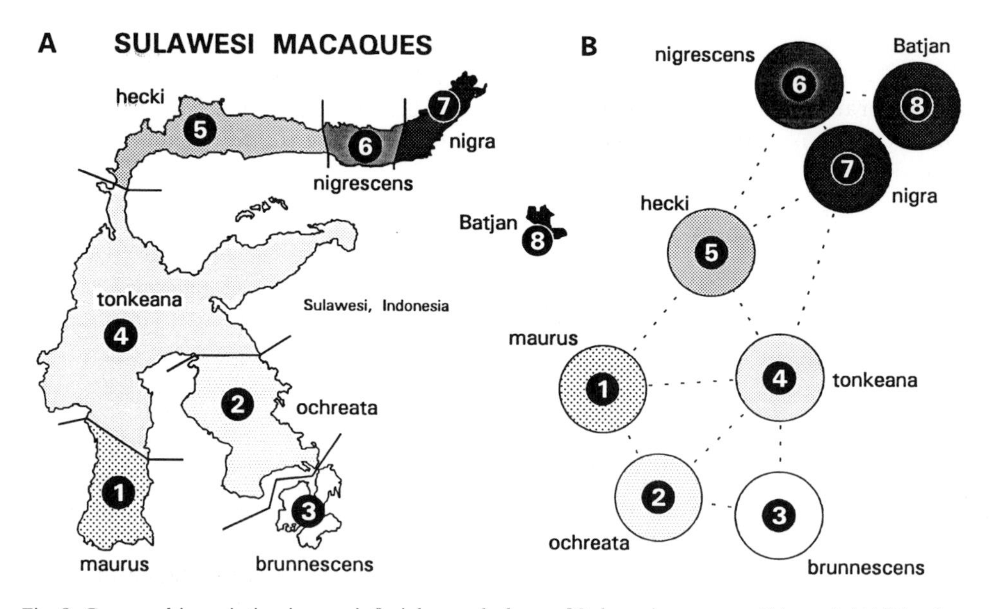

Fig. 8. Geographic variation in craniofacial morphology of Sulawesi macaques (*Macaca*). (A) Distribution of macaques on the island of Sulawesi in Indonesia (subspecies and/or species; adapted from Albrecht, 1978, Fig. 2). A human-introduced population of *nigra* occurs on Batjan (Bacan) Island, 300 km east of Sulawesi (the position in the figure is not to scale). Markers indicate the subspecies/species samples used in the multivariate analysis. (B) Canonical variate analysis of sex-averaged generalized distances, calculated from 24 craniofacial variables, among eight Sulawesi macaques, including the Batjan sample (adapted from Albrecht, 1978, Tables X–XI, Figs. 23–24). The plot summarizes the information in all canonical variates by displaying the full separations among neighboring populations (dotted lines represent actual generalized distances; distances between groups not joined by dotted lines are not to scale). The shaded circles are the usual 90% confidence ellipses based on the pooled within-group dispersion.

phy of the island and correlates with quantitative analyses of both external features (Albrecht, 1978) and genetic traits (Kawamoto *et al.*, 1982). However, due to a lack of specimens from boundary regions, studies of museum materials are inadequate for determining biological relationships between neighboring populations. The possibilities are that morphologically distinct neighbors are characterized by allopatry, parapatry, sympatry, or zones of intergradation, with each possibility implying something different about speciation among the Sulawesi macaques. As outlined by Albrecht (1978), the actual situation is probably some combination of these alternatives, a suggestion supported by the fieldwork of Hamada *et al.* (1987; parapatry at the *hecki-nigrescens* and *nigrescens-nigra* boundaries), Supriatna *et al.* (1990; hybridization at the *maurus-tonkeana* boundary), and Watanabe *et al.* (1991; narrow hybrid zone at the hecki-tonkeana boundary). The differences between these studies and the preliminary observations of Groves (1980; hybridization at the *nigrescens-nigra* boundary, narrow sympatry at the *maurus-tonkeana* boundary, and wide zone of intergradation at the *hecki-tonkeana* boundary) emphasize the problematic nature of assessing speciation even when done in the field.

The skull of the ancestral Sulawesi macaque probably resembled *maurus*, whose craniofacial structure is the least differentiated among Sulawesi macaques and the most like the generalized form of non-Sulawesi species. Subsequent differentiation occurred in two lineages, each characterized by an increasingly derived skull: (1) *maurus* → *tonkeana* → *hecki* → *nigrescens* → *nigra* to the north; and (2) *maurus* → *ochreata* → *brunnescens* to the southeast, with *maurus* eventually isolated in the southwest. This remarkable radiation is not surprising, given the depauperate mammal fauna of Sulawesi and the geotectonic history of the island, which provided a complex series of isolation events conductive to speciation.

Guenons (Cercopithecus)

This genus, the most widespread nonhuman primate in Africa, has as many as 80 subspecies, 19 species, 8 species groups, and 4 subgenera if the talapoin, swamp, and patas monkeys are considered congeneric (Napier, 1981; Lernould, 1988). No comparative work has been done that details multivariate craniometric variation in different guenons. This is surprising, given the taxonomic diversity, the availability of large museum collections, and Verheyen's (1962) tantalizing univariate results showing substantive morphological differences. The one multivariate study of guenons that has been reported deals with nine field measurements of body proportions taken on several hundred animals from Kenya (Turner *et al.*, 1988). Four demes of *C. aethiops pygerythrus* (interdemic $\bar{D}$ = 1.47) are more similar than three demes representing different subspecies of *C. mitis* (inter-subspecific $\bar{D}$ = 3.50); the largest distances are between demes of the two species (interspecific $\bar{D}$ = 4.82).

Savanna Baboons (Papio)

The five taxa generally recognized are distributed allopatrically throughout most of sub-Saharan Africa (*anubis, cynocephalus, hamadryas, papio,* and *ursinus;* see Jolly, this volume). The forms represent distinct morphotypes in external features with many subtle variations. Different authorities treat the

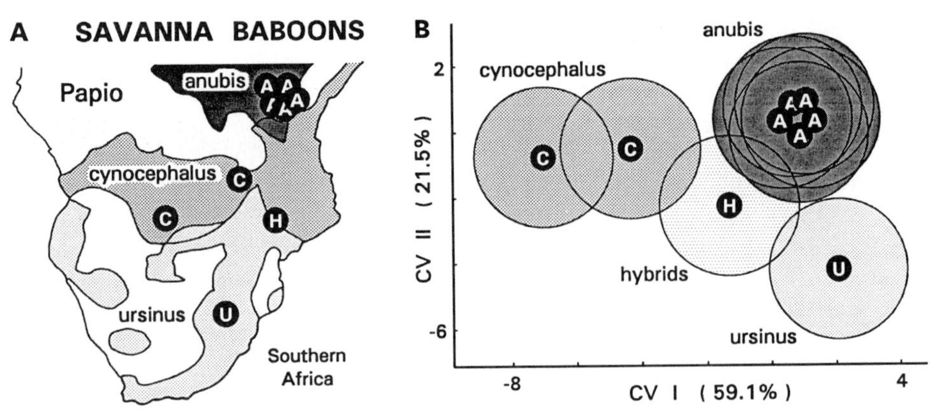

Fig. 9. Geographic variation in savanna baboons (*Papio*). (A) Geographic distribution of baboons in southern Africa (these are considered subspecies or species by different workers; adapted from Hayes *et al.*, 1990, Figs. 1 and 2). Markers indicate demes used in multivariate analysis: A = *anubis*, C = *cynocephalus*, U = *ursinus*, and H = suspected hybrids between *cynocephalus* and *ursinus*. (B) Canonical variate analysis based on 21 maxillary tooth variables for 10 demes of males representing three taxa of savanna baboons from southern Africa (adapted from Hayes *et al.*, 1990, Fig. 5).

savanna baboons as subspecies and/or species, depending on the significance assigned to narrow hybrid zones between neighboring forms (only *papio* and *anubis* are not known to hybridize). Hayes *et al.* (1990, expanding on the early work of Booth and Freedman, 1970) studied multivariate odontometric variation in the three savanna baboons living south of the equator (21 dimensions of maxillary teeth on 257 males and 186 females representing five demes of *anubis*, two demes of *cynocephalus*, one deme of *ursinus*, and one deme of suspected *cynocephalus-ursinus* hybrids) (Fig. 9). The canonical variate ordination for males shows the three forms as dentally distinct, with little overlap in the distributions of their demes. The five *anubis* samples from southwest Kenya cluster tightly together; the two *cynocephalus* samples are more loosely associated, perhaps reflecting the greater geographic distance between them. The small sample of hybrids is intermediate between the presumed parental forms on the first two canonical variates, but later axes emphasize the distinctive morphology of this sample from the *cynocephalus-ursinus* border area in Malawi. The pattern for females differs in having a wider spread of *anubis* demes overlapping with *ursinus* (not shown). The morphological affinity of *anubis* and *ursinus*, for both males and females, is contrary to zoogeographic expectations, given the intervening distribution of *cynocephalus*. These odontometric results suggest how valuable a full study of *Papio* could be if based on osteometric as well as odontometric data collected from all available museum materials, including specimens from zones of intergradation.

Hominoids

Gibbons (Hylobates)

There is considerable controversy over the taxonomy and classification of gibbons (Groves, 1972, 1984; Creel and Preuschoft, 1976, 1984; Marshall and Sugardjito, 1986). Three species are distinctive at a subgeneric rank

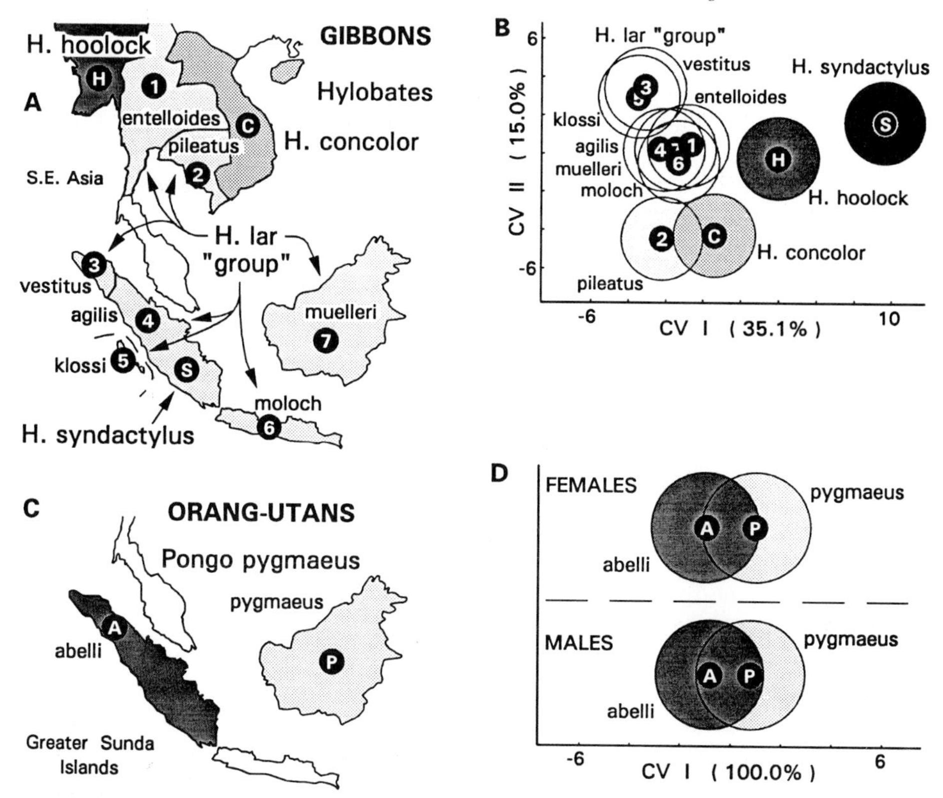

Fig. 10. Geographic variation in gibbons (*Hylobates*) and orang-utans (*Pongo*). (A) Distribution of gibbons in Southeast Asia with shading indicating subgeneric affiliations (range of *H. syndactylus* in Sumatra and West Malaysia not shaded and gibbons from Thai-Malay Peninsula not shown; adapted from Creel and Preuschoft, 1976, Fig. 1). Markers indicate subspecies and species samples used in multivariate analysis (taxonomic designations from Marshall and Sugardjito, 1986): subgenus *Nomascus* (C = *H. concolor* with all subspecies combined), subgenus *Bunopithecus* (H = *H. hoolock* with both subspecies combined), subgenus *Symphalangus* (S = *H. syndactylus syndactylus*), subgenus *Hylobates* (1 = *H. lar entelloides;* 2 = *H. i. vestitus;* 3 = *H. pileatus;* 4 = *H. agilis agilis* and *H. a. unko;* 5 = *H. klossi;* 6 = *H. moloch;* 7 = *H. a. albibarbis, H. muelleri muelleri, H. m. abbotti,* and *H. m. funereus*). (B) Canonical variate analysis based on principal coordinate analysis of sex-averaged generalized distances calculated from 32 landmarks and 15 variables for 10 samples of gibbons (distances from Creel and Preuschoft, 1976, Table 1). (C) Distribution of subspecies of *Pongo pygmaeus* in Indonesia. Markers indicated subspecies samples used in multivariate analysis. (D) Sex-specific canonical variates based on principal coordinate analyses of generalized distances, calculated from 32 skull variables, between the two subspecies of orang-utans (distances from Jacobshagen, 1979, Figs. 1 and 2).

(*H. hoolock, H. concolor,* and *H. syndactylus,* the siamang from Sumatra and West Malaysia, which is the only species sympatric with other gibbons). The remaining forms are often lumped together as the *H. lar* group (subgenus), and there are varying opinions about the number of distinct forms, about their taxonomic ranks, and about their relationships. Marshall and Sugardjito (1986) illustrated the considerable variation in coat color and pattern that confused previous taxonomists who worked with the *H. lar* group (especially troublesome have been *H. lar, H. agilis,* and *H. muelleri*).

Creel and Preuschoft (1976) studied multivariate morphometric differences based on 32 landmarks and 15 linear dimensions on skulls of 217 females and

276 males grouped into 10 geographic samples (Fig. 10A,B). Our principal coordinate analysis of their generalized distance matrix shows the craniometric distinctiveness of the subgeneric forms *H. hoolock, H. concolor,* and *H. syndactylus* (*H. concolor* is well separated from *H. pileatus* on the third and higher axes). Within the *H. lar* group, there is a central cluster formed by samples #1, #4, #6, and #7, while samples #2, #3, and #5 are outliers. The trouble with this early analysis is that some samples of the *H. lar* group contain mixtures of subspecies and species now considered to be distinct (e.g., sample #7 includes *H. agilis albibarbis* and three subspecies of *H. muelleri*).

Subsequently, Creel and Preuschoft (1984) refined their craniometric analyses by subdividing their original groups into samples that better reflect the geographic variants of the *H. lar* group. They found a general pattern of morphometric resemblances among geographic neighbors that did not always honor species boundaries; they suggested this implies gibbon systematics have still not been fully resolved. For example, *H. a. albibarbis* resembles *H. muelleri* more closely than it does conspecifics from either Sumatra or West Malaysia. While the 1984 analyses are to be preferred, they were not portrayed here because they either used size-corrected data or focused on subsets of data that prevented comparisons across the full diversity of gibbons.

Orangutans (Pongo)

The two living subspecies are distinguished by a variety of morphological and genetic characters (Groves, 1986; Courtenay *et al.,* 1988). Jacobshagen (1979) performed sex-specific multivariate analyses using 32 craniofacial dimensions on 29 males and 18 females in attempting to distinguish between *P. pygmaeus pygmaeus* from Borneo and *P. p. abelli* from Sumatra (Fig. 10C,D). The published canonical variate plots show the distributions of individual specimens for the two subspecies to be nonoverlapping and widely separated. These are, however, inconsistent with the generalized distances of D = 1.92 between females and D = 1.54 between males of the the two subspecies. Such small values imply considerable overlap in the population distributions not evident from the canonical variate plots. The canonical variate results are probably an artifact related to having more variables than specimens in each of the analyses (e.g., the popular BMD07M program gives just such misleading results). Thus, additional multivariate work is needed to substantiate the reality of craniometric distinctions within and among orangutans.

Chimpanzees (Pan)

While there is considerable controversy over the distinctiveness of the various chimpanzees, few would debate aligning the three forms of *P. troglodytes,* the common chimpanzees, as a natural group distinct from *P. paniscus,* the pygmy chimpanzee (Groves, 1986). The more reliable distinguishing characters seem to be beard form and face color; most other features tabulated by Groves (1986) are less convincing without extensive studies of variability. Shea and Coolidge (1988) used 17 measurements on 360 crania to investigate morphometric differences among chimpanzees (Fig. 11A,B). Their canonical variate analysis con-

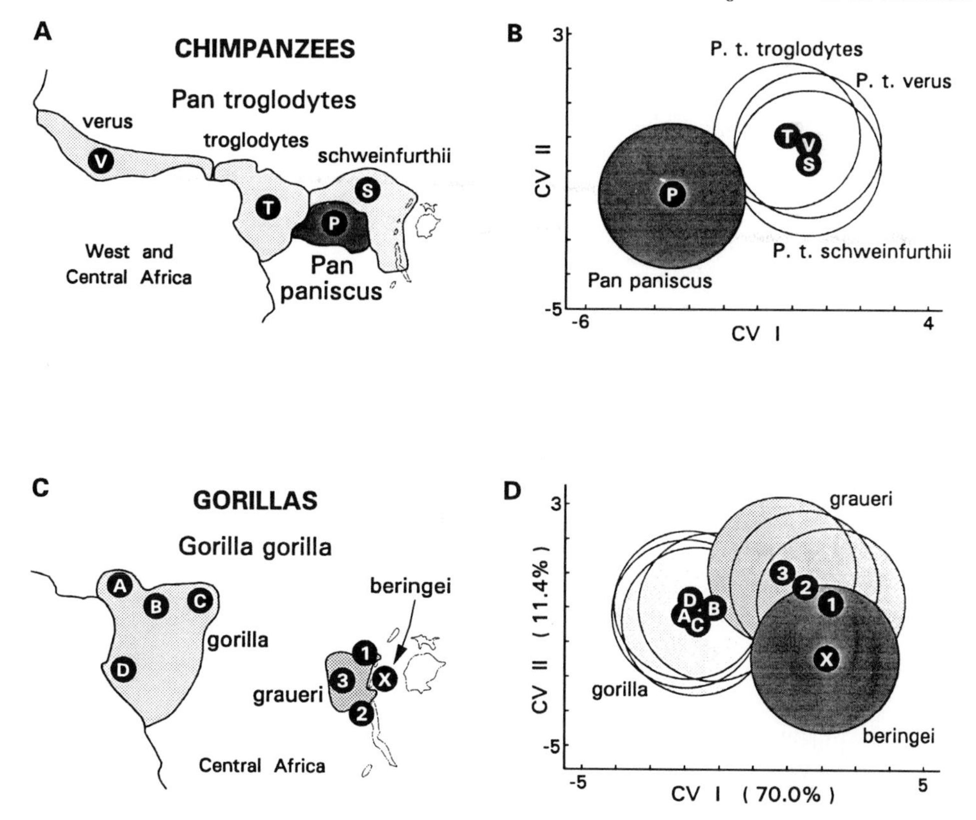

Fig. 11. Geographic variation in chimpanzees (*Pan*) and gorillas (*Gorilla*). (A) Distribution of *P. paniscus* and subspecies of *P. troglodytes* in western and central Africa (adapted from Shea and Coolidge, 1988, Fig. 1). Markers indicate subspecies samples used in multivariate analysis. (B) Canonical variate analysis of combined-sex samples of chimpanzees based on 17 craniofacial variables for four samples representing all forms of chimpanzees (adapted from Shea and Coolidge, 1988, Fig. 4; percentages of explained variation not available). (C) Distribution of subspecies of *G. gorilla* in central Africa (adapted from Groves, 1970, Fig. 1). Markers indicate approximate location of demes used in the multivariate analysis: A–D = *G. g. gorilla,* X = *G. g. beringei,* 1–3 = *G. g. graueri.* (D) Canonical variate analysis based on principal coordinate analysis of sex-averaged generalized distances for eight demes representing all forms of gorilla (distances from Groves, 1970, Table 1, represent averages for separate analyses of six mandibular and 10 cranial variables).

firms the distinctiveness of the pygmy chimpanzee. The comparative homogeneity across the three subspecies of common chimpanzees, compared to variation within other ape species, was interpreted as possibly reflecting a recent differentiation of *P. troglodytes.* The canonical variates, however, were based on combined-sex samples of specimens lumped by subspecies without regard for geographic variation. It is impossible to know how such groupings of specimens, known to be sexually dimorphic and potentially geographically variable, affect canonical variates and generalized distances, methods that rely on interpreting between-group differences as a function of within-group distributions. More detailed studies of geographic variation in chimpanzees are needed, given their wide distribution across central Africa and their importance as models for understanding human evolution. The preliminary report by Shea *et al.* (this volume) on new craniometric data for chimpanzees confirms the distinctiveness of *P. paniscus* and reports that the necessary geographic analyses are in progress.

Gorillas (Gorilla)

The three subspecies have a disjunct distribution in central Africa (Groves, 1970, 1986). Facial features, hair length, and postcranial proportions are usually used to differentiate the subspecies. However, the best evidence comes from the craniometric work by Groves (1970), especially with regard to establishing *G. gorilla graueri* as a second subspecies of eastern gorilla distinct from *G. g. beringei,* the mountain gorilla of the Virunga Volcanoes. Groves distilled his original data set of 45 measurements on several hundred specimens of 19 geographically restricted samples into generalized distances among eight demes of gorillas. These distances represent averages between samples of different demes for separate analyses of males and females using either six mandibular or 10 cranial variables.

Our principal coordinate analysis of Groves' "average" distance matrix shows the demes in three clusters corresponding to the three gorilla subspecies (Fig. 11C,D). The average distance among demes of the same subspecies ($\bar{D}$ = 1.76; 1.37–2.29) is half that among among demes of different subspecies ($\bar{D}$ = 3.51; 2.01–4.46). The geographically intermediate *G. g. graueri* is also craniometrically intermediate. Within *G. g. graueri,* samples from the mountains west of the Rift Lakes (deme #1 from Mt. Tshiaberimu and deme #2 from the Itombwe Mountains) are closer to the neighboring *G. g. beringei* ($\bar{D}$ = 2.21), while deme #3 from the eastern Congo lowlands is closer to *G. g. gorilla* from the lowlands of western Africa ($\bar{D}$ = 3.10).

Modern Humans (Homo sapiens)

Studies of geographic variation among local populations of humans are too numerous to review here. We present only one example from the classic study by Howells (1973) (Fig. 12). He measured 57 linear dimensions and 13 angles on the skulls of 834 males and 818 females representing 17 local populations from different parts of the world. Multivariate analysis revealed regional differentiation into geographic clusters. Interdemic distances among males are $\bar{D}$ = 2.71 (2.07–3.31) for Europe, $\bar{D}$ = 4.85 (3.20–6.96) for Africa, $\bar{D}$ = 3.53 (2.88–3.96) for Australasia, and $\bar{D}$ = 5.26 (3.28–6.47) for North and South America (female $\bar{D}$'s are slightly smaller). These are low compared to $\bar{D}$ = 6.46 (2.97–11.11) for all other distances among males of different regions.

Albrecht (1980c) demonstrated size to be an important factor in this craniometric analysis, explaining about 60% of the among-group variation on the first two canonical axes. He also showed that size is highly correlated with the temperature variables that Guglielmino-Matessi *et al.* (1979) related to Howells' (1973) multivariate results. Thus, geographic variation in human skull morphology is a multifactorial result of (1) an ecogeographic size cline, partly attributable to temperature differences, (2) shape differences, partly attributable to

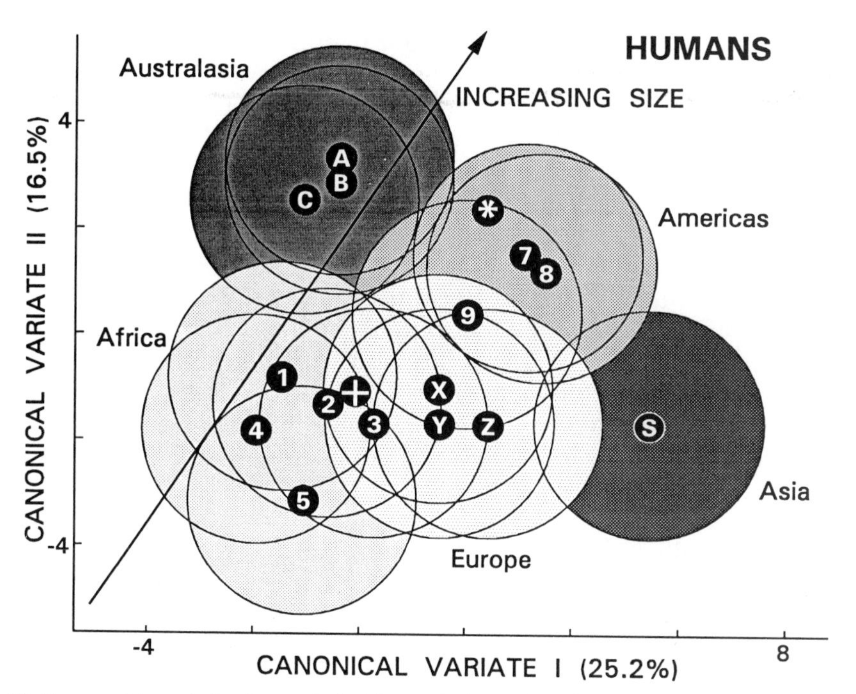

Fig. 12. Geographic variation in craniofacial morphology of modern humans (*Homo sapiens*). Canonical variate analysis based on 70 craniofacial variables for males of 17 local populations from throughout the world (adapted from Howells, 1973, Table 4). Markers indicate demes: Australasia (A = New Guinea, B = Tasmania, C = South Australia), Asia (S = Siberia), Europe (X = Hungary, Y = Norway, Z = Austria), Africa (1 = Zulu, South Africa, 2 = Kenya, 3 = Egypt, 4 = Mali, 5 = Bushmen, South Africa), Americas (7 = Greenland, 8 = South Dakota, 9 = Peru), and two miscellaneous samples plotted without circles (+ = Andaman Islands, * = Hawaii). The size gradient was identified using the method described by Albrecht (1979).

humidity effects, and (3) regional differentiation, reflecting zoogeographic relationships and genetic affinities.

Summary and Synthesis

Geographic Variation in Craniodental Morphology

Our review of multivariate studies demonstrates how common geographic variation is in the skulls and teeth of living primates. Every analysis published to date has revealed biologically and statistically significant amounts of craniodental variation when comparisons are made within and among closely related primate species. This includes size and shape differences for primates of all kinds—prosimians, New World monkeys, colobines, cercopithecines, apes, and humans. These results refute the assertions of those who claim "cranioskeletal differences between primate subspecies of the same species tend to be tiny, if observable at all" (Tattersall, 1986, p. 167) and "sister-species within modern primate genera usually differ little in their skeletal and dental anatomies" (Kimbel, 1991, p. 362).

Geographic effects generally have been ignored or discounted in primatology. Doing so may jeopardize the results and interpretations of studies that are otherwise well formulated. For example, Corner and Richtsmeier's (1991) recent investigation of craniofacial growth in *Cebus apella* divided specimens by sex and dental age without controlling for the origins of specimens acknowledged to be from widely different parts of the species range. Geographic variation was recognized as an important factor in cross-sectional growth studies but was assumed to be unimportant in this primate. Given the consistency of multivariate results reviewed above, however, there is almost certainly considerable geographic variation in the skull of *C. apella,* a species having, by one account (Hill, 1960), 16 subspecies distributed disjunctly across major parts of South America. It is impossible to know how the detailed descriptions of craniofacial growth in *C. apella* may have been affected by using samples of mixed demes and subspecies.

The idea of sampling appropriately to control for geographic variation is no different procedurally, and no less important, than controlling other levels of variation outlined in Figure 1. No one would consider investigating ontogenetic changes without controlling for individual, sexual, and interspecific variation, nor would anyone investigate sexual dimorphism without controlling for ontogenetic, individual, and interspecific variation, and so on. The necessity of such controls is accepted because ontogenetic, intrasexual, intersexual, and interspecific variation are well known. Given the prevalence and magnitude of geographic craniodental variation demonstrated here, it should become standard procedure for primatologists also to consider geographic differences as a significant source of variation, especially as questions focus on progressively finer levels of morphological differentiation.

Congruence between Craniodental Morphology and Taxonomy

Craniodental differences among living primates are generally congruent with taxonomic relatedness. Thus, following our hierachy of variation (Fig. 1), individuals are morphologically most similar on average to other individuals of the same sex of the same deme, demes to other demes of the same subspecies, subspecies to other subspecies of the same species, and species to other species of the same species group, subgenus, or genus. This hierarchy of relatedness is best demonstrated by black-and-white colobus monkeys, whose craniodental variation has been characterized comprehensively in terms of morphological and taxonomic coverage. At the level of individual variation, 95.3% of males and 97.0% of females are most closely associated with their own deme. At higher levels, morphometric distances among colobus demes increase in an orderly fashion, as shown in Fig. 13: (1) interdemic $\bar{D}$ = 3.42 (0.46–6.07) for 19 comparisons among demes of the same subspecies; (2) intersubspecific $\bar{D}$ = 6.45 (2.02–9.71) for 180 comparisons among demes of different subspecies within the same species; and (3) interspecific $\bar{D}$ = 9.39 (6.10–14.37) for 329 comparisons among demes of different species in the subgenus.

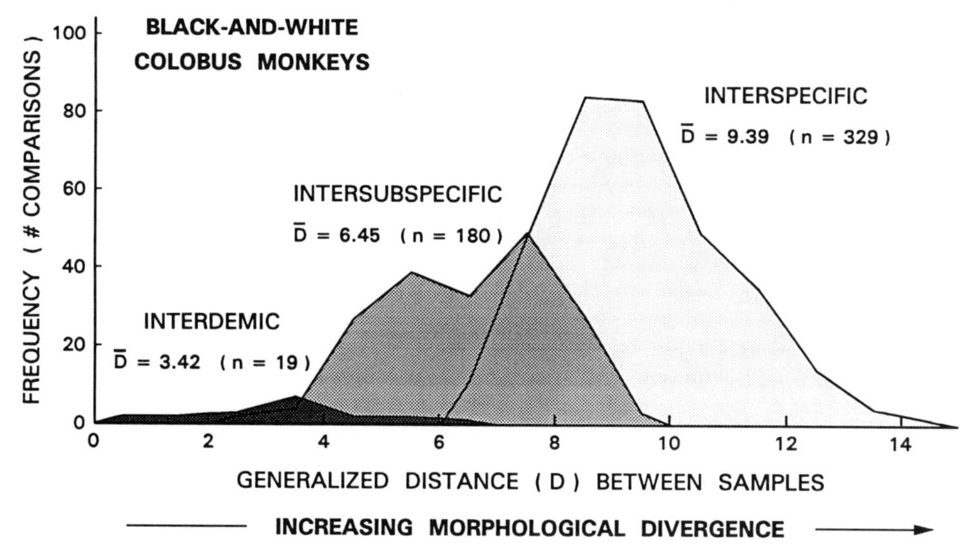

Fig. 13. Frequency distribution of generalized distances (D) among 33 samples of black-and-white colobus monkeys (subgenus *Colobus*). The distances are from the same multivariate analysis of 33 demes of males shown in Fig. 5 (data from Hull, 1979, Table 7). Generalized distances are categorized by rank of comparisons: "interdemic" between demes of the same subspecies, "intersubspecific" between demes of different subspecies in the same species, and "interspecific" between demes of different species. The average generalized distance (D̄) and number of comparisons are given for each category.

The accord between craniodental morphology and taxonomic relatedness is equally evident from the canonical variate plots for *Lepilemur* (Fig. 2), *Saguinus* (Fig. 3), *Ateles* (Fig. 5), non-Sulawesi species of *Macaca* (Fig. 7, Table 1), *Papio* (Fig. 9), *Pan* (Fig. 11), and *Gorilla* (Fig. 11). It would be informative to see how widespread this pattern of craniodental differences is among other taxonomically diverse primate genera. Among the best test cases for detailed, comprehensive studies would be *Galago, Callithrix, Alouatta, Cercopithecus,* and *Presbytis.*

While this pattern of morphologic–taxonomic concordance appears to be a general characteristic of polytypic primate species, the degree of morphometric differentiation varies from taxon to taxon (here, we reiterate the cautions expressed earlier about comparing generalized distances across studies). Interdemic distances within one taxon (e.g., D̄ = 3.42 for black-and-white colobus demes) may equal subspecific distances in another primate (e.g., D̄ = 3.51 for gorilla subspecies), and subspecific differences may equate with specific differences (e.g., in Fig. 11, subspecific distances within *Gorilla gorilla* approximate the divergence of *pan paniscus* from *P. troglodytes*). Even among closely related primates, as demonstrated well by the *Colobus* data (Fig. 13), there is substantial variation and considerable overlap in generalized distances among demes, subspecies, and species. For example, a D = 8.0 between two colobus samples probably rules out interdemic variation within one subspecies but is comparable with either intersubspecific or interspecific differentiation. This means it is impossible to assign taxonomic rank on the basis of generalized distances alone.

Theoretically, there is no necessary, predictable, or required relationship between evolutionary divergence and morphological change. Among primates, perhaps the best example is provided by the broad pattern of craniofacial differentiation in macaques (Albrecht, 1978; Fig. 7). Non-Sulawesi macaques are characterized by conservative changes in skull morphology (mostly size related) despite a long evolutionary history (about 5 million years), subtantial taxonomic variety (numerous subspecies, 12 species, and 4 species groups), wide geographic range (across southern Asia and a relict species in northwestern Africa), and considerable habitat diversity (deciduous temperate forests to evergreen tropical forests). In contrast, the seven Sulawesi macaques display extremes in skull shape suggestive of generic differences among other primates, despite their relatively recent evolution (about 1 million years), close taxonomic affinities (some combination of subspecies and species), restricted distribution (about 2% of the genus range), and homogeneity of environment (equatorial forests). Nevertheless, within each species group, the correlation between morphological and taxonomic affinities of macaques is high, as previously discussed and as documented in Table 1.

Those who emphasize the uncoupling of morphologic and taxonomic differentiation in primates err by generalizing from theoretical possibilities and ignoring contradictory empirical evidence about what actually occurs (e.g., Tattersall, 1986, 1991; Kimbel, 1991). It may be that speciation can take place with little or no morphological differentiation, but in primates this seldom if ever occurs. There are no documented cases of sibling species in primates, despite this being the most extensively studied of all groups of animals. This is not surprising since "sibling species are apparently rarest in organisms . . . that are most dependent on vision in the recognition of epigamic characters" (Mayr, 1963, p. 58).

Indeed, visually oriented, diurnal species of primates display diagnostic characters of external appearance, especially in the head and tail regions, since head-to-head and head-to-tail meetings predominate in social interactions (nocturnal primates, which place more emphasis on olfaction and vocalization, are generally less distinctive in external appearance). Some would refer to these color differences and display structures as premating isolating mechanisms, while others would call them components of specific mate-recognition systems (although, Jolly, this volume, suggests distinguishing features of some baboons may not be effective in this way). As we demonstrated with craniodental morphometrics, primate species and subspecies are also usually differentiated in their skeletal morphology. Moreover, whether considering external or internal features, primates are usually characterized by patterns of morphological variation corresponding to a hierarchial classification from individuals through subgenera (Fig. 1). In summary, while there is no *necessary* relationship between morphologic and taxonomic diversity, the overwhelming weight of empirical observations substantiates the opposite, that is, the evolution of primate diversity at the species level and below has resulted in a structured hierarchy of variation.

We acknowledge the potential circularity of demonstrating a pattern of correspondence between morphology and taxonomy when the taxonomic status of primates is assigned primarily on the basis of morphological similarity in the first place. However, the circularity is illusory since the alpha taxonomy of living primates relies heavily on external morphology, with little emphasis on craniodental features as key characters. More generally, in our view, primate taxonomy primate taxonomy ultimately reflects a consensus hypothesis about actual biological entities and their relationships. The reality of primate taxa is constantly being refined through a long, iterative history of hypothesis testing involving all the available evidence from many fields of inquiry accumulated by hundreds of primatologists from different schools of evolutionary thought. We argue that it is both scientifically valid and biologically valuable to compare individual aspects of an animal's biology, such as craniodental morphometrics, with this consensus taxonomy. Those who disagree are challenged to propose an alternative classification for primates, not based on morphology, that can be used as the basis for testing hypotheses about patterns of biological diversity.

Correlates of Geographic Variation

The geographic variation seen in living primates is related to a number of factors differentially affecting various primates. The primary explanations are levels of interbreeding (gene flow), historical patterns of dispersal and differentiation (zoogeography and historical biogeography), and adaptation to local environmental conditions (ecogeographic variation). Every case of geographic variation reviewed here involves some combination of these factors.

The ordering of morphological resemblances among neighboring demes implies that different levels of gene flow affect the differentiation of primates. It is certainly true that local populations usually cluster closely together in those craniodental studies incorporating multiple samples of polytypic primates (Fig. 2 for *Lepilemur,* Fig. 5 for *Saimiri* and *Ateles,* Fig. 6 for *Colobus,* Fig. 7 for *Macaca,* Fig. 9 for *Papio,* and Fig. 11 for *Gorilla*). Presumably, this homogeneity reflects a communality in the gene pool of interbreeding local populations. Sometimes gene flow is implied when a deme is both morphologically and geographically intermediate between its neighbors. This is most clearly seen in the ring-like arrangement of *Lepilemur* demes (Fig. 2) and the intermediates between named forms of *Ateles* (Fig. 5). In other cases, there seems to be no clear association. Some caution is advised at this level of resolution since demes are being characterized by the usually small samples available in museum collections from scattered localities (i.e., not strictly a geneticist's deme). Few if any primates will ever be represented by requisite sample sizes from enough contiguous localities to make rigorous, definitive investigations of population-level phenomena.

The morphology of demes and subspecies often reflects zoogeographic and/or historical biogeographical relationships. Humans provide a good example of morphological similarities among geographic clusters of closely related primates (Fig. 12). In some cases, primates have nearly the same spatial arrangement in canonical variate data space as they do on a map of their distribution. This resemblance is especially strong in (1) the ringlike distribution of *Lepilemur* around the coastal periphery of Madagascar (Fig. 2), (2) the linear ordering of *Colobus polykomos* subspecies in the narrow band of forest along the coast of western Africa (Fig. 6), and (3) the divergence of Sulawesi macaques from a central population along the arms of the island (Fig. 8).

The inference is that evolution and dispersal in these groups was strongly shaped by regional geography. This means, for example, on zoogeographic grounds alone, that *L. mustelinus* is likely to have a closer affinity with *L. microdon* than *L. leucopus, C. p. polykomos* with *C. p. dollmani* than *C. p. vellerosus,* and *M. nigra* with *M. nigrescens* than *M. brunnescens.* The situation may be complicated when the presumed evolution among closely related primates was not constrained by physical geography to simple pathways.

Recent work by Cheverud and Moore (1990) (*Saguinus* in Fig. 3) and Froehlich *et al.* (1991) (*Ateles* in Fig. 5) demonstrates the insights possible when morphology and historical biogeography are integrated. In *Saguinus fuscicollis,* facial morphometrics are correlated with presumed lines of phylogenetic descent but not geographic contiguity. This is because the formation of subspecies apparently involved a complex series of invasions, isolations, and differentiations, resulting in contiguous races not necessarily being most closely related in terms of their evolutionary history. In *Ateles,* craniodental morphology reflects a complex interplay involving dispersals from Pleistocene refugia, adaptation to terra firma areas around seasonal swamp forests, and isolation by river and habitat barriers.

Some of the craniodental differences among demes of living primates can be explained by ecological and environmental effects. Ecogeographic variation, especially in size, usually becomes evident when studies of sufficient detail are made of primates whose distributions encompass a range of conditions. Latitudinal size clines are known for *Callithrix jacchus* (Albrecht, 1982), *Macaca arctoides* (Albrecht, 1980a), *M. assamensis* (Albrecht, 1980a; Fooden, 1982), *M. fascicularis* (Aimi *et al.,* 1982; Albrecht, 1983; Fooden, 1991), *M. fuscata* (Hamada *et al.,* 1987), *M. mulatta* (Albrecht, 1978), *M. nemestrina* (Fooden, 1975; Albrecht, 1978), *M. radiata* (Fooden, 1981), and *Presbytis entellus* (Fig. 14). In these cases, size usually increases with increasing latitude (Bergmann's rule). Climatic effects on size are known for *Homo sapiens* (e.g., Guglielmino-Matessi *et al.,* 1979), *Papio* (Popp, 1983; Vitzhum, 1986), and *Cercopithecus aethiops* and *C. mitis* (Vitzhum, 1986). In these cases, temperature is more highly correlated with size than is humidity.

Habitat effects are known for *Callithrix jacchus* (Albrecht, 1982), nearly all kinds of Malagasy prosimians (Albrecht *et al.,* 1990), and in many instances of human variation (e.g., Passarello and Vecchi, 1979; Hyndman *et al.,* 1989). In these cases, larger primates are usually found in lusher, more productive habitats. Ecogeographic size differences can be extreme (e.g., Hanuman langurs in Fig. 14), and they can be pervasive (the Malagasy prosimians). On the other hand, not all primates respond in the same way to changes in their physical and biotic environments. For example, toque macaques of Sri Lanka are unaffected by the same extremes of habitat associated with substantial size differences among prosimians of Madagascar. Also, while most primates increase in size with increasing latitude, others decrease (Sundaic pigtail macaques). This diversity in response should be expected for a character such as body size that both affects and is influenced by so many aspects of an animal's existence.

Implications for Interpreting Fossils

The study of geographic variation in living primates has broad and important implications for studying fossils. (1) Geographic variation is common in

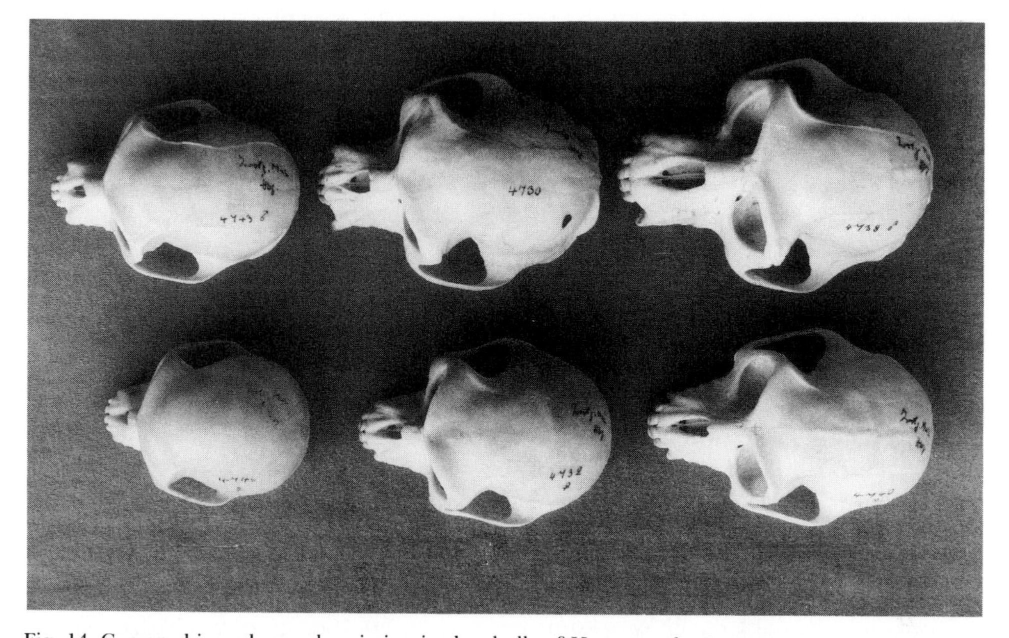

Fig. 14. Geographic and sexual variation in the skulls of Hanuman langurs (*Presbytis entellus*). Specimens are arranged with males on top, females on bottom, *P. e. achates* from North Kanara in southwest India (15°N) on left, *P. e. entellus* from Madhya Pradesh in central India (22°N) in middle, and *P. e. schistacea* from Uttar Pradesh in northern India (28°N) on right. Greatest skull lengths are 95.0 mm for the smallest female and 136.9 mm for the largest male. All are adult specimens from the Zoological Institute and Museum, Hamburg University.

both the internal and external features of living primates. It was, therefore, probably common in fossil primates. (2) Geographic variation in skeletal features represents a significant source variation among living primates. This variability in size and shape should be taken into account by primate paleontologists interested in distinguishing intraspecific and interspecific differences among closely related fossils. (3) Geographic variation in living primates is characterized by a general correspondence between morphological and taxonomic differentiation among demes, subspecies, and species. This means paleoprimatologists may be able to sort out hierarchial variability among fossil specimens at the species level and below. However, the magnitude of morphometric separation varies from species to species in living primates, which means taxonomic ranks of fossils may never be known with certainty. (4) Geographic variation results from interactions among a multitude of geographic, environmental, and biotic effects in living primates. Similar explanations are likely to apply to fossils.

Fossil studies have focused on the role of individual variation and sexual dimorphism in controversies about the numbers and identities of paleospecies. Not enough attention has been given to the possibility that morphological differences among primate fossils might be attributable to geographic variation. The combined effects of intraspecific variation within living primates can be so remarkable that paleontologists should be wary of being led astray by variation within fossil assemblages. The Hanuman langur, *Presbytis entellus,* provides an informative example of what geographic variation might look like in a fossil species (Fig. 14). The subspecies of Hanuman langurs are distributed across the

Indian subcontinent from Sri Lanka in the south to the Himalayas in the north. Females from the north (bottom right in Fig. 14) and south (bottom left) of India appear visually to be about as different as are males and females in general (top vs. bottom rows). Indeed, if canine dimorphism is ignored, the female from the north might be mistaken for a male from the south. These skulls represent a single species whose size follows a classic latitudinal cline (Bergmann's rule) from small langurs in the south to large ones in the north, possibly enhanced by altitudinal effects on body size for animals living at high elevations in the Himalayan mountains.

Now, consider what interpretations might be made if these were skulls of extinct animals from fossil localities in different parts of India. It would be easy to imagine a primate paleontologist assigning them to several different species, if not different genera for the extreme forms. Hanuman langurs provide a dramatic but not exceptional example of intraspecific variability in primate craniofacial morphology. Similar photographic examples of geographic variants can be found in Fooden (1991, Fig. 6) for living *Macaca fascicularis phillipinensis* and in Godfrey *et al.* (1990, Fig. 2) (see our Fig. 15) for the recently extinct *Archaeolemur.* These examples do not include the additional potential source of anagenetic variation through vast stretches of time. Consequently, if the langur skulls were a fossil sample representing, say, 200,000 years of evolution, variation might be even greater than that presently observed. The same could be said for the *Archaeolemur* skulls, which probably capture only several thousand years of recent evolution.

The importance of subspecific variation in paleoanthropology is sometimes disparaged. For example, Kimbel (1991, this volume) questioned the usefulness of subspecies and claimed that their recognition is an impediment to discovering the pattern of phylogenetic relationships in the fossil record. Our view is quite different (we concur with Jolly's, this volume, perspective on the importance of subspecies). We follow the usual definition—a subspecies is a collection of phenotypically similar demes, inhabiting a geographic subdivision of the range of a species, which differ taxonomically from other such collections

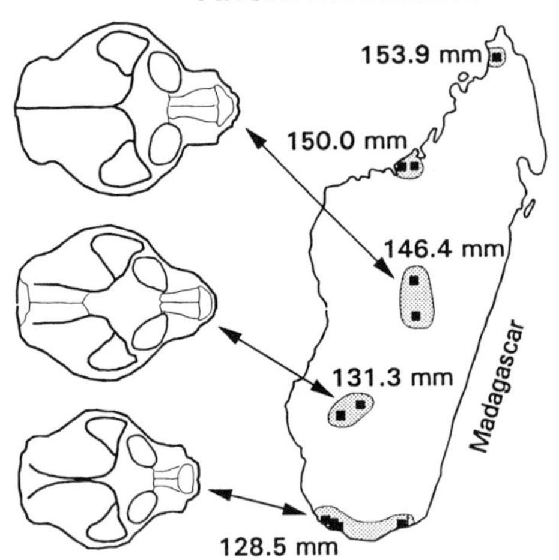

Fig. 15. Geographic variation in skull length of subfossil *Archaeolemur* from Madagascar. Average skull lengths are given for specimens from different parts of Madagascar (adapted from Albrecht *et al.*, 1990, Table II and Fig. 4). Dorsal views or representative skulls from the south, southwest, and central highlands are shown (adapted from Godfrey *et al.*, 1990, Fig. 2).

of demes within a species (e.g., see Mayr, 1969, p. 41). Many taxonomists, such as Simpson (1961, pp. 172–173), have concluded that subspecies "are 'real' in that they correspond with groupings in nature and not solely in the taxonomist's mind." Indeed, living primate subspecies are natural biological groupings, geographically defined and differentiated on the basis of empirical characters as is readily evident from reading any authoritative treatment of primate taxonomy.

Clearly, given the evidence of craniodental variation in living primates, it is now untenable to argue that subspecies are not potentially recognizable in the skeletal remains that comprise the fossil record. Fossil specimens are members of geographic populations (demes and subspecies) frozen in time, just as they represent individuals, sexes, and species. Finally, the pattern of phylogenetic relationships among fossil species cannot be determined successfully without first making decisions about whether the taxa analyzed are species or not. Such decisions require an appreciation of geographic variation as one possibility explaining the presence of distinct fossil morphotypes. To ignore demic and subspecific variation is to run the risk of misidentifying intraspecific groups (i.e., subspecies) for the proper units (species) of phylogenetic analysis.

The argument has been made that, for fossils, "where distinct morphs can readily be identified it would seem most productive to assume they represent species unless there is compelling evidence to believe otherwise" (Tattersall, 1986, p. 168). There is, however, little basis for following this recommendation, given the weight of our findings about craniodental variation. Just as in living primates, distinct morphs of fossils may indicate differences at any level of the hierarchy, including sexes, geographic demes, subspecies, species, or higher levels.

The vagaries of the fossil record seriously confine the paleontologist interested in taxonomic questions. Compared to the studies of living primates reviewed here, fossil samples are usually small, localities are scarce, specimens are fragmentary, and the time dimension is substantial. Nevertheless, when analyzing existing fossil materials, paleoanthropological studies must strengthen their research designs if progress is to be made in sorting out different levels of variation. There are two aspects to consider: (1) selecting the composition of the fossil sample to be studied, and (2) selecting appropriate samples of modern primates to serve as comparative analogs (see Shea *et al.*, this volume, for further discussion on this latter point). With respect to both, the sampling strategy must be appropriate to the research question being asked. For example, if the research involves characterizing sexual dimorphism in a fossil taxon, then it would be ideal to use specimens discovered at a single locality (presumably, representing one time frame). Otherwise, there is no way of knowing the extent to which geographic variation confounds estimates of sexual differences. Likewise, comparative samples of living primates should be chosen to reflect demic levels of sexual variation as closely as possible. Jenkins and Albrecht (1991) used just such a sampling strategy to control for extraneous sources of variation, such as taxonomic and ecogeographic size variation, in their study of sexual dimorphism in Malagasy prosimians.

If, however, the research question is whether fossils from widely spread localities are representative of a single species, then comparative samples should encompass the intraspecific diversity of living primate species. Basing comparison on a single deme or subspecies as representative of an entire species of living primate would underestimate the magnitude of possible species-level differentiation by excluding an important source of variability, geographic variation. The best strategy, in this case, is to carefully construct comparative samples so as to represent all levels of variation from intraspecific through interspecific. This approach provides ranges of overall variation and, at the same time, potentially allows finer grained interpretations about the precise meaning of variation seen in the fossils under study.

Our point is that comparative studies of fossils should differ little from experimental science in their ideal research design; both require careful controls on the different sources of variation involved in the problem. For the paleoprimatologist, this may involve more time, more travel, more museums, more measuring, and more attention to locality data. The result, however, will be an appropriate foundation for rigorous analysis that will be based on larger, more numerous samples that more accurately reflect the nature of morphometric variation within and among species of both living and fossil primates. We join Shea *et al.* (this volume) in their plea for strengthening the comparative framework upon which paleoanthropological studies are based.

Recent studies of *Archaeolemur* illustrate how an understanding of geographic variation can enhance understanding of the biology of fossil primates (Fig. 15; Godfrey and Petto, 1981; Albrecht *et al.*, 1990; Godfrey *et al.*, 1990). The dozens of specimens recovered from fossil localities in all but the eastern tropical regions makes *Archaeolemur* the most common and most widespread of the recently extinct, giant prosimians of Madagascar. The two species generally recognized are the relatively gracile *A. majori* from the south and southwest, and the more robust *A. edwardsi* from the central highlands. Using data on skull length for 31 specimens, Albrecht *et al.* (1990) showed that animals from the arid south (average skull length = 128.5 mm), dry southwest (131.3 mm), and the former bush-woodland-forest mosaic of the now unihabitable central highlands (146.4 mm) follow the same ecogeographic gradient of increasing size as almost all other living and extinct Malagasy prosimians. Two specimens from the northwest (150.0 mm) and one from the north (153.9 mm) are contradictory to the general trend among Malagasy prosimians of finding the largest members of closely related sister taxa in the central highlands.

Godfrey *et al.* (1990; and Godfrey and Petto, 1981) used odontometric measurements on 85 mandibles and 73 maxillae from 18 fossil localities to investigate ecoclinal variation in more detail than attempted by Albrecht *et al.* (1990). Their regression analyses revealed that geographic variation of size differences in *Archaeolemur* is best explained by differences in niche productivity, as inferred from rainfall and floral richness. Thermoregulatory factors were less important, with the possible exception of altitudinal effects on body size of animals that once inhabited the central mountainous regions of Madagascar. Thus, the recognized species may represent nothing more than ecoclinal variants. Taxonomic revision is needed to address this issue and to accommodate the more recently discovered specimens recovered from northwestern and northern Madagascar (see Godfrey *et al.*, 1990). Hopefully, such revisions will include multivariate analyses of craniodental morphology to understand better the role of geographic variation in the differentiation of these fossils.

We conclude by reiterating the prevalence of geographic variation in the craniodental features of living primates. There is no reason to believe primates of the past were any different in this regard. Furthermore, geographic variation impacts the design, results, interpretations, and conclusions of neontological studies just as much as do ontogenetic, individual, sexual, and interspecific variation. Again, there is no reason to believe paleontological studies are any less affected. Since geographic differences exist and have such important implications, it seems obvious that primatologists must learn to contend routinely with geographic variation in all studies, whether these be of living or fossil species. Recognizing geographic variation as an integral part of the hierarchy of biological variability will lead to better understanding of current problems as well as new insights about the biology of primates.

Acknowledgments

We thank Gerald Bales (University of Southern California), Bruce Gelvin (California State University, Northridge), Laurie Godfrey and Michael Sutherland (University of Massachusetts), Steven Hartman (University of New England), and Brian Shea (Northwestern University) for their helpful comments on improving our work. Reviews by Colin Groves (Australian National University) and two anonymous referees were valuable in revising the manuscript. Bill Kimbel and Lawrence Martin should receive special commendations for their efforts in organizing the symposium and editing this volume. Funding support for this research was provided by a NSF grant to GHA (BNS-83078900) and a NSF predoctoral fellowship to JMAM.

References

Aimi, M., Bakar, A., and Supriatna, J. 1982. Morphological variation of the crab-eating macaque, *Macaca fascicularis* (Raffles, 1821), in Indonesia. *Kyoto University Overseas Research Report of Studies on Asian Non-Human Primates,* Kyoto University Primate Research Institute, **2:**51–56.

Albrecht, G. H. 1976. A Multivariate Analysis of the Craniofacial Morphology of the Sulawesi Macaques (Primates: Cercopithecidae). Doctoral Dissertation, University of Chicago, Chicago.

Albrecht, G. H. 1977. Methodological approaches to morphological variation in primate populations: The Celebesian macaques. *Yrbk. Phys. Anthropol.* (1976) **20:**290–308.

Albrecht, G. H. 1978. The craniofacial morphology of the Sulawesi macaques: Multivariate approaches to biological problems. *Contr. Primat.* **13:**1–151.

Albrecht, G. H. 1979. The study of biological versus statistical variation in multivariate morphometrics: The descriptive use of multiple regression analysis. *Syst. Zool.* **28:**338–344.

Albrecht, G. H. 1980a. Latitudinal, taxonomic, sexual, and insular determinants of size variation in pigtail macaques, *Macaca nemestrina. Int. J. Primatol.* **1:**141–152.

Albrecht, G. H. 1980b. Multivariate analysis and the study of form, with special reference to canonical variate analysis. *Am. Zool.* **20:**679–693.

Albrecht, G. H. 1980c. Size variation in the skull of modern human populations. *Am. J. Phys. Anthropol.* **52:**199.

Albrecht, G. H. 1982. The relationships between size, latitude, and habitat in the South American primate, *Callithrix jacchus. Am. J. Phys. Anthropol.* **57:**166.

Albrecht, G. H. 1983. Geographic variation in the skull of the crab-eating macaque, *Macaca fascicularis* (Primates: Cercopithecidae). *Am. J. Phys. Anthropol.* **60:**169.

Albrecht, G. H. 1993. Assessing the affinities of fossils using canonical variates and generalized distances, *Hum. Evol.* (in press).

Albrecht, G. H., Jenkins, P. D., and Godfrey, L. R. 1990. Ecogeographic size variation among the living and subfossil prosimians of Madagascar. *Am. J. Primatol.* **22:**1–50.

Bernstein, I. S. 1966. Naturally occurring primate hybrid. *Science* **154:**1559–1560.

Booth, S. N., and Freedman, L. 1970. Multivariate discriminant analysis applied to cranial features of *Papio ursinus* and *P. cynocephalus*. *Folia Primatol.* **12:**296–304.

Cheverud, J. M., and Moore, A. J. 1990. Subspecific morphological variation in the saddle-back tamarin (*Saguinus fuscicollis*). *Am. J. Primatol.* **21:**1–15.

Coimbra-Filho, A. F., and Mittermeier, R. A. 1973. New data on the taxonomy of the Brazilian marmosets of the genus *Callithrix* Erxleben, 1777. *Folia Primatol.* **20:**241–264.

Corner, B. D., and Richtsmeier, J. T. 1991. Morphometric analysis of craniofacial growth in *Cebus apella*. *Am. J. Phys. Anthropol.* **84:**323–342.

Courtenay, J., Groves, C., and Andrews, P. 1988. Inter- or intra-island variation? An assessment of the differences between Bornean and Sumatran orang-utans, in: J. H. Schwartz (ed.), *Orangutan Biology*, pp. 19–30. Oxford University Press, New York.

Creel, N. and Preuschoft, H. 1976. Cranial morphology of the lesser apes. *Gibbon and Siamang* **4:**219–303.

Creel, N., and Preuschoft, H. 1984. Systematics of the lesser apes: A quantitative taxonomic analysis of craniometric and other variables, in: D. J. Chivers, H. Preuschoft, W. Y. Brockelman, and N. Creel (eds.), *The Lesser Apes*, pp. 562–613. Edinburgh University Press, Edinburgh.

Drenhaus, U. 1975. Ein Beitrag zur Taxonomie der Lemuriformes Gregory 1915 unter besonderer Berücksichtigung der Gattung Lepilemur I. Geoffroy 1851. Doctoral Dissertation, Christian-Albrechts-Universität, Kiel.

Endler, J. A. 1977. *Geographic Variation, Speciation, and Clines*. Princeton University Press, Princeton, NJ.

Endler, J. A. 1989. Conceptual and other problems in speciation, in: D. Otte and J. A. Endler (eds.), *Speciation and its Consequences*, pp. 625–648. Sinauer Associates Publishers, Sunderland, MA.

Fa, J. E. 1989. The genus *Macaca:* a review of taxonomy and evolution. *Mammal Rev.* **19:**45–81.

Fooden, J. 1964. Rhesus and crab-eating macaques: intergradation in Thailand. *Science* **143:**363–365.

Fooden, J. 1969. Taxonomy and evolution of the monkeys of the Celebes (Primates: Cercopithecoidae). *Bibl. Primatol.* **10:**1–148.

Fooden, J. 1971. Report on primates collected in western Thailand January–April, 1967. *Fieldiana: Zool.* **59:**1–62.

Fooden, J. 1975. Taxonomy and evolution of liontail and pigtail macaques (Primates: Cercopithecidae). *Fieldiana: Zool.* **67:**1–169.

Fooden, J. 1976. Provisional classification and key to living species of macaques (Primates: *Macaca*). *Folia Primatol.* **25:**225–236.

Fooden, J. 1979. Taxonomy and evolution of the *sinica* group of macaques: I. Species and subspecies accounts of *Macaca sinica*. *Primates* **20:**109–140.

Fooden, J. 1980. Classification and distribution of living macaques (*Macaca* Lacépède, 1799), in: D. G. Lindburg (ed.), *The Macaques: Studies in Ecology, Behavior and Evolution*, pp. 1–9. Van Nostrand Reinhold, New York.

Fooden, J. 1981. Taxonomy and evolution of the *sinica* group of macaques: 2. Species and subspecies accounts of the Indian bonnet macaque, *Macaca radiata*. *Fieldiana: Zool.*, New Series **9:**1–52, 1981.

Fooden, J. 1982. Taxonomy and evolution of the *sinica* group of macaques: 3. Species and subspecies accounts of *Macaca assamensis*. *Fieldiana: Zool.* New Series **10:**1–52.

Fooden, J. 1983. Taxonomy and evolution of the *sinica* group of macaques: 4. Species account of *Macaca thibetana*. *Fieldiana: Zool.*, New Series **17:**1–20.

Fooden, J. 1986. Taxonomy and evolution of the *sinica* group of macaques: 5. Overview of natural history. *Fieldiana: Zool.*, New Series **29:**1–22.

Fooden, J. 1988. Taxonomy and evolution of the *sinica* group of macaques: 6. Interspecific comparisons and synthesis. *Fieldiana: Zool.*, New Series **45:**1–44.

Fooden, J. 1990. The bear macaque, *Macaca arctoides:* a systematic review. *J. Hum. Evol.* **19:**607–686.
Fooden, J. 1991. Systematic review of Philippine macaques (Primates, Cercopithecidae: *Macaca fascicularis* subspp.). *Fieldiana: Zool.*, New Series **64:**1–44.
Fooden, J., Guoqiang, Q., Zongren, W., and Yingxiang, W. 1985. The stumptail macaques of China. *Am. J. Primatol.* **8:**11–30.
Froehlich, J. W., Supriatna, J., and Froehlich, P. H. 1991. Morphometric analyses of *Ateles:* Systematic and biogeographic implications. *Am. J. Primatol.* **25:**1–22.
Fujita, K. 1987. Species recognition by five macaque monkeys. *Primates* **28:**353–366.
Godfrey, L. R., and Marks, J. 1991. The nature and origins of primate species. *Yrbk. Phys. Anthropol.* **34:**39–68.
Godfrey, L. R., and Petto, A. J. 1981. Clinal size variation in *Archaeolemur* spp. on Madagascar, in: B. Chiarelli and R. L. Corruccini (eds.), *Primate Evolutionary Biology,* pp. 14–34. Springer-Verlag, New York.
Godfrey, L. R., Sutherland, M. R., Petto, A. J., and Boy, D. S. 1990. Size, space, and adaptation in some subfossil lemurs from Madagascar. *Am. J. Phys. Anthropol.* **81:**45–66.
Gould, S. J., and Johnston, R. F. 1972. Geographic variation. *Ann. Rev. Ecol. Syst.* **3:**457–498.
Gower, J. C. 1966a. A Q-technique for the calculation of canonical variates. *Biometrika* **53:**588–590.
Gower, J. C. 1966b. Some distance properties of latent root and vector methods used in multivariate analysis. *Biometrika* **53:**325–338.
Groves, C. P. 1970. Population systematics of the gorilla. *J. Zool., Lond.* **161:**287–300.
Groves, C. P. 1972. Systematics and phylogeny of gibbons. *Gibbon and Siamang* **1:**1–89.
Groves, C. P. 1980. Speciation in *Macaca:* the view from Sulawesi, in: D. G. Lindburg (ed.), *The Macaques: Studies in Ecology, Behavior and Evolution,* pp. 84–124. Van Nostrand Reinhold Company, New York.
Groves, C. P. 1984. A new look at the taxonomy and phylogeny of the gibbon, in: D. J. Chivers, H. Preuschoft, W. Y. Brockelman, and N. Creel (eds.), *The Lesser Apes,* pp. 542–561. Edinburgh University Press, Edinburgh.
Groves, C. P. 1986. Systematics of the great apes, in: D. R. Swindler and J. Erwin (eds.), *Comparative Primate Biology, Volume 1: Systematics, Evolution, and Anatomy,* pp. 187–217. Alan R. Liss, New York.
Guglielmino-Matessi, C. R., Gluckman, P., and Cavalli-Sforza, L. L. 1979. Climate and evolution of skull metrics in man. *Am. J. Phys. Anthropol.* **50:**549–564.
Hamada, Y., Watanabe, T., Suryobroto, B., and Iwamoto, M. 1987. Morphological studies of the Sulawesi macaques: morphological specializations in the black ape (*Macaca nigra*) with ecological and sociological consideration. *Kyoto University Overseas Research Report of Studies on Asian Non-Human Primates* (Kyoto University Primate Research Institute) **6:**31–47.
Hayes, V. J., Freedman, L., and Oxnard, C. E. 1990. The taxonomy of savannah baboons: An odontomorphometric analysis. *Am. J. Primatol.* **22:**171–190.
Hershkovitz, P. 1975. Comments on the taxonomy of Brazilian marmosets (*Callithrix,* Callitrichidae). *Folia Primatol.* **24:**137–172.
Hershkovitz, P. 1977. *Living New World Monkeys (Platyrrhini) with an Introduction to Primates,* Vol. 1. University of Chicago Press, Chicago.
Hershkovitz, P. 1984. Taxonomy of squirrel monkeys genus *Saimiri* (Cebidae, Platyrrhini): A preliminary report with description of a hitherto unnamed form. *Am. J. Phys. Anthropol.* **7:**155–210.
Hill, W. C. O. 1960. *Primates. Comparative Anatomy and Taxonomy. IV. Cebidae.* Part A. Edinburgh University Press, Great Britain.
Howells, W. W. 1973. Cranial variation in man. A study by multivariate analysis of patterns of differences among recent human populations. *Papers of the Peabody Museum of Archaeology and Ethnology,* Vol. 67.
Hull, D. B. 1979. A craniometric study of the black and white *Colobus* Illiger 1811 (Primates: Cercopithecoidea). *Am. J. Phys. Anthropol.* **51:**163–182.
Hyndman, D. C., Ulijhszek, S. J., and Lourie, J. A. 1989. Variability in body physique, ecology, and subsistence in the Fly River Region of Papua New Guinea. *Am. J. Phys. Anthropol.* **79:**89–101.
Jacobshagen, B. 1979. Morphometric studies in the taxonomy of the orang-utan (*Pongo pygmaeus,* L. 1760). *Folia Primatol.* **32:**29–34.
Jenkins, P. D. 1987. Catalogue of Primates in the British Museum (Natural History) and Elsewhere in the British Isles, Part IV: Suborder Strepsirrhini, Including the Subfossil Madagascan Lemurs and Family Tarsiidae. British Museum (Natural History), London.

Jenkins, P. D., and Albrecht, G. H. 1991. Sexual dimorphism and sex ratios in Madagascan prosimians. *Am. J. Primatol.* **24:**1–14.

Jungers, W. L., and Rumpler, Y. 1976. Craniometric collaboration of the specific status of *Lepilemur septentrionalis,* an endemic lemur from the north of Madagascar. *J. Hum. Evol.* **5:**317–321.

Kawamoto, Y., Takenaka, O., and Brotoisworo, E. 1982. Preliminary report on genetic variations within and between species of Sulawesi macaques. *Kyoto University Overseas Research Report of Studies on Asian Non-Human Primates* **2:**23–37.

Kimbel, W. H. 1991. Species, species concepts and hominid evolution. *J. Hum. Evol.* **20:**355–372.

Lernould, J-M. 1988. Classification and geographical distribution of guenons: a review, in: A. Gautier-Hion, F. Bourlière, and J.-P. Gautier (eds.), *A Primate Radition: Evolutionary Biology of the African Guenons,* pp. 54–78. Cambridge University Press, Cambridge.

Marshall, J., and Sugardjito, J. 1986. Gibbon systematics, in: J. Erwin (ed.), *Comparative Primate Biology, Volume 1, Systematics, Evolution, and Anatomy,* pp. 137–186. Alan R. Liss, New York.

Masters, J., and Lubinsky, D. 1988. Morphological clues to genetic species: Multivariate analysis of greater galago sibling species. *Am. J. Phys. Anthropol.* **75:**37–52.

Mayr, E. 1963. *Animal Species and Evolution.* Belknap Press of Harvard University Press, Cambridge, MA.

Mayr, E. 1969. *Principles of Systematic Zoology.* McGraw-Hill, New York.

Mittermeier, R. A., and Coimbra-Filho, A. F. 1981. Systematics: species and subspecies, in: A. F. Coimbra-Filho and R. A. Mittermeier (eds.), *Ecology and Behavior of Neotropical Primates,* Vol. 1, pp. 29–110. Academia Brasileira de Ciências, Rio de Janeiro.

Mittermeier, R. A., Rylands, A. B., and Coimbra-Filho, A. F. 1988. Systematics: species and subspecies—an update, in: R. A. Mittermeier, A. B. Rylands, A. F. Coimbra-Filho, and G. A. B. Fonseca (eds.), *Ecology and Behavior of Neotropical Primates,* Vol. 2, pp. 13–75. World Wildlife Fund, Washington, D.C.

Moore, J. A. 1991. Science as a way of knowing. VII. A conceptual framework for biology. Part III. *Am. Zool.* **31:**349–470.

Napier, P. H. 1976. Catalogue of Primates in the British Museum (Natural History), Part I: Families Callitrichidae and Cebidae. British Museum (Natural History), London.

Napier, P. H. 1981. Catalogue of Primates in the British Museum (Natural History) and Elsewhere in the British Isles, Part II: Family Cercopithecidae, Subfamily Cercopithecinae. British Museum (Natural History), London.

Napier, P. H. 1985. Catalogue of Primates in the British Museum (Natural History) and Elsewhere in the British Isles, Part III: Family Cercopithecidae, Subfamily Colobinae. British Museum (Natural History), London.

Nash, L. T., Bearder, S. K., and Olson, T. R. 1989. Synopsis of *Galago* species characteristics. *Int. J. Primatol.* **10:**57–80.

Natori, M. 1990. Numerical analysis of the taxonomic status of *Callithrix kuhli* based on measurements of the postcanine dentition. *Primates* **31:**555–562.

Natori, M., and Hanihara, T. 1988. An analysis of interspecific relationships of *Saguinus* based on cranial measurements. *Primates* **29:**255–262.

Natori, M., and Hanihara, T. 1989. Interspecific variations of *Saguinus* in dental measurements and its systematic relationships. *Primates* **30:**265–268.

Otte, D., and Endler, J. A. (eds.). 1989. *Speciation and Its Consequences.* Sinauer Associates Publishers, Sunderland, MA.

Passarello, P., and Vecchi, F. 1979. Morphologic variability of human African populations south of the Sahara. *J. Hum. Evol.* **8:**467–474.

Pocock, R. I. 1925a. Additional notes on the external characters of some platyrrhine monkeys. *Proc. Zool. Soc. Lond.* **97:**27–47.

Pocock, R. I. 1925b. The external characters of the catarrhine monkeys and apes. *Proc. Zool. Soc. Lond.* **97:**1479–1579.

Popp, J. L. 1983. Ecological determinism in the life histories of baboons. *Primates* **24:**198–210.

Reyment, R. A., Blackith, R. E., and Campbell, N. A. 1984. *Multivariate Morphometrics.* Academic Press, New York.

Shea, B. T., and Coolidge, Jr., H. J. 1988. Craniometric differentiation and systematics in the genus *Pan. J. Hum. Evol.* **17:**671–685.

Simpson, G. G. 1961. *Principles of Animal Taxonomy.* Columbia University Press, New York.

Supriatna, J. Froehlich, J. W., Costin, C., Southwick, C., Sugardjito, J., and Erwin, J. 1990. Interspecies interactions between *Macaca maurus* and *Macaca tonkeana* in Sulawesi Selatan. Abstracts of XIII Congress of the International Primatological Society, Nagoya and Kyoto, Japan, July 18–24, p. KYT 23 AM.

Tattersall, I. 1982. *The Primates of Madagascar.* Columbia University Press, New York.

Tattersall, I. 1986. Species recognition in human paleontology. *J. Hum. Evol.* **15:**165–175.

Tattersall, I. 1991. What was the human revolution? *J. Hum. Evol.* **20:**77–83.

Thorington, R. W., Jr., 1985. The taxonomy and distribution of squirrel monkeys (*Saimiri*), in: L. A. Rosenblum and C. L. Coe (eds.) *Handbook of Squirrel Monkey Research,* pp. 1–33. Plenum Publishing, New York.

Turner, T. T., Maiers, J. E., and Mott, C. S. 1988. Population differentiation in *Cercopithecus* monkeys, in: A. Gautier-Hion, F. Bourlière and J.-P. Gautier (eds.), *A Primate Radiation: Evolutionary Biology of the African Guenons,* pp. 140–149. Cambridge University Press, Cambridge.

Verheyen, W. N. 1962. Contribution à la craniologie comparée des primates, Les genres *Colobus* Illiger 1811 et *Cercopithecus* Linne 1758. *Musée Royal de L'Afrique Centrale—Tervuren, Belgique Annales—Serie in 8°—Sciences Zoologiques*—n. 105. pp. 1–255.

Vitzthum, V. J. 1986. The Role of Ecological Factors in Odontometric Variability and its Implication for Body Size Adaptations in Cercopithecidae. Doctoral Dissertation, University of Michigan, Ann Arbor.

Watanabe, K., Lapasere, H., and Tantu, R. 1991. External characteristics and associated developmental changes in two species of Sulawesi macaques, *Macaca tonkeana* and *M. hecki,* with special reference to hybrids and the borderland between the species. *Primates* **32:**61–76.

Watanabe, T. 1982. Mandible/basihyal relationships in red howler monkeys (*Alouatta seniculus*): a craniometrical approach. *Primates* **23:**105–129.

Speciation and Morphological Differentiation in the Genus *Lemur*

7

IAN TATTERSALL

Introduction

The event of speciation, the establishment of definitive genetic isolation between sister populations, remains the "black box" of genetics, and by extension of the systematic sciences. What we know about this event serves principally to underline how difficult it is to know whether definitive genetic disruption between closely related populations has actually occurred. It is fairly well agreed that in mammals speciation appears to require the physical division of a parental population. But the genetic mechanisms underlying such events may take place on one or more of many different genomic levels, gross karyotypic and below. Moreover, speciation is not simply a passive correlate of morphological differentiation: Speciation corresponds to no specifiable degree of morphological shift.

These two factors lead to particular problems when we move from consideration of the process of speciation to the practical systematic recognition of its product: species. Nonetheless, it is evident that if species are derived from preexisting species, then those physical characteristics that distinguish daughter species cannot initially be of greater magnitude than those arising within the

IAN TATTERSALL • Department of Anthropology, American Museum of Natural History, New York, New York 10024.

Species, Species Concepts, and Primate Evolution, edited by William H. Kimbel and Lawrence B. Martin. Plenum Press, New York, 1993.

parent species in the course of normal geographic differentiation. While this latter observation suggests strongly that in rejecting typology paleoanthropologists have overenthusiastically embraced the lumping ethic, it also underlines the difficulties attendant on making species discriminations in the fossil record.

Clearly we cannot rely on speciation theory to predict the amount of morphological differentiation we should expect to see between two closely related species, living or fossil. In the case of living organisms, of course, we have other attributes than morphology to guide us in differentiating species, although even here the pitfalls are legion (see, e.g., Tattersall, 1989). But as paleontologists we can hardly abandon the attempt to divide up the fossil record into its basic systematic units simply because it offers us only morphology. This leaves us with only one option, namely to base such attempts upon empirical observations of morphological distinctions among closely related living species.

Obviously, if geographic differentiation is a process unrelated to speciation, we cannot expect such observations to yield hard and fast rules applicable to the fossil record. Nonetheless, geographic differentiation is unquestionably the mechanism that underlies the morphological differences that distinguish related species: It is the engine of morphological change and adaptation in evolution. Appraisal of living primates in this light at the very least offers us a way of establishing limits within which the morphology of species belonging to various groups may be expected generally to vary. In this contribution I thus look at craniodental morphology in one speciose strepsirhine primate genus in an attempt to determine the pattern and degree of morphological differentiation within and among its contained species, with an eye to the potential application of such data to the fossil record of this group. To avoid any suspicion of circularity, I begin by examining the nonmorphological grounds that currently exist for determining the status of the taxa involved.

The Species and Subspecies of the Genus Lemur

The systematics of the subfamily Lemurinae has had an eventful recent history (e.g., Simons and Rumpler, 1988; Groves and Eaglen, 1988), which it is unnecessary to recapitulate here. I follow here the nomenclature adopted by Tattersall and Schwartz (1991) in our recent restudy of relationships within this group. This involves a return to the status current a couple of decades ago, in which the genus *Lemur* embraces the species *L. catta, L. mongoz, L. macaco, L. variegatus, L. fulvus, L. coronatus,* and *L. rubriventer.* Four of these species are monotypic, but *L. variegatus* contains the subspecies *L. v. variegatus* and *L. v. ruber,* and *L. macaco* contains the subspecies *L. m. macaco* and *L. m. flavifrons.* The highly differentiated *L. fulvus* embraces six subspecies: *L. f. fulvus* (including *L. f. mayottensis*), *L. f. albifrons, L. f. rufus, L. f. collaris, L. f. sanfordi,* and *L. f. albocollaris.* All of these species and subspecies are differentiated from one another in external features (pelage coloration, facial markings, tufting of ears, and so forth), although females of *L. f. albifrons* and *L. f. sanfordi* may strongly resemble each other. Many also show marked sexual dichromatism, coupled with the absence of sexual dimorphism either in body size or in the morphology of the hard tissues.

In the present context the initial questions we need to ask about the genus *Lemur* are (1) whether all of the species just listed do indeed represent genetically disjunct populations and (2) whether the subspecies listed are really nondisjunct genetically, despite their morphological differentiation. The attempt to answer these questions dramatically underlines the practical difficulties that exist in implementing both the biological and the recognition species concepts, particularly at the current state of our knowledge of the biology of these populations. The essential problem lies, of course, in recognizing when genetic disjunction has occurred. The simplest expedient is to conclude that it has taken place when free gene flow between two sympatric or parapatric populations is significantly impeded (whatever significance constitutes in this connection). Where obvious premating barriers exist there is of course no problem in determining this, and aficionados of the recognition approach may with equal reason look for the constituents of the specific-mate recognition system (SMRS) involved (e.g., Paterson 1978, 1985). In addition, there will not be any difficulties beyond the possible absence of any natural or artificial experiments in cases where the barriers to gene flow are postmating but involve absolute impediments to zygote formation or to the carrying of fetuses to term. Where restricted gene flow involves no more than the reduced fertility of hybridizing pairs or the decreased viability of offspring, however (and here the consideration of significance enters), practical problems multiply. This is particularly the case where a lack of natural sympatry or parapatry makes it necessary to depend on inferences from captive interbreeding. An equal practical difficulty then applies if we are seeking to identify SMRSs, for captive conditions often distort normal patterns of reproductive behavior.

Frequently, then, we are obliged to infer cases of significantly reduced genetic compatibility from field information on distributions and population interactions. Time and again, as in the cases of virtually all of the lemurs under consideration here, we find that such information is simply not available (and is increasingly unlikely to be as habitats and populations disappear). The upshot is that we will probably never be absolutely certain of the status of some of these recognizably distinct morphs. What follows provides some basic rationale for the current recognition of species vs. subspecies within the genus *Lemur*. It should be noted that the geographical distributions given employ the zoological equivalent of the "ethnographic present"; for justification and for more detailed distribution information, see Tattersall (1982).

The ruffed lemur, *Lemur variegatus*, is found in a 500-mile long strip of Madagascar's eastern rainforest, in different parts of which it occurs sympatrically with five subspecies of *L. fulvus* (themselves parapatric or allopatric to each other), and throughout which it is also sympatric with *L. rubriventer*. The two subspecies of *L. variegatus* probably hybridize naturally in the area between the Vohimaro and Antainambalana rivers. In the north of Madagascar, *L. coronatus* is sympatric with *L. f. sanfordi*, while in a small area of the northwest *L. macaco* it is sympatric with an isolate of *L. f. fulvus*. Further south, *L. mongoz* occurs in sympatry with another population of *L. f. fulvus*, while to the west of the Betsiboka River it co-occurs with *L. f. rufus*. Finally, in the south and southwest, *L. catta* is widely sympatric with *L. f. rufus*. In sum, while all of the subspecies of *L. fulvus* appear to be allopatric or parapatric with regard to each other, each one is sympatric in at least part of its range with another species of genus *Lemur*. *Lemur variegatus*, on the other hand, is the only species of the genus that is found in the same area as both *Lemur fulvus* and another *Lemur* species. In all cases of sympatry between forms recognized here as separate species, behavior supports such recognition; even in those few cases where individuals of different species have been noted to interact socially, they have not been observed to do so reproductively. Little if anything is known about what happens at the mutual boundaries of parapatric *L. fulvus* subspecies, though the existence of many museum specimens of uncertain affinities (invariably poorly documented!) suggests that natural hybridization occurs at least in some places.

While there are no credible reports of reproductive activity in captivity between *L. variegatus* (with a chromosome complement of 2N = 46) and any other species, certain of its congeners will indulge in interspecific reproductive behaviors. *Lemur catta* (2N = 56) has never been recorded to produce offspring with any other species in its genus, but hybridization between certain other *Lemur* species has resulted in offspring. Occasionally such offspring are fertile, although whether ever fully fertile remains doubtful. Thus recent experience at the Duke University Primate Center (DUPC) shows that the crosses *L. f. fulvus* (2N = 60) × *L. mongoz* (2N = 60) and *L. f. fulvus* × *L. macaco* (2N = 44) can produce fertile offspring (E. L. Simons, personal communication). According to Ratomponirina *et al.* (1988), however, the frequency of trivalent formation in the resulting karyotypes suggests a lowered fertility in the second generation, and in the case of *L. f. fulvus* × *L. macaco*, Hamilton and Buettner-Janusch (1977) were also unable to produce any viable second-generation hybrids. In other instances progeny have been totally infertile, as were every one of the 10 offspring (all males) produced at the DUPC by a female *L. rubriventer* (2N = 50) mated with *L. f. fulvus* males (E. L. Simons, personal communication). Recently, Ratomponirina *et al.* (1988) have reported total sterility in the offspring of *L. macaco macaco* × *L. f. albocollaris* and *L. f. fulvus* × *L. coronatus* crosses. Thus, while far from all of the interspecific crosses potentially possible within the genus *Lemur* have yet been tried, curently available hybridization information supports the legitimacy of the seven separate species listed earlier.

Karyotypically, the subspecies of *L. fulvus* present an interesting case. Four of the subspecies (*fulvus, rufus, albifrons, sanfordi*) share the same diploid number of 60 and show similar chromosome morphology, as demonstrated by their G-, Q-, and R-banding patterns (e.g., Rumpler, 1975; Hamilton and Buettner-Janusch, 1977). The two remaining subspecies, however, have different chromosome complements: *L. f. albocollaris* has a diploid number of 48, while *L. f. collaris* is said by Hamilton *et al.* (1980) to be karyotypically heteromorphic, with 2N = 52, 51, and 50, although Rumpler (1990) quotes 2N = 52 as the norm for this subspecies. Fully fertile hybrids result, as expected, from crosses among all of the 2N = 60 *fulvus*. Further, hybridization between *L. f. fulvus* and both *L. f. collaris* and *L. f. albocollaris* yields fertile offspring with normal male spermatogenesis (Rumpler, 1990), as would be expected if all belonged to the same species, while at the same time emphasizing that diploid number and fertility are not necessarily related. But crosses between *L. f. collaris* and *L. f. albocollaris* yield sterile progeny (Rumpler, 1990), with meiotic chromosomes arranged in long chains of six.

As it happens, these latter two subspecies occupy adjacent ranges in the ring of *L. fulvus* populations that surrounds the periphery of Madagascar (Fig. 1). This suggests, at first glance, that *L. fulvus* might represent a classic example of "circular overlap," in which "a chain of intergrading subspecies forms a loop or overlapping circle, of which the terminal forms no longer interbreed" (Mayr, 1942, p. 180). But on close examination many examples of such ring species have proven more complex than originally met the eye, and it appears that *L. fulvus* is another such case. Two of the subspecies of *L. fulvus*, *L. f. fulvus* and *L. f. rufus*, have substantial areas of distribution in both the eastern and western coastal areas of Madagascar, and a small isolate of the former is said to survive in the Ambohitantely Special Reserve (Nicoll and Langrand, quoted in Harcourt and Thornback, 1990), right in the center of the island (Fig. 1). It is clear from these distributions, as well as from the occurrence of subfossil sites in central Madagascar, that at some time in the recent past forest corridors linked the coastal regions of the island across the center, permitting faunal interchange. This, of course, immediately removes the conditions theoretically necessary for the development of a ring species.

For the moment, then, the differentiation of the subspecies of *L. fulvus* poses something of a mystery. Nonetheless, it is clearly mandatory to regard the various 2N = 60 populations of *L. fulvus* as representing a single species, and it seems premature to exclude *collaris* and *albocollaris* from that species when there is no impediment to genetic exchange (unless conceivably in the form of behavioral differences in the wild, but this is unknown) between either and the 2N = 60 forms. *Lemur fulvus collaris* and *L. f. albocollaris* resemble each other rather closely in external characters, and it is preferable to conclude that we are witnessing here the results of a recent event of karyotypic innovation (or succession thereof) that has yet to result in full speciation. Karyotypic polymorphisms are, it should be noted, known in various other Malagasy primate species (e.g., Rumpler, 1990). In sum, we seem to be on respectably firm ground in recognizing the four monotypic and three polytypic species of genus *Lemur* listed earlier.

Morphological Variation among Lemur Species

There is no shortage of variation in morphological detail among these seven species (and within the polytypic ones). Schwartz and I (Tattersall and Schwartz, 1991) recently studied 77 skulls of individuals positively identified to these species and subspecies by associated skins, and we were able to identify 37 characters of the skull and dentition that are consistent within the taxa we studied (which also include *Lepilemur* and *Hapalemur*, both used as outgroups; with only one specimen of *L. m. flavifrons* available, we made no subspecies distinction within *L. macaco*) but that vary among them. The bulk of these characters are dichotomous but some show up to five distinct states.

In most cases morphological distinctions between different states seem subjectively to be quite small (though consistent), as in the curvature of the posterior

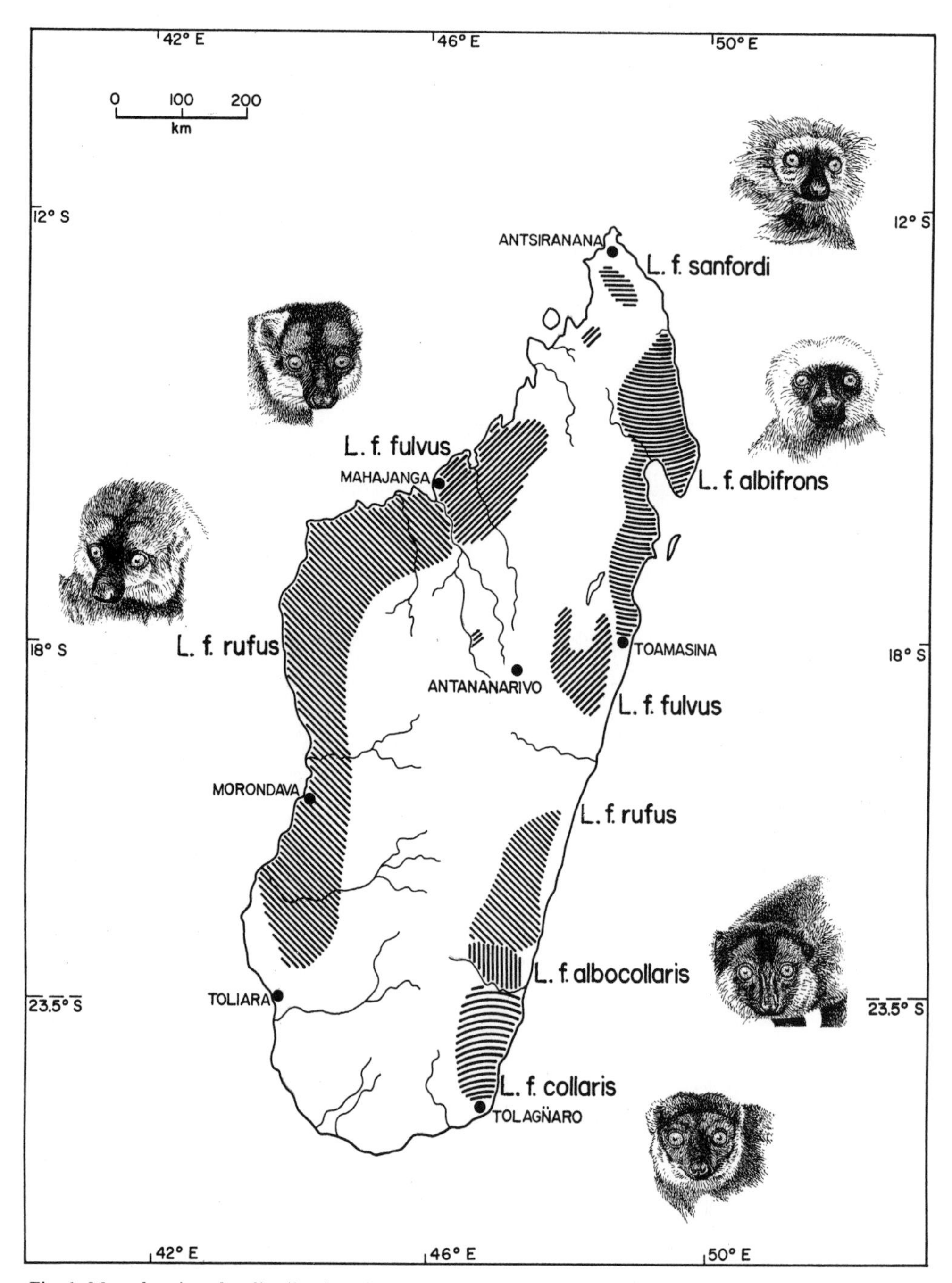

Fig. 1. Map showing the distributions in Madagascar of the various subspecies of *Lemur fulvus*. From Tattersall (1992).

portion of the palatine inferior to the orbitosphenoid (flat or slightly concave to laterally swollen), or in the placement of the M_1 entoconid (lingual or centrally shifted). Some characters, however, show substantial morphological variation. These include the morphology of the cranial base posterior to the hard palate, which ranges from the primitive flat plane to the pronounced excavation seen in *L. rubriventer,* and the formation in certain taxa of a distinct, encased, and approximately spherical bony "balloon" within the posterior part of the paranasal fossa (a space that is itself absent in some forms). Tables 1 and 2 summarize the characters and states identified and the distributions of states among the various taxa examined.

With this much variation to work with one might expect to encounter little difficulty in finding apomorphies of the craniodental system by which to recognize species. The reality turns out to be otherwise, however, for the discordance in character state distributions among the various specific and subspecific taxa involved in truly remarkable, as scrutiny of Tables 1 and 2 will indicate. This discordance is powerfully reflected in the difficulties Schwartz and I encountered in deriving from these data a consensus cladogram summarizing the phy-

Table 1. Cranial Characters/States and Their Distributions[a]

	1	2	3	4	5	6	7	8	9	10	11	12	13	14
Lepilemur	0	9	0	0	0	0	0	0	0	0	0	0	0	0
Hapalemur	0	0	0	0	0	0	0	0	0	0	0	0	0	0
V. v. variegata	0	0	0	1	1	2	0	0	1	1	1	0	0	0
V. v. rubra	1	0	0	1	1	2	0	0	1	1	1	0	0	0
L. catta	1	0	0	1	1	2	0	0	1	1	1	0	0	0
L. mongoz	1	1	1	1	2	2	1	0	3	1	1	0	4	2
L. macaco	2	1	1	1	2	2	1	1	2	1	1	0	3	1
L. rubriventer	2	1	1	1	3	2	0	1	3	1	1	1	2	3
L. coronatus	1	3	0	1	1	1	0	1	1	1	1	0	1	1
L. f. fulvus	1	0	1	1	1	1	0	1	1	1	1	0	1	1
L. f. albifrons	1	2	0	1	1	2	1	1	2	1	1	1	3	1
L. f. rufus	1	1	1	1	2	1	0	1	2	1	1	0	1	1
L. f. collaris	2	0	1	1	1	1	0	1	2	1	1	0	2	1
L. f. albocollaris	1	0	1	1	1	1	0	1	2	1	1	0	1	1
L. f. sanfordi	1	2	1	1	1	1	1	1	9	1	1	0	1	2

[a]Characters/states are: 1, Position of the nasopalatine foramen: posteromedial (0), anteromedial (1), or posterocentral (2); 2, size of nasopalatine foramen: small (0), moderate (1), large (2), or multiple (3); 3, contour of medial orbital wall: concave (0) or swollen laterally (1); 4, contour of prenasopalatine portion of the palatine bone: swollen laterally (1) or unswollen (0); 5, posterior expansion of the prenasopalatine: unextended posteriorly, maxilla widely exposed in orbital floor (0), slightly extended posteriorly, partly obscuring the maxilla posteriorly (1), extends to the edge of the temporal fossa (2), or intrudes into fossa (3); 6, paranasal sinus absent (0), present between P^3 and M^2 (1), or present along the entire length of the nasal fossa (2); 7, posterior portion of the palatine flat or slightly concave (0) or swollen laterally (1); 8, orbitosphenoid concave (0) or outwardly swollen (1); 9, in the floor of the orbit the maxilla is exposed extensively (0), moderately (1), slightly (2), or not at all (3); 10, infraorbital foramen is in line with or posterior to lacrimal foramen (0) or anterior to it (1); 11, maxilla exposed (0) or not exposed (1) in medial orbital wall; 12, encased and approximately spherical bony "balloon" developed within the posterior part of the paranasal fossa (1) or absent (0); 13, frontal sinus absent (0), slight (1), slight to moderate (2), moderate (3), or extensive (4); 14, cranial base behind the hard palate flat (0), forms a steep plane in the sphenoid medially and the palatine laterally (1), or plane commences more posteriorly, at about the level of the wing of the internal pterygoid plate (2), this depression is extremely marked, with a posterior excavation towards the occipital producing paired "post-choanal pits" (3); (9) denotes character variable or missing. Data from Tattersall and Schwartz (1991).

Table 2. Dental Characters/States and their Distributions[a]

	15	16	17	18	19	20	21	22	23	24	25	26	27	28	29	30	31	32	33	34	35	36	37
Lepilemur	9	0	0	0	0	0	3	9	0	3	0	0	0	1	1	1	0	3	1	2	1	2	0
Hapalemur	0	0	0	1	0	0	3	9	0	3	1	0	2	1	0	0	0	1	1	2	0	0	0
V. v. variegata	0	0	2	2	2	0	3	3	0	3	2	0	0	0	2	1	9	0	0	1	1	2	1
V. v. rubra	0	0	2	2	2	0	3	3	0	3	2	0	0	0	2	1	9	0	0	1	1	2	1
L. catta	1	0	0	0	1	0	3	2	0	3	0	0	0	1	0	1	0	0	1	2	0	0	0
L. mongoz	1	0	1	0	3	1	3	4	2	3	0	0	1	1	1	1	9	2	1	2	0	2	0
L. macaco	0	0	2	0	3	3	1	4	1	1	0	0	1	1	2	9	0	1	0	1	9	1	0
L. rubriventer	0	0	1	3	3	2	3	4	3	3	2	0	1	0	2	9	0	1	0	1	9	1	0
L. coronatus	0	0	1	3	3	3	3	4	1	3	3	0	1	1	2	9	9	1	0	1	9	1	0
L. f. fulvus	1	1	1	0	3	2	3	4	1	3	2	0	1	0	2	9	1	1	0	1	9	1	0
L. f. albifrons	0	1	1	0	3	2	1	4	3	3	2	1	1	1	2	9	1	1	0	1	9	1	0
L. f. rufus	1	1	1	0	3	2	1	4	3	3	0	0	1	1	2	9	1	1	0	1	9	1	0
L. f. collaris	0	1	1	0	3	2	0	4	3	2	0	0	1	1	2	9	1	1	0	1	9	1	0
L. f. albocollaris	1	1	1	0	3	2	2	4	1	2	2	0	1	1	2	9	1	1	0	1	9	1	0
L. f. sanfordi	0	1	1	0	3	2	1	4	1	3	2	0	1	1	2	9	1	1	0	1	9	1	0

[a]Characters/states are 15, P^4 has a distolingual protocone fold-like feature (0) or a postprotocrista (1); 16, P^4 unadorned lingually (0) or bears a protostyle (1); 17, preprotocrista of P^4 runs from the protocone to the parastylar region (0), slightly anterior to the paracone (1), or to the paracone (2); 18, M^1 has a postprotocrista (0), a protocone fold (2), a protocone fold and a postprotocrista (3), or lacks both (1); 19, lingual cingulum of M^1 absent (0), ledge-like (1), expanded anteriorly (2), or elaborated into styles (3); 20, M^1 protostyle and hypostyle confluent (1), separated by a crease (2), broadly separated (3), or styles lacking (0); 21, M^1 has both paraconule and metaconule (0), metaconule only (1), paraconule only (2), or lacks conules (3); 22, M^2 lingual cingulum absent (0), ledge-like (1), expanded anteriorly (2), or has style(s) (3); 23, M^2 lacks styles (0), has a protostyle (1), a confluent protostyle and hypostyle (2), or the protostyle and hypostyle are broadly separated (3); 24, M^2 has paraconule and metaconule (0), metaconule only (1), paraconule only (2), or no conules (3); 25, M^2 has a postprotocrista (0), no cristae (1), a protocone fold (2), or a postprotocrista and a protocone fold (3); 26, P_3 metaconid may be absent (0), or present (1); 27, P_4 metaconid absent (0), present (1), or [*Hapalemur*] present but with a distinctive morphology (2); 28, M_1 metastylid present (1) or absent (0); 29, lingual notch on M_1 narrow or pinched (0), broad (1), or absent (2), with a crest connecting the two cusps; 30, M_1 lingual notch, if present, opens centrolingually (0) or posterolingually (1); 31, entoconid of M_1, if present, is lingually placed (0) or centrally shifted (1); 32, paracristid of M_1 confluent with metaconid (0), disjunct from this cusp (1), truncated (2), or anteriorly directed (3); 33, metastylid on M_2 present (1) or absent (0); 34, lingual notch on M_2 narrow or pinched (0), broad (1), or absent (2), with a crest connecting the two cusps; 35, lingual notch of M_2, if present, opens centrolingually (0) or posterolingually (1); 36, lower molar entoconids may be cusplike (0), crestlike (1), or absent (2); 37, talonid basins of the lower molars relatively short (0) or distally elongate (1); (9) denotes character variable or missing. Data from Tattersall and Schwartz (1991).

logenetic relationships of the taxa involved. A heuristic (nonexhaustive) search using the quantitative parsimony procedure PAUP (Swofford, 1989), with state 0 specified as ancestral in the case of all 37 characters (and including a hypothetical ancestor with all states 0), yielded no fewer than 80 different trees, each with the minimum length of 113. Only by eliminating all characters with consistency indices (Swofford, 1989) indicated by PAUP to be below 0.6 (almost half of them) were we able to extract a single most parsimonious tree on the basis of this data set, and among other shortcomings this tree failed to recognize *L. fulvus* as a monophyletic assemblage.

The mechanism behind this confusing situation is clearly homoplasy (parallelism). Eaglen (1983) and Groves and Eaglen (1988) have previously pointed to the extensive parallelism that distinguishes this group and have noted that this extends to complexes of characters other than those examined here. Many of the character states that distinguish the various taxa must have evolved multiple times, independently. Exactly which these are depends on the precise geometry of evolutionary relationships that we recognize within the genus *Lemur,* which, as noted, is difficult to determine on this data set precisely because of the homoplasy involved.

On the basis of the undefinitive sample tree shown in Fig. 2, such independent acquisitions would include lateral swelling of the posterior portion of the palatine in *L. mongoz, L. macaco,* and the *L. f. sanfordi/albifrons* pair; posterior expansion of the prenasopalatine in the *L. rubriventer/mongoz/macaco* ancestor and in *L. f. rufus;* the formation of a full paranasal sinus in *L. catta, L. variegatus,* the *L. rubriventer/L. mongoz/L. macaco* clade, and *L. f. albifrons;* the acquisition of a postnasopalatine "balloon" in *L. rubriventer* and *L. f. albifrons,* the development of a moderate degree of frontal sinus inflation in *L. macaco* and *L. f. albifrons;* the development in both *L. mongoz* and *L. f. sanfordi* of a steep plane descending from the roof of the nasal fossa to terminate near the wing of the internal pterygoid plate; the presence of a postprotocrista on P^4 in *L. catta, L. mongoz, L. f. fulvus, L. f. albocollaris,* and *L. f. rufus;* broad separation of the protostyle and hypostyle in M^2 of *L. rubriventer, L. f. albifrons,* and *L. f. rufus/collaris,* and in M^1 of *L. coronatus* and *L. macaco;* secondary redevelopment of the postprotocrista in *L. mongoz, L. macaco,* and *L. f. rufus/collaris;* the presence of M^1 metaconule in *L. macaco* and the ancestor of *L. fulvus* exclusive of *L. f. fulvus;* the absence of lower molar entoconids in *L. variegatus* and *L. mongoz;* P^4 preprotocrista running to the paracone in *L. variegatus* and *L. macaco;* and many others as well. Using an alternative cladogram would substitute some of these characters for others but would do nothing to reduce their number. Examples of all of these characters are illustrated by Tattersall and Schwartz (1991).

Perhaps this rampant homoplasy is less surprising on reflection than at first sight. After all, we are dealing with taxa that are extremely closely related, and it seems reasonable to expect that the more similar the starting points are, the more likely it is that similar end points will be obtained. Further, the concomitant of this high rate of homoplasy is a high rate of autapomorphy. And it is, of course, autapomorphies that offer us our only practical basis for species recognition in the fossil record. Even if similar autapomorphic states are present in one or other of the outgroup genera, it is fair to include them in a consideration of morphology relative to speciation, as long as they are unique within the genus.

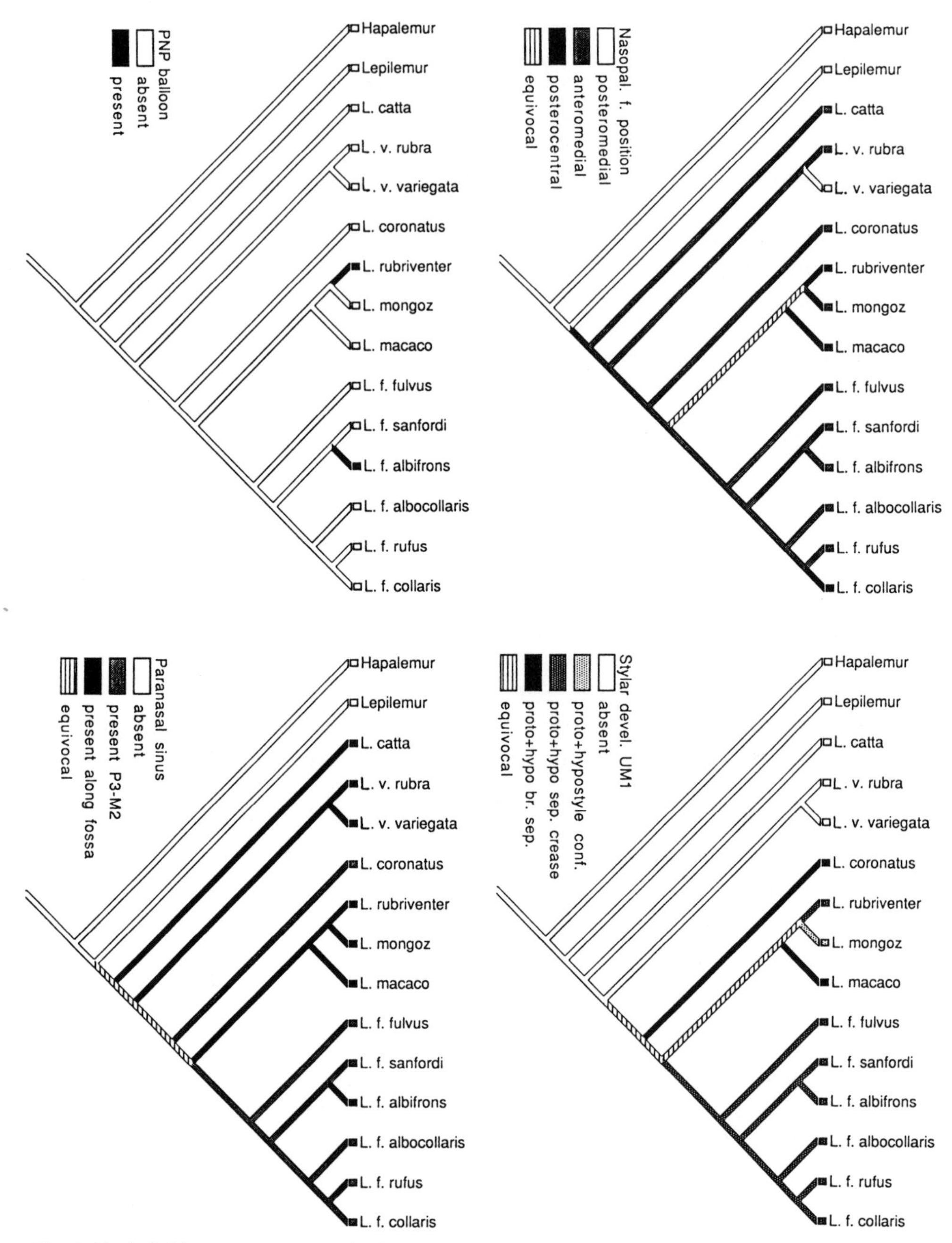

Fig. 2. Undefinitive trees, generated using the MacClade computer procedure (Maddison and Maddison, 1989) to show some typical discordant character distributions under one possible phylogenetic arrangement. Treelength 112; Consistency Index 0.63. This figure is not intended to represent a statement of relationships; it simply shows one among a set of trees that minimize homoplasy in our craniodental data set and is supplied purely for illustrative purposes. The four characters whose state distributions are represented here are, clockwise from top left: position of the nasopalatine foramen; stylar development of the upper first molar; condition of the paranasal sinus; presence or absence of postnasopalatine "balloon."

This is because among lemurs, as among mammals in general, it is the genus that is the true *Gestalt* category at the osteological and dental levels: Genera are in practice morphs that are instantly recognizable one from the other, while distinction among species of the same genus on the basis of hard tissues generally requires close scrutiny. Why exactly this should be is a fascinating question that may have as much to do with human perception as with the evolutionary process [because genera are arbitrarily delimited, if (we hope) monophyletic, assemblages of species]; but the upshot is that since we have little difficulty in discriminating between genera, it is only with intrageneric character state distributions that we need to be concerned in the present context. Among the species under consideration here, only *L. variegatus* comes close to fulfilling the *Gestalt* criterion, due in great part to its relatively large size and its pronounced craniofacial elongation and to the unusual anterior skewing of the lingual portions of its upper molars. However, it is not clear on the basis of this data set whether the ruffed lemur or the ringtailed lemur is the phylogenetic outlier within this group, which is why Schwartz and I (Tattersall and Schwartz, 1991) have suggested retaining the generic appellation *Lemur* for all of the species involved, thus abandoning the recent use of genus *Varecia* for the ruffed lemur. Below I list those few apomorphies in this data set that uniquely characterize the various species:

1. *Lemur variegatus:* protocone fold and preprotocrista on M^1; lingual cingulum of M^1 expanded anteriorly; lingual cingulum present on M^2, with stylar elaboration; lingual notch of M_2 opens posterolingually; and lower molar talonid basins are distally elongate. The nasopalatine foramen is posteromedial in *L. v. variegatus,* but anteromedial in *L. v. ruber,* which has considerably smaller cheek teeth.
2. *Lemur catta:* preprotocrista of P^4 runs from the protocone to the parastylar region; lingual cingulum of M^1 and M^2 ledge-like; lingual notch of M_1 pinched; lower molar entoconids cusplike.
3. *Lemur mongoz:* extensive inflation of frontal sinus; protostyle and hypostyle confluent on M^1; lingual notch of M_1 broad.
4. *Lemur macaco:* metaconule only on M^2.
5. *Lemur rubriventer:* prenasopalatine greatly extended posteriorly, intruding into temporal fossa; "post-choanal pits" in cranial base behind hard palate.
6. *Lemur coronatus:* nasopalatine foramen multiple.
7. *Lemur fulvus:* protostyle on lingual margin of P^4; protostyle and hypostyle of M^1 separated by a crease; entoconid of M^1 centrally shifted. Among the subspecies of *L. fulvus, L. f. albifrons* is unique in having a metaconid on P_3 and *L. f. albocollaris* in having a paraconule only on M^1.

Discussion

Although each species within the genus *Lemur* (unlike each subspecies within *L. fulvus*) thus has at least one skeletodental apomorphy by which it can uniquely be recognized, the list is composed of far fewer such diagnostic features than might have been expected given the large number of characters examined. It is, further, noteworthy that the preponderance of diagnostic features is dental rather than cranial, and that most of these dental features would probably be considered rather minor by a systematist who did not already know that they were the keys to species identification. Of course, many more *combinations* of particular character states are unique for various species; for example, only *L. mongoz* has both no exposure of the maxilla in the floor of the orbit and posterior extension of the prenasopalatine just to the edge of the temporal fossa, and only *L. rubriventer* has both an anteromedial nasopalatine foramen and a flat contour of the posterior portion of the palatine. But the issue I wish specifically to address here is whether a paleontologist would be able to recognize, on the basis of hard parts alone, the same species that we accept on the basis of external characters, distributions, and breeding performance. The critical question thus becomes, "Does the amount of homoplastic noise in the craniodental data set overwhelm the phylogenetic signal?"

The answer to this question is almost certainly "Yes". It would probably be difficult to find a curator who would permit the removal of identifying labels from a large series of lemur skulls, as the appropriate practical experiment demands (if indeed it would be valid at all). But there is little doubt that a paleontologist, even one confronted with complete skulls and dentitions of all of the species under consideration here, would hesitate to recognize more than three, or perhaps four, species in this assemblage. Because of its large size and other attributes mentioned earlier, *L. variegatus* would readily be recognized as separate from the others, and its subspecies might be distinguished on relative dental size. Among the remaining species, *L. catta* might be recognized as distinct from the rest on the basis of the relatively small size of its dentition and its laterally compressed cusps, and *L. rubriventer* might be discriminated on the basis of its highly modified cranial base. The remaining species, however, would pose an enormous puzzle, for although many morphological variations can be identified in the assemblage, homoplasy ensures that they are found in many different combinations. I doubt strongly that a paleontologist would be able to sort out the few unique autapomorphies (or, to use an alternative terminology, diagnostic characters) listed above from the "noise" created by the many parallel acquisitions. The morphological noise-to-signal ratio is simply too great. Instead, the temptation would be to see interspecific variations simply as part of a complex set of intraspecific differences, differences that could then comfortably be ignored because, as we all know, intraspecific variants are at least potentially ephemeral.

One other factor that I find particularly interesting is that some variations of the kind that might most readily catch the eye of a paleontologist are found among the subspecies of *L. fulvus*. Alone among these subspecies, *L. f. sanfordi* shows an excavation of the cranial base comparable to *L. mongoz*, while only *L. f. albifrons* shows an inflation of the frontal sinus comparable to that seen in *L. macaco*. Such observations fit in with two of the most important lessons that result from examination of this data set, namely, (1) that empirical observation supports the inference from theory that very closely related species are unlikely to be distinguished by morphological character shifts greater than those that can arise within a single geographically differentiating species, and (2) that homoplasy is most likely to occur extensively—and to be hardest to recognize—among closely related forms. The third major lesson is that under reasonable current assumptions the craniodental variation seen within the genus *Lemur* underestimates the number of species involved and misidentifies species and subspecies boundaries.

It is possible that a partial and fragmentary sampling of these species in the fossil record might make it possible for a dedicated splitter to recognize (though almost certainly inaccurately) as many species as actually exist. But a more comprehensive fossil representation and a full appreciation of the complex co-occurrence of the various character states would rapidly demonstrate the difficulty of maintaining species discriminations on hard-part morphology and would promote the dismissal of genuine autapomorphies (parallel and unique alike) simply as aspects of complex intraspecific variation. It is true, of course, that the taxa studied here, while all sympatric in the collections of the American Museum of Natural History, would under no circumstances all be found together in the same fossil assemblage; but even though site-by-site analyses would pose a simpler situation, cross-site comparisons would almost certainly be confounded in a similar way.

Summary

Genetically, karyotypically, and behaviorally distinct, the seven species (four monotypic, two with two subspecies, and one with six subspecies) that belong to the strepsirhine genus *Lemur* (including *Varecia*) are also readily differentiable on the basis of external features. Diagnosis on the basis of traits that would be expected to be preserved in the fossil record, however, is another question. Although a great deal of morphological variation exists within the genus *Lemur*, the remarkably extensive homoplasy in craniodental features revealed in the data set of Tattersall and Schwartz (1991) makes species diagnosis marginal in most cases. Further, many of the most striking cranial autapomorphies of the various *Lemur* species have also been evolved independently within the complex of *L. fulvus* subspecies.

Together, these observations support the inference from theory that very closely related species will not show significantly more hard-tissue differentiation than may be accumulated among geographical variants of the same species. They also demonstrate (unsurprisingly, since similar starting points are most likely to produce similar end points) that homoplasy is likely to occur most extensively between closely related forms. They suggest, in addition, that a paleontologist working with hard-tissue characters alone would underestimate the actual species diversity within the genus *Lemur*.

Acknowledgments

I would like to thank Jeffrey Schwartz for allowing me to use our joint data on lemur craniodental morphology in this new context. Judith Masters, Bill

Jungers, and Bob Eaglen made comments that improved the manuscript, though all will still doubtless recoil from much that is left.

References

Eaglen, R. H. 1983. Parallelism, parsimony and the phylogeny of Lemuridae. *Int. J. Primatol.* **4:**249–274.

Groves, C. P., and Eaglen, R. H. 1988. Systematics of the Lemuridae (Primates, Strepsirhini). *J. Hum. Evol.* **17:**513–538.

Hamilton, A. E., and Buettner-Janusch, J. 1977. Chromosomes of Lemuriformes, III. The genus *Lemur:* karyotypes of species, subspecies and hybrids. *Ann. N.Y. Acad. Sci.* **293:**125–159.

Hamilton, A. E., Tattersall, I,. Sussman, R. W., and Buettner-Janusch, J. 1980. Chromosomes of Lemuriformes, VI. Comparative karyology of *Lemur fulvus:* a G-banded karyotype of *Lemur fulvus mayottensis* (Schlegel, 1866). *Int. J. Primatol.* **1:**81–93.

Harcourt, C., and Thornback, J. 1990. *Lemurs of Madagascar and the Comoros. The IUCN Red Data Book.* IUCN, Gland, Switzerland and Cambridge, England.

Maddison, W., and Maddison, D. 1989. MacClade, test version 2.97.36. Computer program distributed by authors.

Mayr, E. 1942. *Systematics and the Origin of Species.* Columbia University Press, New York.

Paterson, H. 1978. More evidence against speciation by reinforcement. *S. Afr. J. Sci.* **74:**369–371.

Paterson, H. 1985. The recognition concept of species, in: E. S. Vrba (ed.), *Species and Speciation.* Transvaal Museum Monograph 4, pp. 21–29. Transvaal Museum, Pretoria.

Ratomponirina, C., Brun, B., and Rumpler, Y. 1988. Synaptonemal complexes in Robertsonian translocation heterozygotes in lemurs. *Kew Chromosome Conferences,* Vol. III, pp. 65–73. HMSO, London.

Rumpler, Y. 1975. The significance of chromosomal studies in the systematics of the Malagasy lemurs, in: I. Tattersall and R. W. Sussman (eds.), *Lemur Biology,* pp. 25–40. Plenum Press, New York.

Rumpler, Y. 1990. Systématique des lémuriens, in: J. J. Roeder and J. R. Anderson (eds.), *Primates: Recherches Actuels,* pp. 13–22. Masson and Cie. Paris.

Simons, E. L., and Rumpler, Y. 1988. *Eulemur:* New generic name for species of *Lemur* other than *Lemur catta. C. R. Acad. Sci. Paris.,* ser III **307:**547–551.

Swofford, D. 1989. PAUP (Phylogenetic Analysis Using Parsimony), version 3.0A. Computer program distributed by author.

Tattersall, I. 1982. *The Primates of Madagascar.* Columbia University Press, New York.

Tattersall, I. 1989. The roles of ecological and behavioral observation in species recognition among primates. *Hum. Evol.* **4:**117–124.

Tattersall, I. 1992. Systematic vs. ecological diversity: the example of the Malagasy primates, in: N. Eldredge, (ed.), *Systematics, Ecology and the Biodiversity Crisis,* pp. 25–39. Columbia University Press, New York.

Tattersall, I., and Schwartz, J. H. 1991. Phylogeny and nomenclature in the "*Lemur*-group" of Malagasy strepsirhine primates. *Anthrop. Pap. Am. Mus. Nat. Hist.* **69:**1–18.

Squirrel Monkey (Genus *Saimiri*) Taxonomy

8

A Multidisciplinary Study of the Biology of Species

ROBERT K. COSTELLO, C. DICKINSON, A. L. ROSENBERGER, S. BOINSKI, AND FREDERICK S. SZALAY

Introduction

Systematists are more than ever faced with the task of interpreting the biological validity of species from data representing many different biological components. How data, such as molecular, biochemical, chromosomal, behavioral, and morphological, may be interpreted objectively within a consistent paradigm is yet without consensus. Characters have often presided over their interpretation, and it is not unknown for technique to have usurped method. The aim of this study is to assess within a paradigm the contributions of a variety of data toward understanding the species taxonomy of the genus *Saimiri*. This is done under the explicit premises entailed in Mayr's biological species concept (BSC; see Mayr,

ROBERT K. COSTELLO • Doctoral Program in Anthropology, The Graduate Center of the City University of New York, New York, New York 10021. C. DICKINSON • Department of Anthropology, University of Chicago, Chicago, Illinois 60637. A. L. ROSENBERGER • U.S. National Zoological Park, Smithsonian Institution, Washington, D. C. 20008. S. BOINSKI • Laboratory of Comparative Ethology, National Institutes of Health—Animal Center, Poolesville, Maryland 20837, and Department of Anthropology, University of Florida, Gainesville, Florida 32611. FREDERICK S. SZALAY • Department of Anthropology, Hunter College, New York, New York 10021.

Species, Species Concepts, and Primate Evolution, edited by William H. Kimbel and Lawrence B. Martin. Plenum Press, New York, 1993.

1942, 1963, 1982), and as expanded by Bock (1986; see also Szalay, this volume). A few remarks are made concerning Paterson's (1980, 1985) species recognition concept with regard to *Saimiri* taxonomy.

Seven prominent works have been published this century that have dealt with *Saimiri* taxonomy, including Elliot (1913), Lonnberg (1940), von Pusch (1942), Cabrera (1958), Hill (1960), Hershkovitz (1984), and Thorington (1985). As a result, one to seven species and up to 17 subspecific taxa have been recognized. Constructing testable hypotheses for the species status of one or another group of *Saimiri* is generally arbitrary, that is, whether a single-species hypothesis or a 1000-species hypothesis is used (neither of which may be appropriate) depends largely on preexisting notions formed from reading literature, examining specimens, etc. There is no default null hypothesis for species. Null hypotheses are probablistic and not necessarily biological. For example, Thorington (1985) studied the geographic pattern of coat color variation to construct hypotheses of species that he then tested with a multivariate craniometric analysis. Hershkovitz (1984) alternatively preferred the congruence of karyology and coat color to determine species (hypotheses). The hypotheses considered in this study are similarly treated, that is, a variety of evidence is evaluated for congruence with each species hypothesis. However, species hypotheses need not be constructed *de novo,* as though this is the first look at the genus. The species hypotheses addressed in this study have been advanced by others (most recently, Hershkovitz, 1984; Thorington, 1985; Ayres, 1985), and the ones selected for testing are regarded as those most important to understanding the number of species in the genus.

There are two parts to the approach that follows. The first is a review of literature, by categories of evidence, which is relevant to our current understanding of the number of species of squirrel monkeys, with discussions on the efficacy of the evidence under the BSC paradigm. The second part is an assessment of new evidence, collected in a series of preliminary studies for the determination of species in *Saimiri.* Four species hypotheses are explicitly put to test by the evidence: *S. boliviensis* (Group), *S. ustus, S. vanzolinii,* and *S. oerstedii* are each treated as species and tested by the evidence. In light of the new evidence, a fifth hypothesis emerges by default, that South American *Saimiri* is one species.

Review of the Current Status of Saimiri

Geographic Distribution

The geographical distribution of *Saimiri* is basically the Amazon basin, with extensions north into the Guyanas and south into Paraguay. A disjunct population also occurs nearly 1000 km to the north in western Panama and Costa Rica (Fig. 1). The genus is widespread in tropical lowland rainforest along river courses. On the southern continent, collecting localities are densely distributed in an hour-glass-shaped swatch of territory (see Hershkovitz, 1984, Fig. 3), with the major east-west axis situated just below the equator. They are abundant along the Rio Amazonas and its continuation, the Rio Solimões, and for varied distances along many tributaries feeding the great river system. In the west, *Saimiri* fans out to higher latitudes following the central arc of the Andes, reaching as far south as 20°S near the Bolivia–Argentina border and in Paraguay. In the east, more museum specimens come from areas north of the Amazon river, from Guyana, Surinam, and French Guyana. Fewer come from the intervening states of Brazil, and some derive from areas in the east below the Rio Amazonas. Localities also occur north and south of the Rio Amazonas in the central region of the basin, but these are scattered and large areas remain to be sampled. Squirrel monkeys are absent from higher altitudes (above 300 m) and more arid regions surrounding the greater Amazon basin.

The geographic distribution of South American *Saimiri* appears to consist mainly of parapatric populations, recognized as either species or subspecies. Populations are referred to *sensu lato,* briefly postponing the taxonomic labeling that follows. An exception to parapatry is a single area of (inferred) sympatry involving *sciureus* and *madeirae,* designated by both Hershkovitz (1984) and Thorington (1985), and located between the Rio Madeirae and Rio Tapajós, just south of the Rio Amazonas. Hershkovitz (1984) proposed a second area of sympatry between *boliviensis* and *sciureus* along the Rio Tapiche. According to Thorington (1985) and DaSilva *et al.* (1992), however, the two forms in question intergrade.

As far as (putative) species are concerned, only *sciureus* inhabits the forest on the north side of the Rio Amazonas (following Hershkovitz, 1984; Thorington, 1985), with the exception of the limited north bank distribution of *vanzolinii* (Ayres, 1985) (as shown as an isolated *boliviensis* in Fig. 1). This species (*sciureus*) also inhabits both banks of the Amazon along the eastern and western reaches of that river; thus, with the exception of *vanzolinii,* parapatry of proposed species occurs only along southern tributaries. The distribution of *vanzolinii* is confined to a small string of islands situated within the Rio Solimões and is distributed along its left (north) bank just above Tefé. This region is close to the boundaries of three other *Saimiri* populations, including *madeirae, boliviensis,* and *sciureus.* The *boliviensis* group appears in the upper reaches of the drainage along the Rio Yucayali and Rio Tapiche, and extends south along the Rio Mamoré/Guapore. Parapatry between *boliviensis* and *madeirae* is reported to occur along the Rio Purús and Rio Guapore/Mamoré (Hershkovitz, 1984; Thorington, 1985). The eastern boundary of *madeirae* is not agreed on (see Species Classification, below); therefore, the geographic relationship to *sciureus* cannot be determined from published reports. Very little information is available for the southern ranges and boundaries of groups.

Species Classification

There is no consensus on species or subspecies, partly due to the lack of significant samples from critical geographic areas, and partly to individual differences of interpretation of diverse data. This is likely to be related to different underlying species concepts. Furthermore, with the exception of Hershkovitz (1984) and Thorington (1985), analyses have not been methodologically explained.

Ayres (1985) named a new species of *Saimiri, S. vanzolinii,* based on pelage

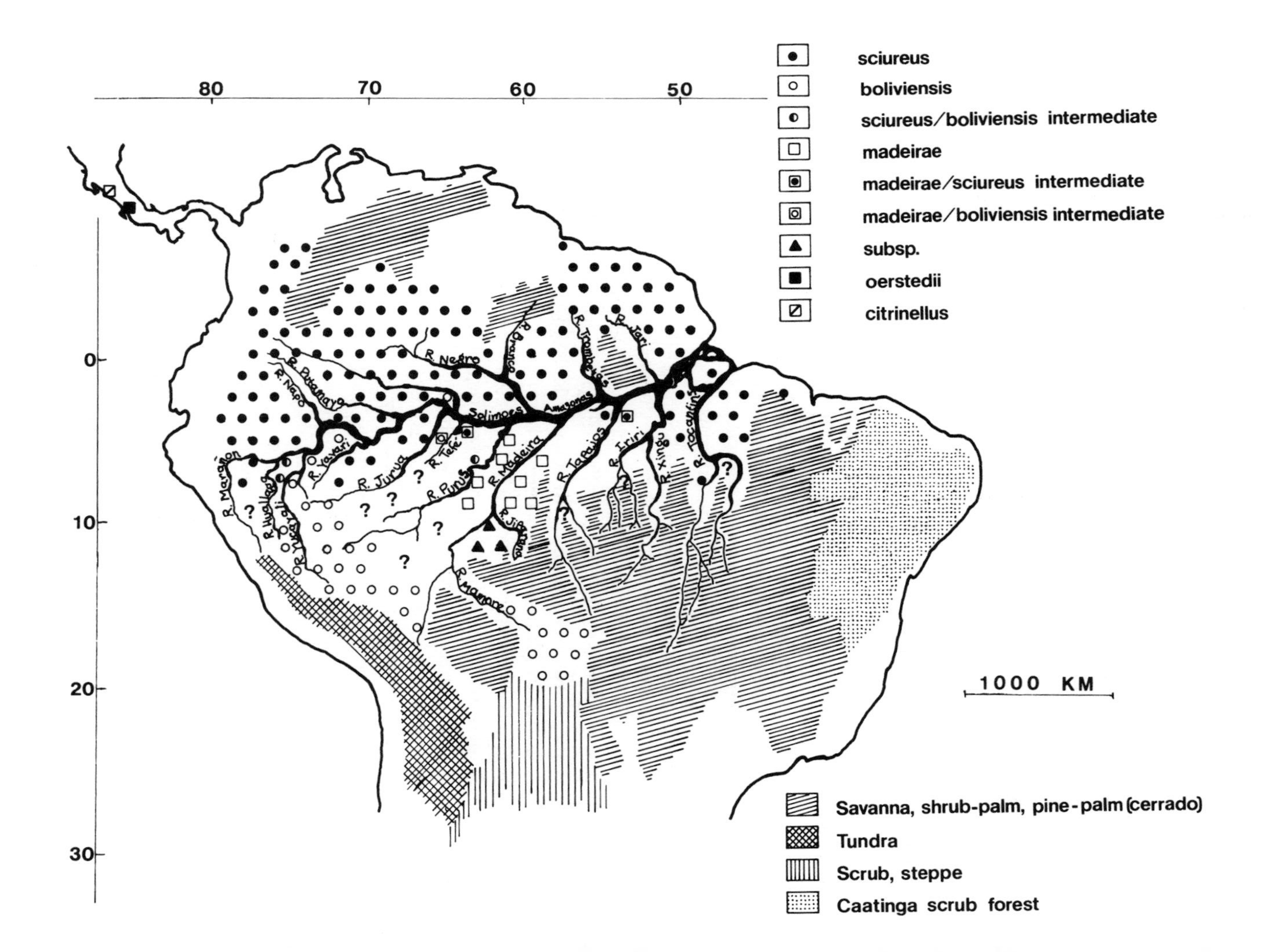
sciureus
boliviensis
sciureus/boliviensis intermediate
madeirae
madeirae/sciureus intermediate
madeirae/boliviensis intermediate
subsp.
oerstedii
citrinellus
80
70
60
50
0
10
20
30
R. Negro
R. Branco
R. Trombetas
R. Jari
Solimoes
Amazonas
R. Napo
R. Putumayo
R. Tapajos
R. Iriri
R. Xingu
R. Tocantins
R. Madeira
R. Purus
R. Tefé
R. Jurua
R. Javari
R. Marañon
R. Huallaga
R. Ucayali
R. Mamore
1000 KM
Savanna, shrub-palm, pine-palm (cerrado)
Tundra
Scrub, steppe
Caatinga scrub forest

coloration and a restricted geographic distribution covering only young and low flooded forest, that is, the *varzea.* The diagnosis of *vanzolinii* was strengthened by chromosome differences (Yonenaga-Yassuda and Chu, 1985) and the discovery of *sciureus* and *madeirae* populations situated parapatrically to *vanzolinii* without any evidence of intergradation. Hershkovitz (1985) considers *vanzolinii* as a subspecific member of *boliviensis.*

Hershkovitz (1984) focused on developing a key to the species and subspecies, groups based largely on their karyotypes, geographic distribution and pelage coloration, but he also included laboratory hybridization, behavior, and craniodental and body measurement. His scheme recognized four species (*boliviensis, ustus, sciureus,* and *oerstedii*), arranged into two species-groups (the Roman- and Gothic-arch forms). The *S. boliviensis* Group included *boliviensis* only; the *S. sciureus* Group included *sciureus, oerstedii,* and *ustus* (= *madeirae* of Thorington, 1985). He noted (p. 300) that the "true taxonomic status (of *oerstedii*) awaits better knowledge of its karyotype." This reflected Hershkovitz's suspicion that *Saimiri* was introduced by people to Central America in Recent time, and the priority given to karyotype in species discrimination. Hershkovitz's "Group" divisions were originally devised by MacLean (1964) on the basis of facial appearance, especially the shape of the whitish eye patch.

Thorington (1985) provided a detailed assessment of the geographic variation in skins and skulls, including univariate and multivariate analyses of 19 craniodental measurements. He recognized two South American species, *S. sciureus* and *S. madeirae,* lumping Roman-arch *boliviensis* together with Gothic arch *sciureus.* To Thorington, *oerstedii* represents a subspecies of *sciureus.* Although he recognized the strong color differentiation of the red-back *oerstedii,* he found them to be no more different cranially from *S. sciureus sciureus* than the latter is from *boliviensis.* Thorington's second distinct species, *S. madeirae,* refers to the same population that Hershkovitz named *S. ustus* and that he located in central Brazil below the Rio Amazonas, bounded west and east by two major rivers, the Rio Purús/Rio Guapore and Rio Tapajós, respectively. Hershkovitz (1984) recognized the distribution of this population extending further east to the Rio Xingu-Iriri. Thorington considered *Saimiri* from east of the Rio Tapajós to be *sciureus.* Thorington's and Hershkovitz's attribution of species status to the *madeirae* form is based upon several findings: (1) inferred sympatry with other *Saimiri* and (2) a strong phenotypic boundary along the species' southwestern limit at the Bolivia–Brazil border.

Why do Hershkovitz (1984) and Thorington (1985) refer to essentially the same population as *S. ustus* and *S. madeirae,* respectively? *Saimiri ustus* was assigned by Geoffroy (1844) to a specimen obtained in Portugal. Geoffroy's animal is illustrated with an olive crown; burnt-orange dorsum (thus its name, "*Saimiri ustus* a dos brule"); yellow-orange feet, hands, and forearms; and a hairy, not naked, or tufted, ear. The specimen locality is unknown. Hershkovitz (1984)

Fig. 1. The geographic distribution of *Saimiri* in Central and South America. Upper left and clockwise are the two forms of Central American *S. oerstedii, oerstedii,* and *citrinellus.* All South American forms are members of *S. sciureus.* No judgment is made on the subspecific status of these geographic forms. Above the Amazon river is *sciureus;* below the Amazon river, from east to west is *sciureus,* the naked-eared *madeirae,* and the dark skull-capped *boliviensis,* with *peruviensis* directly above.

followed Cabrera (1958) when he accepted *ustus* as a valid name for monkeys with naked ears and fulvous hands, and he restricted the type locality to Humaitá (Thomas, 1908). However, the specimen named by Geoffroy does not fit either Hershkovitz's or Thorington's descriptions and, as Thorington (1985) has suggested, it could not have come from anywhere near the type locality. Thomas (1908) named *S. madeirae* based on five specimens collected along the west bank of the Rio Madeira, at Humayta (= Humaitá). His description agrees perfectly with both Hershkovitz's and Thorington's, though he failed to describe the condition of nearly hairless ears. The name *ustus* is best treated as a junior synonym of *sciureus,* from which Geoffroy's animal is nearly indistinguishable, except for the lack of a tuft of hair on the ear. Naked-ear squirrel monkeys of the type described by both Hershkovitz and Thorington hereafter are referred to as *madeirae.*

The confounding state of taxonomy makes it cumbersome to adopt a neutral, yet useful, nomenclature. We refer to geographic groups throughout this report by the last name of a binomen or trinomen, without any intentions as to the species or subspecies status of the groups. For purposes of style and grammar, genus and species are used to begin sentences. Table 1 and Fig. 1 indicate in detail the geographic groups used and some traits that distinguish those groups from each other.

Pelage Coloration

Saimiri species taxonomy has largely been driven by geographic differences of coat coloration. The taxonomic value of patterns of pelage variation is particularly difficult to interpret in *Saimiri,* an animal that is not known to have geographic variation of premating behaviors (discussed under Behavior, below). Differences of sexual dichromatism between geographic groups that are otherwise similar in coat coloration, and that are undisputed members of the same species, suggest some discontinuity in gene flow, coupled with minor differences in sexual selection, or drift. For example, both male and female *oerstedii* from Costa Rica have black crowns, whereas males and females from the nearest geographic population, *citrinellus* of Panama, have gray and black crowns, respectively (Hershkovitz, 1984). Coat color patterns can be used, however, not only to define sympatry and parapatry, but to infer barriers to the introgression of genes, and alternatively, to recognize intermediates and to infer gene flow. A lack of intermediates from areas of sympatry and parapatry is negative evidence for barriers to gene flow or else may be due to sampling error.

The taxonomic groups deserving closest attention are "*boliviensis,*" "*vanzolinii,*" "*madeirae,*" and "*oerstedii*" (see Table 1). Samples of *boliviensis* (including *peruviensis*) *vanzolinii,* and *oerstedii* are distinguished from *madeirae* and *sciureus* by having black, rather than gray or olive, crowns. *Saimiri boliviensis* and *vanzolinii* are further distinguished by the Roman-arch pattern of supercilliary hairs. This is in contrast to the peaked, Gothic-arch pattern of *oerstedii, madeirae,* and *sciureus.* Of the Roman-arch squirrel monkeys, *vanzolinii* is characterized by a continuous, dark dorsal stripe that runs from crown to tip of tail. *Saimiri oerstedii* is the only group with an entirely red back, and it is the only black-crowned group

Table 1. Pelage Variation and Social Organization

	boliviensis	*madeirae*	*oerstedii*	*sciureus*	*vanzolinii*
Coat color patterns					
Supercilliary Patch	Roman	Gothic	Gothic	Gothic	Roman
Crown	Black, tapers caudally	Bluish-gray	Black	Olive to olive-gray	Black, tapers caudally
Dorsum	Gray, black, orange and agouti	Burnt-orange, black dorsal stripe	Red-orange, long hair	Gray, olive, or burnt-orange	Gray/black, with black stripe
Forelimb	Yellow hand and forearm, gray shoulder	Fulvous hands or hands and forearm, shoulder gray	Orange-yellow hand and forearm, olive-gray shoulder	Orange-yellow hand and forearm, olive-gray shoulder	Yellow hand and forearm, gray shoulder
Behavior					
	Sexually segregated, dominance hierarchy, male emigration	?	Sexually integrated, no dominance hierarchy, female philopatry	Sexually integrated, dominance hierarchy	?

within the Gothic-arch division. *Saimiri madeirae* is distinguished from *sciureus* by a gray crown, naked ears, and fulvous hands, whereas *sciureus* has an olive to olive-gray crown, and the tufted ears and yellow forearms in common with every other group.

Thorington's (1985) analysis of *boliviensis* concluded that there is clinal variation with intergradation between *boliviensis* and *sciureus* in Peru (Fig. 1) and an abrupt western border separating *boliviensis* from *madeirae* along the Rio Guapore/Mamoré and along the Rio Purús. Hershkovitz (1984) does not acknowledge this intergradation.

Specimens of *madeirae* and *sciureus* from around the Rio Tapajós have a mix of traits. Along the east side of the river, naked-ear monkeys share coat color features with *sciureus*. These animals have olive-gray crowns and yellow forearms, in contrast to the gray crown and fulvous hands of *madeirae*, near Humaitá (Thorington, 1985). That this suggests intergradation between *madeirae* and *sciureus* was not overlooked by Thorington (1985), yet the apparent sympatry between them convinced both Thorington and Hershkovitz that the two do not likely interbreed. Thorington also mentioned *madeirae* specimens from Calama (the southern portion of their range) with yellow forearms and gray crowns without suggesting intergradation.

The apparent intergradation along the Tapajós river and on the right (west) bank of the Rio Madeira, at Calama, is equivocal with regard to the species status of *madeirae*. There is inferred sympatry between the Rio Tapajós and Rio Madeira, without intergradation, and there is evidence that may be interpreted as intergradation with *sciureus* east of the Rio Tapajós, and at Calama, in the southwest quadrant of the group's distribution. Either *madeirae* is polytypic, with subspecies located east and south, those enigmatic specimens are intermediates, they represent other species, or the populations are polymorphic.

Cranio-Dental Studies

Niche separation, that is, differential use of the environment, is one common component of species, and species adaptations that vary geographically may reflect differences of the environment, not changes in niche. Cranio-dental features may, therefore, reflect form-functional differences in populations or species-specific adaptations. Establishing such characters in no way implies they are acquired at "speciation," or that they represent species. It only refers to the fact that species' niches evolve, and adaptive, form–function correlates may be discovered.

Dental evidence relevant to the taxonomy of *Saimiri* is limited to a few studies. Orlosky (1973) argued for species status of *oerstedii* based on the presence of statistically significant differences in traditional dental dimensions when compared to a sample of South American *sciureus.* Galliari and Colillas (1985) recorded dental eruption sequences of *boliviensis* and compared their results to those of a similar study of Colombian *sciureus* by Long and Cooper (1968). Their findings reveal significant differences in the sequence and timing of eruption, specifically in the sequences of premolar eruption and the relatively delayed eruption of M_3 in *boliviensis.* Based on this evidence, Galliari and Colillas (1985) suggested that the two groups are separate species. A perfectly valid alternate interpretation of their work suggests the differences are interpopulational, and the notion of species-specific eruption sequences should not be generally applied. Rosenberger *et al.* (1991) showed that standard dimensions of P_4-M_2 did little to segregate the populations recognized by Hershkovitz (1984) and Thorington (1985).

Thorington (1985) used univariate and multivariate cranial analysis to test species and subspecies hypotheses. When cranial differences were found concordant with pelage differences, taxa were upheld or renamed. Thorington's analysis is not unequivocal, and this is shown in the indeterminancy of the *madeirae–sciureus* comparison. With regard to skull length, the sample he used from the Rio Tapajós is not a cranial intermediate between *madeirae* to the west and *sciureus* to the east. Another sample (*madeirae*) from between the Tapajós and Madeira rivers is more similar to *sciureus* from the same locality than to any other *sciureus* sample, and it falls nearly midway between *sciureus* to the east and *madeirae* from further west. Other "single taxon" samples were separated by skull size, such as a northern sample of *boliviensis* from a southern *boliviensis* group. The deciding principle employed by Thorington to discriminate taxa based on differences of size and proportion is whether there are geographically intermediate samples that are also morphologically intermediate.

Based on the craniodental evidence, a stronger case can be made for the synonymy of *boliviensis* and *sciureus* than for the autonomy of *madeirae.* Cranial evidence for *oerstedii* does not point strongly one way or the other.

Chromosomes

Chromosome studies of squirrel monkeys began in the late 1950s (Bender and Mettler, 1958). While all squirrel monkeys examined to date possess a diploid chromosome number of 44, animals from certain geographic regions have been found to vary in the number of acrocentric (V-shaped) vs. metacentric (X-shaped) chromosomes they possess (Table 2; Jones *et al.*, 1973; Ma *et al.*, 1974; Jones and Ma, 1975; Hershkovitz, 1984; Assis and Barros, 1987; Lacy *et al.*, 1988). This variability has been attributed to pericentric inversions occurring in chromosome pairs 15 and 16 (Ma *et al.*, 1974; Lau and Arrighi, 1976; Moore *et al.*, 1990).

Among captive animals of unknown origin, karyotyping alone is not always a conclusive means of differentiating geographically separated populations within *Saimiri* because certain geographically separated populations possess the same ratio of acrocentric to metacentric chromosomes (e.g., *oerstedii, madeirae,* and *peruviensis*) and because this ratio can vary among second and later generation hybrids (Lacy *et al.*, 1988; VandeBerg *et al.*, 1990b). Populations that are generally agreed to belong to the same species show interpopulational variance in acrocentric and biarmed pairs equal to those occurring between groups that are more separated geographically and that show greater differences in other features (e.g., *boliviensis* and *peruviensis* subspecies are as different from one another as, for example, *boliviensis* from *oerstedii*).

Recently, Moore *et al.* (1990) compared several homologous chromosomes among the *boliviensis* Group (*boliviensis* and *peruviensis*) and *sciureus* in order to document differences in the distribution of C-banding patterns and nucleolar organizer regions. Most significant was their finding that only *sciureus* possesses a terminal C-band in chromosome 5.

Chromosomal traits may have causal relevance as postmating isolating *effects,* though there is little evidence to suggest a necessary correlation with speciation (e.g., Sites *et al.*, 1987; Carson, 1987). The number of species documented for intraspecific variation in numbers of chromosomes is steadily increasing (e.g., Hamilton *et al.*, 1980; Rumpler, 1975; Hershkovitz, 1983), and the usefulness of karyotypes in discriminating taxa is not general and remains more of a case-by-case procedure (clusters of Hawaiian *Drosophila* species have identical banding sequences for major chromosomal elements; Carson, 1987). One con-

Table 2. ***Saimiri*** **Chromosome Data**

Taxon	2n	Number of acrocentric autosome pairs	Number of biarmed autosome pairs	Terminal C-band in chromosome 5	Ref.[a]
sciureus	44	7	14	Present	1,2,4
macrodon	44	6	15	?	1,2
oerstedii	44	5	16	?	1
madeirae	44	5	16	?	3
boliviensis	44	6	15	Absent	4
peruviensis	44	5	16	Absent	1,2,4

[a]1, Jones *et al.* (1973); 2, Ma *et al.* (1974); 3, Assis and Barros (1987); 4, Moore *et al.* (1990).

clusion is that inversions are not necessary for speciation. The application of inversions to the taxonomy of groups has been without a thorough understanding of the meiotic effects of inversions on the process of speciation (e.g., Sites *et al.*, 1987) and has resulted in the naming of "chromosomal species." Also the theoretical works of Barton (1979) and Spirito *et al.* (1983) suggest that more than one or two chromosomal rearrangements would be required to effect (reduce) appreciably gene flow in hybrid zones. Until generalizations can be supported on this subject, inversions and other chromosomal polymorphisms should be treated as the symptoms of mating isolation (biological or geographical), not the causes of it.

The notoriously high rates of abortion, stillbirth, and infant mortality among captives (Kaplan, 1977; Wolf *et al.*, 1975; Dukelow, 1982; VandeBerg *et al.*, 1990b) might be indirect evidence that infertile hybrid matings are a more widespread phenomenon than previously recognized (Moore *et al.*, 1990). There are also numerous instances of anomalous karyotypes in the hybrid offspring of parents from geographically separated sources (Moore *et al.*, 1990). To the best of our knowledge, no controlled breeding experiments have been performed to determine the fertility of hybrid crosses between geographically separated *Saimiri* populations. Until *in vitro* cytological studies of meiosis for both sex combinations are executed and establish the frequency of nondisjunction, numbers of chiasmata, and also the degree to which recombination takes place, we cannot say there are genomic postmating isolating mechanisms between parapatric or sympatric samples.

Biochemistry

Protein electrophoresis has played a role in squirrel monkey systematics, and biochemical markers may someday provide an alternative to karyotyping as a means of assessing the affinities of individual animals. Recent studies (DaSilva *et al.*, 1987a,b, 1992; Lacy *et al.*, 1988; VandeBerg *et al.*, 1990a,b) have compared a number of distinct geographic and chromosomal forms using standard electrophoretic methods. VandeBerg *et al.* (1990a,b) describe the first study of biochemical genetic markers in a relatively large number of pedigreed individuals of *boliviensis, peruviensis,* and *sciureus.* Twenty-six blood proteins were surveyed, 14 of which were polymorphic. Allelic frequencies at 13 of these variable loci showed statistically significant differences among the three populations and were useful in detecting hybrids. For example, the distribution of two alleles at the *ADA* locus consistently distinguished *sciureus* from *boliviensis* and *peruviensis* samples. Other alleles were restricted to one or two groups.

In a preliminary survey, DaSilva *et al.* (1987b) found genetic distances to be lowest among two geographic samples of *sciureus* from the Rio Jari and the Rio Tocantins, tributaries located on opposite banks of the lower Amazon, and a Peruvian *sciureus* population (*S. s. macrodon* of Hershkovitz)—some of the most widely separated populations within South America. These authors also report low genetic distances between *peruviensis* and *boliviensis.* Finally, they suggest that *madeirae* from central Brazil south of the Rio Amazonas is closer to *sciureus* than to *boliviensis* or *peruviensis,* but their results do not agree with another preliminary study based on karyotyping (Assis and Barros, 1987). In yet another report, Lacy *et al.* (1988) made a preliminary electrophoretic survey of *boliviensis* from Peru and of *sciureus* from Colombia and the Guyanas. Their results did not support any previously proposed classificatory scheme.

One study (DaSilva *et al.*, 1992) sampled 49 animals from Peruvian Amazonia in the area where *macrodon* and *peruviensis* overlap in their distributions. This is in the area where Hershkovitz thought the two groups were sympatric. Along the Rio Ucayali, eight of the animals collected were identified as Roman-arch. Six of those showed biochemical admixture. Another 14 Gothic-arch squirrel monkeys also showed admixture. In all, 22 of the 49 (45%) showed some indication of admixture.

Sampling may play an enormous role in the results gained by electrophoretic studies and many scenarios could be described. If there are hybrid zones, such as the areas described by DaSilva *et al.*, (1992) and Thorington (1985), and one or more exist along the borders of a geographic group, then specimens originating in, near, or away from a zone may effect the results. DaSilva *et al.* (1987b) concluded that *madeirae* is closer to *sciureus* than it is to *boliviensis* or *peruviensis*. Was their *madeirae* sample from near the Rio Tapajós, and thus geographically close to *sciureus* where introgression may be taking place, or was that sample from west of the Rio Purús, in the vicinity of *boliviensis* populations? Lack of specific geographical information for samples used in species-level biochemical studies lessens the usefulness of these contributions for discriminating parapatric species.

Biochemical studies give weak support for a Roman-arch group and a Gothic-arch group, beyond which little can be inferred or taxonomically resolved.

Behavior

Does *Saimiri* exhibit any survival or reproductive behaviors that may function to isolate groups? Three aspects of behavior that may be directly related to (strategies of) reproduction and niche are social organization, vocalization, and feeding behavior (Table 1, Figs. 2 and 3, and Table 3, respectively). Both captive and wild studies provide information on these areas.

Geographic differences of behavior among squirrel monkeys have emerged from observations on captive animals (e.g., Mendoza *et al.*, 1978; Martau *et al.*, 1985). Some of the variations appear to reflect population-specific, and inferred, genotypic differences, such as aspects of social behavior (Mendoza and Mason, 1989), stereotyped displays (Maclean, 1964), and vocalizations (Winter, 1969).

An extensive series of captive studies is available on social behavior and its physiological basis for two forms of *Saimiri: sciureus* and *boliviensis*. Observations indicate (1) *boliviensis* has a sexually segregated social organization in which males and females remain spatially separated outside of the breeding season, whereas *sciureus* males and females are integrated throughout the year. (2) In *boliviensis* social groups, independent linear dominance hierarchies are present within each sex. Among *sciureus* groups, a single linear hierarchy includes both males and females. Male and female social relationships in the two groups ap-

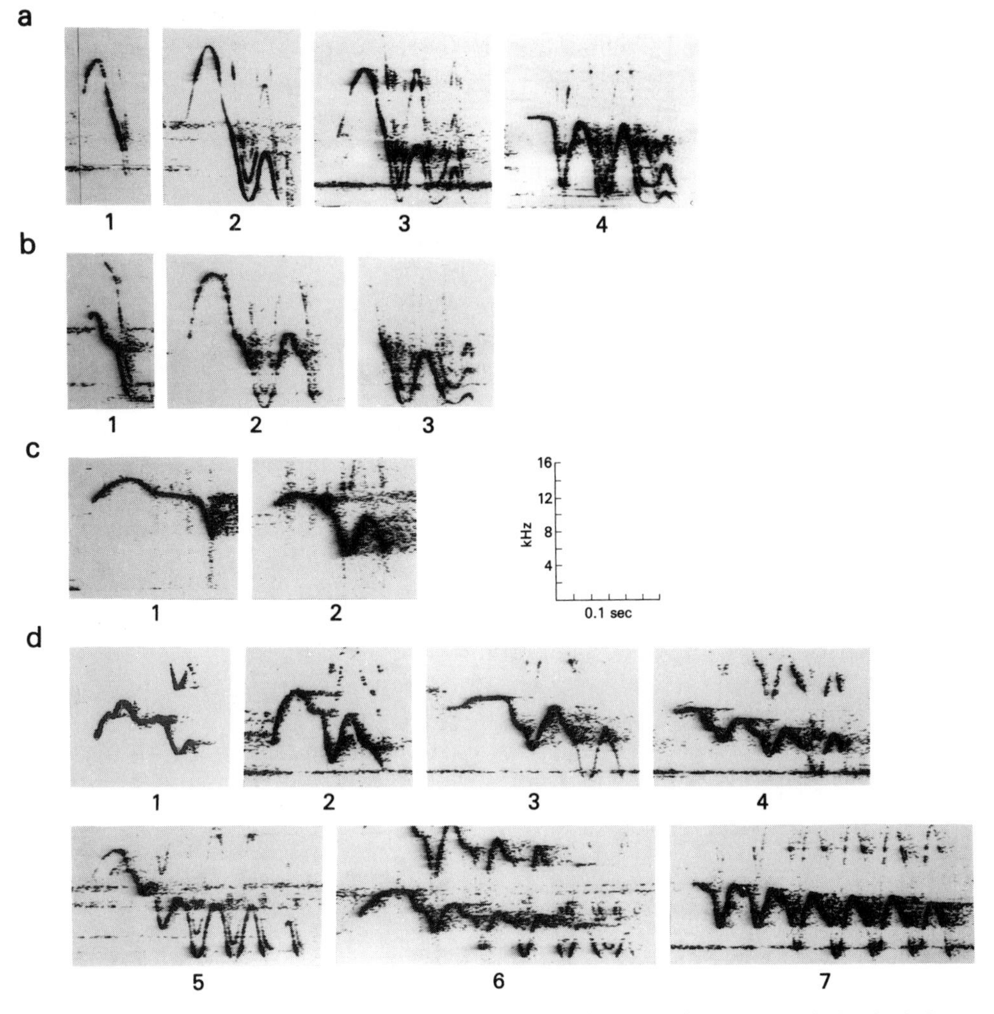

Fig. 2. The major categories of calls produced by *S. oerstedii* in Costa Rica: (a) smooth chuck, (b) bent mast chuck, (c) peep, and (d) twitters. Subtypes within each call category are designated by numbers.

pear to be regulated by two distinct mechanisms, dominance relationships in *sciureus* (Mendoza *et al.*, 1978) and sexual segregation in *boliviensis*. (3) Female *boliviensis* have higher female cortisol titers, longer and more sustained adrenal responses to stress, are dominant over males, and more actively initiate agonistic interactions compared to *sciureus* (Coe *et al.*, 1985).

Unfortunately, the degree to which these data can be applied to the taxonomy of *Saimiri* is limited. Much of the information on *Saimiri* behavior derives from studies in which the exact geographic source of the captive animals is not known.

Field studies of *oerstedii* in Costa Rica and *boliviensis* in Peru reveal strong differences in the social behavior of these groups from each other and particularly among females (Mitchell *et al.*, 1991). In *oerstedii*, female relationships appear to be undifferentiated. There are no female–female alliances and no female dominance hierarchies. Females also disperse from the troop in which they are born. On the other hand, female *boliviensis* in Peru exhibit differentiated female relationships, a female dominance hierarchy, and female philopatry. An important evolutionary factor thought to be responsible for the disparity in the social relationships of female squirrel monkeys in Costa Rica and Peru relates to differences in the distribution of fruit resources (Mitchell *et al.*, 1991). In Peru, the fruit patches are large, and it may be advantageous for females to acquire

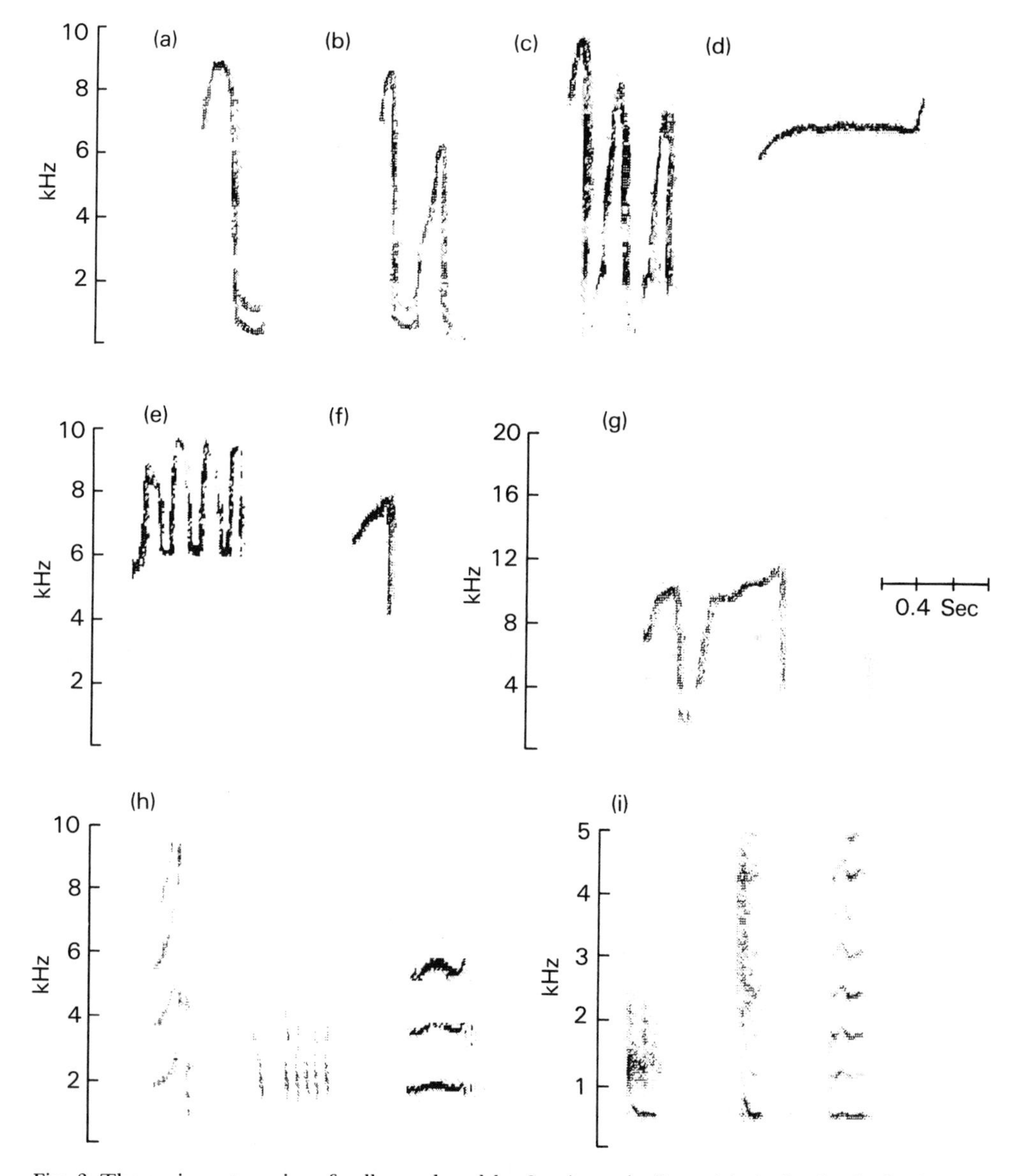

Fig. 3. The major categories of calls produced by *S. sciureus* in Peru: (a) single chuck, (b) double chuck, (c) multiple chuck, (d) peep, (e) tweet, (f) peep-chuck, (g) tweet-chuck, and (h, i) examples of six "mother" calls. Mother calls are vocalizations that appear to be directed exclusively by mothers to their infants and are not found among *S. oerstedii* in Costa Rica.

and to defend them aggressively; hence the dominance hierarchies. In Costa Rica, fruit patches are generally small and unlikely to satiate even one individual. There is no direct competition for food, and dominance hierarchies have not evolved as a consequence of aggressive female interactions.

The limited field information on male squirrel monkey behavior describe differences between *oerstedii* and *boliviensis* (Boinski, 1987b; Mitchell, 1990; Boinski and Mitchell, 1992). Costa Rican *oerstedii* males usually remain in their natal troop until full adulthood. There is only one observed instance of male migration. Aggression among *oerstedii* males within a troop has never been observed, though subadult and young males engage in extensive play and rough play.

S. boliviensis monkeys from Peru emigrate from their natal troop before maturity and form all-male subgroups. The few adult males that remain with the females and young throughout the year have overwhelmingly agonistic interactions with each other and with the females. Aside from the brief mating season, females behave aversively toward males.

Males of the genus, without known exceptions, also undergo annual male fatting (Dumond and Hutchinson, 1967), a reproductive physiology unique among primates.

The vocal behavior of squirrel monkeys has been studied in the laboratory and more recently in the field (Boinski and Newman, 1988; Boinski, 1991; Boinski and Mitchell, 1992). *Saimiri* have easily identifiable calls; nearly all have a simple, pure-tone frequency structure that ranges up to about 12 kHz. Most vocalizations fall into two categories, chucks and peeps (see Figs. 2 and 3), at least based on the phonemic typology of a human scrutinizing a sonogram. Yet contrary to other aspects of their behavior (particularly foraging), vocalizations are geographically differentiated, making it unlikely that geographically separated populations of squirrel monkeys would completely recognize each other's vocal repertoire (Snowdon *et al.*, 1985; Boinski, 1991; Boinski and Mitchell, 1992). Nearly every population of squirrel monkeys studied to date has a set of unique vocalizations, albeit these are not commonly produced (Boinski and Mitchell, 1992). Squirrel monkeys do not produce mating calls, and the likelihood that vocalizations function to preclude hybridization appears to be small (e.g., DaSilva *et al.*, 1992).

However, there are indications that suggest vocalizations may play a role in the postmating success of hybrid individuals. Data from captive animals provide evidence for the genetic basis of the acoustic structure of squirrel monkey vocalizations (Winter *et al.*, 1973; Newman and Symmes, 1982). At least some vocalizations of newborn individuals closely approach adult versions of the same call, indicating that the role of learning is small. The offspring of crosses between *boliviensis* × *sciureus* have structurally intermediate vocalizations. *S. boliviensis* and *sciureus* do not respond to the calls of a separated, distressed infant of the other group but do respond to the "separation peeps" from infants of their own group (Snowdon *et al.*, 1985). Additional differences between the vocal behavior of *sciureus* and *boliviensis* in captivity include the rate at which females exchange chucks and the pattern of extremely slight structural modifications that can occur in these chucks, depending on whether the chucks initiate or end a series (Biben *et al.*, 1986; Masataka and Biben, 1987).

Table 3. Group Feeding Statistics

Trait	*oerstedii* (Costa Rica; Boinski, 1987, 1988)	*boliviensis* (Peru, Terborgh, 1983)	*boliviensis* (Peru, Mitchell *et al.*, 1981)	*sciureus* (Colombia, Klein and Klein, 1975)
Group size	35–65	35	45–75	25–35
Home range	200 ha	>250 ha	250–500 ha	65–130 ha
Density	0.36/ha	0.50/ha	0.60/ha	0.50–0.80/ha
Time allocated to foraging insects	45–65%	50–75%		
Time allocated to rest	5–10%	11%		

Squirrel monkey populations in the wild glean foliage on terminal twigs; forage on soft, berry-like, fruits; and prey on arthropods, particularly caterpillars and grasshoppers (Mitchell *et al.*, 1991; Janson and Boinski, 1992). Occasionally nectar is eaten, but there are no well-documented instances of squirrel monkeys eating nonreproductive plant tissue, such as leaves, bark, or stems. Variations reported from different field studies reflect site differences in the abundance and distribution of food sources, as foraging techniques and food preferences appear identical across sites (Janson and Boinski, 1992). In light of this homogeneity (Table 3), and the close similarity of body size and morphology across populations, it is intriguing to ask whether sympatry could exist in *Saimiri*, as it does in *Cebus* and *Saguinus*, for example. If sympatry does exist, what are the niche differences, if any?

There are no obvious "challenging" behavioral patterns or displays that are likely to inhibit matings between individuals from geographically isolated populations. Wild individuals observed in both Costa Rica and Peru commonly copulate with no evident preparatory interactions, no stereotypical mating displays, and no vocalizations specific to mating (Boinski, 1992; Mitchell, personal communication). In fact, males and females, both in captivity and the wild, typically lack any premating affiliative behaviors apart from the limited association and occasional olfactory investigations that occur just prior to copulation (Mendoza *et al.*, 1991). For squirrel monkeys, it may be extremely difficult to identify even tentatively any effective behavioral reproductive isolating mechanisms. Olfactory signaling may as yet offer an area where barriers can be identified.

The behavioral evidence does show (1) the same feeding behavior is shared by all squirrel monkeys; (2) nontrivial differences in the social organization of three groups, *boliviensis*, *sciureus*, and *oerstedii*, each appearing under different ecological conditions, to optimize their feeding strategy; (3) a genetic basis for the acoustic structure of vocalizations, at the deme or population level, which may act to diminish hybrid viability; and (4) no known premating behaviors that could possibly function in the reproductive isolation of groups.

We cannot say there is more than one mate-recognition system present within the genus. Yet the differences in vocalizations and social behavior, in captivity, suggest a genetic basis for both, a basis that differs between the three groups. Ecological differences may indeed explain why social behaviors differ, and those behaviors may have their bases fixed within the group genomes. All these may well be clinal across territory that is not yet sampled or adequately sampled, where geographic groups come into contact.

Methods and Materials

The "species question" of primary importance with regard to parapatric and sympatric areas in South America is whether there is (indirect) evidence for gene flow across parapatric regions and between forms thought to be sympatric with one another. Character studies are often the basis on which we make inferences about introgression. Each character provides information on the phenotypic differences that have accrued in the forms of *Saimiri.* These may reveal clues about the processes that transform phenetic groups into cladistic ones. The allopatrically distributed Central American *oerstedii* is another matter. We can only assess differences for their significance to the biological validity of the group as a species, knowing there has been geographic separation for some indeterminate amount of time. In this sense, characters are markers that offer indirect evidence of the natural status of a group. To these two ends, we have measured and described characters of *sciureus, boliviensis, madeirae, vanzolinii,* and *oerstedii.* We do not treat the subspecific status of groups.

Geographic Distribution and Pelage Coloration

Skins were examined at three Brazilian museums, Museu Paraense "Emilio Goeldi," Belém (MG); Museu Nacional, Rio de Janeiro (MN); Museu de Zoologia, São Paulo (MZ); and at the American Museum of Natural History, New York (AMNH) and the Field Museum of Natural History, Chicago, to assess coat color variation and patterns of geographic distribution. Specimens from critical areas were recorded on 8-mm videotape and reviewed extensively.

Dentition

A sample of specimens from the American Museum of Natural History was selected to represent as broad a geographic distribution as possible, and to include the major phenetic and taxonomic groups, *boliviensis, sciureus, madeirae,* and *oerstedii.* A minimum of 15 specimens, equally apportioned to sex, when possible, was measured from each major geographic group, for a total of 110 specimens.

We investigated topographic relations of crown morphology and standard tooth-size measures by digitizing the crown surface in two dimensions (see, for example, Fink, 1990; MacLeod, 1990; Rohlf, 1990 for an overview on digitizing). Gathering coordinate data from corresponding landmarks on homologous features allows questions to be asked about topographic regions (O'Higgins, 1989). We do not refer to these landmarks as homologous; they are defined geometrically, by least radius or curvature of a surface or by the intersection of two or more surface features, for example. Traditional maximum and minimum dimensions taken on teeth often do not represent such corresponding points on homologous features, and regardless of the extensive naming of tooth surface features, the actual boundaries of homologous features are undefined. Coordinate data allow exact topographic relations of crown morphology to be quantified. Although not executed for this study, three-dimensional (3-D) digitizing of crown surfaces is feasible and has been done by others (for example, Kanazawa *et al.*, 1983a, 1988; Hartman, 1986, 1989; Richmond, 1987). A significant constraint of 3-D methods is the restriction of samples to nearly unworn teeth and the virtual elimination of M1 from the analysis because of the early eruption of that tooth in the eruption sequence (Hartman, 1989). In two dimensions (2-D), landmark locations may be interpreted on moderately worn teeth without serious compromise of the measure. In 3-D, landmarks must be similarly interpreted on worn teeth; however, 3-D distances are affected by the differential wear of tooth areas or by shape differences (and 3-D metrical analysis is generally one- or two-dimensional).

Landmarks were preassigned to the crown surfaces of all lower first molars before measurement (Fig. 4). The spatial positions of all landmarks were recorded as Cartesian coordinates using a video-digitizing system interfaced with a desktop computer. Linear dimensions were calculated from the X, Y data. Specimens (complete lower jaws) were placed on a stack of flat plastic shims on a stage without further manipulation. The shims were used to maintain the focal distance from the lens by adding or subtracting them, thus avoiding changes in scale with lens adjustments. Any orienting procedure is arbitrary (Ramaekers, 1975; Kanazawa *et al.*, 1988; Hartman, 1989), and this one was selected for ease and repeatibility. A wear-graded series of lower first molars, representing pristine to obliterated crowns with slight increases in wear between adjacent teeth, was studied to determine the effect of wear on the apparent locations of landmarks.

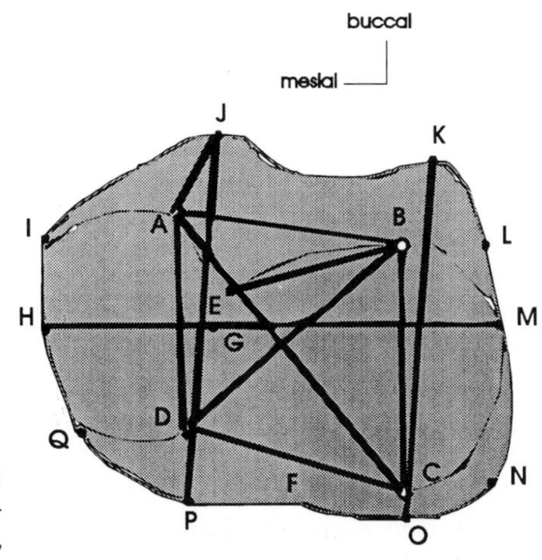

Fig. 4. Landmarks assigned to M_1 and the 11 length variables used for two-dimensional analysis. A, protoconid; B, hypoconid; C, entoconid; D, metaconid; AJ, maximum buccal flare; BE, chord length of cristid obliqua.

A number of specimens were remeasured on different days without reference to the state of wear of those specimens. Coordinates were plotted on graph paper at lens magnification and compared. Differences from day to day were negligible.

Principle components, Pearson correlation coefficients, and t- and f-tests were performed, and simple statistics were calculated using SAS. Frequencies of variables significant at the 0.05 level were determined to assess the relative contribution of variables in the discrimination of groups. In addition, angles between cusps were calculated to compare the topographic relationships of cusps between groups. All variables used in the analysis were checked for approximation to normal distribution. Eleven out of all possible variables were chosen for analysis.

Biochemistry

We have compared squirrel monkeys from four geographic areas using high resolution two-dimensional electrophoresis (2DE; O'Farrell, 1975). This method separates proteins using two independent, intrinsic parameters: protein isoelectric point, a charge parameter, and protein molecular weight. When very sensitive methods are used to stain the proteins separated by the 2DE procedure, hundreds of proteins can be resolved in a single gel, producing tissue-specific patterns that are highly reproducible (Harrison *et al.*, 1992).

Blood samples were studied from six *boliviensis*, four *peruviensis*, and four *sciureus* squirrel monkeys. The majority of these samples were provided by the Department of Comparative Medicine, University of South Alabama College of Medicine; two of the six *boliviensis* were obtained from the Delta Regional Primate Research Center. A single individual in the collection of the Santa Ana Zoo, which we believe represents *oerstedii* on the basis of phenotype, was also examined.

Whole blood was collected in EDTA and the plasma fraction was separated by centrifugation. Plasma proteins were denatured and made soluble in four volumes of a mixture containing 2% sodium dodecyl sulfate (SDS), 1% dithiothreitol (DDT), and 10% glycerol in 40 mM cyclohexylaminoethane sulfonate (CHES, pH 9.5) ("SDS mix" of Tollaksen *et al.*, 1984). Sample aliquots were heated for 5 min in a 95 °C water bath and then stored at −75°C until needed.

High resolution 2DE (O'Farrell, 1975) was performed, with modifications (Daufeldt and Harrison, 1984), using the ISO-DALT™ multiple gel casting and electrophoresis system (N. G. Anderson and N. L. Anderson, 1978; N. L. Anderson and N. G. Anderson, 1978; Tollaksen *et al.*, 1984). Charge (Hickman *et al.*, 1980) and molecular weight standards (Edwards *et al.*, 1982) were added to each gel to facilitate alignment of protein spot patterns between gels. A modification (Harrison *et al.*, 1992) of the ammoniacal silver staining method of Guevara *et al.* (1982) was used to visualize the separated proteins.

A 2DE gel was run for each of the 15 individual squirrel monkeys in our sample (e.g., Fig. 5). In addition, each possible pair-wise combination of the four taxonomic samples was run together on a single gel (termed *coelectrophoresis*), using one individual to represent each group. Forty-eight of the best resolved

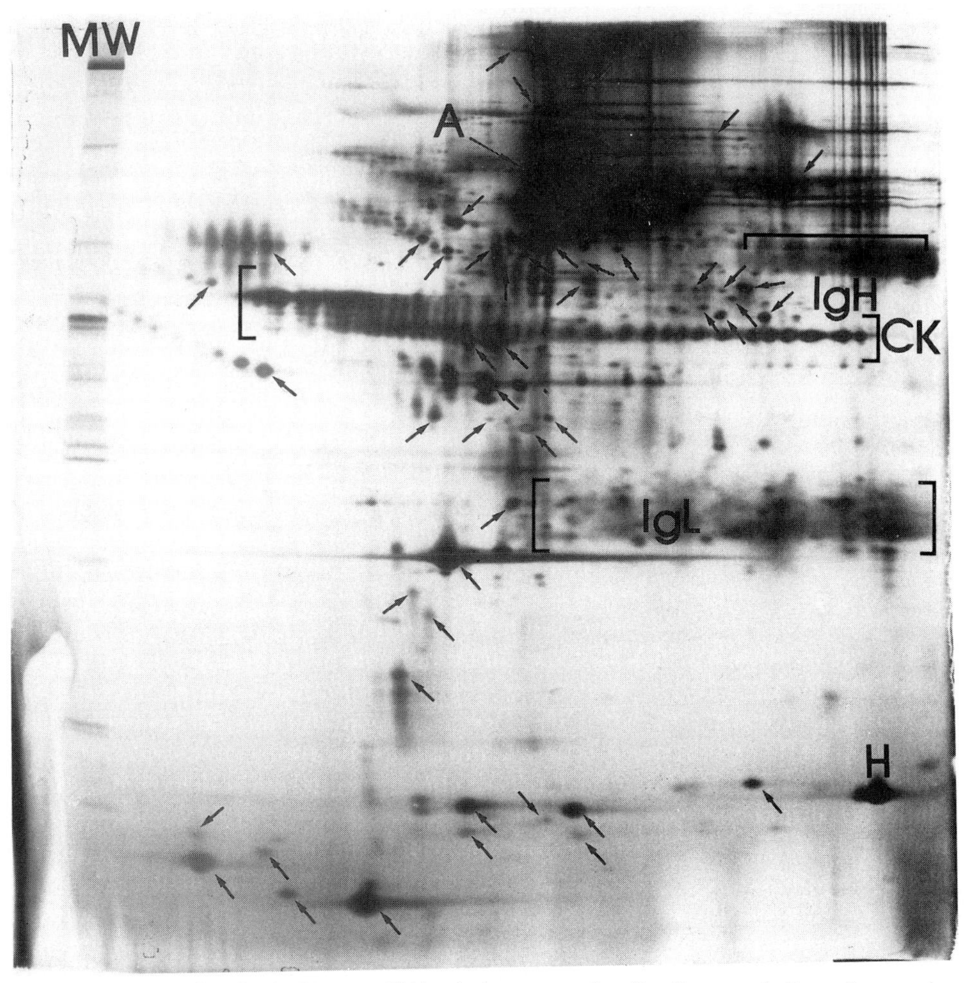

Fig. 5. Two-dimensional gel of *S. oerstedii* blood plasma proteins. Small arrows indicate the proteins (n = 48) compared among squirrel monkey populations for differences in electrophoretic mobility. A, albumin; MW, molecular weight standards; CK, creatine kinase charge standards; IgH, immunoglobulin heavy chains region; IgL, immunoglobulin and light chains region; H, hemoglobin chain.

proteins (Fig. 5) were surveyed for differences in electrophoretic mobility among the 15 squirrel monkeys in our sample. The 2DE gels, and the archival "XRD images" (Harrison, 1984) made from them, were scored manually over a light box. Protein spot patterns were aligned using the internal charge and weight standards and the constellation of invariant proteins in the immediate vicinity of the proteins being compared. Reference was made to the co-electrophoresis gel patterns in an effort to verify or to rule out apparent mobility differences initially detected during gel-to-gel comparisons.

Genotypic data (not shown) were analyzed with the help of the BIOSYS-1 (Swofford and Selander, 1989) and NTSYS-pc (Rohlf, 1988) software packages. A contingency chi-square test was performed to test the significance of inter-populational differences in allelic frequency values (Workman and Niswander,

1970; Swofford and Selander, 1981). A number of similarity and distance coefficients were applied to the gene frequency data, including Nei's (1978) unbiased genetic identity and genetic distance, and a modified Roger's distance (Wright, 1978). Coefficients were clustered using the UPGMA algorithm of Sneath and Sokal (1973).

New Evidence Bearing on Species of Saimiri: Data and Discussion

Geographic Distribution and Pelage Coloration

Specimens collected since 1984 from Brazilian Amazonia provide important information from "species" boundary areas south of the Rio Amazonas. These specimens are from (1) the area of Tefé, where *vanzolinii, sciureus,* and *madeirae* occur (Ayres, 1985) and (2) the left and right banks of the Rio Jiparaná, a tributary of the Rio Madeirae, near the probable southern extent of *madeirae's* distribution. Additional comments are given to specimens collected from complicated areas bearing on the validity of *Saimiri madeirae,* the area of inferred sympatry, and the eastern border of that group along the Rio Tapajós.

Two specimens from Tefé labeled *S. ustus,* collected in 1985, are in the "Emilio Goeldi" collection and are of interest for their mix of traits. One, MG13210 (male), is from the left bank of the Rio Tefé, at the mouth of the Rio Bauana (Ponta da Castanha: see Ayres, 1985, Map 1). This specimen has the burnt-orange dorsum and fulvous hands typical of *madeirae,* a grayish preauricular patch typical of *sciureus*, but unknown in *madeirae* specimens from near Humaitá, and ear hair intermediate between naked and tufted (Fig. 6).

The second specimen, MG13209 (female), also from the west bank of the river, at Lago Boia, does not have an equivalently burnt-orange dorsum as the first, and is more evenly gray/black and orange, similar to some *boliviensis,* but with a darker gray tail, approximating a condition intermediate between *vanzolinii* and other *Saimiri.* The hands are fulvous and the forearms have slightly

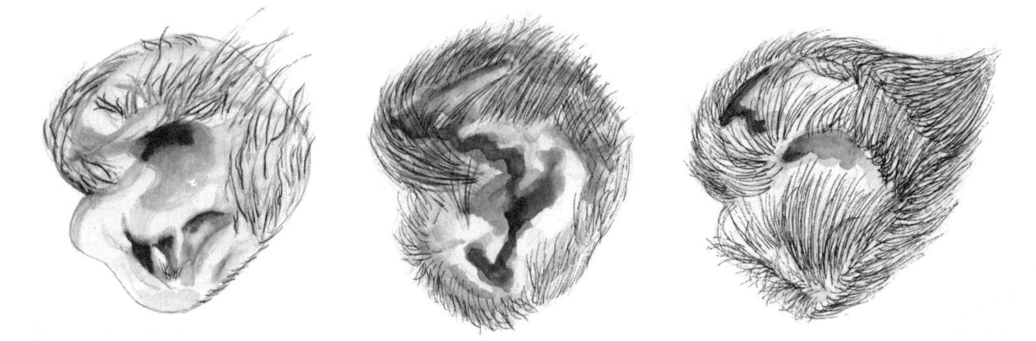

Fig. 6. Expressions of the variation of ear hair: from left to right; naked-eared squirrel monkey c.f. *madeirae;* intermediate expression of hair typical of samples from the border areas of *madeirae;* tufted ear typical of all South American *Saimiri* except *madeirae.*

more of this color, but are not covered with orange or yellow, like all other *Saimiri*. The crown is mottled olive and dark gray to black, with black bars along the margin near the ears and black on the forehead, with a nearly complete Roman-arch pattern. The preauricular patch is uncharacteristically mottled for *Saimiri*, and ear hair is even greater than the first specimen, although not tufted.

MG13210 is likely a hybrid, or a generation or two removed from a hybrid parent, of *sciureus* and *madeirae*. MG13209 is probably the offspring of *madeirae* crossed with *vanzolinii* or *boliviensis*. We have seen only one other specimen, from Fordlândia, that has a mottled crown, mostly black with less olive. That specimen is an interesting mix of *sciureus* and *madeirae* features, and mottling may occur in some specimens as a result of hybridization. The geographic proximity to *vanzolinii* and the darker tail may implicate *vanzolinii* admixture; however, the distribution of *boliviensis* in the region is unknown. Neither specimen can be neatly pigeon-holed into any taxon, unless one arbitrarily selects a single character for that purpose.

A few more specimens from Tefé are in the collection of the Museu de Zoologia (MZ19013, MZ19014, MZ19015, collected in 1984). These are all right bank (east) collections from Vila Vale, near the confluence of the Rio Tefé with the Rio Solimões. These specimens more closely approximate *madeirae* from Humaitá, with a few exceptions. The skull cap is not the consistent bluish-gray found in *madeirae* from Humaitá and the ears show more hair. These specimens may be viewed as weak indicators of gene exchange with *sciureus* from the left bank of the Rio Tefé, and perhaps from small groups of *sciureus* that have crossed the river. A similar situation may occur with *sciureus* having crossed the Rio Solimões. Alternatively, there may be an east-west cline, without intergradation, where central populations of *madeirae* with naked ears, fulvous hands, and blue-gray crowns shift toward olive-gray crowns and hairier ears as they approach the eastern and western boundaries. To accept the latter, one must also accept that the morphological direction of the cline is twice toward the condition of *madeirae's* neighbors.

There was extensive collecting recently at a number of sites along the upper reaches of the Rio Madeira, including the hydroelectric site of Samuel on the Rio Jamari, Rondônia ("Emilio Goeldi," Museu Nacional, 1988), the Rio Madeirae at Humaitá, and a Madeira tributary, the Rio Jiparaná (Ferrari and Lopes, 1992). Opposite Humaitá, on the right bank of the Rio Madeirae, specimens are typical *madeirae* as described by Thomas (1908). The same is true of specimens taken on the right bank of the Rio Jiparaná. No monkeys were collected on the left bank of the Rio Madeirae, but a household pet squirrel monkey from that side is reported to be identical to right bank monkeys (Ferrari and Lopes, 1992). However, specimens from the left bank of the Rio Jiparaná and from Samuel, which is situated between the Rio Jiparaná and the Rio Madeirae, differ in some features. Most striking is the combination of all golden hands and forearms, together with naked ears. All specimens examined are consistent for these features, and none have exclusively fulvous hands. Features are so consistent that we consider this form to be a new subspecies. There is yet no information on the southern boundary of this phenotype and whether or not it is contiguous with *boliviensis*.

Between the Rio Madeira and Rio Tapajós, some specimens have a mix of traits and others are either *sciureus* or *madeirae*. Three specimens from the Rio Arapiuns (MG), are *sciureus* without any *madeirae*-like states of coat color characters; additionally, the orange-yellow on the feet extends onto the leg. We know of no new specimens that can shed light on the area of inferred sympatry.

A number of specimens are known from east of the Rio Tapajós. Three specimens from Fordlândia (MZ) are remarkable in having very little hair on their ears, fulvous hands and forearms, and olive crowns, with the crown color extending onto the dorsum in a broad stripe, specimens that in every way except the ears, are like *sciureus*. One specimen, a female, has the black and olive mottled crown mentioned above. From Bom Jardim (MZ) are specimens intermediate for hand-forearm color, the orange being restricted to the distal portion of the forearm above the hand. Ears are variable for hair, some completely denuded, others intermediate (e.g., Fig. 6). The next major river to the east is the Rio Iriri/Xingú, and *Saimiri* reported from along the Iriri are referred to as *sciureus* (Martins *et al.*, 1986). Because of the mosaic nature of characters where *sciureus* and *madeirae* occur, however, survey observations without specimen samples are not the strongest evidence for the recognition of boundaries.

Pelage variation and zoogeography are still equivocal. New specimens from the western range of *madeirae* weaken the hypothesis that *madeirae* and *sciureus* are separate species. Other specimens weaken the species status of *vanzolinii;* they include one probable hybrid from Tefé, samples that bear remarkable resemblance to *vanzolinii* from the nearest *boliviensis* sample along the Rio Jurua and its tributary the Rio Eiru, and nearly identical *boliviensis* from around Envira. *Saimiri vanzolinii,* as remarked by Hershkovitz (1985), is best considered a member of *boliviensis* until further evidence indicates otherwise.

Pelage variation is distributed in zoogeographic patterns, with South American groups of *Saimiri* distributed along and between the major rivers. All groups, though, are not contained by river boundaries, such as *sciureus* north and south of the Rio Amazonas. The parapatric distribution of populations may indeed indicate either insufficient niche differences for the successful coexistence of those groups or nonadaptive stochastic differences between them. Traits may have accumulated through episodic periods of isolation.

Dental

Table 4 lists Pearson correlation coefficients (and the significance probabilities calculated for N = 111 under the null hypothesis that the correlation is zero) of 11 variable means (listed in Table 5) for the total *Saimiri* sample. Some variables with high correlation values are easily understood as the alternate sides of the same triangle (Fig. 4). The variable pairs AD/AC, BC/AC, CD/AC, AB/BD, AD/BD, BC/BD, and CD/BD are of this type and range from 0.45 to 0.71 (R = 0.0001). The two diagonals, AC and BD, also have a high correlation (0.44). Variable BE, the chord measure of the cristid obliqua, has higher correlation coefficients with AB and BD. That crest is anchored anteriorly near the protoconid base (near landmark A) and distally at the hypoconid (landmark B). Thus BE is the alternate side of triangles ABE and DBE. AD/BC (0.48) refer to the distances between the anterior trigonid cusps (protoconid-metaconid) and

Table 4. Pearson Correlation Coefficients

PROB > |R| UNDER HO: RHO = 0 N = 111

	AB	AC	AD	AJ	BC	BD	BE	CD	HM	JP	KO
AB	1.00										
AC	0.38	1.00									
	.0001										
AD	0.05	0.66	1.00								
	.6102	*.0001*									
AJ	0.23	−.021	−0.32	1.00							
	.0145	*.0290*	*.0005*								
BC	0.18	0.60	0.48	0.09	1.00						
	.0586	*.0001*	*.0001*	*.3256*							
BD	0.46	0.44	0.46	0.24	0.71	1.00					
	.0001	*.0001*	*.0001*	*.0119*	*.0001*						
BE	0.67	0.39	0.19	0.04	0.37	0.50	1.00				
	.0001	*.0001*	*.0490*	*.7139*	*.0001*	*.0001*					
CD	0.31	0.59	0.05	0.07	0.23	0.45	0.27	1.00			
	.0010	*.0001*	*.5828*	*.4505*	*.0171*	*.0001*	*.0040*				
HM	0.35	0.35	0.04	0.06	0.21	0.20	0.39	0.33	1.00		
	.0001	*.0002*	*.6739*	*.5282*	*.0259*	*.0396*	*.0001*	*.0003*			
JP	0.36	0.41	0.32	0.34	0.49	0.48	0.30	0.21	0.28	1.00	
	.0001	*.0001*	*.0006*	*.0003*	*.0001*	*.0001*	*.0014*	*.0289*	*.0025*		
KO	0.31	0.33	0.22	0.31	0.59	0.55	0.34	0.19	0.24	0.65	1.00
	.0010	*.0004*	*.0198*	*.0009*	*.0001*	*.0001*	*.0003*	*.0480*	*.0119*	*.0001*	

the posterior talonid cusps (hypoconulid-entoconid). The relatively high correlation coefficients that remain to be accounted for relate maximum buccolingual dimensions of the tooth, JP, KO, to each other, and to the talonid cusps, AD, BC, and the diagonal BD. Notably, the trigonid cusps dimension, AD, is not as highly correlated (although highly significant) with the maximum trigonid width, JP (0.32), as it is with the talonid cusps, BC (0.48). This appears related to the fact that there is an inverse correlation between AJ, the distance from the protoconid, A, to the maximum buccal flare of the tooth, J (an indirect measure of

Table 5. Variable Means

Variable	*oerstedii*	*sciureus*	*madeirae*	*boliviensis*	*peruviensis*
AB	1.42	1.51	1.44	1.50	1.43
AC	2.25	2.30	2.25	2.29	2.21
AD	1.58	1.54	1.53	1.53	1.49
AJ	0.68	0.77	0.74	0.82	0.82
BC	1.62	1.74	1.72	1.78	1.75
BD	1.82	1.95	1.90	1.99	1.97
BE	1.03	1.09	1.03	1.10	1.03
CD	1.23	1.28	1.20	1.36	1.26
HM	2.83	2.73	2.65	2.74	2.73
JP	2.61	2.73	2.73	2.71	2.72
KO	2.42	2.58	2.56	2.58	2.74

cingulum development), and AD, the trigonid cusp distance, and AC, the diagonal. There is a very slight tendency for tooth width to be maintained independent of intercusp spread.

T- and f-tests were performed for the 11 variables (Fig. 7A) with representative samples of the geographical groups. The inset box shows the order of placement of variables being compared. More variables were significant for *oerstedii* ($p > 0.05$), 24, compared to all other groups. *Oerstedii* also registered the greatest number of significantly different variables (nine) when compared to just one other group, *sciureus*. *Sciureus* follows with a total of 18, seven of which result from the comparison with *madeirae*. *Madeirae* has 17 total significant variables; *peruviensis* has 12, and *boliviensis* has 9. The groups showing fewest significant variables are *boliviensis* and *peruviensis* when each are compared to *sciureus*. A histogram (Fig. 7B) shows the frequencies at which variables were involved in discriminating groups. The three highest variable frequencies—JP, BD, and KO—reflecting trigonid width, hypoconid-metaconid distance, and talonid breadth, respectively, also have relatively high correlation values to each other (0.48–0.65). Variable AD, the protoconid-metaconid distance, does not discriminate.

Figure 8 shows the mean topographic positions of cusps and Table 6 is a list of the mean angles formed between cusps. In Fig. 8, all lengths between cusps are scaled to the longest base line, BC, which is made equal for all groups. Notable differences between group patterns in Fig. 8 are dependent on two variables: size and shape, and the distance between cusps and their angles. Because all group means for variables are scaled to an equal baseline BC, any other lengths that are equal between two groups will not appear equal on the diagram if their lengths for BC are not also equal. The diagram is used only to depict angular, and not length, relationships.

From Fig. 8, we see that in *peruviensis* the protoconid (A) and metaconid (B) are shifted lingually relative to other topographies. The relations between cusps

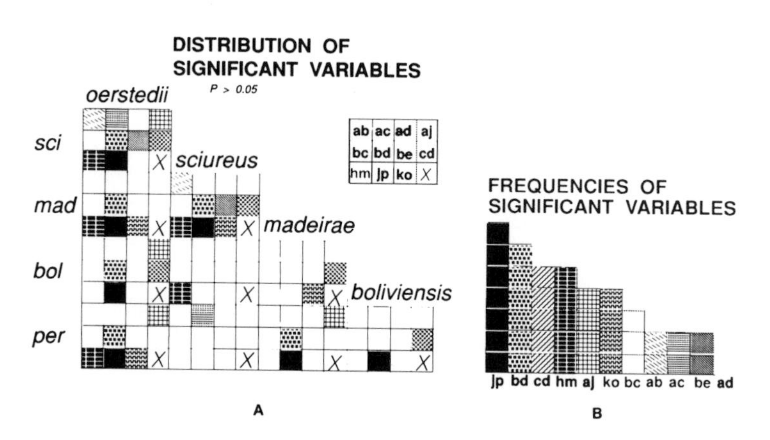

Fig. 7. (**A**) Distribution of significant variables calculated from the t-test, $p < 0.05$. The box with variables ab - ko shows the order of variables that correspond to patterns within the matrix. X fills an empty box. Samples of taxa are compared two by two. (**B**) The histogram shows the frequencies for each variable in the discrimination of all pair-wise samples.

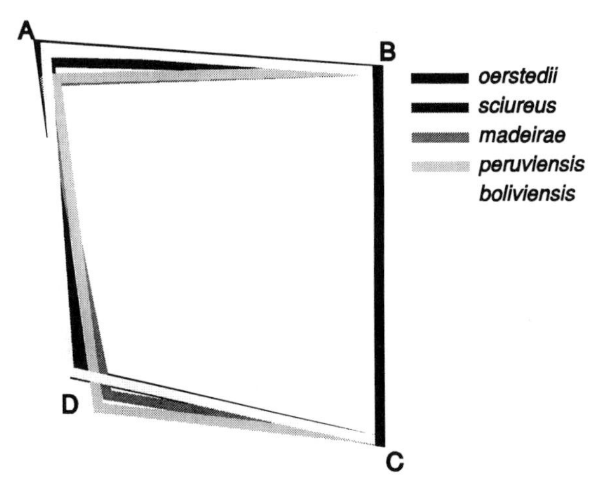

Fig. 8. Mean topographic relations of cusps among samples scaled to the longest base (BC) and superimposed.

approach a rhombus in shape, with alternating acute and oblique angles. A shift of the angles of the rhombus, without changing distances between cups, will not affect the statistical analysis of lengths. Thus, *sciureus* and *peruviensis* forms appear visually different, yet may be discriminated by only one distance, the diagonal AC.

The two forms considered sympatric in their distributions south of the Rio Amazonas, *madeirae* and *sciureus,* were similarly compared. The specimens used were collected from the area along the Rio Madeira, at Borba and Rosarinho. These samples compared more closely than any other two. This may, however, be due to sampling error, since samples were $n = 3$ for *madeirae* and $n = 5$ for *sciureus.* Figure 7A shows seven variables that discriminate these two forms. The sample used for Fig. 7A included localities where the two forms are not known to be sympatric.

This 2-D study of *Saimiri* M_1 reveals that two molar patterns are differentiable. One is found in the Central American *oerstedii* and the other in South American *Saimiri,* with differential expression among geographic groups of the latter. The Central American molar pattern is distinguished from the South American pattern by (1) less buccal flare (and less buccal cingulum) at the trigonid, resulting in a narrower tooth anteriorly without reduction of protoconid-metaconid distance; (2) a narrower talonid basin with hypoconid-entoconid distance reduced; (3) a reduced posterior tooth width; and (4) greater molar length.

The South American molar pattern is simply a broader and shorter tooth

Table 6. Molar Cusp Angles

Taxon	A	B	C	D
oerstedii	78.0	94.6	76.0	111.4
sciureus	84.8	91.2	77.0	107.0
madeirae	80.6	88.9	80.0	110.5
boliviensis	82.1	94.7	76.4	106.8
peruviensis	85.3	88.9	82.8	103.0

that has greater buccal flare and talonid cusps spread farther apart. This pattern is expressed least in *sciureus,* to an intermediate extent in *boliviensis,* and is greatest in the *peruviensis.* Whether or not molar differences and the grading of this tooth have adaptive significance, are the result of genetic drift, or represent a cline without interruption is presently a matter of speculation. Ecological differences in the habitat structure do occur (Mitchell *et al.,* 1991), and they may reflect selective variables driving regional adaptive, not niche, differences (*sensu* Bock and von Wahlert, 1965).

The dental evidence may not be appropriate for the discrimination of species in *Saimiri.* These data are limited and quite preliminary. However, they do suggest that *oerstedii* is more distinct, thus more distant, from any South American form of the genus than any of these are from each other. In addition, they do not support any multiple species hypotheses within South America.

Biochemical

Twelve of the 48 proteins surveyed showed evidence of polymorphism. Gene frequency data for *boliviensis, peruviensis,* and *sciureus,* along with the genotype of the single *oerstedii* we studied, are presented in Table 7. Unbiased average heterozygosity values (Nei, 1978) were 3.3% in *boliviensis,* 4.5% in *peruviensis,* and 5.3% in *sciureus.*

The contingency chi-square analysis suggests statistically discernible differences in gene frequencies between the sample populations we examined exist in only 5 of 12 variant proteins at the 5% error level (Table 8). Genetic identity and distance coefficients are presented in Table 9.

The main relevance of the protein data is with the genetic distances inferred from it. The genetic distances are so close between the groups that only *oerstedii* may approach a minimum distance adequate for the recognition of species [see VandeBerg (1990b) for species and subspecies comparisons]. These results, though plagued by an inadequate sample, indicate that *oerstedii* has undergone genome evolution to a greater extent than any other group. What amount of time is involved is an open question, but the suggestion by Hershkovitz (1984) that these animals were introduced by humans in Recent time is not supported.

We hope to have compensated for the small sample to some extent by examining a fairly large number of genetic loci ($n = 48$) (Nei and Roychoudhury, 1973; Nei, 1978; Gorman and Renzi, 1979; but see Archie *et al.,* 1989). However, there are serious problems other than sample size that render these findings tentative. (1) The autosomal codominant nature of the apparently polymorphic proteins we examined has yet to be confirmed; such confirmation will require study of a much larger sample of pedigreed individuals. (2) Due to the nature of the staining technique, the assessment of homology for the 48 proteins identified for study is tentative. (3) These data do not objectively identify the groups as "real."

Few biochemical markers have been found that fit preconceived notions of the number of taxa thought to exist. There are many alleles shared between two or more geographic groups. Who is to say which markers define groups and

Table 7. Gene Frequencies at 12 Variable Loci for Four Squirrel Monkey Samples

Locus and alleles	*boliviensis* N = 6	*peruviensis* N = 4	*sciureus* N = 4	*oerstedii* N = 1
1A	0.083	0.000	0.000	0.000
1B	0.917	1.000	1.000	1.000
7A	0.000	0.000	0.250	0.000
7B	1.000	1.000	0.750	1.000
8A	1.000	1.000	0.750	1.000
8B	0.000	0.000	0.250	0.000
9A	0.000	0.000	0.000	1.000
9B	1.000	1.000	1.000	0.000
10A	0.000	0.125	0.000	0.000
10B	1.000	0.875	1.000	1.000
15A	1.000	1.000	0.750	1.000
15B	0.000	0.000	0.250	0.000
17A	0.500	0.875	0.875	0.000
17B	0.500	0.125	0.125	1.000
18A	0.083	0.500	1.000	0.000
18B	0.000	0.000	0.000	1.000
18C	0.750	0.000	0.000	0.000
18D	0.000	0.125	0.000	0.000
18E	0.000	0.375	0.000	0.000
18F	0.167	0.000	0.000	0.000
19A	1.000	0.750	1.000	1.000
19B	0.000	0.125	0.000	0.000
19C	0.000	0.125	0.000	0.000
43A	0.250	0.625	0.000	0.000
43B	0.750	0.000	0.000	0.000
43C	0.000	0.375	1.000	1.000
44A	1.000	1.000	0.500	1.000
44B	0.000	0.000	0.500	0.000
46A	1.000	1.000	0.750	1.000
46B	0.000	0.000	0.250	0.000

which play an important role in isolating species? We performed the protein study by grouping specimens into taxonomic categories and then tested for significant differences between categories. Essentially, this begs the questions of species. By contrast, the dental data were treated locality by locality before any localities were combined, though circularity may still creep into the testing of groups. We do not know if, for example, all the *boliviensis* or *sciureus* samples would have clustered together if a discriminant function analysis were performed on all the individual animals comprising the groups. We have completed some work along this line and the groups mostly hold together, but there were aberrant individuals.

Table 8. Contingency Chi-squared Analysis at 12 Polymorphic Loci

Locus	Number of Alleles	Chi square	D.F.	p
1	2	1.552	3	0.67038
7	2	5.893	3	0.11694
8	2	5.893	3	0.11694
9	2	30.000	3	0.00001
10	2	2.845	3	0.41617
15	2	5.893	3	0.11694
17	2	8.625	3	0.03474
18	6	65.769	15	0.00001
19	3	5.893	6	0.43530
43	3	32.704	6	0.00001
44	2	12.692	3	0.00535
46	2	5.893	3	0.11694

Animals originating from between the Rio Tapajós and Rio Xingú, for example, are considered to belong to *madeirae* by Hershkovitz (1984) and *sciureus* by Thorington (1985). Depending on which author is followed, comparisons of animals from between the Tapajos and Xingu with other taxa of *Saimiri* will be interpreted differently.

A conservative conclusion of this and other studies (e.g., VandeBerg *et al.*, 1987, 1988, 1990a,b) is only that differences between squirrel monkeys from different geographic localities have been detected. Our electrophoretic study does not support species differences between South American groups, and it minimally supports species distinction of *oerstedii*. Results from different electrophoretic studies do not all give a consistent picture. Again, sampling error may be at work here; the geographic origin of specimens will likely effect results, even though specimens used from study to study appear to be members of the same taxonomic group. Not only would large samples of a group be desirable, so would samples over the range of the group.

Table 9. Genetic Identity and Distance Coefficients Derived from 2DE Analysis of 48 Plasma Proteins in Four Samples of *Saimiri*

	Squirrel Monkeys			
Sample	*boliviensis*	*sciureus*	*peruviensis*	*oerstedii*
boliviensis	*****	0.958	0.981	0.941
sciureus	0.213	*****	0.981	0.934
peruviensis	0.151	0.155	*****	0.941
oerstedii	0.244	0.261	0.247	*****

Above diagonal: Nei's (1978) unbiased genetic identity.
Below diagonal: Modified Rogers distance (Wright, 1978).

Conclusions

Patterns of pelage variation, geographic distribution, and survival and reproductive behaviors, together with the biochemical and dental evidence reported here, consistently indicate there are few differences between samples of South American *Saimiri* that support multiple species hypotheses. The evidence is similarly concordant in showing greater differences between Central and South American *Saimiri* than exist between South American groups. Based on this evidence, we consider the genus *Saimiri* to contain two nominal species, the Central American *S. oerstedii* Reinhardt, 1872, and the South American *S. sciureus* Linnaeus, 1758. Thorington (1985, p. 21) postulated, "it is not obvious that *oerstedii* is sufficiently different from the South American squirrel monkeys to be considered a distinct species," and "that it is no more distinct in other ways from *Saimiri sciureus sciureus* than the latter is from *Saimiri sciureus boliviensis,* and that by analogy, it, too, should be considered a subspecies of *Saimiri sciureus.*" The data we provide on blood proteins and molar patterns demonstrate greater differences than previously reported. Although this study is preliminary, we are encouraged by the consistency with which the various data direct us to our conclusions, and we regard this as a strength of the biological species concept. Not one of the preliminary studies is convincing on its own; together, however, they each corroborate the hypotheses we have accepted.

Under Paterson's (1980, 1985) species recognition concept, we must conclude that one, rather than two, species of *Saimiri* exists. We were unable to find any significant differences in the mate-recognition system (as defined by Patterson) of *S. oerstedii* and *S. sciureus.* Yet appreciable genomic, behavioral, pelage, and dental differences have accrued with *extensive* isolation of *S. oerstedii.* The apparently gradual buildup of differences in multiple systems, without a "niche shift," appears to be the effect of random genetic change or assortative mating (e.g., see Froehlich *et al.*, 1991, on species of *Ateles*). Some threshold must finally be reached where these differences have an irreversible effect on the outcome of mating. Such studies are not available for *Saimiri.*

Summary

Two species taxa are recognized, the Central American *S. oerstedii* and the South American *S. sciureus.* The previously named *S. madeirae* and *S. boliviensis* are regarded as geographic forms of *S. sciureus.*

Two molar patterns distinguish South and Central American *Saimiri.* The molar pattern shared by South American squirrel monkeys appears to grade clinally ("*sciureus*" → "*boliviensis*" → "*peruviensis*"). Samples of sympatrically distributed "*madeirae*" and "*sciureus*" are compared and found closer morphologically than any samples compared between parapatric groups.

Genetic distances computed from protein frequency data show (1) slight differences between South American geographic groups and (2) a greater distance between Central American monkeys and any of those groups. The degree to which *S. oerstedii* may be genetically distinct is minimal for species recognition. Karyology is so far minimally informative at the species level.

Patterns of pelage variation and geographic distribution indicate hybridization where groups meet. There is evidence to suggest intergradation between "*madeirae*" and "*sciureus*," "*vanzolinii*" and "*madeirae*" (this report), and "*boliviensis*" and "*sciureus*" (Thorington, 1985; DaSilva *et al.*, 1992). Populations of "*boliviensis*" geographically near to "*vanzolinii*" share coat patterns with the former.

Behavioral evidence does not identify any premating behaviors that could function to isolate any one group of *Saimiri* from any other. All squirrel monkeys share the same feeding strategy, diet, and breeding systems, and differ in social organization only in conjunction with changes in forest structure and the distribution of food resources.

ACKNOWLEDGMENTS

We wish to thank Drs. Guy Musser, Stephen Ferrari, Fernando Novaes, Alfredo Langutth, and Paulo E. Vanzolini for their help and permission to use museum collections; Dr. Lawrence Williams, Department of Comparative Medicine, University of South Alabama College of Medicine, Dr. Marion Ratterree of the Delta Regional Primate Research Center, and Dr. Connie Sweet of the Santa Ana Zoo for providing blood samples. The rabbit psoas muscle molecular weight markers were generously provided by the Molecular Anatomy Program, Argonne National Laboratory. We thank Dr. Harold H. Harrison and Ms. Kathy Miller, MT (ASCP) for their kind assistance and for use of the facilities of the Molecular Pathology Laboratory, University of Chicago Medical Center. Special thanks to Wilson McCord and Maribel Miranda for their artistic assistance, and Lodewijk Werre, Doug Broadfield, and Gavin Naylor for their help. This study was supported in part by NIH grant #RR01254, NIH grant #RR0016427, the L.S.B. Leakey Foundation, the Wenner-Gren Foundation for Anthropological Research, and PSC CUNY doctoral grant #661222.

References

Anderson, N. G., and Anderson, N. L. 1978. Analytical techniques for cell fractions XXI. Two-dimensional analysis of serum and tissue proteins: multiple isoelectric focusing. *Anal. Biochem.* **85:**331–340.

Anderson, N. L., and Anderson, N. G. 1978. Analytical techniques for cell fractions XXII. Two-dimensional analysis of serum and tissue proteins: multiple gradient-slab gel electrophoresis. *Anal. Biochem.* **85:**341–354.

Archie, J. W., Simon, C., and Martin, A. 1989. Small sample size does decrease the stability of dendrograms calculated from allozyme-frequency data. *Evolution* **43:**678–683.

Assis, M. F. L. and Barros, R. 1987. Karyotype pattern of Saimiri *ustus. Int. J. Primatol.* **8:**552 (abstr.).

Ayres, J. M. 1985. On a new species of squirrel monkey, genus *Saimiri,* from Brazilian Amozonia (Primates, Cebidae), Papeis Avulsos Zool., S. Paulo **36:**147–164.

Barton, N. 1979. Gene flow past a cline. *Heredity* **43:**333–339.

Bender, M. A. and Mettler, L. E. 1958. Chromosome studies of primates. *Science* **128:**186–190.

Biben, M., Symmes, D., and Masataka, N. 1986. Temporal and structural analysis of vocal exchanges in squirrel monkeys (Saimiri *sciureus*). *Behaviour* **98:**259–273.

Bock, W. J. 1986. Species concepts, speciation, and macroevolution, in: K. Iwatsuki, P. H. Raven, and W. J. Bock (eds.). *Modern Aspects of Species,* pp. 31–57. University of Tokyo Press, Tokyo.

Bock, W. J. and von Wahlert, G. 1965. Adaptation and the form-function complex. *Evolution* **19:**269–299.

Boinski, S. 1987a. Habitat use by Costa Rican squirrel monkeys (*Saimiri oerstedi*). *Folia Primatol.* **49:**151–167.

Boinski, S. 1987b. Mating patterns in squirrel monkeys (*Saimiri oerstedi*): Implications for seasonal sexual dimorphism. *Behav. Ecol. Sociobiol.* **21:**13–21.

Boinski, S. 1988. Sex differences in the foraging behavior of squirrel monkeys in a seasonal habitat. *Behav. Ecol. Sociol.* **23:**177–167.

Boinski, S. 1991. The coordination of spatial position: a field study of the vocal behavior of adult female squirrel monkeys. *Animal Behav.* **41:**89–102.

Boinski, S. 1992. Olfactory communication among squirrel monkeys: a field study. *Folia Primatol.* (in press).

Boinski, S., and Newman, J. D. 1988. Preliminary observations on squirrel monkey (*Saimiri oerstedi*) vocalizations in Costa Rica. *Am. J. Primatol.* **14:**329–343.

Boinski, S., and Mitchell, C. L. 1992. Ecological and social factors affecting vocal behavior of adult female squirrel monkeys. *Ethology*

Cabrera, A. 1958. Catalogo de los Mamiferos de America del Sur, Rev. Mus. Argent. Cienc. Nat., Bernardino Rivadavia, Cienc. Zool. **4:**i–iv, 1–307.

Carson, H. L. 1987. Tracing ancestry with chromosomal sequences. *TREE* **2:**203–207.

Coe, C. L., Smith, E. R., and Levine, S. 1985. The endocrine system of the squirrel monkey, in: L. A. Rosenblum and C. L. Coe (eds.), *Handbook of Squirrel Monkey Research,* pp. 191–218. Plenum Press, New York.

DaSilva, B. T. F., Sampaio, M. I. C., Schneider, H., Schneider, M. P. C., Villavicencio, H., Montoya, H., Encarnacion, F., and Salzano, F. M. 1987a. Genetic variability in two species of *Saimiri* from Peruvian Amazonia. *Int. J. Primatol.* **8**(5)**:**528 (abstr.).

DaSilva, B. T. F., Sampaio, M. I. C., Schneider, H., Schneider, M. P. C., Villavicencio, H., Montoya, H., Encarnacion, F., and Salzano, F. M. 1987b. Preliminary analysis of genetic distance between squirrel monkeys. *Int. J. Primatol.* **8**(5)**:**528 (abstr.).

DaSilva, B. T. F., Sampaio, M. I. C., Schneider, H., Schneider, M. P. C., Montora, E., Encarnacion, F., and Salzano, F. M. 1992. Natural hybridization between *Saimiri* taxa in the Peruvian Amazonia. *Primates* **33**(1)**:**107–113.

Daufeldt, J. A. and Harrison, H. H. 1984. Quality control and technical outcome of ISO-DALT two-dimensional electrophoresis in a clinical laboratory setting. *Clin. Chem.* **30:**1972–1980.

Dukelow, W. R. 1982. The squirrel monkey. in: J. Remfrey (ed.), *Universities Federation for Animal Welfare Handbook,* pp. 151–176. MTP Press, London.

Dumond, F. V. and Hutchinson, T. C. 1967. Squirrel monkey reproduction: The "fatted" male phenomenon and seasonal spermatogenesis. *Science* **158:**1067–1070.

Edwards, J. J., Tollaksen, S. L., and Anderson, N. G. 1982. Proteins of human urine. III. Identification and two-dimensional electrophoretic map positions of some major urinary proteins. *Clin. Chem.* **28:**941–948.

Elliot, D. G. 1913. *A Review of the Primates,* Vol. 1, American Museum of Natural History, New York.

Ferrari, S., and Lopes, C. 1992. New data on the distribution of primates in the region of the confluence of the Jiparaná and Madeira rivers in Amazonas and Rondônia, Brazil. *Goeldiana Zool.* **11:**1–12.

Fink, W. L. 1990. Data acquisition for morphometric analysis in systematic biology, in: F. J. Rohlf and F. L. Bookstein (eds.), *Proceedings of the Michigan Morphometrics Workshop.* Special publication number 2. The University of Michigan Museum of Zoology, Ann Arbor, MI.

Froehlich, J. W., Supriatna, J., and Froehlich, P. H. 1991. Morphometric analyses of *Ateles*: systematic and biogeographic implications. *Am. J. Primatol.* **25:**1–22.

Galliari, C. A. and Colillas, O. J. 1985. Sequences and timing of dental eruption in Bolivian captive-born squirrel monkeys (*Saimiri sciureus*). *Am. J. Primatol.* **8:**195–204.

Geoffroy Saint-Hilaire, I. 1844. Description des mammiferes nouveaux ou imparfaitement connus de la collection du Museum D'Histoire Naturelle, et remarques sur la classification et les caracteres des mammiferes. *Arch. Mus.,* Paris: 5–42.

Gorman, G. C., and Renzi, J. R., Jr. 1979. Genetic distance in heterozygosity estimates in electrophoretic studies: Effects of sample size. *Copeia* **2:**242–249.

Guevara, J., Johnston, D. A., Ramagli, L. S., Martin, B. A., Capetillo, S., and Rodriquez, L. V. 1982. Quantitative aspects of silver deposition in proteins resolved in complex polyacrlyamide gels. *Electrophoresis* **3:**197–205.

Hamilton, A. E., Tattersall, I., Sussman, R. W., and Buettner-Janusch, J. 1980. Chromosomes of Lemuriformes, VI. Comparative karyology of *Lemur fulvus:* a G-banded karyotype of *Lemur fulvus mayottensis* (Schlegel, 1966). *Int. J. Primatol.* **1:**81–93.

Harrison, H. H. 1984. Improved record keeping and photography of silver-stained two-dimensional electrophoretograms by way of "XRD images." *Clin. Chem.* **30:**1981–1984.

Harrison, H. H., Miller, K. L., Dickinson, C., and Daufeldt, J. A. 1992. Quality assurance and reproducibility of high resolution two-dimensional electrophoresis and silver-staining in polyacrylamide gels. *Am. J. Clin. Pathol.* **97:**66–74.

Hartman, S. E. 1986. A Stereophotogrammetric Analysis of Occlusal Morphology of Extant Hominoid molars. PhD Dissertation, State University of New York, Stony Brook.

Hartman, S. E. 1989. Stereophotogrammetric analysis of occlusal morphology of extant hominoid molars: phenetics and function. *Am. J. Phys. Anthropol.* **80:**145–166.

Hershkovitz, P. 1983. Two new species of night monkeys, genus *Aotus* (Cebidae, Platyrrhini): a preliminary report on *Aotus* taxonomy. *Am J. Primatol.* **4:**209–243.

Hershkovitz, P. 1984. Taxonomy of squirrel monkeys genus *Saimiri* (Cebidae, Platyrrhini): A preliminary report with description of a hitherto unnamed form. *Am. J. Primatol.* **6:**257–312.

Hershkovitz, P. 1985. A preliminary taxonomic review of the South American bearded saki monkeys genus *Chiropotes* (Cebidae, Platyrrhini), with the description of a new species. *Fieldiana* **27**(1363)**:**1–46.

Hickman, B. J., Anderson, N. L., Willard, K. E., and Anderson, N. G. 1980. Internal charge standardization for two-dimensional electrophoresis, in: B. J. Radola (ed.), *Electrophoresis '79.* Walter de Gruyter & Co, Berlin.

Hill, W. C. O. 1960. *Primates, Comparative Anatomy and Taxonomy. IV. Cebidae.* Part A. Interscience, New York.

Janson, C. H. and Boinski, S. 1992. Morphological and behavioral adaptations for foraging in generalist primates: the case of the cebines. *Am. J. Phys. Anthropol.* **88:**483–498.

Jones, T. C., Thorington, R. W., Hu, M. M., Adams, E., and Cooper, R. W. 1973. Karyotypes of squirrel monkeys (*Saimiri sciureus*) from different geographic regions. *Am. J. Phys. Anthropol.* **38:**269–278.

Jones, T. C. and Ma, N. S. F. 1975. Cytogenetics of the squirrel monkey (*Saimiri sciureus*). *Fed. Proc.* **34:**1646–1650.

Kaplan, J. N. 1977. Breeding and rearing squirrel monkeys (*Saimiri sciureus*) in captivity. *Animal Sci.* **27:**557–567.

Klein, L. L. and Klein, D. J. 1975. Social and ecological contrasts between four taxa of neotropical primates, in: R. Tuttle (ed.), *Socioecology and Psychology of Primates,* pp. 59–88. The Hague, Mouton.

Kanazawa, E., Sekikawa, M., and Ozaki, T. 1983a. Three-dimensional measurement of the occlusal surface of upper first molars in a modern Japanese population. *Acta Anat.* **116:**90–96.

Kanazawa, E., Sekikawa, M., and Ozaki, T. 1983b. Three-dimensional measurement of the occlusal surface of upper molars in a Dutch population. *J. Dent. Res.* **63:**1298–1301.

Kanazawa, E., Morris, D. H., Sekikawa, M., and Ozaki, T. 1988. Three-dimensional measurement of the occlusal surface of the upper first molar in South African samples. *J. Anthrop. Soc. Nippon* **96**(4)**:**405–415.

Lacy, R. C., Foster, M. L., and the Primate Department Staff. 1988. Determination of pedigrees and taxa of primates by protein electrophoresis. *Int. Zoo Yrbk.* **27:**159–168.

Lau, Y.-F. and Arrighi, F. E. 1976. Studies of the squirrel monkey, *Saimiri sciureus,* genome. I. Cytological characterizations of chromosomal heterozygosity. *Cytogene. Cell Genetics* **17:**51–60.

Linneaus, C. 1758. *Systema Naturae,* 10th ed.

Long, J. O. and Cooper, R. W. 1968. Physical growth and dental eruption in captive-bred squirrel monkeys *Saimiri sciureus* (Leticia, Colombia), in: L. A. Rosenblum and R. W. Cooper (eds.), pp. 193–205. Academic Press, New York.

Lonnberg, E. 1940. Notes on some members of the genus *Saimiri, Ark. Zool.* **32A**(21)**:**1–18.

Ma, N. S. F., Jones, T. C., Thorington, R. W., and Cooper, R. W. 1974. Chromosome banding patterns in squirrel monkeys (*Saimiri sciureus*). *J. Med. Primatol.* **3:**120–137.

Maclean, P. D. 1964. Mirror display in the squirrel monkey. *Science* **146:**950–952.

MacLeod, N. 1990. Digital images and automated image analysis systems, in: F. J. Rohlf and F. L. Bookstein (eds.), *Proceedings of the Michigan Morphometrics Workshop.* Special publication number 2. The University of Michigan Museum of Zoology, Ann Arbor, MI.

Martau, P. A., Caine, N. G., and Candland, D. K. 1985. Reliability of the emotions profile index, primate form with *Papio hamadryas, Macaca fuscata,* and two *Saimiri* species. *Primates* **26:**501–505.

Martins, E. S., Ayres, J. M., and Ribeiro do Valle, MB. 1986. On the status of *Ateles belzebuth marginatus* with notes on other primates of the Iriri river basin. *Prim. Conserv.* **9:**87–91.

Masataka, N. and Biben, D. 1987. Temporal rules regulating vocal exchanges of squirrel monkeys (*Saimiri sciureus*). *Behaviour* **101:**311–319.

Mayr, E. 1942. *Systematics and the Origin of Species.* Columbia University Press, New York.

Mayr, E. 1963. *Animal Species and Evolution.* Harvard University Press, Cambridge.

Mayr, E. 1982. *The Growth of Biological Thought.* The Belknap Press of Harvard University Press, Cambridge.

Mendoza, S. P., and Mason, W. A. 1989. Behavioral and endocrine consequences of heterosexual pair formation in squirrel monkeys. *Physiol. Behav.* **46:**597–603.

Mendoza, S. P., Lowe, E. L., and Levine, S. 1978. Social organization and social behavior in two subspecies of squirrel monkeys (*Saimiri sciureus*). *Folia Primatol.* **30:**126–144.

Mendoza, S. P., Lyon, D. M., and Saltzman, W. 1991. Sociopsychology of squirrel monkeys. *Am. J. Primatol.* **23:**37–54.

Mitchell, C. L. 1990. The Ecological Basis for Female Social Dominance: A Behavioral Study of the Squirrel Monkey (*Saimiri sciureus*) in the Wild. Ph.D. thesis, Princeton University.

Mitchell, C. L., Boinski, S., and van Schaik, C. P. 1991. Competitive regimes and female bonding in two species of squirrel monkey (*Saimiri oerstedi* and *S. sciureus*). *Behav. Ecol. Sociobiol.* **28:**55–60.

Moore, C. M., Harris, C. P., and Abee, C. R. 1990. Distribution of chromosomal polymorphisms in three subspecies of squirrel monkeys (genus *Saimiri*). *Cytogen. Cell Gen.* **53:**118–122.

Nei, M. 1978. Estimation of average heterozygosity and genetic distance from a small number of individuals. *Genetics* **89:**583–590.

Nei, M. and Roychoudhury, A. K. 1973. Sampling variances of heterozygosity and genetic distance. *Genetics* **76:**379–390.

Newman, J. D. and Symmes, D. 1982. Inheritance and experience in the acquisition of primate vocal behavior, in: C. T. Snowdon, C. H. Brown, and M. R. Peterson (eds.), *Primate Communication,* pp. 259–278. Cambridge University Press, Cambridge.

Newman, J. D., Smith, J. H., and Talmadge-Riggs, G. 1983. Structural variability in primate vocalizations and its functional significance: an analysis of squirrel monkey chuck calls. *Folia Primatol.* **40:**114–124.

O'Farrell, P. H. 1975. High-resolution two-dimensional electrophoresis of proteins. *J. Biol. Chem.* **205:**4007–4021.

O'Higgins, P. 1989. Developments in cranial morphometrics. *Folia Primatol.* **53:**101–124.

Orlosky, F. 1973. Comparative Dental Morphology of Extant and Extinct Cebidae. University Microfilms, Ann Arbor, MI.

Paterson, H. E. H., 1980. A comment on 'mate recognition systems.' *Evolution* **34:**330–331.

Paterson, H. E. H. 1985. The recognition concept of species, in: E. S. Vrba (ed.), *Species and Speciation,* pp. 21–29. Transvaal Museum Monograph No. 4. Transvaal Museum, Pretoria.

Ramaekers, P. 1975. Using polar coordinates to measure variability in samples of Phenacolemur: a method of approach, in: F. S. Szalay (ed.), *Approaches to Primate Paleobiology. Contrib. Primatol.* **5:**106–135 (Karger, Basel).

Reinhardt, J. 1872. Et Bidrag til Kunskab om Aberne i Mexico og Centralamerika. Videnskapabelige Meddeleser, Naturhistorisk Forening Kobenhavn, series 3(4–6):150–158.

Richmond, S. 1987. Recording the dental cast in three dimensions. *Am. J. Orthod.* **92:**199–206.

Rohlf, F. J. 1988. *NTSYS-pc: Numerical Taxonomy and Multivariate Analysis System* (version 1.40). Exeter Publishing Company, Setauket, NY.

Rohlf, F. J. 1990. An overview of image processing and analysis techniques for morphometrics, in: F. J. Rohlf and F. L. Bookstein (eds.), *Proceedings of the Michigan Morphometrics Workshop.* Special publication number 2. The University of Michigan Museum of Zoology, Ann Arbor, MI.

Rosenberg, A. L., Hartwig, W. C., Takai, M., Setoguchi, T., and Shigehara, N. 1991. Dental variability in *Saimiri* and the taxonomic status of *Neosaimiri fieldsi,* an early squirrel monkey from La Venta, Colombia. *Int. J. Primatol.* **12**(3):1–11.

Rumpler, Y. 1975. The significance of chromosomal studies in the systematics of the Malagasy lemurs, in: I. Tattersall and R. W. Sussman (eds.), *Lemur Biology,* pp. 25–40. Plenum Press, New York.

Sites, J. W., Jr., Porter, C. A., and Thompson, P. 1987. Genetic structure and chromosomal evolution in the *Sceloporus grammicus* complex. *Natl. Geogr. Res.* **3**(3):343–362.

Sneath, P. H. A. and Sokal R. R. 1973. *Numerical Taxonomy.* WH Freeman & Company, San Francisco.

Snowdon, C. T., Coe, C. L., and Hodon, A. 1985. Population recognition of infant isolation peeps in the squirrel monkey. *Anim. Behav.* **33:**1145–1151.

Spirito, F., Rossi, C., and Rizzoni, M. 1983. Reduction of gene flow due to partial sterility of heterozygotes for a chromosome mutation I. Studies on a "neutral" gene not linked to the chromosome mutation in a two population model. *Evolution* **37:**785–797.

Swofford, D. L., and Selander, R. B. 1981. BIOSYS: A fortran program for the comprehensive analysis of electrophoretic data in population genetics and systematics. *J. Hered.* **72:**281–283.

Swofford, D. L., and Selander, R. B. 1989. BIOSYS-1: A computer program for analysis of allelic variation in population genetics and biochemical systematics. *Illinois Natural History Survey,* Champaign, IL.

Szalay, F. S. and Bock, W. J. 1990. Evolutionary theory and systematics: relationships between process and patterns. *Z. Zool. Syst. Evolut.-Forsch.* **29:**1–39.

Terborgh, J. 1983. *Five New World Primates: A Study in Comparative Ecology.* Princeton University Press, Princeton, NJ.

Thomas, O. 1908. Four new Amazonian monkeys. *Ann. Mag. Nat. Hist.* Series 8, **2:**88–91.

Thorington, R. W. 1985. The taxonomy and distribution of squirrel monkeys (*Saimiri*), in: L. A. Rosenblum and C. L. Coe (eds.), *Handbook of Squirrel Monkey Research,* pp. 1–33.

Tollaksen, S. L., Anderson, N. L., and Anderson, N. G. 1984. *Operation of the ISO-DALT system, 7th ed.,* U.S. Department of Energy Publication ANL-BIM-48-1, Argonne, IL.

VandeBerg, J. L., Cheng, M.-L., Moore, C. M., and Abee, C. R. 1987. Genetics of squirrel monkeys (genus *Saimiri*): Implications for taxonomy and research (abstr.). *Int. J. Primatol.* **8:**423.

VandeBerg, M. L., Cheng, M.-L., Moore, C. M., and Abee, C. R. 1988. Genetic distances among squirrel monkey populations and taxonomic implications (abstr). *Isozyme Bull.* **21:**181.

VandeBerg, J. L., Aivaliotis, M. J., Williams, L. E., and Abee, C. R. 1990a. Biochemical genetic markers of squirrel monkeys and their use for pedigree validation. *Biochem. Gen.* **28:**41–55.

VandeBerg, J. L., Williams-Blangero, S., Moore, C. M., Cheng, M.-L., and Abee, C. R. 1990b. Genetic relationships among three squirrel monkey types: Implications for taxonomy, biomedical research, and captive breeding. *Am. J. Primatol.* **22:**101–111.

von Pusch, A. 1942. Die Arten der Gattung *Cebus. Z. Saugetierkhd.* **16**(vol. for 1942)**:**183–237.

Winter, P. 1969. Dialects in squirrel monkeys: vocalizations of the roman arch type. *Folia Primatol.* **10:**216–229.

Winter, P., Handley, P., Ploog, D., and Schott, D. 1973. Ontogeny of squirrel monkey calls under normal conditions and under acoustic isolation. *Behavior* **47:**230–239.

Wolf, R. H., Harrison, R. M., and Martin, T. W. 1975. A review of reproductive patterns in New World monkeys. *Lab. Animal. Sci.* **25:**814–821.

Workman, P. L. and Niswander, J. D. 1970. Population studies on southwestern Indian tribes. II. Local genetic differentiation in the Papago. *Am. J. Hum. Gen.* **22:**24–49.

Wright, S. 1978. *Evolution and the Genetics of Populations, Vol. 4. Variability Within and Among Natural Populations.* University of Chicago Press, Chicago.

Yonenaga-Yassuda, Y. and Chu, T. H. 1985. Chromosome banding patterns of *Saimiri vanzolinii* Ayres, 1985 (Primates, Cebidae). *Papéis Avulsos Zool.* **36**(15)**:**165–168.

Measures of Dental Variation as Indicators of Multiple Taxa in Samples of Sympatric *Cercopithecus* Species

9

DANA A. COPE

Introduction

Recognition of species in the fossil record is a critical issue for students of primate evolution. Since dental remains provide the largest samples for assessing intra- and interspecific variation, dental variation in fossil mammals is frequently compared with data for recent related species when investigating systematic diversity (Gingerich, 1974, 1979; Gingerich and Shoeninger, 1979; Kay, 1982a,b; Kay and Simons, 1983; Kelley, 1986; Kimbel and White, 1988; Martin and Andrews, 1984; Simpson, 1941a, and many others). The coefficient of variation (CV) has been the most frequently used statistic in a majority of these studies.

The applicability of this measure of relative variation to fossil samples is strengthened by observations that indicate a consistent pattern, whereby first and second molars in primates and other mammals tend to exhibit a low degree of size variation (within a definite range) compared to other teeth (Gingerich,

DANA A. COPE • Department of Sociology and Anthropology, College of Charleston, Charleston, South Carolina 29424.

Species, Species Concepts, and Primate Evolution, edited by William H. Kimbel and Lawrence B. Martin. Plenum Press, New York, 1993.

1974, 1979; Gingerich and Shoeninger, 1979; Gingerich and Winkler, 1979; Simpson *et al.*, 1960). However, the significance of comparisons of variation in living species with that in fossil samples thought to represent distinct taxa remains unclear. For example, if dental dimensions in a fossil sample are within the range of variation in recent species, does this constitute "proof" that only one taxon is present? Conversely, do estimates of variation beyond that documented for living species indicate that a sample consists of more than one species (Kelley, this volume)? If the variation in a fossil sample exceeds single-species' maxima derived from extant reference data, how much greater must the variation be to indicate multiple taxa?

It has recently been argued that the CV is of dubious utility because pooled samples of two different species of the same genus may have a CV of similar magnitude to that of a single species (Martin, 1983; Plavcan, 1989) and that confidence limits at small sample sizes do not permit a distinction between single- and pooled-species CVs at smaller sample sizes (Plavcan, 1989), even if the pooled-species CV is larger than those observed in extant species. It has also been suggested that for fossil samples biased by size and sexual composition, the range expressed as a percentage of the mean and the maximum/minimum index of a sample might be a useful alternatives (Martin and Andrews, 1984). Although the use of range-based statistics in this way has been severely criticized (Simpson, 1941b, 1947), Martin and Andrews (1984) point out that as few as two specimens may be used to test a single-species hypothesis. Martin and Andrews (1984) discount elevated CVs for M_1 (n = 5) and P4 (n = 7) in the Ravin de la Pluie hominoid sample (Kay and Simons, 1983) as evidence of more than one species. The large values are attributed to the small sample size. Sokal and Braumann's (1980) study of the sampling distribution of the CV indicates that (on average) it tends to be underestimated at small sample sizes. However, their results also show that the standard error increases, suggesting that the largest possible estimates of a population CV will come from smaller samples. This factor must be considered when faced with a fossil sample in which there are only one or two variables exhibiting unusually high variation.

Given these apparent problems with the CV, some systematists have shown preferences for other measures of relative variation or discounted the assumption that fossil species should exhibit no more variation than living ones. For example, Martin and Andrews (1984) cite range statistics (based on larger reference sample sizes than those of the fossil sample) as evidence that the variation in the Ravin de la Pluie hominoid sample is within the normal single-species range for hominoids. Kelley (1986) has pooled some specimens formerly assigned to *Proconsul africanus* into *Proconsul nyanzae* on the basis of anterior dental morphology and variation, but argued that the resulting elevated CVs for posterior dental dimensions reflect a greater degree of sexual dimorphism than that present in living catarrhines. Although CVs are not reported, a similar argument is made (Kelley, this volume) with regard to variation in the Lufeng hominoid material. In another interesting case, Kimbel and White (1988) acknowledge that the presence of a single species should be viewed as a null hypothesis and conclude, on the basis of a number of factors, that this hypothesis cannot be rejected for the material generally attributed to *A. afarensis*. However, the CVs for this sample are high [even when calculated without the correction factor suggested by Sokal and Rohlf (1981)]. This is especially true considering that some of the reference data to which they are compared seem to have unusually high values for single-species samples.

Clearly, there is a need for experimentation to determine what conclusions can or cannot be drawn from comparisons of dental variation in fossil and recent species. The purpose of this study is to clarify what is to be gained from such comparisons and to assess the utility of three statistics of relative variation (the coefficient of variation, the range as a percentage of the mean, and the maximum/minimum index) in detecting the presence of more than one species in morphologically homogeneous dental samples. This is best achieved by an examination of recent sympatric primate species, where the size, as well as the sexual and taxonomic composition of samples, can be controlled. Furthermore, the pooled-species samples analyzed here consist of different species collected in the same geographic areas, so it is possible to determine the actual effect of combining sympatric populations of different species in a single sample. Finally, by determining the effect of sample size on statistics of relative variation, an assessment of their utility with respect to small samples, such as those characteristic of fossil assemblages, may be achieved.

Species of the genus *Cercopithecus* are ideal for these analyses because large samples of sympatric populations of different species are available in museum collections. Although it is possible to show significant differences in a number of mean tooth dimensions of sympatric species of *Cercopithecus,* the magnitudes of such differences are not great and the ranges of variation overlap considerably (Cope, 1989). As is the case for most mammalian genera, sympatric species of *Cercopithecus* differ mostly in size (Gingerich, 1979; Greenfield, 1979). Thus, this study simulates the effect of multiple, morphologically homogenous species in a single fossil assemblage and compares intra- and interspecific variation in a context similar to that which may be encountered by a paleontologist. In addition, single- and pooled-species samples available from Cameroon are sufficiently large that a series of randomly obtained, hypothetical "fossil" samples of various sizes can be generated. This methodology allows a comparison of the ability of various measures of relative variation to distinguish between single- and pooled-species samples at a given sample size. It also allows estimation of the probability that a measure of variation of a given magnitude and at a given sample size could have been obtained from a sample of only one species. Such an estimate is important, since it potentially provides an opportunity to determine an objective "critical value" for rejection of the null hypothesis that the sample is composed of only one species. As indicated above, the absence of such a criterion has been a problem when the variation in a fossil sample is simply compared to that observed in extant reference data.

Materials and Methods

Samples and Measurements

A series of 34 dental measurements, primarily buccolingual breadths or mesiodistal lengths, were obtained for 13 *Cercopithecus* species or subspecies samples. (The various museum collections are listed in the acknowledgements). Measurements were taken to the nearest tenth of a millimeter. Talon(id) and trigon(id) breadths were taken on all molars. Mesiodistal lengths of incisors were not taken, as rapid wear in *Cercopithecus* has a significant influence on this dimension. Canine lengths were taken as maximum length and canine breadths were taken perpendicular to the maximum length (see Cope, 1989, for a complete description of measurements). Mesiodistal measurements were designed in such a way as to allow moderately worn teeth to be included. If wear had significantly influenced a dimension, it was not included in the sample. Measurements were taken on the left side of the dentition. When teeth from the left side could not be used, measurements of the right antimere were substituted if available. A total of 46 specimens were measured twice. Estimates of average error range from approximately 1% to 3%. Therefore, measurement error apparently does not significantly influence the results.

Table 1 shows the samples used in the analyses, and Fig. 1 illustrates the geographic areas from which the specimens were collected. Only species and subspecies samples from geographically restricted populations were used, to minimize the effect of intraspecific geographic variation. In some cases, more than one population of the same species was included in the analysis, since these provide independent estimates of maximum single-species variation in geo-

Table 1. *Cercopithecus* Dental Samples and Sample Identification Codes Used in Subsequent Tables

Locality and samples	Code	Females	Males	Total
Kisangani area				
C. wolfi denti	wd	11	11	22
C. ascanius schmidti	as	16	22	38
C. mitis stuhlmani	ms1	13	12	25
C. lhoesti lhoesti	ll	10	11	21
C. neglectus[a]		5	5	10
Pooled-species total		55	61	116
Bolobo area				
C. wolfi wolfi	ww	10	11	21
C. ascanius whitesidei[b]	aw	14	14	28
C. neglectus	Cn	12	19	31
Pooled-species total		36	44	80
Southern Cameroon				
C. pogonias grayi	pg	28	37	65
C. cephus cephus	cc	33	46	79
C. nictitans nictitans	nn	52	56	108
Luebo area				
C. ascanius katangae	ak	24	20	44
Kapatagat/Mt. Elgon area				
C. mitis stuhlmani	ms2	17	15	32
Mt. Kenya area				
C. mitis kolbi	mk	10	13	23
Total sample		255	292	547

[a]Not used in the determination of single-species maximum values because of small sample size.

[b]It has been suggested that specimens from this area might be placed in a new unnamed subspecies (Gautier-Hion *et al.*, 1988, Fig. 5)

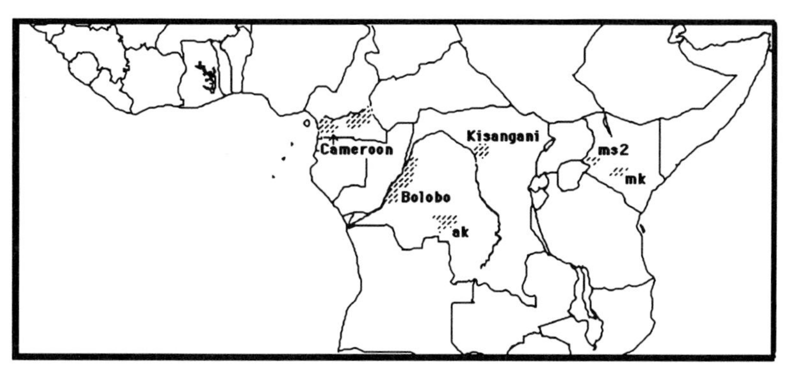

Fig. 1. Map of the central portion of Africa, showing locations of samples. Sympatric samples of multiple species are identified by area and single-species samples by code (see Table 1).

graphically restricted samples. Thus, there are 13 single-species samples used as reference data in this study, but only eight *different* species are represented. The sex of each specimen was determined according to museum records and examination of the canines and P_3, which in all species of *Cercopithecus* exhibit clearly distinct male and female morphs (Cope, 1989).

Analyses of the Original Data

All 13 species or subspecies samples were compared to determine maximum single-species values for the CV, the range as a percentage of the mean (R%), and the maximum/minimum index ($I_{max/min}$). The maximum estimates of relative variation could then be compared to those derived from pooled-species samples of sympatric taxa from south of the Sanaga River in Cameroon (three species), the Bolobo area of Western Zaire (three species), and the area just east of Kisangani (north of the Zaire River) in Eastern Zaire (five species). Note that in the latter case, a *C. neglectus* sample of 10 individuals (five males and five females) included in the pooled-taxa Kisangani sample was not used to establish maximum single-species reference values because of small sample size, although for most variables its inclusion would not have altered the resulting maximum values. Because single-sex samples have been reported sometimes to have a higher CV than combined-sex samples (Martin, 1983; Martin and Andrews, 1984), both single- and combined-sex values of the CV, R%, and $I_{max/min}$ were calculated for each species sample. The maximum single-species values in some cases were those of either all-male or all-female samples. Ninety-five percent confidence limits for each sample CV (calculated *without* the correction factor for small samples, according to Sokal and Rohlf, 1981) were also compared. Single-sex values for $I_{max/min}$ never exceeded those of mixed-sex samples. When single-sex $I_{max/min}$ reference maxima equaled those of combined-sex samples, the tabled values shown are those of single-sex samples, since it is important to note the smallest sample capable of providing a maximum estimate of single-species variation.

It was also of interest to understand the degree to which differences in the mean dental dimensions of sympatric *Cercopithecus* species influenced the magnitude of pooled-species CVs (see Plavcan, 1989, this volume). Therefore, single-classification analysis of variance and the Student–Newman–Kuels *a posteriori* test (Sokal and Rohlf, 1981) for significant differences in mean dental dimensions between species were applied to each geographic sample of sympatric species. This allowed for a comparison of pooled-species CV values with the pattern of differences in means between species that might naturally occur in pooled assemblages.

Simulation of Variation in Relation to Sample Size

The applicability of estimates of variation to fossil samples was addressed by randomly generating hypothetical "fossil" distributions from the large Cameroon sample of 252 individuals representing three species. Twenty single-species samples were drawn from a large population of both sexes of *C. nictitans* (n = 108) at samples sizes of 50, 45, 40, 35, . . . down to a sample size of five individuals. All individuals had an equal probability of being sampled during each simulation at a given sample size. A similar methodology was followed in deriving samples of the same sizes from the total Cameroon sample, which included *C. pogonias* and *C. cephus,* in addition to *C. nictitans.* In this case, the sampling process followed two steps. First, 28 individuals were sampled from each sex of each species, since the sample of 28 females of *C. pogonias* represented the smallest single-sex sample from Cameroon. Then a pooled-species sample of desired size was randomly derived from these individuals. This ensured an experimental design in which there was equal probability of sampling any species or sex in drawing each pooled-species simulation sample. Only results for M_2 are reported for reasons of space. Larger samples of second molars were available than for first molars, and results for M_2 are typical of those teeth in which variation appears to reflect the presence of multiple species.

Results

Range-Based Measures of Relative Variation

In comparisons of the three measures of variation for single- and pooled-species samples, range-based statistics for pooled-species samples more frequently exceed single-species reference maxima than do coefficients of variation. This gives an initial impression that the two range statistics are superior to the CV in detecting the presence of multiple taxa in a sample. Table 2 provides a comparison of non-Cameroon maximum single-species values for R% with the maximum values derived from a comparison of the generally larger single- and pooled-species values from Cameroon. Out of 34 total variables, pooled-species values from Cameroon, Bolobo, and Kisangani exceed single-species maxima for 26, 14, and 28 dimensions, respectively. Comparing the Cameroon and the non-

Table 2. Single-Species Maxima and Pooled Sympatric Species Values of the Range/Mean × 100 (R%)[a,b]

	Maximum single-species value from outside Cameroon			Cameroon maximum single-species R%			Cameroon pooled-species R% (3 taxa)		Bolobo pooled-species R% (3 taxa)		Kisangani pooled-species R% (5 taxa)	
Variable	R%	Code	*N*	R%	Code	*N*	R%	*N*	R%	*N*	R%	*N*
M_3 length	32.7	akF	24	34.3[c]	nnC	106	*45.6*	249	*39.2*	80	*49.0*	110
M_3 post. br.	38.0[c]	akC	44	36.6	pgC	65	*47.5*	247	36.0	79	*44.7*	110
M_3 ant. br.	33.4[c]	akC	44	31.1	nnF	49	*37.7*	249	*33.6*	80	*42.0*	110
M_2 length	27.7	ms1C	25	32.8[c]	nnC	108	*40.1*	252	*36.7*	80	*42.0*	116
M_2 post. br.	32.4	akC	44	37.9[c]	ccC	79	36.1	252	30.9	80	*46.4*	115
M_2 ant. br.	32.9	akF	24	43.8[c]	ccC	79	41.8	252	32.6	80	38.5	116
M_1 length	25.6[c]	akC	39	25.3	nnC	95	*35.9*	231	29.9	70	*44.9*	107
M_1 post. br.	30.9	awC	28	35.5[c]	nnF	46	35.5	240	34.5	78	*47.4*	110
M_1 ant. br.	26.7	awC	27	39.6[c]	nnF	45	*44.7*	234	38.1	74	37.5	109
P_4 length	36.8	ms1C	25	48.1[c]	pgC	62	*48.1*	240	45.2	61	*51.9*	108
P_4 breadth	37.7[c]	asF	15	33.1	ccC	78	*40.1*	240	27.0	68	*49.0*	110
P_3 length	75.4[c]	akC	43	75.2	pgC	65	*80.7*	250	*93.4*	76	*89.8*	115
P_3 breadth	45.6	asF	16	46.7[c]	ccC	78	*48.7*	249	42.9	72	*62.1*	111
C_1 length	63.2[c]	ms2C	32	57.8	ccC	78	*66.8*	249	*67.8*	79	*82.3*	113
C_1 breadth	62.1	wwC	21	63.3[c]	nnC	108	*73.2*	252	*66.9*	79	*74.7*	114
I_2 breadth	36.7	asC	33	54.4[c]	nnC	103	*60.8*	237	33.3	74	52.0	104
I_1 breadth	39.2	akC	33	41.2[c]	ccC	51	*60.6*	183	35.3	67	37.1	93
M^3 post. br.	50.7	akC	43	52.6	pgF	28	*59.4*	247	51.3	76	*60.2*	104
M^3 ant. br.	29.7	CnC	28	33.0[c]	pgF	28	*40.3*	247	*38.7*	76	*39.1*	105
M^2 length	25.7	CnC	31	28.3[c]	nnC	108	*38.4*	251	*32.8*	80	*43.2*	116
M^2 post. br.	32.3	akC	24	49.5[c]	nnF	52	*52.5*	252	34.6	80	36.2	116
M^2 ant. br.	38.5[c]	CnC	31	34.2	ccC	79	32.7	252	*39.7*	80	38.2	116
M_1 length	39.5[c]	CnC	31	32.0	nnC	108	35.8	251	*43.0*	80	*41.2*	115
M^1 post. br.	29.3	akF	24	34.9[c]	ccC	78	*36.2*	249	279	79	*42.6*	114
M^1 ant. br.	28.0	akF	24	33.1[c]	ccC	77	32.1	250	*33.1*	79	*42.8*	114
P^4 length	34.7	akC	40	43.9[c]	nnF	52	*51.1*	249	41.0	78	*54.2*	112
P^4 breadth	31.0	asC	37	31.3[c]	nnC	108	*34.5*	251	29.4	77	*55.2*	113
P^3 length	36.3	akC	41	43.0[c]	nnC	100	*53.4*	227	*48.7*	74	*59.0*	106
P^3 breadth	43.1[c]	asF	15	32.6	nnC	104	38.4	243	32.0	75	*55.4*	108
C^1 length	82.4[c]	ms2C	30	81.5	nnC	107	*97.2*	247	74.7	76	*84.3*	106
C^1 breadth	59.3	asC	30	64.6[c]	nnC	107	*70.3*	247	*79.8*	75	*84.5*	109
I^2 breadth	51.7[c]	asF	13	40.3	nnC	96	47.2	197	25.7	57	*73.4*	93
I^1 breadth	36.5	ms2C	31	44.1[c]	pgC	43	*46.3*	183	29.8	57	*45.0*	77

[a]A pooled-species value in italics exceeds the single-species reference maximum value.
[b]Species codes are listed in Table 1. M = all-male sample; F = all-female sample; C = combined-sex sample.
[c]Maximum single-species R%.

Cameroon maximum single-species values allows an assessment of the effect of sample size on measures of relative variation. The larger Cameroon samples provide 23 reference maximum values, while non-Cameroon samples account for only 11. In addition, the pooled three-species Bolobo sample (which is of a size comparable to single-species samples from Cameroon) exceeds reference maxima only about half as often as the larger pooled three-species sample from Cameroon. This suggests a significant effect of sample size on R% values. Therefore, although observed pooled-species values for R% do frequently exceed the reference maximum, there is some evidence indicating that larger samples have a greater probability of yielding a reference maximum value (23 of the single-species maximum values derive from the larger Cameroon samples). Only three single-species maximum values are from samples of less than 25 individuals. The pooled-species values from Cameroon are, of course, from the largest samples of all, and it should be noted that they exceed reference maximum values almost as frequently as the five-species sample from Kisangani (although Kisangani pooled-species values do exceed those of Cameroon in 23 out of 34 cases). However, the failure of R% in a fossil sample to exceed a reference maximum does not constitute "proof" that only one taxon is present, since small samples (such as those commonly available from the fossil record) are likely to produce smaller values for range-based statistics. In addition, for a substantial number of dental dimensions, pooled-species values do not exceed the reference maximum for single-species samples (8 for Cameroon, 20 for Bolobo, and 6 for Kisangani).

The pattern is much the same for $I_{max/min}$, shown in Table 3. In this case, values for pooled-species samples from Cameroon, Bolobo, and Kisangani exceed reference maxima in 26, 13, and 27 cases, respectively. Cameroon species account for 22 reference maxima, non-Cameroon species for nine maxima, and in three cases both Cameroon and non-Cameroon samples provide equivalent reference maxima. Again, the larger Cameroon pooled-species sample exceeds the reference maximum nearly as often as the five-species Kisangani sample (although Kisangani values are greater than those of Cameroon for 24 variables).

The Coefficient of Variation

The results of this study also indicate that if a sample CV value does not exceed a single-species reference maximum, this does not prove that only one taxon is present. Table 4 provides pooled-species and maximum single-species CV values. Out of 34 dental dimensions considered in this study, 23 yield CVs that are smaller than the reference maxima in the Cameroon pooled-species sample and 18 that are smaller in the Bolobo sample. Only eight variables yield CVs smaller than the reference maximum in the five-species Kisangani sample. The fact that many CV values for pooled-species samples do not exceed reference maxima should not be unexpected, since closely related primate species do not differ substantially in all or even most characters (Tattersall, 1986, this volume).

Sources and patterns of intraspecific variation may make some variables unsuitable for detecting a pooled-species sample. For example, the variation in canine and P_3 dimensions in *Cercopithecus* is more a reflection of the sexual rather than the taxonomic composition of a sample (Cope, 1988, 1989). In the case of guenon species, CVs for lengths of posterior teeth appear to be the best indicators of multiple taxa in a sample, since they consistently exceed reference maximum values in all three pooled-species samples by approximately 1% or more. Other dimensions sometimes exceed reference maxima, but by much smaller amounts, begging the issue of establishing criteria for rejecting the hypothesis that only one species is present (see below). For posterior teeth, *Cer-*

Table 3. Single-Species Maxima and Pooled Sympatric Species Values of the Maximum/Minimum Index ($I_{max/min}$)[a,b]

	Maximum single-species $I_{max/min}$ from outside Cameroon			Cameroon maximum single-species $I_{max/min}$			Cameroon pooled-taxa $I_{max/min}$ (3 species)		Bolobo pooled-taxa $I_{max/min}$ (3 species)		Kisangani pooled-taxa $I_{max/min}$ (5 species)	
Variable	Index	Code	*N*	Index	Code	*N*	Index	*N*	Index	*N*	Index	*N*
M_3 length	1.36	ms1C	25	1.41[c]	nnC	106	*1.57*	249	*1.47*	80	*1.64*	110
M_3 post. br.	1.45[c]	akC	44	1.45[c]	pgC	65	*1.63*	247	1.43	79	*1.56*	110
M_3 ant. br.	1.38[c]	akC	44	1.37	nnF	49	*1.47*	249	*1.40*	80	*1.51*	110
M_2 length	1.33	ms1C	25	1.40[c]	nnC	108	*1.49*	252	*1.45*	80	*1.52*	116
M_2 post. br.	1.37	akC	44	1.43[c]	ccM	46	1.43	252	1.36	80	*1.59*	115
M_2 ant. br.	1.36	akC	44	1.53[c]	ccC	79	1.53	252	1.39	80	1.48	116
M_1 length	1.27	ms1C	25	1.29[c]	nnC	95	*1.43*	231	*1.35*	70	*1.54*	107
M_1 post. br.	1.36	awC	28	1.42[c]	nnC	98	1.42	240	1.42	78	*1.59*	110
M_1 ant. br.	1.31	awC	27	1.46[c]	nnF	45	*1.55*	234	*1.47*	74	1.45	109
P_4 length	1.43	ms1C	25	1.57[c]	pgC	62	*1.62*	240	1.57	61	*1.66*	108
P_4 breadth	1.46[c]	asF	15	1.39	nnC	105	*1.48*	247	1.32	68	*1.65*	110
P_3 length	2.09[c]	awC	43	2.04	pgC	65	*2.28*	250	*2.46*	76	*2.51*	115
P_3 breadth	1.57[c]	asF	16	1.56	nnC	106	*1.60*	249	1.54	72	*1.90*	111
C_1 length	1.91[c]	ms2C	32	1.81	nnC	108	*1.95*	249	*1.97*	79	*2.28*	113
C_1 breadth	1.78	wwC	21	1.89[c]	nnC	108	*2.04*	252	*1.92*	79	*2.13*	114
I_2 breadth	1.47	asC	33	1.64[c]	nnC	103	*1.74*	237	1.39	74	*1.70*	104
I_1 breadth	1.50	akC	33	1.55[c]	ccC	51	*1.61*	183	1.43	67	1.46	93
M^3 length	1.37	ms2C	32	1.51[c]	ccC	79	*1.63*	250	*1.55*	77	*1.53*	105
M^3 post br.	1.67	akC	43	1.72[c]	pgF	28	*1.86*	247	1.70	76	*1.83*	104
M^3 ant. br.	1.36	CnC	28	1.40[c]	pgF	28	*1.51*	247	*1.49*	76	*1.47*	105
M^2 length	1.30	CnC	31	1.33[c]	nnC	108	*1.46*	251	*1.38*	80	*1.54*	116
M^2 post. br.	1.36	ms1C	25	1.57[c]	nnF	52	*1.65*	252	1.41	80	1.42	116
M^2 ant. br.	1.49[c]	CnC	31	1.40	ccC	79	1.40	252	1.49	80	1.46	116
M_1 length	1.58[c]	CnC	31	1.39	nnC	108	1.42	251	1.58	80	1.51	115
M^1 post. br.	1.33	akC	42	1.40[c]	ccF	46	*1.43*	249	1.33	79	*1.53*	114
M^1 ant. br.	1.33	akC	42	1.38[c]	ccC	77	1.38	250	1.38	79	*1.53*	114
P^4 length	1.42	ms1C	25	1.54[c]	nnF	52	*1.64*	249	1.50	78	*1.71*	112
P^4 breadth	1.38[c]	asC	37	1.38[c]	nnM	56	*1.41*	251	1.34	77	*1.74*	113
P^3 length	1.43	akC	41	1.51[c]	nnC	100	*1.66*	227	*1.63*	74	*1.79*	106
P^3 breadth	1.54[c]	asF	15	1.38	pgC	62	1.47	243	1.38	75	*1.75*	108
C^1 length	2.17[c]	ms2C	30	2.17[c]	nnC	108	*2.50*	247	2.07	76	*2.33*	106
C^1 breadth	1.85	asC	37	1.88[c]	nnC	107	*1.94*	247	*2.13*	75	*2.33*	109
I^2 breadth	1.65[c]	asF	13	1.48	nnC	96	1.58	197	1.29	57	*2.04*	93
I^1 breadth	1.42	ms2C	31	1.60[c]	pgM	25	*1.63*	183	1.35	57	1.55	77

[a]A pooled-species value in italics exceeds the single-species reference maximum value.
[b]Species codes are listed in Table 1. M = all-male sample; F = all-female sample; C = combined-sex sample.
[c]Maximum single-species $I_{max/min}$.

copithecus species of similar body size sometimes differ significantly in mean breadths of posterior teeth, with some species having relatively broader posterior teeth for their body size (see below). In contrast to mesiodistal lengths, however, breadths of posterior teeth almost never exhibit multimodal distributions in pooled-species samples (Cope, 1989). Perhaps lengths better reflect body size differences between sympatric *Cercopithecus* species. Regardless of the possible reasons, CVs of mesiodistal lengths do appear to be consistently greater in pooled-taxa samples than in single-species or single-sex samples (Table 4). Despite this, Table 5 shows that the confidence limits calculated for reference maximum CVs tend to overlap those of pooled-species samples.

The one possible exception to the general trend for mesiodistal lengths is the pattern for M^1. The large maximum single-species CV here (9.33) is most likely an aberrant value, since the next highest CV is 6.99 from a combined-sex

Table 4. Coefficient of Variation (CV) in Single- and Pooled Species Samples[a,b]

	Non-Cameroon maximum single-species CV			Cameroon maximum single-species CV			Cameroon pooled-species CV (3 taxa)		Bolobo pooled-species CV (3 taxa)		Kisangani pooled-species CV (5 taxa)	
Variable	CV	Code	*N*	CV	Code	*N*	CV	*N*	CV	*N*	CV	*N*
M_3 length	6.63	awF	14	7.52[c]	pgF	28	*8.87*	249	*9.62*	80	*11.34*	110
M_3 post. br.	8.36[c]	wdC	21	7.83	pgF	28	*8.72*	247	*8.51*	79	*10.76*	110
M_3 ant. br.	6.83	ms1F	13	7.37[c]	pgF	28	*7.67*	249	*7.65*	80	*8.66*	110
M_2 length	6.23[c]	ms1C	25	6.15	ccF	33	*8.21*	252	*8.53*	80	*10.78*	116
M_2 post. br.	7.45	ms1C	25	7.84[c]	ccM	46	7.71	252	6.19	80	*9.90*	115
M_2 ant. br.	7.27	awM	14	7.42[c]	ccM	46	7.03	252	7.19	80	*8.63*	116
M_1 length	5.81[c]	ms1M	12	5.47	nnC	95	*7.45*	231	*7.94*	70	*10.83*	107
M_1 post. br.	6.95	ms1F	13	7.00[c]	ccF	31	6.54	240	*7.36*	78	*9.76*	110
M_1 ant. br.	6.88[c]	ms1F	13	6.68	pgF	28	6.37	234	*7.85*	74	*9.27*	109
P_4 length	9.31[c]	ms1C	12	8.58	pgF	25	*9.80*	240	*10.13*	61	*12.32*	108
P_4 breadth	8.67[c]	asF	15	8.02	pgF	27	7.42	247	6.42	68	*9.34*	110
P_3 length	20.46[c]	CnC	29	17.91	nnC	107	19.17	250	*22.45*	76	*22.02*	115
P_3 breadth	10.08[c]	asF	16	8.88	nnC	106	9.22	249	*10.87*	72	*12.79*	111
C_1 length	19.89[c]	ms2	32	16.66	nnC	108	15.16	249	19.11	79	*20.15*	113
C_1 breadth	20.74[c]	wwC	21	16.36	nnC	108	15.80	252	18.27	79	17.17	114
I_2 breadth	8.81[c]	asF	14	7.86	nnM	54	7.48	237	7.54	74	*9.07*	104
I_1 breadth	9.07[c]	asF	13	8.44	ccF	21	6.87	183	6.88	67	7.43	93
M^3 length	8.08[c]	ms1M	12	7.32	pgF	28	*9.20*	250	*9.28*	77	*9.56*	105
M^3 post br.	11.97[c]	akM	20	10.04	pfF	28	9.92	247	11.81	76	11.36	104
M^3 ant. br.	7.12	am1M	12	7.63[c]	ccM	44	*7.68*	247	*7.81*	76	*8.21*	105
M^2 length	6.47[c]	ms1C	25	6.25	pgF	28	*7.88*	251	*7.65*	80	*10.40*	116
M^2 post. br.	7.48	awF	14	7.69[c]	nnF	52	7.49	252	7.13	80	7.44	116
M^2 ant. br.	7.76[c]	CnC	31	6.59	ccM	46	6.60	252	7.32	80	7.66	116
M_1 length	9.33[c]	CnF	12	5.58	nnC	108	7.59	251	9.16	80	*10.86*	115
M^1 post. br.	7.67[c]	mkC	22	6.97	ccM	46	6.55	249	6.81	79	*8.59*	114
M^1 ant. br.	6.82	CnC	30	7.42[c]	ccM	45	6.28	250	7.30	79	*8.21*	114
P^4 length	7.97[c]	ms1C	25	7.77	nnF	52	*9.54*	249	*10.54*	78	*13.02*	112
P^4 breadth	7.44[c]	asF	16	6.77	ccM	46	7.00	251	7.06	77	*9.61*	113
P^3 length	8.94[c]	ms2M	14	8.41	ccM	42	*9.92*	227	*10.46*	74	*11.69*	106
P^3 breadth	9.79[c]	asF	15	7.27	ccF	31	7.05	243	6.88	75	9.53	108
C^1 length	25.54[c]	ms2C	30	20.26	nnC	107	19.64	247	22.74	76	22.27	106
C^1 breadth	16.86[c]	mkC	21	13.95	nnC	107	13.32	247	*17.39*	75	*16.96*	109
I^2 breadth	11.85[c]	asF	13	8.38	ccM	34	8.92	197	6.60	57	10.12	93
I^1 breadth	10.53[c]	akM	16	8.82	pgF	18	7.31	183	7.16	57	7.28	77

[a]A pooled-species value in italics exceeds the single-species reference maximum value.
[b]Species codes are listed in Table 1. M = all-male sample; F = all-female sample; C = combined-sex sample.
[c]Maximum single-species CV.

Table 5. 95% Confidence Limits of CV for Mesiodistal Lengths of Posterior Teeth[a,b]

	Cameroon single species maximum	Non-Cameroon single species maximum	3 species from Bolobo pooled	3 species from Cameroon pooled	5 species from Kisangani pooled
M_3	5.46–9.58	3.92–9.34	8.10–11.14	8.08–9.66	*9.83–12.85*
M_2	6.29–7.70	3.70–8.75	7.19–9.87	7.49–8.93	*9.38–12.18*
M_1	4.68–6.26	2.84–8.38	*6.60–9.28*	*6.77–8.13*	*9.36–12.30*
P_4	6.08–11.08	5.13–13.49	8.30–11.96	8.92–10.68	10.66–13.98
M^3	5.31–9.33	4.52–11.64	7.78–10.78	8.39–10.01	8.25–10.87
M^2	4.53–7.95	4.58–8.36	6.44–8.86	7.19–8.57	*9.05–11.75*
M^1	4.79–6.37	5.14–13.52	7.71–10.61	6.92–8.26	9.44–12.28
P^4	6.26–9.28	5.64–10.30	8.85–12.23	8.70–10.38	*11.30–14.74*
P^3	6.56–10.26	5.29–12.59	8.74–12.18	9.01–10.83	10.10–13.28

[a]Sample sizes and species identification codes are in Table 4.
[b]Confidence limits in italics do not overlap those of a single-species maximum CV.

sample of the same species, and the other estimates are between 4 and 6. In addition, simulations indicate that small samples can occasionally yield CVs more than twice the value of the population from which they are drawn, although the probability of this occurring is very low (see below). The value of 9.33 can thus be seen as an extreme case relative to the other 38 values used in determining a reference maximum.

Comparison of the larger Cameroon samples with the other reference samples yields a pattern opposite that of range-based statistics. Only 8 of 34 reference maxima derive from the Cameroon samples. Very low and very high values are more common for smaller samples, so it may be that there is less of a systematic effect of sample size on the CV than on range-based statistics. In the latter case, lower values tend to occur most often in smaller samples, and larger values in bigger samples. Maximum CV values derived from a number of relatively small reference samples may, in fact, overestimate true single-species variation. This would be the opposite of the bias of range statistics, which would underestimate variation for smaller samples. The simulations discussed below are intended to clarify this phenomenon.

Clearly, most CVs for mesiodistal lengths of posterior teeth do reflect the presence of pooled species in a sample (Table 4). If one can assume that single-species populations are similar in variation, then a reference maximum is probably an overestimate of the true variation in a species. Therefore, it is not surprising that the confidence limits for reference maxima frequently overlap with those of a multiple-species sample (Table 5).

The CV confidence limits for posterior tooth lengths of *most* single-species samples, however, do not overlap those of pooled-species samples. Figure 2 shows CV estimates and 95% confidence limits of the CV for M_2 length in single- and pooled-species samples of both sexes for the three geographic samples for which multiple sympatric species are available. In most cases, 95% confidence limits of the CV for single-species samples do not overlap with those of pooled-species samples. The two exceptions involve a sample of only 10 individuals and

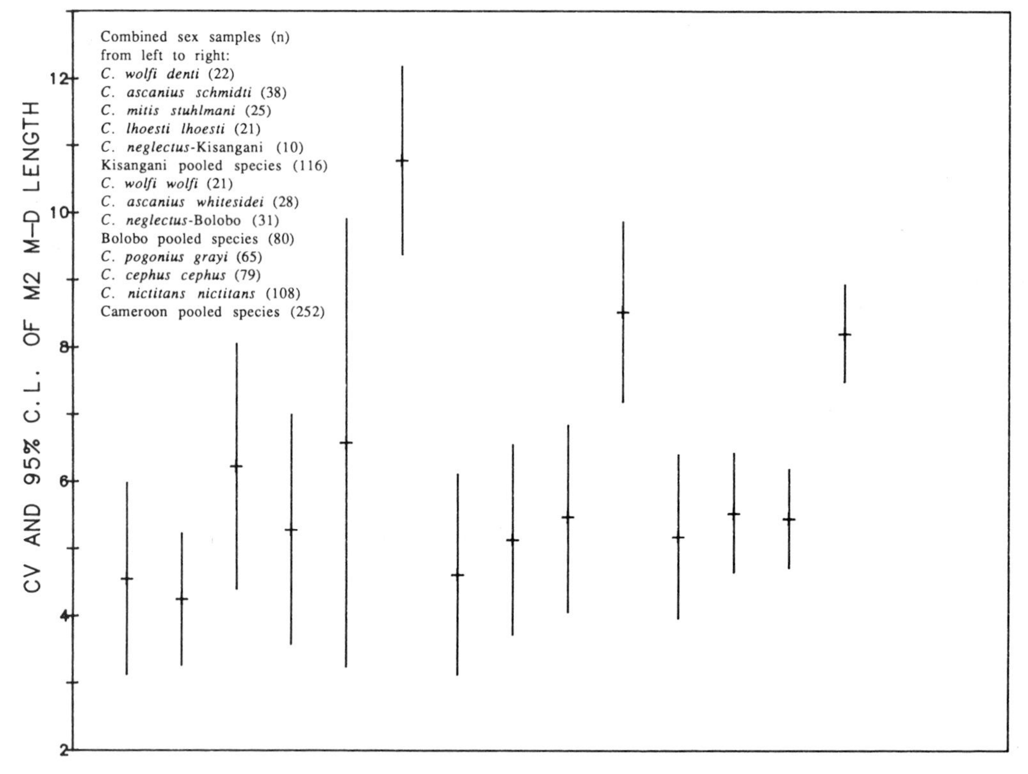

Fig. 2. Coefficient of variation of M_2 length (horizontal lines) and 95% confidence limits (vertical lines) for single- and pooled-species samples, males and females combined.

the reference maximum value. All other single-species samples appear to have significantly smaller CVs than pooled-species samples. These results suggest that a single-species maximum CV based on a number of comparative samples of varying size and sexual composition is likely to fall in the upper part of a range of potential estimates. Thus, it may be viewed as being closer to the highest range of possible CV estimates than to the actual population value of the CV. This is especially true if reference samples of less than 25 individuals are used. At the sample sizes usually available from the fossil record, it is unrealistic to expect the 95% confidence limits of a reference maximum not to overlap those of a pooled-species CV. Other means must be employed to determine criteria for falsification of the null hypothesis (see below).

Sexual Dimorphism and the CV of Posterior Teeth

The sexual composition of samples has only a slight effect on the CV for posterior teeth, because sexual dimorphism for these dimensions is low (Cope, 1988, 1989; Gingerich and Schoeninger, 1979; Kay, 1982a; Plavcan, this volume). Figure 3 shows 95% confidence limits of the CV for M_2 mesiodistal length for Cameroon single- and pooled-species samples of all possible sexual compositions. The results are representative for those of the other two geographic samples and mesiodistal lengths of other posterior teeth. Females tend to be somewhat more variable than males (because of smaller means), but these differences are not significant. However, irrespective of sexual composition, the CVs for pooled-species samples are generally significantly larger than those for single-species samples. Note the great similarity of the CV in the pooled-species samples, regardless of sexual composition.

The Magnitude of the CV in Relation to Differences in Means between Sympatric Species

Although the number of taxa in a pooled-species sample may be expected to influence the magnitude of the CV (i.e., Kisangani tends to have the largest values and more frequently exceeds reference maxima, see Table 4 and Fig. 2), the relationship is not a simple one because of the patterns of tooth-size differences among sympatric species. Both the Cameroon and Bolobo samples can be said to include two "small" and one "large" species, while Kisangani has two

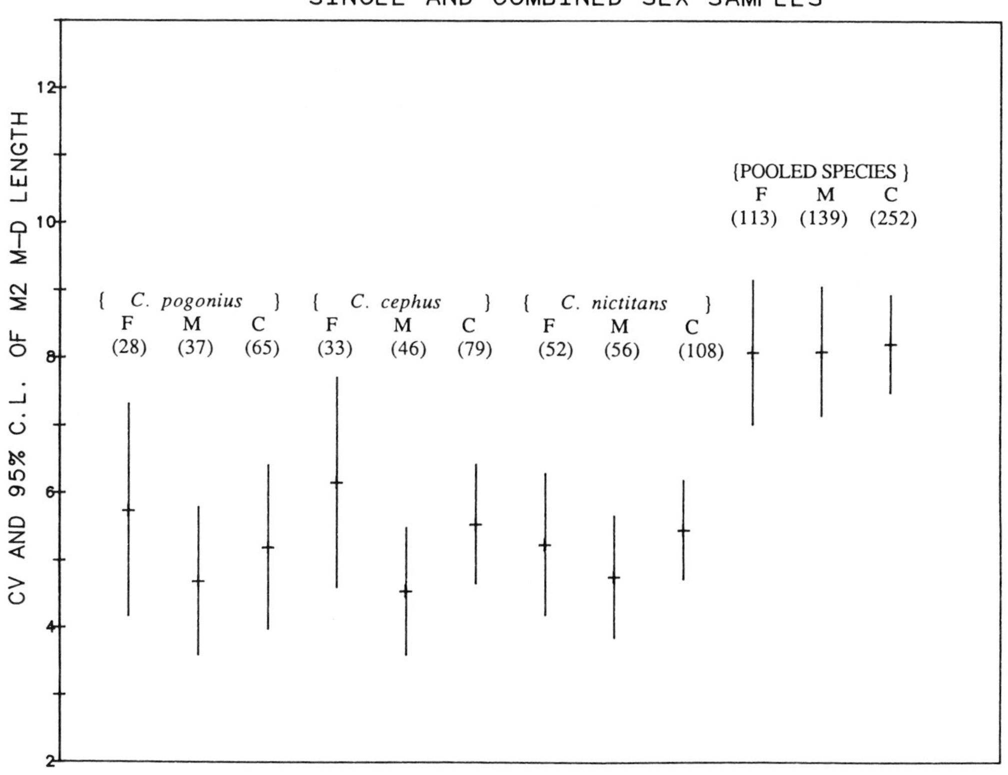

Fig. 3. Coefficient of variation of M_2 length (horizontal lines) and 95% confidence limits (vertical lines), Cameroon samples. F = female, M = males, C = combined-sex values.

"small" and three "large" species (for a detailed listing and discussion of differences in mean dental metrics, see Cope, 1989). Table 6 provides pooled-species CVs for pairs of Cameroon species. Table 7 provides a summary of differences in mean tooth size among these taxa (based on single-classification analysis of variance and the Student–Newman–Kuels test for significant differences between means). A comparison of means of dental metrics for the three species reveals a generally consistent pattern, wherein mesiodistal lengths tend to be significantly greater in *C. nictitans* and smallest in *C. cephus* (Cope, 1989). *C. pogonias* mean tooth lengths are usually intermediate and not significantly different from those of *C. cephus* (except for M^2 and M^1 length, where combined-sex samples for all three species are significantly different). These data are consistent with records of differences in body weight taken in the field: *C. nictitans* is the heavier species, while *C. pogonias* is only slighter heavier on average than *C. cephus* (Gautier-Hion, 1980).

As is the case for the molars, mean dimensions of premolars for *C. nictitans* are significantly larger than those of the other two taxa. *C. cephus* is generally intermediate for premolar size and *C. pogonias* is the smallest. For length of P_4 the difference between *C. cephus* and *C. pogonias* is significant. *C. pogonias* has relatively broad posterior teeth compared to their breadth, with means often not significantly different from those of *C. nictitans,* although means for M_2 posterior breadth, M^2 posterior breath, and M^1 posterior breadth are significantly larger in *C. pogonias* than in both of the other two Cameroon species.

As a result of the pattern of tooth size differences among *Cercopithecus* species, it can be shown that the degree to which a pooled-species CV is elevated is not simply the result of an additive effect of increasing the number of species in a sample. Instead, the magnitude of a pooled-species CV is a reflection of the average difference between means within a sample (see also Plavcan, this volume), regardless of the number of species present. In samples balanced for taxonomic composition (Cope and Lacy, in press), the correlation between the elevation of the CV (expressed as the percent difference between pooled species and reference maximum CV) and the average difference among means in the pooled sample is very high (r is just under 0.90). In the present study, different pooled-species samples differ in the proportions of sympatric species represented, since all measured specimens have been included in the calculation of pooled-species CVs. Still, the effect of average differences between species means in pooled-species samples on the magnitude of the CV can be clearly seen.

A comparison of Cameroon pooled-species values in Tables 4 and 6 shows the way in which the CV of pooled sympatric species depends on the pattern of differences in mean dental dimensions shown in Table 7. The pooled sample of *C. nictitans* and *C. cephus* actually yields greater CV values than the three-species Cameroon sample for 24 out of 34 dimensions, reflecting the inclusion in a sample of what are generally the largest and smallest species from that geographic area. The CVs of this sample exceed reference maximum values in 10 of 34 variables, all cases in which the three-species Cameroon sample CV also exceeds the reference maximum. The fact that the CV for P_4 length exceeds the reference maximum in the three-species sample, but not in the combined *C. nictitans/C. cephus* sample, makes sense if one considers that the *C. cephus* mean is significantly larger than the *C. pogonias* mean for this variable. For this variable,

Table 6. Coefficients of Variation for Pooled Pairs of Cameroon Species[a]

	Reference maximum[b]		*C. nictitans* & *C. cephus*		*C. nictitana* & *C. pogonias*		*C. pogonias* & *C. cephus*	
Variable	CV	*N*	CV	*N*	CV	*N*	CV	*N*
M_3 length	7.52	28	*9.02*	184	*8.57*	171	6.15	143
M_3 post. br.	8.36	21	*8.88*	182	8.34	168	7.57	144
M_3 ant. br.	7.37	28	*7.88*	184	6.75	170	*7.68*	144
M_2 length	6.23	25	*8.34*	187	*7.78*	173	5.39	144
M_2 post. br.	7.84	46	7.42	187	6.66	173	*8.68*	144
M_2 ant. br.	7.42	46	7.14	187	5.97	173	*7.52*	144
M_1 length	5.81	12	*7.43*	169	*7.48*	157	4.92	136
M_1 post. br.	7.00	31	6.58	175	6.18	163	6.72	142
M_1 ant. br.	6.88	13	6.56	170	6.08	159	6.22	139
P_4 length	9.31	12	8.40	178	*10.62*	165	7.66	137
P_4 breadth	8.67	15	7.21	183	7.48	169	6.74	142
P_3 length	20.46	29	19.32	185	19.09	172	16.90	143
P_3 breadth	10.08	16	9.37	184	9.27	171	8.07	143
C_1 length	19.89	32	16.84	186	16.11	171	12.61	141
C_1 breadth	20.74	21	16.17	187	16.38	173	12.87	144
I_2 breadth	8.81	14	7.92	175	7.27	165	6.53	134
I_1 breadth	9.07	13	7.15	131	6.30	132	7.13	103
M^3 length	8.08	12	*9.51*	185	*8.62*	171	7.01	144
M^3 post br.	11.97	20	10.22	182	9.44	170	9.13	142
M^3 ant. br.	7.63	44	*8.07*	182	6.32	170	*7.78*	142
M^2 length	6.47	25	*8.18*	187	*7.29*	172	5.59	143
M^2 post. br.	7.69	52	7.31	187	6.94	173	*7.90*	144
M^2 ant. br.	7.76	31	6.40	187	5.72	173	7.33	144
M_1 length	9.33	12	8.06	187	6.90	172	4.98	143
M^1 post. br.	7.67	22	6.45	185	6.00	171	7.09	142
M^1 ant. br.	7.42	45	6.29	185	5.43	173	6.96	142
P^4 length	7.97	25	*9.05*	187	*9.85*	170	6.39	141
P^4 breadth	7.44	16	7.29	186	5.80	173	7.08	143
P^3 length	8.94	14	*9.87*	173	*9.71*	154	7.97	127
P^3 breadth	9.79	15	7.22	181	6.92	166	6.73	139
C^1 length	25.54	30	20.42	185	19.78	169	16.68	140
C^1 breadth	16.86	21	13.78	185	13.67	169	11.00	140
I^2 breadth	11.85	13	8.97	156	8.75	137	7.73	101
I^1 breadth	10.53	16	6.96	140	7.24	128	7.87	98

[a]A pooled-species value in italics exceeds the single-species reference maximum.
[b]Single species reference maximum from Table 4.

the average difference between means for the pooled *C. nictitans*/*C. cephus* sample is less than the difference in means for the pooled *C. nictitans*/*C. pogonias*/*C. cephus* sample.

The *C. nictitans*/*C. pogonias* pooled-sample CV values also reflect patterns of tooth-size differences among species. In this case, only seven values are greater than the reference maximum. All of these cases are lengths of posterior teeth that also exceed single-species variation in the total Cameroon sample. The pooled *C. nictitans*/*C. pogonias* sample CVs are greater than those of the three-species Cameroon sample in only nine variables. This is because for most variables, the *C. nictitans*/*C. pogonias* sample does not include the smallest species. In

Table 7. Results of Analysis of Variance and SNK Procedure on the Combined Sex Cameroon Samples[a]

Variable	Means	Variable	Means	Variable	Means	Variable	Means
M_3 length	C P N C P N * *	P_4 length	P C N P C * N * *	M^3 length	C P N C P N * *	P^4 length	P C N P C N * *
M_3 post. breadth	C P N C P N * *	P_4 breadth	P C N P C N * *	M^3 post. breadth	C P N C P * N * *	P^4 breadth	C P N C P * N * *
M_3 ant. breadth	C P N C P * N *	P_3 length	P C N P C N * *	M^3 ant. breadth	C P N C P * N *	P^3 length	P C N P C N * *
M_2 length	C P N C P N * *	P_3 breadth	P C N P C N * *	M^2 length	C P N C P * N * *	P^3 breadth	C P N C P N * *

M_2 post. breadth	C N P	C_1 length	C P N	M^2 post. breadth	C N P	C^1 length	C P N
	C		C		C		C
	N *		P		N *		P
	P * *		N * *		P * *		N * *
M_2 ant. breadth	C N P	C_1 breadth	C P N	M^2 ant. breadth	C N P	C^1 breadth	C P N
	C		C		C		C
	N *		P		N *		P
	P *		N * *		P *		N * *
M_1 length	P C N	I_2 breadth	C P N	M^1 length	C P N	I^2 breadth	P C N
	P		C		C		P
	C		P		P *		C
	N * *		N * *		N * *		N * *
M_1 post. breadth	C P N	I_1 breadth	n.s.	M^1 post. breadth	C N P	I^1 breadth	n.s.
	C				C		
	P				N *		
	N * *				P * *		
M_1 ant. breadth	C P N			M^1 ant. breadth	C N P		
	C				C		
	P				N *		
	N * *				P *		

[a]Species are arranged from the smallest mean on the left and top to largest mean on the right and bottom C = *C. cephus*; P = *C. pogonias*; N = *C. nictitans*.
An asterisk indicates a significant difference (at $p < 0.05$) between two means based on the SNK procedure; n.s. indicates a nonsignificant ($p > 0.05$) F-ratio.

one case (M_1 length), the value of the CV for the pooled *C. nictitans*/*C. cephus* sample is very similar to that of the *C. nictitans*/*C. pogonias* sample, reflecting the fact that the mean and standard deviations for *C. pogonias* (5.31, 0.257) are virtually the same as those of *C. cephus* (5.33, 0.267).

In four cases in which the CV of the pooled *C. nictitans*/*C. pogonias* sample exceeds that of the three-species sample (P_4 length, P_4 breadth, P^4 length, P^3 breadth), *C. pogonias* has a smaller mean than *C. cephus*. Thus, the CV values for the pooled *C. nictitans*/*C. pogonias* samples for these variables reflect the combination of the Cameroon species with the largest and smallest means.

The other four dimensions in which the *C. nictitans*/*C. pogonias* pooled sample has larger CVs than the three-species sample are those of the canines. Although the *C. pogonias* sample does have larger combined-sex means than *C. cephus* for these dimensions, it has a greater proportion of females than the *C. cephus* sample (see Table 1). These dimensions are highly sexually dimorphic, and the standard deviations of females are less than those of males. Therefore, the proportion of females has a great influence on the mean in relation to the standard deviation, and thus the magnitude of the CV.

The *C. pogonias*/*C. cephus* pooled-species sample exceeds the reference maximum in only five dimensions, all breadths of posterior teeth. In four of these cases, the CV of the three-species sample also exceeds the reference maximum. In the other case (M^2 posterior breadth), the greater value is the result of *C. pogonias* having a significantly larger mean value than *C. nictitans*. Thus, for this variable, the paired sample of *C. pogonias*/*C. cephus* includes the largest and smallest species from the area. Eleven variables in this paired-species sample yield CVs higher than those of the total Cameroon sample. Most of these (eight) are breadths of posterior teeth, for which *C. pogonias* is as large as, or larger than, *C. nictitans* in mean dimensions.

It should be noted that while pooled samples (of two or three species), including *C. nictitans*, exhibit a number of posterior tooth lengths that exceed reference maxima by values greater than 1%, none of the pooled *C. pogonias*/*C. cephus* samples yield a CV exceeding single-species variation by such a margin. This is despite the fact that these taxa do differ significantly in a number of dental dimensions. Differences between means must be substantial in order for a pooled-species CV to be recognized as such. Thus, it is unlikely that a sample including these latter two taxa could be recognized as including pooled species through an analysis of the CV, as has been noted by Plavcan (1989, this volume) in an independent study including many of the same specimens of the same species.

Simulation of Hypothetical "Fossil" Samples

The examination of three statistics of variation as indicators of multiple species based only on the original samples still leaves some uncertainty about the utility of these statistics at small sample sizes such as those characteristic of the fossil record. There also remain some questions regarding the use of a reference maximum as a "critical value" above which a single-species hypothesis may be rejected. The simulated values of these statistics, randomly drawn from larger populations, clarify these issues considerably.

Single-species and pooled-species samples drawn from larger populations clearly indicate that the CV is the most robust measure of variation in smaller samples, such as those commonly available from the fossil record. Figures 4–6 illustrate the distribution of values obtained at each sample size for R%, $I_{max/min}$, and the CV, respectively. Both R% and $I_{max/min}$ distributions systematically shift downward with decreasing sample size. Even when samples are drawn from a pooled-species population, values rarely exceed the single-species reference maximum (horizontal line) in samples of 20 or less. For R%, only 0–20% of the pooled-species estimates exceed the reference maximum at sample sizes of 20 or less, while only 30–65% of the pooled-species estimates for sample sizes of greater than 20 exceed this value. For $I_{max/min}$, only 0–15% of the estimates for samples of 20 or less exceed the reference maximum, while 25–65% do for samples of greater than 20. Thus, range-based measures of variation cannot be used to support a hypothesis that only one species is present in a sample, because

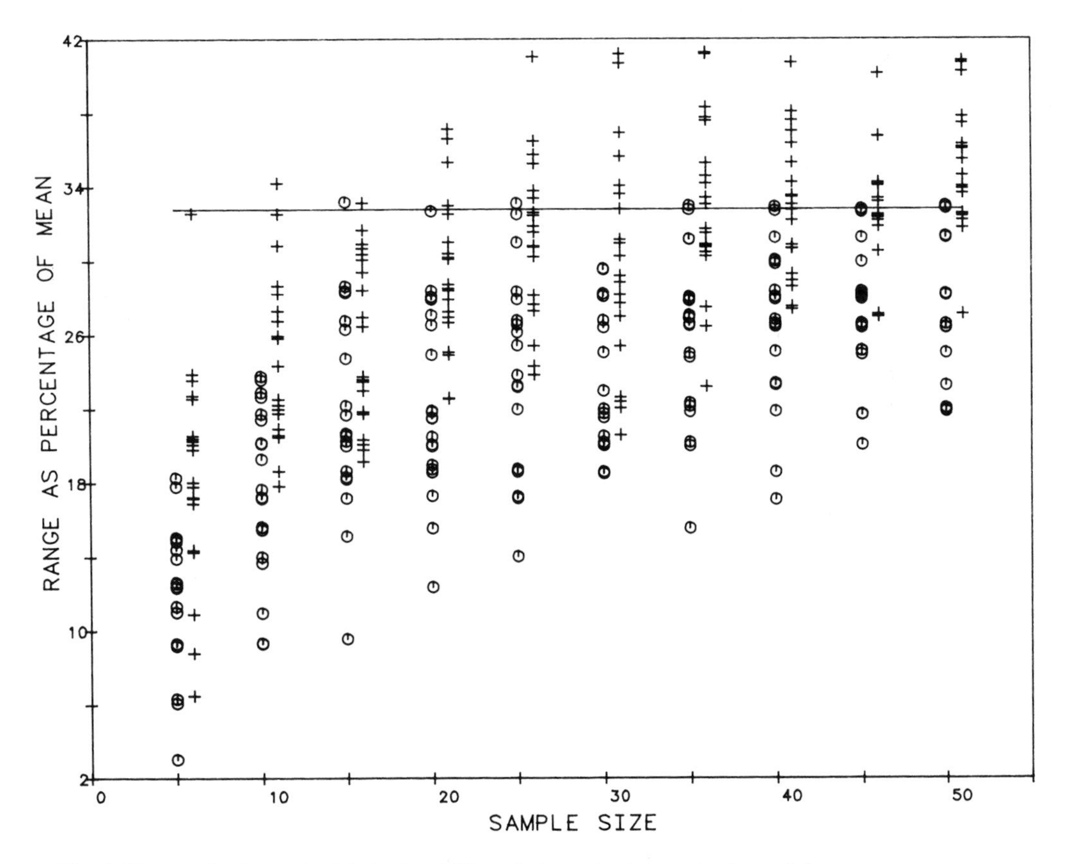

Fig. 4. Twenty single-species (circles) and 20 pooled-species (crosses) values of the range as a percentage of the mean (R%) of M_2 length at various sample sizes, obtained randomly from a sample of three sympatric species from Cameroon. The actual pooled-species value is 40.13; the horizontal line represents the single-species reference maximum of 32.79, which is the value for *C. nictitans*.

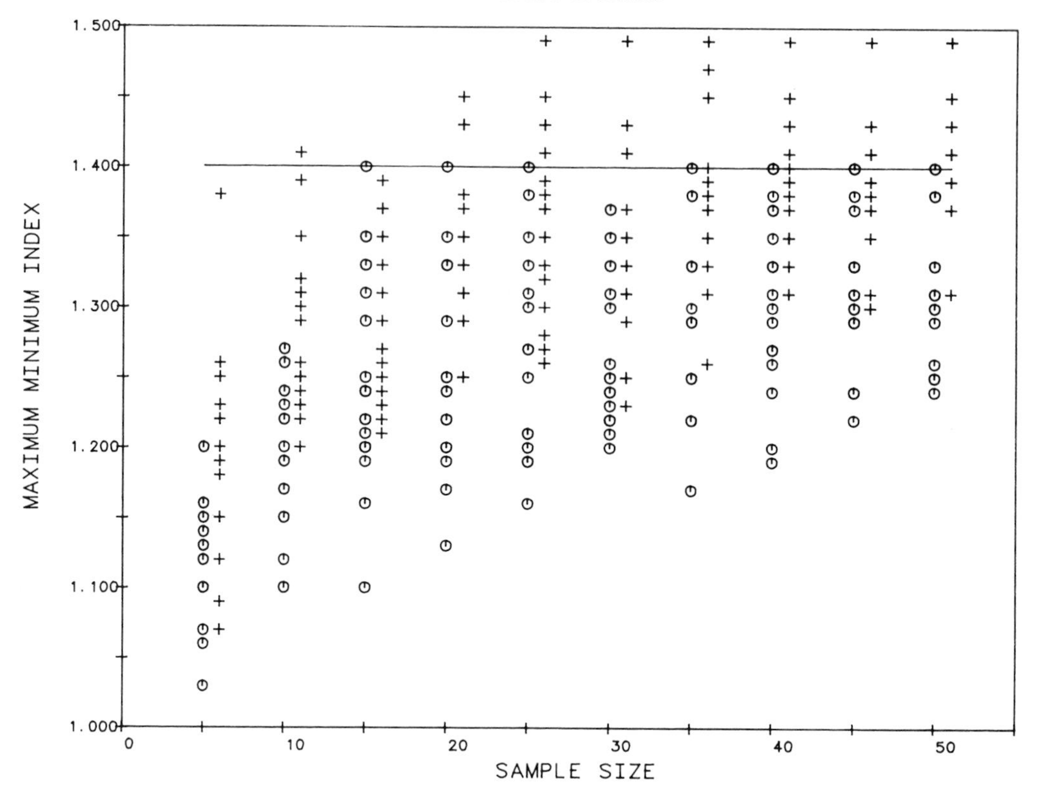

Fig. 5. Twenty single-species (circles) and 20 pooled-species (crosses) values of the maximum/minimum index ($I_{max\ min}$) of M_2 length at various sample sizes, obtained randomly from a sample of three sympatric species from Cameroon. The actual pooled-species value is 1.49; the horizontal line represents the single-species reference maximum of 1.40, which is the value for *C. nictitans*.

of the high probability of making a type II error (which in this context means failure to reject a false single-species hypothesis). However, if the value of a range-based statistic in a small sample does exceed a reference maximum, the probability is very high that more than one species is present, since single-species samples are strongly subject to the same bias toward smaller values at smaller sample sizes (see Martin and Andrews, this volume).

The CV appears to provide the best means of detecting multiple species in small samples. Figure 6 shows that, on average, CVs for each sample size are relatively stable, although there is a slight downward shift of values in samples under 10. The major trend with decreasing sample size is a broadening of the range of values obtained. In contrast to range-based statistics, the central tendencies of the CV sampling distributions remain much the same at decreasing sample sizes. Note that values above the reference maximum can be obtained from a single-species sample but that values greater than 1.25% above the reference maximum are quite rare.

This observation suggests that in the case of the present data, use of a reference maximum as a critical value would lead to an unacceptable level of type I errors (in this case, meaning a false rejection of a single-species hypothesis). Table 8 provides estimated probabilities (based on smaller samples randomly drawn from the large *C. nictitans* sample) of obtaining a CV estimate of a given magnitude beyond the reference maximum from a single-species population for each sample size. An additional 20 single-species samples were drawn at sample sizes of 5–20 to see if the trend toward a greater range of CV estimates in smaller samples was accurate. The results are similar to those originally obtained. The estimated probabilities of obtaining CV estimates over a given value based on 40 samples drawn from a single-species population are shown for sample sizes of 20 or less, while probabilities given for sample sizes of 25–50 are based on 20 randomly drawn samples. Note that one estimate of 11.28 (at $n = 5$) was obtained when drawing additional samples, suggesting that aberrantly high CV values derived from single-taxa samples are possible, although the probability of such an outcome is less than 0.05.

The data in Table 8 show that use of the current reference maximum CV of 6.23 for M_2 as a critical value for rejection of a single-species hypothesis would lead to an unacceptable rate of type I errors at smaller sample sizes. Rather, the

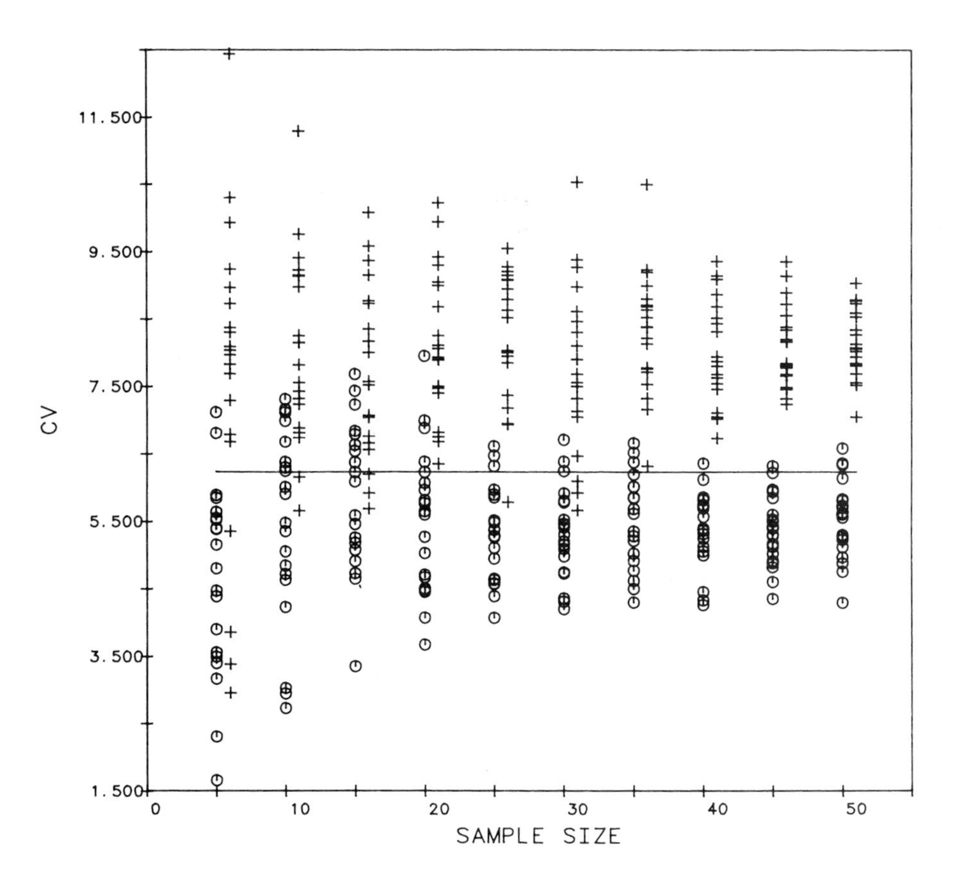

Fig. 6. Twenty single species (circles) and 20 pooled-species (crosses) values of the coefficient of variation (CV) of M_2 length at various sample sizes, randomly drawn from a sample of three sympatric species from Cameroon (actual pooled species CV = 8.21; actual *C. nictitans* CV = 5.46). The line represents the maximum single-species value (6.23) obtained from reference data (see Table 4).

Table 8. Estimated Probabilities (Based on Simulations) That a Sample CV of a Given Magnitude Is Derived from a Single-Species Sample (M_2 Mesiodistal Length)[a,b]

N	>6.23	>6.48	>6.73	>6.98	>7.23	>7.48	>7.73	>8.23	>8.73	....	>11.48
5	0.125	0.100	0.100	0.050	0.025	0.025	0.025	0.025	0.025	0.025	<0.025
10	0.400	0.275	0.255	0.175	0.075	0.025	<0.025				
15	0.325	0.200	0.175	0.125	0.100	0.025	<0.025				
20	0.275	0.225	0.200	0.125	0.050	0.050	0.050	<0.050			
25	0.150	0.100	<0.050								
30	0.150	0.050	<0.050								
35	0.150	0.100	<0.050								
40	0.050	<0.050									
45	0.050	<0.050									
50	0.150	0.050	<0.050								

[a]Actual CV of population = 5.46 (N = 108); single species reference maximum = 6.23.
[b]Estimates for sample sizes 5–20 are based on 40 simulations, while those for sample sizes of 25–50 are based on 20 simulations.

simulations suggest that a CV of 7.48 or greater has a probability of 0.05 or less of occurring at all sample sizes. As sample size increases from 5 to 50, the suggested critical values are lower, reflecting the decreasing range of estimates obtained in larger samples. How often do pooled-species samples exceed this value? In statistical terms, this would be the power of the test. Based on the simulations, 60–70% of the estimates are greater at sample sizes of 20 or less, while in samples of greater than 20 individuals, 65–95% exceed this value. Thus, although pooled-species CVs do not always exceed the point at which the null hypothesis can be rejected, they do so much more frequently than range statistics. Note the much more extensive overlap in pooled-species and single-species sampling distributions of range-based statistics compared to considerably less overlap in the pooled-species and single-species sampling distributions of the CV (see Figs. 4–6). Similarly derived critical values can be established for range-based statistics (and should be if these statistics are going to be used), but the more extensive overlap in single- and pooled-species values results in substantially less power to test the null hypothesis.

Discussion

The observation that pooled-species variation need not always exceed single-species maxima underscores the fact that the presence of a single species in a sample must be considered as the null hypothesis and cannot be proven through studies of relative variation in dental metrics. The ability to falsify the null hypothesis for a fossil sample depends on three factors: (1) that pooled species in a sample differ significantly in some of the metric characters available for study, (2) that the sample accurately represents the variation present in the source population(s), and (3) that adequate criteria exist for determining the magnitude of variation required to falsify the null hypothesis. The first factor, for the most part, is not under the control of the paleontologist, because it depends on whether closely related species differ in characteristics of body parts recovered from the fossil record. The second is partially beyond control but is also a matter of determining the appropriate measure of variation, given the sample size available in the fossil record. The third factor, closely related to the second, requires an estimation of the probability of deriving a given magnitude of variation from a single-species sample. If these problems can be overcome, the issue becomes one of power; that is, do pooled-species samples frequently exceed a given magnitude of variation that single-species samples rarely do? In more formal terms, can one minimize the probability of making type I and type II errors when testing the null hypothesis? To my knowledge this question has never been specifically addressed, but through the random generation of hypothetical "fossil" samples drawn from larger populations, our ability to understand and avoid type I and type II errors can be improved (Cope and Lacy, 1992).

The CV best addresses these factors for several reasons. First, and perhaps most importantly, the CV is simply a much more accurate estimate of the relative variation in the source population than any range-based statistic. This greater robusticity extends to sample sizes as small as five. In samples of $n = 5$ or greater, it is unlikely that range statistics will provide any information on relative variation that is not better represented by the CV.

Because the use of a correction factor for CVs based on small samples (Sokal and Rohlf, 1981) increases the estimate of variation, it should be avoided when testing a single-species hypothesis. Although, on average, the CV is moderately underestimated in small samples, the largest possible estimates derived from a population will also come from smaller samples. Since the correction factor is added to the conventional sample CV, it further distorts inaccurately high estimates resulting from sampling error. This will increase the probability of making a type I error when testing a single-species hypothesis. If a number of relatively small samples are used in establishing reference maximum values, then the use of the correction factor for estimating their CVs will greatly increase the probability of making a type II error when assessing a fossil sample.

Conclusions

The results of this study indicate that no measure of relative variation can be used to prove that a fossil sample consists of only one species. The presence of a single species must be viewed as the null hypothesis and can only be falsified, not proven. Since closely related species do not differ in all or even necessarily most characters, pooled-taxa samples should not be expected to exhibit inflated variation for all or even most variables.

There are two problems with using range-based statistics to test a single-species hypothesis. One is the low probability of exceeding reference maximum values in small samples, such as those commonly available from the fossil record. This is true even when one is sampling multiple species in which the actual range of the pooled populations does exceed the single-species maximum. Secondly, the extensive overlap in the sampling distributions of single- and pooled-species samples indicates that even when sample sizes of reference and fossil samples are comparable, the power of a test based on range statistics is much lower than that for the CV.

If a small fossil sample does yield a range based statistic exceeding the reference maximum (see Martin and Andrews, this volume), then the probability is high that more than one species is represented if reference data are composed of samples comparable in size to (or larger than) the fossil sample. This is because range-based measures of variation systematically reduce with decreasing sample size, resulting in a bias against demonstrating the presence of multiple species in a sample.

The CV has been shown to be significantly greater in most cases for mesiodistal dimensions of posterior teeth in pooled-species samples than in single-species samples. In addition, the probability of obtaining estimates of the CV from pooled-species samples that are higher than single-species reference maxima is much greater in smaller samples than is the case for range-based measures of variation.

Although it is difficult to show the statistical significance of these differences at smaller sample sizes, the use of maximum single-species CV values derived from a number of reference samples and random simulations appears to be a relatively conservative way of establishing a threshold for distinguishing CV estimates derived from single- and pooled-species samples. For posterior tooth dimensions, the sexual composition of samples has only a minor effect in *Cercopithecus* and probably most catarrhines, an advantage since the sex of specimens in a fossil sample is unknown. A CV of posterior tooth dimensions in single-species samples exceeding the reference maximum by 1.25% appears to have a probability of 0.05 or less of occurring at most sample sizes. Dimensions exhibiting high sexual dimorphism should be avoided, as the CVs are sensitive to fluctuations in the sex ratio of samples. Correction factors for the CV of small samples are useful for describing the probable variation of a small fossil sample, but should be avoided when establishing reference maxima or actually testing for evidence of multiple species.

The degree to which these results apply to other primate genera remains to be seen. For example, would simulations on samples from other genera yield a similar value by which the CV must exceed the reference maximum in order to reject the null hypothesis? In this study, the reference data and the samples being "tested" are of the same genus. Despite generally remarkable consistency, coefficients of variation for single-species samples of hominoids and cercopithecoids reported in the literature occasionally vary widely (compare, for example, values in Kay, 1982a, and Martin and Andrews, 1984) due to the size and geographic distribution of samples. The reference maxima used here would in many cases be greater if values reported in the literature were used. It would seem that the approach of using multiple, geographically restricted, biological populations (even if that means using more than one population of the same species) in establishing reference maxima might help to standardize the results. If a large series of comparative samples is used, one would be justified in omitting extreme outliers (if based on sample sizes of less than 25) as reference maximum values, making the test of a fossil sample more powerful. Of course, fossils sampled over long time periods may well exceed the variation present in geographically restricted recent samples from museum collections, due to anagenesis (Kelley, 1986; Plavcan, this volume). This factor can only be controlled by determining the time span over which a fossil sample has accumulated. Studies of other fossil mammals in well-controlled stratigraphic contexts might shed some light on this question. Increasing the number of reference samples and/or broadening their phylogenetic composition will tend to increase the ultimate reference maximum, making the test of a single-species hypothesis more conservative.

Thus, if a large number of samples of recent species that are closely related to the fossil sample in question are available, one can more powerfully test a single-species hypothesis by concentrating only on these taxa as reference data. On the other hand, if such samples are not available, then the use of a wide variety of comparative species provides a more conservative and, under such conditions, reliable test.

The use of reference maxima to identify variables in fossil samples that show high relative variation for a single species is supported by this study. However, the data indicate that reference maxima should not always be regarded as "critical values" for rejecting the null hypothesis of a single species. The results of simulating CV values at various sample sizes indicates that false rejections of a single species hypothesis would consequently occur at unacceptably high rates (10–40% of the time at sample sizes of 20 or less). Conversely, use of reference maxima as critical values for range-based statistics will usually result in high rates of type II errors (incorrect failure to reject a single-species hypothesis). In the present study, the single-species sample with the largest sample size (and presumably the sample providing the best estimate of true population variation) was used to model single-species variation at a number of different samples sizes. The results show clearly that any "critical value" of variation for a single species is strongly dependant on sample size (regardless of the statistic used to measure that variation). The best approach is to simulate the sampling distribution of the statistic (CV, R%, $I_{max/min}$) at the same sample size as the fossil assemblage of interest. This allows the estimation of the probability that a given estimate of sample variation could be derived from a single-species population. Thus, one minimizes the rate of type I errors but at the same time removes the need to use a statistical test that involves comparing the 95% confidence limits of both fossil and reference samples. This approach, while minimizing false rejections of a single-species hypothesis, increases the power (ability to detect actual pooled-species samples) of the test.

The underlying assumption behind this approach to assessing systematic diversity in morphologically homogeneous assemblages is that fossil species within a restricted geographic area and time span should be about as variable as recent species. Kelley (this volume) strongly objects to the application of this kind of methodology to hominoids. He questions not only the use of cercopithecoid species as relevant data but also the basic assumption that fossil species are comparable in variation to recent ones. However, the observed variation for numerous living mammalian species suggests a pattern so widespread that (for practical purposes) the assumption does not seem unreasonable. No alternative scientific hypothesis testing procedure has yet been proposed. The kind of approach suggested here is intended to supplement, rather than to replace, other means of determining the number of species present in a fossil sample (such as the study of qualitative morphology).

The use of the coefficient of variation is not a panacea for all taxonomic problems in the fossil record. This study supports the idea that when a fossil sample shows consistently elevated CVs for a number of dental dimensions, this is compelling evidence that more than species is present. On the other hand, one or two variables showing elevated CVs might simply be the result of sampling error. However, the CV can be used to attempt to falsify the proposition that only a single species is present in a sample with a reasonable expectation of success in the event that the null hypothesis is indeed false.

ACKNOWLEDGMENTS

I am grateful to Bill Kimbel and Lawrence Martin for inviting me to participate in this symposium and for their thoughtful comments regarding this chapter. Gene Albrecht and Bill Jungers provided important information about primate collections and useful discussions of primate variation. I would also like to thank Peter Andrews, Claud Bramblett, Colin Groves, John Kappelman, Robert Malina, Joe Miller, Michael Plavcan, and Bernard Wood for their comments. This project was supported by NSF grant BNS-8615824. Supplementary funds were provided by the Graduate School of the University of Texas at Austin. I wish to thank the following institutions and persons for permitting access to the specimens under their care: American Museum of Natural History, Wolfgang Fuchs; Field Museum of Natural History, Bruce Patterson; the National Museum of Natural History, Richard Thorington; the Powell-Cotton Museum, Birchington, Kent, UK, Derek Howlett; the British Museum (Natural History), London, UK, Paula Jenkins; the Odontological Museum, Royal College of Surgeons, London, UK, Caroline Grigson; and the Musee Royal de L'Afrique Centrale, Tervuren, Belgium, D. Thys van den Audenaerde, M. Louette and D. Meirte. Marc Colyn of the Station Biologique de Paimfont Plélan-le-Grand, France, kindly allowed me access to his personal collection of *Cercopithecus* from East of Kisangani.

References

Cope, D. A. 1988. Dental variation in three sympatric species of *Cercopithecus*. (abstract) *Am. J. Phys. Anthropol.* **75:**198–199.

Cope, D. A. 1989. Systematic Variation in *Cercopithecus* Dental Samples. Ph.D. Thesis, University of Texas, Austin.

Cope, D. A., and Lacy, M. G. 1992. Falsification of a single species hypothesis using the coefficient of variation: A simulation approach. *Am. J. Phys. Anthropol.* **89:**359–378.

Gautier-Hion, A. 1980. Seasonal variations of diet related to species and sex in a community of *Cercopithecus* monkeys. *J. Anim. Ecol.* **49:**237–269.

Gautier-Hion, A., Bourlière, F., Gautier, J.-P., and Kingdon, J. (eds.) 1988. *A Primate Radiation: Evolutionary Biology of the African guenons.* Cambridge University Press, Cambridge.

Gingerich, P. D. 1974. Size variability of the teeth in living mammals and the diagnosis of closely related sympatric fossil species. *J. Paleont.* **48:**895–903.

Gingerich, P. D. 1979. Paleontology, phylogeny and classification: an example from the mammalian fossil record. *Syst. Zool.* **28:**451–464.

Gingerich, P. D., and Shoeninger, M. J. 1979. Patterns of tooth size variability in the dentition of primates. *Am. J. Phys. Anthropol.* **51:**457–466.

Gingerich, P. D., and Winkler, D. A. 1979. Patterns of variation and correlation in the dentition of the red fox, *Vulpes vulpes. J. Mammal.* **60:**691–704.

Greenfield, L. O. 1979. On the adaptive pattern of "*Ramapithecus.*" *Am. J. Phys. Anthropol.* **50:**527–548.

Kay, R. F. 1982a. *Sivapithecus simonsi,* a new species of miocene hominoid, with comments on the phylogenetic status of the ramapithecinae. *Int. J. Primatol.* **3:**113–173.

Kay, R. F. 1982b. Sexual dimorphism in ramapithecinae. *Proc. Natl. Acad. Sci. USA* **79:**209–212.

Kay, R. F., and Simons, E. L. 1983. A reassessment of the relationship between later Miocene and subsequent Hominoidea, in: R. L. Chiochon and R. S. Courrucini (eds.), *New Interpretations of Ape and Human Ancestry,* pp. 577–623. Academic Press, New York.

Kelley, J. 1986. Species recognition and sexual dimorphism in *Proconsul* and *Rangwapithecus. J. Hum. Evol.* **15:**461–495.

Kimbel, W. H., and White, T. D. 1988. Variation, sexual dimorphism and the taxonomy of *Australopithecus,* in: F. E. Grine (ed.), *Evolutionary History of the "Robust" Australopithecines,* pp. 175–192. Aldine de Gruyter, New York.

Martin, L. B. 1983. The Relationships of the Later Miocene Hominoids. Ph.D. Thesis, University of London.

Martin, L. B., and Andrews, P. 1984. The phyletic position of *Graecopithecus freybergi* Koenigswald. *Cour. Forsch. Inst. Senckenberg* **69:**25–40.

Plavcan, J. M. 1989. The coefficient of variation as an indicator of interspecific variability in fossil assemblages (abstract). *Am. J. Phys. Anthropol.* **78:**285.

Simpson, G. G. 1941a. The species of *Hoplophoneus. Am. Mus. Novitates.* **1123:**1–21.

Simpson, G. G. 1941b. Range as a zoological character. *Am. J. Sci.* **239:**785–804.

Simpson, G. G. 1947. Note on the measurement of variability and on relative variability of teeth of fossil mammals. *Am. J. Sci.* **245:**522–525.

Simpson, G. G., Roe, A., and Lewontin, R. C. 1960. *Quantitative Zoology.* Harcourt, Brace and Co., New York.

Sokal, R. R., and Braumann, C. A. 1980. Significance tests for coefficients of variation and variability profiles. *Syst. Zool.* **34:**449–456.

Sokal, R. R., and Rohlf, F. J. 1981. *Biometry.* W. H. Freeman and Co., San Francisco.

Tattersall, I. 1986. Species recognition in human paleontology. *J. Hum. Evol.* **15:**165–175.

Catarrhine Dental Variability and Species Recognition in the Fossil Record

10

J. MICHAEL PLAVCAN

Introduction

Debate about species recognition in the primate fossil record is pervasive. Numerous studies have come to different conclusions regarding the sexual and taxonomic composition of, for example, *Proconsul* samples from East Africa (Kelley, 1987; Pickford, 1986), *Sivapithecus* from Asia (Kay, 1982a,b; Wu and Oxnard, 1983), hominoids from Rain Ravine (Kay, 1982a, Martin and Andrews, 1984), *Australopithecus afarensis* (Cole and Smith, 1987; Johanson *et al.*, 1982; Kimbel *et al.*, 1985; Kimbel and White, 1988; Olson, 1981, 1985; Senut and Tardieu, 1985), *A. africanus* (Zwell and Pilbeam, 1972), and *Homo habilis* (Wood, 1985, this volume), to name a few. Since the majority of fossil specimens consist of teeth, a good deal of this debate centers on the interpretation of dental variation.

Closely related extant mammalian species can be difficult to recognize on the basis of dental crown morphology alone (Bown and Rose, 1987; Gingerich, 1974, 1985; Kay, 1982a,b; Tattersall, this volume), so postcanine tooth size is often used to help discriminate species in the fossil record. Gingerich (1974)

J. MICHAEL PLAVCAN • Department of Biological Sciences, University of Cincinnati, Cincinnati, Ohio 45221.

Species, Species Concepts, and Primate Evolution, edited by William H. Kimbel and Lawrence B. Martin. Plenum Press, New York, 1993.

pointed out that often closely related extant mammalian species that occur sympatrically differ in body size. Since postcanine tooth size is strongly correlated with body size in interspecific comparisons (Gingerich *et al.*, 1982; Gould, 1975), it is assumed that body-size differences between species are reflected by postcanine tooth-size differences. When dentally similar species differing in postcanine tooth size are mixed together in the fossil record, sample variability increases beyond that characteristic of extant, single species (Gingerich, 1974; Kay, 1982a). Kay (1982b, p. 114) underscores the assumption that ". . . variability within and between extinct species is of the same kind as, and does not differ in magnitude from, that seen in closely allied extant species." Thus if dental variability in a sample of fossils is greater than that seen in any living species, more than one species is assumed to be present. This protocol is based on comparison of fossil sample variation to intraspecific variability in extant species. As pointed out by Tattersall (1986), comparison of variability in fossil samples to patterns of interspecific variability in extant species has been neglected.

Excessive variation in tooth size in fossil samples may arise from four sources: the mixing of species, geographic variation, temporal variation, and sexual dimorphism (Kay, 1982a). Geographic and temporal variation presumably can be controlled by restricting a sample to limited localities and time horizons. Sexual dimorphism can be controlled by taking advantage of the fact that, even in the most dimorphic primates, dimorphism in upper and lower P4–M2 tooth size does not substantially increase sample variation (Plavcan, 1990). Thus, tests of a multiple species hypothesis usually focus on variation in P4–M2.

Most studies have used simple univariate and bivariate statistics to quantify dental metric variation in fossil samples. Multivariate statistics are potentially extremely useful, but they require a complete matrix of measurements. Since most fossil specimens are extremely fragmentary, sample sizes available for a multivariate analysis are usually prohibitively small.

The coefficient of variation (or CV) has been suggested to be the best univariate technique for quantifying tooth-size variation in fossil samples and testing the multiple-taxon hypothesis (Cope, 1989, this volume; Fleagle *et al.*, 1980; Gingerich, 1974; Kay, 1982a,b). Cope has demonstrated that among several different statistics (including frequency distributions, range as percentage of the mean, ratio of largest to smallest specimen) used for this purpose, the CV is the best, particularly at the small sample sizes so characteristic of the fossil record. The use of CVs for testing a multiple-taxon hypothesis has been criticized on the grounds that, occasionally, low CVs may be derived from artificially constructed, mixed-species samples (Cope, 1989, this volume; Kelley, 1987; Martin, 1983; Martin and Andrews, 1984; Plavcan, 1989, 1990; Vitzthum, 1990). It now seems clear that, while low CVs cannot be used to prove that only one species is present in a sample, high CVs should indicate the presence of more than one species in a sample.

Bivariate comparisons of tooth size are also popular for comparing variation in fossil samples to that of extant species (e.g., Kay, 1982b; Pickford, 1986). Such comparisons allow a simultaneous assessment of variation in both tooth size and proportions. Length and breadth dimensions of postcanine teeth in a fossil sample are plotted and compared to plots of similar tooth dimensions from an extant species. Discrete clusters of data points are usually interpreted as separate species, though sexual dimorphism may be suspected if the clusters fall within the total range of variation of the extant species. Normally, patterns of tooth size that may be formed when closely related species are mixed together in one sample are not considered.

This study compares patterns of tooth-size variation that may arise from intraspecific variation to those that may arise from interspecific variation. This is done to determine whether there are distinct patterns or degrees of tooth-size variation that can be used to recognize whether a sample is composed of a single species, which may or may not be sexually dimorphic, or more than one species. The emphasis is on CVs and bivariate analysis of tooth size, these being the most common methods for quantifying tooth size variation in fossils.

Materials and Methods

Samples

Dental metric data were gathered for 34 species of catarrhine primates (Table 1). Only wild-shot individuals of known sex and locality were selected from museum collections. Specimens from as small a geographic range as possible were selected, and individuals were carefully sorted by subspecies (for detailed information about specimen locality, see Plavcan, 1990). Classification of species and subspecies follows Napier (1981, 1985). Because inferences about patterns of interspecific variation are dependent on the taxonomic status of the samples used, care was taken to avoid comparing samples of dubious taxonomic status. It should be kept in mind that the variation within most single "species" used in this analysis, in fact, represents variation within subspecies of each species, and thus does not include geographic variability. Total intraspecific variation, inclusive of all populations and subspecies, may or may not be greater than that reported here for each species. This fact tends to make the results of this study more conservative, since subtle differences between locally sympatric species might be swamped by total intraspecific variation.

Variation in tooth size was examined in samples composed of one species (Table 1), and artificial samples composed of pairs of congeneric species whose geographic ranges overlap in the wild (Table 2). Except for the great apes, all mixtures of species were restricted to subspecies whose ranges overlap, and therefore might occur together in the fossil record. Thus, for example, *Colobus angolensis cottoni,* which occurs in the Ituri forest region of northeast Zaire, was mixed with *C. badius oustaleti,* which also occurs in the Ituri forest region of northeast Zaire, rather than *C. badius badius,* which occurs in Sierra Leone, Liberia, and Western Ivory Coast.

Because the great apes are used for comparison to most hominoid and hominid fossil samples, it is of interest to examine mixes of all species, even though congeneric species do not naturally occur in sympatry (only the gorilla and the chimpanzee actually have overlapping ranges). The lack of diversity

Table 1. Samples Used to Model Variability, Showing Means and Standard Deviations of CVs (in Parentheses) from Iterative Sampling Experiments[a]

Species	*N*	MD	BL	
			Trigonid	Talonid
Colobus angolensis cottoni	35	4.62(0.52)	4.92(0.67)	4.15(0.34)
Colobus badius badius	49	5.42(0.92)	5.02(0.88)	4.99(0.66)
Colobus badius oustaleti	46	5.14(0.74)	4.57(0.62)	4.53(0.50)
Colobus guereza caudatus	45	3.09(0.43)	4.57(0.65)	4.42(0.79)
Colobus polykomos polykomos	50	3.80(0.58)	4.65(0.54)	4.41(0.69)
Colobus satanas	44	3.24(0.48)	4.08(0.50)	4.48(0.56)
Colobus verus	46	3.56(0.42)	4.40(0.63)	4.67(0.53)
Cercopithecus aethiops hilgerti	33	5.16(0.36)	5.16(0.45)	5.77(0.70)
Cercopithecus ascanius whitesidei	51	4.81(0.63)	5.62(0.54)	5.68(0.69)
Cercopithecus cephus cephus	40	4.57(0.55)	5.66(0.67)	6.61(0.57)
Cercopithecus lhoesti lhoesti	27	—	—	—
Cercopithecus mona	30	5.84(0.53)	7.16(0.84)	6.28(0.64)
Cercopithecus neglectus	41	6.90(0.64)	7.28(0.62)	6.76(0.57)
Cercopithecus nictitans nictitans	45	5.32(0.63)	5.84(0.76)	5.83(0.73)
Cercopithecus pogonias grayi	38	5.00(0.66)	6.11(0.52)	6.25(0.51)
Cercocebus agilis agilis	24	—	—	—
Cercocebus albigena aterrimus	48	5.14(0.63)	6.04(0.61)	5.79(0.52)
Cercocebus torquatus atys	39	4.92(0.56)	5.46(0.67)	5.68(0.79)
Macaca fascicularis fascicularis	27	—	—	—
Macaca mulatta mulatta	50	6.84(0.95)	7.32(0.70)	7.60(0.81)
Macaca nemestrina nemestrina	34	5.24(0.54)	6.65(0.58)	6.86(0.77)
Macaca sinica	41	5.39(1.25)	5.60(0.54)	5.72(0.74)
Hylobates hoolock	52	4.46(0.50)	4.13(0.49)	4.90(0.92)
Hylobates lar carpenteri	40	5.16(0.57)	5.43(0.81)	7.32(1.37)
Hylobates syndactylus syndactylus	55	5.20(0.70)	5.44(0.86)	4.96(0.84)
Hylobates klossi	21	—	—	—
Gorilla gorilla gorilla	40	5.70(0.70)	6.68(0.86)	5.84(0.97)
Pan paniscus	38	4.88(0.58)	5.81(0.50)	5.93(0.64)
Pan troglodytes troglodytes	39	4.15(0.40)	4.48(0.46)	4.55(0.36)
Pongo pygmaeus pygmaeus	40	6.02(0.55)	6.71(0.86)	6.21(0.89)
Presbytis cristata pyrrhus	40	3.57(0.42)	3.88(0.35)	4.03(0.50)
Presbytis melalophos chrysomelas	31	5.22(0.81)	4.64(0.47)	4.52(0.72)
Presbytis obscura obscura	45	4.62(0.65)	4.48(0.45)	4.62(0.51)
Presbytis rubicunda chrysea	33	2.38(0.29)	3.18(0.43)	2.70(0.27)

[a]Sampling was for 50 iterations using 20 individuals in each iteration. Sex ratio was not controlled.

among the great apes, and the fact in the past hominoids may have been more diverse, greatly reduces the utility of these comparisons.

Variation in tooth size among several subspecies of seven species was also examined (Table 3). For *Presbytis entellus,* samples of seven subspecies from almost the entire range in India was available (Fig. 1), allowing a limited examination of how subspecific variation in tooth size is associated with geographic distance in this species. It must be pointed out that the representation of subspecies in each of these samples is limited, and conclusions need to be reinforced by more detailed analysis using more complete data sets.

Analyses

Measurements

The greatest lengths and breadths of all teeth on one side of the dentition were measured for each specimen. All measurements were taken with a Wild M5 binocular microscope with a calibrated reticle in the eyepiece. Measurements were converted to hundredths of millimeters. For each mandibular molar tooth, two breadth measurements were taken: one across the trigonid basin and one across the talonid basin. When a tooth was unmeasurable, its antimere, if present, was used. Specimens showing strong tooth wear were not measured. Details of measurement technique and protocol are available in Plavcan (1990).

The Coefficient of Variation

An iterative sampling procedure was used to generate CVs from single- and mixed-species samples. For each single-species or mixed-species sample, a computer program randomly selected a specified number of individuals and calculated CVs for all dental dimensions. Fifty iterations were performed for each mixed- and single-species sample at $n = 10$ and $n = 20$. In other words, for each single-species sample, such as *Cercopithecus pogonias,* the computer randomly chose 20 specimens (without replacement) and calculated the CV of each tooth dimension, then repeated this procedure 49 times more. For each mixed-species sample, such as a combination of *C. pogonias* and *C. cephus,* the computer randomly chose 10 individuals of *C. pogonias* and 10 individuals of *C. cephus,* calculated the CV of each tooth dimension, then repeated this procedure 49 times more. Finally, for each single- and mixed-species sample, the means and standard deviations of the CVs so generated for each tooth dimension were calculated.

Sex ratio was not controlled, because it has a trivial effect on the magnitude of the CV in upper and lower P4–M2, even in the most dimorphic species (Plavcan, 1990). For mixed-species samples, equal numbers of individuals from each species were chosen, maximizing CVs in these samples and biasing the results in favor of the ability of the CV to identify mixed-species samples. While paleontological samples may be composed of several hundred specimens (for example, Bown and Rose, 1987, on omomyid primates from the Bighorn Basin), a sample size of 20 would be considered large in many cases. For example, among fossil *Proconsul* and *Rangwapithecus* examined by Kelley (1987), sample sizes of canine teeth ranged from 6 to 21. Kay (1982a,b) examined a total sample (before sorting specimens by species) of Siwalik hominoid postcanine teeth, ranging in sample size from 11 to 35, while the Rain Ravine hominoid sample examined in the same report ranged from 3 to 7. Kimbel and White (1988) reported sample sizes ranging from 3 to 30 for four species of *Australopithecus.*

Comparison of subspecies within the same species was carried out in the same manner as for comparisons between different species. CVs generated from pairwise mixtures of subspecies were compared to CVs generated from the single-subspecies samples. In each case, 50 iterations were performed for each

Table 2. Mixes of Species to Model Variability, in Samples Composed of Two Closely Related, Sympatric Species Followed by Means and Standard Deviation (in Parentheses) of Mandibular M_1 CVs from Experimental Runs

Species	MD	BL	
		Trigonid	Talonid
Colobus			
C. angolensis cottoni + *C. badius oustaleti*	5.16(0.91)	4.97(0.85)	4.52(0.42)
C. angolensis cottoni + *C. guereza caudatus*	4.91(0.68)	5.90(0.93)	6.07(0.95)
C. guereza caudatus + *C. badius oustaletti*	4.56(0.71)	6.13(0.83)	6.47(0.68)
C. b. badius + *C. p. polykomos*	6.35(0.98)	8.80(0.88)	9.20(0.79)
C. verus + *C. b. badius*	11.43(1.07)	10.63(0.98)	9.78(0.96)
C. verus + *C. p. polykomos*	14.95(0.80)	17.27(0.92)	16.97(0.85)
Cercopithecus			
C. ascanius whitesidei + *C. l. lhoesti*	11.12(0.62)	11.01(1.65)	11.70(1.49)
C. ascanius whitesidei + *C. neglectus*	9.59(1.04)	9.36(1.41)	8.91(1.33)
C. l. lhoesti + *C. mona*	8.74(0.74)	10.10(1.78)	9.35(1.52)
C. l. lhoesti + *C. neglectus*	6.60(0.99)	8.16(1.47)	8.24(1.11)
C. mona + *C. neglectus*	7.94(0.79)	8.46(1.19)	6.91(1.06)
C. c. cephus + *C. n. nictitans*	6.46(0.80)	6.13(0.93)	6.52(0.86)
C. c. cephus + *C. pogonias grayi*	4.64(0.71)	5.75(0.84)	6.15(0.88)
C. n. nictitans + *C. pogonias grayi*	7.08(0.83)	6.29(0.69)	6.27(0.78)
Cercocebus			
C. albigena aterrimus + *C. a. agilis*	6.67(0.85)	10.61(0.92)	9.61(0.95)
C. albigena aterrimus + *C. torquatus atys*	9.66(1.01)	10.99(1.19)	10.68(1.25)
C. torquatus atys + *C. a. agilis*	6.07(0.92)	6.20(1.10)	6.27(0.98)
Macaca			
M. f. fascicularis + *M. n. nemestrina*	12.17(0.98)	12.70(1.02)	15.37(1.23)
Presbytis			
P. melalophos chrysomelas + *P. cristata pyrrhus*	10.97(0.64)	7.82(0.66)	7.48(0.64)
P. melalophos chrysomelas + *P. o. obscura*	6.89(0.77)	4.60(0.66)	4.59(0.74)
P. melalophos chrysomelas + *P. rubicunda crysea*	4.12(1.29)	4.24(0.58)	3.66(0.77)
P. rubicunda crysea + *P. cristata pyrrhus*	11.09(0.48)	6.52(0.54)	7.44(0.49)
P. rubicunda crysea + *P. o. obscura*	6.64(0.66)	4.09(0.40)	3.76(0.45)
P. cristata pyrrhus + *P. o. obscura*	6.50(0.82)	7.55(0.63)	7.37(0.75)
Hylobates			
H. hoolock + *H. lar carpenteri*	7.30(0.80)	6.31(0.75)	7.13(1.89)
H. s. syndactylus + *H. klossi*	15.76(0.87)	16.71(1.18)	16.50(1.17)
Great apes			
G. g. gorilla + *P. paniscus*	23.61(0.86)	22.41(1.11)	21.98(1.23)
G. g. gorilla + *P. p. pygmaeus*	10.48(0.90)	9.39(0.90)	8.04(1.04)
G. g. gorilla + *P. t. troglodytes*	17.90(0.78)	18.73(1.09)	16.47(1.10)
P. paniscus + *P. p. pygmaeus*	15.40(1.30)	15.82(1.58)	16.73(1.38)
P. paniscus + *P. t. troglodytes*	7.74(0.96)	6.52(1.10)	8.00(0.92)
P. p. pygmaeus + *Pan t. troglodytes*	9.60(0.14)	12.08(0.18)	11.14(0.18)

mixed- and single-subspecies sample for n = 10 and n = 20. The exception to this was *P. entellus,* in which the small sample sizes restricted all comparisons to n = 10. Furthermore, because of the small sample sizes available for several subspecies of *P. entellus,* only 10 iterations were carried out for each single- and mixed-subspecies sample.

Analysis of CVs was performed on untransformed data. It is important to note that log transformation, while normalizing the distribution of data, also normalizes variances. [This holds as long as the variance scales to the square of the mean. When variance scales in proportion to the mean itself, log transformation actually results in relatively smaller variances with increasing means (Bryant, 1986). This fact should also be kept in mind for the results of the bivariate comparisons reported below.] Calculating the CV of log-transformed data actually reintroduces a size effect, invalidating comparisons of variation between large and small species. This is because normalization of variances removes the well-known size affect on variance, such that standard deviations of large objects are comparable to those of small objects. Since the CV is the sample standard deviation divided by the sample mean, CVs calculated from log-transformed data will tend to actually get smaller for larger species!

Table 3. Samples Used to Investigate Subspecific Variation, Showing Means and Standard Deviations (in Parentheses) of CVs from Iterative Sampling Experiments[a]

			BL	
Species	*N*	MD	Trigonid	Talonid
Cercopithecus aethiops				
hilgerti	33	5.12(0.82)	5.15(0.88)	5.70(1.19)
sabaeus	21	6.76(0.86)	8.06(2.72)	6.17(1.20)
C. mitis				
kolbi	20	5.13(0.78)	5.94(1.27)	5.75(0.85)
stuhlmanni	19	7.02(2.05)	6.71(0.94)	6.85(1.19)
C. pogonias				
grayi	38	4.84(1.25)	6.27(1.00)	6.15(1.05)
pogonias	15	3.16(0.58)	5.78(0.65)	5.03(0.79)
Colobus guereza				
caudatus	45	3.04(0.74)	4.29(1.07)	4.19(1.08)
occidentalis	19	4.85(1.20)	6.73(1.48)	7.74(1.80)
Pan troglodytes				
troglodytes	39	4.25(0.82)	4.42(0.85)	4.62(0.54)
schweinfurthei	15	4.23(0.74)	5.10(1.40)	4.57(1.00)
Macaca mulatta				
mulatta	50	6.66(1.53)	7.34(1.31)	7.52(1.31)
villosa	13	9.97(1.04)	5.86(0.76)	4.99(0.54)
Presbytis entellus[b]				
achates	11	4.90	6.89	5.94
ajax	13	3.39	3.95	4.92
entellus	16	5.66	6.59	5.04
hector	10	5.26	4.15	3.28
priam	9	5.67	8.62	9.82
thersites	32	4.66(0.62)	4.79(0.77)	4.94(1.02)

[a]Sampling was for 50 iterations using 10 individuals in each iteration. Sex ratio was not controlled.

[b]Sample size of all subspecies of *P. entellus* except *P. e. thersites* were too small for random sampling experiments; therefore, only one CV for each subspecies sample was calculated.

Fig. 1. Approximate geographic location within India of the six subspecies of *Presbytis entellus* used in this study.

Bivariate Tooth-Size Plots

Bivariate comparisons of postcanine tooth-size dimensions were carried out for each single- and mixed-species sample. Particular attention was paid to those samples in which the CV could not identify the presence of two species. Sample size was not controlled in the bivariate comparisons. Comparisons were made between dimensions of the same tooth (such as M_1 length vs. M_1 breadth) and between dimensions of different teeth in the same jaw.

All bivariate comparisons were made with ln-transformed data. All plots were scaled to the same dimensions so that the relative scatter of points could be compared (again, under the assumption that variance increases with the square of the mean). Ln-transformation does not change the pattern of dispersal of points in bivariate plots but is essential if species of different sizes are compared. It is well known that standard deviations increase with increasing size (Simpson *et al.*, 1960; Sokal and Rohlf, 1981). Comparison of the scatter of points in a bivariate plot can be deceptive if this phenomenon is not taken into consideration. For example, Pickford (1986) compared untransformed M1 dimensions of *Proconsul nyanzae* to a sample of *Pan troglodytes,* which appears considerably larger than all but the largest fossil specimens. While, visually, the scatter of points seemed somewhat larger in the fossils, in fact the relative variation of the fossil sample was even larger than it appeared in the plot because of the correlation between dispersion and size.

Results

The Coefficient of Variation

Within single-species samples, CVs are consistently lowest in the region of P4–M2 (Fig. 2). The effect of sexual dimorphism in these teeth is trivial in comparison to the overall range of CVs found within single-species samples. Even a taxon characterized by relatively large postcanine tooth size dimorphism, such as *Pongo,* while showing somewhat inflated CVs in these teeth compared to most other taxa, still produces CVs well within the range of CVs generated from other single-species samples. Among all teeth, M_1 usually shows the lowest CVs, the value rarely exceeding 8.0, even with sample sizes of 10 or less. For this reason, most of the following analyses focus on M_1.

The CVs of mixed-species samples almost completely overlap those generated from the single-species samples in all tooth dimensions. While the total range of CVs generated from the mixed-species samples greatly exceeds the total range of CVs generated from the single-species samples in all postcanine tooth-size dimensions, a surprisingly large number of mixed-species CVs fall within the range of single-species CVs (Fig. 3).

The magnitude of CVs generated from mixed-species samples is, of course, a simple function of average tooth-size differences between the species. When species differ greatly in tooth size, as in the mixture of *Macaca fascicularis* and *M. nemestrina* (Fig. 4, top), CVs of postcanine tooth dimensions are consistently higher than those generated from any single-species sample. When species are nearly identical in tooth size, as in pairwise mixtures of *Cercopithecus pogonias, C. nictitans,* and *C. cephus,* single- and mixed-species CVs are very similar (Fig. 4, bottom). In this latter case mixtures of *C. nictitans* with either of the other two *Cercopithecus* species actually results in inflated CVs (Cope, this volume). However, the slightly elevated CVs from these mixtures all fall well within the range of single-species CVs derived from all catarrhines.

There are some taxonomic differences in molar tooth-size variability among single-species samples (Table 1). In general, species with greater postcanine dimorphism show higher CVs, but the magnitude of the effect is minimal in comparison to the overall range of CVs (Plavcan, 1990). The great apes, in particular *Gorilla* and *Pongo,* show high postcanine CVs. Cercopithecines tend to show higher postcanine tooth-size dimorphism, and thus higher average postcanine CVs, than colobines.

The only group in which most mixtures of species produce high postcanine CVs are the great apes (Table 2). However, these samples are extremely artificial, since most do not actually occur sympatrically in the wild. Furthermore, these species can be easily recognized on the basis of postcanine tooth morphology, rendering CVs of little additional benefit in recognizing distinct species. Mixtures of *Pongo* and *Pan* consistently produce enormous postcanine CVs, but I seriously doubt that any paleontologist would mix specimens from Sumatra with specimens from Liberia. Of the great apes, only *Pan* and *Gorilla* actually occur in regional sympatry, and mixtures of these species produce very high postcanine

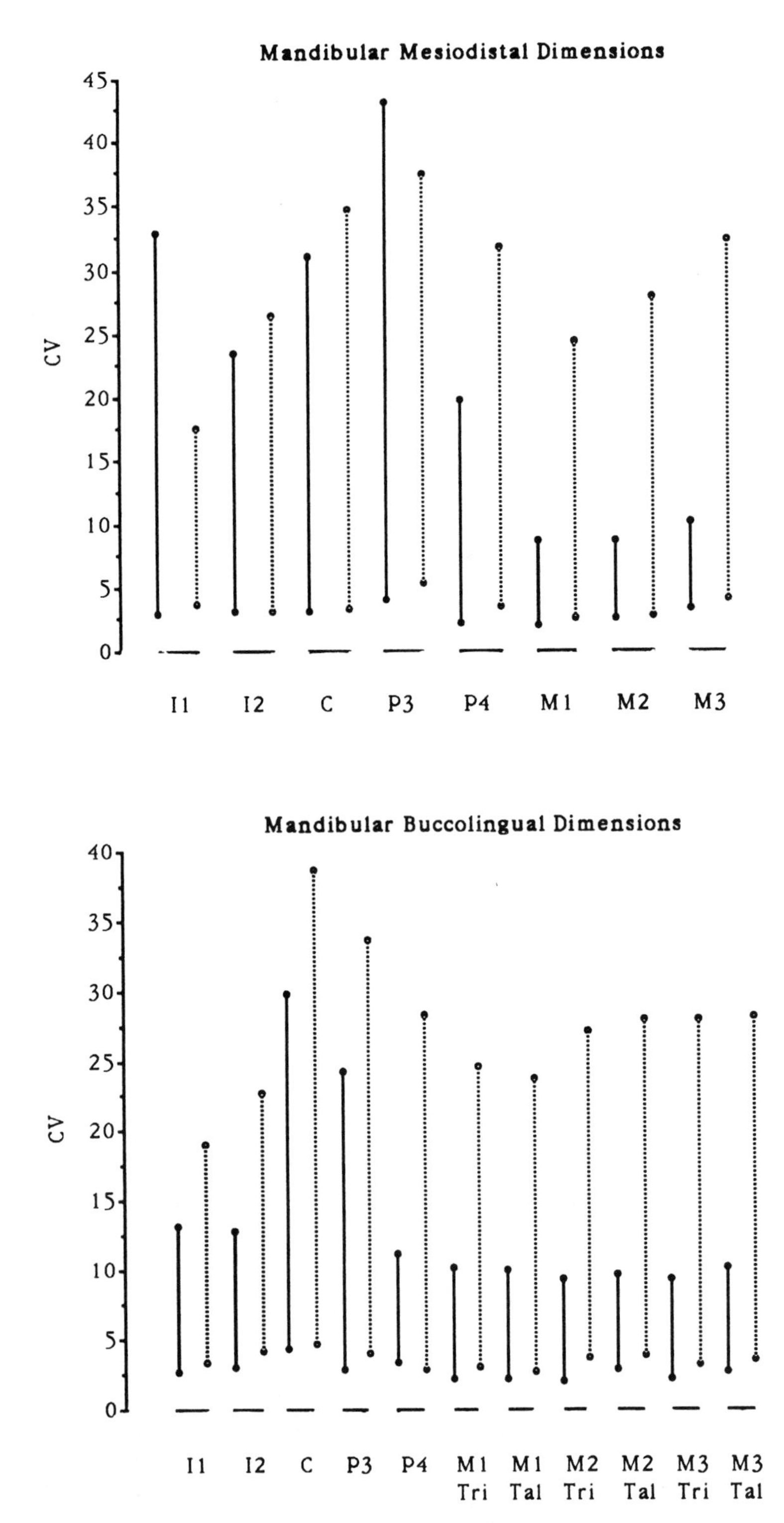

Fig. 2. Range of CVs generated for all single-species samples (solid lines) compared to the range of CVs generated for all mixed-species samples (dotted lines) for each measurement of the mandibular dentition at $n = 20$.

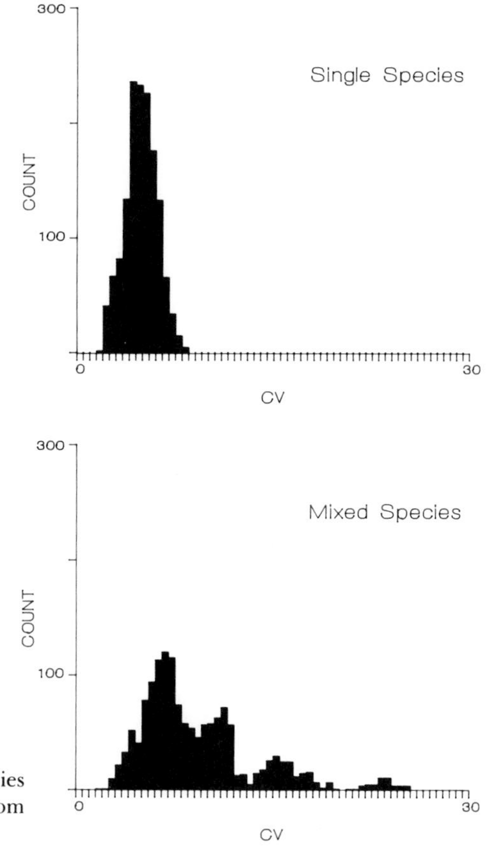

Fig. 3. Frequency histograms of all single-species CVs (top panel) and all mixed-species CVs (bottom panel) generated for $n = 20$.

CVs. However, these species differ enough in tooth morphology such that tooth size would not be needed to distinguish between them.

Mixtures of *Presbytis entellus* from the full extent of its range in India, Pakistan, Bangladesh, and Nepal produce enormous CVs (Table 4). For example, mixtures of *P. e. thersites,* from Sri Lanka and *P. e. ajax* from Kashmir produce CVs that are on average greater then 13.00 for each lower first molar tooth dimension, well above the CVs generated for single-species samples. However, mixtures of *P. entellus* populations from restricted parts of its range produce CVs entirely within the range of single-species CVs. In particular, those populations that are relatively near one another, such as *P. e. ajax* and *P. e. hector,* or *P. e. achates* and *P. e. priam* (Fig. 1), produce CVs well within the overall anthropoid range of single-species CVs. This is because there is a geographic cline in tooth size within this species, such that subspecies from Sri Lanka and the south of India (such as *P. e. thersites*) are much smaller than more northerly populations (such as *P. e. ajax*). Mixtures of subspecies within *Cercopithecus aethiops, C. mitis, C. pogonias, Colobus guereza, Macaca mulatta,* and *Pan troglodytes* did not produce postcanine CVs markedly inflated beyond those of the single-subspecies samples from which they were composed, though standard deviations of the CVs generated from the mixed-subspecies samples tend to be higher than those generated from the single-subspecies samples.

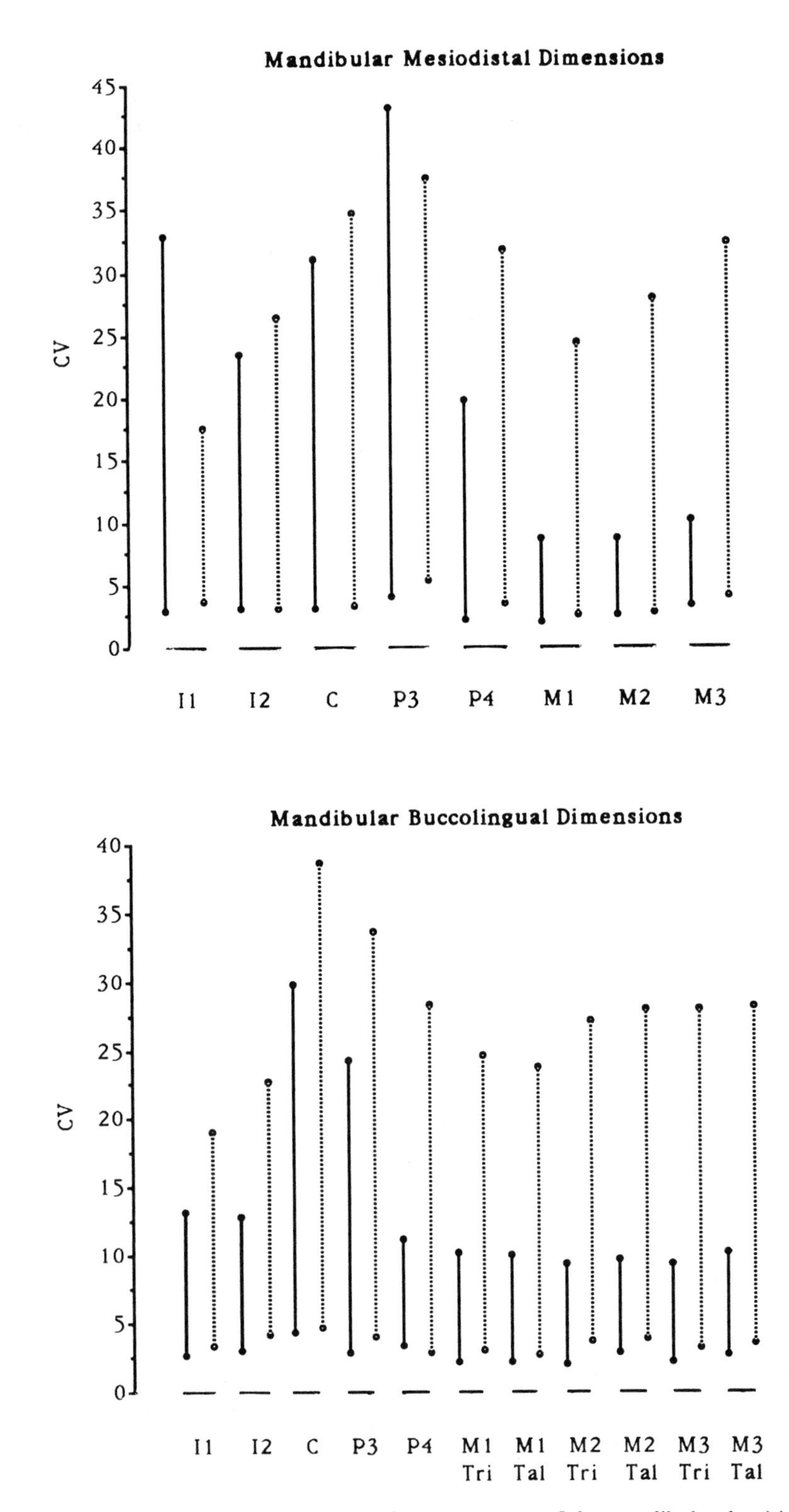

Fig. 4. Comparison of the ranges of CVs of each measurement of the mandibular dentition generated from mixed-species samples (dotted lines) and single-species samples (solid lines) of *Macaca fascicularis* and *M. nemestrina* (top panel), and *Cercopithecus pogonias, C. cephus,* and *C. nictitans* (bottom panel). Combinations of the latter set of species were all pairwise.

Table 4. Mixes of Subspecies Used to Investigate Subspecific Variation, Showing Means and Standard Deviations (in Parentheses) of CVs from Iterative Sampling Experiments[a]

Species	MD	BL	
		Trigonid	Talonid
Cercopithecus aethiops hilgerti + *C. a. sabaeus*	7.52(1.49)	6.80(1.74)	7.42(1.76)
C. mitis kolbi + *C. m. stuhlmani*	5.52(1.86)	6.14(1.42)	6.28(1.68)
C. pogonias grayi + *C. p. pogonias*	4.04(1.04)	5.96(1.24)	6.09(1.32)
Colobus guereza caudatus + *C. g. oustaleti*	4.92(1.30)	6.76(1.82)	5.84(1.92)
Pan troglodytes troglodytes + *P. t. schweinfurthi*	4.27(0.90)	4.94(1.05)	4.54(0.83)
Macaca mulatta mulatta + *M. m. villosa*	7.93(1.44)	6.93(1.60)	6.99(1.46)
Presbytis entellus			
P. e. ajax + *P. e. hector*	3.82(1.10)	4.20(0.66)	4.38(0.79)
P. e. ajax + *P. e. achates*	10.41(0.87)	11.55(1.23)	10.01(1.11)
P. e. ajax + *P. e. entellus*	7.20(1.13)	7.63(1.05)	6.63(0.73)
P. e. ajax + *P. e. priam*	11.09(1.33)	10.87(1.35)	11.08(1.50)
P. e. ajax + *P. e. thersites*	13.03(1.04)	14.42(0.92)	13.31(1.39)
P. e. achates + *P. e. entellus*	5.78(0.89)	7.50(1.36)	7.24(0.83)
P. e. achates + *P. e. hector*	10.12(0.97)	10.42(1.10)	9.14(0.68)
P. e. achates + *P. e. priam*	4.57(0.58)	8.02(0.87)	7.15(1.29)
P. e. achates + *P. e. thersites*	5.70(1.36)	6.14(0.65)	6.68(1.08)
P. e. entellus + *P. e. hector*	8.20(0.85)	8.01(1.23)	4.74(1.33)
P. e. entellus + *P. e. priam*	6.22(0.75)	7.66(0.97)	8.48(1.04)
P. e. entellus + *P. e. thersites*	7.39(1.05)	7.64(1.14)	9.04(1.40)
P. e. hector + *P. e. priam*	10.83(1.05)	10.16(1.29)	9.74(1.18)
P. e. hector + *P. e. thersites*	12.91(0.62)	12.63(0.90)	12.31(0.80)
P. e. priam + *P. e. thersites*	5.29(1.24)	8.25(1.17)	9.61(1.27)

[a]Sampling was for 50 iterations (except for combinations of *P. entellus*) using five individuals from each subspecies for each iteration. Sex ratio was not controlled.

Bivariate Plots

Figure 5 compares sexual dimorphism in M1 in *Pongo pygmaeus* to that in *Presbytis cristata*. The orangutan has the greatest postcanine dimorphism found in any catarrhine, and probably in any primate (Plavcan, 1990). Male and female dispersions partially overlap, and the distribution of both sexes is nearly continuous. The sample of *P. cristata* shows a degree of molar dimorphism found in most catarrhines. Females are slightly smaller than males, but the distributions overlap to such a degree that males and females could not be sorted without prior knowledge of sex.

Most mixed-species samples that are indistinguishable by the CV show slight differences in tooth size and shape in bivariate comparisons. Figures 6–8 demonstrate three situations found in bivariate plots of P4–M2 tooth-size dimensions of mixed-species samples from which low CVs were generated. Each figure shows two versions of the same plot, one of which indicates the identity of each specimen by species. Figure 6 shows a mixture of *Colobus angolensis* and *C. badius*. While there is a slight difference between the two species in tooth proportions, one cannot tell that two species are present in the sample when the data points are not identified as to taxon. The total dispersion of points is intermediate between that of *P. pygmaeus* and *P. cristata* (Fig. 5). This pattern is common

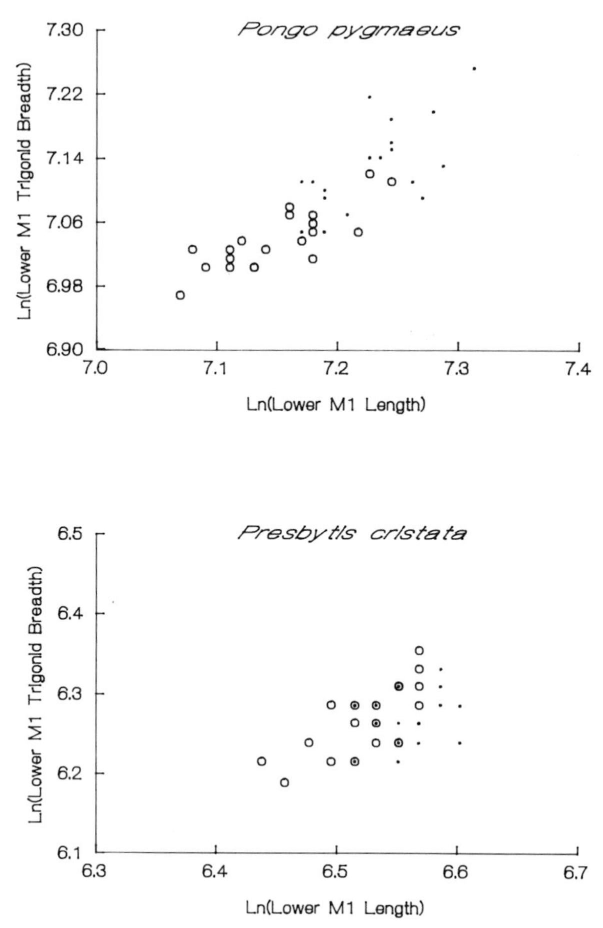

Fig. 5. Bivariate plots of ln-transformed lower first molar dimensions of *Pongo pygmaeus* (top) and *Presbytis cristata* (bottom) demonstrating sexual dimorphism. Sexual dimorphism in the postcanine dentition of *Pongo* is as great as or greater than that in any other catarrhine. The pattern of *P. cristata* is by far the most common among catarrhines (■, males; ○, females).

among the mixed-species samples. Figure 7 shows a similar plot for a combination of *P. cristata* and *P. obscura*. While the species differ in tooth size, the distribution of points is continuous, and the total range of variation is comparable to that of *Pongo pygmaeus* (Fig. 5). On the basis of tooth size alone, one would be hard pressed to decide whether the sample was composed of two species, or a single, sexually dimorphic species. Along with the pattern shown in Figure 6, this pattern is common among the mixed-species samples. Figure 8 displays a similar plot for a mixture of *P. cristata* and *P. melalophos*. While the total range of variation slightly exceeds that of *Pongo pygmaeus* (Fig. 5), there is a distinct difference in tooth proportions that is never apparent with sexual dimorphism. The CVs of the tooth dimensions of this combination of species fall at the upper end of the single-species range.

Plots of species showing large postcanine tooth size differences usually show two distinct clusters of points with no overlap (Fig. 9). In such cases CVs accu-

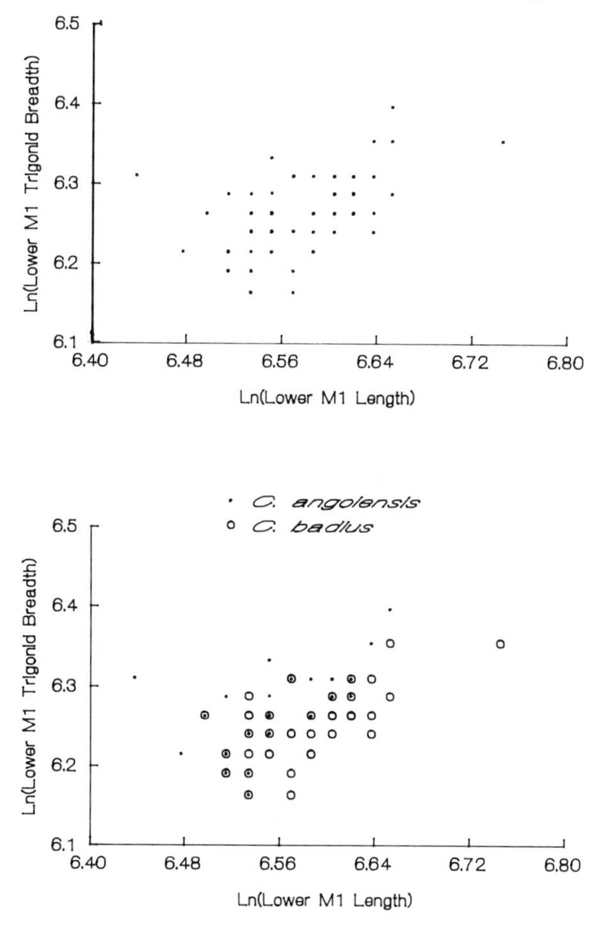

Fig. 6. Two bivariate plots of the same ln-transformed lower molar dimensions of a mixed-species sample of *Colobus angolensis* and *C. badius*. The top panel shows the data points without regard to species; the bottom panel shows the same plot with the data points identified by species.

rately identify the presence of two species, and individuals can be easily and accurately sorted by species on the basis of tooth size alone.

Discussion

The results of this study clearly demonstrate that closely related species that have overlapping geographic ranges—and thus might occur together in the fossil record—are occasionally virtually indistinguishable by postcanine tooth size. A number of combinations of species used in this analysis would be classified as a single species on the basis of this character were they found together in the fossil record. This strongly implies that diversity occasionally will be undersampled in the fossil record when species are recognized on the basis of postcanine tooth size and stresses the importance of using other characters, such

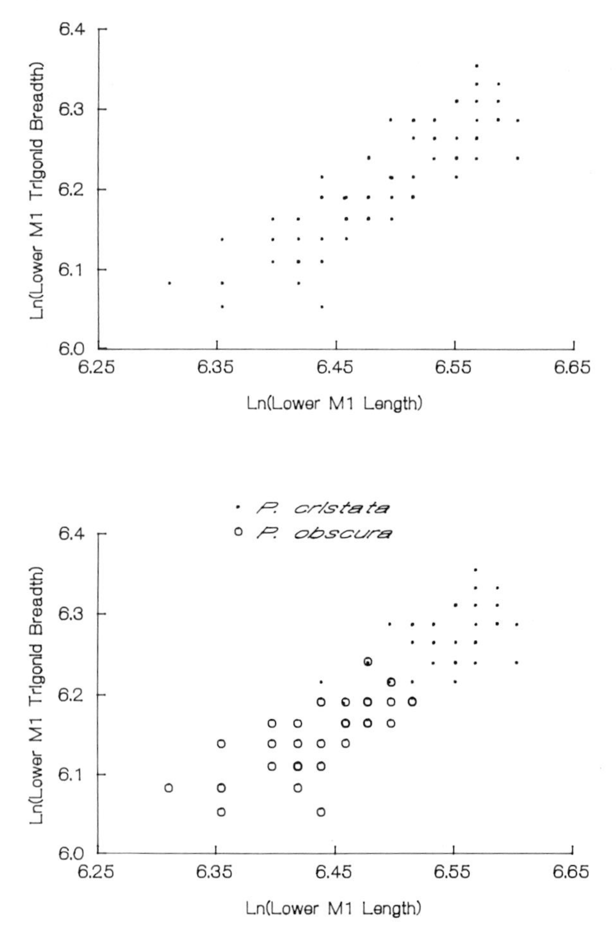

Fig. 7. Two bivariate plots of the same ln-transformed lower molar dimensions of a mixed-species sample of *Presbytis cristata* and *P. obscura*. The top panel shows the data points without regard to species; the bottom panel shows the same plot with the data points identified by species.

as those of the anterior dentition (Bown and Rose, 1987; Kelley, 1987) for species recognition whenever possible. The degree of undersampling depends on how commonly species that are similar in both dental morphology and tooth size actually occur in the same geographic area.

The CV is of limited use for identifying mixed-species samples in the fossil record. While low CVs offer no positive evidence of the species composition of a sample, high postcanine CVs should suggest the presence of more than one species. Even so, CVs should be used with caution. Sexual dimorphism, geographic variation, and presumably temporal variation within a single lineage can inflate sample variability. Sexual dimorphism in upper and lower P4–M2 of catarrhine species tends to be very low, and even in the most dimorphic species it has only a limited effect on postcanine CVs (Plavcan, 1990). CVs for more dimorphic teeth, such as P_3 or the canines (Leonard and Hegmon, 1987; Kelley, this volume), should not be used to test a multiple-species hypothesis, since sample variation due to dimorphism is often as great as that due to interspecific differences in tooth size (Plavcan, 1990).

Postcanine CVs can be inflated by geographic variation among populations of a single species, though the generality of such a conclusion must be considered tenuous because of the limited data for subspecies within species available for this study. Among the species examined in this study, only within *Presbytis entellus* does mixture of subspecies inflate sample variation substantially beyond that seen in single-species samples. But even within this species, inflated sample variation only occurs when populations separated by great distances are mixed together. This implies that restriction of specimens by locality should minimize the chances that geographic variants of a single species would be recognized as two separate species.

That geographic variants of a single species can differ substantially in tooth size implies that temporal variation can also result in increased sample variation.

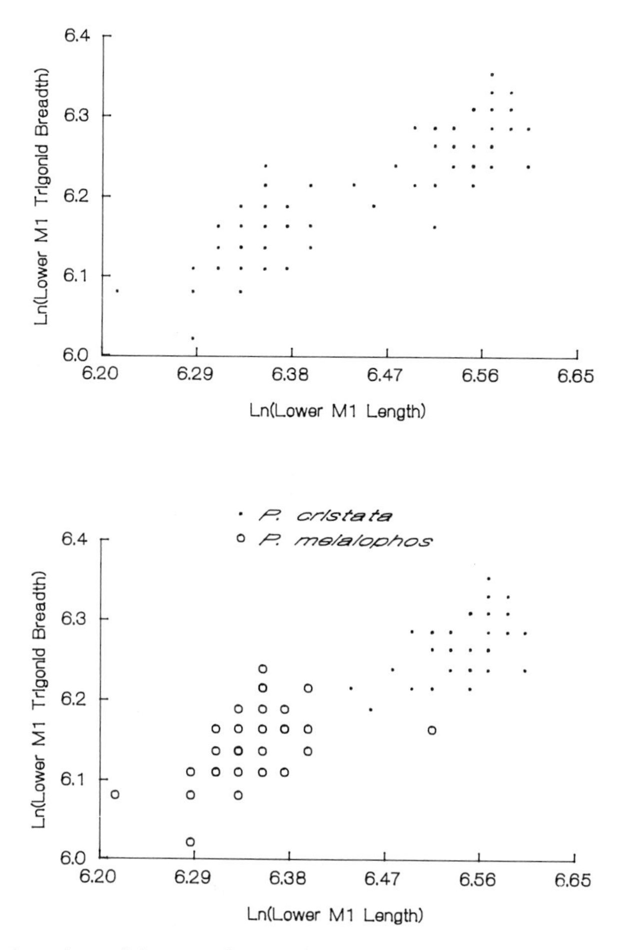

Fig. 8. Two bivariate plots of the same ln-transformed lower molar dimensions of a mixed-species sample of *Presbytis cristata* and *P. melalophos*. The top panel shows the data points without regard to species; the bottom panel shows the same plot with the data points identified by species.

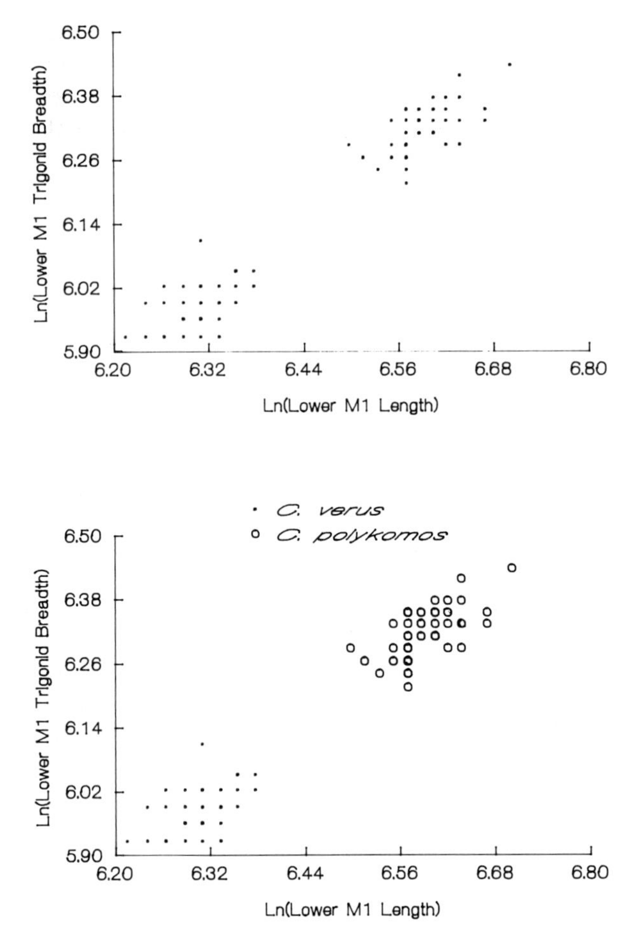

Fig. 9. Two bivariate plots of the same ln-transformed lower molar dimensions of a mixed-species sample of *Colobus verus* and *C. polykomos.* The top panel shows the data points without regard to species; the bottom panel shows the same plot with the data points identified by species.

To control for this possibility, geologic horizon should be restricted as carefully as possible. If this cannot be done, then intraspecific variation cannot be ruled out as a possible cause of increased sample variability.

While CVs offer a quantitative method of comparing sample variation tooth dimensions, bivariate plots allow examination of individual data points, along with a subjective assessment of patterns of tooth-size variation. Because bivariate plots can detect differences in tooth proportions, they potentially can reveal subtle differences in tooth size that univariate statistics cannot. However, the interpretation of the significance of patterns of variation in such plots can be highly subjective. For example, Pickford (1986) rejects the hypothesis that the East African sample of *Proconsul* contains two sexually dimorphic species of similar size (Greenfield, 1972) on the basis that bivariate plots of canine tooth size form two clusters of points (this is the same type of evidence used by Kelley (this volume) to assert that the Lufeng sample of hominoids is composed of only a single species).

As Pickford points out (1986, p. 156) ". . . size was the main criterion for sorting *Proconsul* species." However, the tooth-size distribution presented as evidence of a single dimorphic species could easily be generated from either a single sexually dimorphic species or a mixture of two dimorphic species. That ". . . the dentition is also more variable than in living apes. . ." (Pickford, 1986) suggests, according to the results of the present study, that in fact two species are present. Without further morphological evidence, the hypothesis that the sample is composed of a single dimorphic species cannot be tested against the hypothesis that more than one species is present.

In the same study, Pickford (1986) argues that continuity in the clustering of data points suggests that only a single species is present. But, as shown above, this pattern is easily reproduced when closely related species are pooled in a sample, and by itself offers no evidence whatsoever about either the taxonomic or sexual composition of the sample. Conversely, Kay (1982b) uses discontinuity in the scatter of points of *Sivapithecus* from the Siwalik Hills to assign specimens to three species. Just as a continuous distribution of points can be generated from either a single, sexually dimorphic species or a mixture of closely related species, so too can a discontinuous distribution be generated from either small samples of a single, sexually dimorphic species (such as *Pongo pygmaeus,* illustrated in Fig. 10), or a mixture of closely related species. While Kay convincingly demonstrates that the Siwalik Hills sample is probably composed of more than one species on the basis of the high variation in tooth size (though stratigraphic control is poor and the possibility of a geographic or temporal component to variation cannot be ruled out), the assignment of particular specimens to one or the other species by size alone is unwarranted. The assumption inherent in this practice—that closely related, sympatric species should usually show discrete distributions in tooth size—is simply not true.

Claims that sexual dimorphism in the postcanine dentition of particular fossil taxa exceed that in any living species (Kelley 1986, 1991, this volume;

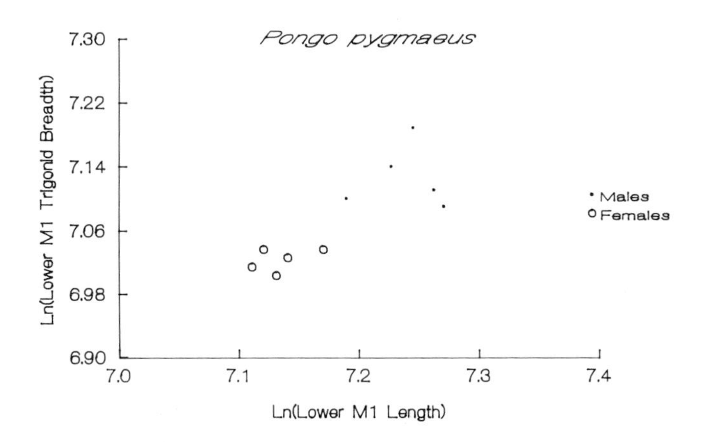

Fig. 10. Plot of ln-transformed lower molar tooth dimensions of a randomly drawn sample of 10 *Pongo pygmaeus* (five males and five females) demonstrating a discrete tooth size distribution of males and females resulting from random sampling. The computer generated this figure in the fourth attempt of random sampling from the distribution presented in Fig. 5.

Pickford, 1986) are difficult to falsify and should be approached with caution. The fossil record cannot be interpreted only with reference to itself. Therefore, we must consistently rely on neontological models to interpret variation in the fossil record (Kay, 1982b; Kay and Cartmill, 1977). To justify the claim that the biological parameters of fossil species differ from those of living species on the basis of fossil evidence is circular, since the proof rests on the data that are being questioned. On this basis, if tooth-size variation in a fossil sample exceeds that in all living species, we have little choice but to infer that the sample is composed of more than one species, rather than two sexes, unless there is other, strong evidence to support the dimorphism hypothesis. This is not to say that extinct taxa were never more dimorphic (in the postcanine dentition) than living taxa, nor does it preclude the demonstration of extreme dimorphism in fossil taxa. Rather, it is very difficult to identify such dimorphism.

The practice of rigorously interpreting the fossil record with reference to the present has been criticized on the grounds that it condemns the past to be a replica of the present (Kelly, this volume). Few would argue with the idea that many species in the past differed from their living descendants. The problem is testing hypotheses about how extinct species differed from those living today. With the exception of anatomically unique features that have no analog among extant species, virtually all that is known or inferred about extinct species derives from comparison between fossils and modern species at some level. The rigorous demonstration of a unique phenomenon in the fossil record (such as extreme dimorphism) should rest on evidence that a particular neontological model could not apply to the extinct species. For example, Kelley could easily demonstrate that the Lufeng hominoids probably had greater *canine* dimorphism than any living species. Assuming that Kelley's assignation of sex to each specimen is correct (a procedure that is based on the assumption that the morphology of male and female canines in the extinct species mirrors that of living species), it could be demonstrated that no matter how many species are in the sample, the canine dimorphism of all would have to be greater than that of any living species. The same logic applies to the inferences about dimorphism in the molar teeth. A careful demonstration that dimorphism in the postcanine teeth must be greater than that seen in any living species, no matter how many species are in the sample, would support Kelley's interpretation. Such a demonstration rests on the accurate identification of the sex of individual specimens. As presented in this volume, though, the pattern of postcanine tooth size variation of the Lufeng hominoids is consistent with both the single- and multiple-species hypotheses (in particular, the hypothesis that there are two dimorphic species of similar, but not identical, body size), and, strictly speaking, there is no absolute way to choose between one or the other hypothesis. However, the level of tooth-size dimorphism proposed by Kelley is nearly twice as great as that of any living mammal. This suggests that until the multiple-species hypothesis can be ruled out with confidence, it should represent the working hypotheses.

Kelley (this volume) discusses several reasons why the data from my study do not apply to the hominoid fossil record. He argues that cercopithecoids are biologically different from hominoids and that the large body size of hominoids precludes the existence of two species of similar body size in sympatry. To sustain this argument, Kelley must assert that all fossil apes are biologically similar to the living apes. Of course, this imposes a neontological model of the present onto the past, producing a world that is an exact replica of the present (Kelley, this volume), and we are left trying to decide which neontological model is most reasonable for application to the hominoid fossil record. That fossil apes were biologically similar to modern apes is simply a hypothesis, not evidence that the results of this study do not apply to fossil apes.

This study confirms the observation that morphological differentiation (at least in tooth size) does not necessarily correlate with speciation (see also Cope, this volume; Tattersall, this volume; Groves, this volume). At least in some cases, closely related species are difficult to distinguish on the basis of tooth morphology alone (Gingerich, 1974, 1985; Kay, 1982a,b; Tattersall, this volume). Enormous geographic variation in tooth size may be found among populations of a single species (as in *Presbytis entellus*), while congeneric, sympatric species can be virtually indistinguishable in tooth size (as among several *Cercopithecus, Colobus,* and *Presbytis* species).

The ability to test a taxonomic hypothesis is limited by the available evidence. While currently debated species concepts emphasize reproductive or genetic aspects of populations rather than morphological discontinuity, in practice most systematists recognize neontological species by morphology. Neontologists face similar problems as paleontologists in species recognition, but have available for study evidence from behavior, ecology, morphology, and genetics to test the hypothesis that a sample of organisms represents a single species. For example, were modern savannah baboons known only from morphology, there probably would be little argument that hamadryas and olive baboons comprise discrete species easily recognized by morphological characters. However, field studies have revealed a degree of reproductive continuity between the two groups that suggest (by the criteria of the biological species concept) that they be subsumed within a single species (Thorington and Groves, 1970; Jolly, this volume). Such a test can never be achieved in the fossil record. Thus, lacking other criteria for recognition, a species in the fossil record is practically recognized as a cluster of morphologically similar individuals in which variation does not exceed that of similar recognized extant species (Simpson, 1961). Such a "paleospecies" is thus simply a hypothesis that a group of individuals constitutes a sample from a single, biological species.

Anagenetic change (temporal variation) in an evolving lineage is difficult to accommodate within any species concept and renders any attempt at standardization of species recognition at some point arbitrary. For example, while some advocate the concept of paleospecies (Bown and Rose, 1987; Gingerich, 1985; Simpson, 1961) to recognize time-successive populations of an evolving lineage, others criticize the practice as arbitrarily delineating boundaries between populations of a single "evolutionary" species using breaks in the fossil record (Wiley, 1981). Even the latter approach (recognition of evolving lineages as a single species), though, involves arbitrary systematic practice since it, too, relies on obvious breaks in the fossil record to delineate species and creates variation in "species" that has no modern analogue.

Paleontologists are above all interested in documenting tempo and mode in evolution. Necessarily, the first step in this process involves the identification of clusters of individuals that are morphologically similar within the bounds of variation of modern species. Only after this is done can relations between samples be meaningfully assessed. The hypothesis that a temporal succession of samples represents a single evolving lineage only has meaning after the phylogenetic relationships between stratigraphically successive groups have been assessed. Since an extinct species is only a taxonomic hypothesis, sinking successive populations into a single extremely variable species can only obscure patterns of variation and confuse meaningful tests about evolutionary change. On the other hand, it must be noted that recognizing a cluster of individuals as representative of a species does not imply that phyletic transformation of species cannot or does not occur.

After all this, can tooth-size variability be used to assist in species recognition in the fossil record? The answer is a qualified "yes." While it is easy to demonstrate that congeneric species may be indistinguishable by tooth size alone, postcanine dental variation in extant species is consistently limited. When variation in a fossil sample exceeds that characteristic of living species, then more than one species is probably present in the sample, even if individuals cannot be sorted on the basis of tooth size alone. The difficulty of assigning individuals to a particular species imposed by fragmentary evidence should not be used as justification for claiming that an extremely variable sample is composed of only one species. Conversely, if variation in a fossil sample is consistent with that of extant species, then only one species can be recognized, with the caveat that the grouping may be revised as further morphological evidence becomes available. Since teeth can vary between species in proportions as well as size, it is well to consider both variation within single tooth dimensions as well as covariation between different traits. Above all, interpretations of patterns of variation in fossils must be based on comparison to both intraspecific and interspecific variability of extant species in as many anatomical features as possible. By so doing, taxonomic hypotheses based on the meaning of sample variability can be tested much more rigorously.

Summary

Tooth size is commonly used to assist in the recognition of species in the fossil record. While tooth-size variation is usually compared to intraspecific variation in extant species, rarely is interspecific variation considered. The results of this study show that tooth size can be used to recognize the presence of more than one species in a sample when variation exceeds that in any living species. However, closely related extant species with overlapping geographic ranges commonly do not differ enough in tooth size to be recognized as distinct species, implying that diversity will be underestimated in the fossil record when tooth size is used to recognize species. This result supports observations that, in general, morphological differentiation does not necessarily parallel speciation and underscores the fact that species in the fossil record represent taxonomic hypotheses. Only when this is recognized can hypotheses of the tempo and mode of evolution be clearly formulated within the limits of paleontological data.

ACKNOWLEDGMENTS

I thank Bill Kimbel and Lawrence Martin for their kind invitation to participate in this symposium. Richard F. Kay, Gene Albrecht, Bill Kimbel, Lawrence Martin, Dana Cope, Jay Kelley, Rick Madden, Anna Yoder, and above all, Matt Cartmill, provided encouragement and helpful discussions. Comments from two anonymous reviewers greatly improved the manuscript. The following curators kindly provided access to the specimens in their care: Guy Musser (American Museum of Natural History), Paul Jenkins [British Museum (Natural History)], Bruce Latimer (Cleveland Museum of Natural History), Hans-Walter Mittmann (Landessammlungen fur Naturkunde Karlsruhe), A. W. Crompton (Museum of Comparative Zoology at Harvard), Michel Tranier (Museum National D'Histoire Naturelle, Paris), Renata Angermann (Museum fur Naturkunde der Humboldt-Universitat), Danny Mierte (Musee Royal de L'Afrique Centrale), H. Schaeffer (Naturhistorisches Museum, Basel), Adam Stanczak (Naturhistoriska Riksmuseet, Stockholm), Gerhard Storch (Natur-Museum Senkenberg, Frankfurt), Derrick Howlett (Powell-Cotton Museum), Caroline Grigson (Odontological Museum, Royal College of Surgeons), Chris Smeenk (Rijksmuseum van Natuurlijke Historie, Leiden), F. Dieterlen (Staatliches Naturkundemuseum, Stuttgart), Elisabeth Lang (Musee Zoologique de l'Univeriste Louis Pasteur et de la Ville Strasbourg), Richard Thorington (United States National Museum), Peter Schmid (University of Zurich), and Richard Kraft (Zoologische Staatssammlung, Munich). This research was supported by NSF grant BNS 83-00646 to Richard F. Kay, a Sigma-Xi Grant-in-Aid, and an NSF Dissertation Improvement Grant to JMP.

References

Bown, T. M., and Rose, K. D. 1987. Patterns of dental evolution in early Eocene anaptomorphine primates (Omomyidae) from the Bighorn Basin, Wyoming. Paleont. Soc., Mem. 23. *J. Paleontol.* **61(Suppl. 5).**

Bryant, E. H. 1986. On use of logarithms to accommodate scale. *Syst. Zool.* **35:**552–559.

Cole, T. M., and Smith, F. H. 1987. An odontometric assessment of variability in *Australopithecus afarensis. Hum. Evol.* **2:**221–234.

Cope, D. A. 1989. Systematic Variation in Cercopithecus Dental Samples. Ph.D. Dissertation, University of Texas at Austin.

Fleagle, J. G., Kay, R. F., and Simons, E. L. 1980. Sexual dimorphism in early anthropoids. *Nature* **287:**328–330.

Gingerich, P. D. 1974. Size variability of the teeth in living mammals and the diagnosis of closely related sympatric fossil species. *J. Paleontol.* **48:**895–903.

Gingerich, P. D. 1985. Species in the fossil record: concepts, trends, and transitions. *Paleobiology* **11:**27–41.

Gingerich, P. D., Smith, B. H., and Rosenberg, K. 1982. Allometric scaling in the dentition of primates and prediction of body weight from tooth size in fossils. *Am. J. Phys. Anthropol.* **58:**81–100.

Gould, S. J. 1975. On the scaling of tooth size in mammals. *Am. Zool.* **15:**351–362.

Greenfield, L. O. 1972. Sexual dimorphisms in *Dryopithecus africanus. Primates* **13:**395–410.

Kay, R. F. 1982a. Sexual dimorphism in Ramapithecinae. *Proc. Natl. Acad. Sci. USA* **79:**209–212.

Kay, R. F. 1982b. *Sivapithecus simonsi,* a new species of Miocene hominoid, with comments on the phylogenetic status of the Ramapithecinae. *Int. J. Primatol.* **3:**113–173.

Kay, R. F., and Cartmill, M. 1977. Cranial morphology and adaptations of *Palaecthon nacimienti* and other Paromomyidae (Plesiadapoidea, ?Primates), with a description of a new genus and species. *J. Hum. Evol.* **6:**19–35.

Kelley, J. 1987. Species recognition and sexual dimorphism in *Proconsul* and *Rangwapithecus. J. Hum. Evol.* **15:**461–495.

Kelley, J. 1991. Taxonomic implications of sexual dimorphism in *Lufengpithecus. Am. J. Phys. Anthropol.* **Suppl. 12:**103.

Kimbel, W. H., and White, T. D. 1988. Variation, sexual dimorphism and the taxonomy of *Australopithecus,* in: F. E. Grine (ed.), *Evolutionary History of the "Robust" Australopithecines,* pp. 175–192. Aldine de Gruyter, Hawthorne, New York.

Kimbel, W. H., White, T. D., and Johanson, D. C. 1985. Craniodental morphology of the hominids from Hadar and Laetoli: evidence of "*Paranthropus*" and *Homo* in the mid-Pliocene of eastern Arica?, in: E. Delson (ed.), *Ancestors: The Hard Evidence,* pp. 120–137. Alan R. Liss, New York.

Johanson, D. C., Taieb, M., and Coppens, Y. 1982. Pliocene hominids from the Hadar Formation, Ethiopia (1973–1977): stratigraphic, chronologic and paleoenvironmental contexts, with notes on hominid morphology and systematics. *Am. J. Phys. Anthropol.* **57:**373–402.

Leonard, W. R., and Hegmon, M. 1987. Evolution of P3 morphology in *Australopithecus afarensis. Am. J. Phys. Anthropol.* **73:**41–63.

Martin, L. B. 1983. The Relationships of Later Miocene Hominoids. Ph.D. Dissertation, University of London.

Martin, L. B., and Andrews, P. 1984. The phyletic position of *Graecopithecus freybergi* Koenigswald. *Cour. Forsch. Inst. Senkenberg* **69:**25–40.

Napier, P. H. 1981. Catalogue of Primates in the British Museum (Natural History) and Elsewhere in the British Iles. Part II: Family Cercopithecidae, Subfamily Cercopithecinae. British Museum (Natural History), London.

Napier, P. H. 1985. Catalogue of Primates in the British Museum (Natural History) and Elsewhere in the British Iles. Part III: Family Cercopithecidae, Subfamily Colobinae. British Museum (Natural History), London.

Olson, T. R. 1981. Basicranial morphology of the extant hominoids and Pliocene hominids: the new material from the Hadar Formation, Ethiopia and its significance in early human evolution and taxonomy, in: C. B. Stringer (ed.), *Aspects of Human Evolution,* pp. 99–128. Taylor and Francis, London.

Olson, T. R. 1985. Cranial morphology and systematics of the Hadar Formation hominids and "*Australopithecus*" *africanus,* in: E. Delson (ed.), *Ancestors: The Hard Evididence.* pp. 102–119. Alan R. Liss, New York.

Pickford, M. 1986. Sexual dimorphism in *Proconsul,* in: M. Pickford and B. Chiarelli (eds.), *Sexual Dimorphism in Living and Fossil Primates,* pp. 133–170. Il Sedicesimo, Firenze.

Plavcan, J. M. 1989. The coefficient of variation as an indicator of intra- and interspecific variability in fossil assemblages. *Am. J. Phys. Anthropol.* **78:**285.

Plavcan, J. M. 1990. Sexual Dimorphism in the Dentition of Extant Anthropoid Primates. Ph.D. Dissertation, University Microfilms International, Ann Arbor, MI.

Senut, B., and Tardieu, C. 1985. Functional aspects of PlioPleistocene hominid limb bones: implications for taxonomy and phylogeny, in: E. Delson (ed.), *Ancestors: The Hard Evidence,* pp. 193–201. Alan R. Liss, New York.

Simpson, G. G. 961. *Principles of Animal Taxonomy.* Columbia University Press, New York.

Simpson, G. G., Roe, A., and Lewontin, R. C. 1960. *Quantitative Zoology.* Harcourt, Brace and Co., New York.

Sokal, R. R., and Rohlf, R. J. 1981. *Biometry.* W. H. Freeman and Company, San Francisco.

Tattersall, I. 1986. Species recognition in human paleontology. *J. Hum. Evol.* **15:**165–175.

Thorington, R. W., and Groves, C. P. 1970. An annotated classification of the Cercopithecoidea, in: J. R. Napier and P. R. Napier (eds.), *Old World Monkeys: Evolution, Systematics, and Behavior,* pp. 629–647. Academic Press, New York.

Vitzthum, V. J. 1990. Odontometric variation within and between taxonomic levels of Cercopithecidae: implications for interpretations of fossil samples. *Hum. Evol.* **5:**359–374.

Wiley, E. O. 1981. *Phylogenetics: The Theory and Practice of Phylogenetic Systematics.* John Wiley and Sons, New York.

Wood, B. 1985. Early *Homo* in Kenya, and its systematic relationships, in: E. Delson (ed.), *Ancestors: The Hard Evidence,* pp. 206–214. Alan R. Liss, New York.

Wu, R., and Oxnard, C. E. 1983. *Ramapithecus* and *Sivapithecus* from China: some implications for higher primate evolution. *Am. J. Primatol.* **5:**303–344.

Zwell, M., and Pilbeam, D. 1972. The single species hypothesis, sexual dimorphism, and variability in early hominids. *Yrbk. Phys. Anthropol.* **16:**69–79.

Multivariate Craniometric Variation in Chimpanzees 11

Implications for Species Identification in Paleoanthropology

BRIAN T. SHEA, STEVEN R. LEIGH, AND COLIN P. GROVES

> A basic problem in taxonomy at and below the species level lies in distinguishing variability within a population from that among populations. Methods for comparing populations which use the variance within samples as the yardstick for measuring the variance between two or more samples offer one way to deal with this problem. . .
>
> Creel and Preuschoft, 1976, p. 220

Introduction

The relevant literature in evolutionary biology on the formation, definition, and recognition of species is immense, to say the least. In reality, however, the composition and predominant emphasis of the symposium and present volume,

---

BRIAN T. SHEA • Departments of Cell, Molecular and Structural Biology/Anthropology, Northwestern University, Chicago, Illinois 60611. STEVEN R. LEIGH • Department of Anthropology, Northwestern University, Evanston, Illinois 60628. *Present address:* Department of Anatomical Sciences, State University of New York at Stony Brook, Stony Brook, New York 11794. COLIN P. GROVES • Department of Prehistory and Anthropology, Australian National University, Canberra, ACT 2601, Australia.

Species, Species Concepts, and Primate Evolution, edited by William H. Kimbel and Lawrence B. Martin. Plenum Press, New York, 1993.

entitled *Species, Species Concepts, and Primate Evolution,* present a much more restricted focus. We take this emphasis to be on the relationships between processes of speciation and the morphological patterns that may or may not accompany species formation in neontological groups, with the application of this knowledge to the primate fossil record in order to adequately identify and delimit extinct species in a variety of paleoanthropological investigations.

It is unfortunate but decidedly true that the study of species from this perspective is characterized mostly by what we *cannot* directly investigate and therefore analyze in a scientific sense. We begin with the proposition that examining the morphological results or correlates of speciation in museum collections of the bones of extant or extinct taxa cannot directly inform us in any way concerning the dynamic evolutionary *processes* related to speciation. In general, *patterns* of results may be utilized to test for the presence or operation of particular *processes* explicated directly using other approaches (assuming there is a firm basis for directly linking the two), but such patterns permit no direct observation of these underlying processes. In other words, to directly examine processes and mechanisms of speciation we need to investigate the factors of mating patterns, gene flow, adaptation, drift, etc. in populations of (rapidly reproducing) living organisms in their natural habitats and relate these to the cessation of successful genetic exchange among groups [in the traditional "biological species concept" of Mayr (1970) and/or the "recognition concept" of Paterson (1985)].

The inability to discern historical processes and mechanisms from current static patterns is but one significant limitation. A second layer of obfuscation emerges in that, quite simply, there is of course no theoretically predictable or necessary relationship between genetic and morphological divergence. This is a very important, if unsubtle, point, and it poses only major pragmatic difficulties for the neontologist, since he or she can at least adopt an empirical stance and crosscheck patterns of intra- and interspecific variability against geographical, behavioral, and genetic data relating to the ability of particular groups to successfully interbreed in the wild. In other words, we can utilize morphology to create groupings that are essentially *hypotheses* about species distinctions, and these must be tested against the genetic and behavioral data of the Biological Species Concept or the Recognition Concept. For the paleontologist, the situation is much worse, since sorted patterns can never be crosschecked against such information, and therefore we are left with groupings that are essentially untestable, save through indirect means.

Given the previous considerations, we believe the inescapable conclusion is that the only scientific approach to delimiting species in the fossil record is to develop a *comparative, empirical* matrix derived from living taxa. As Delson (1990, p. 141) has argued, ". . . the problem of central importance today is to develop empirical approaches to delineating species in the fossil record (or in museum collections generally) that correspond to those few that can be defined through direct application of the biological species concept." This decidedly empirical approach is the only rigorous set of criteria that we can utilize.

If this tells us *what* to do, it certainly does not prescribe *how* to do it. Opinions here may vary, but ultimately any empirical approach must be based in some way on Darwin's central insight, that is, the tremendous potential and biological significance of variation *within* species. In other words, attempts to recognize species as distinct from other taxonomic units, such as subspecies or local populations, must be able to successfully partition between-group variation from within-group variation. Statistical and even nonquantitative approaches to this issue are well known, and a number of papers in this volume, in addition to our own contribution, can be consulted for theoretical and methodological discussions (see especially, Albrecht and Miller).

For the paleoanthropologist, certain specific guidelines can be developed to supplement this general approach. It seems likely that the most relevant context in which to test the degree and nature of variation in extinct forms is one of phylogenetic proximity, though ecological similarity may also be heavily weighted. For early hominids, the most relevant taxa are first extant *Homo sapiens* and *Pan troglodytes/Pan paniscus,* and then the remaining hominoids. The view proposed here echoes that of Delson (1990, p. 143):

> Clearly, there must be some morphological or taxonomic baseline employed in any empirical discussion of species recognition. For paleoanthropology, especially in the later phases of human evolution, the obvious standard is variation in living people and the taxonomic level of separation accepted for modern populations.

As an example of the utility of this empirical and scientific approach, Delson (1990) goes on to prescribe a specific criterion regarding the taxonomic status of the Neanderthals in the growing debate over Middle and Upper Pleistocene human evolution. He suggests that variation among the fossil hominid specimens be assessed for a variety of characters and then compared to the observed variation in populations of *Homo sapiens* across Eurasia. A decision regarding specific or subspecific status would be based on this comparison of potential between-group variation with known within-group variation.

It is important to stress that since our approach here is entirely empirical and not all extant groups are represented by comparable amounts of morphological and genetic diversity, in some cases we will be able to make specific distinctions with greater confidence and scientific rigor than others. For example, the empirical base for species within, say, *Cercopithecus, Papio,* or *Macaca,* is stronger than for the living large-bodied hominoids, which are represented by only five quite distinct species, which for these purposes also comprise four separate genera.

Selected Examples

A brief review of but several corrective applications of this simple empirical recipe in the field of paleoanthropology will hopefully suffice to make a point. The following examples provide quite different approaches and morphological foci, but they all share the common goal of establishing interspecific differences in fossil collections relative to a comparative matrix of both interspecific and intraspecific variation in living relatives.

The first example is taken from discussions of the systematics of *Australopithecus afarensis.* It will be recalled that in an early publication, Johanson and colleagues argued for multiple distinct taxa, including a species of early *Homo* (e.g., Johanson and Taieb, 1976). Following a protracted collaboration and debate with Tim White, Johanson and his colleagues ultimately concluded that this fossil collection represented but a single variable species, and one exhibiting morphological variation resulting from at least the influences of time, geography, allometry, and sexual dimorphism. We are fortunate, in this case, to have the popular book *Lucy* by Johanson and Edey (1981), for it catalogs with refreshing candor the personal transformation that Johanson underwent as he shifted his systematic conclusions. The key scientific component of this decision was based on the rigorous and tedious comparison of morphological variation within a moderate sample of *Pan troglodytes* skulls, and the juxtaposition of this to the patterns of variation observed in the fossil hominid collection. This was a relatively straightforward and simple empirical procedure, and it involved no statistically complex machinations of continuous quantitative characters (though it certainly might have). If the ultimate result of this procedure caused Johanson to balk at the necessity of reversing himself in print, at least the general approach that led to the new conclusion was one that he could embrace wholeheartedly. This is because his dissertation work involved an exhaustive study of both qualitative and quantitative variation among 826 specimens of the subspecies of *P. troglodytes* (*P.t. verus, P.t. troglodytes,* and *P.t. schweinfurthii*) and the purportedly distinct species, *P. paniscus,* the pygmy chimpanzee (Johanson, 1974a). One interesting result of this study relevant to our own contribution was the conclusion that the pygmy chimpanzee represented a distinct species and not merely a fourth subspecies of *P. troglodytes* (Johanson, 1974a,b). Whether one accepts the particular conclusion concerning a single early species labeled *Australopithecus afarensis* or not (and at different points, various workers have accepted multiple taxa here), the essential point relates to the rigorous process utilized to reach it, that is, the careful comparison of patterns of variation within and between related species groups.

Our second example is a recent contribution to the growing debate over the number of distinct species possibly contained within the group of fossils commonly classified as *Homo habilis.* Support for a multispecies interpretation has been based primarily on two regions, craniofacial morphology and endocranial volume, with Stringer (1986), Lieberman *et al.* (1988), and others suggesting that there is too much variation in these features to be encompassed by a single species. A recent rigorous analysis by Miller (1991) has attacked this conclusion from the perspective of brain size variability. Miller systematically examined the effects of temporal variation, sample size, sample choice, sex ratio, geographic variation, and measurement technique on coefficients of variation (CV) in endocranial volume for extant hominoid species. Subsequent to various statistical and comparative analyses, he concluded that there is no good reason to argue that published CVs in the range of 12.7 for *H. habilis* are persuasive evidence for multiple species, as Stringer (1986) and others have suggested. Among other important points (such as the demonstration that CVs for volumetric measurements like cranial capacity cannot be assessed using a "yardstick" derived from linear comparisons), Miller illustrates the large confidence intervals derived in extant hominoids for small samples such as the *H. habilis* collection.

We chose this study as a good example of an essentially quantitative and statistical approach to comparing within- and between-group patterns of variation. We strongly feel that these types of rigorous and replicable comparisons are needed to replace the frequent culling of literature data to provide what appears to be largely *post hoc* support for conclusions reached on other grounds. We do not necessarily believe, nor does Miller (1991), that this work effectively eliminates the *possibility* that the *H. habilis* material, in fact, comprises more than a single species [indeed, one of us—Groves (1989)—has explicitly argued for multiple species here]. What we do believe is that this type of approach needs to be applied to many morphological regions and data sets before any final answer will be forthcoming. For example, Lieberman and co-worker's (1988) claim that facial dimorphism between KNM-ER 1813 and 1470 is too great for a single hominoid species is an important conclusion (and one based on direct analogy to levels of variation in extant hominoids), but this issue requires careful reexamination (Miller, 1990).

Our last example involves a nonmetric character, that is, the nature of the articulations between the nasal bone and the adjoining frontal and maxillary bones. We chose this example of work by Eckhardt (1987), since it deals with a qualitative character of the type often utilized by paleoanthropologists, and the claim is sometimes made that heavily quantitative investigations of continuous morphometric features (e.g., Groves, 1970; Creel and Preuschoft, 1976; Albrecht, 1978) are of little direct relevance to the traditional practice of systematics in fragmentary fossil forms. Eckhardt's (1987) study emerged as a test of claims by Olson (1978, 1985a,b) that (1) specific patterns of nasal region articulation were diagnostic of extant hominoid and fossil hominid taxa and (2) two immature Plio-Pleistocene hominids (AL 333-105 and Taung) could thereby be reliably grouped with distinct lineages leading to *Paranthropus* and *Homo,* respectively. Olson's work distinguished three purportedly distinct patterns, characteristic of extant great apes, *H. sapiens,* and the paranthropines. Eckhardt's reanalysis of the extant great apes demonstrated extensive polymorphism in the nasal region within every taxon of extant pongid, including several clear examples of the supposed paranthropine pattern in *P. troglodytes.* This finding is all the more striking considering that Eckhardt sampled only moderate numbers of crania from two museums. Obviously, these between-species patterns are not as distinct and diagnostic as claimed by previous authors, and therefore their utility in cladistic analyses, and especially classifying individual specimens, is highly suspect. Parenthetically, we might add that Kiessling and Eckhardt (1990) have recently demonstrated that the configuration of the palatine fenestrae, claimed by Schwartz (1983) to separate species of extant hominoids, and then proposed by him to be a synapomorphy linking *Pongo* and *Homo,* in fact exhibits considerable within-taxon variation.

Although the preceding example involves qualitative, nonmetric traits, the process of comparison and analysis is essentially identical to that in a quantitative study, that is, between-species distinctions can only be reliably established relative to within-species variation. Various recent statistical approaches to the analysis of categorical variables make the approximation to classical between- vs. within-group contrasts even greater.

Within- and Between-Group Cranial Variation in Chimpanzees

The work we describe here results from an investigation of patterns of craniometric variation in populations and subspecies of common (*P. troglodytes*) and pygmy (*P. paniscus*) chimpanzees. It can be seen as an extension of previous work on gorillas by Groves (1970) and on different samples of chimpanzees by Shea and Coolidge (1988). We do not undertake in this paper a complete analysis of the interrelationships among local populations; that work is forthcoming (Shea *et al.*, manuscript in preparation). Rather, we focus here on the traditionally defined geographical subspecies with two primary goals in mind. The first is to contribute to the long-standing need for such an investigation of within- and between-species cranial variation in what is likely to be our closest living relative. The second is to utilize this particular analysis as an example of the more general issue of basing species distinctions on careful comparisons of within-species patterns of variation.

We understand that the data set we analyze here is in many respects quite different from those generally utilized by paleoanthropologists to make taxonomic and systematic assessments. For example, our basic data are quantitative craniometrics, our specimens are virtually complete, and we have data on sex, dental age, and locale of origin. Paleoanthropological systematists generally work with qualitative features derived from fragmentary specimens of uncertain sex, age, and (sometimes) specific origin. These disparities might be quite important if our goal here was direct extrapolation of our results to hominid systematics, but that is emphatically not our primary purpose. Rather we are trying to demonstrate the conceptual organization and partial execution of an approach to craniometric variation in the genus *Pan* that integrates analyses of both within-group and between-group patterns of variation and covariation, specifically to assess the hypothesis that the pygmy chimpanzee likely represents a distinct species of *Pan,* and not merely a fourth subspecies of *P. troglodytes.*

In stressing this aspect of the present contribution, we in no way wish to undermine the argument that in theory the results of analyses of patterns of within- and between-group variation in chimpanzees are directly applicable to an understanding of the hominid fossil record. To the contrary, we are appalled that decades of work in a field where practitioners often literally outnumber the primary objects of study have not yielded more comprehensive investigations of the factors generating variation in qualitative or quantitative characters in our wide-ranging, closest relatives, the chimpanzees. The preceding three examples from paleoanthropological debates provide ample evidence of the valuable perspective that even a moderate emphasis on intra- and interspecific variation in living hominoids can provide.

Materials and Methods

This paper presents some preliminary analyses and results of a larger study specifically designed to examine between-population variation within the chimpanzees. Appendices I and II list the locality names (arranged by currently accepted major subgroupings of the chimpanzees), sample size by sex, and measurement definitions. A map illustrating the approximate location for these samples is also included.

All measurements were taken by Colin Groves and have frequently been utilized in previous studies of hominoid cranial variation and systematics (e.g., Groves, 1970, 1986; Groves *et al.*, 1992). We had originally planned to incorporate a large sample of *P. t. verus* (Frankfurt collection) measured by Daris Swindler, but analysis of comparable measurements on common samples demonstrated some statistically significant differences between Groves and Swindler, so we have not included these additional specimens. Unfortunately, this means that our sampling of populations within the commonly accepted subspecies of *P. t. verus* is weaker than we would like.

For the purposes of this paper, our analytical approach is restricted to discriminant and principal component analyses. Detailed presentations of the theory and method of these multivariate statistical procedures can be found in a variety of texts and in such papers as Albrecht (1978), Creel and Preuschoft (1976), and Groves (1970). Briefly, multiple discriminant analysis (or canonical variates analysis) produces linear functions of weighted variables that maximize between-group variance while minimizing within-group variance, therefore distinguishing multiple groups to the greatest extent possible relative to the dispersion within the groups. Discriminant functions themselves consist of vectors of weights by which individual variable values are multiplied to yield the specimen's score on that particular function; centroid values indicate the position of the group means on each function. Correlations between variable weights and the positions of groups along the discriminant functions are utilized to interpret the patterns of variation. Mahalanobis distances among individuals or centroids are used to map within- and between-group dispersion, and to produce classification functions that assign group association of individual specimens. The Mahalanobis distances may also serve as raw data for multivariate clustering procedures. In this paper, we use multiple discriminant analysis to test for the presence and degree of multivariate separation among groups of *P. paniscus, P. troglodytes verus, P. t. troglodytes*, and *P. t. schweinfurthii.*

We complete discriminant analyses on both raw data and on adult cranial values size-corrected by deriving residuals relative to least-squares bivariate regressions on basicranial length. This allows us to "extract" shape differences among crania of different sizes that result from differential positioning along a common vector of allometric growth, and to focus on those shape differences (regression residuals) that result from departures from this common trend.

Our second multivariate technique is principal components analysis (PCA). PCA involves the rigid, geometric rotation of original axes to a new position such that the first axis or component is positioned to account for the maximum amount of variation, subsequent components being orthogonal and accounting for progressively less of the total variance. PCA is particularly well suited to situations where it is desirable to preserve Euclidean distances among specimens and to emphasize the major axis of variation. When ontogenetic samples are available for closely related groups, the first and subsequent components derived from a covariance matrix of logarithmic values can often be interpreted as axes summarizing the variance due to ontogenetic scaling (first component) and that resulting from slope and position differences in allometric trajectories (second and subsequent components). Shea (1985) provides additional details on the relationships between bivariate allometry and Jolicoeur's (1963) multivariate generalization of allometry. In this study, we use principal components analysis of ontogenetic and adult data to assess the amount of variance in adult skull form that results from (growth) allometric factors, particularly in comparing *P. paniscus* to the various groups of *P. troglodytes*.

We utilized the statistical software package SYSTAT/SYGRAPH (version 4.0; Wilkerson, 1988a,b) for the discriminant function and principal components analyses reported here.

Results

Figure 1 is a plot of 90% confidence ellipsoids about group centroids on discriminant functions I and II, derived from an analysis of 13 cranial dimensions on adult females of the three commonly accepted subspecies of *P. troglodytes* and the purportedly distinct *P. paniscus*. The three groups of *P. troglodytes* cluster together on both discriminant functions I and II, but *P. paniscus* is quite distinct from the *P. troglodytes* on function I, which accounts for 81.2% of the total variance. Table 1 records the classification accuracy in the groups based on this analysis; percentage correct classification averages around 75% for the *P. troglodytes* subspecies but is 100% for *P. paniscus*. These results indicate a marked separation of adult *P. paniscus* from the three groups of common chimpanzees, though also suggesting the possibility that differences in overall skull size may be an important factor in producing this separation. Table 5 presents data on intergroup centroid distances and indicates that the *P. t. verus* to *P. t. schweinfurthii* value is the greatest among the three subspecies of *P. troglodytes*, which agrees with their geographic distribution. Of course, the greatest intergroup centroid distances are observed in comparisons of *P. paniscus* versus all the subspecies of *P. troglodytes*. Comparable analyses on males yield very similar results to those described here for females.

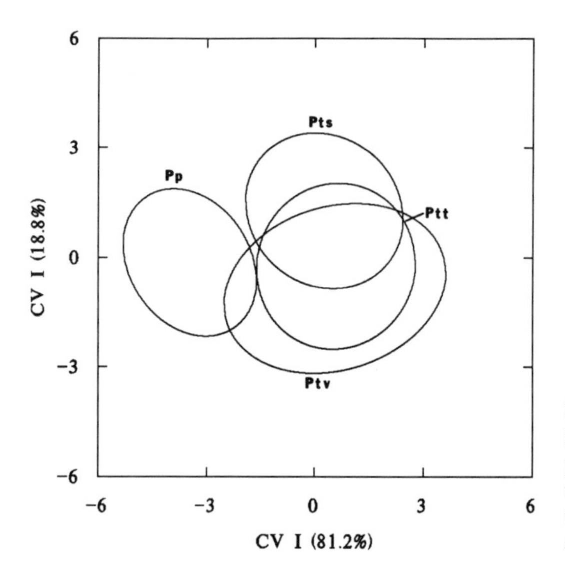

Fig. 1. A plot of 90% probability ellipsoids from a multiple discriminant function analysis of non-size-corrected adult females, with *P. paniscus* included. Note the separation of the pygmy chimpanzees from the subspecies of *P. troglodytes*.

Table 1. Classification Accuracy for the Discriminant Analysis Completed on Non-Size-Corrected Females with *P. paniscus* Included

	Ptv	Ptt	Pts	Pp	Total	% Correct
Ptv	14	3	0	0	17	82
Ptt	22	79	22	2	125	63
Pts	4	5	30	0	39	77
Pp	0	0	0	26	26	100

Taxon membership (rows) by predicted membership (columns).

Figure 2 is a plot of the 90% ellipsoids for the three subspecies of female *Pan troglodytes,* with pygmy chimpanzees deleted from the analysis. Once again, these are non-size-corrected adult values for the 13 cranial dimensions. There is still significant discrimination (though a number of the univariate F tests are no longer significant, whereas they all are when *P. paniscus* is included). Table 2 reveals that the classification accuracy for this analysis does not change for the subspecies of *P. troglodytes* when the divergent *P. paniscus* is deleted from the overall sample. There is no marked or clear patterning to the group dispersion within *P. troglodytes* in this case, e.g., a geographical gradient with *P. t. verus* and *P. t. schweinfurthii* most separated. In fact, the data in Table 5 on intergroup centroid distances reveal that the eastern *P. t. schweinfurthii* is morphometrically more distinct from the centrally located *P. t. troglodytes* than it is from the extreme western group, *P. t. verus.* Males show a similar pattern to that described here for females.

Figure 3 illustrates the results of a discriminant analysis run on adult female cranial values that have been size-corrected using bivariate ontogenetic regressions. Note the much greater approximation of the *P. paniscus* 90% probability

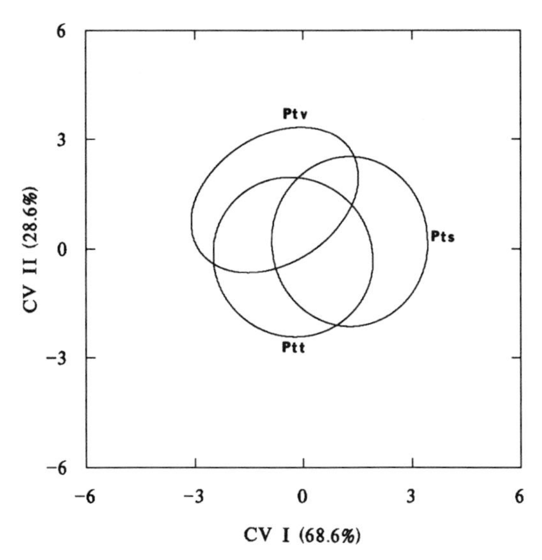

Fig. 2. A plot of 90% probability ellisoids from a multiple discriminant function analysis of non-size-corrected adult females of *P. troglodytes.*

Table 2. Classification Accuracy for the Discriminant Analysis Completed on Non-Size-Corrected Females with *P. paniscus*

	Ptv	Ptt	Pts	Total	% Correct
Ptv	14	3	0	17	82
Ptt	23	79	23	125	63
Pts	3	6	30	39	77

Taxon membership (rows) by predicted membership (columns).

ellipsoid to those of the *P. troglodytes* subspecies than is the case with the non-size-corrected data (Fig. 1). Distances among the group centroids (Table 5) show that the degree of distinctiveness of *P. paniscus* is much diminished, particularly relative to *P. t. troglodytes* and *schweinfurthii.* Following size correction, the distance between the *P. t. verus* and *P. t. schweinfurthii* centroids is only slightly less than that between the *P. paniscus* and *P. t. schweinfurthii* centroids. Classification accuracy drops somewhat (Table 3) when the data are size-corrected, though the bulk of the *P. paniscus* sample is still correctly classified. The variables most responsible for producing the between-group differences on discriminant function I are interorbital breadth and bicanine breadth, indicating some relative narrowness in these regions in the pygmy chimpanzee. Coolidge (1933) long ago stressed the narrow interorbital region and smaller palatal and canine dimensions of the pygmy chimpanzee.

Table 4 presents the classification accuracy for the three subspecies of *P. troglodytes* after regression-correction and with *P. paniscus* excluded from the analysis. The percentage of correct classifications drops substantially for *P. t.*

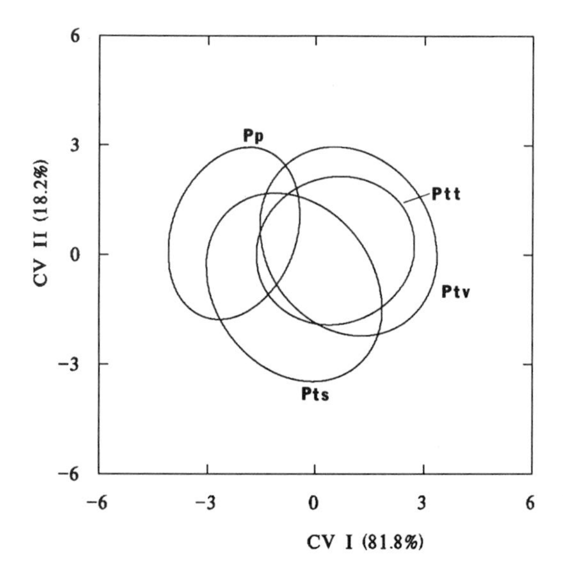

Fig. 3. A plot of 90% probability ellisoids from a multiple discriminant function analysis of regression size-corrected adult females, with *P. paniscus* included.

Table 3. Classification Accuracy for the Discriminant Analysis Completed on Regression-Corrected Females with *P. paniscus* Included

	Ptv	Ptt	Pts	Pp	Total	% Correct
Ptv	14	3	0	0	17	82
Ptt	25	72	20	8	125	58
Pts	4	4	24	7	39	62
Pp	0	0	1	25	26	96

Taxon membership (rows) by predicted membership (columns).

Table 4. Classification Accuracy for the Discriminant Analysis Completed on Regression-Corrected Females without *P. paniscus*

	Ptv	Ptt	Pts	Total	% Correct
Ptv	14	3	0	17	82
Ptt	28	24	7	59	41
Pts	29	6	73	108	66

Taxon membership (rows) by predicted membership (columns).

troglodytes and *P. t. schweinfurthii* following size correction. Data from Table 5 reveal that the two morphometrically most divergent subspecies are now the eastern and western groups.

Table 6 summarizes changes in classification accuracy before and after size correction, both with and without *P. paniscus* included. In all cases where the classificatory precision changes (12 of 14 comparisons), it decreases following size correction, and often by a substantial margin. This indicates that a significant component of the distinctiveness of *P. paniscus* and the *P. troglodytes* subspecies is due to allometric factors, which are at least partially controlled for through such regression-correction procedures.

The intergroup centroid distances summarized in Table 5 supplement the classificatory changes by demonstrating that morphometric relationships are altered following size correction. For example, in the non-size-corrected analysis, *P. paniscus* most closely approximates *P. t. verus* among the common chimpanzee groups, while subsequent to regression correction, it is closest to *P. t. schweinfurthii.* In the analyses where only the *P. troglodytes* subspecies were included, the two most divergent groups prior to size correction are *P. t. troglodytes* and *P. t. schweinfurthii;* subsequent to regression adjustment, *P. t. verus* and *P. t. schweinfurtii* are most distinct.

Principal components analysis of the combined ontogenetic and adult data provide a means of directly assessing and comparing multivariate allometric trajectories and their concordance and/or divergence (see Shea, 1985). Figure 4 is a plot of the component scores for the mean values (four age categories, two

Table 5. Intergroup Centroid Mahalanobis Distances

	Ptv	Ptt	Pts	Pp
Non-size-corrected centroid distances, *P. paniscus* included				
Ptv	0			
Ptt	1.270	0		
Pts	2.276	1.564	0	
Pp	3.015	4.048	3.985	0
Regression-corrected centroid distances, *P. paniscus* included				
Ptv	0			
Ptt	0.443	0		
Pts	1.965	1.524	0	
Pp	3.178	2.848	2.229	0
Non-size-corrected centroid distances, *P. troglodytes*				
Ptv	0			
Ptt	1.656	0		
Pts	2.141	2.624	0	
Regression-corrected centroid distances, *P. troglodytes*				
Ptv	0			
Ptt	1.416	0		
Pts	2.166	1.590	0	

sexes) in the four groups. Principal component I is clearly a vector of allometric growth, arranging the skulls according to overall size, explaining the predominant amount of overall variation, and yielding variable loadings that accurately capture the allometric size-related shape changes during ontogeny. The truncation of the *P. paniscus* dispersion accurately reflects its smaller overall size and the allometric underpinnings of a significant portion of its cranial shape differences relative to the *P. troglodytes* specimens. There is no significant separation of *P. paniscus* from the combined or individual samples of *P. troglodytes* on

Table 6. Summary Table of Classification Accuracies (Percentage Correct Classification) for Males/Females of Each Group

	With *P. paniscus*		Without *P. paniscus*	
	Non-size-corrected	Regression-corrected	Non-size-corrected	Regression-corrected
Ptv	87/82	56/82	67/82	61/82
Ptt	62/63	60/58	64/63	63/41
Pts	78/77	70/62	83/77	70/66
Pp	100/100	94/96		

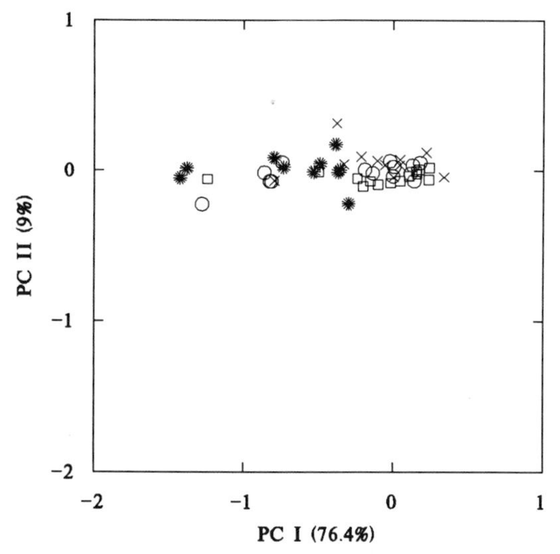

Fig. 4. A plot of principal component scores for age/sex group means in the four chimpanzee samples. Symbols: stars, *P. paniscus;* circles, *P. t. verus;* squares, *P. t. troglodytes,* x's, *P. t. schweinfurthii.* Note the size distribution along the first principal component, with adult *P. paniscus* crania falling in with juvenile *P. troglodytes* crania. The second principal component effects no statistically significant separation of the groups.

component II, as would be expected if patterns of multivariate growth allometry differed substantially between the two purported species (Shea, 1985).

Since the overall discriminatory power drops when *P. paniscus* is deleted from the analysis or when all the groups are size corrected, we were interested in seeing whether individual contrasts were successful in distinguishing the various pairs of the three purported subspecies groupings of *P. troglodytes.* Results of these pairwise comparisons revealed significant discrimination for both males and females in each of the *Ptt/Ptv, Ptt/Pts,* and *Ptv/Pts* discriminant analyses, both prior and subsequent to size correction.

In all the preceding analyses, sexes were treated separately. Separate runs for each sex were completed in the multiple discriminant analyses, and in the principal components analysis all sexes and ages were included, though not pooled. Results of the principal components analysis illustrated in Figure 4 demonstrate that including both sexes in the analysis does not erase the distinctiveness of adult *P. paniscus* from the various *P. troglodytes,* even where this difference is primarily driven by size and allometrically related shape. We have also completed some multiple discriminant analyses on pooled sex samples. Although some details of variable loadings and classification change subtly, the basic patterns of marked divergence of *P. paniscus* from all *P. troglodytes,* plus more limited but still statistically significant separation among the subspecies of *P. troglodytes,* are not altered.

Discussion

There are several key concepts around which we will concentrate our discussion of the results presented here. These are (1) the issue of establishing species distinctions relative to within-species patterns of variation, (2) the importance of size correction in such studies, and (3) the taxonomic implications of our ability to establish significant morphometric and morphological differences among groups of specimens. We will follow this portion of the discussion with a consideration of some related general issues in paleoanthropological systematics.

Our results demonstrate that adult *P. paniscus* are clearly and fairly uniformly distinguishable from adult *P. troglodytes,* particularly when subspecific and interpopulational variation within the latter is considered. In other words, the differences *between* the two "types" of chimpanzee are significantly greater than the considerable variation cataloged *within* the broadly ranging *P. troglodytes.* There is some previous work utilizing different samples and measurements that supports this conclusion (Shea and Coolidge, 1988), though most previous studies of craniometric differences between pygmy and common chimpanzees (e.g., Shea, 1982, 1984; Cramer, 1977; Fenart and Deblock, 1973; Creel, 1986) have contrasted the former with either a single subspecies or much more restricted sampling of *P. troglodytes.*

For a number of years, most primatologists have agreed with a separation of pygmy and common chimpanzees at the species level, although there have been some notable dissenters at various times, including one of us (Schwarz, 1934; Schultz, 1954; Horn, 1979; Groves, 1972). Our analysis of quantitative cranial variation supports the separation of the pygmy chimpanzee at the species level, however. Other body systems (e.g., Shea, 1983a, 1984, 1985; Jungers and Susman, 1984; Zihlman, 1984; Zihlman and Cramer, 1978; Morbeck and Zihlman, 1989; Dahl, 1985), socioecological adaptations (e.g., Wrangham, 1986), genetics (e.g., Stanyon *et al.,* 1986), and behavioral differences (e.g., de Waal, 1989) perhaps provide even more convincing evidence for a specific separation of the pygmy chimpanzee.

We reach this taxonomic conclusion in spite of the fact that our own analyses demonstrate that much (though certainly not all) of the morphometric differentiation between the pygmy and common chimpanzee crania disappears following (ontogenetic) allometric size correction. We wish to stress that it is extremely important to determine the component of morphological differences that results from size differences, utilizing a clearly articulated theory and method of "size correction." However, even the demonstration that all morphological and/or morphometric differences could be attributed to overall size differences through the pervasive effects of allometry need not require a taxonomic reassessment of subspecific propinquity. This is simply because size and allometric effects are not merely troublesome obfuscation to be removed to yield a clearer view of "true" biological or phylogenetic distinctions; indeed, size differences have so many pervasive effects on myriad biological systems and levels (e.g., Lindstedt and Calder, 1981) that they should be central to systematic conclusions in and of themselves. In addition, in the present case previous work has shown both that pygmy chimpanzee crania can be distinguished from those of *P. troglodytes* (*troglodytes*), even after size and allometric factors have been controlled (Shea, 1982, 1984) and that *P. paniscus* exhibits some fundamental reorganizations and shifts among major body regions that result in a clearly different anatomical pattern from what is seen in any of the common chimpanzees (Shea, 1983a). In particular, at common sizes *P. paniscus* appears to have relatively smaller facial complexes and relatively longer hindlimbs than *P. troglodytes* (though morphological differences are predominantly ontogenetically scaled within these respective regions).

The importance of size correction (utilizing ontogenetic allometries as the criterion of subtraction) is also nicely illustrated by the fact that after such procedures, crania of *P. paniscus* emerge as morphometrically closest to *P. t. schweinfurthii,* rather than to *P. t. verus,* as in the non-size-corrected analyses. Based on geographical factors, we might expect the maximum distance between *P. paniscus* and *P. troglodytes* to occur in *P. t. verus.* Coolidge (1933) originally suggested that the pygmy chimpanzee most closely approached the eastern variant of *P. troglodytes,* in contradistinction to Schwarz (1934), who argued for greatest morphological proximity to *P. t. troglodytes.* We should stress that we do not imply anything regarding the historical origin of *P. paniscus* by these results, but merely refer to the relative degrees of morphometric distinctiveness.

The impact of size correction is also demonstrated in the analyses excluding *P. paniscus.* Here, a somewhat unexpected pattern (maximum divergence between two neighboring groups in the non-size-corrected analyses) is replaced by a result where geographic and biological distance are correlated in the regression-corrected analysis. It appears that the primary differences driving the divergence of *P. t. troglodytes* and *P. t. schweinfurthii* in the non-size-corrected analysis are due to size and allometric factors, since the morphometric distance decreases substantially in the regression-corrected analysis, allowing the eastern-western contrast to emerge as primary (see Table 5). This finding is supported by results demonstrating that body size and skeletal size in *P. t. schweinfurthii* (especially females) is substantially smaller than *P. t. troglodytes* (Jungers and Susman, 1984; Morbeck and Zihlman, 1989).

Quite apart from any implications for our specific example here, we reiterate our contention that detailed and proper assessment of allometric factors is vital to any comparisons of this nature. There is an increasing awareness of the theoretical and methodological reasons for basing systematic assessments on ontogenetic as well as adult comparisons. Shea (1984) provides one systematic approach to the investigation of adult morphological differences in terms of allometric and ontogenetic factors. The case described earlier regarding mandibular proportions and the systematic assessment of the Laetoli/Hadar hominids provides an example of the potential significance of these factors for paleoanthropology, although it should be pointed out in passing that no rigorous comparative analyses of such investigations have ever been published. We have only the anecdotal accounts provided in Johanson and Edey (1981) to assure us that allometric factors account for the shape divergence observed in "Lucy" and other small individuals of *A. afarensis.* Indeed, in general, paleoanthropologists lag far behind primatologists and other mammalogists in the application of such approaches to questions of adaptation, taxonomy, and the factors affecting the rate and direction of evolutionary change.

The third primary point that we would like to stress is that our results clearly reveal the ability to discriminate in statistically significant fashion among the three primary groups (or multiple local populations) or *P. troglodytes,* even subsequent to the removal of *P. paniscus* from the analysis. All pairwise comparisons for both sexes yield significant discrimination, in spite of the considerable overlap among these groups. This causes us to wonder what some paleontologists might do with comparable results derived from fossil collections, whether these distinctions are based on the continuous, quantitative variation that we have assessed here or more qualitative but still quite variable features. Tattersall (1986, p. 168) argues that "it is critical to avoid relegating distinct morphological variants observed in the fossil record to the status of subspecies," advocating that "where distinct morphs can be readily identified it would seem most productive to assume that they represent species." Kimbel (1991) essentially echoes these views. Our analyses of variation within *P. troglodytes* certainly establish "distinct morphological variants," even though these are clearly intraspecific groups of populations and not species. Albrecht and Miller's (this volume) extensive compilation and review of primate intraspecific and interspecific craniodental variation provides many additional examples and relevant discussion. We now turn to more general discussion of these issues, particularly as they relate to the hominid fossil record.

Why Intraspecific Variation Must Be Investigated

Our conception, like that of many other evolutionary biologists, is that the species represents the lowest and most fundamental level of biological organization that is relevant to phylogenetic reconstruction and the determination of evolutionary patterns. As Kimbel and Rak (this volume) point out, the species is at once a genealogical and taxonomic unit, while other classificatory levels are not. Analysis of evolutionary processes and patterns should ideally be based on species units. Indeed, it is vital that we not misidentify distinct species as mere subspecies, precisely because the former are the primary units of evolutionary change (Tattersall, 1986).

But, by the same argument, surely it is also vital that we not misidentify mere subspecies (or other intraspecific variants) as distinct species. On further reflection, it is apparent that these two cases of potential misidentification are clearly not symmetric or equally problematic excesses in terms of their implications. The more strongly we believe that species and subspecies represent vastly different units in terms of evolutionary significance, the more imperative it is that we not misidentify intraspecific differences as species distinctions, even if the cost is the "lumper's" erroneous reduction of the number of true species.

A simple example will illustrate this point. Assume, in a given extant group, that we in fact have three distinct species, each containing three distinguishable major variants or subspecies. A "lumper" might identify but two highly variable species, while a "splitter" might come up with six or more distinct species. If our understanding of evolutionary processes and patterns requires that we work with specific, rather than subspecific, distinctions, then the lumper will probably provide us with a more accurate picture of the world. His or her distinctions, though admittedly underestimating diversity, will at least truly reflect species differences. The splitter will develop patterns and infer processes based on erroneous distinctions, for half of his or her "species" differences are nonexistent and misleading, being based on artificial and misleading units of evolutionary change.

Consider a hypothetical example from the hominid fossil record. Again assume we have three distinct species in successive time periods, each containing three subspecies (though with fossils this is truly a hypothetical, we could never test and therefore "know" it). Our three species are in reality characterized by a progressive increase in both the mean overall body size and the relative brain size, or encephalization quotients, yielding an interspecific allometric slope significantly greater than 0.75. As is often the case, the subspecies within each species exhibit brain/body scaling around 0.2–0.4, so that, relative to each species mean, the smaller variant will have relatively larger brain sizes and the larger variant will have smaller encephalization quotients (see Gould, 1975; Jerison, 1973; Shea, 1983b; Riska and Atchley, 1985). Our lumper would construct a pattern of increased encephalization, even though this would be based on only two rather than the three actual species. The splitter might produce a pattern of three cases of increased encephalization and three cases of decreased encephalization. The latter would do greater damage to the actual "truth" in this case.

If the *consequences* of misidentifying the proper taxonomic and evolutionary level of groups are not equal, we must stress again that the *likelihood* of this occurring is also not comparable in the extant vs. extinct groups. The determination of legitimate species distinctions is often difficult in living forms, but in addition to morphological features, evolutionary biologists can incorporate data on genetics, ecology, social organization, mating behavior, and specific mate-recognition systems, and geographical, temporal, and behavioral isolating mechanisms. This array of information provides a potentially rigorous test of species-distinct morphs utilizing the biological species concept. One nice example of such is the buttressing of morphological data with evidence from karyotypic discontinuities, geographical distribution, ecological specializations, social behavior, and breeding data by those claiming species distinction between *Galago crassicaudatus* and *G. garnetti* (Masters, 1988a,b). While species recognition may often be difficult in extant forms, it is decidedly more of a problem when dealing with extinct variants. Not only are purported fossil species not directly testable according to criteria of the biological species concept, but the broad range of information available for living forms is nonexistent in the case of extinct groups.

Given the necessary and virtually exclusive reliance on morphological patterns or features, plus our inability to directly test for the presence of isolating mechanisms, etc., the most rigorous course open to the paleobiologist is to indirectly test patterns and distributions within the fossil sample against those derived from careful analysis of *both species and subspecies* of close phylogenetic relatives and groups occupying similar ecological and adaptive zones. Since we can never know whether our fossil samples are subdivided properly, this approach is our most replicable procedure and the one that suggests the greatest probability of correctness.

An alternative way to conceptually approach and summarize this issue is to consider types I and II errors in hypothesis testing. For a reminder, type I errors are made when the tested hypothesis is falsely rejected; type II errors occur when the tested hypothesis is not rejected when it is actually false (Hays, 1973). One generally compensates for the possibility of type I error by setting a very small p value, so the probability of falsely rejecting is decreased. This is done when it is deemed extremely important not to falsely reject the null hypothesis, as, for example, when we hypothesize that an experimental medicine is unsafe due to negative side effects on the population. Here we wish to be overcautious in our statistical testing due to the potential negative implications for significantly large numbers of people, so we would set a very small p value, thus ensuring that we are very unlikely to falsely reject this null hypothesis and erroneously conclude that the medicine is indeed safe when it will likely have adverse effects for some people. Therefore, increasing the power of avoiding type I error is accomplished by setting very low *p* values for rejection, and it is done when specific circumstances warrant particular concern for falsely rejecting a stated hypothesis. The likelihood of committing a type II error is decreased through such means as altering sample sizes and other aspects of the test procedures (Hays, 1973). In many situations in evolutionary biology, the frequent assumption of minimizing type I errors might not be defensible, and indeed this is why *p* values are sometimes set at levels of 0.05 or even higher.

As noted above, we strongly feel that it is important to avoid granting species status to morphological variants in the fossil record that in fact do not represent distinct species. We argue this because it would elevate an ephemeral and arbitrary taxonomic unit of possibly no real evolutionary significance (cf. Tattersall, 1986; Kimbel, 1991) to a position as a fundamental unit of evolutionary pattern and process, which would then become reified in systematic discussions as a distinct species, and which would be relatively immune to any further testing as to its validity, given its extinct status (presumably we haven't all forgotten *Ramapithecus*?). Thus, we could state our null hypothesis as something like the following: "These two or (or more) variants represent subspecific or other intraspecific levels of variation if the degree and kind of both quantitative and qualitative morphological variation in the purportedly distinctive features does not significantly exceed that determined for the sample of extant relatives deemed most appropriate for comparison."

Further building on the application of types I and II errors, we note that we are significantly more concerned with avoiding type I errors in fossil forms as compared to extant groups, particularly because of the lack of direct testability of nonmorphological aspects of species distinctiveness, and due to the paucity of supplemental data on other biological and behavioral systems available in the fossil record. As we also noted above, making a type II error (in this case, accepting fossil morphs as subspecies when they are in fact distinct species) is more acceptable to us in theory, since it is unlikely to alter our reconstruction and understanding of primary evolutionary patterns. We believe that an accurate representation of the these patterns is more important than the recognition and naming of all individual species that might be present in the fossil record (though, of course, that number can never be known with certainty in any case). For the morphologically oriented neontologist working on systematic revisions (e.g., Albrecht, 1978; Groves, 1970, 1986), we would be more comfortable with increasing the likelihood of Type I error somewhat, since we would be confident that rejection of the null hypothesis would generate a new and alternative hypothesis of species distinctiveness, which could and would be subject to much additional testing utilizing other morphological data, plus information from genetics, behavior, and ecology. Again, the case cited above concerning morphospecies of greater galagos serves as an example of the possibility and relevance of such crosscheck testing, as does the work of Jolly (this volume) on morphological, genetic, and ecological variation in *Papio*.

Our decisions regarding statistical rigor and systematic approaches intersect directly in the realm of the types of evolutionary units and the amount of biological information available in extant vs. extinct organisms. Our "take-home" message is simple—*the more unimportant and irrelevant one thinks patterns of intraspecific variation are for understanding evolution on the species level, the more imperative it is not to misidentify subspecies as species in the untestable fossil record.* Individuals maintaining this view should logically be spending even more time and effort understanding the processes and patterns of intraspecific variation in living forms than those colleagues who believe that such studies may actually tell us something about speciation processes and events themselves (see below). The relevance of the preceding discussion to the data and morphological problem focused on in this paper is clear, that is, the specific distinctiveness of *P. paniscus* on a morphological level is supported when examined against a thorough investigation of intraspecific variation within the wide-ranging *P. troglodytes*, and this conclusion derived from morphological comparison is heavily buttressed by a variety of data from genetics, ecology, social organization, and so forth.

Tattersall's (1986) apparent distinction between a focus on intraspecific as opposed to interspecific variation is problematic in several regards. First of all, implicit in the one type of analysis is the other, e.g., a catalog of interspecific features presumably demonstrates or assumes that the between-species distinctions are not encompassed by the within-species pattern of either species. Secondly, Tattersall (1986), Kimbel (1991), and others argue that application of studies of intraspecific variation in extant forms misses the boat for one or more of the following reasons (see especially Tattersall, 1986, pp. 166–167): (1) speciation may take place in the absence of morphological divergence, because distinguishing features are often related to soft tissue characters and not bony morphology; and (2) in closely related species, morphological variation for the vast majority of anatomical characters will overlap substantially or totally, while a few key features differentiate the species. In regard to the latter point, this may certainly be true, but this in no way whatsoever undermines a careful examination of both intraspecific and interspecific patterns in extant forms as a model for interpreting the fossils. Presumably, the "differentiating" features hypothesized here would stand apart from the "nondifferentiating," intraspecific features in relevant extant models. If they did not, in the absence of any other special convincing evidence, species designation would be an unsupported and unscientific supposition, though of course it *could* be true. "A species may exist independent of our ability to recognize or adequately define it," as Kimbel (1991, p. 362) assures us. However, as scientists we are not professionally interested in truth or potential existence, but rather testability, replicability, demonstrability, and falsifiability. We concur with Kimbel (1991, p. 362) that "the fact that the recognition of species is almost always based on some aspect of the morphology of organisms does not mean that our theoretical concept of species should be purely morphological," but only in the case of extant groups, where additional nonmorphological information can be attained and tested. In the case of fossil taxa, we basically have only morphology, and a pitifully small window into that system, as has been recognized by many other paleobiologists (e.g., Rose and

Bown, 1986). Paleoanthropologists have many wonderful assets in the fossil and archaeological record of millions of years of ape and human evolution, but they must appreciate the information they do not have access to, and where they simply must rely on data from living models to make scientific and probabilistic hypotheses and predictions about their extinct forms.

The first issue raised above, i.e., that morphological divergence does not coincide with speciation and genetic differentiation in any necessary and predictable way, is a theoretical postulate that we must and do accept, as noted in our Introduction. However, we understand relatively little about what drives this potential dissociation (or lack thereof) and whether we might expect differential relationships between morphology and speciation in particular taxa or ecological situations, as Jackson and Cheetham (1990) have noted. In other words, this raises the empirical issue of how true this statement is for living primates where species designations are testable and relatively robust. Albrecht and Miller's (this volume) comprehensive summary and analysis of multivariate intra- and interspecific differentiation in craniodental features of a wide variety of primates provides a very important perspective on this issue. They have shown that in all cases examined in extant primates, purported subspecies or other intraspecific groupings (and, of course, species) can be significantly separated using multivariate, quantitative techniques. Arguments presupposing the widespread existence of morphologically indistinguishable sibling species in extant primates currently have no support.

Is Intraspecific Variation Relevant to Speciation?

Both Tattersall (1986) and Kimbel (1991) express apparent frustration at approaches focusing on the identification and explanation of patterns of intraspecific variation. Stressing the fact that there is no necessary relationship between morphological divergence and speciation, Tattersall (1986) suggests that closely related species will often overlap considerably or even totally in most osteological characters, yet be characterized by clear differences in a few key features. Tattersall rejects any strong focus on intraspecific variation, arguing that the most commonly asked question, that is, whether the limits of variation we see in our fossil assemblages exceed the limits of variation we see in samples drawn from extant primate species, is inappropriate. He suggests that this approach should be replaced by a focus on between-species variation, for "hand in hand with this concern for [intraspecific] variation has gone the triumph of the lumping ethic which, in laudably rejecting the typological thinking that bedevilled paleontology in the earlier years of this century, has in certain cases proceeded with an equal lack of realism to the opposite extreme." These arguments serve as a prelude for Tattersall's conclusion that the Middle-to-Late Pleistocene hominid fossil record provides evidence of multiple species (though he does not tell us exactly how many).

Kimbel's (1991) recent discussion of species concepts and the hominid fossil record makes many similar points. He stresses that closely related species will often overlap considerably in a majority of their morphological features. He clearly feels that the study of intraspecific patterns of variation in extant forms bears little or no relation to the establishment of morphological species in the fossil record, since there is no basis to believe that "subspecies have intrinsic evolutionary significance (i.e., as incipient species)" (Kimbel, 1991, p. 362). Like Tattersall, Kimbel complains of what he perceives to be current "preoccupation with ever-expanding envelopes of intraspecific variation," and he warns that "to treat morphologically discrete clusters in the fossil record as subspecies confuses an arbitrary tool of taxonomy with evolutionary units, and thus thwarts the first task of the paleoanthropologist, which is to discover the pattern of phylogenetic relationships in the hominid fossil record" (Kimbel, 1991, p. 363). Kimbel's view of subspecific variation requires (in nearly tautalogical fashion) that morphological distinctions observed in the fossil record are evidence of distinct species. We base this assessment on the following quote (Kimbel, 1991, p. 363):

> Since significant morphological diversification among early hominids could hardly have proceeded without the geographical isolation and subsequent expansion of differentiated local populations, it is far more likely that the fossil record of this diversity reflects the successful establishment of new species than the maintenance of stable local populations through geological time. It is speciation that "provides morphological changes with enough permanence to be registered in the fossil record." (Futuyma, 1987, p. 467)

Several important points need to be stressed in relation to the arguments of Tattersall (1986) and Kimbel (1991). The first is that while we also believe that patterns of intraspecific variation are often ephemeral, constantly shifting, and frequently unrelated to actual processes of speciation, we are certainly hesitant to baldly state that this is always the case. Many authorities (e.g., Endler, 1977, 1989; Bock, 1979, 1986; Wright, 1982; Templeton, 1982; Charlesworth *et al.*, 1982) present broader and less restrictive depictions of the nature of speciation, and it is perhaps no accident that none of these authorities are bound to the fossil record for their primary interpretations. In this volume, Jolly, in particular, has discussed infraspecies groups and their potential relation to speciation.

Kimbel (1991, p. 362) paraphrases Mayr (1982, no page citation given) as claiming that maintaining the view that "subspecies taxa have intrinsic evolution significance (i.e., as incipient species) is no longer acceptable." We find that this brief paper by Mayr, entitled "Of What Use Are Subspecies," says nothing of the sort. In it, Mayr reviews the history of the use of the subspecies category as a taxon. He notes that this somewhat typological approach to intraspecific variation was at once related to two important biological meanings: (1) the "incipient species" of Darwin, or *a stage in the speciation process* (our italics); and (2) evidence of adaptive response of species to local climatic conditions. In the conclusion of his paper, Mayr (1982, p. 595) argues, and we firmly agree, that these two important biological phenomena (i.e., incipient speciation and local adaptation) are best studied with a populational approach, rather than by utilizing rigid subspecies classifications. Most significant for an understanding of Mayr's views of speciation and its possible relationship to intraspecific variation (at least as stated in this 1982 paper), we note that Mayr (1982, p. 594) argued that "subspecies are not types, [but] populations or groups of populations," and he went on to caution that "the designation of "incipient species" [is] true only of isolates but not of contiguously distributed continental subspecies." Mayr (1982, p. 595) cites arguments by Lack as supporting the notion that "subspecies that [are] incipient species had to acquire ecological differences before they could invade each other's geographic ranges. That different subspecies differ in their ecological requirements had been known to perceptive naturalists since Darwin's time." This recognition of *some* (though not all) intraspecific groups as *potentially* incipient species seemingly contradicts Kimbel's (1991, p. 362) paraphrase of Mayr (1982). Finally, we point out that Mayr (1982, p. 594) favorably describes his own participation in taxonomic revisions of a plethora of species designations made by prior ornithologists (for instance, of 607 species of North American birds alone, 315 taxa that originally had been described as full species were subsequently reduced to subspecies status). He concludes, in words that should serve as an admonition to paleoanthropologists, that "the newly recognized polytypic species were much more distinct, real entities of nature than the purely morphologically defined species of the 1880's. Morphological difference was replaced as species criterion by reproductive isolation" (Mayr, 1982, p. 594).

The diversity of opinion among evolutionary biologists regarding the relationships between intraspecific variation or units (demes, populations, subspecies) and species-level differences is further revealed by contrasting Kimbel (1991) and Tattersall (1986). Contrary to Kimbel's extreme statement regarding the lack of any "intrinsic evolutionary significance (i.e., as incipient species)" of subspecies, Tattersall accepts that some subspecies may reflect incipient speciation, for he writes (1986, p. 168) that "only with speciation, with the development of reproductive isolation, does the identity of *what began as a subspecies* become definitely established" (italics added). Jolly (this volume) advocates a reconsideration of the subspecies (in certain cases and according to specified criteria) as incipient species and legitimate evolutionary units.

We close this section by reiterating that by no means do we intend to suggest that subspecies are necessarily or even frequently incipient species, and therefore primary units of evolutionary change. We merely emphasize that the relationship between subspecific and specific levels of classification should remain an open question, to be addressed through data collection and hypothesis testing (e.g., see Liem, 1984, and Smith, 1991, for discussion of possible processes of sympatric speciation). We simply do not wish to see real possibilities excluded *a priori* by theoretical necessities (of cladism and/or punctuated equilibrium), declamatory fiat, or preferred interpretations of fossil assemblages.

Special Issues with the Fossil Record

Since we have stressed the inadequacies of the fossil record in providing direct and reliable tests of species distinctions, we should also point out its unique strength, that is, time depth. In this regard, we concur, to some extent, with both Tattersall (1986) and Kimbel (1991) in their suggestions that the maintenance of significantly different morphological variants across long periods of time in the fossil record is more likely to be indicative of species- rather than subspecies-level distinctions. Undoubtedly, as we have seen in recent discussions of the later stages of human evolution, there is much room for debate and disagreement in terms of what would constitute "significant morphological differences" or "long periods of time." We also stress in passing that there is considerable diversity of opinion regarding the validity and meaning of subspecific variation in the fossil record. Mayr (1970, p. 210–213) provides some relevant discussion, and Krishtalka and Stucky (1985) offer a detailed analysis of within- and between-species patterns of morphological change in the early Eocene artiodactyl *Diacodexis*.

An example of the application of subspecies or intraspecific level variation to hominid taxonomy and phylogeny is presented by Tobias (1980). He suggested that the Laetoli and Hadar assemblages are morphologically distinct and should be labeled *A. africanus afarensis* and *A. africanus aethiopicus*, respectively. Tobias (1980) views both of these subspecies as conspecific with *A. africanus transvaalensis*. No one has followed this proposal, to our knowledge, and we certainly do not advocate its acceptance here. The point we would like to stress, however, is that both Tobias (1980) and White (1985) concur is seeing the Laetoli and Hadar fossils as statistically and qualitatively distinct in *some* features and to *some* degree. Also central to our point is that both Tobias (1980) and White (1985) reach their (admittedly differing) taxonomic conclusions by carefully placing these Laetoli-Hadar differences in a context of intraspecific and interspecific patterns of variation, using extant pongids, extinct hominids, and extant humans as primary guidelines. This is particularly true of White's (1985) detailed, point-by-point comparison of the Laetoli and Hadar dental samples. Without the comparative matrix of intraspecific and interspecific variation (especially in living apes and humans), we would be left with the Laetoli and Hadar collections representing "morphologically discrete clusters in the fossil record" (Kimbel, 1991, p. 363). Yet surely we wouldn't want to therefore simply *assume* distinctions at the species level. The process works effectively in both directions when applied to the fossil record.

Current Debates in Paleoanthropology

As noted in our introduction, the primary focus of this edited volume is on the theory and method of species recognition as it can be applied to the primate fossil record. This focus is easily blurred in a field of investigation and a fossil record that all would agree have frequently become distorted and twisted with contentious debate and posturing for the significant media and general interest that understandably surrounds the issue of human origins and evolution. It is frequently observed (usually by those outside the field) that the practitioners of human paleontology seem to outnumber the objects of study, and some suspect that the multiplication of taxa often has much to do with this very human factor. This is simply not as much the case in the study of, say, fossil turtles. We make this point in order to raise the possibility that a concern for the general process of how we go about recognizing and testing species distinctions in the fossil record has been, to some extent, sacrificed to the context of heated and ongoing debates in paleoanthropology. We refer here to issues such as variation within *H. habilis,* Middle Pleistocene hominid fossils, the status of Neandertals, and the "out-of-Africa" hypothesis. We note that both Tattersall (1986) and Kimbel (1991, this volume) engage the general issues of species recognition through explicit consideration of portions of the hominid fossil record.

We wish to stress that, in spite of the fact that one of us has recently entered the morass of hominid systematics and produced a view somewhat divergent from current consensus (Groves, 1989), on the whole we have no particular bone to pick with consensus interpretation of hominid classification. Rather, our concern is with the general analytical process through which these conclusions regarding species recognition are reached. In that light, we must bemoan the fact that if only a small portion of the time, arguments, and analyses by paleoanthropologists over the past several decades had been dedicated to extensive and rigorous analysis of within- and between-group patterns of both quantitative and qualitative variation in modern humans, great apes, and other primates, hominid systematics would be in much more robust shape than it is today. We emphatically do not share the predictions of despair by some (e.g., Tattersall, 1986; Kimbel, 1991) that such an approach would lead us to a reincarnation of the 1960s and 1970s version of the "Michigan school," single-species lumping, which never afflicted much of the rest of the field, in any case. Significantly, that perspective was driven by a largely indefensible theoretical postulate that multiple species of hominid could not coexist; it was not the outcome of extensive analyses of real patterns of intra- and interspecific variation within extant humans, apes, and other primates.

Our advocacy of a particular analytical approach based on patterns of both intra- and interspecific variation in extant relatives of known taxonomic status should definitely not be interpreted as emanating from a single species interpretation of Middle and Upper Pleistocene hominid variation. We distance ourselves from that particular conclusion because it is no more based in the kind of careful comparative matrix that we propose than are the conclusions of its detractors. This has recently been pointed out in a letter to *Science* by Van Vark and Bilsborough (1991), who criticized unsupported statements about degrees of morphological diversity in some fossil hominids in relation to modern humans. The multivariate distance data (based on Howells' 1973 monograph) presented by Van Vark and Bilsborough (1991) were certainly informative and quite corroborating, but it is somehow symptomatic of the field that after decades that have witnessed no major synthetic analyses juxtaposing hominid fossil variation to that present in modern *Homo,* the great apes, or other relevant groups, we must go to the Letters column in *Science* for a sketchy discussion of such issues and data analysis.

These examples lead us to raise a frustrated complaint to hominid and primate paleontologists, and also provide a rationale for the study and results that we have briefly presented here on one of our closest living relatives. In a word, we are staggered that Delson must suggest in 1990 the need for a rigorous comparison of Neandertal variation with that seen in modern humans. Or that, in spite of Howells' (1973, 1989) extensive and exemplary quantitative studies, no comparable investigations of relevant nonmetric morphological features have been undertaken by paleoanthropologists. We wish we could legislate a remedy to this unfortunate and inexcusable situation, but that will only eventuate when individual researchers and their students acknowledge the need for such data and set about the admittedly arduous task of gathering it.

How would we propose such an investigation be set up? First of all, rigorous consideration of *both* intraspecific and interspecific patterns in relevant extant species must be complemented by detailed analysis of *both* quantitative and qualitative morphological features. Relevant extant taxa are generally determined to be the phylogenetically closest relatives to the fossil taxa in question, though Tattersall (1986) raises the possibility that in certain cases this might not be appropriate, and ecologically more similar models might be utilized. This is a good idea, as long as the linkage is reliably established. We would suspect that the more information we can gather on patterns of intra- and interspecific variation in extant forms, reflecting phylogenetic and/or adaptive proximity, the better our assessments of the fossils will be. But some formalized context of extant models must be used and explicitly presented. Albrecht and Miller (this volume) provide a more detailed, multilevel approach focusing on extant species as models, identification of variation due to several important sources (e.g., sexual, geographic, temporal), and types of analyses that might be executed. We would stress, in addition, that the general principles of approach behind such a research program can be applied to either quantitative morphometric features or qualitative morphological distributions.

In sum, we propose that paleoanthropology would benefit tremendously if a great deal more extensive background work were undertaken on the patterns of variation in the relevant qualitative or quantitative features, *before* species groupings were determined, character states delimited, polarities estimated, and phylogenies reconstructed. This would probably greatly reduce or clarify the number of equally parsimonious phylogenies, the requisite frequency of homoplasy, and the debates over basic units of analysis.

A Paleontological Example

An instructive example of how knowledge of extant patterns and relationships can inform the fossil record is provided by Jackson and Cheetham's (1990) study of morphospecies in Cheilostome bryozoans. These are a diverse group of predominantly sessile invertebrates that inhabit the ocean floor and that have a fossil record stretching back almost 100 million years. These authors utilized evidence from breeding experiments and protein electrophoresis to show that in extant forms the recognized morphospecies have heritable morphologies, are genetically distinct, and there is no evidence of morphologically cryptic species. They carefully assessed within-species compared to between-species variation in a group of morphological features in three extant genera. Having directly analyzed the extent to which morphospecies correspond to biological species in this group, they then felt confident in sorting fossil morphs into distinct species. Their analysis supported previously demonstrated patterns, which they conclude are suggestive of punctuated speciation in fossil cheilostomes. Interestingly, they conclude that "the widely supposed lack of correspondence between morphospecies and biospecies . . . may result as much out of uncritical acceptance of outdated, subjectively defined taxa as from any fundamental biologic differences between the two kinds of species" (Jackson and Cheetham, 1990, p. 582). This issue is obviously one that requires a great deal of empirical investigation in a wide variety of taxa. As noted above, Albrecht and Miller (this volume) have made an important contribution to primate biology in this regard.

The key point for us in the Jackson and Cheetham (1990) paper on marine invertebrates is obviously not one of phylogenetic proximity to fossil or extant humans (though some of these may also be predominantly sessile and without backbones), or concern for any particular interpretation of pattern in terms of tempo and mode of evolutionary change. Rather, we find this to be an exemplary case study of how knowledge gained from careful analysis of relevant extant forms can be profitably applied to the fossil record to determine basic evolutionary units prior to making assessments of phylogenetic patterns. This general approach can and should be readily applied to varying degrees in assessing morphospecies in primate paleontology and paleoanthropology.

Conclusions

Our investigation of craniometric variation in chimpanzees nicely illustrates an example of establishing species differences relative to a context of significant intraspecific variability. Our use of the traditional subspecies of *P. troglodytes* should not be taken as indicative of a belief that such necessarily represent primary evolutionary units; rather, they are utilized here as a tool to establish the breadth and patterning of variance for a particular morphological system. We are currently undertaking a more detailed investigation of quantitative craniometric variation among local populations within both *P. paniscus* and *P. troglodytes* in an attempt to more fully examine the hierarchy of inputs (cf. Albrecht and Miller, this volume) producing the observed patterns in these taxa.

We have utilized this preliminary analysis of craniometric variation within the genus *Pan* as a vehicle for focusing on a number of key issues related to the recognition and analysis of species within the fossil record. Our primary point is neither novel nor complex. We believe that species distinctions among extinct morphs can never be directly tested, and therefore should be rigorously based in a comparative matrix of both intraspecific and interspecific variation in extant forms, where hypotheses of species status can be more scientifically assessed. The construction of this comparative matrix should perhaps entail more effort and application than the study of the fragmentary fossils themselves. It should include data on both quantitative and qualitative traits, particularly those features that workers believe may be novel and derived. The approach should encompass a multilevel analysis of factors producing intraspecific and interspecific variation (see Albrecht and Miller, this volume) in an extant comparative sample comprising both close relatives and ecologically or evolutionarily similar species. Although it is frustrating how little effort is often expended by paleoanthropologists in undertaking such a foundation prior to engaging in contentious systematic and interpretive debates, we intend any criticism herein as constructive and stress that we are aware that the present brief analysis of chimpanzee craniometrics represents but a small portion of the necessary examination of even that single genus.

There are those who disparage as misleading and disorienting attempts to interpret variability in fossil forms in a rigorous context of extant intraspecific and interspecific variation. We have attempted to discuss here some of the significant theoretical and methodological problems that can arise from a failure to undertake such detailed comparative investigations. A single example, briefly alluded to above, may make the point much more effectively than any abstract argument, however. Though it hardly seems necessary, we remind the reader that a few, scrappy gnathic remains of Miocene hominoids were interpreted for many years by general consensus (with some notable exceptions) in paleoanthropology as representing a new and distinct genus (*Ramapithecus*), one sharing with true hominids some of the key derived and differentiating features of our lineage. The paleobiological and evolutionary implications of this conclusion were immense, and this episode awaits the detailed perspective of a true historian of science from outside the field.

Paleoanthropologists should take as a grave admonition the fact that current consensus now places this group of fossils comfortably within a single species of *Sivapithecus* (e.g., Kay, 1982), entirely erasing the characters previously interpreted as indicating a distinct genus. This error could have been avoided if the purportedly distinctive characters had been rigorously assessed in the context of relevant extant forms, particularly *Pongo* and the other large-bodied hominoids (indeed, a few lonely voices tried to raise these issues, e.g., Greenfield, 1979). We believe it is very important not to repeat these types of mistakes, and it is with this central goal in mind that we have presented the views expressed here. We do not believe that this is an isolated or extreme case, unlikely to have been duplicated elsewhere in paleontological systematics or to potentially reoccur in future such analyses. We worry that it may be all too common where the phylogenetic species concept is applied without extensive grounding in an analysis of both intraspecific and interspecific morphological variation in extant models.

Acknowledgments

We would like to thank Bill Kimbel and Lawrence Martin for the invitation to submit a chapter to this volume. We greatly appreciate the helpful comments of Bill Kimbel, Gene Albrecht, and several anonymous reviewers on previous drafts of the paper. The senior author wishes to stress that, in spite of the collaborative component of this multiauthored work, no one should assume that all of the specific views and opinions stated herein are necessarily shared by the other authors.

References

Albrecht, G. H. 1978. The craniofacial morphology of the Sulawesi macaques. *Contr. Primatol.* **13:**1–151.

Bock, W. J. 1979. The synthetic explanation of macroevolutionary change—a reductionistic approach. *Bull. Carnegie Mus.* **13:**20–69.

Bock, W. J. 1986. Species concepts, speciation, and macroevolution, in: K. Iwatsuki, P. H. Raven, and W. J. Bock (eds.), *Modern Aspects of Species,* pp. 31–57. University of Tokyo Press, Tokyo.

Charlesworth, B., Lande, R., and Slatkin, M. 1982. A neo-Darwinian commentary on macroevolution. *Evolution* **36:**474–498.

Coolidge, H. J., Jr. 1933. *Pan paniscus,* pygmy chimpanzee from south of the Congo River. *Am. J. Phys. Anthrop.* **18:**1–59.

Cramer, D. L. 1977. Craniofacial morphology of *Pan paniscus. Contr. Primatol.* **10:**1–64.

Creel, N. 1986. Size and phylogeny in hominoid primates. *Syst. Zool.* **35:**81–99.

Creel, N., and Preuschoft, H. 1976. Cranial morphology of the lesser apes. *Gibbon and Siamang* **4:**219–303.

Dahl, J. F. 1985. The external genitalia of female pygmy chimpanzees. *Anat. Rec.* **211:**24–28.

Delson, E. 1990. Commentary on: Species and species concepts in paleoanthropology, in: M. K. Hecht (ed.), *Evolutionary Biology at the Crossroads,* pp. 141–145. Queens College Press, New York.

Endler, J. A. 1977. *Geographic Variation, Speciation, and Clines.* Princeton University Press, Princeton, NJ.

Endler, J. A. 1989. Conceptual and other problems in speciation, in: D. Otte and J. A. Endler (eds.), *Speciation and its Consequences,* pp. 625–648. Sinauer Associates, Sunderland, MA.

Eckhardt, R. B. 1987. Hominoid nasal region polymorphism and its phylogenetic significance. *Nature* **328:**333–335.

Fenart, R., and Deblock, R. 1973. *Pan paniscus-Pan troglodytes.* Craniometrie. Etude comparative et ontogenique selon les methodes classiques et vestibulaire. *An. Mus. R. Afr. Cent. Tervuren* **I:**1–593.

Futuyma, D. J. 1987. On the role of species in anagensis. *Am. Nat.* **130:**465–473.

Gould, S. J. 1975. Allometry in primates, with emphasis on scaling and the evolution of the brain. *Contrib. Primatol.* **5:**244–292.

Greenfield, L. O. 1979. On the adaptive pattern of *Ramapithecus. Am. J. Phys. Anthropol.* **50:**527–548.

Groves, C. P. 1970. Population systematics of the gorilla. *J. Zool. Lond.* **161:**287–300.

Groves, C. P. 1972. Phylogeny and classification of primates, in: R. N. T-W Fiennes (ed.), *Pathology of Simian Primates,* Vol. 1, pp. 11–57. Karger, Basel.

Groves, C. P. 1986. Systematics of the great apes, in: D. S. Swindler and J. Erwin (eds.), *Comparative Primate Biology, Vol I: Systematics, Evolution and Anatomy,* pp. 187–217. Alan R. Liss, New York.

Groves, C. P. 1989. *A Theory of Human and Primate Evolution.* Oxford University Press, New York.

Groves, C. P., Westwood, C., and Shea, B. T. 1992. Unfinished business: Mahalanobis and a clockwork orang. *J. Hum. Evol.* **22:**327–340.

Hays, W. L. 1973. *Statistics for the Social Sciences,* 2nd ed. Holt, Rinehart and Winston, New York.

Horn, A. D. 1979. The taxonomic status of the Bonobo chimpanzee. *Am. J. Phys. Anthrop.* **51:**273–282.

Howells, W. W. 1973. *Cranial Variation in Man.* Papers of the Peabody Museum, Harvard, no. 67, 259 pp.

Howells, W. W. 1989. *Skull Shapes and the Map.* Papers of the Peabody Museum, Harvard, no. 79, 189 pp.

Jackson, J. B. C., and Cheetham, A. H. 1990. Evolutionary significance of morphospecies: a test with cheilostome Bryozoa. *Science* **248:**579–583.

Jerison, H. J. 1973. *Evolution of the Brain and Intelligence.* Academic Press, New York.

Johanson, D. C., and Taieb, M. 1976. Plio-Pleistocene hominid discoveries in Hadar, Ethiopia. *Nature* **260:**293–297.

Johanson, D. C., and Edey, M. 1981. *Lucy: The Beginnings of Humankind.* Simon and Schuster, New York.

Johanson, D. C. 1974a. An Odontological Study of the Chimpanzee with Some Implications for Hominoid Evolution. PhD dissertation, University of Chicago, Chicago, IL.

Johanson, D. C. 1974b. Some metric aspects of the permanent and deciduous dentition of the pygmy chimpanzee. *Am. J. Phys. Anthropol.* **41:**39–48.

Jolicoeur, P. 1963. The multivariate generalization of the allometry equation. *Biometrics* **19:**497–499.

Jungers, W. L., and Susman, R. L. 1984. Body size and skeletal allometry in African apes, in: R. L. Susman (ed.), *The Pygmy Chimpanzee,* pp. 131–178. Plenum Press, New York.

Kay, R. F. 1982. *Sivapithecus simonsi,* a new species of Miocene hominoid, with comments on the phylogenetic status of the Ramapithecinae. *Int. J. Primatol.* **3:**113–173.

Kiessling, E., and Eckhardt, R. B. 1990. Palatine fenestrae: windows on hominoid variation and its interpretation. *Am. J. Phys. Anthropol.* **81:**249.

Kimbel, W. H. 1991. Species, species concepts and hominid evolution. *J. Hum. Evol.* **20:**355–372.

Krishtalka, L., and Stucky, R. K. 1985. Revision of the Wind River faunas, early Eocene of central Wyoming. Part 7. Revision of *Diacodexis* (Mammalia, Artiodactyla). *Ann. Carnegie Mus.* **54:**413–486.

Lieberman, D. E., Pilbeam, D. R., and Wood, B. A. 1988. A probabilistic approach to the problem of sexual dimorphism in *Homo habilis:* A comparison of KNM-ER 1470 and KNM-ER 1813. *J. Hum. Evol.* **17:**503–511.

Liem, K. F., and Kaufman, L. S. 1984. Intraspecific macroevolution: functional biology of the polymorphic cichlid species *Cichlasoma minckleyi,* in: A. A. Echelle and I. Kornfield (eds.), *Evolution of Fish Species Flocks,* pp. 203–215, University of Maine Press, Orano.

Linstedt, S. L., and Calder, W. A. III. 1981. Body size, physiological time, and longevity of homeothermic animals. *Q. Rev. Biol.* **56:**1–16.

Masters, J. 1988a. Morphological clues to genetic species: Multivariate analysis of greater galago sibling species. *Am. J. Phys. Anthropol.* **75:**37–52.

Masters, J. 1988b. Speciation in the greater galagos (Prosimii:Galaginae): Review and synthesis. *Biol. J. Linn. Soc.* **34:**149–174.

Mayr, E. 1970. *Population, Species and Evolution.* Belknap Press, Cambridge, MA.

Mayr, E. 1982. Of what use are subspecies? *The Auk* **99:**593–595.

Miller, J. A. 1990. *Homo habilis:* the sexual dimorphism hypothesis reconsidered. *Am. J. Phys. Anthropol.* **81:**269.

Miller, J. A. 1991. Does brain size variability provide evidence of multiple species in *Homo habilis? Am. J. Phys. Anthropol.* **84:**385–398.

Morbeck, M. E., and Zihlman, A. L. 1989. Body size and proportions in chimpanzees, with special reference to *Pan troglodytes schweinfurthii* from Gombe National Park, Tanzania. *Primates* **30:**369–382.

Olson, T. R. 1978. Hominid phylogenetics and the existence of *Homo* in Member 1 of the Swartkrans Formation, South Africa. *J. Hum. Evol.* **7:**159–178.

Olson, T. R. 1985a. Cranial morphology and systematics of the Hadar Formation, Ethiopia and its significance in early human evolution and taxonomy, in: E. Delson (ed.), *Ancestors: The Hard Evidence,* pp. 99–128. Alan R. Liss, New York.

Olson, T. R. 1985b. Taxonomic affinities of the immature hominid crania from Hadar and Taung. *Nature* **316:**539–540.

Paterson, H. E. 1985. The recognition concept of species, in: E. S. Vrba (ed.), *Species and Speciation,* Transvaal Museum Monograph No. 4, pp. 21–29. Transvaal Museum, Pretoria.

Riska, B., and Atchley, W. R. 1985. Genetics of growth predict patterns of brain-size evolution. *Science* **229:**668–671.

Rose, K. D., and Bown, T. M. 1986. Gradual evolution and species discrimination in the fossil record. *Contr. Geol. Univ. Wyoming,* Spec. Pap. **3:**119–130.

Schultz, A. H. 1954. Bemerkungen zum Variabilitat und Systematik der Schimpansen. *Saugetierk. Mitt.* **2:**159–163.

Schultz, A. H. 1969. *The Life of Primates.* Universe Books, New York.

Schwartz, J. H. 1983. Palatine fenestrae, the orangutan and hominoid evolution. *Primates* **24:**231–240.

Schwarz, E. 1934. On the local races of the chimpanzee. *Ann. Mag. Nat. Hist. Lond.* **13:**576–583.

Shea, B. T. 1982. Growth and Size Allometry in the African Pongidae: Cranial and Postcranial Analyses. PhD dissertation. Duke University, Durham, North Carolina.

Shea, B. T. 1983a. Paedomorphosis and neoteny in the pygmy chimpanzee. *Science* **222:**521–522.

Shea, B. T. 1983b. Phyletic size change and brain/body scaling: a consideration based on the African pongids and other primates. *Intl. J. Primatol.* **4:**33–62.

Shea, B. T. 1984. An allometric perspective on the morphological and evolutionary relationships between pygmy (*Pan paniscus*) and common (*Pan troglodytes*) chimpanzees, in: R. L. Susman (ed.), *The Pygmy Chimpanzee: Evolutionary Biology and Behavior,* pp. 89–130. Plenum Press, New York.

Shea, B. T. 1985. Bivariate and multivariate growth allometry: statistical and biological considerations. *J. Zool. Lond.* **206:**367–390.

Shea, B. T., and Coolidge, N. J., Jr. 1988. Craniometric differentiation and systematics in the genus *Pan. J. Hum. Evol.* **17:**761–685.

Simons, E. L., and Pilbeam, D. 1965. Preliminary revision of the Dryopithecinae. *Folia Primat.* **3:**81–152.

Smith, T. B. 1991. A double-billed dilemma. *Nat. Hist.* January, 1991:14–19.

Stanyon, R., Chiarelli, B., Gottlieb, K., and Patton, W. H. 1986. The phylogenetic and taxonomic status of *Pan paniscus:* a chromosomal perspective. *Am. J. Phys. Anthropol.* **69:**489–498.

Stringer, C. B. 1986. The credibility of *Homo habilis,* in: B. Wood, L. Martin, and P. Andrews (eds.), *Major Topics in Primate and Human Evolution,* Cambridge University Press, Cambridge, U.K.

Tattersall, I. 1986. Species recognition in human paleontology. *J. Hum. Evol.* **15:**165–175.

Templeton, A. R. 1982. Genetic architectures of speciation, in: C. Barigozzi (ed.), *Mechanisms of Speciation,* pp. 105–121. Alan R. Liss, New York.

Tobias, P. V. 1980. "*Australopithecus afarensis*" and *A. africanus:* critique and an alternative hypothesis. *Palaeont. Afr.* **23:**1–17.

van Vark, G. N., and Bilsborough, A. 1991. Letters: "Shaking the family tree." *Science* **253:**834.

Waal, F. de. 1989. *Peacemaking Among Primates.* Harvard University Press, Cambridge, MA.

White, T. D. 1985. The hominids of Hadar and Laetoli: an element-by-element comparison of the dental samples, in: E. Delson (ed.), *Ancestors: The Hard Evidence,* pp. 138–152. Alan R. Liss, New York.

Wilkerson, L. 1988a. *SYSTAT: The System for Statistics.* SYSTAT, Inc., Evanston, IL.

Wilkerson, L. 1988b. *SYGRAPH.* SYSTAT, Inc., Evanston, IL.

Wrangham, R. W. 1986. Ecology and social relationships in two species of chimpanzee, in: D. I. Rubenstein and R. W. Wrangham (eds.), *Ecological Aspects of Social Evolution: Birds and Mammals.* pp. 352–378. Princeton University Press, Princeton, NJ.

Wright, S. 1982. Character change, speciation, and the higher taxa. *Evolution* **36:**427–443.

Zihlman, A. L. 1984. Body build and tissue composition in *Pan paniscus* and *Pan troglodytes,* with comparisons to other hominoids, in: R. L. Susman (ed.), *The Pygmy Chimpanzee: Evolutionary Biology and Behavior.* pp. 179–200. Plenum Press, New York.

Zihlman, A. L., and Cramer, D. L. 1978. Skeletal differences between pygmy (*Pan paniscus*) and common (*Pan troglodytes*) chimpanzees. *Folia Primatol.* **29:**86–94.

Appendix I

Locality List (See Accompanying Figure): Sample Size for Adult Males and Females

	Male	Female
P. t. verus		
1. Guinea, Sierra Leone, Liberia	14	17
2. Ivory Coast, Ghana, Togo	4	2
3. Lagos, Benin (City)	2	1
P. t. troglodytes		
4. Cameroun/Nigeria border (Cross River district)	7	8
5. Mt. Cameroun and Bamenda Highlands	5	5
6. Edea/Ongue (lower Sanaga)	3	3
7. Kribi/Bipindi (Cameroun Coast	9	16
8. Efulen/Ebolowa (inland of coast)	4	3
9. Yaounde/Akonolinga	10	7
10. Lomia/Dja River	8	8

Appendix I. (*Continued*)

	Male	Female
11. Batouri and upper Sanaga	7	54
12. Southeast Cameroun (middle Sangha)	4	1
13. Rio Muni and borders	2	2
14. Gabon estuary	5	4
15. Lambarene/Mimongo	5	3
16. Makokou/Belinga (northeast Gabon)	2	3
17. Mambili/Ouesso (northeast Congo)	5	6
18. Brazzaville	1	—
19. Sette Camma/Fernan Vaz (southern Gabon coast)	7	19
20. Mayombe	8	7
P. t. schweinfurthii		
21. Lisala region (between Oubangui and Zaire rivers)	6	6
22. Uele River	14	14
23. Kisangani district	4	4
24. Ituri/Lake Albert	8	5
25. Rutshuru/Toro/Ankole	4	7
26. Entebbe	—	—
27. Rwanda	3	2
28. Burundi	—	2
29. Kivu/Maniema	3	3
30. Fizi/Boko	—	1
31. Moba	1	2
32. Kibwesa	1	—
P. paniscus		
33. Mbandaka/Bolobo	3	1
34. Befale/Lopori/Wamba	14	21
35. Lomela/Lubefu	—	4
36. Kasai	—	1

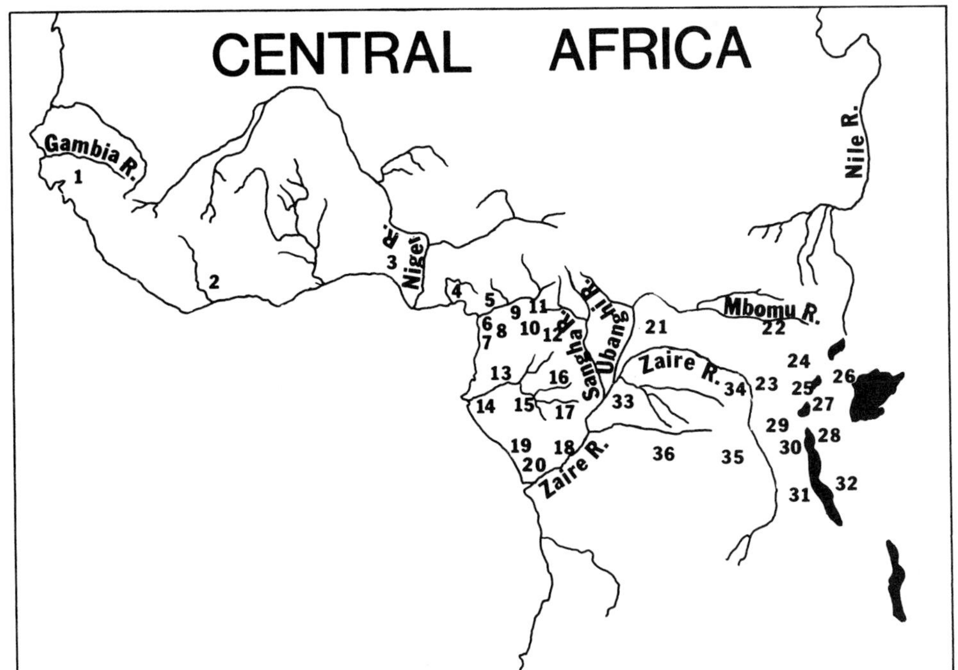

Appendix II

Cranial measurements
Prosthion-opisthocranion length
Glabella-opithocranion length
Bizygomatic breadth
Bieuryonic breadth
Postorbital breadth
Biorbital breadth
Orbital breadth
Interorbital breadth
Bicanine breadth
Nasion-prosthion length
Nasion-basion length
Basion-bregma length
Prosthion-staphylion length

Species and Species Recognition in the Primate Fossil Record

3

Species Concepts and Species Recognition in Eocene Primates

12

KENNETH D. ROSE AND
THOMAS M. BOWN

Introduction

The problem of how to apply the species concept to fossils is not new (see, e.g., Trueman, 1924; Simpson, 1943, 1951; Sylvester-Bradley, 1956; Imbrie, 1957), nor has it been resolved. Despite heated debate and voluminous literature on the subject, a consensus seems more elusive today than several decades ago. Part of the reason for this discord may be the varied approaches, philosophies, and perspectives—some of which have led to new views on the nature of species—of participants in the debate, as well as different systematic objectives and questions being asked. Theoreticians, historians, and philosophers of science have joined paleontologists and neontologists in contributing to the debate. As will be seen, theoretical arguments often conflict with more operational approaches offered by paleontologists working with real specimens.

The most common problem encountered by most paleontologists is to decipher how many species exist in a sample from a single locality and horizon. This problem has usually been approached by assessing the amount of morphological and metric variation in the sample and comparing this to established

KENNETH D. ROSE • Department of Cell Biology and Anatomy, The Johns Hopkins University School of Medicine, Baltimore, Maryland 21205. THOMAS M. BOWN • Branch of Paleontology and Stratigraphy, U.S. Geological Survey, Denver, Colorado 80225.

Species, Species Concepts, and Primate Evolution, edited by William H. Kimbel and Lawrence B. Martin. Plenum Press, New York, 1993.

limits in extant biological species. If the acceptable range for a single species is exceeded, the conclusion that more than one species was present is considered justified, although overlapping ranges may still render the boundary between them ambiguous. Because this is a nondimensional example (restricted to one time and place), various species concepts tend to coincide, since lineages, reproductive populations, and morphological clusters all yield the same result. Nevertheless, it must be remembered that conventional species criteria (reproductive, genetic) are unavailable in fossil samples and must be inferred.

Even among field paleontologists, however, relatively few have been faced with the dilemma of how to recognize and to name species in relatively complete stratigraphic sequences. Can new species originate in an unbranched lineage (i.e., anagenetic or phyletic speciation) or only through cladogenesis? In either case, are species boundaries perceived to be sharp (as indicated by the punctuated equilibrium model), or fuzzy and perhaps even arbitrary (as suggested by gradualism)? Or do both situations exist? In fact, the incompleteness of most fossiliferous sections has historically constrained most field paleontologists from pursuing more theoretical approaches to species constructs. Most paleontological species, by virtue of their morphological and temporal distance from one another, have been perceived to be distinct and taxonomically secure.

Recognition of species in the fossil record depends to a large extent on what is meant by the terms *species* and *speciation,* and these definitions are in turn colored by one's systematic and evolutionary philosophy. Often what is meant by a species in the fossil record—sometimes called *paleospecies* or *chronospecies*—is an approximation to the Mayr-Dobzhansky biological species concept. The *evolutionary species* (*sensu* Simpson or Wiley), a rather different concept, designates a lineage. Whether or not such a lineage can or should be subdivided is an issue of contention; nevertheless, subdivision into successive species (which then really approach biological species, at least at any one horizon) has been common (e.g., Gingerich, 1976, 1980).

Morphological breadth has been the conventional criterion for discriminating species in the fossil record. However, some systematists caution against reliance on morphology for identifying species, citing sibling species and polytypic species as potential problems ("degree of morphological distinctness is not a suitable criterion, as proven by sibling species and conspicuous morphs" Mayr, 1982, p. 273). Instead they insist that monophyly and recency of common ancestry, that is, evolutionary continuity, be the only criteria. Such information is rarely available, however, and must itself be inferred ultimately from morphological evidence.

By all accounts speciation is the process resulting in the formation of new species; it is usually considered to involve splitting of a lineage (cladogenesis). According to adherents of the punctuated equilibrium model, it also involves an abrupt morphological shift followed by a much longer period of stasis (Eldredge and Gould, 1972). But many workers—principally paleontologists dealing with empirical records—admit the possibility of the origin of new species within an unbranched lineage (usually by continual and gradual structural change), that is, phyletic or anagenetic speciation (e.g. Simpson, 1961, p. 201; Raup and Stanley, 1978; Harris and White, 1979; Enay, 1983; Jaeger, 1983; Chaline, 1983; Gingerich, 1983, 1985; Bown and Rose, 1987; Van Valen, 1988; Hulbert and Mac-

Fadden, 1991). The distinction is of utmost importance, for if one excludes the possibility of anagenetic origin of species *by definition*, then new species can only be recognized at branching events (see also Tintant, 1983).

Raup and Stanley (1978, p. 104) justified recognition of new species through phyletic transition when it is judged that "the accumulated differences are such that the later populations *would* be reproductively isolated from the initial population *if* they were living at the same time. At this point a new species has formed by phyletic transition." This is no doubt the basis on which paleospecies are often regarded as generally equivalent to biospecies. However, other authors (for example, those adopting the evolutionary or cladistic species concepts) reject the notion that new species can arise anagenetically, regardless of the extent of morphological divergence, because to them an evolutionary species *is* a lineage and it can only give rise to another such species or lineage by splitting (e.g., Wiley, 1978; Hull, 1979).

It is sometimes claimed that branching produces species by a different mechanism from anagenesis. If this could be demonstrated it might be considered support for restricting the term *speciation* to cladogenetic events. (Alternatively, it might simply show that there are at least two ways of producing species.) Some authors (e.g., Martin, 1990), however, contend that the evidence supports no such difference: Whether there is only one lineage involved or diverging lineages, evolutionary change in each lineage occurs in exactly the same manner. The only difference is the resulting number of lineages. Hence anagenetic speciation should be acknowledged. We agree.

Consider a case in which an original population is split into two populations. Over time, each of the isolated populations evolves anagenetically and gradually, diverging from the parent population. The evolutionary process in each isolated lineage is the same as in the parent population and must be the same whether or not the other lineage existed. At least one (perhaps both) of the daughter lineages must be a new species, but which one? At what point in either of the now separate lineages has a new species evolved, and how are we to recognize this? Is the species boundary discrete and nonarbitrary? If sufficient evolution occurs in one or both lineages to require recognition of a new species, whatever the criteria (reproductive, morphological), is this transition any different from an equal degree of evolution within a single phyletic lineage? If one of the daughter lineages was not preserved in the fossil record (or had not existed at all), should that alter the species status of the other, which underwent the same process and extent of evolutionary change from the parent population in either case?

Recently, arguments have erupted concerning how closely classification should reflect phylogeny (or the cladogram). Many cladists contend that they should correspond precisely. But classification relies on discontinuities between categories, whereas phylogeny incorporates continuity and discontinuity (Martin, 1990). We agree with Fox (1986, p. 76) that "it is impossible to classify in a non-arbitrary way continuous phenomena in a classification made up of discontinuous categories having well-marked boundaries." The primary purpose of classification is to promote communication (e.g., Gingerich, 1979b), whereas phylogeny is the actual history of a natural process, evolution. Of course, phylogenetic relationships should provide the foundation for an evolutionary classification. Phylogeny should remain dynamic as new evidence accrues, but stability in classification is desirable until sufficient evidence justifies a change (see also Simpson, 1961).

In this chapter, we briefly review some current species concepts, with particular attention to proposals and examples of their applicability to the fossil record. Subsequently, we consider alternative ways that species may be designated in some especially problematical cases of primates from the densely fossiliferous early Eocene Willwood Formation of Wyoming, and show how the same fossil record is subject to varying systematic classifications depending on the perspective of the taxonomist.

Species Concepts

There has been considerable debate, often acrimonious, over the proper definition of species and what should be the underlying emphasis in such a definition—for example, genetic, reproductive, or evolutionary continuity; ecological niche; or phenetic similarity.

Most species concepts are related to one of two differing evolutionary philosophies, the traditional evolutionary systematics of Simpson and Mayr, and the cladistic method of Hennig. Cladists generally argue that classification should be an exact reflection of phylogeny, whereas more traditional evolutionary systematists accept that classification must be consistent with phylogenetic history but they should not necessarily correspond precisely. Both philosophies aver that monophyly should be the basis of any phylogeny and classification, but to evolutionary systematists monophyly indicates a single origin, whereas most cladists insist that it also includes all descendants (holophyly). Cladistic classifications generally admit only monophyletic groups (in the holophyletic sense; e.g., Smith and Patterson, 1988; DeQueiroz and Donoghue, 1988). Evolutionary systematists, on the other hand, acknowledge the existence, indeed importance, of paraphyletic groups (primitive or ancestral taxa left after subsets with derived traits are separated out; e.g., Simpson, 1961; Mayr, 1969; Van Valen, 1978; Carroll, 1988; see also Harrison, this volume).

The objection to paraphyletic groups is rooted in the cladistic attempt to make classification a mirror of phylogeny, which it cannot and should not be (see also Ashlock, 1979; Gingerich, 1979b). The biological species concept is more compatible with horizontal classifications, the cladistic concept with vertical classifications. The tenet of punctuated equilibrium, that species are as distinct from their progenitors at their origin as they will ever be, is more in accord with cladistic classifications; for it suggests that species are discrete, easily classifiable units, whether viewed horizontally or vertically. By this model species should be easy to recognize in the fossil record. Simpson (1961) favored a compromise between horizontal and vertical classifications, and his evolutionary species concept is consistent with this.

Biological Species

For many years the prevailing species view has been Mayr's (e.g., 1963) biological species concept (BSC), which stresses genetic and reproductive continuity among members, and reproductive isolation between species (see also Dobzhansky, e.g., 1970). It is therefore sometimes called the *isolation concept* of species. The BSC is based on what are now terminal taxa (extant species) (Fig. 1). Biological species are rarely recognized by the reproductive criteria of the definition; instead we rely on phenetic characters to discriminate them: "A morphological species . . . is an inference as to the most probable limits of the biological (genetic) species" (Dobzhansky, 1970, p. 360). Thus the BSC, or an approximation to it, has generally proven satisfactory for application to most fossils as well, because of their limited occurrences.

[Among recently proposed species concepts purported to be preferable to the BSC are the mate-recognition concept ["that most inclusive population of individual biparental organisms which share a common fertilization system" (Paterson, 1985, p. 25)] and the cohesion concept ("the most inclusive population of individuals having the potential for phenotypic cohesion through intrinsic cohesion mechanisms," such as gene flow and genetic drift; Templeton, 1989, p. 12). Both of these concepts have been criticized for their dependence on particular processes of speciation, among other problems, and neither is significantly different from the BSC (e.g., Mayr, 1988a; Chandler and Gromko, 1989). Moreover, neither concept is any more readily applicable to the fossil record; therefore, we do not consider them further in this essay. Other more serious challenges to the BSC are discussed below.]

According to Mayr (1982, p. 273), "The biological species concept, expressing a relation among populations, is meaningful and truly applicable only in the nondimensional situation. It can be extended to multidimensional situations [i.e., extended in space and time] only by inference." Thus considerable debate has focused on whether paleontological species are equivalent to biological species (*sensu* Mayr), as well as on practical problems of applying biological species to the fossil record (e.g., Imbrie, 1957; Maglio, 1971; Gould, 1983; Jaeger, 1983; Tattersall, 1986; Jackson and Cheetham, 1990). In view of various other species

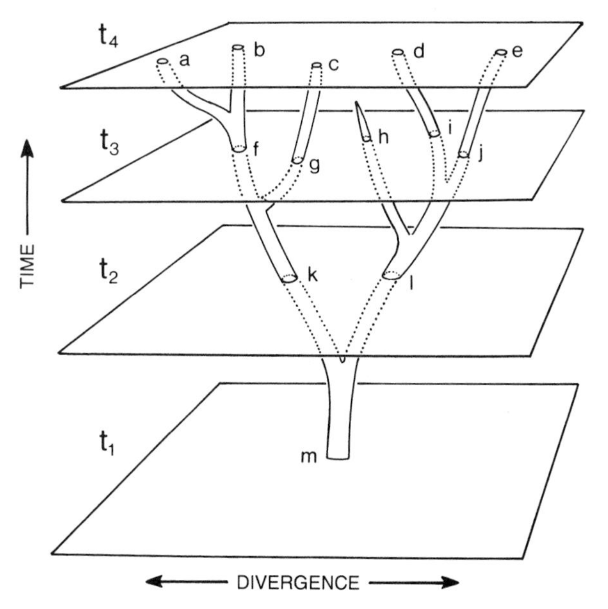

Fig. 1. Hypothetical clade, evolving gradually, cut at four time planes. Time t_4 shows extant (biological) species (a–e); earlier horizons depict contemporaneous paleontological species that are approximately equivalent to biological species (f–m). Lineages or evolutionary species (e.g., mlh, mlid, mlje) followed back in time overlap extensively unless arbitrary boundaries are drawn. The horizontal axis, divergence, usually equals morphology. [Modified after Newell (1956), and Simpson (1961).]

concepts being advocated at present, some might argue that this equivalency does not matter, since the BSC is no longer desirable. Many will disagree with Mayr's (1988b, p. 328) contention that the "biological species concept continues to be what most biologists have in mind when talking about species." Nevertheless, we believe this applies to most paleontologists as well.

Evolutionary Species

Simpson (1961, p. 153) proposed the evolutionary species concept (ESC) as a means of involving evolution in the species concept, thereby extending the biological species concept in time. He defined an evolutionary species as "a lineage (an ancestral-descendant sequence of populations) evolving separately from others and with its own unitary evolutionary role and tendencies." According to Mayr (1982, p. 294), however, this is the definition "of a phyletic lineage, but not of a species." Simpson clearly understood the dilemma of unbounded evolutionary species, which resulted from equating the species with a lineage (Figs. 1 and 2). Hence he reasoned that subdivision of such a lineage into successional species (= chronospecies or paleospecies) was necessary for classification, and that this should be based on morphological variation consistent with that in biological species (Simpson, 1961, p. 165):

> If you start at any point in the sequence and follow the line backward through time, there is no place where the definition [of an evolutionary lineage] ceases to apply. You never leave an uninterrupted, separate, unitary lineage and therefore never leave the species with which you started unless some other criterion of definition can be brought in. If the fossil record were complete, you could start with man and run back to a protist still in the species *Homo sapiens*. Such classification is manifestly both useless and somehow wrong in principle. Certainly the lineage must be chopped into segments for purposes of classification, and this must be done arbitrarily, because there is no nonarbitrary way to subdivide a continuous line.

The necessity for such arbitrary division in the face of relatively complete fossil records has been widely acknowledged (e.g., Mayr, 1982; Gingerich, 1985, 1987; Rieppel, 1986; Rose and Bown, 1986; Bown and Rose, 1987; Martin, 1990). Implicit in this view is acceptance of phyletic speciation. Despite arbitrary

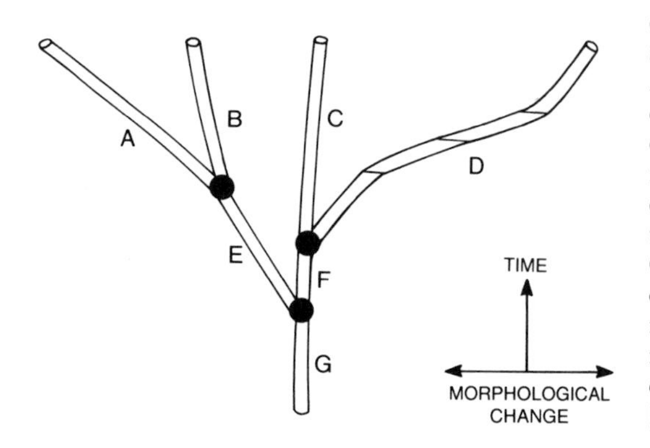

Fig. 2. Hypothetical clade illustrating cladistic species A–G, equivalent to lineage segments between nodes (terminations of A–D are also equivalent to biological species). According to Wiley's ESC, E, F, and G do not necessarily become extinct at nodes; for example, EA or EB might be a single evolutionary species (as is D) regardless of the extent of morphological divergences. GFC is a single species according to most concepts other than cladistic species. Segments in D are paleontological or chronospecies with arbitrary boundaries (in lineage GFD), according to Simpson's ESC. [Modified after Bonde (1981).]

boundaries, successional species themselves are not arbitrary; they have prescribed limits, that is, morphological breadth consistent with extant biological species. Thus successional species are in this sense objective (Simpson, 1943; Newell, 1956; Maglio, 1971; Fox, 1986; Wei, 1987), because diagnoses and characterizations can be provided regardless of where the boundaries are placed. Even this is questionable, however, where there are relatively complete records showing continual change (Bown and Rose, 1987).

Does arbitrary division of lineages preclude the segments from being "real" species? Who can say, for what constitutes a "real" species is itself moot. However, the absence of sharp boundaries between successive species is insufficient to reject them as real, for even at cladogenesis lineages may diverge gradually and continuously, lacking sharp boundaries between species.

Wiley (1978, 1981) espoused the ESC in slightly modified form and argued forcefully that an unbranched lineage should not be subdivided (see also Hennig, 1966; Hull, 1979; Frost and Hillis, 1990), regardless of how much morphological divergence has occurred. This poses obvious problems for paleontological application. Like Hennig, Wiley would generally delimit species by cladogenetic ("speciation") events, but Wiley rejects Hennig's principle that a species necessarily becomes extinct when it splits to give rise to new species. Thus Wiley's ESC allows survival of a parent species after giving rise to a daughter species. According to Wiley (1981, p. 25) an ancestral species may become extinct at cladogenesis if it is "subdivided in such a way that neither daughter species has the same fate and tendencies as the ancestral species," but how this can be determined is not explained.

Bock (1979), while accepting the BSC in nondimensional situations, also subscribed to the evolutionary species concept (which he aptly called a *phyletic lineage*). Although Bock (1979, pp. 28–29) considered lineage splitting as the only source of new species, and acknowledged that at any one time a phyletic lineage consists of only one species, he perceptively reasoned that samples of the lineage viewed at different times "are not different species *nor are they the same species*" (emphasis added; see Figs. 1 and 2). Consequently, he concluded that the cladistic view of species as lineage segments between branching events "is simply not the same as the biological species concept."

Certainly the number of lineages that existed is of interest, but is this what we really wish to indicate by species? Several difficulties arise in applying a lineage species concept to the fossil record. If an evolutionary species is the lineage of ancestral-descendant populations, then at what point does it no longer have "its own unitary evolutionary role and tendencies"? Indeed, as Mayr (1982, p. 294) asked, how could this ever be determined? The question is further complicated by Simpson's stipulation that the role can change within a species. Moreover, where does an evolutionary species begin? Branching lineages lead back eventually to a single ancestral lineage (which must, therefore, be part of all descendant evolutionary species), and all descendants (no matter how many lineages) are in reproductive and evolutionary continuity with the ancestor; but all cannot be the same species because branching has occurred. Thus reproductive continuity is not a useful criterion for identifying a single evolutionary species (Reif and Brettreich, 1985)! Unless arbitrary species boundaries are imposed, the ESC violates its own definition, because each evolutionary species cannot evolve separately or have its own evolutionary role and tendencies. Simpson (1961, pp. 164–165) recognized this paradox, and this is why he believed that arbitrary segmentation is inescapable (see also Reif and Brettreich, 1985).

The problem was perhaps best summarized by Reif and Brettreich (1985, pp. 424–425), who concluded that there is no objective way to determine if successive samples represent the same species or not, and that

> . . . species can only be recognized objectively within a time plane, if at all. . . If the biological species criterion is applicable in a given higher taxon, a "horizontal" (i.e., within a time-plane) delimitation of species is possible at any time. A "vertical" delimitation (of evolutionary species) requires an arbitrary definition in addition to the biological species criterion.

Cladistic Species

Ridley (1989, p. 12) recently suggested that noncladistic application of the BSC may underestimate the scope of a species and overestimate the true number of species, defining "arbitrarily limited, rather than cladistic, branch units." Like Hennig (1966) and Wiley (1978, 1981), Ridley considers that an unbranched lineage represents a single species, notwithstanding the extent of morphological evolution.

The cladistic species concept (Hennig, 1966; Bonde, 1981; Ridley, 1989; see also Westoll, 1956) maintains that a parent species must become extinct at a cladogenetic event, producing two new daughter species, regardless of how much or how little change occurred between parent and each "new" species. Thus species are delineated as those groups of individuals between two branching events (Fig. 2). It has been argued that this is the only nonarbitrary way to delineate species in an evolving continuum. Perhaps so, but Mayr (1982, p. 293) contends that to require extinction of the parent species, by definition, is theoretically wrong, because speciation in a peripheral isolate "[has] no effect on the main body of the species, which continues its evolutionary life without changing its species status, since it is unaffected by the budding off of a peripheral daughter species" (see also Hull, 1979; Mayr, 1988b). As aforementioned, according to Wiley (1981) the parent species may either persist or become extinct at speciation.

Ridley (1989, p. 5) optimistically asserted that, in theory, application of a cladistic species concept solves the species problem: "the only problem will be practical questions of whether we can recognize [species]." But this is no small problem for the practicing paleontologist! Any theoretical unit that cannot be made operational is unsatisfactory. Use of a cladistic species concept is inherently circular and impractical, for it requires *a priori* knowledge of phylogeny before a species can be recognized. But to reconstruct phylogeny requires analysis of relationships among species, which in turn necessitates that we have first discriminated species.

According to the cladistic species concept, discrimination of a distinct species may not require any morphological differences from a parent species, but only the existence of a "sister" species. Therefore, discovery of fossils extraneous to a specified lineage, but indicating that the lineage had split, could dictate its recognition as a distinct species, even if the latter cannot be separated morphologically from the parent species. As a result, taxonomic stability and utility would be sacrificed for the sake of a putatively more objective species definition. The paradox of the cladistic species concept is, of course, that even cladistic phylogeny reconstruction must be based ultimately on morphological characters; but if this evidence is irrelevant to species identification, how can phylogeny be reconstructed?

Even if placing boundaries at nodes were theoretically more objective, practical recognition of cladistic species so defined is no more objective than if their boundaries were arbitrarily drawn between nodes. This is because cladistic species are discriminated by their position in a diagram (i.e., between nodes in a cladogram), rather than by any real (e.g., phenetic) criteria that can be observed in fossil or recent specimens. In fact, designation of a branching "event" (= node on a cladogram) as the lower boundary of a species is a convention and is no less arbitrary than subdivision of an evolving lineage at convenient stratigraphic levels (both practices may prove useful and necessary), for (as discussed above) the lineage is continuous back through time to and beyond the branching "event" (Simpson's paradox).

Phylogenetic Species

The phylogenetic species concept (PSC) has been recommended by some authors as preferable to the biological or other species concepts, because it follows "from the perspective of the *results* of evolution rather than from . . . the processes thought to produce those results" (Cracraft, 1983, p. 169), and it is purportedly based on evolutionary units (e.g., Eldredge and Cracraft, 1980; Cracraft, 1983, 1987; McKitrick and Zink, 1988; Echelle, 1990). Cracraft (1983, p. 170) defined a phylogenetic species as "the smallest diagnosable cluster of individual organisms within which there is a parental pattern of ancestry and descent." As Ereshefsky (1989) pointed out, this definition is flawed because it is too ambitious for consistent applicability to real situations (for instance, it could refer to local demes or to morphological variants in fossil samples), and it defines away subspecies (some present subspecies would become full species while others would be subsumed; see also Echelle, 1990).

The PSC prescribes that "species are now defined in terms of diagnostic characters" (Cracraft, 1983, p. 171), not by reproductive isolation. In this respect it conflicts not only with the BSC but also with Wiley's ESC, according to which the extent of phenetic change is not a valid criterion for species discrimination. Mayr (1988b) suggested that this kind of species definition approximates a typological concept in practice. The PSC may be useful for discriminating species that evolved by the punctuated equilibrium model, but it offers no guidelines for dealing with gradually evolving sequences of species. For these reasons, the PSC fails to provide a preferable alternative to the BSC or to Simpson's version of the ESC for application to the fossil record.

Other Species Concepts

Van Valen (1976, p. 233) proposed the ecological species concept as a modification of Simpson's ESC, emphasizing the ecological niche: "A species is a lineage . . . which occupies an adaptive zone minimally different from that of any other lineage in its range and which evolves separately from all lineages outside its range." [More recently Van Valen (1988) has adopted a more eclectic concept that includes characteristics of the biological, evolutionary, and ecological species concepts.]

De Queiroz and Donoghue (1988), De Queiroz and Gauthier (1990), and Mishler and Donoghue (1982) advocated a species concept based on monophyly (in the restricted sense of holophyly) rather than reproductive criteria. Adherence to this concept requires that species names be applied only to monophyletic populations; a population considered ancestral to any other would not be recognized as a separate species or as any other monophyletic taxon. (This conflicts with the cladistic species concept of Hennig, Bonde, and Ridley discussed above.) According to De Queiroz and Donoghue (1988, p. 332),

> The ancestral population of a monophyletic group recognized as a genus is part of that genus but not of any less inclusive monophyletic taxon. . . . Although not assigning all organisms or populations to taxa of species rank violates a longstanding convention, this alone is insufficient grounds for rejecting a definition of the species category based on monophyly.

Nonetheless, few paleontologists would find it very practical to relegate most fossils only to a higher taxon without species assignments. Moreover, as fossil records improve and relationships become better documented, this convention would require that we abandon well-established species found to be ancestral to any later taxa! Thus, in the interest of preserving a monophyletic species concept, consistency and stability in taxonomy again would be lost.

Donoghue (1985) proposed that populations whose monophyletic status is uncertain (for which neither monophyly or paraphyly is demonstrable) be recognized as "metaspecies" and their names designated by an asterisk; however, "demonstrably paraphyletic groups would not be recognized as taxa" (De Queiroz and Donoghue, 1988). Following this procedure, such familiar species as *Australopithecus africanus* or *Homo habilis* might be considered metaspecies, but more likely would not be recognized as taxa at all! In contrast, Van Valen (1988, p. 52) explicitly accepted different species of genera such as *Australopithecus*, even within the same lineage, "because of their adaptive differences as reflected in the total phenotype."

Like the cladistic species concept, discrimination of monophyletic species requires prior knowledge of phylogeny rather than use of reproductive or phenetic criteria. For this reason, and the obvious drawback that ancestral populations cannot be accorded species names, it seems to be a less than optimal solution of the species problem for paleontology.

Ideally there would be a universal and invariant concept of species, applicable for sexual and asexual species, and in multidimensional as well as nondimensional situations. But such a concept may be unattainable, even theoretically. Mishler and Donoghue (1982, p. 500) suggested that "The search for a universal species concept, wherein the basal unit in evolutionary biology and in taxonomy is the same, is misguided." Many other systematists have also reached the conclusion that different concepts *should* apply in different situations (e.g., De Queiroz and Donoghue, 1988; Stebbins, 1987; Endler, 1989; Westergaard, 1989). In application to real cases, many different species concepts are currently advocated, in most cases with justification.

Species in Paleontology

Although it is important for a species concept to have a sound theoretical basis in evolutionary biology, whatever concept is adopted must allow delimitation and recognition of species in the fossil record. However, most theoretical concepts are difficult to apply in practice (Endler, 1989).

A frequent concern of adherents to the BSC is whether paleontological species, based on phenetic criteria, constitute a reasonable approach to the actual number of species. Some paleontologists argue that morphological evidence available in fossils may not reveal specific differences that exist (e.g., among sibling species; see Jaeger, 1983). Although morphological differences and reproductive isolation do not always coincide (e.g., Cope, Tattersall, this volume), the basic assumption that they usually do underlies the recognition of virtually all paleospecies; morphology still offers the best approximation to biological species in the fossil record (e.g., Dobzhansky, 1970).

Accepting that morphological evidence generally provides the criteria by which we *discriminate* species in the fossil record, there are two types of situations that pose particular problems for recognition of fossil samples as species: similar or overlapping morphologies in contemporaneous (often sympatric) samples, and overlapping morphologies in temporally successive samples. In both cases species boundaries may be ambiguous. In contemporaneous samples it is first necessary to determine how many lineages are present. This is usually considered in terms of the biological species concept, but can only be dealt with in phenetic terms. It is arbitrary to the extent that the paleontologist must decide how much morphological variation is too much to have represented a single species and, then, which characters to use to separate the sample into more than one species (e.g., Gingerich, 1974; see also Cope, Plavcan, this volume). In many cases, nonoverlapping or clearly bimodal character or size distributions (if inconsistent with sexual or other dimorphism) render such a decision clear-cut. Frequently, however, morphological overlap is extensive, yet ranges exceed acceptable standards, or statistical tests indicate more than one species (as shown by many contributors to this volume). In such instances, if two (or more) lineages are indicated, both must be considered distinct species, even if the boundary between them is obscure. We presented examples of this situation in the early Eocene omomyid primates *Tetonius* and *Absarokius* from the Bighorn Basin, Wyoming (Bown and Rose, 1987).

Temporally successive samples that include evolutionary intermediates pose the greatest problem for delimiting species, partly because of differential evolutionary rates: Change in morphology is not necessarily accompanied by reproductive isolation, and isolation may be acquired with minimal morphological divergence (e.g., Mayr, 1988b). Moreover, morphological characters often change in mosaic fashion. It is such evolutionary intermediates that we focus on in this section.

Gaps in the fossil record, usually representing stratigraphic breaks, often provide convenient boundaries between morphological paleospecies. If stratigraphic gaps were filled in by fossils, one of two patterns should be evident (Fig. 3): either an abrupt morphological break at the inception of a new species (according to the punctuated equilibrium model), or gradual transformation from parent species to daughter species (whether or not splitting occurred). In the former instance, species boundaries would be relatively nonarbitrary and readily designated; in the latter, they would be nebulous and arbitrary. Therefore, in a gradually evolving lineage, use of stratigraphic gaps to delineate species is no less arbitrary than any other arbitrary line drawn in a continuous record.

How can such a continuous sequence be classified? Should it be divided into species and, if so, how? Simpson (1943, p. 171) was one of the first to grapple with this dilemma, noting the existence of continuous sequences of apparently gradually changing populations:

> which change so much that every taxonomist places the earliest and latest members of the line in different species or genera. Clearly a species as a subdivision of such a temporal, or vertical, succession is quite a different thing from a species as a spatial, or horizontal, unit and cannot be defined in the same way.

Simpson further wondered "whether such subdivisions should be called species and whether vertical classification should not proceed on an entirely different plan from the basically and historically horizontal Linnaean system." At that time he opted to name the successive samples as subspecies in a chronocline,

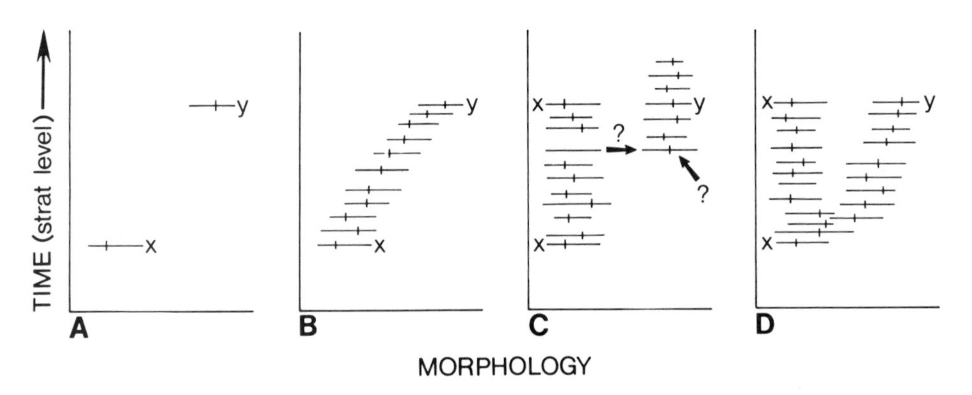

Fig. 3. Effect of completeness of the fossil record and evolutionary philosophy on the interpretation of two fossil samples. Samples differing significantly in morphology and age (A) are usually considered distinct, but discovery of intermediates suggesting gradual transition from x to y (B) makes species discrimination problematic. According to some species concepts (see text), the lineage is regarded as a single species unless evidence of punctuation, immigration, or cladogenesis exists (C,D). But x and y are equally distinct in all four examples and are no more similar to each other simply because transitional forms are known. [From Rose and Bown (1986).]

but subsequently he regarded them as successive or successional species (Simpson, 1961; see discussion of ESC above).

While we do not advocate subdivision of anagenetic lineages where morphological change is minimal, we concur with Simpson that it is both useful and necessary to recognize as species successive samples that show a significant degree of morphological distinction from other samples, for the extent of difference may equal or surpass that in diverging lineages undergoing allopatric speciation. Following widespread opinion, subdivision of unbranched lineages should be based on comparable variation to that observed among extant biological species (e.g., Simpson, 1951, 1961; Sylvester-Bradley, 1956; Imbrie, 1957; Raup and Stanley, 1978; Gingerich, 1985; Mayr, 1988b); that is, at any time plane (geological horizon) Mayr's BSC is theoretically applicable. Thus successive species should be morphologically and ecologically distinct and, by inference, would have been unable to interbreed had they coexisted (Raup and Stanley, 1978).

Some authors reject the notion that morphology and reproductive isolation are useful criteria for species discrimination. For example, Ridley (1989, pp. 11–12) asserted that for this scenario "the correct, cladistic interpretation is obviously that the lineage is all one species . . . whether the earlier and later forms could interbreed is no more relevant than whether they have the same ecological adaptations." But this interpretation could result in cladistic species that far exceed in morphological breadth most other paleontological as well as extant species. Because he equated lineages with species, Ridley argued that species diversity in the fossil record has been overestimated by others. In contrast, Tattersall (1986) and Hecht (1983) contend that closely allied species may differ only slightly and often by anatomy not available in fossils; hence, species diversity in the fossil record has probably been underestimated.

In gradually evolving situations where the record is sufficiently dense and continuous, the boundaries between successive species are necessarily arbitrary (e.g., Newell, 1956; Rhodes, 1956; Simpson, 1961; Gingerich, 1983, 1985, 1987; Rieppel, 1986; Rose and Bown, 1986; Martin, 1990). Even Mayr, an adherent of the cladogenetic view of speciation, has admitted (1982, p. 295),

> It is only in the cases of a sequence of ancestor-descendant species which transform gradually into each other in a single phyletic lineage that a sharp delimitation between temporal species taxa is impossible. Here biological evolution fails to accommodate the wishes of the taxonomist. Fortunately, the fossil record is more accommodating. Its deficiencies usually provide sufficient gaps in the lineages to permit a delimitation of vertical species taxa, as artificial as this may be. It seems that we will have to accept this compromise solution since the evidence does not seem to support the claim of some proponents of the theory of "punctuated equilibria" that there is never any phyletic speciation and that all new species originate in founder (or refuge) populations or even by saltations.

The difficulty of species demarcation is not restricted to cases of single phyletic lineages, but also obtains in instances of apparent sympatric gradual cladogenesis [e.g., the Eocene primates *Absarokius* (Bown and Rose, 1987) and *Cantius* (Harrisville-Wolfe, 1991), the radiolarian *Eucyrtidium* (Prothero and Lazarus, 1980)].

Inspired by the controversy over rates and patterns of evolution, a large literature on individual cases has appeared in recent years. These studies have documented that gradual evolution—whether by anagenesis or cladogenesis—of what had been regarded as different species has occurred in many different zoological groups, thus necessitating arbitrary imposition of species boundaries. Without the much improved fossil records that have linked together successive species, there would be no ambiguity about their distinction.

For example, Sheldon (1987) found that previously named "end members" of several different lineages of trilobites were linked by extensive series of intermediates, rendering boundaries between species nebulous. There are now many cases of apparent gradual anagenesis linking series of previously recognized species of planktonic foraminiferans (e.g., Malmgren and Kennett, 1981; Bolli, 1986; Wei, 1987; Wei and Kennett, 1988). Malmgren *et al.* (1984) documented the transition from one foraminiferan species to another, without lineage splitting, during an apparently short but observable interval ("punctuated gradualism"). But the seemingly punctuated nature of that transition now appears to have been an artifact of sediment compression; hence, the transition was in fact much slower and more gradual and the boundary between species even more ambiguous (MacLeod, 1991). Chaline and Laurin (1986) reported that five previously named species of European arvicolid rodents actually represented successive species of a single lineage, which had evolved gradually through phyletic speciation. Improved records of microchoerine primates from the upper Eocene of France document the gradual cladogenesis between *Necrolemur antiquus* and *Microchoerus* cf. *erinaceus* (Godinot, 1985), and new specimens of the early Tertiary plesiadapiforms *Carpolestes* and *Phenacolemur* bridge gaps between long-known species (Rose, 1981; Gingerich, 1987). These examples demonstrate that, as fossil records improve, species boundaries often become more equivocal and arbitrary. While increasing rapidly, cases like these are still relatively rare.

The most celebrated examples in the mammalian fossil record come from the early Eocene Willwood Formation of the Bighorn Basin, Wyoming, a series of densely fossiliferous floodplain paleosols stacked about 700 m thick and spanning 3 million years (Bown and Kraus, 1981, 1987; Bown *et al.*, in press; Wing *et al.*, 1991). Gingerich (e.g., 1976, 1980) and others have documented numerous cases of gradual transformation between successional mammalian species, and primates have figured prominently in these studies.

Eocene Primates from the Bighorn Basin

Cantius

Gingerich and Simons' (1977) stratophenetic study of the adapid primate *Cantius* (then referred to *Pelycodus*) from the Bighorn Basis charted changes in molar size and other morphological traits (such as hypocone and mesostyle expression) through the Willwood Formation. They demonstrated that temporally successive samples of *Cantius*, which had conventionally been treated as the distinct species *C. ralstoni* and *C. trigonodus* based on morphological differences, were linked by intermediate stages, thus producing a continuum of intergrading successional samples (see also Bookstein *et al.*, 1978; Rose and

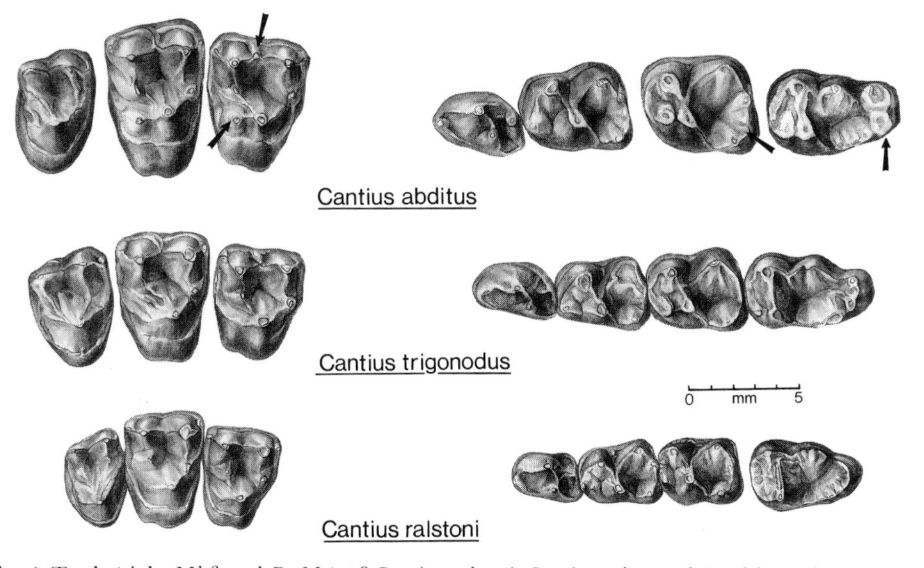

Fig. 4. Teeth (right M^{1-3} and P_4-M_3) of *Cantius ralstoni, C. trigonodus,* and *C. abditus,* showing differences in size and structure by which they have been differentiated. Arrows indicate particular areas of interest (hypocone and mesostyle on uppers, entoconid notch of lowers, and M_3 heel). Intermediates now known display continuous gradation in these traits suggesting that these species are segments of a gradually evolving continuum. [*C. ralstoni* upper teeth based on AMNH 16089 (holotype) and AMNH 16070, lowers drawn from AMNH 16093 supplemented by USGS 13586; *C. trigonodus* based on AMNH 15017 (holotype); *C. abditus* based on USGS 6328, lowers supplemented by UM 66000 (holotype).]

Bown, 1984b; Harrisville-Wolfe, 1991). Thus what had long been recognized as distinct species apparently evolved through gradual anagenesis. Gingerich and Simons named two new species, *C. mckennai,* linking the two previous species, and *C. abditus,* a larger and more derived species emerging from *C. trigonodus** (Fig. 4). Whether or not one agrees with the naming of additional species in this lineage, there has been universal agreement among paleontologists that more than one species should be recognized. As Gingerich (1987, p. 1058) observed, "Documentation of a continuous series of specimens connecting species of *Cantius* does not thereby make all species the same; they are as different as they were in Matthew's time, but intermediates are now present as evidence of transition." (see Fig. 3; see Rose and Bown, 1986, p. 121, for a similar view.)

All four *Cantius* species were considered to represent arbitrarily bounded segments along a single unbranched lineage. By a cladistic or evolutionary species concept, however, this would be considered a single species. But if *C. trigonodus* persisted after giving rise to the new species *C. abditus* (as suggested by Harrisville-Wolfe, 1991 and unpublished manuscript), cladogenesis must have occurred, and we are compelled to recognize both species, even though the precise boundary between them must be arbitrarily imposed (because we can

*Beard (1988) synonymized *C. trigonodus* with the San Juan Basin species *C. frugivorus.* We are not convinced that these widely separated forms are conspecific, and to avoid confusion here we retain the traditional nomenclature for the Bighorn Basin species.

never pinpoint the initial divergence) and their morphological characteristics and differences have not changed.

Omomyids

We have elsewhere presented evidence for gradual transformation within lineages of early Eocene omomyids from the Bighorn Basin (Rose and Bown, 1984a, 1986; Bown and Rose, 1987), and these have been considered convincing examples of anagenetic change or even phyletic speciation (Gould, 1988; Jones, 1988; Mayr, 1988b). Here we will focus instead on species discrimination and nomenclature in these continuously evolving sequences.

Our analysis was based on a modification of Gingerich's (1979a) stratophenetic method, whereby we analyzed tooth dimensions and structural characters in a stratigraphic context. This method has been criticized (e.g., Forey, 1982) for a too literal reading of the stratigraphic record, in which successive samples are linked regardless of morphology. This is, of course, an exaggeration; these studies require detailed knowledge of the teeth and groups in question, and in many cases teeth compared stratophenetically are in fact essentially identical except in size. Nonetheless, teeth in closely related species often do differ in characteristics other than size, and it is clearly advantageous to examine all available traits. In so doing, one would expect to find stasis in most features, accompanied by change in others (e.g., Rose and Bown, 1986; Bown and Rose, 1987; also compare Gingerich, 1985, with Krishtalka and Stucky, 1985, on *Diacodexis*).

Teilhardina. European early Sparnacian *Teilhardina belgica* is one of the most primitive known omomyids, which occupies a phylogenetic position in or very near the ancestry of North American early Wasatchian *T. americana* (Fig. 5). Neither species has any evident autapomorphies, although they are separated by a geographic and slight morphological gap presumed to coincide with a brief temporal hiatus. (Thus *T. americana* is derived relative to *T. belgica,* but its derived traits are also present in its immediate descendants.) Samples occurring in a 150-m interval of the Willwood Formation, beginning with *T. americana,* show progressive changes, including loss of P_1, increase in size of I_1, reduction in the canine, basal inflation of cheek teeth, and lower and broader P_{3-4} with increasing elevation of the metaconid (Fig. 6). These characters are brought to extreme in the samples we called *T. crassidens;* smaller size and still lower P_{3-4} with widely open trigonids characterize *T. tenuicula.* A series of stratigraphic and morphological intermediates showing a wide range of morphology seems to document the branching of the two later species and provides a more or less continuous gradation from *T. americana* to both species (Figs. 6 and 7), but neither this split, nor the probable branching of *Tetonius* from *Teilhardina,* can be placed precisely. The dichotomy between *T. tenuicula* and *T. crassidens* appears to have resulted from separation of a peripheral isolate within the same basin. A species of *Teilhardina,* most likely *T. crassidens,* probably gave rise to the basal species of *Anemorhysis.*

Where should the species boundaries be drawn in this clade? Both daughter species are morphologically and cladistically distinct. But if either *T. tenuicula* or *T. crassidens* were unknown, should we no longer consider the other species

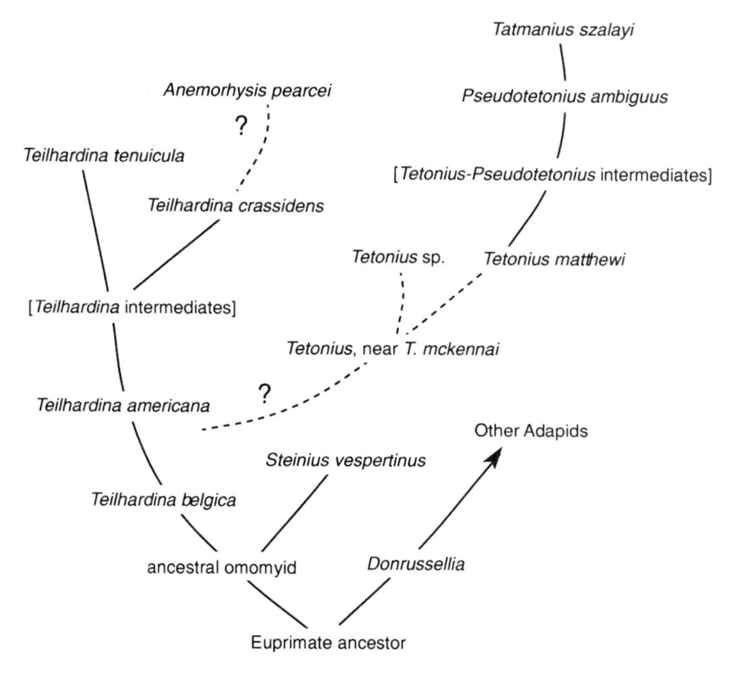

Fig. 5. Phylogenetic tree of some omomyid primate species, mostly from the early Eocene of the Bighorn Basin; *Teilhardina belgica* and *Donrussellia* are European.

distinct only because it no longer has a sister lineage? In our study (Bown and Rose, 1987) we opted to restrict each species to a relatively uniform morphology (approximating the phenotypic breadth of extant primate species) that could be readily diagnosed from others, situating arbitrary boundaries in stratigraphic intervals where fossils were sparse. This conforms to Simpson's ESC. The phylogenetic species concept would presumably recognize at least as many species, and their boundaries would remain problematic. Applying Wiley's ESC, one could designate as evolutionary species the lineages *T. americana* (or an earlier form) through *T. tenuicula,* or through *T. crassidens* (and species of *Anemorhysis*) (see Fig. 5)—thus resulting in only two species—but the beginning of either evolutionary species would be arbitrary, and neither could be discriminated by its morphology. A cladistic species concept would extend *T. tenuicula* and *T. crassidens* earlier in time into the group designated as intermediates (Bown and Rose, 1987) and impose their lower boundary (equivalent to the termination of *T. americana*) at the cladogenetic "event." Of course, this cannot be known, so the boundary in practice would have to be drawn arbitrarily anyway. Finally, according to De Queiroz and Donoghue's (1988) monophyletic species concept, if *T. belgica, T. americana,* and *T. crassidens* prove to be ancestral to later taxa, none should be recognized at the species level.

Tetonius–Pseudotetonius–Tatmanius. Five different species names are applied to parts of this lineage in the Bighorn Basin: *Tetonius homunculus, Tetonius* sp., *T. matthewi, Pseudotetonius ambiguus,* and *Tatmanius szalayi* (Bown and Rose, 1987, 1991). *Tetonius homunculus* was used for the type specimen for technical reasons and can be ignored here (it represents an undeterminable part of the *T. mat-*

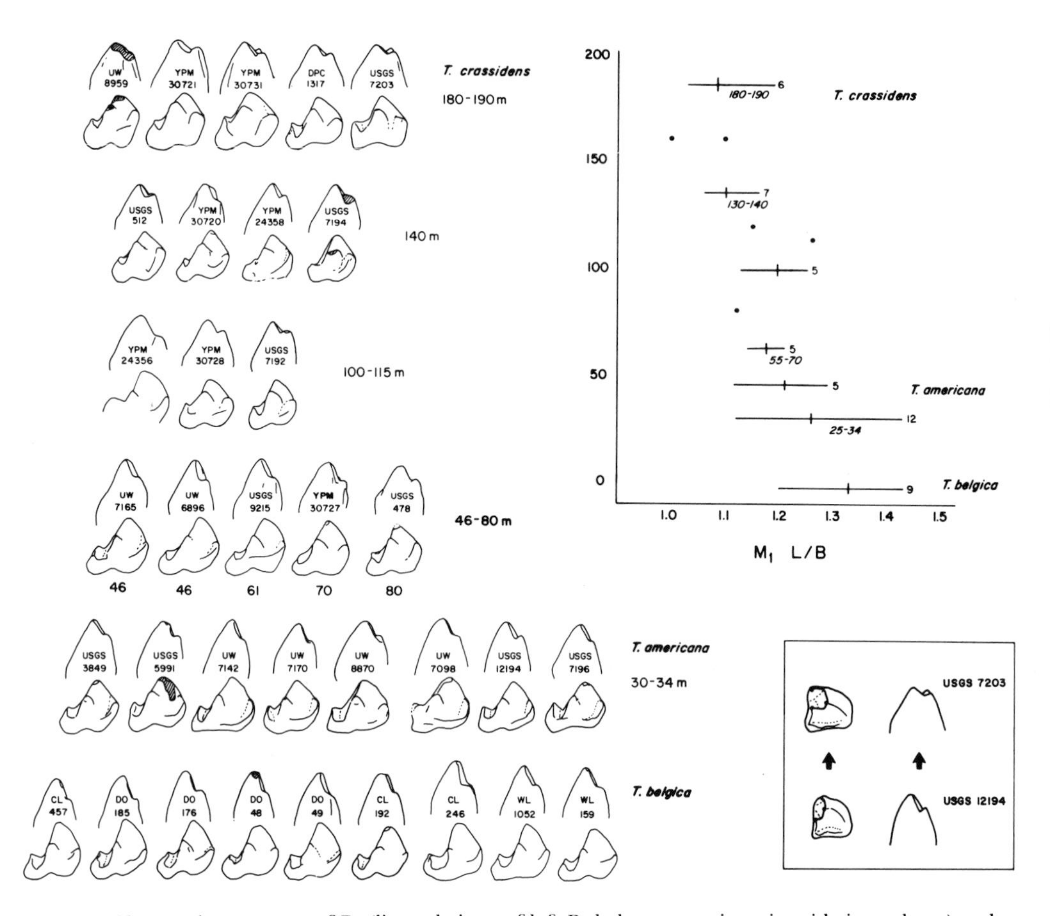

Fig. 6. Changes in structure of P_4 (lingual views of left P_4 below, posterior trigonid views above) and shape of M_1 in the lineage from *Teilhardina belgica* to *T. crassidens*. Meter levels are measured from base of the Willwood Formation in the southern part of the Bighorn Basin. *Teilhardina belgica* is from the basal Sparnacian of Belgium, arbitrarily placed here at the 0-level, but probably older than *T. americana* in any case. . [From Bown and Rose (1987).]

thewi–P. ambiguus lineage). The simplest interpretation of the evolution of these omomyids is that the remaining four species constitute a lineage in the sequence listed, which also corresponds to their chronological occurrences and increasingly derived morphology (Fig. 5). Alternatively, *Tetonius* sp. may represent a short-lived offshoot. *Tetonius mckennai,* known from a different basin, is a plausible structural ancestor for the lineage and was reasonably derived from a primitive species of *Teilhardina*.

Features that accumulated gradually in the line from *Tetonius matthewi* to *Pseudotetonius ambiguus* mostly relate to compression of the anterior dentition: hypertrophy of I_1, loss of P_2, and reduction of intervening teeth to P_4 (including loss of one root of P_3). The same traits, together with reduction of P_4 and *its*

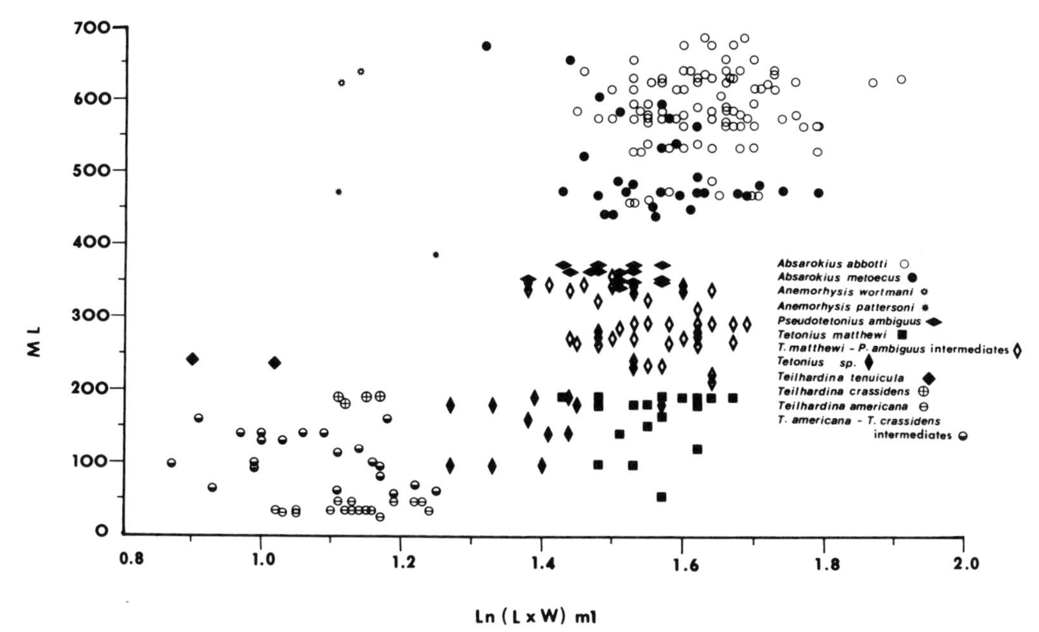

Fig. 7. Plot of M_1 area in omomyids from the southern part of the Bighorn Basin, according to stratigraphic level (ML) in the Willwood Formation. This kind of plot displays the distribution of specimens in the section and suggests a first-order separation of taxa, but many other aspects must be examined in order to discriminate paleontological species and to propose phylogenetic relationships. [From Bown and Rose (1987).]

roots, characterize *Tatmanius szalayi*, which appears to be derived from *P. ambiguus* (Bown and Rose, 1991). These successional species, and several unnamed intermediate stages, evolved without evidence of branching; yet together they encompass far greater morphological diversity than related modern primate species (Figs. 5 and 8).

The size range in early samples of *Tetonius* exceeds normal ranges within comparable biological species. To underscore the probability that more than one sympatric species (i.e., more than one lineage) is represented, we listed smaller specimens whose P_4 fell below the size range of the next successive sample—an arbitrary guideline—as *Tetonius* sp. (Bown and Rose, 1987). Our reluctance to apply a formal species name reflects uncertainty that more than one species was present. The sample is not obviously bimodal, and consistent morphological criteria to separate *Tetonius* sp. and *T. matthewi* are lacking. The specimens co-occur, intergrade completely, and (from what is currently known) can be separated only arbitrarily. Nonetheless, the broad size range of the entire sample, the abrupt disappearance of the smaller specimens above 190 m, and possible proportional differences of P_4 suggest that two lineages may well have coexisted, with only the larger one in continuity with later samples. If so, then both are species by all concepts, even if the boundary between them must be drawn arbitrarily. Alternatively, if evidence for branching is denied and comparable variation to biological species is rejected as a species criterion, then a reasonable case can be made for a single variable species. In that case, one could regard

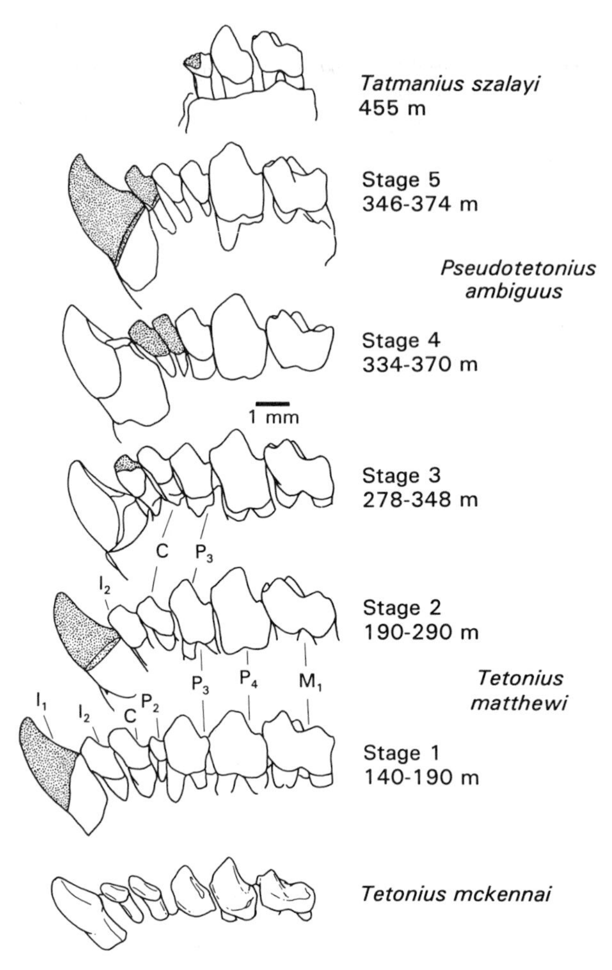

Fig. 8. Transformation of the lower dentition (left I_1-M_1, buccal view) in the *Tetonius matthewi-Pseudotetonius ambiguus* lineage. Stages 1 and 5 have been considered different species since 1915, but transitional forms now mandate an arbitrary species boundary. Possible extensions of the lineage (more primitive *Tetonius mckennai* and more derived *Tatmanius szalayi*) are also depicted; for these two, gaps in the fossil record still provide convenient species boundaries. Alternatively, if all of these belong to the same unbranched lineage, they represent a single cladistic or evolutionary species. [Modified from Bown and Rose (1987, 1991).]

Tetonius sp., *T. matthewi,* and *P. ambiguus* as parts of the same unbranched lineage, that is, a single evolutionary or cladistic species.

Tatmanius szalayi, probably a descendant of *P. ambiguus,* is, for now at least, a valid species by virtue of its stratigraphic and morphological isolation, comparable to what separated *Pseudotetonius ambiguus* from *T. matthewi* (then called *T. homunculus*) in 1974 before intensive collecting filled in the gaps. If this hiatus were similarly filled, *Tatmanius szalayi,* too, would be subsumed under the same evolutionary species unless evidence of branching could be determined.

Species in this clade have been delineated based on the recognition criteria (morphology) of the BSC or, where continuous lineages exist, Simpson's ESC.

We combined the latter with numbered stages (which overlap in part with species) in order to reflect progressively derived morphology (Fig. 8). Thus stage 1 specimens (excluding smaller specimens grouped as *Tetonius* sp.) and contemporaneous stage 2 individuals were assigned to *Tetonius matthewi,* whereas stage 5 and contemporaneous stage 4 specimens were allocated to *Pseudotetonius ambiguus.* All stage 3 and most stage 2 and 4 specimens were considered intermediates without formal assignment. The species boundaries were drawn arbitrarily at stratigraphic horizons rather than at morphological (stage) boundaries, because the latter overlap and transitional specimens exist between stages as well. Hence, through most of the *Tetonius–Tatmanius* clade we believe only a single species existed at any temporal horizon. The lineage consisted of variable species with arbitrary boundaries, an interpretation that is, contrary to Krishtalka's contention (this volume), decidedly *non*typological.

Several alternative taxonomic treatments of this lineage are possible, however. For instance, our stages might be equated with species. This interpretation, which might result from application of the phylogenetic species concept, would significantly increase the number of species. In fact, rigidly applying the PSC's "smallest diagnosable cluster" could result in even more species than numbered stages. Our stages would be the same as species also if they had resulted from a series of closely spaced cladogenetic events, after which the primitive stage persisted for a short time together with the derived and eventually prevailing stage. This interpretation would justify recognition of distinct species for each stage; but it would result in as many as four contemporaneous, sympatric species (for example, at about 340 m; see Bown and Rose, 1987: Fig. 68), differing for the most part in relatively trivial characters that are usually associated with intraspecific variability. Similar cases among extant mammals are rare, and we believe it is much more probable that the record here is of a single lineage composed of a succession of variable species. This is supported by the fact that our five stages are not distinct, but rather were arbitrarily discriminated, with transitional forms bridging them as well. Our position is most accurately represented by using Simpson's ESC, with arbitrary time-parallel species boundaries indicating that at no time during the unbranched lineage is more than one species in existence (see also Gingerich, 1979b).

Wiley's ESC focuses on lineages too, thus its application to the Bighorn Basin omomyids in Fig. 5 also avoids the risk of overestimating the number of species (i.e., lineages) at any one time; but it neglects morphological diversity and encounters other difficulties inherent to its definition. For example (Figs. 5 and 9), we might recognize the following as evolutionary species: (1) *Teilhardina belgica* through *Anemorhysis pearcei,* (2) *Teilhardina tenuicula* (or alternatively *T. belgica* through *T. tenuicula*), (3) *Tetonius mckennai* through *Tatmanius szalayi* (missing data between *Teilhardina* and *Tetonius* make this a convenient branch), and (4) *Tetonius* sp. Each of these is a lineage or, more precisely, part of a lineage (as can be seen in the phylogenetic tree but is not obvious in the cladogram). Several other alternative evolutionary species can be envisioned (for example, *Teilhardina belgica* through *Tatmanius szalayi*). The problem with the ESC, as already discussed, is that lineages "overlap eventually as they are followed backward in time" (Simpson, 1961, p. 164), hence the beginning of an evolutionary species must be arbitrarily imposed. We may choose to recognize it at a branching

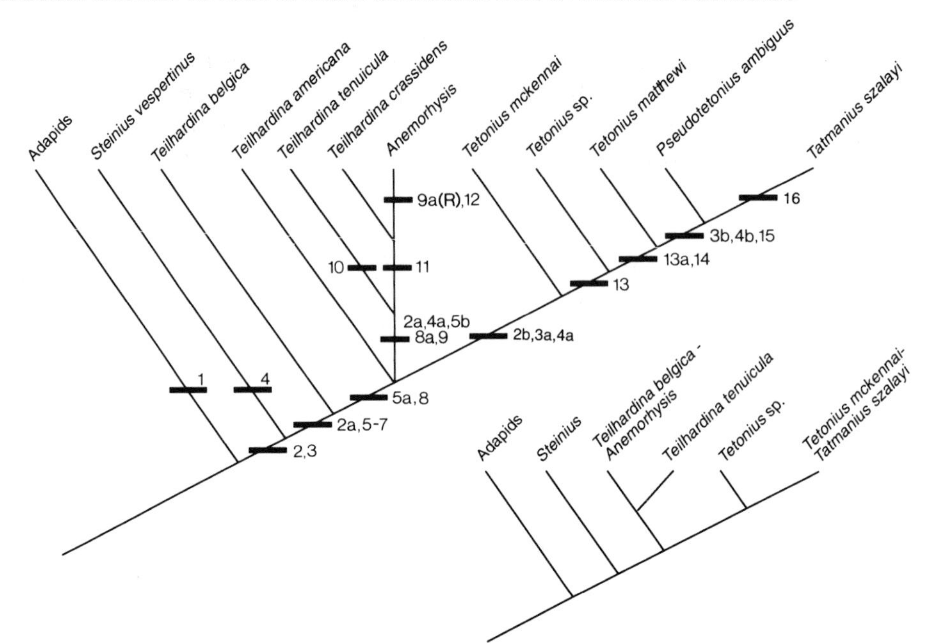

Fig. 9. Two of many possible cladograms of the taxa in Figure 5; bars indicate derived traits (listed below). Note that in the larger cladogram several species lack autapomorphies because they are probably ancestral to later taxa. The smaller cladogram illustrates the collapse of several taxa according to a strict ESC (*sensu* Wiley); the resulting taxa are also arbitrarily bounded. Because of the difficulty of indicating gradually evolving traits on a cladogram, a phylogenetic tree may be a preferable way to depict ancestral-descendant relationships when a dense fossil record exists. Here letters are used to indicate such traits. Probable derived characters: (1) tarsal features (see Dagosto, 1988; Gebo, 1988); (2) P_1 small, (a) P_1 very small or absent, (b) P_1 always absent; (3) P_2 l-rooted, (a) P_2 very small or absent, (b) P_2 always absent; (4) $I_1 > I_2$, (a) I_1 noticeably enlarged, (b) I_1 very large; (5), (a, b) Cheek teeth progressively relatively broader; (6) premolars becoming relatively shorter; (7) M_3 slightly reduced relative to $M_{1\text{-}2}$; (8) P_4 metaconid slightly elevated, (a) P_4 metaconid markedly elevated; (9) P_3 with metaconid, (a) P_3 metaconid lost (reversal); (10) $P_{3\text{-}4}$ low crowned, with open trigonids; (11) P_4 trigonid closed; (12) P_4 with broader, basined talonid; (13) slight overall size increase, (a) further size increase; (14) P_4 larger relative to M_1; (15) I_2, C, P_3 reduced and compacted, P_3 with bilobed or single root; (16) P_4 compacted and reduced relative to M_1 and with bilobed root.

"point," but where gradual transition and reproductive continuity occurred, this is completely arbitrary, for no such "point" exists. For instance, unless we prescribe such an arbitrary definition, *Teilhardina belgica* would be a member of all four evolutionary species. Furthermore, *T. belgica* is indicated as the beginning of each evolutionary species only because of a convenient gap in the record preceding it; but as the record improves, the beginning of each species could be the ancestral omomyid, the ancestral euprimate, or, as Simpson wrote, a protist!

The monophyletic species concept leads to one of two possible interpretations of this clade, neither very satisfactory from our perspective: (1) Species of *Tetonius* and *Pseudotetonius* might be regarded as metaspecies if neither their monophyly nor paraphyly can be demonstrated, or (2) if we accept their probable ancestor-descendant relationships, then none of them can be recognized as species.

Although the sequence discussed here (at least beginning with *Tetonius matthewi*) appears to be a single, unbranched lineage, the lack of fossil evidence for cladogenesis is not evidence that lineage splitting did not take place. In fact, numerous splitting "events" may well have occurred, but evidence of them may not have been preserved or discovered. If Simpson (1953) was correct that phyletic evolution probably does not continue long without branching, it is likely that many branches off the main lineage existed. In fact, short-lived branches may be included within samples already analyzed. Their existence would justify at least as many cladistic species names as there were cladogenetic events. Successive waves of immigration, replacing the existing species with a slightly more derived population, would yield a similar pattern in the fossil record. Because they intergrade morphologically, these species, in practice, would still require arbitrarily drawn boundaries.

The critical importance of a stratigraphic framework for understanding the pattern of evolution (Bown and Rose, 1987; MacLeod, 1991) is exemplified by these studies of Bighorn Basin primates, and this is true regardless of the species concept adopted.

Nomenclature of Evolutionary Intermediates

Although Linnaean binominals have been successfully applied to most extant populations, it has long been recognized that the temporal component, with its concomitant phenotypic change, presents difficulties for application to some paleontological samples (e.g., Trueman, 1924). We previously suggested that the discontinuous nature of species names may be mistaken to imply discrete, static species separated by "punctuations" where neither has been demonstrated (Bown and Rose, 1987; see also Krishtalka and Stucky, 1985; Sheldon, 1987). For example, Stanley (1982) cited Schankler's (1980) stratigraphic distribution of mammalian species from the Willwood Formation as evidence of stasis and, implicitly, sharp boundaries (punctuations) between successive species. But Schankler's account presented only occurrence data and gave no indication of modes of evolution. In fact, many of the mammalian lineages concerned have been shown to have evolved gradually and continuously in both anagenesis and cladogenesis (e.g., Gingerich, 1976, 1980, 1985; Gingerich and Simons, 1977; Rose and Bown, 1984a, 1986; Bown and Rose, 1987).

What Stanley observed was stasis in taxonomic names, not in successive populations (see also Maynard Smith, 1981; Carroll, 1988; Krishtalka, this volume). As noted above, these relatively continuous fossil records pose particular problems of species recognition and taxonomy, for no discrete boundaries exist between successive species; hence, conventional binominals may be inappropriate or inadequate. As Thomas (1956, p. 26) observed, "the 'frozen mechanics' of the Linnaean system are incompatible with this concept of the gradual modification of species in time." Nevertheless, comparatively few paleobiologists have been confronted by actual cases of intermediates that could not be adequately classified using the Linnaean system.

When they are found, how should evolutionary intermediates be dealt with taxonomically? Maynard Smith (1987, p. 516) recently asserted "Communication requires that we give names to fossils, and convenience suggests that they had better be Linnaean names." Most solutions proposed do, in fact, employ Linnaean names or a modification thereof.

Chaline (1983) proposed a conservative approach, maintaining the binominal system for segments of evolving lineages that are sufficiently distinct (essentially Simpson's recommendation). For intermediate stages he suggested using trinominals. This practice would allow the taxonomist to follow the code of nomenclature, but would result in the use of subspecies names for temporal subdivisions, to which some would object. Much earlier, however, Simpson (1943) used conventional trinominals (subspecies) to indicate stages in a chronocline of the early Tertiary phenacodontid *Ectocion osbornianum.*

Krishtalka and Stucky (1985) preferred "informal lineage segments" to designate anagenetic changes in the primitive artiodactyl *Diacodexis secans,* reserving subspecies for geographic variants. They introduced potential confusion, however, by referring to the segments with a hyphenated compound species name, which has the appearance of a formal Linnaean name and could be mistaken for a subspecies (e.g., *D. secans-primus*). Only the hyphen sets their usage apart from Simpson's (1943) subspecies chronocline. Moreover, Crusafont-Pairo and Reguant (1970) employed identical hyphenated compound species names in a different sense to indicate intermediate forms.

Less orthodox was Tobias' (1969) use of two generic names to indicate transitional species, e.g., *Australopithecus/Homo habilis,* but it still involved Linnaean names. Somewhat different was our own informal designation of transitional samples using the earlier and later species names (e.g., *Tetonius matthewi–Pseudotetonius ambiguus* intermediates). This usage was an intentional effort to avoid formal taxonomic assignments for intermediates (Bown and Rose, 1987). Although this is an admittedly cumbersome designation (illustrating how unsatisfactory the Linnaean system is for describing evolutionary intermediates), it accurately describes the inferred evolutionary position of the specimens so assigned.

Others have contended that, while it is necessary to recognize morphological change, this should not necessarily involve Linnaean species names (e.g., Fox, 1986; Maglio, 1971). Maglio defended the use of informal stage numbers or names for successive segments of well-documented lineages where "morphological overlap is great" (see also Harris and White, 1979). As discussed above, we also employed stage numbers, in conjunction with Linnaean binominals, to designate arbitrary segments in the *Tetonius matthewi–Pseudotetonius ambiguus* phyletic lineage (Rose and Bown, 1984, 1986; Bown and Rose, 1987). In his detailed study of Ordovician trilobites from Wales, however, Sheldon (1987, p. 561) found that because of "intermediate morphologies and temporary trend reversals, practical taxonomic subdivision of each lineage proved impossible."

Westergaard (1989) recently proposed an admittedly radical method of classification, based essentially on the cladistic species concept. This system would classify all organisms (not just intermediates) in a cladospecies and a cladosubspecies, the names of which would be modified by a series of letter and number prefixes and suffixes to indicate whether they are sexual species, stem-taxa, holophyletic or paraphyletic, or anagenetically evolved, as well as the time of their origin and extinction, and other such data. The proposal is so impractical that it is unlikely to acquire much popularity.

Suffice it to say that there is still no consensus on how to refer to evolutionary intermediates, formally or informally, with modified Linnaean names, or by some other system. Nevertheless, evolutionary (and morphological, temporal) intermediates do exist, punctuationists notwithstanding, and the various attempts to deal with them testify to the need for a satisfactory means of designating them (see also Stucky and Krishtalka, 1990).

Discussion and Conclusions

> There is no way of eliminating the species problem in paleontology; all general methods of defining and recognizing fossil species must be both subjective and arbitrary. (Raup and Stanley, 1978, p. 110)

We reluctantly agree with Raup and Stanley's pessimistic assessment. Obviously the concept of species is not the same to all systematists. How species are discriminated and named in the fossil record depends on several interrelated issues: in particular, the investigator's systematic philosophy and perception of evolutionary patterns and phylogeny, and the questions being explored (see also Stebbins, 1987; Smith and Patterson, 1988). Viewpoints are diverse, and this is unlikely to change very soon. Whether or not the species is the "real" unit of evolution, and whether species (except in the nondimensional sense) can be objectively defined or recognized probably cannot be answered (see also Eldredge and Gould, 1972). Consequently the persistence of more than one species concept seems inescapable, for ". . . no one species concept can meet the needs of all comparative biologists" (De Queiroz and Donoghue, 1988, p. 317). Indeed, as Endler (1989, p. 632) observed, "It is unproductive, and often positively misleading, to apply one species concept to all species or to answer all questions."

Each species concept suits the purpose of its proponents. Each has its weaknesses and none may be inherently superior to the others. Which concept is "closer to biological reality" remains controversial. Much confusion and argument might be avoided if systematists would indicate the species concept they adopt and how they apply species names to taxa under study (Mishler and Donoghue, 1982). This is particularly important in paleontology, where concepts of species, as we have seen, differ dramatically and can potentially result in substantial taxonomic differences. But above all, a species concept must have practicability. For paleontologists, it must be a concept that can be applied to fossil samples.

For this reason we favor the evolutionary species concept as outlined by Simpson. As paleontologists, we regard Simpson's ESC to be preferable to a monophyletic species concept that would exclude paraphyletic and ancestral species. Simpson's ESC is probably a better reflection of biological reality than the phylogenetic species concept, which is typological and tends to overestimate the actual number of species. It also seems to be a closer approximation to biological and evolutionary reality than is the cladistic species concept, because it does not require extinction of an ancestral species at cladogenesis. In addition, it is more practical than Wiley's ESC in allowing subdivision of unbranched lineages—admittedly arbitrarily—that have undergone substantial phenetic evolution. Both the cladistic and evolutionary species concepts are unsatisfactory for practical classification: The cladistic species concept requires recognition of a new species name (at cladogenesis) for a persisting ancestral species, even if indistinguishable from its predecessor, whereas Wiley's ESC might require phenetically and statistically different samples to be called the same species if they are parts of the same unbranched lineage. Although proponents have claimed that both the cladistic species concept and Wiley's ESC are more objective than Simpson's ESC, we have argued above that they trade one arbitrary boundary for another.

The imposition of arbitrary species boundaries in phyletic continua has been considered a particularly serious flaw in Simpson's ESC, but it results from the nature of the evolutionary record and is inevitable if there is to be any hope of consistent classification. Using Simpson's method, species are characterized by comparable, limited morphological breadths and are in this respect objective, although their boundaries may need to be drawn arbitrarily; phylogeny need not be known in detail to assign a sample. In fact, this is no different from the way most paleontological and extant species are recognized in practice. Despite criticism of the BSC, many systematists continue to employ the extent of phenotypic variation in recent populations as a criterion of species discrimination. If this is accepted, why should not the same criterion be applied to a temporal series? A drawback of Wiley's ESC and the monophyletic and cladistic species concepts is that the phylogenetic pattern, rather than phenotypic attributes, is considered the basis for delineating species; yet morphological traits must be used initially to reconstruct the phylogenetic pattern. Obvious problematic situations may result.

For example, without imposing arbitrary species boundaries, we are compelled to include the entire range of morphologies in the lineage from *Tetonius matthewi* to *Pseudotetonius ambiguus* (or perhaps *T. mckennai* through *Tatmanius szalayi*) in a single species (Fig. 8). By the same reasoning, *Homo habilis-H. erectus-H. sapiens,* which has been considered an unbranched lineage (e.g., Tobias, 1985), would be collapsed into a single species. Although this practice may accurately reflect lineage diversity, it conceals significant evolutionary change and constructs a species so broad in morphological scope (and inferred reproductive criteria) that it is no longer comparable to other paleontological or neontological species (the overwhelming number of which have been based on the BSC). As a consequence, diagnoses would become difficult or impossible to construct.

A secondary result would be the demise of species-based biostratigraphy. If the fossil record were typically as dense as in the Bighorn Basin, however, paleontologists would no doubt discover that many well-established species are actually segments of gradually evolving anagenetic lineages, raising the dilemma of how or whether such species should be discriminated. The fact is that many species names currently in use probably represent such segments of less well-documented lineages, and it should be realized that *apparent* taxonomic diversity (as it has been inferred from species names, not only from lineage splitting), as well as morphological diversity, has significantly increased through gradual phyletic evolution. In fact, Chaline (1983) reported that 53% of the speciations in 38 lineages of rodents he studied were anagenetic.

It would be easier to reject the possibility of anagenetic speciation if the process of change that takes place in a phyletic lineage were fundamentally different from that occurring in two lineages during or following cladogenesis, but this has not been demonstrated. Indeed, it is even quite possible that apparent anagenetic speciation is actually the result of cladogenesis where one branch was short-lived or not preserved after the split. According to many biologists, a morphologically distinct population is a separate species only if a sister lineage exists. But how long must a sister lineage persist to justify recognition of one or both branches as species? A million generations? One thousand? One hundred? Surely divided populations of the parent species are not yet distinct species, but at what point should we acknowledge their specific status?

It is evident that a single evolving lineage cannot be distinguished from one that involved a split *by the record of that transforming lineage alone,* because the actual transformation is the same whether a sister lineage existed or not. Therefore, anagenetic change should, in our view, be recognized with a new species name when it has accumulated to a degree at least as great as between extant species. Lineage diversity, however, should be recognized by some other means (e.g., by evolutionary lineages designated by the first and last species in the lineage, see below and Fig. 9; substantial parts of such lineages may overlap others because the "first" species is necessarily arbitrary).

These considerations underscore the difference in meaning and objectives of different species concepts. The importance of determining the number of lineages at a given time or through time is undeniable, but it is not the only objective of systematists and evolutionary biologists. Evolutionary species as lineages—as Simpson realized—are not equivalent to what biologists have long recognized as species (see also Mayr, 1982; Bock, 1979). To be useful, a classification must allow separation of clusters that differ by an acceptable, limited degree; taxonomic groups based on similarity are useful (see Ashlock, 1979). At the same time, such a classification cannot be based solely on phenetic characters; there is an underlying assumption of genetic/reproductive continuity, because phylogeny should be the basis for such classification. But a classification that would separate identical clusters into different species and group different (primitive and derived) ones into the same is not useful (see Ashlock, 1979). Thus, recognition of four evolutionary species for Bighorn Basin omomyids may be an accurate reflection of phylogeny or terminal taxic diversity; but it is an inadequate reflection of morphological (and chronotaxic) diversity through time, for which many more clusters (successional species) are necessary.

The Linnaean binominal system, despite its shortcomings, remains the best and most stable way to recognize related clusters of fossils with a prescribed morphological breadth. Linnaean names have been widely applied to such clusters in the belief that they approximate "biological species," although this is rarely if ever demonstrable. Many are decidedly *not* the equivalent of evolutionary species (lineages), which are also important to recognize and study. For these, the names *lineage* or *phyletic lineage* may be preferable to *species* (see also Bock, 1979; Mayr, 1982, p. 294). We suggest that they be designated not by a single name but by the beginning species (which is an arbitrary choice if its origin was gradual) and the terminal species, for example, *Teilhardina americana–Anemorhysis pearcei* lineage.

Classification cannot and should not reflect phylogeny exactly (see also Van Valen, 1978; Ashlock, 1979; Gingerich, 1979b); its objectives are not the same. Stability in classification is desirable; views of phylogeny should change with new data, but these changes should not always require altering the classification. Research on evolutionary rates and modes, morphological diversity, biostratigraphy, and other problems requires continued recognition of morphological clusters that approximate biological species. Only this practice can maintain a view of species that is relatively comparable through time and space.

ACKNOWLEDGMENTS

We are grateful to Bill Kimbel and Lawrence Martin for inviting us to participate in this symposium. In adition, we thank David Archibald, Norman MacLeod, Alan Walker, David Weishampel, and especially Lawrence Witmer for lively discussion and debate on the nature of species. We also thank Chris Beard, Bill Kimbel, David Krause, Lawrence Martin, Ian Tattersall, and Lawrence Witmer for helpful comments on the manuscript. Needless to say, this does not imply endorsement of our views. Elaine Kasmer prepared the figures. This research has been supported by NSF grants BSR-8500732 and BSR-8918755.

References

Ashlock, P. D. 1979. An evolutionary systematist's view of classification. *Syst. Zool.* **28:**441–450.

Beard, K. C. 1988. New notharctine primate fossils from the Early Eocene of New Mexico and southern Wyoming and the phylogeny of Notharctinae. *Am. J. Phys. Anthrop.* **75:**439–469.

Bock, W. J. 1979. The synthetic explanation of macroevolutionary change—a reductionistic approach. *Bull. Carnegie Mus. Nat. Hist.* **13:**20–69.

Bolli, H. M. 1986. Evolutionary trends in planktic Foraminifera from Early Cretaceous to Recent, with special emphasis on selected Tertiary lineages. *Bull. Cent. Rech. Explor.-Prod. Elf-Aquitaine* **10:**565–577.

Bonde, N. 1981. Problems of species concepts in palaeontology. *Inter. Symp. Concpt. Meth. Paleo. Barcelona, 1981:*19–34.

Bookstein, F. L., Gingerich, P. D., and Kluge, A. G. 1978. Hierarchical linear modeling of the tempo and mode of evolution. *Paleobiology* **4:**120–134.

Bown, T. M., and Kraus, M. J. 1981. Lower Eocene alluvial paleosols (Willwood Formation, northwest Wyoming, U.S.A.) and their significance for paleoecology, paleoclimatology, and basin analysis. *Palaeogeog., Palaeoclim., Palaeoecol.* **34:**1–30.

Bown, T. M., and Kraus, M. J. 1987. Integration of channel and floodplain suites, I. Developmental sequence and lateral relations of alluvial paleosols. *J. Sed. Pet.* **57:**587–601.

Bown, T. M., and Rose, K. D. 1987. Patterns of dental evolution in early Eocene anaptomorphine primates (Omomyidae) from the Bighorn Basin, Wyoming. *Paleont. Soc. Mem.* 23 (*J. Paleont.* 61, suppl. to no. 5):1–162.

Bown, T. M., and Rose, K. D. 1991. Evolutionary relationships of a new genus and three new species of omomyid primates (Willwood Formation, lower Eocene, Bighorn Basin, Wyoming). *J. Hum. Evol.* **20:**465–480.

Bown, T. M., Rose, K. D., Wing, S. L., and Simons, E. L. 1992. Distribution and stratigraphic correlation of fossil mammal and plant localities in the upper Paleocene-lower Eocene Fort Union, Willwood, and Tatman Formations, southern Bighorn Basin, Wyoming. *U.S. Geol. Surv. Prof. Paper.* (in press).

Carroll, R. L. 1988. *Vertebrate Paleontology and Evolution.* W. H. Freeman and Company, New York.

Chaline, J. 1983. Les rôles respectifs de la spéciation quantique et diachronique dans la radiation des arvicolidés (Arvicolidae, Rodentia), conséquences au niveau des concepts. *Colloques Int. C.N.R.S.* **330:**83–89.

Chaline, J., and Laurin, B. 1986. Phyletic gradualism in a European Plio-Pleistocene *Mimomys* lineage (Arvicolidae, Rodentia). *Paleobiology* **12:**203–216.

Chandler, C. R., and Gromko, M. H. 1989. On the relationship between species concepts and speciation processes. *Syst. Zool.* **38:**116–125.

Cracraft, J. 1983. Species concepts and speciation analysis. *Curr. Ornithol.* **1:**159–187.

Cracraft, J. 1987. Species concepts and the ontology of evolution. *Biol. Philos.* **2:**329–346.

Crusafont-Pairo, M., and Reguant, S. 1970. The nomenclature of intermediate forms. *Syst. Zool.* **19:**254–257.

Dagosto, M. 1988. Implications of postcranial evidence for the origin of euprimates. *J. Hum. Evol.* **17:**35–56.

De Queiroz, K., and Donoghue, M. J. 1988. Phylogenetic systematics and the species problem. *Cladistics* **4:**317–338.

De Queiroz, K., and Gauthier, J. 1990. Phylogeny as a central principle in taxonomy: Phylogenetic definitions of taxon names. *Syst. Zool.* **39:**307–322.

Dobzhansky, T. 1970. *Genetics of the Evolutionary Process.* Columbia University Press, New York.

Donoghue, M. J. 1985. A critique of the biological species concept and recommendations for a phylogenetic alternative. *Bryologist* **88:**172–181.

Echelle, A. A. 1990. In defense of the phylogenetic species concept and the ontological status of hybridogenetic taxa. *Herpetologica* **46:**109–113.

Eldredge, N., and Cracraft, J. 1980. *Phylogenetic Patterns and the Evolutionary Process.* Columbia University Press, New York.

Eldredge, N., and Gould, S. J. 1972. Punctuated equilibria: an alternative to phyletic gradualism, in: T. J. M. Schopf (ed.), *Models in Paleobiology,* pp. 82–115. Freeman, Cooper, San Francisco.

Enay, R. 1983. Spéciation phylétique dans le genre d'ammonite téthysien *Semiformiceras* Spath, du Tithonique inférieur des chaines bétiques (Andalousie, Espagne). *Colloques Int. C.N.R.S.* **330:**115–123.

Endler, J. A. 1989. Conceptual and other problems in speciation, in: D. Otte and J. A. Endler (eds.), *Speciation and its Consequences,* pp. 625–648.Sinauer Associates, Sunderland, MA.

Ereshefsky, M. 1989. Where's the species? Comments on the phylogenetic species concepts. *Biol. Philos.* **4:**89–96.

Forey, P. L. 1982. Neontological analysis versus palaeontological stories, in: K. A. Joysey and A. E. Friday (eds.), *Problems of Phylogenetic Analysis,* pp. 119–157. Systematics Association Special Volume, Vol. 21, Academic Press, New York.

Fox, R. C. 1986. Species in paleontology. *Geosci. Can.* **13:**73–84.

Frost, D. R., and Hillis, D. M. 1990. Species in concept and practice: herpetological applications. *Herpetologica* **46:**87–104.

Gebo, D. L. 1988. Foot morphology and locomotor adaptation in Eocene primates. *Folia Primatol.* **50:**3–41.

Gingerich, P. D. 1974. Size variability of the teeth in living mammals and the diagnosis of closely related sympatric fossil species. *J. Paleont.* **48:**895–503.

Gingerich, P. D. 1976. Paleontology and phylogeny: patterns of evolution at the species level in early Tertiary mammals. *Am. J. Sci.* **276:**1–28.

Gingerich, P. D. 1979a. The stratophenetic approach to phylogeny reconstruction in vertebrate paleontology, in: J. Cracraft and N. Eldredge (eds.), *Phylogenetic Analysis and Paleontology,* pp. 41–77. Columbia University Press, New York.

Gingerich, P. D. 1979b. Paleontology, phylogeny, and classification: an example from the mammalian fossil record. *Syst. Zool.* **28:**451–464.

Gingerich, P. D. 1980. Evolutionary patterns in early Cenozoic mammals. *Ann Rev. Earth Planet. Sci.* **8:**407–424.

Gingerich, P. D. 1983. Origin and evolution of species: evidence from the fossil record. *Colloques Int. C.N.R.S.* **330:**125–130.

Gingerich, P. D. 1985. Species in the fossil record: concepts, trends, and transitions. *Paleobiology* **11:**27–41.

Gingerich, P. D. 1987. Evolution and the fossil record: patterns, rates, and processes. *Can. J. Zool.* **65:**1053–1060.

Gingerich, P. D., and Simons, E. L. 1977. Systematics, phylogeny, and evolution of early Eocene Adapidae (Mammalia, Primates) in North America. *Contrib. Mus. Paleont. Univ. Mich.* **24:**245–279.

Godinot, M. 1985. Evolutionary implications of morphological changes in Palaeogene primates. *Spec. Papers Palaeontol.* **33:**39–47.

Gould, S. J. 1983. Dix-huit points au sujet des équilibres ponctués. *Colloques Int. C.N.R.S.* **330:**39–41.

Gould, S. J. 1988. Trends as changes in variance: a new slant on progress and directionality in evolution. *J. Paleont.* **62:**319–329.

Harris, J. M., and White, T. D. 1979. Evolution of the Plio-Pleistocene African Suidae. *Trans. Am. Phil. Soc.* **69**(2)**:**1–128.

Harrisville-Wolfe, C. 1991. Phyletic gradualism in a sample of the Early Eocene adapid *Cantius* (Mammalia, Primates) from the Willwood Formation in the Bighorn Basin, Wyoming. *Am. J. Phys. Anthropol.* **Suppl. 12:**89–90.

Hecht, M. K. 1983. Microevolution, developmental processes, paleontology and the origin of vertebrate higher categories. *Colloques Int. C.N.R.S.* **330:**289–294.

Hennig, W. 1966. *Phylogenetic Systematics.* University of Illinois Press, Urbana.

Hulbert, R. C., Jr., and MacFadden, B. J. 1991. Morphological transformation and cladogenesis at the base of the adaptive radiation of Miocene hypsodont horses. *Am. Mus. Novitates* **3000:**1–61.

Hull, D. L. 1979. The limits of cladism. *Syst. Zool.* **28:**416–440.

Imbrie, J. 1957. The species problem with fossil animals, in: E. Mayr (ed.), *The Species Problem,* pp. 125–153. AAAS Publ. 50, Washington, D.C.

Jackson, J. B. C., and Cheetham, A. H. 1990. Evolutionary significance of morphospecies: a test with cheilostome Bryozoa. *Science* **248:**579–583.

Jaeger, J.-J. 1983. Équilibres ponctués et gradualisme phylétique: un faux débat? *Colloques Int. C.N.R.S.* **330:**145–153.

Jones, J. S. 1988. Gaps in fossil teeth: saltations or sampling errors? *TREE* **3:**208–213.

Krishtalka, L., and Stucky, R. K. 1985. Revision of the Wind River faunas, early Eocene of central Wyoming. Part 7. Revision of *Diacodexis* (Mammalia, Artiodactyla). *Ann. Carnegie Mus.* **54:**413–486.

MacLeod, N. 1991. Punctuated anagenesis and the importance of stratigraphy to paleobiology. *Paleobiology* **17:**167–188.

Maglio, V. J. 1971. The nomenclature of intermediate forms: an opinion. *Syst. Zool.* **20:**370–373.

Malmgren, B. A., and Kennett, 1981. Phyletic gradualism in a Late Cenozoic planktonic foraminiferal lineage: DSDP Site 284, southwest Pacific. *Paleobiology* **7:**230–240.

Malmgren, B. A., Berggren, W. A., and Lohmann, G. P. 1984. Species formation through punctuated gradualism in planktonic Foraminifera. *Science* **225:**317–319.

Martin, R. D. 1990. *Primate Origins and Evolution.* Princeton University Press, Princeton, NJ.

Maynard Smith, J. 1981. Macroevolution. *Nature* **289:**13–14.

Maynard Smith, J. 1987. Darwinism stays unpunctured. *Nature* **330:**516.

Mayr, E. 1963. *Animal Species and Evolution.* Belknap Press, Cambridge.

Mayr, E. 1969. *Principles of Systematic Zoology.* McGraw-Hill, New York.

Mayr, E. 1982. *The Growth of Biological Thought.* Belknap Press, Cambridge.

Mayr, E. 1988a. The why and how of species. *Biol. Philos.* **3:**431–441.

Mayr, E. 1988b. *Toward a New Philosophy of Biology.* Belknap Press, Cambridge.

McKitrick, M. C., and Zink, R. M. 1988. Species concepts in ornithology. *Condor* **90:**1–14.

Mishler, B. D., and Donoghue, M. J. 1982. Species concepts: a case for pluralism. *Syst. Zool.* **31:**491–503.

Newell, N. D. 1956. Fossil populations, in: P. C. Sylvester-Bradley (ed.), *The Species Concept in Palaeontology,* pp. 63–82. Syst. Assoc. (London) Publ. 2.

Paterson, H. E. H. 1985. The recognition concept of species, in: E. S. Vrba (ed.), *Species and Speciation,* pp. 21–29. Transvaal Museum Monograph no. 4, Pretoria.

Prothero, D. R., and Lazarus, D. B. 1980. Planktonic microfossils and the recognition of ancestors. *Syst. Zool.* **29:**119–129.
Raup, D. M., and Stanley, S. M. 1978. *Principles of Paleontology,* 2nd ed. W. H. Freeman and Co., San Francisco.
Reif, W.-E., and Brettreich, J. 1985. Definitions of evolutionary species: a discussion. *N. Jb. Geol. Palaont. Mh.,* **7:**421–426.
Rhodes, F. H. T. 1956. The time factor in taxonomy, in: P. C. Sylvester-Bradley (ed.), *The Species Concept in Palaeontology,* pp. 33–52. Syst. Assoc. (London) Publ. 2.
Ridley, M. 1989. The cladistic solution to the species problem. *Biol. Philos.* **4:**1–16.
Rieppel, O. 1986. Species are individuals. A review and critique of the argument. *Evol. Biol.* **20:**283–317.
Rose, K. D. 1981. The Clarkforkian Land-Mammal Age and mammalian faunal composition across the Paleocene-Eocene boundary. *Univ. Mich. Papers Paleont.* **26:**1–197.
Rose, K. D., and Bown, T. M. 1984a. Gradual phyletic evolution at the generic level in early Eocene omomyid primates. *Nature* **309:**250–252.
Rose, K. D., and Bown, T. M. 1984b. Early Eocene *Pelycodus jarrovii* (Primates: Adapidae) from Wyoming: phylogenetic and biostratigraphic implications. *J. Paleont.* **58:**1532–1535.
Rose, K. D., and Bown, T. M. 1986. Gradual evolution and species discrimination in the fossil record. *Contrib. Geol., Univ. Wyoming, Spec. Pap.* **3:**119–130.
Schankler, D. M. 1980. Faunal zonation of the Willwood Formation in the central Bighorn Basin, Wyoming. *Univ. Mich. Papers Paleont.* **24:**99–114.
Sheldon, P. R. 1987. Parallel gradualistic evolution of Ordovician trilobites. *Nature* **330:**561–563.
Simpson, G. G. 1943. Criteria for genera, species, and subspecies in zoology and paleozoology. *Ann. N.Y. Acad. Sci.* **44:**145–178.
Simpson, G. G. 1951. The species concept. *Evolution* **5:**285–298.
Simpson, G. G. 1953. *The Major Features of Evolution.* Simon and Schuster, New York.
Simpson, G. G. 1961. *Principles of Animal Taxonomy.* Columbia University Press, New York.
Smith, A. B., and Patterson, C. 1988. The influence of taxonomic method on the perception of patterns of evolution. *Evol. Biol.* **23:**127–216.
Stanley, S. M. 1982. Macroevolution and the fossil record. *Evolution* **36:**460–473.
Stebbins, G. L. 1987. Species concepts: semantics and actual situations. *Biol. Philos.* **2:**198–203.
Stucky, R. K., and Krishtalka, L. 1990. Revision of the Wind River faunas, early Eocene of central Wyoming. Part 10. *Bunophorus* (Mammalia, Artiodactyla). *Ann. Carnegie Mus.* **59:**149–171.
Sylvester-Bradley, P. C. (ed.) 1956. The Species Problem in Palaeontology. *Syst. Assoc. (London) Publ.* **2:**1–145.
Tattersall, I. 1986. Species recognition in human paleontology. *J. Hum. Evol.* **15:**165–175.
Templeton, A. R. 1989. The meaning of species and speciation: a genetic perspective, in: D. Otte and J. A. Endler (eds.), *Speciation and its Consequences,* pp. 3–27. Sinauer Associates, Sunderland, MA.
Thomas, G. 1956. The species conflict—abstractions and their applicability, in: P. C. Sylvester-Bradley (ed.), *The Species Concept in Palaeontology,* pp. 17–31. Syst. Assoc. (London) Publ. 2.
Tintant, H. 1983. Cent ans après Darwin, continuité ou discontinuité dans l'évolution. *Colloques Int. C.N.R.S.* **330:**25–37.
Tobias, P. V. 1969. Bigeneric nomina: a proposal for modification of the rules of nomenclature. *Am. J. Phys. Anthropol.* **31:**103–106.
Tobias, P. V. 1985. Punctuational and phyletic evolution in the hominids, in: E. S. Vrba (ed.), *Species and Speciation,* pp. 131–141. Transvaal Museum Monograph no. 4, Pretoria.
Trueman, A. E. 1924. The species concept in palaeontology. *Geol. Mag.* **61:**355–360.
Van Valen, L. 1976. Ecological species, multispecies, and oaks. *Taxon* **25:**233–239.
Van Valen, L. 1978. Why not to be a cladist. *Evol. Theory* **3:**285–299.
Van Valen, L. 1988. Species, sets, and the derivative nature of philosophy. *Biol. Philos.* **3:**49–66.
Wei, K.-Y. 1987. Multivariate morphometric differentiation of chronospecies in the late Neogene planktonic foraminiferal lineage *Globoconella. Marine Micropaleontol.* **12:**183–202.
Wei, K.-Y., and Kennett, J. P. 1988. Phyletic gradualism and punctuated equilibrium in the late Neogene planktonic foraminiferal clade *Globoconella. Paleobiology* **14:**345–363.
Westergaard, B. 1989. A new species concept: cladospecies (and cladosubspecies). Pluralistic realism and biotaxonomy: A radical solution to classification exemplified by hominids, in: G. Giacobini

(ed.), *Hominidae, Proceedings of the 2nd International Congress on Human Paleontology.* pp. 71–74. Jaca Book, Milan.

Westoll, T. S. 1956. The nature of fossil species, in: P. C. Sylvester-Bradley (ed.), *The Species Concept in Palaeontology,* pp. 53–62. Syst. Assoc. (London) Publ. 2.

Wiley, E. O. 1978. The evolutionary species concept reconsidered. *Syst. Zool.* **27:**17–26.

Wiley, E. O. 1981. *Phylogenetics.* John Wiley & Sons, New York.

Wing, S. L., Bown, T. M., and Obradovich, J. D. 1991. Early Eocene biotic and climatic change in interior western North America. *Geology* **19:**1189–1192.

Anagenetic Angst
Species Boundaries in Eocene Primates

13

LEONARD KRISHTALKA

Ideas are like children: there are none so wonderful as your own.

A recent fortune-cookie proverb

Introduction

About 57 million years ago a dramatic event occurred in the evolution of North American mammalian faunas: the simultaneous appearance on this continent of major new groups of mammals, including perissodactyls and artiodactyls, adapid and omomyid primates, and hyaenodontid creodonts (Krishtalka *et al.*, 1987 and references therein; Gingerich, 1989; Krause and Maas, 1990). The event marks the onset of the Wasatchian Land Mammal Age. Subsequently, these and other groups underwent an explosive radiation that is well preserved in Eocene deposits, especially those in intermontane basins in the Western Interior.

The fossil record of this mammalian diversification—skull, dental, and skeletal material—is, for many taxa, stratigraphically dense and geographically widespread. The excellence and wealth of preserved material often permits the precise reconstruction of phylogenetic relationships among individual species and higher clades. In turn, the overlay of the temporal record of these taxa on

LEONARD KRISHTALKA • Section of Vertebrate Paleontology, Carnegie Museum of Natural History, Pittsburgh, Pennsylvania 15213.

Species, Species Concepts, and Primate Evolution, edited by William H. Kimbel and Lawrence B. Martin. Plenum Press, New York, 1993.

their hypothesized relationships reveals the evolutionary patterns and rates of morphological change that were operative during this radiation.

As such, study of Eocene mammalian evolution provides a paleontological test of current models of evolutionary change and suggests novel hypotheses. This has already been applied to some Eocene primates (Gingerich and Simons, 1977; Gingerich, 1979a,b; Rose and Bown, 1986; Bown and Rose, 1987; Swarts, 1989), condylarths (Gingerich, 1974, 1985; Bookstein *et al.,* 1978; West, 1979; Redline, 1990), marsupials (Krishtalka and Stucky, 1983a,b), and artiodactyls (Krishtalka and Stucky, 1985; Stucky and Krishtalka, 1990).

The systematic and evolutionary conclusions that have emerged from these studies reveal that Eocene mammalian paleontology can no longer hide behind geologic convenience, namely, discontinuities and gaps. It is now beset with the same theoretical and practical dilemmas that have long confounded studies of other taxa blessed with a rich record. The dilemmas are actually cause for applause: They are here courtesy of paleontological success. The fossil record of many taxa of Eocene mammals is now so robust that systematists face familiar conundra:

- What is a species of Eocene mammal? How are these species to be delimited?
- Should species be tied to cladogenetic events, that is, restricted to phena arising from lineage splitting?
- Should an anagenetic lineage of successional, morphologically overlapping populations be considered one species or divided arbitrarily into a series of "phyletic species"?
- Do the solutions to these conundra rely on (and reinforce) separate species concepts for paleontology and neontology or a unitary one?
- Do the solutions confirm the fortune-cookie proverb, namely, do they merely reify one's personal view of the evolutionary process? Taxonomic decisions regarding the composition of paleontological species can foreordain the evolutionary pattern and rate, as well as conclusions concerning species turnover and biodiversity (Wiley, 1978; Krishtalka and Stucky, 1985; Smith and Patterson, 1988).

Paleontological Species and Evolutionary Patterns

A general review of theoretical species concepts here would be unoriginal and repetitive simply because the literature on the subject is voluminous, compelling, and lucid (see especially, Simpson, 1951, 1961; Newell, 1956; Sylvester-Bradley, 1956; Wiley, 1978; Hull, 1979; Gingerich, 1985; Donoghue, 1985; Paterson, 1985; Fox, 1986; Rose and Bown, 1986, this volume; Mayr, 1988; De Queiroz and Donoghue, 1988; Ridley, 1989; De Queiroz and Gauthier, 1990; and references therein). Rather, it seems best here for an empirical example from the fossil record to introduce theory, and ecumenical for that example not to be an Eocene primate but another contemporaneous mammal. *Diacodexis secans,* an early and middle Eocene artiodactyl, is an appropriate one because of first-hand experience with the conundra its record presents (Krishtalka and Stucky, 1985, 1986). Those conundra, as well as the species' geographic and stratigraphic occurrence, fit the Eocene primates discussed below.

Diacodexis secans is one of the two earliest known North American artiodactyls and a species basal to the radiation of later members of the order. It occurs over a span of about 7 million years from the earliest Wasatchian (Sandcouleean) through the middle Bridgerian (Blacksforkian) of Wyoming and Colorado. It is one of the most commonly represented taxa in early Eocene and early middle Eocene mammalian faunas in North America.

Each sample of dental remains of *D. secans* from a discrete fossiliferous horizon reveals a normally distributed range of morphological variation involving size, the bunodonty and robusticity of the teeth, length of premolars, and degree of development and expression of particular molar cusps, crests, and basins. When these samples are arranged stratigraphically from oldest to youngest, two morphoclinal patterns emerge:

1. Temporally adjacent samples (from the same basin or different basins) overlap to the degree that they constitute a morphological continuum and cannot be divided objectively into diagnosable phena.
2. The morphological continuum formed by temporally adjacent samples is not one of isomorphy or stasis. Rather, the morphological continuum shows a successional, progressive increase in the frequency and degree of expression of a number of derived dental features. This morphologic change is often mosaic, involving different characters changing at different rates.

As such, the entire temporal record of *D. secans* preserves gradual and continuous morphoclinal change, or classic anagenetic evolution. Moreover, the amount of anagenetic change in the *D. secans* morphocline is impressive and evolutionarily significant: Stratigraphically earliest and latest samples are distinct to the degree that would demand recognition as separate species were they contemporaneous; they show no overlap in that suite of evolving features that describes the morphocline. Indeed, were these two populations contemporaneous, their morphology would imply a splitting of the younger "species" from the older one.

Diacodexis secans is at once both a species *and* an evolving lineage or, as termed by Krishtalka and Stucky (1985), a *species-lineage*. *Diacodexis secans* is a species in the sense that it is a morphologically cohesive unit in space separated by morphologic gaps at any point in its record from penecontemporaneous species of *Diacodexis*. It is a lineage *and* a species in the sense that its fossil record of successive populations exhibits an indivisible (other than arbitrary) morphological continuum through time and, as such, implies reproductive and genetic continuity as strongly as paleontology can. I would like to think that *Diacodexis secans* embodies Simpson's (1951, p. 289) evolutionary species concept (ESC): ". . . a phyletic lineage (ancestral-descendant sequence of interbreeding populations) evolving independently of others and with its own separate and unitary evolutionary role and tendencies. . ."

Were the record of organisms like *D. secans* one of relative morphological stasis, workers of all systematic persuasion would agree that the taxon represents one long-lived species of Eocene artiodactyl. But it isn't and they don't, and so is born a vast literature on species concepts and categories (classes *vs.* individuals) in paleontology. To reiterate, the record of what Krishtalka and Stucky (1985) called *D. secans* does not exhibit morphological stasis. Rather, it preserves significant morphological change through time in the frequency and degree of expression of a suite of dental features; end samples of the species-lineage would be recognized as discrete species were they contemporaneous.

Some authors have chosen to express this pattern of anagenetic change taxonomically by an arbitrary — morphological and/or stratigraphic — subdivision of a species-lineage continuum into a succession of named Linnaean species. Gingerich (1974, 1985) did so for *Hyopsodus,* Gingerich (1979a) and Gingerich and Simons (1977) for notharctine primates, and Bown and Rose (1987) for anaptomorphine primates. The arbitrary subdivisions are called *phyletic species* and the process by which they arose, *phyletic speciation,* or the "origin of new species by phyletic evolution" (Rose and Bown, 1986, p. 122).

Early Eocene Primates

Perhaps the most thoroughly documented example of this practice is the exhaustive analysis by Bown and Rose (1987). One of the anagenetic anaptomorphine lineages they describe is the gradual transition from *Tetonius matthewi* to *Pseudotetonius ambiguus* based on hundreds of specimens from the 140 to 374-m stratigraphic interval of the Willwood Formation in the Bighorn Basin. The morphological transition includes the gradual "loss of P_2, progressive reduction in size of I_2-P_3, and hypertrophy of I_1" (Bown and Rose, 1987, p. 105, Fig. 67).

The nature of the *Tetonius-Pseudotetonius* lineage mirrors what Krishtalka and Stucky (1985) described for *Diacodexis secans.* It exhibits an increase in the frequency and degree of expression of derived dental features along a temporal and morphological continuum. There are no morphological breaks in the sequence and the pattern of change is mosaic. Bown and Rose divided this lineage into five "arbitrary time-successive stages," numbered "1" to "5"; two or more successive stages are often represented in the same sample and/or species because the morphological change is not digital but merely a gradual shift in the normal distribution of derived traits. Thus *Tetonius matthewi* encompasses stage 1 and early stage 2, "*Tetonius-Pseudotetonius* Intermediates" stages 2–4, and *Pseudotetonius ambiguus* stages 4–5. The process that transformed *Tetonius matthewi* through the intermediate taxon (?species) into *Pseudotetonius ambiguus* was "anagenetic speciation" and took slightly more than a million years (Bown and Rose, 1987, p. 126).

Other anagenetic anaptomorphine transitions identified by Bown and Rose in the early Eocene of the Bighorn Basin are *Teilhardina americana-T. americana/T. crassidens* intermediates-*Teilhardina crassidens; Absarokius metoecus-Absarokius abbotti;* and *Absarokius abbotti-Strigorhysis* cf. *bridgerensis.* Turning to notharctine primates, Gingerich and Simons (1977) propose "anagenetic speciation" within an early Eocene lineage of *Cantius: C. ralstoni-C. mckennai-C. trigonodus-C. abditus.*

Anagenetic Angst

By their own admission, advocates of "phyletic species" and "phyletic speciation" within anagenetic lineages have a basic target: punctuated equilibrium. They wish to counter, with the use of taxonomy, the bugaboo that punctuated equilibrium is the evolutionary rule. Although the target may be worthy, the weapon isn't. And their anagenetic angst has caused them to confuse and encumber morphologic change with speciation.

The theory of punctuated equilibrium denies that species undergo major anagenetic change with any appreciable frequency. Its view of evolution is digital: species are essentially static morphologic units during their tenure on Earth; these histories of equilibrium are punctuated with episodes of significant morphologic change, which occur only during cladogenetic events.

However, evolutionary gradualism—significant, incremental and directional morphological change—*both within anagenetic lineages and at cladogenetic events* has been demonstrated in thumping fashion. Besides the aforementioned Eocene mammals, gradualism describes the evolution of late Tertiary and Quaternary voles, muskrats, elephants, pigs, and impalas (Barnosky, 1987 and references therein), rodents (Chaline and Laurin, 1986; Fahlbusch, 1989), and other mammals (Simpson, 1953; Barnosky, 1982; Krishtalka and Stucky, 1983a,b; Endler, 1977), as well as a number of invertebrates, including Ordovician trilobites (Sheldon, 1987). Clearly, punctuated equilibrium does not fit many organisms that have a fossil record sufficiently rich and dense to test the hypothesis. Indeed, the entire chassis of punctuated equilibrium has been brought into question (Brown, 1987).

The rub, however, is advertising. Rose and Bown (1986, p. 121), for example, worry

> . . . some now consider it [gradualism] a relatively rare or insignificant mode of evolution. . . To regard [an anagenetic lineage of] successional forms as a single species simply because intermediates linking them are now known would . . . *obscure important information on evolution.* (emphasis added)

Their concern is McLuhanesque: The species (and species name) is the evolutionary message. Cladogenesis advertises itself (and punctuated equilibrium) with new names for new species, but anagenetic lineages as single species broadcasts nothing about the degree or process (gradualism) of morphological evolution within the lineage. Ergo, the need for "anagenetic species."

The worry expressed by Bown, Rose, Gingerich, and others is legitimate. Too many studies of evolutionary patterns and rates are simplistic: They measure morphologic stasis and change (e.g., Stanley, 1982) by surveying the names of species in the fossil record and their geologic longevity. What these studies are equating, naively, is nomenclatural stasis/change with morphological stasis/change. That these studies see a predominance of stasis and punctuated change is guaranteed. What these studies are measuring, however, is the evolutionary pattern among taxonomic names, not species.

Diacodexis secans, for example, shows the folly of such methodological approaches to a synthetic study of evolutionary patterns. *Diacodexis secans* exhibits, by our taxonomic choice, nomenclatural stasis over 7 million years, but it is most definitely not an example of morphological stasis. Yet this species was cited ["*D. metsiacus*" in Stanley (1982), which also included specimens of another genus, *Antiacodon*] as an example of long-term morphological stasis precisely because its name was static. A similar facile correlation would bypass the significant evolutionary changes that Bown, Rose, Gingerich, and Simons have demonstrated in species of early Eocene notharctine and anaptomorphine primates.

Despite the worthy cause, formal Linnaean "phyletic species" are a poor choice to advertise the major role of gradualism (anagenetic change) in evolution. One problem is that advertising is a fickle business: "Anagenetic species" may send an unintended message. For example, after demonstrating gradualism in trilobites, Sheldon (1987, p. 563) cautioned that ". . . to subdivide a lineage into two or more arbitrary species or subspecies would give a false impression of punctuation and stasis (especially in range charts)."

More important, however, are the fundamental flaws in this taxonomic practice revealed by Newell (1956), Maglio (1971), Wiley (1978), Hull (1979), Eldredge and Cracraft (1980), Krishtalka and Stucky (1985), Fox (1986), and many others. Basically, the erection or retention of "anagenetic taxa" or "phyletic species" implies three untenable propositions:

Implication 1: that "phyletic (anagenetic) species" and "phyletic (anagenetic) speciation" are real biological phenomena discrete from cladogenetic species and cladogenesis, although this is unsupported by neontology. Speciation, no matter the model (allopatric, sympatric, parapatric), is by the evidence and thus by definition cladogenetic: It involves the multiplication of reproductively isolated clades and at least a momentary increase in species diversity. "Phyletic speciation," on the other hand, differs both in process—it does not involve reproductive isolation—and in product: It does not yield a net increase in the number of species at any moment in time. "Phyletic speciation" is merely pseudospeciation, an artifice, much like its twin, pseudoextinction.

Implication 2: that two discrete notions of species (anagenetic and cladogenetic) are identical or even equivalent. To paraphrase Fox (1986; also see Eldredge, this volume), if cladogenetic events delimit "real" species, how can subsequent arbitrary subdivisions of such species (i.e., "anagenetic" or "phyletic" species) also be species? Moreover, such practice and resultant theory are circular and nontestable. The artificial recognition of anagenetic "species" predetermines that there is a process of "anagenetic speciation," and vice-versa.

Implication 3: that there has been taxonomic turnover or a change in taxonomic richness during a period of geologic time, although neither has occurred. This criticism is the flip side to the simplistic correlation of species names/longevity with morphological change in the attempts to demonstrate the prevalence of stasis and punctuation in the history of life. Theoreticians assay the fossil record for species richness and turnover per epoch, subepoch, or land mammal age (e.g., number of anaptomorphines during the early Eocene or Wasatchian) by simply counting the number of named species for that interval. A succession of "anagenetic species" will artificially increase the count and decrease the value of the assay.

"Anagenetic" or "phyletic speciation" is pseudospeciation, literally, speciation in name only. It should not be confused with cladogenesis or lineage bifurcation, which results in a net increase of at least one species in penecontemporaneous taxa and a real increase in species diversity. As Fox (1986) quotes Hull (1979, pp. 431–432), "A continuously evolving lineage should no more be divided into distinct species than an organism undergoing ontogenetic development should be divided into distinct organisms."

Hull's use of the "organism" metaphor is deliberate. He (Hull, 1976), Ghislen (1974), and Eldredge (1989 and references therein; this volume) have argued compellingly that species are individuals—in Eldredge's terminology "spatiotemporal bounded entities"—with a beginning (cladogenetic origin), history, and end (true extinction). During that history, a species may evolve anagenetically, observe stasis, do either at different times, and bud off new species. The species as individual is implicit in Simpson's (1951, 1961) recognition of the evolutionary species—". . . with its own separate and unitary evolutionary role and tendencies. . ."—and is explicit in Wiley's (1978, p. 18) modification of the concept.

Wiley's (1978, p. 21) Corollary 4 also deals with what Rose and Bown (this volume) call Simpson's paradox:

> . . . if you start at any point in the sequence and follow the line backward through time, there is no point where the definition [of the ESC] would cease to apply. . . . If the fossil record were complete you could start with man and run back to a protist still in the species *Homo sapiens*. (Simpson, 1961, p. 165)

Hence, single lineages must be divided, albeit artificially, into "anagenetic species." Not so. The protist-to-*Homo* lineage *would* be interrupted at those cladogenetic events in which the ancestral species became extinct. Also, counters Wiley (1978, p. 21), ". . . there is no doubt that one can run from man to protist in one classificatory taxon, but, in my opinion, that taxon would be Eucaryota, not species *Homo sapiens*."

Interestingly, Simpson (1951, pp. 295–296) had no paradox 10 years earlier, in 1951, when he first defined the evolutionary species and warned against its division into a sequence of "anagenetic species":

> The whole sequence of populations . . . is genetically continuous and it fulfills the conditions of both genetical and evolutionary definitions of a species . . . By these concepts, it is a single taxonomic group, defined as a species.

Simpson felt strongly that evolutionary species and "anagenetic species" were "radically and fundamentally incongruent"; to consider both as species meant "to abandon any evolutionary significance for taxonomy. . ." This conclusion, of course, now provides high irony, seeing that Bown, Rose, and Gingerich cite evolutionary significance as one of the *raisons d'être* for "anagenetic species."

Other Anagenetic Issues

Advocates raise three other rationales for "phyletic species," apart from advertising phyletic gradualism. The first appeals to gaps, the second to typology, and the third asks for comparable morphologic limits in paleontological and neontological species.

1. *Gaps.* As argued by Rose and Bown (1986, p. 121), "gaps in the fossil record are biologically no more meaningful for separating such [paleontological] species than are any other arbitrary guidelines." The logic here is suspect. Gaps in the fossil record as not necessarily arbitrary; gaps *may or may not* prove to be biologically meaningful. We don't know the significance of a particular gap until either it is filled or shown to be real. Until then a gap signals, at the very least, missing evidence and, at the very most, phylogenetic discontinuity, especially if the gap coincides with a discrete morphologic jump.

2. *Typology.* Gingerich (1979b, p. 458) maintains, "The successive species [in an anagenetic lineage] . . . separated by an arbitrary boundary . . . are no less distinctive and real because they are parts of a continuum." The sentiment is echoed by Rose and Bown (1986, p. 121): ". . . species considered distinct by established criteria are no less distinct if evolutionary intermediates can be demonstrated."

This smacks of classic typological thinking. The statements are true only if the "established criteria" willy-nilly exclude new evidence of anagenetic intermediates. And they do not or should not. Otherwise, Recent species would be a joke: Populations thought to be discrete species but subsequently linked by geographically intermediate and interbreeding populations would still have to be separate species because they were originally "considered distinct by established criteria"!

Rose and Bown take typology to a logical extreme with their "phyletic transformation" of one genus, *Tetonius* (and *Absarokius*), into another, *Pseudotetonius* (and *Strigorhysis*). They and Gingerich seem to be prisoners of history: Once named and described, a species is a species is a species is a species! Ditto for genus. Anagenetic intermediates, discovered later, hover in valence between these types.

Wyoming geography suffers from the same typological thinking. In central Wyoming, the Wind River becomes ("speciates anagenetically" into) the Bighorn River at the entrance to the Wind River Canyon. The canyon links the Wind River Basin (where I work) and the Bighorn Basin (where Bown, Rose, and Gingerich work). It is one river (one species) described as two only because different parties explored and named different sections of the river at different times. "Wind River" has priority.

3. *Morphologic limits.* Gingerich (1985, p. 38) claims that, "Inclusion in the same [paleontological] species of organisms differing in size or form by factors greater than those characterizing living species reduces the functional comparability of species." Again, this worry is echoed by Rose and Bown (1986, p. 121): "To regard the successional forms . . . as single species simply because intermediates linking them are now known would greatly amplify the morphologic limits of paleontological species compared to biospecies . . ."

No, not if we stick to comparable measures. The morphologic limits of living species are bounded by space, those of paleontological species *by time as well as space.* "Functional comparability" is not threatened if we compare morphologic limits across space for living species to morphologic limits across space (across a single locality or fossiliferous horizon) for paleontological species.

Obviously, the morphologic range (*through time*) of an anagenetically shifting paleontological species may be amplified compared to the morphologic range (*across space*) of a biospecies. But those ranges or limits are not comparable, nor should they be, as they occur and originated across different dimensions. Were we able to observe living species for a few million years, the morphologic limits through time would then be comparable.

The same principle precludes invoking the rationale of reproductive isolation for divvying up anagenetic lineages into species. Raup and Stanley (1978) approved of "anagenetic species" because end members of a single anagenetic lineage would be reproductively isolated were they contemporaneous. The whole point is that end members of an anagenetic lineage are not contemporaneous and that the morphologic differences between them did not arise through reproductive isolation. Just the opposite! The morphological differences between end members arose through anagenetic change, namely, reproductive and genetic continuity. Basically, as Maglio (1971) emphasized, no matter how one cuts the continuum, one cannot pretend that arbitrary "phyletic species" are equivalent to neontological species.

Other Solutions

How then do we advertise noncladogenetic evolutionary change through time? How do we have taxonomy reflect the kind of anagenetic change in one paleontological species that mimics the morphological effects of speciation in cladogenesis? The need to do so is clear. If classification is to reflect phylogeny, it cannot ignore anagenetic change, especially if one or more species arise from the anagenetic lineage during its geologic extent (e.g., see the unabashedly paraphyletic cladogram in Krishtalka and Stucky, 1985, p. 476; see also Redline 1990, p. 142).

If the answer is not "phyletic species," neither is it the radicle of cladistic and holophyletic species concepts (Donoghue, 1985; De Queiroz and Donoghue, 1988; Ridley, 1989; De Queiroz and Gauthier, 1990). All that these species concepts seem to accomplish is to render nature digital and neat by the fiat of convention, especially those aspects of phylogeny that are analog and messy to classify. Obvious examples of such conventions are the tenets requiring that ancestral taxa suffer pseudoextinction at cladogenetic events, or that taxa ancestral to holophyletic species not be relegated to a species but remain in a generic orbit.

For me, these conventions hint of the Bridge-on-the-River-Kwai syndrome: The General, obsessed with building the bridge, forgets he is in a war. The obsession here seems to be paraphyly. True, paraphyletic taxa make for inelegant cladograms and classificatory conundra. But, for example, if the founder effect (speciation in a peripheral isolate) is a real biological phenomenon, so is paraphyly (see Mayr, 1982; Wiley, 1978): The mainland parental species does not necessarily become extinct when a gravid female departs for an island. The tenet that ancestral species must become extinct (nomenclaturally, at least) at all branching events is an artifice: It merely produces "new cladogenetic species" by pseudoextinction, much as phyletic gradualism produces "new anagenetic species" by pseudospeciation.

There are two ironies here. Both schools, though well intentioned, have come to endorse false processes—pseudospeciation and pseudoextinction. And both schools, though highly critical of one another, have come convergently to advocate pseudospecies. Their solutions may be tidy, but they cause too much collateral damage: They give us an illusory, ideal view of nature.

Paleontologists have seemingly exhausted other taxonomic solutions for a phyletic continuum. Most are cumbersome and some ill advised. Numbered stages, as used by Maglio (1973) and White and Harris (1977) for stratigraphically successive phena of East African elephants and pigs, are simple, straightforward, unambiguous designations, free of other evolutionary allusions. The problem here, frankly, is exposure: "Stages" are not usually picked up by synthetic studies. As stated earlier, evolutionary overviews tend to count species names in lieu of delving into the primary literature. Rose and Bown (1986; also Bown and Rose, 1987) endorsed the stage system but burdened it with their typological philosophy. They superimposed two "phyletic species" and one intermediate taxon on the five stages of the *Tetonius-Pseudotetonius* lineage.

Given the problem of exposure, Krishtalka and Stucky (1985) subdivided the *Diacodexis secans* species-lineage into arbitrary and informal "lineage segments," with a hyphenated trinomial designation. From oldest (Sandcouleean) to youngest (Blacksforkian) these are *D. s.-primus, D. s.-metsiacus, D. s.-kelleyi,* and *D. s.-secans.* Lineage segments, despite the subspecieslike moniker, are not temporal subspecies. Lineage segments also figured in Swarts' (1989) analysis of the Eocene primate *Microsyops* and Redline's (1990) revision of the Eocene condylarth *Hyopsodus.* R. A. Martin (personal communication, 1990), documenting species lineages of Pleistocene muskrats, improved the lineage segment syntax by substituting slashes for hyphens (e.g., *D. s.-metsiacus* becomes *D. s./metsiacus*). But, either way, trinomial lineage segments are unwieldy and do not make for an elegant reflection of evolutionary change.

The use of temporal subspecies (e.g., Beden, 1983 for fossil elephants) compounds the problem by confusing geographical and temporal relationships (Simpson, 1961; Gingerich, 1979b, p. 459; Krishtalka and Stucky, 1985, p. 418). The subspecies concept should be reserved for and not subverted from its formal usage in neontology: designations of spatial patterns within one species. What is needed for the fossil record are designations of temporal patterns within an anagenetic lineage/evolutionary species. Moreover, given the right record, the subspecies concept (as used in neontology) can be applied to spatial patterns in paleontology, even to Eocene mammals (Robinson, 1966; Stucky and Krishtalka, 1990).

Remaining nomenclatural practices for intermediate forms constitute typological variants of "phyletic species" and range from the clumsy to the comical:

1. A "bi-" prefix (Crusafont-Pairo and Reguant, 1970), which, if applied to the *Tetonius–Pseudotetonius* lineage, for example, would result in three "species": *T. matthewi,* the intermediate *T. matthewi bi–P. ambiguus,* and *P. ambiguus*
2. Fractions (see Tobias, 1969), whereby the intermediate form would be designated (with the anaptomorphine example):

$$\frac{\textit{Pseudotetonius ambiguus}}{\textit{Tetonius matthewi}}$$

3. Free-form nomenclatural mongrels (see Tobias, 1969), such as "*Teto-pseudotetonius matthiguus*" for the intermediate form.

Modest Proposals

Humor aside, none of the solutions is wholly satisfactory and the reason is clear. It is not treasonous to admit that Linnaean nomenclature—conceived for a static, typological, and atemporal world—is ill suited to express a paleontological record of morphologic flux through time in single species. Unfortunately, because taxonomic systems were formulated first for living species, we are stuck with the Linnaean binominal. So be it.

But, rather than remaining prisoners of the Recent, let's admit progress. Paleobiology has pushed and extended our knowledge of the species concept into deep time, beyond Linnaeus and beyond the Recent. Species are, to quote Hull (1979, p. 431), "integrated lineages developing continuously through time," or Simpson's (1951) and Wiley's (1978) "evolutionary species."

Evolutionary species do not differ from neontological species in kind, only in degree, namely, geologic extent and thus life history. Indeed, even this degree of difference between paleontological and Recent species begins to blur when faced with evidence of major anagenetic evolution within species of Darwin's finches during real time, the past 40 years (Grant, 1991). Very simply, Recent species represent the current, extant state of their life histories as evolutionary species. Put another way, unless Recent species sprang fresh from the brow of Jove (Fox, 1968) they are merely the Recent records of their tenure as evolutionary species.

Thus, the notion that paleospecies and biospecies require different species concepts or taxonomic treatments is untenable. As Simpson (1951), Wiley (1978), Fox (1986), and Eldredge (this volume) hint, both paleospecies and biospecies are faithful at once to the Biological Species Concept *and* the Evolutionary Species Concept in those parameters (across space, through time) that are applicable. There is no either/or choice: the Evolutionary Species Concept merely extends the Biological Species Concept into time, as Simpson (1951) intended. The same follows for taxonomic treatment: One binominal describes the species-as-individual, whether it refers to the entire anagenetic lineage through time, or a particular, single Eocene record from that anagenetic lineage, or the particular, single record we call the Recent. The taxonomic conventions should differ only where they must reflect different sorts of processes, patterns, and parameters, for example, morphological flux across space (subspecies) or through time (stages, lineage segments).

Two modest proposals seem to follow:

1. Let classification and taxonomy serve their roles. Let them communicate the empirical record rather than impose a particular ideology. Paleontology gives

us the empirical evidence for evolutionary species and life histories in the fossil record of organisms, including that of Eocene anaptomorphines, notharctines, artiodactyls, and other groups. We should express it in a fashion that is neither ambiguous nor loaded.

Anagenetic lineages/evolutionary species are single species that should be divided into arbitrary, successional infraspecific units to reflect the morphological flux through time. The most practical units seem to be either numbered stages or lineage segments. Thus, *Diacodexis secans* would be divided into stages 1 through 4 or, as already done (Krishtalka and Stucky, 1985), four lineage segments. If, in the interests of consistency, there is suddenly a consensus on stages, I won't fight it. In either case, the species *Diacodexis secans* refers at once to its tenure and properties as an evolutionary species through time and to any particular single record of its members during that time.

2. What about the paradox of advertising? Will the message of anagenetic change within evolutionary species be lost? After all, numbered stages and lineage segments don't have the Linnaean imprimatur of a succession of formal "phyletic species."

Well, in current classifications, extinct species are flagged with a dagger (†), a symbol that could, but has not, upset sensibilities. Where warranted, let's then flag evolutionary species as such with a simple, appropriate, and conspicuous mark to send the evolutionary message of a life history of anagenetic morphological transformation. I vote for an "↗," as in ↗*Diacodexis secans*, ↗*Tetonius matthewi*, ↗*Teilhardina americana*, and so on. It will serve to warn theoreticians, authors of synthetic studies, and Whig paleobiologists to look beyond the binominal.

References

Barnosky, A. D. 1982. [Review of] Evolutionary relationships of middle Eocene and younger species of *Centetodon* (Mammalia, Insectivora, Geolabididae) with a description of the dentition of *Ankylodon* (Adapisoricidae), by J. A. Lillegraven, M. C. McKenna and L. Krishtalka. *J. Vert. Paleont.* **2:**261–267.

Barnosky, A. D. 1987. Punctuated equilibrium and phyletic gradualism: some facts from the Quaternary mammalian record, in: H. H. Genoways (ed.), *Current Mammalogy, Vol. 1,* pp. 109–148. Plenum Press, New York.

Beden, M. 1983. Family Elephantidae, in: J. M. Harris (ed.), *Koobi Fora Research Project, v. 2, The Fossil Ungulates: Proboscidea, Perissodactyla, and Suidae,* pp. 40–129. Clarendon Press, Oxford.

Bookstein, F. L., Gingerich, P. D., and Kluge, A. G. 1978. Hierarchical linear modeling of the tempo and mode of evolution. *Paleobiology* **4:**120–134.

Bown, T. M., and Rose, K. D. 1987. Patterns of dental evolution in early Eocene anaptomorphine primates (Omomyidae) from the Bighorn Basin, Wyoming. *J. Paleont.* **61** 5 (Suppl.)**:**1–162.

Brown, W. L. 1987. Punctuated equilibrium excused: the original examples fail to support it. *Biol. J. Linnean Soc.* **31:**383–404.

Chaline, J., and Laurin, B. 1986. Phyletic gradualism in a European Plio-Pleistocene *Mimomys* lineage (Arvicolidae, Rodentia). *Paleobiology* **12:**203–216.

Crusafont-Pairo, M., and Reguant, S. 1970. The nomenclature of intermediate forms. *Syst. Zool.* **19:**254–257.

De Queiroz, K., and Donoghue, 1988. Phylogenetic systematics and the species problem. *Cladistics* **4:**317–338.

De Queiroz, K., and Gauthier, J. 1990. Phylogeny as a central principle in taxonomy: phylogenetic definitions of taxon names. *Syst. Zool.* **39:**307–322.

Donoghue, M. J. 1985. A critique of the biological species concept and recommendations for a phylogenetic alternative. *Bryologist* **88:**172–181.

Eldredge, N. 1989. *Macroevolutionary Dynamics.* McGraw-Hill, New York.

Eldredge, N., and Cracraft, J. 1980. *Phylogenetic Patterns and the Evolutionary Process.* Columbia University Press, New York.

Endler, J. A. 1977. *Geographic Variation, Speciation and Clines.* Princeton University Press, Princeton, NJ.

Fahlbusch, V. 1989. Evolutionary lineages in mammals. *Abh. Naturwiss. Ver. Hamburg* **28:**213–224.

Fox, R. C. 1968. Studies of late Cretaceous vertebrates II. Generic diversity among multituberculates. *Syst. Zool.* **17:**339–342.

Fox, R. C. 1986. Paleoscene #1. Species in paleontology. *Geosci. Can.* **13**(2)**:**73–84.

Ghiselin, M. T. 1974. A radical solution to the species problem. *Syst. Zool.* **23:**536–544.

Gingerich, P. D. 1974. Stratigraphic record of early Eocene *Hyopsodus* and the geometry of mammalian phylogeny. *Nature* **248:**107–109.

Gingerich, P. D. 1979a. Phylogeny of middle Eocene Adapidae (Mammalia, Primates) in North America: *Smilodectes and Notharctus. J. Paleont.* **53:**153–163.

Gingerich, P. D. 1979b. Paleontology, phylogeny and classification: an example from the mammalian fossil record. *Syst. Zool.* **28:**451–464.

Gingerich, P. D. 1985. Species in the fossil record: concepts, trends and transitions. *Paleobiology* **11:**27–41.

Gingerich, P. D. 1989. New earliest Wasatchian mammalian fauna from the Eocene of northwestern Wyoming: composition and diversity in a rarely sampled high-floodplain assemblage. *Univ. Mich. Paper Paleontol.* **28:**1–97.

Gingerich, P. D., and Simons, E. L. 1977. Systematics, phylogeny, and evolution of early Eocene Adapidae (Mammalia, Primates) in North America. *Contrib. Mus. Paleont. Univ. Michigan* **24:**245–279.

Grant, P. R. 1991. Natural selection and Darwin's finches. *Sci. Am.* **265:**82–87.

Hull, D. L. 1976. Are species really individuals? *Syst. Zool.* **25:**174–191.

Hull, D. L. 1979. The limits of cladism. *Syst. Zool.* **28:**416–440.

Krause, D. W., and Maas, M. C. 1990. The biogeographic origins of late Paleocene-early Eocene mammalian immigrants to the western interior of North America. *Geol. Soc. Am. Special Pap.* **243:**71–105.

Krishtalka, L., and Stucky, R. K. 1983a. Revision of the Wind River faunas, early Eocene of central Wyoming. Part 3. Marsupialia. *Ann. Carnegie Mus.* **52:**205–228.

Krishtalka, L., and Stucky, R. K. 1983b. Paleocene and Eocene marsupials of North America. *Ann. Carnegie Mus.* **52:**229–263.

Krishtalka, L., and Stucky, R. K. 1985. Revision of the Wind River faunas, early Eocene of central Wyoming. Part 7. Revision of *Diacodexis* (Mammalia, Artiodactyla). *Ann. Carnegie Mus.* **54:**413–486.

Krishtalka, L., and Stucky, R. K. 1986. Early Eocene artiodactyls from the San Juan Basin, New Mexico, and the Piceance Basin, Colorado, in: K. M. Flanagan and J. A. Lillegraven (eds.), *Vertebrates, Phylogeny and Philosophy,* G. G. Simpson Memorial Volume. pp. 183–196. *Univ. of Wyoming Contrib. Geol.* Special Paper No. 3.

Krishtalka, L., Stucky, R. K., West, R. M., McKenna, M. C., Black, C. C., Bown, T. M., Dawson, M. R., Golz, D. J., Flynn, J. J., Lillegraven, J. A., and Turnbull, W. D. 1987. Eocene (Wasatchian through Duchesnean) biochronology of North America, in: M. O. Woodburne (ed.), *Cenozoic Mammals of North America: Geochronology and Biostratigraphy,* pp. 77–115. University of California Press, Berkeley.

Maglio, V. J. 1971. The nomenclature of intermediate forms: an opinion. *Syst. Zool.,* **20:**370–373.

Maglio, V. J. 1973. Origin and evolution of the Elephantidae. *Trans. Am. Phil. Soc.* **63:**1–49.

Mayr, E. 1982. *The Growth of Biological Thought.* Belknap Press, Cambridge.

Mayr, E. 1988. The why and how of species. *Biol. Philos.* **3:**431–441.

Newell, N. D. 1956. Fossil populations, in: P. C. Sylvester-Bradley (ed.), *The Species Concept in Palaeontology,* pp. 63–82. Systematic Association, London.

Paterson, H. E. H. 1985. The recognition concept of species, in: E. S. Vrba (ed.), *Species and Speciation,* pp. 21–29. Transvaal Museum Monograph 4, Pretoria.

Raup, D. M., and Stanley, S. M. 1978. *Principles of Paleontology*, 2nd ed. W. H. Freeman and Co., San Francisco.

Redline, A. D. 1990. Phylogenetic Pattern and Systematics of Early Eocene *Hyopsodus* (Mammalia, Condylarthra). M.Sc. thesis, Dept. Geol. Planetary Sci., University of Pittsburgh.

Ridley, M. 1989. The cladistic solution to the species problem. *Biol. Philos.* **4:**1–16.

Robinson, P. 1966. Fossil Mammalia of the Huerfano Formation, Eocene of Colorado. *Bull. Yale Peabody Mus. Nat. Hist.* **21:**1–95.

Rose, K. D., and Bown, T. M. 1986. Gradual evolution and species discrimination in the fossil record, in: K. M. Flanagan and J. A. Lillegraven (eds.), *Vertebrates, Phylogeny and Philosophy*, G. G. Simpson Memorial Volume, pp. 119–130. *Univ. of Wyoming Contrib. Geol.* Special Paper 3.

Sheldon, P. R. 1987. Parallel gradualistic evolution of Ordovician trilobites. *Nature* **330:**561–563.

Simpson, G. G. 1951. The species concept. *Evolution* **5:**285–298.

Simpson, G. G. 1953. *The Major Features of Evolution.* Columbia University Press, New York.

Simpson, G. G. 1961. *Principles of Animal Taxonomy.* Columbia University Press, New York.

Smith, A. B., and Patterson, C. 1988. The influence of taxonomic method on the perception of patterns of evolution. *Evol. Biol.* **23:**127–216.

Stanley, S. M. 1982. Macroevolution and the fossil record. *Evolution* **36:**460–473.

Stucky, R. K., and Krishtalka, L. 1990. Revision of the Wind River Faunas, early Eocene of central Wyoming. Part 10. *Bunophorus* (Mammalia, Artiodactyla). *Ann. Carnegie Mus.* **59:**149–171.

Swarts, J. D. 1989. The Evolution of Microsyopinae. Doctoral Dissertation, Dept. Anthropology, University of Pittsburgh.

Sylvester-Bradley, P. C. (ed.). 1956. *The Species Concept in Palaeontology.* Systematic Association, London.

Tobias, P. V. 1969. Bigeneric nomina: a proposal for modification of the rules of nomenclature. *Am. J. Phys. Anthropol.* **31:**103–106.

West, R. M. 1979. Apparent prolonged evolutionary stasis in the middle Eocene hoofed mammal *Hyopsodus. Paleobiology* **5:**252–260.

White, T. D., and Harris, J. M. 1977. Suid evolution and correlation of African hominid localities. *Science* **198:**13–21.

Wiley, E. O. 1978. The evolutionary species concept reconsidered. *Syst. Zool.* **27:**17–26.

Cladistic Concepts and the Species Problem in Hominoid Evolution

14

TERRY HARRISON

Introduction

Over the past 20 years, since the initial application of Hennigian phylogenetic principles to the study of human evolution, the usage of cladistic concepts has become increasingly popular in paleoanthropology (Eldredge and Tattersall, 1975; Delson *et al.*, 1977; Tattersall and Eldredge, 1977; Bonde, 1977; Olson, 1978; White *et al.*, 1981; Skelton *et al.*, 1986; Wood and Chamberlain, 1986; Stringer, 1987; Chamberlain and Wood, 1987; Kimbel *et al.*, 1988; Tobias, 1988; Groves, 1989). The rigorous operational framework, in conjunction with its potential for the application of Popperian deductive reasoning in testing inferences about character states and morphocline polarities upon which phylogenetic hypotheses are based, has made cladistics an attractive methodological approach, even among some of its initial antagonists (Nelson, 1970, 1971a,b); Bonde, 1977; Bock, 1977; Szalay, 1977; Platnick, 1977, 1978, 1979; Platnick and Gaffney, 1977; Patterson, 1978; Mayr, 1968, 1981; but see Cartmill, 1981 for a critique of the utility of Popper's model of scientific enquiry for testing phylogenetic interpretations).

An early area of contention between cladists and more traditional systematists, however, and one that still persists today, is the nature of the relationship between phylogenetic inference and classification (Hull, 1970; Nelson, 1972,

TERRY HARRISON • Department of Anthropology, New York University, New York, New York 10003.

Species, Species Concepts, and Primate Evolution, edited by William H. Kimbel and Lawrence B. Martin. Plenum Press, New York, 1993.

1974a,b; Ashlock, 1974, 1979; Mayr, 1974, 1981; Szalay, 1977; Bock, 1977; Martin, 1981; Szalay and Bock, 1991). Hennig's (1966) scheme of ranking taxa according to their absolute time of divergence has been abandoned as impractical, and the majority of cladists today have adopted more flexible classificatory schemes (Crowson, 1970; Nelson 1971a, 1974a; Delson and Andrews, 1975; Farris, 1976; Løvtrup, 1977; Patterson and Rosen, 1977; Eldredge and Cracraft, 1980; Patterson, 1980; Groves, 1989). Nevertheless, in spite of the burgeoning number of seemingly idiosyncratic variants in classificatory methods that have arisen as a response to these problems, the overriding requirement that taxa *must* be monophyletic* is upheld as a basic tenet of all cladistic classificatory schemes (Hennig, 1966; Ashlock, 1971, 1972, 1979; Bonde, 1977; Eldredge and Cracraft, 1980; Mayr, 1981; Ax, 1987).

As a long-term practitioner of cladistics I have had the opportunity to apply cladistic methodology to a range of problems in primate paleontology (Harrison 1981, 1982, 1986, 1987a,b, 1988). Practical experience has served to reinforce my appreciation, based on purely theoretical considerations, that cladistics is the most appropriate (i.e., biologically the most meaningful and scientifically the most rigorous) method to employ in order to attempt to reconstruct phylogenetic relationships. Recent critiques of the use of cladistics in paleoanthropology (e.g., Habgood, 1989; Trinkaus, 1990) have responded more to weaknesses in its application by practitioners than to problems inherent to the method itself. Nevertheless, I do not mean to imply that the theoretical premise of cladistics is not without its limitations. In fact, it has become increasingly evident to me that a number of major theoretical and conceptual difficulties do arise, especially when Hennigian phylogenetic principles are translated into the taxonomic sphere. That these problems impinge directly or indirectly on the question of the species problem in paleontology makes them pertinent to discuss in the context of the present volume.

The aim of this paper is to examine the nature of the interface between cladistic methods of phylogenetic inference and classification. Cladists contend that classification should represent a precise reflection of phylogenetic affinity, while evolutionary systematists argue that additional information, such as the degree of divergence or level of diversification, may be incorporated in the construction of a classification. However, as has been emphasized repeatedly in the literature, phylogenetic inference and classification are two distinct procedures (Simpson, 1961; Bock, 1977; Martin, 1981). The former is a hypothetical representation of actual evolutionary events, while the latter is a human construct that provides a means of conceptualization, communication, and storage of information about taxa (Simpson, 1961; Bock, 1977). As the two processes can be decoupled, there is no intrinsic reason why a cladistic classification need necessarily reflect perfectly the set of relationships expressed in a cladogram, since the latter by itself is the purest expression of such relationships.

A classification is an essential tool, and like all tools, utility is at a premium.

***Monophyly* is used throughout the text in the sense proposed by Hennig (1966), in which a "monophyletic group is a group of species descended from a single (stem) species, and which includes all species descended from this stem species." The term is used as a direct equivalent of Ashlock's (1971) proposed replacement term, *holophyly*.

With this in mind, Simpson (1961) proposed that classifications should fulfil three important principles: (1) the basis of classification should be the most biologically significant relationships among organisms and should bring in as many of those as is practical, (2) classifications should be consistent with the relationships used as its basis, and (3) classifications should be as stable as possible without contravening the two preceding principles. These would seem to be worthwhile and realistic goals for all systematists, regardless of their philosophical persuasion. In fact, cladists, with their uncompromising concern for establishing vertical classifications, are perhaps in the most advantageous position to succeed in applying these principles. This is because cladists, quite rightly, recognize the paramount importance of phyletic information as the most appropriate biological data to employ in constructing a classification.

As will be seen from the following discussion, however, it is the inflexible application of the concept of monophyly in classification, without due regard to empirically based biological research, that is at the root of a number of taxonomic problems. Cladists who adhere, in all cases, to the principles of strict monophyly when attempting to construct classifications are in danger of undermining the utility of a classification by ignoring the first goal of Simpson's principles, which is that classifications should maximize (or more appropriately optimize) their biologically significant content. As I shall discuss in a later section, in certain circumstances other types of biological information may prove to be more appropriate, or at least *more readily accessible,* than phylogenetic data, and the neglect by cladists of this wider array of biological phenomena may impose unnecessary limitation on the usefulness of their classifications. It is an ironic twist, therefore, that the very strength of cladistics—its purported methodological rigor for reconstructing phylogenetic relationships—is one of its weaknesses when it comes to translating the subtle complexities of paleobiological evidence into a classificatory scheme (see also Bock, 1977).

It should be noted from the outset that this paper is not intended to be a definitive review of the problem. It is simply the presentation of a selection of examples drawn from my own research that serve to illustrate a number of key conceptual difficulties that I have encountered in applying cladistic concepts to taxonomic issues in hominoid evolution. Moreover, I cannot claim that the problems presented here are entirely original, or that I am in a position to offer any real solutions to them, but it is my hope that a reiteration of some of the limitations of cladistic methods of classification, in the context of specific examples, may provoke new attempts by systematists to attain a closer correspondence between what are currently perceived as cladistically correct procedures and observed biological phenomena.

The examples I have chosen are taken primarily from my research on the taxonomy and phylogenetic relationships of the early Miocene catarrhines from East Africa (Harrison 1981, 1982, 1987a,b, 1988). The aims of my research in this area have been threefold: to carry out a detailed taxonomic revision of the fossil primates, to identify their evolutionary relationships, and to make inferences about their natural history. The attainment of these research goals has involved the following discrete procedural steps: (1) the recognition of species groupings based on comparisons of individual specimens, (2) the identification of the species and genus (based on comparison with type specimens) to which these groupings should be assigned, (3) reconstruction of the paleobiology of each species based on the morphology and inferred functional affinities of their craniodental and postcranial remains, (4) an assessment of the cladistic relationships among the different species, and (5) a wider cladistic analysis involving relevant fossil and extant catarrhines in order to determine the higher taxonomic groupings to which the fossil species may be assigned.

Different conceptual problems with the application of cladistic methodology have been encountered at different phases in the analysis, corresponding with increasing taxonomic level. Specific problems are most easily identifiable at three different levels: (1) at the family–group level and above (corresponding to step 5 in the research design outlined above); (2) at the genus–group and species level, or the alpha-taxonomic level (corresponding to step 4); and (3) at the level of species recognition (corresponding to step 1). It is important, however, to note that the problems are not necessarily exclusive to the taxonomic levels identified. The cases presented here were selected merely because they best highlight the deficiencies of the cladistic method, but they presumably represent individual examples of more wide-ranging phenomena that occur throughout the continuum of hierarchical inclusiveness from the individual to higher order taxonomic categories.

The Family-Group Level and Above

Cladistic analyses have proved to be most effective in resolving relationships among the major groups of living and fossil primates (e.g., Delson, 1975; Rosenberger, 1977; Szalay and Delson, 1979; Schwartz and Tattersall, 1985, 1987; Ford, 1986; Schwartz, 1986; Harrison, 1987a; Groves, 1989), and there has been some measure of success among current workers with regard to the relationships among modern primates at the family-group level and above. This success is due largely to the fact that the last common ancestors of the major groups are sufficiently distantly removed in time to allow the recognition of major adaptive patterns that characterize the extant representatives. Under these circumstances, it is a relatively simple task to identify homologies, to weed out homoplasies, and to delineate more precisely robust transformation sequences in reconstructing ancestral morphotypes. Because of this it is much easier to recognize a member of a major group, regardless of the extent of its convergence on the adaptive pattern of a distantly related group. As I shall discuss in the next section, correct determination of character polarity and the identification of homologies vs. homoplasies become progressively more troublesome as one descends the taxonomic hierarchy to the level of the genus or species.

Despite the apparent success in using cladistic methods for resolving the relationships of the major extant groups of primates, problems occur when the analysis is transferred from the phylogenetic to the taxonomic domain. Especially significant is the way in which fossil taxa, particularly stem groups, are treated in relation to extant higher level taxa. The relative ranking of modern higher level taxa is usually defined by their level of species diversity and/or by their degree of adaptive divergence (Simpson, 1961). This implies that each taxonomic category of the same rank represents approximately the same absolute level of diversity or divergence. There is an implicit assumption, for example, that the mammalian families Cervidae, Canidae, and Cercopithecidae, and also, for that matter, their less speciose counterparts, the Antilocapridae, Ailuropodidae, and Hominidae, are in some sense biologically equivalent to one another. The purely practical considerations of this approach, in addition to the retarding influence exerted by the desirability to maintain nomenclatural stability, explains why most cladistic classifications of modern primates do not deviate significantly at the family–group level from Simpson's (1945) revision, based on an entirely different philosophical framework, combining cladistic and gradistic concepts.

However, different criteria are commonly used for ranking fossils, particularly stem groups, in a cladistic classification. As noted by Simpson (1975), among others, the strict requirement of monophyly in cladistic classifications may lead to two hardly distinguishable fossil species being classified in widely different taxonomic categories. This is because fossil species, particularly those comprising stem groups, have a *realized* phylogenetic history, while modern taxa, being the contemporary products of their evolutionary history, have only the *potential* to produce major new taxa.

A simple example may serve to illustrate the problem. The phylogram in Fig. 1 depicts a hypothetical situation in which two modern groups of species, here recognized as two separate families, family A and family B, are derived from fossil sister species of an earlier radiation. Following the logic of doctrinaire cladistic philosophy, species A should be classified in family A, while its sister species, species B, should be classified in a different family, family B. Similarly,

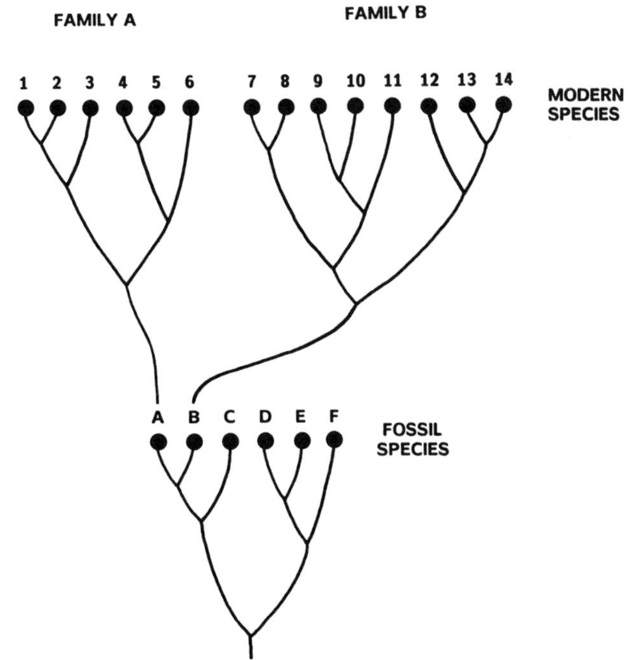

Fig. 1. Hypothetical phylogram illustrating the evolutionary relationships between a stem group and its modern descendants (see text for discussion).

the sister taxon to species A and B, species C, should be classified in a monotypic family, family C. However, species A, B, and C may be very similar to each other, being distinguished by relatively minor morphological difference, such as the structure of the genitalia, pelage color, or molar enamel thickness. After all, it would only require the recognition of a single synapomorphy in each of the fossil species to relate it to a later radiation.

The fossil species in the group depicted in Fig. 1 are clearly related to one another, and they may show a comparable level of adaptive diversity as each of the modern families, but yet they are subdivided into four separate families. In a cladistic scheme, therefore, a paleofamily may not correspond directly with a family composed entirely of modern species. Higher taxa based on paleontological samples will tend to exhibit lower levels of divergence from neighboring groups of equivalent rank and will also tend to be less speciose. In addition, fossil taxa are more likely to be subject to dramatic alterations in their ranking because of the diachronic context and because of the possibility that major new discoveries will influence the level of the diversity. With further, more intensive, research it could be discovered, for example, that each of the individual fossil species in Figure 1 gave rise to a major radiation and that each should be classified in separate superfamilies, or for that matter, even in separate orders.

This type of approach is eminently reasonable from a phylogenetic perspective, but it has less relevance to a paleontologist who is interested in the paleobiology, ecology, and community structure of the individual species. It gives a false impression to those who are less familiar with the biology of the fossil group that they are as different from each other as are representatives of different modern families. Based purely on morphological grounds, and disregarding their relationship to later descendant taxa, we might be tempted, viewing them in a neontological perspective, to consider species A through F as a cluster of closely related species of the same genus. However, in a paleontological scheme, where their descendant relationships may have been realized and subsequently identified by paleontologists, they are given different generic ranks within separate families.

Clearly, there is a discordance here in the way in which cladists view paleontological samples with realized evolutionary history, and modern species that represent the terminal products of this evolutionary history, with only the potential to produce descendant species. An inherent problem in this approach is that paleontologists rarely view the classification of fossil groups in exactly the same way as modern groups. Paradoxically, a paleontologist analyzing species A through F would almost certainly arrive at a very different classificatory scheme from that of a neontologist who was able to study the exact same group of species as living organisms. Similarly, it can be assumed that present classifications of living species will need to be modified in the distant future when the evolutionary history of modern species has been realized. Future (or perhaps I should say futuristic) neontologists may be faced with the perplexing problem of reclassifying species we recognize today as congeneric, such as the closely related Old World monkeys *Cercopithecus nictitans* and *C. mitis,* as members of separate superfamilies or orders, if they happened to give rise to separate adaptive radiations.

When viewed in a diachronic context, taxonomic categories, in a purely cladistic framework, are relative concepts that are determined by descent. They are, therefore, inherently unstable through time. Taxa would become more or less fixed absolutely only in the unlikely event that all lineages derived from a species have become extinct and that all of the constituent species of this group are known to paleontologists. Surely one of the major objectives of paleontology is to understand and to appreciate past life in a way that is congruent with our view of biological systems as seen today. If our basis for communicating about fossil groups, the classificatory system, is not directly equivalent to that used for modern groups, then it seems logical to conclude that we are presenting a distorted and unnecessarily abstracted view of past life that does not correspond closely to empirically based observations of the relationships that are known to exist among living species today.

The early Miocene catarrhines from East Africa provide an excellent example of the problems concerned. The fossil primates represent a diverse community, comprising 10 genera and 13 species (see Table 1; Andrews, 1978; Harrison, 1981, 1982, 1988; Leakey and Leakey, 1986a,b, 1987). Early workers on the fossil catarrhines from East Africa considered them to be hominoids, with the smaller species linked phyletically to the hylobatids and the larger forms to the great apes (Hopwood, 1933; Le Gros Clark and Leakey, 1951; Le Gros Clark and Thomas, 1952; Simons and Pilbeam, 1965; Simons, 1972; Andrews, 1974, 1978; Delson and Andrews, 1975; Andrews and Simons, 1977).

There is very little evidence, however, actually to link any of the early Miocene catarrhines from East Africa to the extant hominoids (see Harrison,

Table 1. A List of Currently Recognized Species of Noncercopithecoid Catarrhines from the Early Miocene of East Africa Arranged According to Estimated Average Body Size

	Body weight[a]	Dietary category[b]
Small species		
Micropithecus clarki	4 kg	Frugivore
Limnopithecus legetet	5 kg	Frugivore/folivore
Limnopithecus evansi	5 kg	Frugivore/(folivore)
Simiolus enjiessi	5 kg	Frugivore/folivore
Kalepithecus songhorensis	6 kg	Frugivore
Dendropithecus macinnesi	8 kg	Folivore/(frugivore)
Medium-sized species		
Nyanzapithecus vancouveringi	10 kg	Folivore
Turkanapithecus kalakolensis	12 kg	Folivore/(frugivore)
Proconsul africanus	15 kg	Folivore/frugivore
Rangwapithecus gordoni	15 kg	Folivore
Large species		
Afropithecus turkanensis	30 kg	Folivore/frugivore
Proconsul nyanzae	30 kg	Folivore/frugivore
Proconsul major	50 kg	Folivore/frugivore

[a]Body weight estimates are based on comparisons of cranial, dental, and postcranial material with modern antropoid primates.

[b]Inferred dietary categories are based on comparisons of cranial and dental material with modern anthropoid primates. Where two categories are separated by a slash, the first is considered to be the primary dietary resource (more than 50% of its diet). A category enclosed by parentheses indicates that it is an important, but relatively minor resource.

1987a, 1988). In fact, assessment of the phylogenetic relationships of the members of this group is severely limited by the lack of adequate material. The dentition of each species is generally well known, and in most cases several good jaw fragments or partial skulls are also represented in the collections. A detailed understanding of the morphology of the dentition and lower face has contributed significantly to the much improved appreciation of the alpha-taxonomy of this group in recent years (Andrews, 1978; Harrison, 1982, 1988; Leakey and Leakey, 1986a,b, 1987). However, important information concerning the morphology of the cranium and postcranium, critical for adequately resolving the relationships of early catarrhine primates, is entirely lacking for most of the species. Good data are only available for *Dendropithecus macinnesi* and *Proconsul africanus* (Harrison, 1987a and references therein).

Comparisons of the partial skeletons of *D. macinnesi* from Rusinga Island, as well as the extensive series of craniodental specimens, provides no clear evidence of close affinities with the modern hominoids. In fact, *Dendropithecus* exhibits a uniform absence of the extensive suite of shared derived traits that unites the extant representatives of the hominoid clade (Harrison, 1982, 1988; Fleagle, 1983; Rose, 1983, in press; Andrews, 1985; Rose *et al.*, 1992). The available evidence therefore suggests that *Dendropithecus*, at least, is not a hominoid in the strict sense and that it is best considered a basal catarrhine that diverged prior to the appearance of the last common ancestor of the hominoids and Old World monkeys (Fig. 2).

In fact, the same conclusion is probably generally applicable to the other small catarrhines from East Africa, such as *Limnopithecus, Micropithecus, Kalepithecus,* and *Simiolus*. There is little or no morphological evidence to support

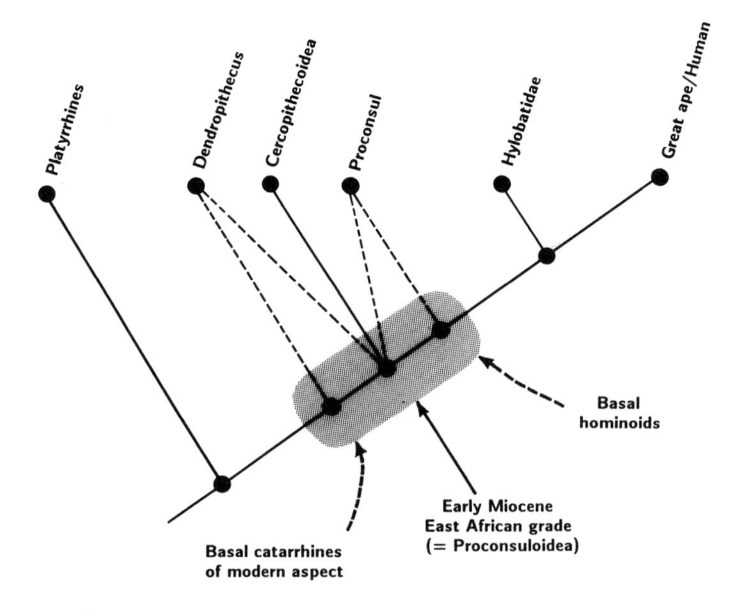

Fig. 2. A cladogram illustrating the inferred phylogenetic relationships of *Dendropithecus* and *Proconsul* to extant anthropoid primates. The shaded area depicts the possible extent of the gradistic zone occupied by the Proconsuloidea.

the contention that any of the small catarrhines (with the possible exception of the oreopithecid *Nyanzapithecus*) share a particularly close phyletic relationship with extant hominoids (Andrews, 1980, 1985; Harrison, 1982, 1986, 1987a,b, 1988; but see Leakey and Leakey, 1987 for an alternative view). The small catarrhines from East Africa probably represent, therefore, a heterogeneous grade of undifferentiated basal catarrhines that branched off prior to the dichotomy of the two modern superfamilies, the Cercopithecoidea and the Hominoidea (Harrison, 1982, 1987a, 1988; Andrews, 1985; Fleagle, 1988; Groves, 1989; Rose *et al.*, 1992).

Opinion is more deeply divided, however, over the relationships of the large catarrhines from East Africa, such as *Proconsul, Rangwapithecus, Afropithecus,* and *Turkanapithecus.* Most workers accept that they are hominoids, probably representing a combination of basal hominoids, preceding the appearance of the last common ancestor of the extant hominoids, and hominoids of modern aspect, appearing subsequent to the last common ancestor of the extant hominoids) (Andrews, 1985; Andrews and Martin, 1987; Andrews, 1988; Tattersall *et al.*, 1988; Delson, 1988; Walker and Teaford, 1989). The best known taxon, *Proconsul,* however, conforms in most respects to the ancestral catarrhine morphotype, and in my view the widespread acceptance of its hominoid status is more the product of uncritical supposition than of careful scrutiny (Harrison, 1987a, 1988; see also Feldesman, 1986).

A few possible synapomorphies in the postcranial skeleton have been identified that could be used to support claims that *Proconsul,* and probably also *Afropithecus,* represent conservative basal hominoids (Rose, 1983, 1992, in press; Walker and Pickford, 1983; Andrews, 1985; Senut, 1989; Rafferty, 1990; Rose *et al.*, 1992). I do not believe, however, that a sufficiently strong case has yet been made to assert with confidence that any of the individual genera of early Miocene catarrhines from East Africa should be linked uniquely with the extant hominoids (Fig. 2).

We may infer from the current evidence that the early Miocene anthropoids from East Africa represent part of a larger radiation of conservative catarrhines that probably includes a number of stem taxa that gave rise to the extant hominoids and Old World monkeys. This general conclusion has led most cladists to recognize at least some level of taxonomic differentiation in the group (Andrews, 1985, 1988; Delson, 1988; Tattersall *et al.*, 1988). For example, a recent classification based on cladistic principles (Tattersall *et al.*, 1988) placed the early Miocene catarrhines in four different families and three superfamilies. Taxonomic differentiation may be defensible in terms of a purely phylogenetic perspective of the group, but what about in terms of adaptive diversity?

When compared with extant primates, the level of morphological and adaptive diversity in the early Miocene catarrhines from East Africa could readily be encompassed by a single superfamily. In fact, the fossil taxa are no more different from each other in their craniodental or postcranial morphology than are modern genera classified together in a single family, such as *Lemur* and *Megaladapis* (in the family Lemuridae), *Aotus* and *Alouatta* (in the family Cebidae), or *Presbytis* and *Papio* (in the family Cercopithecidae). If the early Miocene noncercopithecoid catarrhines from East Africa were classified using the same criteria as neontological taxa, without due regard to their descendant relationships, I have no doubt that most zoologists would confine the entire group to a single family.

Clearly, there are morphological differences that separate the species, and these provide the basis for recognizing a large number of genera. In addition, it is quite evident that the two best-known species, *Dendropithecus macinnesi* and *Proconsul africanus,* are sufficiently distinct from one another to indicate that they occupied quite different adaptive niches (Fleagle, 1983, 1988; Rose, 1983, 1988; Fleagle and Kay, 1985). Nevertheless, when the group is considered as a whole, they appear to span a quite limited range of diversity, especially when allometric and ecological correlates of species that are known to range in estimated average body size from 4 to 50 kg are taken into consideration (Table 1).

As a scientist interested in issues of paleobiology, such as adaptive diversity, ecological relationships, and community structure, as well as phylogeny, I find that a classification based purely on cladistic criteria is, in this case, much less meaningful, and therefore much less useful, than one based on a combination of cladistic and gradistic concepts. It should be stressed that in most cases a classification based entirely on phylogenetic information will also reflect adequately the level of diversity and other paleobiological factors, and in these instances a cladistic classification is most desirable. The discordance between the cladistic and patristic data only becomes a serious concern when attempting to classify members of a stem group, such as the early Miocene catarrhines from East Africa. This is because stem groups usually exhibit low levels of morphological or adaptive differentiation from the last common ancestor, but yet, in a cladistic analysis, they exhibit high levels of taxonomic differentiation due to the strict requirement of monophyly.

Thus, a cladistic classification of stem groups will reflect only phyletic information, and any inferences about the paleobiology of the group that are usually implied by the classification will be entirely spurious. To recognize three separate superfamilies among the early Miocene catarrhines from East Africa may be justifiable on phylogenetic grounds, but it surely gives a distorted impression about their adaptive diversity, especially when it is considered that all modern anthropoids are traditionally included in only three superfamilies.

For these reasons, stem groups should be considered as a special category in cladistic analyses and should be dealt with in a different way from other groups. In order to maximize the biologically meaningful content of a classification, it might be most appropriate, in certain exceptional circumstances, to consider a combination of cladistic and patristic data. Use of a grade concept is applicable, even within the context of a primarily cladistic analysis, because of the practical benefits and flexibility that it may provide in a classificatory scheme. However, it is important to emphasize that the concept should be employed with extreme caution and should never be applied to a group where good evidence provides indisputable support of paraphyly or polyphyly.

In my recent cladistic analyses of the early catarrhine primates from East Africa (Harrison, 1987a, 1988), I included almost all of the genera in a single superfamily, the Proconsuloidea. The taxon is a patristic group comprising undifferentiated basal catarrhines of modern aspect (Harrison, 1988). The group may also contain *unidentified* basal hominoids, which lack, in their preserved parts, clearly recognizable synapomorphies with the living apes. Until better material is available, it seems reasonable, for practical purposes at least, to include all East African early Miocene catarrhines that lack clear-cut affinities with either of the two extant superfamilies in the Proconsuloidea (Harrison, 1988).

I feel that this position is justified given the following philosophical and practical considerations: It is certain that the early Miocene catarrhines from East Africa represent a narrow grade, very close to the initial radiation of all recent catarrhines (see Fig. 2). Their general morphological similarity suggests that the various species represent a closely related group that diverged relatively recently from their last common ancestor. However, to demonstrate clear-cut relationships either among them or with other groups of catarrhines has proved exceedingly difficult. The problem arises mainly because the early Miocene catarrhines from East Africa largely retain the ancestral morphotypic pattern defining the extant catarrhines. It is difficult to establish the affinities of the constituent members of a stem group because of their tendency to retain a high frequency of plesiomorphic traits.

In addition, the paucity of material for most of the species makes adequate assessments of their relationships to one another and to other catarrhine groups rather tenuous. It is this combination of phyletic conservatism and the lack of adequate material that makes it impossible at present to determine, with any degree of reliability, the evolutionary relationships of most members of this group. For this reason, I prefer to recognize Proconsuloidea as a useful "wastebasket" category that is not demonstrably paraphyletic or polyphyletic and that includes all those forms from the early Miocene with uncertain affinities to later catarrhines.

In this sense, it is a temporary clustering of species that maintains its integrity only while inadequate material is available or while clear-cut affinities with other groups of catarrhines cannot to be demonstrated. For example, the early Miocene catarrhine *Nyanzapithecus vancouveringorum* is excluded from the Proconsuloidea because it can be shown to have close phyletic affinities to *Oreopithecus*, an undoubted hominoid from the late Miocene of Italy (Harrison, 1986, 1987b, 1991). It may be presumed that other species will be removed from the Proconsuloidea as their relationships to later catarrhines become more securely established.

I do not believe that this approach is such a radical departure from that already employed by most other cladistically oriented paleoanthropologists; the main difference is that I am explicitly acknowledging its practical application within the methodological context. For example, the genus *Australopithecus* (if we exclude those species commonly included in *Paranthropus*) is regarded by most workers as including two species—*A. afarensis* and *A. africanus*. Recent reviews of the phylogenetic relationships of the australopithecines favor the interpretation that *A. afarensis* represents the sister taxon of all later hominids, while *A. africanus* is the sister taxon of *Homo* (e.g., see the contributions in Grine, 1988). If this is the case, then *Australopithecus* constitutes a nonmonophyletic grade of basal hominids.

In order for the classification to reflect these inferred relationships, *A. africanus*, as the type species of the genus, could be retained in *Australopithecus* or be transferred to the genus *Homo*, while *A. afarensis* would need to be removed from *Australopithecus* and subsequently recognized by the prior name *Prae-*

anthropus africanus (Weinert, 1950). In my view this option may prove to be a necessary and desirable course of action, but I can fully appreciate that the majority of workers might prefer to retain *Australopithecus* as a paraphyletic clustering of stem species (just as I do for the Proconsuloidea), at least until such time as the relationships of the early hominids have been more firmly established.

It is important to recall that phylogenetic inference and classification are two discrete operations with quite different functions and properties. For instance, a cladogram, which represents a hypothesis of inferred phylogenetic relationships, is highly labile and is subject to dramatic alterations with the incorporation of new material and information. On the other hand, if the primary purpose of classification is to provide a basis for communication, then some degree of stability is an important requirement if utility is to be maintained. As stem groups tend to exhibit high levels of conservatism, phylogenetic hypotheses tend to be less robust and are subject to repeated revisions, especially if the individual species are rather poorly known.

With our present limited knowledge about the phylogenetic relationships and paleobiology of the early Miocene catarrhines from East Africa, it is likely that we can entertain a wide array of alternative cladograms. This is a necessary and desirable means of exploring the feasibility of different phylogenetic hypotheses, but it is unlikely that concordant changes in the classificatory schemes will be as well received. Until a consensus is reached about the phylogenetic affinities of the early Miocene catarrhines from East Africa, it seems most practical, and in the interests of taxonomic stability, to retain those species of uncertain status in the taxon Proconsuloidea.

The inclusion of the majority of early Miocene catarrhines from East Africa in a single superfamily more realistically reflects the diversity of the group when compared with modern taxa and avoids the unnecessary oversplitting that tends to occur with stem groups. In addition, discussion of the morphological diversity, ecological differentiation, and community structure of the Proconsuloidea can be more readily related to data derived from neontological studies.

In conclusion, the Proconsuloidea should be retained as a taxonomic grouping because it is, at present, the most meaningful way to deal with the constituent members and to communicate information about them. Given the uncertainties surrounding their precise phyletic relationships, it seems much more reasonable to make use of the wealth of paleobiological information that we have accumulated on the group as the primary basis for their taxonomy. In this way we are able to produce a relatively stable, yet biologically meaningful classification, at least as an interim measure, until such time as the uncertainties surrounding their relationships have been resolved.

In conclusion, it is important to point out that I am not advocating that cladists should "soften" their approach as a general rule in order to adopt a more Simpsonian inclination. It is more a matter of allowing greater flexibility within the confines of general cladistic methodology to accommodate the obvious shortcomings of paleontological data and the inherent limitations of the phylogenetic method. I firmly believe that under ideal circumstances it would be most desirable for *all* classifications to reflect cladistic (i.e., monophyletic) relationships *only* and that we should strive, wherever possible, to achieve this objective. Unfortunately, however, our data are rarely perfect, and I would contend that under certain exceptional circumstances formal cladistic classificatory procedures may provide less worthwhile results than those based on other criteria.

As discussed above, the recognition of monophyletic groups based on inadequately known representatives of a stem group may be beyond the resolution of current applications of cladistic method. In these cases, it would be much better to use the more readily accessible and more inherently stable data on adaptive diversity (data that are no less biologically relevant than cladistic evidence) as a criterion for taxonomic attribution. The recognition of a gradistically defined taxon for poorly known stem groups, such as the Proconsuloidea, is therefore recommended for exceptional cases only, and in these instances such groups should be considered to have integrity only until such time as the cladistic affinities of the included species have been more firmly established.

The Genus Group and Species Levels

Cladistic analyses of fossil primates at the alpha-taxonomic level have so far proved to be much less successful than those dealing with higher-level categories. My own attempts, for example, to resolve the inferred relationships *among* closely related fossil primates, such as the early Miocene catarrhines from East Africa or the Eurasian pliopithecids, have been largely unsuccessful. Perhaps this is hardly surprising when one considers that neontologists, with access to the full range of information from anatomy, ontogeny, behavior, and ecology, have been equally frustrated in their attempts to resolve the relationships among closely related species of modern primates, such as those included in *Cercopithecus* or *Lemur* (Tattersall, 1986, this volume; Tattersall and Schwartz, 1991).

In 1982, I attempted to assess the relationships between the genera of early Miocene catarrhines from East Africa (Harrison, 1982). I constructed a simple cladogram, based primarily on dental characteristics, that purported to illustrate the relationships among the various taxa (Fig. 3 and Table 2). In this scheme, *Limnopithecus, Dendropithecus, Proconsul,* and *Rangwapithecus* are linked at node 2 by a shared derived character complex, the possession of upper and lower molars with large and well-defined occlusal basins, while *Micropithecus* is excluded by the retention of the primitive catarrhine molar pattern, with restricted occlusal basins (Table 2). *Rangwapithecus* and *Proconsul* are further linked at node 4 by several synapomorphies, such as the development of a moderately elongated lower face and a modified molar pattern (see Fig. 3 and Table 2 for details), and are represented as the sister group of *Limnopithecus* and *Dendropithecus.* The latter share a derived molar pattern that includes a more rectangular occlusal outline, higher and more conical cusps, and further development of crests and occlusal basins. This set of inferred relationships was considered at the time to be a tentative phylogenetic statement only, but one that was consistent with the available fossil evidence.

There is a strong impetus among cladists to generate best-fit cladograms as summary statements, just as I did in my 1982 study, in order to provide hypotheses of possible relationships. However, I believe that too few workers are willing

Table 2. A List of the Characters Used to Define the Nodes in Fig. 3

Node	Characters
Node 1:	Nasal aperture ovoid in shape and higher than broad
	Anterior margin of the orbit situated above P^3–P^4
	Palate relatively long and narrow
	Mandibular symphysis with superior transverse torus only, or dominant over the inferior transverse torus
	Mandible moderately deep
	I^1 slightly higher than broad
	Upper and lower canines only moderately bilaterially compressed
	Upper canines with single mesial groove
	Upper premolars with buccal cusp moderately higher than the lingual cusp
	Upper and lower molars with the following features:
	Crowns short and broad, and ovoid in occlusal outline
	Cusps low and conical
	Occlusal crests low and quite rounded
	Relatively restricted occlusal basins
	Lack of secondary wrinkling
	M3 relatively large
	P_4 longer than broad
	Lower incisors relatively high crowned
Node 2:	Upper and lower molars with large and well-defined occlusal basins
Node 3:	Upper and lower molars with the following features:
	Crowns rectangular in occlusal outline
	High conical cusps
	Sharp and well-defined crests
	Large occlusal basins
Node 4:	Position of anterior margin of the orbit above M^1
	Face moderately long
	Molar cusps high voluminous and rounded
Node 5:	Nasal aperture relatively broad
	Palate relatively short and broad
	Mandible relatively deep
	M3 reduced in size
	Incisors large relative to the size of the molars
Node 6:	Nasal aperture relatively very narrow
Node 7:	Inferior transverse torus well developed
	Mandible relatively shallow
	I^1 high and relatively narrow
	Canines bilaterally compressed
	Upper canines in males with double mesial groove
Node 8:	P_4 broader than long
Node 9:	Mandible relatively deep
	I^1 high and narrow
	Buccal cusp of upper premolars only slightly more elevated than the lingual cusp
	Upper molars long and narrow, increasing in size from M1 to M3
	Lower molars very long and narrow
	Molars with secondary wrinkling

Adapted from Harrison (1982).

to concede that some problems are insoluble given the quality of the information that they have at hand. In many cases it is possible to construct a cladogram, even a parsimonious one, that is based on too little information to be a realistic assessment of relationships. In other words, the level of information that is actually required to construct a cladogram is often much less than is required to

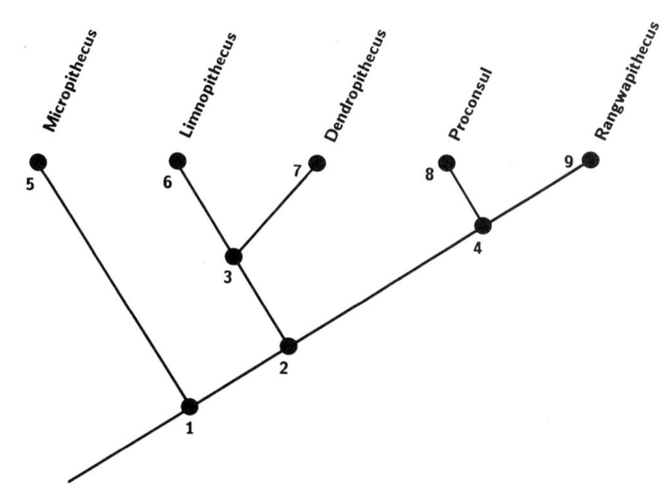

Fig. 3. Cladogram showing the purported evolutionary relationships between some of the genera of early Miocene catarrhines from East Africa, as suggested by Harrison (1982). The characters used to define each node are given in Table 2. [Adapted from Harrison (1982).]

provide inferences about relationships. Cladists commonly ignore the option that no cladogram is preferable to a weakly supported one based on insufficient or conflicting data. This requires that systematists develop a keen appreciation of the threshold at which phylogenetic hypotheses shift from intuitive guesswork to empirically based objectivity.

Recognizing the shortcomings of the data is much less detrimental to future advances in understanding phylogenetic relationships than trying to establish relationships with insufficient support, simply because it is axiomatic that such relationships exist in reality. With hindsight, I now recognize that it is simply not possible, given the quality of the information available, to determine the nature of the relationships among the genera of fossil catarrhines from the Miocene of East Africa. In fact, based on my experience with other fossil groups and my comparative studies of extant primates, I would argue that *most* attempts to assess the relationships among closely related forms are likely to prove inconclusive. If this is true, then the classification of fossil species at the generic level may have to rely much more on the degree of phenetic similarity, rather than on the precise nature of their phylogenetic relationships, as is usually advocated by most cladists.

There are four main reasons why attempts to resolve the relationships among closely related groups of primates may be unsuccessful: (1) Holomorphological insufficiency—there is just not enough morphological information preserved in the fossil record to deduce adequately what relationships exist. (2) The problems of correctly inferring the polarity of character transformations (which, among closely related forms, tend to be based on minor morphological differences, such as the detailed structure of the molars in the case of the East African fossil catarrhines) and the relatively narrow transformational spectrum due to the relative recency of their last common ancestor. (3) It is extremely difficult to establish relationships among closely related members of a stem groups, such as the early Miocene catarrhines from East Africa, because they tend to retain a high frequency of plesiomorphic characters. Furthermore, the majority of well-established derived traits characterizing the species of fossil anthropoids from East Africa are autapomorphous features, which, although useful for determining species and generic distinctions, are of no value for assessing relationships among the taxa. (4) There is an increased tendency for character complexes in closely related forms to be canalized along identical structural pathways as a result of functional convergence, thereby leading to a high incidence of homoplasies that are indistinguishable from homologies. This is probably true for the East African Miocene catarrhines. I now suspect that the set of relationships depicted in Fig. 3, being based primarily on molar and lower facial morphology, is more likely to be reflecting common adaptive patterns due to similar diets and allometric constraints (see Table 1 for information on estimated body sizes and dietary inferences in the fossils) than the degree of phyletic affinity.

In combination, these problems are likely to obscure any attempts to resolve phylogenetic relationships between closely related species. I should point out that these are not problems uniquely associated with cladistics; they are general theoretical problems encountered by all systematists. However, unlike most other systematists, who usually formulate phylogenetic hypotheses in rather general terms, cladists base their assessments on explicitly identified characters and their inferred morphocline polarities. These represent testable statements that are open to critical scrutiny and reevaluation by other workers. As a consequence, I do not attribute my lack of success in attempts to resolve the relationships among the Miocene catarrhines from East Africa as due to the shortcomings of cladistic methods. On the contrary, the methodological framework prescribed by cladists provides an ideal means by which a scientist can recognize that the reconstruction of relationships is not possible given the data available.

Species Recognition

There has been a noticeable trend among hominid paleontologists in recent years to extend the usage of cladistic methods to encompass the problem of species recognition. In such instances, cladistic concepts have been employed as a basis for diagnosing species and for identifying individual specimens that should or should not be included in the hypodigm. Emphasis is placed on identifying the presence of apomorphies for recognizing samples of hominids as distinct species. Good examples of this type of approach have been set forth in recent reviews of the status of *Homo erectus* (Andrews, 1984a, 1984b; Stringer, 1984, in press; Wood, 1984; Groves, 1989).

From their studies of *Homo erectus,* the above authors draw the following general conclusions: (1) that the Asian sample of *Homo erectus,* containing the type material from Indonesia, is united by a series of autapomorphies; (2) that these autapomorphies essentially exclude *Homo erectus sensu stricto* from direct ancestry to *Homo sapiens;* (3) that all or most of these autapomorphies are lacking in some of the earlier specimens from East Africa traditionally included in *Homo erectus;* (4) that the lack of synapomorphies in the combined samples from East Africa and Asia implies that *Homo erectus* is a grade concept; and (5) that the autapomorphies in the Asian sample can be considered sufficient grounds to support the possible exclusion of the East African forms (including KNM-ER 3733, KNM-ER 3883, and OH 9, according to Stringer and Andrews, although Wood and Groves retain OH 9 in *H. erectus sensu stricto*) from the hypodigm of *H. erectus.*

These authors are understandably guarded about formally recognizing a separate species for the East African sample. As Stringer (1984) notes, "we must beware of the position of saying on cladistic grounds that *H. sapiens* did not evolve from *H. erectus* but from a different species showing similar characteristics which lived at the same time!" Nevertheless, Clarke (1990) has recently argued that the African material previously attributed to *H. erectus* should be assigned to a separate species, *Homo leakeyi* Heberer, 1963, based on the type specimen OH 9.

However, identification of the valid species name for the hypodigm comprising the African specimens has proved problematic. Most authors include a combination of specimens from North, South, and East Africa that have been associated with the following species group names: *Telanthropus capensis* Broom and Robinson, 1949; *Atlanthropus mauritanicus* Arambourg, 1954; *Homo leakeyi* Heberer, 1963; *Tchadanthropus uxoris* Coppens, 1965; *Homo erectus olduvaiensis* Tobias, 1968; and *Homo ergaster* Groves and Mazák, 1975. *Telanthropus capensis* is available and has priority over all other names, but if the species is transferred to *Homo,* then the name is preoccupied by *Homo capensis* Broom, 1917, based on the modern-looking Boskop skull.

Groves (1989) has argued that *Atlanthropus mauritanicus* is unavailable because it was proposed as a provisional name only, but this ruling (Article 15 of the *International Code of Zoological Nomenclature*) does not apply to names published before 1961. Nevertheless, as noted by Clarke (1990), *A. mauritanicus* is not available because Arambourg (1954) failed to provide a definition that purports to differentiate it from other taxa [Article 13 (a) (i)]. The name *Homo leakeyi,* accepted as valid by Clarke (1990), is excluded from availability under the provision of Article 15, because Heberer (1963) proposed the name provisionally only (as previously pointed out by Simons *et al.,* 1969, and Szalay and Delson, 1979). An earlier, unrelated reference to *Homo leakeyi* by Paterson (1940), based on the anatomically modern human material from Kanjera, is a *nomen nudum* and is, therefore, unavailable. Similarly, a poorly preserved skull from Tchad, named provisionally as *Tchadanthropus uxoris* by Coppens (1965), is also unavailable.

The subspecies *Homo erectus olduvaiensis* Tobias, 1968, based on the type specimen OH 9, is considered to be the valid species group name for the African sample by Groves (1989). This too, however, can be excluded from availability on the grounds that the name was considered by Tobias (1968) as a conditional proposal only. This leaves *Homo ergaster* as the only possibility as an available name for the African species, although some authors (e.g., Leakey *et al.,* 1978; Stringer, 1986; Groves, 1989) would question whether the type specimen, KNM-ER 992, should really be included in the same species as forms such as OH 9, KNM-ER 3733, and KNM-ER 3883.

A number of authors (Rightmire, 1984, 1987, 1990; Bilsborough and Wood,

1986; Hublin, 1986; Turner, 1986; Turner and Chamberlain, 1989; Habgood, 1989; Bräuer, 1990; Kennedy, 1991) have provided critical responses to this cladistic approach to analyzing *H. erectus.* I must say that as a cladist, which many of the critics are not, I agree with their conclusions. In my view there are good grounds to support the argument that cladistic methods are inappropriate to use in the recognition and definition of species groupings (see also Kimbel and Rak, this volume).

Hennig (1966) was very explicit about the lowest operational unit for cladistic analyses—the species. Species groupings need to be formulated as a prerequisite to cladistic analysis by taking into account all forms of intraspecific variation. Species can only be identified in the fossil record in terms of morphological criteria, and although there are no absolute rules for the recognition of species boundaries, intuitive ideas of the range of well-known extant groups provide models for the inclusion or exclusion of individuals (Hennig, 1966). In other words, fossil species are recognized by employing entirely phenetic concepts, based on analogies derived subjectively from morphological ranges of variation seen in modern species (see also Rose and Bown, this volume).

It is, therefore, important to identify species groupings as an initial step in a cladistic analysis, and not to use cladistics as a method to identify species groupings. After all, cladistics is a method that allows the interpretation of phylogenetic relationships, and phylogenetic relationships cannot be expressed at the infraspecies level. Of course, it is possible to construct detailed cladograms based on subspecies, populations, or individuals of a single species, and to identify synapomorphies and autapomorphies that define the various nodes, but these may contain nothing meaningful about evolutionary relationships.

A detailed cladistic analysis of infraspecific groups could certainly reveal a significant genealogical structure; one that may even indicate that a particular population within a species is the "sister group" of a neighboring species (see the macaque example cited by Jolly, this volume). The reticulate nature of the ancestral-descendant relationships among infraspecies groups, however, rather than the bifurcating relationship among species, serves to confound any attempt at detailed character analysis. This is because species share a common gene pool. Apomorphies, including autapomorphies, being genetically determined, can flow between different populations of the same species. If this is the case, then shared derived traits may not necessarily be the result of common descent, but rather the product of interbreeding. As a consequence, it is not until gene pools become permanently segregated by a speciation event that autapomorphies and synapomorphies can take on any special relevance in phylogenetic interpretations (see Kimbel and Rak, this volume, for further discussion of this issue).

Cladistic analyses below the species level are not comparable to cladistic analyses above the species level. They are, in fact, merely structurally analogous in terms of their operational procedures. As an essential first step in any cladistic analysis of *H. erectus,* it is necessary, therefore, to decide whether the African and Asian samples previously attributed to *Homo erectus* are in fact representative of a single species or not.

One of the key issues in this debate is how one is able to identify which specimens should be assigned to a single species. An extensive series of characters has been provided in the past to distinguish *H. erectus* from modern humans (Weidenreich, 1936, 1943; Le Gros Clark, 1964; Howell, 1978; Rightmire, 1990). However, as noted by Andrews (1984a), Stringer (1984), and Wood (1984), many of these characters are also found in early *Homo,* and thus may be plesiomorphic for the genus. Although the latter authors acknowledge that a combination of plesiomorphies and apomorphies can be used to define a species in relation to its sister species, they have stressed the particular importance of autapomorphies in recognizing species [see Wood (1984) for the most explicit exposition of this type of approach].

It is important to note, however, that a species diagnosis is merely a phenetically derived morphotype that serves to distinguish the species from all other closely related species. The morphocline polarity of individual characters comprising the list of diagnostic features need not be assessed at this stage. This is done only after the species has been recognized, when its phylogenetic relationships need to be determined, and when variations in character states have been carefully considered. Apomorphies can be used to infer relationships at the species level and above, but not to establish the group identity of individuals within a species. Of course autapomorphies, being uniquely derived traits, are especially helpful in distinguishing species, but synapomorphies can be used to distinguish species from more conservative taxa, and plesiomorphies can be used to distinguish species from more derived taxa.

Definitions containing a combination of apomorphies and plesiomorphies are likely to be the most practical, given the mosaic nature of evolution. For this reason a single list of traits is unlikely to distinguish clearly a species, such as *Homo erectus,* from all other closely related species. This is why a differential diagnosis, in which each species is compared on an individual basis with all other related species, is preferable over a single universal diagnosis. Turner and Chamberlain (1989) and Rightmire (1990) have recently provided a revised set of traits that does, in fact, serve to distinguish *H. erectus* from both *H. habilis* and *H. sapiens.*

In their recent critical essay, Turner and Chamberlain (1989) implied that Tattersall's (1986) review of the species problem in hominid evolution provides support for the view that the African and Asian samples of *H. erectus* should be distinguished at the species level. Although Tattersall (1986) does not address this question directly, based on his empirical studies of modern primate taxa, he argues that separate species should be recognized in the fossil record if distinct morphs can readily be identified from their bony anatomy. I sympathize with Tattersall's viewpoint in this regard. The criteria that I have used to assign individual specimens to species groupings among Miocene catarrhines is also based largely on my experience of the degree of morphological variation seen in modern primate taxa.

These same criteria, however, could not be met by some of the currently accepted taxonomic groupings of hominids. For example, I would have to concur with Tattersall that if I found two morphs as different as Neandertals and modern humans in the Miocene there is little doubt that I would recognize them as two distinct species. Nevertheless, I do not believe that this line of reasoning is necessarily pertinent to the question of *Homo erectus.* The distinction here is simply a matter of degree.

Tattersall contends that separate species should be identified where distinct morphs can *readily* be identified. It may be presumed that he is not arguing that any observable difference constitutes sufficient grounds to distinguish morphs as separate species, or we would soon be reduced to the absurd situation that all individuals are distinct morphs that merit species recognition. Clearly, individual specimens that exhibit low levels of morphological variation should be clustered together to form morphs. One or more morphs can then be combined in such a way that the type or degree of variation does not exceed that considered acceptable for a single species. In my view this is the case for *H. erectus*—the Asian material is not sufficiently different (i.e., readily identifiable enough) from the African sample to merit the recognition of a separate species.

Recently a number of authors have carefully reviewed the extent of the difference between the African and Asian *H. erectus* samples and have noted that there is a good deal of individual variation in the characters that have been purported to separate the two samples (Hublin, 1986; Turner and Chamberlain, 1989; Habgood, 1989; Rightmire, 1990; Bräuer, 1990). This is not to deny that there are identifiable differences between the Asian and African material, it is just that they may not be as profound and consistently clear-cut as was initially believed.

Hublin (1986) has even suggested that many of the individual differences between the two groups may be part of a single functional complex, associated with bone hypertrophy of the cranial vault. The degree of variation observable within the entire sample from Asia and Africa may be considered appropriate for a single geographically widespread species. In fact, Rightmire (1990) has commented that the amount of variation observed is, perhaps, not surprising, given the wide geographical and temporal distribution of the two samples. The weight of the evidence would, therefore, favor the assessment of Rightmire that the *H. erectus* samples from Africa and Asia are morphologically consistent with a single polytypic species that is widely distributed through space and time.

The differences between the Asian sample and the African sample may be explained as populational differences, rather than as a consequence of post-speciational divergence. In this case, Andrews' (1984a,b) and Clarke's (1990) phylogenetic conclusions, that the African species is close to the direct ancestry of *Homo sapiens,* while *Homo erectus* from Asia is a more specialized collateral relative, can still be accommodated with only minor modifications (see Fig. 4). If we exclude the possibility that all *H. erectus* populations across the Old World graded imperceptibly into *H. sapiens* (Eldredge and Tattersall, 1975, 1982; Delson *et al.,* 1977), then the presence of unique specializations in the Asian population may indeed make it more likely that the African population is the ancestral population from which *H. sapiens* was derived (Turner, 1986; Rightmire, 1990).

Conclusions

The examples that I have presented above serve to highlight some of the limitations of the cladistic method. Although not all of them specifically address the species problem *per se,* the underlying theoretical, conceptual, and practical

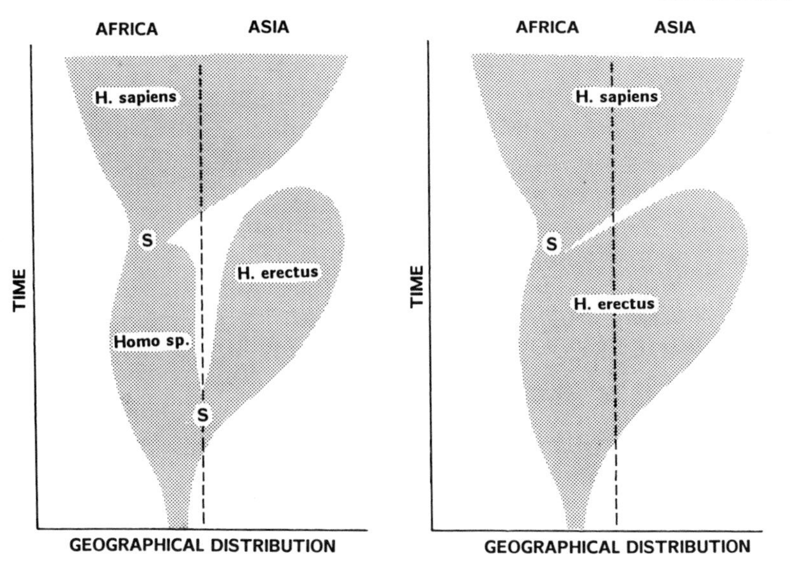

Fig. 4. Phylogenetic tree illustrating the two main competing hypotheses of the relationships between archaic *Homo sapiens sensu lato* and the African and Asian samples of *Homo erectus*. Left: The hypothesis preferred by Andrews (1984a, 1984b) and Clarke (1990). Right: The hypothesis preferred by Rightmire (1984, 1987, 1990), Turner and Chamberlain (1989), and the present author. S denotes a speciation event.

problems that they exemplify do impinge either directly or indirectly on the species concept, and influence the way in which cladists perceive species. The difficulties arise primarily from the fact that the strict operational procedures employed by cladists for inferring phylogenetic relationships tend to pervade all other aspects of their research, including classification and species recognition. A much more flexible approach is required to accommodate the subtle complexities of the natural world.

If our primary objective as cladists is to obtain a better appreciation of phylogenetic affinities and of other biological phenomena, it is clear that utilitarianism must be a guiding principle. If this is so, then systematists should be concerned with applying modified techniques and concepts to suit particular practical problems, rather than dogmatically adhering to rigid procedures that are perceived as theoretically correct.

In this chapter I have identified a number of issues that I feel need more careful consideration by cladists in relation to the species concept in the fossil record. First, diachronicity and descent, unique properties of the paleontological record, create special problems that complicate the direct transferal of concepts and data derived from modern species to the fossil record. However, if systematists are to succeed in making meaningful statements about biological phenomena in the past, it is imperative that they pay close and detailed attention to the use of modern analogues as a basis for determining the limits of paleospecies.

Second, paleontological data are always much less complete than neontological data. It is, therefore, highly unlikely that paleontologists will be able to settle issues that have proven to be beyond the capabilities of neontologists. For example, neontologists have been trying for decades to explain and to deal with the widely different levels of intraspecific and interspecific morphological variation seen in modern taxa, while cladistic analyses of closely related extant species have mostly yielded inconclusive results. This is not intended to cast a shadow of pessimism over the entire analysis of the fossil record, but we need to be willing to acknowledge that some key phylogenetic problems are insoluble given the present limitations of the paleontological evidence.

Finally, cladists should aim to maximize the biological content of their classifications. One way to achieve this would be to make fuller use of paleobiological information, which is generally excluded by cladists on the grounds that it is applicable in establishing patristic affinities only.

It is important to emphasize that I am not trying to advocate a complete reformulation of cladistic methods of classification to include more Simpsonian concepts as a general principle. Clearly cladistic relationships should be given priority in any classification, but where this information is lacking or is only tentatively formulated, patristic information may have greater relevance for assessing taxonomic affinity. Cladists contend that mixed classifications of this kind will necessarily lead to misunderstandings, because the criteria upon which the classification is based are not explicitly expressed. However, phyletic relationships are precisely reflected only in the form of a cladogram. A classification, on the other hand, as an entity fully distinct from the cladogram, can incorporate both cladistic and patristic information because it is constrained merely by its usefulness as a means of communication about species.

Acknowledgments

I would like to thank Bill Kimbel and Lawrence Martin for inviting me to participate in the symposium, "Species, Species Concepts and Primate Evolution" at the 1991 AAPA meetings in Milwaukee and to contribute to this companion volume. Peter Andrews, Clifford Jolly, Bill Kimbel, Lawrence Martin, Chris Stringer, Fred Szalay, and Bernard Wood provided helpful comments and suggestions that greatly improved the clarity and accuracy of the text. I am especially indebted to Peter Andrews, Clifford Jolly, Todd Olson, and Eric Delson for many critical discussions over the past 12 years that have helped to provoke and to shape the ideas presented in this chapter, although I should emphasize that none of them is in any way to be held responsible for the outcome. I would also like to thank my graduate students at New York University, especially Eric Baker, Eugene Harris, Tim Newman, and Bill Sanders, who have had to endure, as a captive audience, my endless monologues (they would contend soliloquies) on the subject of this chapter. In addition, I am grateful to the following colleagues and institutions for allowing me access to material in their care: Richard Leakey, Meave Leakey, and Emma Mbua, National Museums of Kenya; Peter Andrews, Chris Stringer, and Paula Jenkins, The Natural History Museum, London; and Guy Musser and Wolfgang Fuchs, American Museum of Natural History.

References

Andrews, P. J. 1974. New species of *Dryopithecus* from Kenya. *Nature* **249:**188–190.

Andrews, P. J. 1978. A revision of the Miocene Hominoidea of East Africa. *Bull. Br. Mus. Nat. Hist. (Geol.).* **30:**85–224.

Andrews, P. J. 1980. Ecological adaptations of the smaller fossil apes. *Z. Morph. Anthropol.* **71:**164–173.

Andrews, P. J. 1984a. An alternative interpretation of characters used to define *Homo erectus. Cour. Forsch. Inst. Senckenberg* **69:**167–175.

Andrews, P. J. 1984b. The descent of man. *New Scientist* **May 3, 1984:**24–25.

Andrews, P. J. 1985. Family group systematics and evolution among catarrhine primates, in: E. Delson (ed.), *Ancestors: The Hard Evidence,* pp. 14–22. Alan R. Liss, New York.

Andrews, P. J. 1988. Hominoidea, in: I. Tattersall, E. Delson, and J. Van Couvering (eds.), *Encyclopedia of Human Evolution and Prehistory,* pp. 248–255. Garland, New York.

Andrews, P. J., and Martin, L. B. 1987. Cladistic relationships of extant and fossil hominoids. *J. Hum. Evol.* **16:**101–118.

Andrews, P. J., and Simons, E. L. 1977. A new African Miocene gibbon-like genus, *Dendropithecus* (Hominoidea, Primates) with distinctive postcranial adaptations: Its significance to origin of Hylobatidae. *Folia Primatol.* **28:**161–170.

Arambourg, C. 1954. L'hominien fossile de Ternifine (Algerie). *C. R. Seanc. Acad. Sci. Paris* **239:**893–895.

Ashlock, P. D. 1971. Monophyly and associated terms. *Syst. Zool.* **20:**63–69.

Ashlock, P. D. 1972. Monophyly again. *Syst. Zool.* **21:**430–437.

Ashlock, P. D. 1974. The uses of cladistics. *Ann. Rev. Ecol. Syst.* **5:**81–99.

Ashlock, P. D. 1979. An evolutionary systmatist's view of classification. *Syst. Zool.* **28:**441–450.

Ax, P. 1987. *The Phylogenetic System. The Systematization of Organisms on the Basis of their Phylogenesis.* John Wiley, Chichester.

Bilsborough, A., and Wood, B. A. 1986. The nature and fate of *Homo erectus,* in: B. Wood, L. Martin, and P. Andrews (eds), *Major Topics in Primate and Human Evolution,* pp. 295–316. Cambridge University Press, Cambridge.

Bock, W. J. 1977. Foundations and methods of evolutionary classification, in: M. Hecht, P. C. Goody, and B. M. Hecht (Eds.), *Major Patterns in Vertebrate Evolution,* pp. 851–895. Plenum, New York.

Bonde, N. 1977. Cladistic classification as applied to vertebrates, in: M. Hecht, P. C. Goody, and B. M. Hecht (eds.), *Major Patterns in Vertebrate Evolution,* pp. 741–804. Plenum, New York.

Bräuer, G. 1990. The occurrence of some controversial *Homo erectus* cranial features in the Zhoukoudian and East African hominids. *Acta Anthrop. Sinica* **9:**350–358.

Broom, R. 1917. Fossil man in South Africa. *Am. Mus. J.* **17:**141–142.

Broom, R., and Robinson, J. T. 1949. A new type of fossil man. *Nature* **164:**322–323.

Cartmill, M. 1981. Hypothesis testing and phylogenetic reconstruction. *Z. Zool. Syst. Evolut.-Forsch.* **19:**73–96.

Chamberlain, A. T., and Wood, B. A. 1987. Early hominid phylogeny. *J. Hum. Evol.* **16:**119–133.

Clarke, R. J. 1990. The Ndutu cranium and the origin of *Homo sapiens. J. Hum. Evol.* **19:**699–736.

Coppens, Y. 1965. L'hominien du Tchad. *C. R. Acad. Sci. (Paris)* **260D:**2869–2871.

Crowson, R. A. 1970. *Classification and Biology.* Heinemann Educational Books, London.

Delson, E. 1975. Evolutionary history of the Cercopithecidae, in: F. S. Szalay (ed.), *Approaches to Primate Paleobiology,* Contributions to Primatology, Volume 5. pp. 167–217. Karger, Basel.

Delson, E. 1988. Catarrhini, in: I. Tattersall, E. Delson, and J. Van Couvering (eds.), *Encyclopedia of Human Evolution and Prehistory,* pp. 111–116. Garland, New York.

Delson, E., and Andrews, P. J. 1975. Evolution and interrelationships of the catarrhine primates, in: W. P. Luckett and F. S. Szalay (eds.), *Phylogeny of the Primates: A Multidisciplinary Approach,* pp. 405–446. Plenum, New York.

Delson, E., Eldredge, N., and Tattersall, I. 1977. Reconstruction of hominid phylogeny: A testable framework based on cladistic analysis. *J. Hum. Evol.* **6:**263–278.

Eldredge, N., and Cracraft, J. 1980. *Phylogenetic Patterns and the Evolutionary Process: Method and Theory in Comparative Biology.* Columbia University Press, New York.

Eldredge, N., and Tattersall, I. 1975. Evolutionary models, phylogenetic reconstruction, and another look at hominid phylogeny, in: F. S. Szalay (ed.), *Approaches to Primate Paleobiology*, Contributions to Primatology, Volume 5. pp. 218–242. Karger, Basel.

Eldredge, N., and Tattersall, I. 1982. *The Myths of Human Evolution.* Columbia University Press, New York.

Farris, F. S. 1976. Phylogenetic classification of fossils with recent species. *Syst. Zool.* **25:**271–282.

Feldesman, M. R. 1986. The forelimb of the newly "rediscovered" *Proconsul africanus* from Rusinga Island, Kenya: Morphometrics and implications for catarrhine evolution, in: R. Singer and J. K. Lundy (eds.), *Variation, Culture and Evolution in African Populations,* pp. 179–193. Witwatersrand University Press, Johannesburg.

Fleagle, J. G. 1983. Locomotor adaptations of Oligocene and Miocene hominoids and their phyletic implications, in: R. L. Ciochon and R. S. Corruccini (eds.), *New Interpretations of Ape and Human Ancestry,* pp. 301–324. Plenum, New York.

Fleagle, J. G. 1988. *Primate Adaptation & Evolution.* Academic Press, New York.

Fleagle, J. G., and Kay, R. F. 1985. The paleobiology of catarrhines, in: E. Delson (ed.), *Ancestors: The Hard Evidence,* pp. 23–36. Alan R. Liss, New York.

Ford, S. M. 1986. Systematics of the New World monkeys, in: D. R. Swindler and J. Erwin (eds.), *Comparative Primate Biology, Volume 1: Systematics, Evolution and Anatomy,* pp. 73–135. Alan R. Liss, New York.

Grine, F. E. 1988. *Evolutionary History of the "Robust" Australopithecines.* Aldine de Gruyter, New York.

Groves, C. P. 1989. *A Theory of Human and Primate Evolution.* Clarendon Press, Oxford.

Groves, C. P., and Mazák, V. 1975. An approach to the taxonomy of the Hominidae: Gracile Villafranchian hominids of Africa. *Casopis pro Mineralogii Geologii* **20:**225–247.

Habgood, P. J. 1989. An investigation into the usefulness of a cladistic approach to the study of the origin of anatomically modern humans. *Hum. Evol.* **4:**241–252.

Harrison, T. 1981. New finds of small fossil apes from the Miocene locality of Koru in Kenya. *J. Hum. Evol.* **10:**129–137.

Harrison, T. 1982. Small-Bodied Apes from the Miocene of East Africa. Ph.D. Thesis, University of London.

Harrison, T. 1986. New fossil anthropoids from the middle Miocene of East Africa and their bearing on the origin of the Oreopithecidae. *Am. J. Phys. Anthropol.* **71:**265–284.

Harrison, T. 1987a. The phylogenetic relationships of the early catarrhine primates: a review of the current evidence. *J. Hum. Evol.* **16:**41–80.

Harrison, T. 1987b. A reassessment of the phylogenetic relationships of *Oreopithecus bambolii* Gervais. *J. Hum. Evol.* **15:**541–583.

Harrison, T. 1988. A taxonomic revision of the small catarrhine primates from the early Miocene of East Africa. *Folia Primatol.* **50:**59–108.

Harrison, T. 1991. The implications of *Oreopithecus bambolii* for the origins of bipedalism, in: B. Senut and Y. Coppens (eds.), *Origine(s) de la Bipédie Chez les Hominidés, Cahiers de Paléoanthropologie,* pp. 235–244. CNRS, Paris.

Heberer, G. 1963. Uber einen neuen archanthropinen Typus aus der Oldoway-Schlucht. *Z. Morph. Anthropol.* **53:**171–177.

Hennig, W. 1966. *Phylogenetic Systematics.* University of Illinois Press, Urbana.

Hopwood, A. T. 1933. Miocene primates from Kenya. *J. Linn. Soc. London Zool.* **38:**437–464.

Howell, F. C. 1978. Hominidae, in: V. J. Maglio and H. B. S. Cooke (eds.), Evolution of African Mammals, pp. 154–248. Harvard University Press, Cambridge.

Hublin, J. J. 1986. Some comments on the diagnostic features of *Homo erectus. Anthropos* (Brno) **23:**175–187.

Hull, D. L. 1970. Cladism gets sorted out. *Paleobiology* **6:**131–136.

Kennedy, G. E. 1991. On the autapomorphic traits of *Homo erectus J. Hum. Evol.* **20:**375–412.

Kimbel, W. H., White, T. D., and Johanson, D. C. 1988. Implications of KNM-WT 17000 for the evolution of "Robust" *Australopithecus,* in: F. E. Grine (ed.), *Evolutionary History of the "Robust" Australopithecines.* pp. 259–268. Aldine de Gruyter, New York.

Leakey, R. E., and Leakey, M. G. 1986a. A new Miocene hominoid from Kenya. *Nature* **324:**143–146.

Leakey, R. E., and Leakey, M. G. 1986b. A second new Miocene hominoid from Kenya. *Nature* **324:**146–148.

Leakey, R. E., and Leakey, M. G. 1987. A new Miocene small-bodied ape from Kenya. *J. Hum. Evol.* **16:**369–387.

Leakey, R. E., Leakey, M. G., and Behrensmeyer, A. K. 1978. The hominid catalogue, in: M. G. Leakey and R. E. Leakey (eds.), *Koobi Fora Project, Volume 1: The Fossil Hominids and an Introduction to their Context, 1968–1974,* pp. 86–182. Clarendon Press, Oxford.
Le Gros Clark, W. E. 1964. *The Fossil Evidence for Human Evolution.* University of Chicago Press, Chicago.
Le Gros Clark, W. E., and Leakey, L. S. B. 1951. *The Miocene Hominoidea of East Africa.* Fossil Mammals of Africa. No. 1. British Museum (Natural History), London.
Le Gros Clark, W. E., and Thomas, D. P. 1952. *Associated Jaws and Limb Bones of Limnopithecus macinnesi.* Fossil Mammals of Africa. No. 5. British Museum (Natural History), London.
Løvtrup, S. 1977. *The Phylogeny of Vertebrata.* John Wiley and Sons, New York.
Martin, R. D. 1981. Phylogenetic reconstruction versus classification: the case for clear demarcation. *Biologist* **28:**127–132.
Mayr, E. 1968. Theory of biological classification. *Nature* **220:**545–548.
Mayr, E. 1974. Cladistic analysis or cladistic classification? *Z. Zool. Syst. Evolut.-Forsch.* **12:**94–128.
Mayr, E. 1981. Biological classification: Toward a synthesis of opposing methodologies. *Science* **214:**510–516.
Nelson, G. J. 1970. Outline of a theory of comparative biology. *Syst. Zool.* **19:**373–384.
Nelson, G. J. 1971a. "Cladism" as a philosophy of classification. *Syst. Zool.* **20:**373–376.
Nelson, G. J. 1971b. Paraphyly and polyphyly: redefinitions. *Syst. Zool.* **20:**471–472.
Nelson, G. J. 1972. Phylogenetic relationship and classification. *Syst. Zool.* **21:**227–231.
Nelson, G. J. 1974a. Classification as an expression of phylogenetic relationships. *Syst. Zool.* **22:**344–359.
Nelson, G. J. 1974b. Darwin-Hennig classification: A reply to Ernst Mayr. *Syst. Zool.* **23:**452–458.
Olson, T. R. 1978. Hominid phylogenetics and the existence of *Homo* in Member I of the Swartkrans Formation, South Africa. *J. Hum. Evol.* **7:**159–178.
Paterson, T. T. 1940. Geology and early man: II. *Nature* **146:**49–52.
Patterson, C. 1978. Verifiability in systematics. *Syst. Zool.* **27:**218–222.
Patterson, C. 1980. Cladistics. *Biologist* **27:**234–240.
Patterson, C., and Rosen, D. E. 1977. Review of the ichthydectiform and other mesozoic teleost fishes and the theory and practice of classifying fossils. *Bull. Am. Mus. Nat. Hist.* **158:**81–172.
Platnick, N. I. 1977. Cladograms, phylogenetic trees, and hypothesis testing. *Syst. Zool.* **26:**438–442.
Platnick, N. I. 1978. Classifications, historical narratives and hypotheses. *Syst. Zool.* **27:**365–369.
Platnick, N. I. 1979. Philosophy and the transformation of cladistics. *Syst. Zool.* **28:**537–546.
Platnick, N. I., and Gaffney, E. S. 1977. Systematics: a Popperian perspective. *Syst. Zool.* **26:**360–365.
Rafferty, K. L. 1990. The Functional and Phylogenetic Significance of the Carpometacarpal Joint of the Thumb in Anthropoid Primates. MA thesis, New York University.
Rightmire, G. P. 1984. Comparisons of *Homo erectus* from Africa and Southeast Asia. *Cour. Forsch.-Inst. Senckenberg* **69:**83–98.
Rightmire, G. P. 1987. Species recognition and *Homo erectus. J. Hum. Evol.* **15:**823–826.
Rightmire, G. P. 1990. *The Evolution of* Homo erectus. Cambridge University Press, Cambridge.
Rose, M. D. 1983. Miocene hominoid postcranial morphology: Monkey-like, ape-like, neither, or both? in: R. L. Ciochon and R. S. Corruccini (eds.), *New Interpretations of Ape and Human Ancestry,* pp. 405–420. Plenum, New York.
Rose, M. D. 1988. Another look at the anthropoid elbow. *J. Hum. Evol.* **17:**193–224.
Rose, M. D. in press. Locomotor anatomy of Miocene hominoids, in: D. Gebo (ed.), *Postcranial Adaptation in Nonhuman Primates.* Northern Illinois Press, DeKalb.
Rose, M. D. 1992. Kinematics of the trapezium-1st metacarpal joint in extant anthropoids and Miocene hominoids. *J. Hum. Evol.* **22:**255–266.
Rose, M. D., Leakey, M. G., Leakey, R. E. F., and Walker, A. C. 1992. Postcranial specimens of *Simiolus enjiessi* and other primitive catarrhines from the early Miocene of Lake Turkana, Kenya. *J. Hum. Evol.* **22:**171–237.
Rosenberger, A. L. 1977. *Xenothrix* and ceboid phylogeny. *J. Hum. Evol.* **6:**461–481.
Schwartz, J. H. 1986. Primate systematics and a classification of the order, in: D. Swindler and J. Erwin (eds.), *Comparative Primate Biology, Volume 1: Systematics, Evolution and Anatomy,* pp. 1–41. Alan R. Liss, New York.
Schwartz, J. H., and Tattersall, I. 1985. Evolutionary relationships of living lemurs and lorises (Mammalia, Primates) and their potential affinities with European Eocene Adapidae. *Anthropol. Pap. Am. Mus. Nat. Hist.* **60:**1–100.

Schwartz, J. H., and Tattersall, I. 1987. Tarsiers, adapids and the integrity of the Strepsirhini. *J. Hum. Evol.* **16:**23–40.

Senut, B. 1989. *Le Coude des Primates Hominoïdes: Anatomie, Fonction, Taxonomie, Évolution.* Cahiers de Paléoanthropologie. CNRS, Paris.

Skelton, R. R., McHenry, H. M., and Drawhorn, G. R. 1986. Phylogenetic analysis of early hominids. *Curr. Anthropol.* **27:**21–43.

Simons, E. L. 1972. *Primate Evolution.* MacMillan, New York.

Simons, E. L., and Pilbeam, D. R. 1965. Preliminary revision of the Dryopithecinae (Pongidae, Anthropoidea). *Folia Primatol.* **3:**81–152.

Simons, E. L., Pilbeam, D., and Ettel, P. C. 1969. Controversial taxonomy of fossil hominids. *Science* **166:**258–259.

Simpson, G. G. 1945. The principles of classification and a classification of mammals. *Bull. Am. Mus. Nat. Hist.* **85:**1–350.

Simpson, G. G. 1961. *Principles of Animal Taxonomy.* Columbia University Press, New York.

Simpson, G. G. 1975. Recent advances in methods of phylogenetic inference, in: W. P. Luckett and F. S. Szalay (eds.), *Phylogeny of the Primates: A Multidisciplinary Approach,* pp. 1–31. Plenum, New York.

Stringer, C. B. 1984. The definition of *Homo erectus* and the existence of the species in Africa and Europe. *Cour. Forsch.-Inst. Senckenberg* **69:**131–143.

Stringer, C. B. 1986. The credibility of *Homo habilis,* in: B. Wood, L. Martin, and P. Andrews (eds.), *Major Topics in Primate and Human Evolution,* pp. 266–294. Cambridge University Press, Cambridge.

Stringer, C. B. 1987. A numerical cladistic analysis for the genus *Homo. J. Hum. Evol.* **16:**135–146.

Stringer, C. B. in press. "*Homo erectus*" et "*Homo sapiens* archaique": Peut on définir *Homo erectus?* in: J. J. Hublin and A. M. T. Tillier (eds.), *Aux origines d'*Homo sapiens. Presses Universitaires de France, Paris.

Szalay, F. S. 1977. Ancestors, descendants, sister groups and testing of phylogenetic hypotheses. *Syst. Zool.* **26:**12–18.

Szalay, F. S., and Bock, 1991. Evolutionary theory and systematics: relationships between process and patterns. *Z. Zool. Syst. Evolut.-Forsch.* **29:**1–39.

Szalay, F. S., and Delson, E. 1979. *Evolutionary History of the Primates.* Academic Press, New York.

Tattersall, I. 1986. Species recognition in human paleontology. *J. Hum. Evol.* **15:**165–176.

Tattersall, I. 1992. in: W. H. Kimbel and L. B. Martin (eds.), *Species, Species Concepts and Primate Evolution.* Plenum, New York.

Tattersall, I., and Eldredge, N. 1977. Fact, theory and fantasy in human paleontology. *Am. Sci.* **65:**204–211.

Tattersall, I., and Schwartz, J. H. 1991. Phylogeny and nomenclature in the *Lemur*-group of Malagasy strepsirhine primates. *Anthrop. Pap. Am. Mus. Nat. Hist.* **69:**1–18.

Tattersall, I., Delson, E., and Van Couvering, J. 1988. *Encyclopedia of Human Evolution and Prehistory.* Garland, New York.

Tobias, P. V. 1968. Middle and early Upper Pleistocene members of the genus *Homo* in Africa, in: G. Kurth (ed.), *Evolution and Hominization,* pp. 176–194. G. Fischer, Stuttgart.

Tobias, P. V. 1988. Numerous apparently synapomorphic features in *Australopithecus robustus, Australopithecus boisei* und *Homo habilis:* Support for the Skelton-McHenry-Drawhorn hypothesis, in: F. E. Grine (ed.), *Evolutionary History of the "Robust" Australopithecines,* pp. 293–308. Aldine de Gruyter, New York.

Trinkaus, E. 1990. Cladistics and the hominid fossil record. *Am. J. Phys. Anthropol.* **83:**1–11.

Turner, A. 1986. Species, speciation and human evolution. *Hum. Evol.* **1:**419–430.

Turner, A., and Chamberlain, A. 1989. Speciation, morphological change and the status of African *Homo erectus. J. Hum. Evol.* **18:**115–130.

Walker, A., and Pickford, M. 1983. New postcranial fossils of *Proconsul africanus* and *Proconsul nyanzae,* in: R. L. Ciochon and R. S. Corruccini (eds.), *New Interpretations of Ape and Human Ancestry,* pp. 325–351. Plenum, New York.

Walker, A., and Teaford, M. 1989. The hunt for *Proconsul. Sci. Am.* **260:**76–82.

Weidenreich, F. 1936. The manibles of *Sinanthropus pekinensis:* A comparative study. *Palaeont. Sin.* Series D, **7:**1–162.

Weidenreich, F. 1943. The skull of *Sinanthropus pekinensis:* a comparative study of a primitive hominid skull. *Palaeont. Sin.* New Series D, **10:**1–484.

Weinert, H. 1950. Uber die neuen Vor- und Fruhmenschenfunde aus Afrika, Java, China und Frankreich. *Z. Morphol. Anthropol.* **42:**113–148.
White, T. D., Johanson, D. C., and Kimbel, W. H. 1981. *Australopithecus africanus:* its phyletic position reconsidered. *S. Afr. J. Sci.* **77:**445–470.
Wood, B. A. 1984. The origin of *Homo erectus. Cour. Forsch.-Inst. Senckenberg* **69:**99–111.
Wood, B. A., and Chamberlain, A. T. 1986. *Australopithecus:* grade or clade?, in: B. Wood, L. Martin, and P. Andrews (eds.), *Major Topics in Primate and Human Evolution,* pp. 220–248. Cambridge University Press, Cambridge.

Species Discrimination in *Proconsul* from Rusinga and Mfangano Islands, Kenya

15

M. F. TEAFORD, A. WALKER, AND G. S. MUGAISI

Introduction

Until the 1930s, the Kenyan islands of Rusinga and Mfangano were unknown to the paleontological community. In 1931–1932, L. S. B. Leakey and D. G. MacInnes discovered the first fossils at these sites, including a hominoid that eventually came to be known as *Proconsul* (Hopwood, 1933; Leakey, 1943; MacInnes, 1943). These discoveries prompted further expeditions and additional discoveries that ultimately led to a series of papers (e.g., LeGros Clark and Leakey, 1951; LeGros Clark, 1952; Napier and Davis, 1959) that firmly established the genus *Proconsul* as one of the best known fossil hominoids. Sixty years after the initial discoveries, the amount of fossil material from Rusinga and Mfangano islands has grown dramatically, yet investigators are still in disagreement over the number of species of *Proconsul* represented at these sites.

From one perspective, there is only one species of *Proconsul* on Rusinga and Mfangano, and any variation in size among specimens is thought to reflect

M. F. TEAFORD AND A. WALKER • Department of Cell Biology and Anatomy, The Johns Hopkins University School of Medicine, Baltimore, Maryland 21205 G. S. MUGAISI • National Museums of Kenya, Nairobi, Kenya.

Species, Species Concepts, and Primate Evolution, edited by William H. Kimbel and Lawrence B. Martin. Plenum Press, New York, 1993.

sexual dimorphism. This interpretation was put forth by MacInnes (1943), who referred a variety of *Proconsul* material to one species, *P. africanus.* Shortly afterwards, this viewpoint fell into disfavor as Le Gros Clark and Leakey (1950, 1951) suggested that additional material warranted the recognition of two species of *Proconsul* from Rusinga and Mfangano: a smaller, *Colobus*-sized species, *Proconsul africanus,* and a larger, chimpanzee-sized species, *Proconsul nyanzae.* For the next 20 years, most publications on the topic (e.g., Simons and Pilbeam, 1965; Pilbeam, 1969) merely sanctioned the two-species hypothesis. As a result, when Greenfield (1972) noted that *P. africanus* seemed to be represented only by females, few people appeared to take notice, and most subsequent publications (e.g., Andrews, 1978; Bosler, 1981) retained the two-species perspective.

More recently, Kelley (1986) and Pickford (1986) have resurrected the single-species hypothesis, claiming that the canines of *Proconsul* (from Rusinga and Mfangano) can be easily categorized into male and female morphotypes, and the variation in postcanine tooth size can be encompassed in one dimorphic species rather than two less-dimorphic species. Both Kelley and Pickford admit that this single species would be more variable in dental dimensions than any living catarrhine (Fig. 1), but as Kelley suggests, perhaps that ought to be expected given the "time-accumulated nature" of most fossil samples (Kelley, 1986, p. 484). Taking this a step further, Kelley (1986) has suggested that marked sexual dimorphism, complete with all its biological implications, might be characteristic of *Proconsul.* This would seem to be twisting the facts to fit the situation, however, for if the excessive variation in the *Proconsul* sample were due to time averaging, then any measure of sexual dimorphism computed for that sample would not be analogous to similar measures computed for modern biological species. Instead, Kelley's high measures of sexual dimorphism would be nothing more than artifacts of time averaging.

While it is indeed tempting to use a single species hypothesis to sidestep such problems as the assignment of isolated teeth to species within the Rusinga/Mfangano *Proconsul* sample, two key factors argue against it. First, as men-

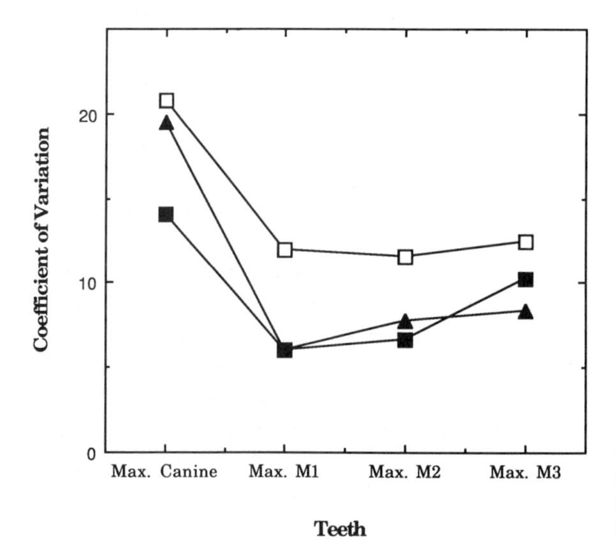

Fig. 1. CVs for bucco-lingual breadths of maxillary teeth in *Proconsul* (□), *Pan* (■), and *Gorilla* (▲). *Proconsul* measurements from Andrews (1978).

tioned above, it would leave us with a paleospecies that is more variable than any species of modern catarrhine. We would be choosing to ignore all of our modern primate analogues by using paleospecies to establish the limits of variation of paleospecies. Without a living reference group against which to calibrate our paleontological interpretations, any degree of variation might be deemed acceptable if the situation warranted it, and as Martin (1991) has noted, none of these alternative interpretations would be falsifiable.

The second problem with the *Proconsul* single-species hypothesis is that it is based solely on analyses of teeth. As Greenfield (1972), Andrews (1978), and Bosler (1981) have all noted, the distribution of male and female *Proconsul* canines within the Rusinga–Mfangano samples is puzzling. The postcanine teeth are all morphologically similar, and the ranges of measurements for most teeth show just enough overlap (between the larger and smaller *Proconsul* individuals) to make the identification of isolated specimens difficult (Pickford, 1986). When one also considers the fact that the sample of palates and mandibles from Rusinga and Mfangano is still surprisingly limited and that much of the *Proconsul* dental sample is composed of isolated teeth, another possibility emerges: Perhaps teeth are relatively poor indicators of taxonomic affinities in *Proconsul* (Teaford *et al.,* 1988). From this perspective, the historical "swapping" of dental specimens between taxa may not be evidence in support of the single-species hypothesis (contra. Pickford, 1986), but instead evidence that we should look elsewhere in the *Proconsul* sample for taxonomically relevant distinctions. The obvious place to look would be in the postcranium.

Despite the growing amount of work done on modern primate postcrania, we still know surprisingly little about patterns of postcranial variation, especially between taxa, and how postcranial variation might compare with craniodental variation. Most recent work (e.g., Wood, 1976; Oxnard, 1983; Leutenegger and Larson, 1985; McHenry, 1986) has focused on sexual dimorphism in the postcranium. Thus, the questions asked and the methods used have been different from those used in defining species units in paleontological interpretations. As might be expected, the bulk of that work (e.g., Gingerich, 1974; Gingerich and Schoeninger, 1979; Kay, 1982a,b; Kay and Simons, 1983; Martin, 1983; Martin and Andrews, 1984; Kelley, 1986; Kelley and Xu, 1991) has involved teeth, the most common elements in the primate fossil record. Still, there seems to be an underlying assumption in some of these analyses that teeth are developmentally more stable than bone; for example, "once formed, . . . teeth do not grow during the remainder of an animal's life" (Gingerich, 1974, p. 895). Due to the vast potential of bone for remodeling, there is a tendency to think of postcranial skeletal elements as more prone to morphological variation than are teeth. This is simply not the case. Some postcranial elements show more variation than others. But so do some teeth. In fact, in studies of a wide range of mammals, Yablokov has noted that one of the most striking aspects of the postcranial skeleton is its "extremely low amount of variability" (1974, p. 43). The most important point to keep in mind in analyses of postcranial variation is that comparisons should be confined to functional complexes, organ systems, etc. In other words, "the variability for organs of different systems, and different measurable parameters within a system of organs, differ considerably from each other" (Yablokov, 1974, p. 75; see Tague, 1991, for examples from the primate pelvis). Taken to an extreme, comparisons of dental variation in one species with calcaneal variation in another would be fruitless. However, comparisons of calcaneal variation among taxa might lead to some interesting insights.

Recent expeditions to Rusinga and Mfangano (Beard *et al.*, 1986; Walker *et al.*, 1986; Teaford *et al.*, 1988; Walker and Teaford, 1988, 1989; Ward, 1991; Ward *et al.*, 1991) have more than doubled our samples of most postcranial elements of *Proconsul.* Initial analyses of that material have reaffirmed that the larger form of *Proconsul* from Rusinga and Mfangano is primarily an "upscaled" version of the smaller (Walker and Pickford, 1983; Beard *et al.*, 1986; Pickford, 1986). However, estimates of body weight based on the new material have yielded body weight differences of 3.4 : 1 between the larger and smaller *Proconsul* (Ruff *et al.*, 1989). Kelley (this volume) has claimed that these estimates are too high, because they do not include a range of body weights for the smaller species. From Kelley's perspective, the actual figure might be more "conservatively" estimated at 2.6 : 1. Unfortunately, we have no fossils to bracket the range of body weights exhibited by the smaller *Proconsul* species. Thus, as Kelley admits, his arguments are speculative. Even if his hypothesized ratios based on unknown fossils prove to be true, the fact remains that, no matter which primate species he chooses to examine (even *Mandrillus,* for which body size data are sketchy at best), he cannot come up with a ratio as large as his modified ratio from Ruff *et al.* (1989). Moreover, if we only have three genera of modern great apes to compare with the fossils, and if two of those modern genera (*Gorilla* and *Pongo*) are much larger than the fossil hominoids under consideration, then how can Kelley possibly eliminate the third genus (*Pan*) from consideration because it is "highly unusual"? Given the estimates of body size for *Proconsul,* it would seem far more reasonable to use the chimpanzee, or even the gibbon, as the model of body size dimorphism in *Proconsul,* rather than focus on the extreme values documented for the gorilla, orang, and baboon. If by chance the true body-size differences between large and small *Proconsul* turn out to be closer to those presented by Ruff *et al.*, then they could not be accommodated within a single species of modern land mammal. So the question remains, if the differences in body weight are so large, why are the teeth so similar?

One possibility is that the specimens traditionally referred to *P. africanus* simply have relatively large teeth, while those referred to *P. nyanzae* have relatively small teeth. This might blur the size-related dental distinctions between these taxa through a narrowing of the ranges of standard dental measurements and thus decrease values for traditional measures of variation (e.g., coefficient of variation).

Unfortunately, it is difficult to estimate the relative size of the dentition of *Proconsul.* As is so often the case at Miocene sites, the association between cranial and postcranial remains at most *Proconsul* sites is fairly poor, to say the least. Even at the newly described Kaswanga Primate Site (Walker and Teaford, 1988), surprisingly few cranial remains have been recovered. We are left with nothing more than ratios of tooth size to jaw size as indicators of relative tooth size. Still, available evidence supports the idea that the two forms of *Proconsul* on Rusinga and Mfangano differed in relative tooth size. For instance, five large specimens and five small specimens are complete enough to allow the measurement of the

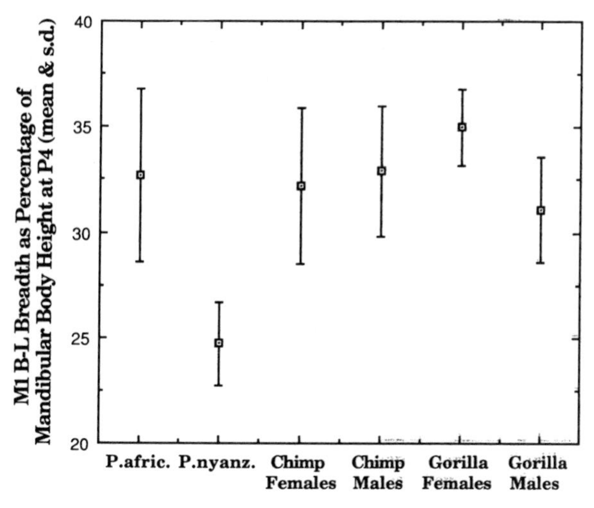

Fig. 2. Bucco-lingual breadth of M_1 as a percentage of mandibular body height at P4 for samples of *Proconsul* and modern chimp and gorilla. Measurements of chimp and gorilla from Pilbeam (1969).

buccolingual breadth of M_1 and mandibular body height at P_4. If the former is divided by the latter, results suggest that specimens traditionally referred to *P. nyanzae* have relatively smaller M_1s than do specimens traditionally referred to *P. africanus* ($p < .01$; Mann-Whitney test) (see Fig. 2). It should be emphasized, however, that these results are merely suggestive, as the small sample sizes and poor correlations between variables for *P. nyanzae* preclude other statistical analyses (e.g., regression).

With all of this in mind, the purpose of this paper is to begin to examine variation in the postcranium of *Proconsul* to see if patterns of postcranial variation might be similar to or different from those exhibited by the dentition, and also to see how patterns of postcranial variation in *Proconsul* compare with those for modern primates. In the process, we hope to shed new light on the question of the number of species of *Proconsul* represented at the Miocene sites on Rusinga and Mfangano.

Materials and Methods

Materials

While the sample of *Proconsul* postcranial material has grown dramatically in the last 10 years, we still have few complete specimens of certain postcranial elements, most notably long bones of the forelimb. One area in which sample sizes are beginning to approach respectability, however, is the ankle. The *Proconsul* specimens used in this study are listed in Table 1. Due to the fragmentary nature of some specimens, not all measurements could be taken for each specimen. Thus, seven or eight *Proconsul* calcanei were used in these analyses, as were 10 or 11 *Proconsul* tali. Modern comparative samples included 40 gorillas (20

Table 1. *Proconsul* Specimens Used in This Study

Specimens used by Harrison and/or Langdon	New specimens from Rusinga and Mfangano
Tali (N = 16)	
KNM-RU 1743	KNM-RU 5940
KNM-RU 1744	KNM-RU 5945
KNM-RU 1745	KNM-RU 14230
KNM-RU 1896	KNM-MW 13142
KNM-RU 2036	KNM-MW 17380
KNM-RU 3105	KPS T10
KNM-RU 4347	KPS T22
KNM-RU 5872	KPS T46
Calcanei (N = 8)	
KNM-RU 1755	KNM-RU 18387
KNM-RU 2036	KNM-MW 13142
KNM-RU 5872	KPS T13
	KPS T19 & T20
	KPS T48

males and 20 females) and 38 chimpanzees (19 males and 19 females) from the Cleveland Museum of Natural History and 49 baboons (26 males and 23 females) from a new collection of *Papio cynocephalus* at the National Museums of Kenya.

Measurements

Three factors entered into the selection of measurements for this study. First, could the measurements be taken on most of the fossil specimens? For example, maximum length of the calcaneus could not be used because most fossil calcanei are missing their heel processes. Second, to facilitate comparisons, had the measurements been used in previous analyses of Miocene material? Third, could the measurements provide indications of the size of the bone in question and also at least one of its articular facets?

With these factors in mind, the following measurements were recorded:

Talus

1. breadth of talar body, excluding lateral malleolar facet (Harrison, 1982, measurement #5)
2. maximum height of talus (Harrison, 1982, measurement #16).
3. width of posterior subtalar facet (Harrison, 1982, measurement #13)

Calcaneus

1. maximum breadth of calcaneus (Rose, 1986, measurement #2)
2. maximum depth of calcaneus (Rose, 1986, measurement #3)
3. width of posterior subtalar facet (Rose, 1986, measurement #9)
4. width of sustentacular subtalar facet (Rose, 1986, measurement #11)

Data Analysis

Investigators have generally taken one of two approaches to analyses of variation in fossil hominoid assemblages. Either they have split their sample into males and females, and looked at the degree of dimorphism within the sample, or they have lumped all specimens together and looked at the degree of variation within their samples (e.g., Kay, 1982a,b; Kay and Simons, 1983; Kelley and Xu, 1991; Martin, 1983; Martin and Andrews, 1984; Kelley, 1986; McHenry, 1986; Pan *et al.,* 1989). In all cases, the standards of comparison were samples of modern primates. For example, if the fossils showed less dimorphism than modern primates, then the fossil sample was thought to have included only one species (e.g., McHenry, 1986; Pan *et al.,* 1989). If the fossils showed greater variation than that in modern primates, then the fossil sample was thought to have included more than one species (Kay, 1982a,b; Kay and Simons, 1983). Kelley (1986), Kelley and Xu (1991), and Pickford (1986) have recently challenged the latter approach by suggesting that some paleospecies might be more variable than modern biological species.

Other discussions and debates surrounding this work have focused on methods of data analysis, specifically, methods for documenting variation. Some investigators (e.g., Kay 1982a,b; Kay and Simons, 1983) have advocated the use of the coefficient of variation (CV), and others (e.g., Martin, 1983; Martin and Andrews, 1984; Pan *et al.,* 1989) have advocated the use of additional, range-based measures, that is, using the range as a percentage of the mean and the maximum/minimum index. Recent work by Cope (1989, this volume) and Plavcan (1989, this volume) has helped to clarify the issues in this debate by showing that the single-species hypothesis should always be treated as the null hypothesis. As a result, it can never be proven, only disproven. This still leaves the coefficient of variation as a useful tool that can frequently detect multiple taxa from pooled samples when the range-based statistics cannot (Cope, 1989, this volume; Plavcan, 1989, this volume). It also shows, however, that Simpson's criticisms of range-based statistics (Simpson, 1947; Simpson *et al.,* 1960) must be kept in proper perspective. Yes, range-based statistics are unlikely to falsify single-species interpretations, especially when sample sizes are small. As the range-based statistics can be calculated for extremely small samples, however, and as they do tend to underestimate the amount of variation in small samples (Martin, 1983), they still have their advantages over CVs in certain situations. As Cope has noted, "if [one of these range-based estimates] in a small sample does exceed the single species maximum, it strongly suggests that more than one species is present, since both methods are highly biased against such an outcome" (1989, p. 190). For these reasons, all three measures of variation (the coefficient of variation, the range as a percentage of the mean, and the maximum/minimum index) were used in this study. For all samples of fossil material, CVs were *not* corrected for small sample sizes, as suggested by Simpson *et al.* (1960). This was to avoid further increases in CV values that might already be unusually high due to the small sample sizes (Cope, this volume).

Results

For each measurement of the calcaneus and each measurement of the talus, the *Proconsul* sample from Rusinga and Mfangano showed greater variation than did the gorilla, chimpanzee, or baboon samples (see Table 2 and Figs. 3–5). The coefficients of variation for the *Proconsul* sample are generally two times as large as those documented for the modern primates. The range-based measures of variation consistently show comparable differences, although these are not as dramatic as those provided by the coefficients of variation. This is probably to be expected given the disparity in sample size between the fossil and comparative samples.

Discussion

Overall, the results for the postcranial material are encouraging and intriguing, especially as compared with measures of dental variation (Kelley, 1986) (compare Figs. 1, 4 and 5). For every variable measured, the *Proconsul* sample showed greater variation than the gorilla sample, which generally showed greater variation than the baboon sample, which usually showed greater variation than the chimpanzee sample. The differences between gorilla and baboon, on the one hand, and chimpanzee, on the other, probably reflect the already well-documented differences in sexual dimorphism between these species (Schultz, 1969; Wood, 1976; Senut, 1986).

The differences between *Proconsul* and the modern primates are more problematic. Those based on coefficients of variation are generally far more dramatic than those derived from the range-based statistics. Given the small sample sizes for the *Proconsul* material (n = 7–11), this is probably just further evidence that the range-based statistics underestimate variation in small samples (Simpson, 1947, Simpson *et al.*, 1960; Cope, 1989; Plavcan, 1989). Still, the only measurements for which the range-based statistics for *Gorilla* and *Proconsul* approach each other are those for the height of the talus and calcaneus. This might well be due to unusual variation in the *Gorilla* sample, as the tali and calcanei of some male gorillas are surprisingly flat, whereas those of other male gorillas are relatively high.

The fact remains that, without exception, the range-based statistics for *Proconsul* exceed those for *Gorilla, Papio,* and *Pan.* In the face of such small *Proconsul* samples, this is strong evidence in support of the idea that more than one species of *Proconsul* is represented in the material collected from Rusinga and Mfangano.

Results based on the coefficient of variation provide resounding support for this idea. As noted earlier, CVs will vary from anatomical element to anatomical element and from measurement to measurement within anatomical elements. Still, the CVs for the *Proconsul* material are far larger than those obtained in most investigations of mammalian postcrania (e.g., Tague, 1991; Yablokov, 1974), where CVs of 3–15 are more the norm for single species. The CVs for the

Table 2. Measures of Variability for Ankle Measurements

	Coefficient of variation	Range as a percentage of mean	Maximum divided by minimum
Talus			
Breadth			
Proconsul (*n* = 11)	25.0	65.9	1.85
Gorilla (*n* = 40)	10.4	38.4	1.48
Pan (*n* = 38)	7.4	28.5	1.32
Papio (*n* = 49)	10.5	35.2	1.43
P. africanus (*n* = 8)	11.4	37.8	1.43
P. nyanzae (*n* = 3)	10.7	19.1	1.22
Height			
Proconsul (*n* = 11)	24.6	51.5	1.69
Gorilla (*n* = 40)	12.3	48.3	1.62
Pan (*n* = 38)	7.3	27.2	1.31
Papio (*n* = 49)	8.7	36.5	1.43
P. africanus (*n* = 6)	2.5	6.6	1.07
P. nyanzae (*n* = 5)	2.5	6.4	1.07
Width of post. subtalar facet			
Proconsul (*n* = 10)	26.1	79.2	2.21
Gorilla (*n* = 40)	14.2	61.3	1.80
Pan (*n* = 38)	8.6	41.2	1.49
Papio (*n* = 44)	10.0	42.9	1.54
P. africanus (*n* = 5)	12.0	30.9	1.37
P. nyanzae (*n* = 5)	16.8	40.8	1.51
Calcaneus			
Width			
Proconsul (*n* = 8)	22.3	52.2	1.65
Gorilla (*n* = 40)	11.8	40.6	1.52
Pan (*n* = 36)	8.0	30.6	1.36
Papio (*n* = 49)	9.5	35.5	1.44
P. africanus (*n* = 6)	8.7	21.6	1.24
P. nyanzae (*n* = 2)		Sample too small	
Height			
Proconsul (*n* = 7)	23.6	57.6	1.75
Gorilla (*n* = 40)	13.3	52.5	1.69
Pan (*n* = 37)	6.7	25.9	1.30
Papio (*n* = 49)	11.6	49.9	1.67
P. africanus (*n* = 5)	8.4	20.7	1.23
P. nyanzae (*n* = 2)		Sample too small	
Width of post. subtalar facet			
Proconsul (*n* = 8)	25.2	67.9	1.87
Gorilla (*n* = 40)	12.2	48.3	1.61
Pan (*n* = 36)	9.5	35.5	1.42
Papio (*n* = 49)	10.2	48.5	1.61
P. africanus (*n* = 6)	19.3	44.8	1.52
P. nyanzae (*n* = 2)		Sample too small	
Width of susten. facet			
Proconsul (*n* = 8)	27.5	71.7	1.98
Gorilla (*n* = 40)	13.2	49.2	1.66
Pan (*n* = 37)	11.4	42.6	1.54
Papio (*n* = 49)	11.0	42.2	1.51
P. africanus (*n* = 6)	16.3	40.9	1.47
P. nyanzae (*n* = 2)		Sample too small	

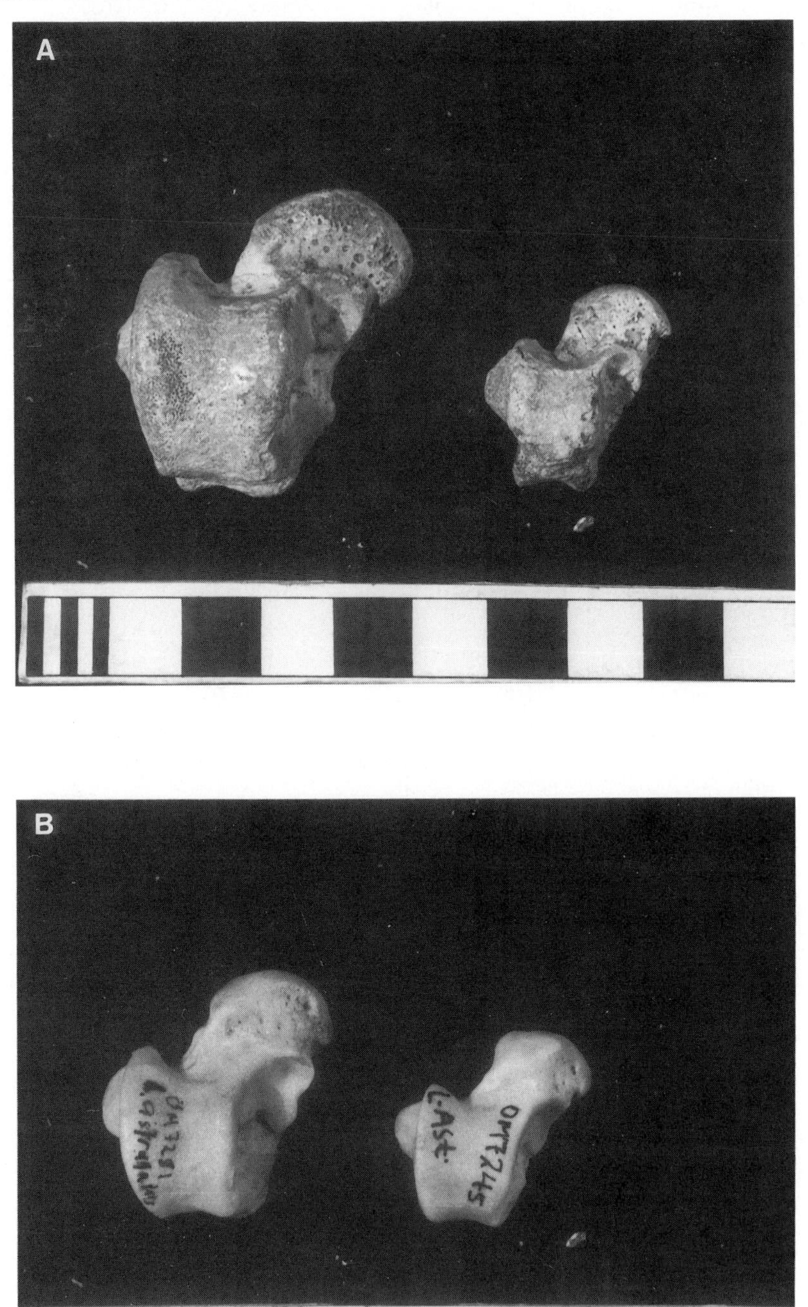

Fig. 3. Comparison of large and small tali in *Proconsul* and *Papio cynocephalus*. (A) *Proconsul* specimens from Rusinga traditionally assigned to *P. nyanzae* (KNM-RU 1743 on left) and *P. africanus* (KNM-RU 1744 on right). (B) Specimens of *Papio cynocephalus;* male (OM 7281) on left and female (OM 7245) on right.

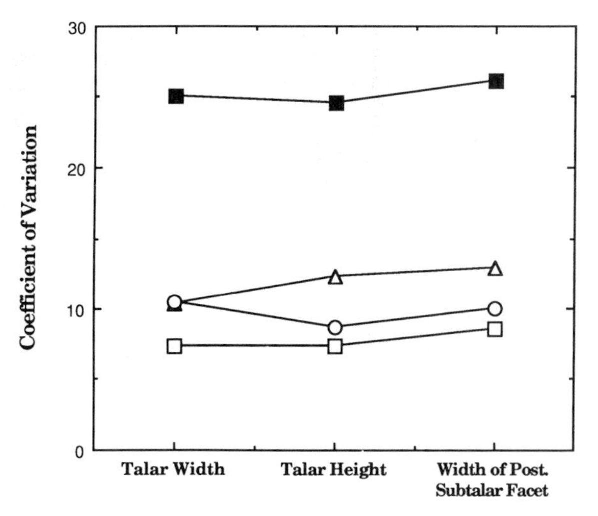

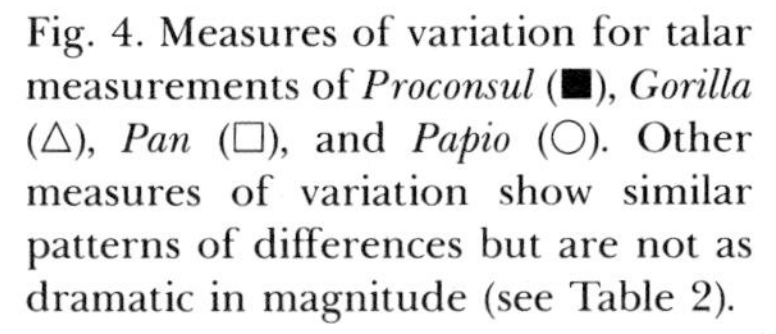
Fig. 4. Measures of variation for talar measurements of *Proconsul* (■), *Gorilla* (△), *Pan* (□), and *Papio* (○). Other measures of variation show similar patterns of differences but are not as dramatic in magnitude (see Table 2).

Proconsul postcranial material are also far larger than those obtained for most *Proconsul* teeth, reaffirming that the postcanine dentition, in particular, is probably of little use in sorting through species of *Proconsul* (Pickford, 1986; Teaford *et al.*, 1988).

Taken together, the coefficients of variation and the range-based statistics paint such a convincing picture of more than one species of *Proconsul* on Rusinga and Mfangano that one cannot help but wonder, how could there be anything *less* than two species of *Proconsul* at these sites?

There are four potential problem areas with these analyses. First, many of the *Proconsul* ankle bones used in this study (15 out of 22 to be exact) were found unassociated with teeth. As a result, other than initial morphological assessments in which the material was identified as primate, the assignment of specimens to taxa has been based primarily on differences in size. This raises the possibility

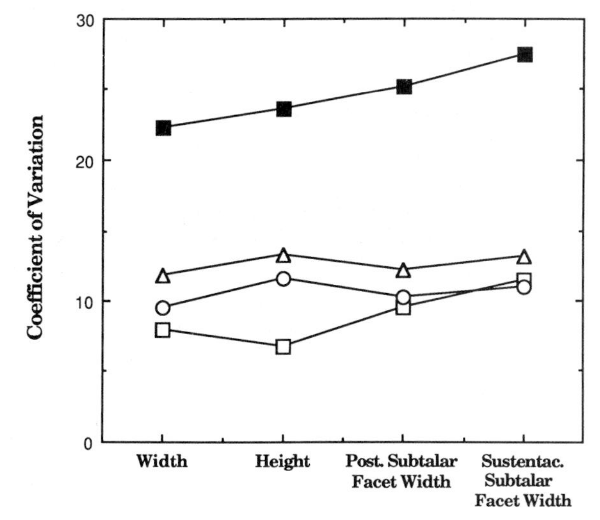

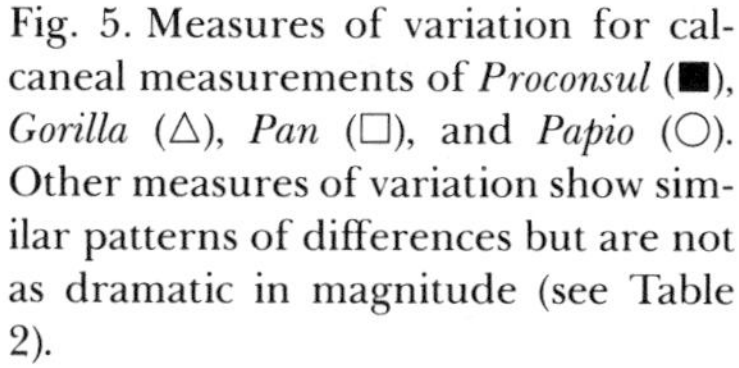
Fig. 5. Measures of variation for calcaneal measurements of *Proconsul* (■), *Gorilla* (△), *Pan* (□), and *Papio* (○). Other measures of variation show similar patterns of differences but are not as dramatic in magnitude (see Table 2).

that specimens from genera other than *Proconsul* have been included in the sample. Based on our current knowledge of the distribution of Miocene hominoids in western Kenya, the most likely candidates to be included in this sample are specimens of *Dendropithecus* (Andrews and Simons, 1977) and *Nyanzapithecus* (Harrison, 1986). *Dendropithecus* is smaller than the smaller species of *Proconsul* from Rusinga and Mfangano.

Based on size and morphological differences (e.g., shape and orientation of the medial malleolar facet), two new, isolated, talar specimens were thought to be attributable to *Dendropithecus* and thus were not included in the present analyses. *Nyanzapithecus* from Rusinga is unknown postcranially. Based on its dental measurements (Harrison, 1986), it is also smaller than the smaller species of *Proconsul* from Rusinga and Mfangano. Again, the elimination of the two smaller new specimens from our sample would seem to counter this potential problem. It would also leave the range-based portion of these analyses relatively conservative by effectively decreasing the variation of measurements. Coefficients of variation, however, might be slightly increased due to the smaller sample size (Martin, 1983).

A second potential problem with the present study concerns the matter of sample size. Previous work (Martin, 1983; Cope, 1989) has shown that the different measures of variation used in this study are affected differently by variations in sample size. Thus, while range-based statistics generally underestimate variation in small samples ($n < 10$), CVs from small samples can overestimate variation (Martin, 1983; Cope, 1989, this volume). In the present study, sample sizes for *Proconsul* are much smaller than those for the modern comparative samples of *Pan, Papio,* and *Gorilla.*

Could the large CVs for the small *Proconsul* samples merely be statistical artifacts? To answer this question, 100 random samples were drawn from each of the modern comparative samples used in this study (i.e., 100 from *Pan,* 100 from *Papio,* and 100 from *Gorilla*). Each random sample had eight individuals, and the sex of each individual was disregarded in the random selection process. Thus some samples had a disproportionate number of males and some had a disproportionate number of females. Each measure of variation used in this study was computed for the smaller random samples.

As can be seen from the ranges of values in Table 3, some of the measures of variation for the random samples were certainly larger than those obtained in the original analyses. Some, however, were also smaller. This is reflected in the fact that the mean coefficients of variation for these random samples are quite close to those obtained for the larger original samples (compare Tables 2 and 3). By contrast, the mean values for the range-based statistics from the random samples are noticeably lower than the values obtained for the larger original samples. This is further evidence that range-based measures of variation change in response to changes in sample size (Simpson, 1947; Simpson *et al.,* 1960; Martin, 1983; Cope, 1989).

The key point to be derived from the analysis of these random samples is that none of the resultant measures of variation equaled the values originally obtained for *Proconsul* (see Table 3). In light of these results, it is extremely unlikely that the high measures of variation for the *Proconsul* ankle bones are statistical artifacts.

Table 3. Maximum, Minimum, Mean, and Standard Deviation of Measures of Variability for Ankle Measurements: Computed from 100 Random Samples of *Gorilla*, *Pan*, and *Papio*

	Coefficient of variation	Range as a percentage of mean	Maximum divided by minimum
Talus			
Breadth			
Proconsul (n = 11)	25.0	65.9	1.85
Gorilla (n = 100 random samples, 8 individuals in each)			
max.	15.2	39.7	1.48
min.	4.8	14.9	1.16
mean	10.1	27.9	1.33
s.d.	2.2	6.0	.08
Pan (n = 100 random samples, 8 individuals in each)			
max.	10.5	28.1	1.32
min.	3.6	10.7	1.11
mean	7.1	20.0	1.22
s.d.	1.6	4.5	.06
Papio (n = 100 random samples, 8 individuals in each)			
max.	13.8	36.5	1.43
min.	4.6	12.7	1.14
mean	10.3	27.8	1.32
s.d.	1.8	4.8	.06
Height			
Proconsul (n = 11)	24.6	51.5	1.69
Gorilla (n = 100 random samples, 8 individuals in each)			
max.	17.9	50.6	1.62
min.	5.5	16.4	1.18
mean	12.1	35.0	1.42
s.d.	2.8	8.3	.12
Pan (n = 100 random samples, 8 individuals in each)			
max.	11.1	28.0	1.31
min.	2.1	6.9	1.07
mean	7.0	19.6	1.21
s.d.	1.8	5.2	.06
Papio (n = 100 random samples, 8 individuals in each)			
max.	11.7	36.4	1.43
min.	3.1	8.0	1.08
mean	8.4	24.2	1.27
s.d.	1.9	5.7	.07
Width of post. subtalar facet			
Proconsul (n = 10)	26.1	79.2	2.21
Gorilla (n = 100 random samples, 8 individuals in each)			
max.	20.3	65.2	1.80
min.	6.7	18.8	1.20
mean	13.9	39.8	1.49
s.d.	3.2	10.8	.15
Pan (n = 100 random samples, 8 individuals each)			
max.	11.6	31.9	1.37
min.	4.0	9.9	1.10
mean	7.4	21.7	1.24
s.d.	1.7	5.2	.07

(*continued*)

Table 3. (*Continued*)

	Coefficient of variation	Range as a percentage of mean	Maximum divided by minimum
Papio (n = 100 random samples, 8 individuals in each)			
max.	15.9	45.3	1.54
min.	5.5	12.3	1.13
mean	10.0	28.0	1.33
s.d.	2.4	7.6	.10
	Calcaneus		
Width			
Proconsul (n = 8)	22.3	52.2	1.65
Gorilla (n = 100 random samples, 8 individuals in each)			
max.	15.8	41.3	1.52
min.	5.8	18.0	1.20
mean	11.7	31.1	1.37
s.d.	2.0	5.4	.08
Pan (n = 100 random samples, 8 individuals each)			
max.	13.0	30.9	1.36
min.	3.2	8.2	1.08
mean	7.7	21.8	1.25
s.d.	1.9	5.4	.07
Papio (n = 100 random samples, 8 individuals in each)			
max.	12.7	35.9	1.44
min.	6.2	15.5	1.17
mean	9.5	25.4	1.30
s.d.	1.5	4.5	.06
Height			
Proconsul (n = 7)	23.6	57.6	1.75
Gorilla (n = 100 random samples, 8 individuals in each)			
max.	19.0	55.9	1.69
min.	7.1	21.8	1.22
mean	13.1	37.1	1.45
s.d.	2.6	8.0	.11
Pan (n = 100 random samples, 8 individuals each)			
max.	9.1	26.9	1.30
min.	3.8	10.8	1.11
mean	6.7	19.0	1.21
s.d.	1.3	3.8	.05
Papio (n = 100 random samples, 8 individuals in each)			
max.	18.2	50.6	1.67
min.	7.4	21.6	1.24
mean	11.6	32.4	1.39
s.d.	2.2	6.5	.10
Width of post. subtalar facet			
Proconsul (n = 8)	25.2	67.9	1.87
Gorilla (n = 100 random samples, 8 individuals in each)			
max.	16.7	59.9	1.61
min.	6.9	19.4	1.21
mean	11.8	33.6	1.40
s.d.	2.3	8.0	.11

(*continued*)

Table 3. (*Continued*)

	Coefficient of variation	Range as a percentage of mean	Maximum divided by minimum
Pan (*n* = 100 random samples, 8 individuals each)			
max.	14.7	35.8	1.42
min.	5.0	14.9	1.16
mean	9.3	26.2	1.30
s.d.	2.0	5.2	.07
Papio (*n* = 100 random samples, 8 individuals in each)			
max.	15.3	47.9	1.61
min.	5.2	13.1	1.14
mean	9.9	27.8	1.33
s.d.	2.2	7.4	.10
Width of susten. facet			
Proconsul (*n* = 8)	27.5	71.7	1.98
Gorilla (*n* = 100 random samples, 8 individuals in each)			
max.	18.1	49.7	1.66
min.	7.3	19.7	1.22
mean	12.9	36.7	1.45
s.d.	2.7	8.3	.13
Pan (*n* = 100 random samples, 8 individuals each)			
max.	15.1	42.9	1.54
min.	5.2	13.7	1.15
mean	10.9	29.9	1.36
s.d.	2.0	5.7	.08
Papio (*n* = 100 random samples, 8 individuals in each)			
max.	16.1	43.3	1.51
min.	3.5	10.7	1.11
mean	10.8	30.6	1.36
s.d.	2.3	6.4	.08

Measures for *Proconsul* sample included for comparison.

A third potential problem with the present study is that it is based solely on analyses of the talus and calcaneus. Are these results representative of what might be found using other postcrania? For instance, the Miocene hominoids might have been part of an adaptive radiation that involved adaptations to new substrates that, in turn, led to strong selective pressures affecting the feet. In that case, would analyses of other parts of the skeleton (such as the forelimb) yield similar results? This is one of the questions we will be trying to answer as we work through analyses of other *Proconsul* skeletal material.

The fourth potential problem with this study concerns the "time-averaged nature of fossil assemblages" (Kelley, 1986, p. 484). How much of the variation in the Rusinga and Mfangano sample can be attributed to evolutionary changes that occurred during the deposition of the deposits at these sites? As Kelley (1986) has correctly intimated, fossil samples can rarely be viewed as populations in the strict ecological sense. Still, the fossiliferous deposits of the Rusinga Group are about as close as we can get to a geological instant in the Miocene. As Drake *et al.* (1988) have noted, the best dates for the Hiwegi Formation consistently fall at 17.9 Ma ± 0.1. Most of the deposits are mudstones and siltstones that were laid down in low-energy depositional environments. However, this does not necessarily mean the deposits took a long time to accumulate. For example, over 4 m of deposits were laid down at the R114 site around a large tree (Walker *et al.*, 1986; Walker and Teaford, 1989). When all of this is coupled with the fact that there are no discernible evolutionary trends in the associated fauna at the *Proconsul* sites, then the chances of significant evolutionary changes within the *Proconsul* sample would appear to be slim indeed.

Taking this a step further, if time is but a minor complicating factor in the interpretation of the *Proconsul* material, then why focus on extreme values of variation in our comparisons with modern primate material? If *Proconsul* was about the size of a modern pygmy chimpanzee, then why should we expect it to have the sexual dimorphism of a gorilla? The *Proconsul* ankle measurements show that, if the sample is split into the traditional small and large species groupings, the resultant measures of variation are, for the most part, similar to those for the modern primates (see Table 2 and Fig. 6). The only exception involves measures of talar height, which show unusually low measures of variation for each *Proconsul* group. Given the results from the other measurements and the rest of this study, and given the small sample sizes for these *Proconsul* groupings, the low measures of variation are probably best viewed as artifacts of small sample size.

From this perspective, *Proconsul* would still exhibit at least as much sexual dimorphism as the chimpanzee (Ruff *et al.*, 1989). Of course, if the two species of *Proconsul* from Rusinga and Mfangano islands each had the sexual dimorphism of a chimpanzee, then the question remains, why are there only two groups of canines for the entire *Proconsul* sample at these sites (Bosler, 1981; Kelley, 1986; Pickford, 1986)? Perhaps the combination of a megadont dentition in the smaller species and a microdont dentition in the larger species has left us with a fossil sample representing two dimorphic species whose tooth sizes overlap (Martin and Andrews, this volume) despite the pronounced size differences in the postcrania.

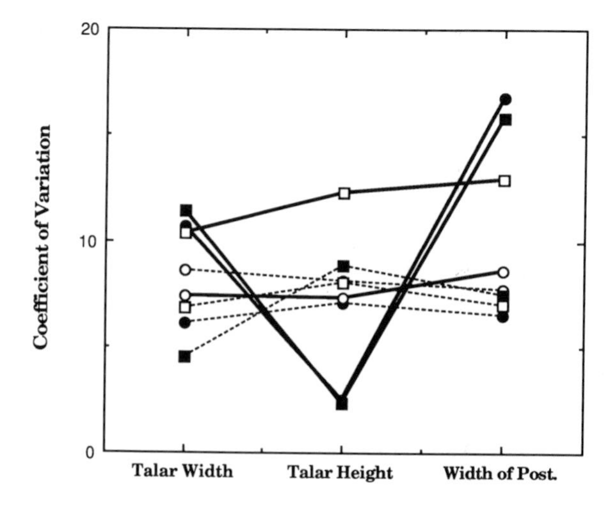

Fig. 6. CVs for talar measurements. *Proconsul* sample is split into traditional species groupings, labeled *P. africanus* and *P. nyanzae* (solid symbols, solid lines); mixed-sex samples of gorilla and chimpanzee are represented by open symbols and solid lines; single-sex samples of gorilla and chimpanzee are represented by dotted lines.

Uniformitarianism, Operational Limits, and Falsifiability

> The past history of our globe must be explained by what can be seen to be happening now.
>
> Hutton (1795)

Kelley (this volume) has taken us to task, claiming that our strict reliance upon modern analogues puts unrealistic limitations on our interpretations—in effect, ruling out the possibility of many evolutionary novelties. Kelley claims that this "misuse" of uniformitarianism guarantees "a world of the past that is an exact replica of the present." The present authors have no quarrels with uniformitarianism. Our disagreement is with Kelley's interpretation of it. As he notes in this volume, "uniformitarianism is about processes, not outcomes or events." The clear implication is that, from a uniformitarian perspective, processes remain relatively stable, while outcomes or events can be very different. We feel that the basic philosophical difference between our papers can be traced to the fact that Kelley has failed to maintain a distinction between processes, on the one hand, and outcomes or events, on the other.

To get to the heart of this debate, we must return to basic biology. Each of us exhibits a unique combination of morphological features and behavioral patterns. All of these traits, including those cited by Kelley (i.e., tooth morphology, body size, and reproductive behavior) are recognizable characteristics of *individuals*. As such, these traits have complex evolutionary histories, but each of them can also be viewed as evolutionary outcomes or events at our particular instant in time.

If we look at the fossil record, we frequently find evolutionary novelties, that is, combinations of morphology and behavior that cannot be found today. Much of the recent research of the present authors has helped to document such novelties in the fossil record (see, e.g., Beard *et al.*, 1986; Solounias *et al.*, 1988; Teaford *et al.*, 1988; Walker and Pickford, 1983; Walker and Teaford, 1989; Ward *et al.*, 1991). However, while the traits in question may differ, we feel that the underlying *processes* that led to their development remain the same: evolution by natural selection, genetic drift, etc. It is these processes, not the resultant morphology and behavior, that remain relatively constant in the eyes of uniformitarianism.

Clearly, variation lies at the heart of these processes. Without variation, natural selection and genetic drift would not exist. Without natural selection and genetic drift, the process of evolution would be radically different. If the amount of variation in a fossil sample is tied closely to the process of evolution as we know it, then the amount of variation within a given sample must be treated differently than individual traits, such as the size of the teeth. If we cannot use modern analogues to set testable limits of variation for prehistoric species, then we must also accept the fact that the process of evolution as we know it has changed. At the very least, this would mean that our prehistoric "species" are *not* biological species and cannot be treated as such in paleobiological interpretations (e.g., Kelley, 1986). It would also mean that we could never falsify even highly speculative or questionable hypotheses (Martin, 1991).

In the face of these possibilities/implications, we would seem to have but two alternatives: either use the present as a means to interpret the past or change our species concepts. There are plenty of available substitutes for the biological species concept (see Rose and Bown, this volume, for a review). However, if we opt for one of these substitutes, we must do so very carefully, for it may leave paleobiological interpretation in indecipherable chaos!

Conclusions

This study is the first attempt to analyze postcranial variation in the Miocene hominoid *Proconsul*. *Proconsul* ankle bones from the Miocene sites on Rusinga and Mfangano Islands, Kenya, have been shown to be more variable than might be expected based on previously published analyses of dental remains. The *Proconsul* postcranial sample is far more variable than postcranial samples of modern baboons, chimpanzees, or gorillas. This leaves us with little alternative but to insist that there were indeed two species of *Proconsul* at these sites.

Acknowledgments

We wish to thank Bill Kimbel and Lawrence Martin for inviting us to participate in the symposium, "Species, Species Concepts, and Primate Evolution" at the 1991 AAPA Meetings in Milwaukee. We thank the Government of Kenya for permission to carry out research in Kenya and the Governors of the National Museums of Kenya. We thank Bw. Kamoya Kimeu and his team for invaluable help in the field and at the National Museum in Nairobi. We also thank the following individuals and institutions for allowing access to skeletal material in their care: Bruce Latimer, Cleveland Museum of Natural History; Meave Leakey, Emma Mbua, and Nina Mudida, National Museums of Kenya. Special thanks go to Jay Kelley, Lawrence Martin, Ken Rose, Mike Rose, Bob Tague, and Dave Weishampel for their insightful comments on the manuscript. We also wish to thank Rose Keller for her help in data analysis. This work was supported by the National Museums of Kenya and by NSF grants 8418567, 8803570, and 8904327.

References

Andrews, P. J. 1978. A revision of the Miocene Hominoidea of East Africa. *Bull. Br. Mus. (Nat. Hist.).* **30:**85–224.

Andrews, P. J., and Simons, E. L. 1977. A new African Miocene gibbon-like genus, *Dendropithecus* (Hominoidea, Primates) with distinctive postcranial adaptations: its significance to the origin of Hylobatidae. *Folia Primatol.* **28:**161–169.

Beard, K. C., Teaford, M. F., and Walker, A. 1986. New wrist bones of *Proconsul africanus* and *P. nyanzae* from Rusinga Island, Kenya. *Folia Primatol.* **47:**97–118.

Bosler, W. 1981. Species groupings of Early Miocene dryopithecine teeth from East Africa. *J. Hum. Evol.* **10:**151–158.

Cope, D. A. 1989. Systematic Variation in *Cercopithecus* Dental Samples, Ph.D. Dissertation, University of Texas, Austin.

Drake, R. E., Van Couvering, J. A., Pickford, M. H., Curtis, G. H., and Harris, J. A. 1988. New chronology for the Early Miocene mammalian faunas of Kisingiri, Western Kenya. *J. Geol. Soc., Lond.* **145:**479–491.

Gingerich, P. D. 1974. Size variability of the teeth of living mammals and the diagnosis of closely related sympatric fossil species. *J. Paleont.* **48:**895–903.

Gingerich, P. D., and Schoeninger, M. J. 1979. Patterns of tooth size variability in the dentition of primates. *Am. J. Phys. Anthropol.* **51:**457–466.

Greenfield, L. O. 1972. Sexual dimorphism in *Dryopithecus africanus. Primates* **13:**395–410.

Harrison, T. 1982. Small-Bodied Apes from the Miocene of East Africa, Ph.D. Dissertation, University College London.

Harrison, T. 1986. New fossil anthropoids from the Middle Miocene of East Africa and their bearing on the origin of the Oreopithecidae. *Am. J. Phys. Anthropol.* **71:**265–284.

Hopwood, A. T. 1933. Miocene primates from Kenya. *J. Linn. Soc. Lond.* **38:**437–464.

Hutton, J. 1795. *Theory of the Earth.* William Creech, Edinburgh.

Kay, R. F. 1982a. Sexual dimorphism in Ramapithecinae. *Proc. Natl. Acad. Sci.* **79:**209–212.

Kay, R. F. 1982b. *Sivapithecus simonsi,* a new species of Miocene hominoid, with comments on the phylogenetic status of the Ramapithecinae. *Int. J. Primatol.* **3:**113–173.

Kay, R. F., and Simons, E. L. 1983. A reassessment of the relationship between later Miocene and subsequent Hominoidea, in: R. L. Ciochon and R. S. Corruccini (eds.), *New Interpretations of Ape and Human Ancestry,* pp. 577–624. Plenum Press, New York.

Kelley, J. 1986. Species recognition and sexual dimorphism in *Proconsul* and *Rangwapithecus. J. Hum. Evol.* **15:**461–495.

Kelley, J., and Xu, Q. 1991. Extreme sexual dimorphism in a Miocene hominoid. *Nature* **352:**151–153.

Leakey, L. S. B. 1943. A Miocene anthropoid mandible from Rusinga, Kenya. *Nature* **152:**319–320.

Le Gros Clark, W. E. 1952. Report on fossil hominoid material collected by the British-Kenya Miocene Expedition, 1949–1951. *Proc. Zool. Soc. Lond.* **122:**273–286.

Le Gros Clark, W. E., and Leakey, L. S. B. 1950. Diagnoses of East African Miocene Hominoidea. *Q. J. Geol. Soc. Lond.* **105:**260–262.

Le Gros Clark, W. E., and Leakey, L. S. B. 1951. The Miocene Hominoidea of East Africa, in: *Fossil Mammals of Africa,* British Museum of Natural History, London, **1:**1–117.

Leutenegger, W., and Larson, S. 1985. Sexual dimorphism in the postcranial skeleton of New World primates. *Folia Primatol.* **44:**82–95.

MacInnes, D. G. 1943. Notes on the East African Miocene primates. *J. E. Afr. Uganda Nat. Hist. Soc.* **17:**141–181.

Martin, L. B. 1983. The Relationships of the Later Miocene Hominoidea, Ph.D. Dissertation, University College, London.

Martin, L. B. 1991. Teeth, sex and species. *Nature* **352:**111–112.

Martin, L., and Andrews, P. 1984. The phyletic position of *Graecopithecus freybergi* Koenigswald. *Cour. Forsch.-Inst. Senckenberg* **69:**25–40.

McHenry, H. M. 1986. Size variation in the postcranium of *Australopithecus afarensis* and extant species of Hominoidea. *Hum. Evol.* **1:**149–156.

Napier, J. R., and Davis, P. R. 1959. The forelimb skeleton and associated remains of *Proconsul africanus,* in *Fossil Mammals of Africa,* Vol. 16, pp. 1–70. British Museum of Natural History, London.

Oxnard, C. E. 1983. Sexual dimorphism in the overall proportions of primates. *Am. J. Primatol.* **4:**1–22.

Pan, Y., Waddle, D. M., and Fleagle, J. G. 1989. Sexual dimorphism in *Laccopithecus robustus,* a Late Miocene hominoid from China. *Am. J. Phys. Anthropol.* **79:**137–158.

Pickford, M. 1986. Sexual dimorphism in *Proconsul,* in: M. Pickford and B. Chiarelli (eds.), *Sexual Dimorphism in Living and Fossil Primates,* pp. 133–170. Il Sedicesimo, Firenze.

Pilbeam, D. R. 1969. Tertiary Pongidae of East Africa: Evolutionary relationships and taxonomy. *Bull. Peabody Mus. Nat. Hist. (Yale Univ.)* **31:**1–185.

Plavcan, J. M. 1989. The coefficient of variation as an indicator of interspecific variability in fossil assemblages. *Am. J. Phys. Anthropol.* **78:**285.

Rose, M. D. 1986. Further hominoid postcranial specimens from the Late Miocene Nagri Formation of Pakistan. *J. Hum. Evol.* **15:**333–367.

Ruff, C. B., Walker, A., and Teaford, M. F. 1989. Body mass, sexual dimorphism and femoral proportions of *Proconsul* from Rusinga and Mfangano Islands, Kenya. *J. Hum. Evol.* **18:**515–536.

Schultz, A. H. 1969. *The Life of Primates.* New York: Universe Books.

Senut, B. 1986. Long bones of the primate upper limb: monomorphic or dimorphic? *Hum. Evol.* **1:**7–22.

Simons, E. L., and Pilbeam, D. R. 1965. Preliminary revision of the Dryopithecinae (Pongidae, Anthropoidea). *Folia Primatol.* **3:**81–152.

Simpson, G. G. 1947. Note on the measurement of variability and on relative variability of teeth of fossil mammals. *Am. J. Sci.* **245:**522–525.

Simpson, G. G., Roe, A., and Lewontin, R. C. 1960. *Quantitative Zoology.* Harcourt, Brace & World, New York.

Solounias, N., Teaford, M. F., and Walker, A. 1988. Interpreting the diet of extinct ruminants: the case of a non-browsing giraffid. *Paleobiology* **14:**287–300.

Tague, R. G. 1991. Commonalities in dimorphism and variability in the anthropoid pelvis, with implications for the fossil record. *J. Hum. Evol.* **21:**153–176.

Teaford, M. F., Beard, K. C., Leakey, R. E., and Walker, A. 1988. New hominoid facial skeleton from the Early Miocene of Rusinga Island, Kenya, and its bearing on the relationship between *Proconsul nyanzae* and *Proconsul africanus. J. Hum. Evol.* **17:**461–477.

Walker, A., and Pickford, M. 1983. New postcranial fossils of *Proconsul africanus* and *Proconsul nyanzae,* in: R. L. Ciochon and R. S. Corruccini (eds.), *New Interpretations of Ape and Human Ancestry,* pp. 325–352. Plenum Press, New York.

Walker, A., and Teaford, M. F. 1988. The Kaswanga Primate Site: an Early Miocene hominoid site on Rusinga Island, Kenya. *J. Hum. Evol.* **17:**539–544.

Walker, A., and Teaford, M. F. 1989. The hunt for *Proconsul. Sci. Am.* **260:**76–82.

Walker, A., Teaford, M. F., and Leakey, R. E. 1986. New information concerning the R114 *Proconsul* site, Rusinga Island, Kenya, in: J. Else and P. Lee (eds.), *Primate Evolution* pp. 143–149. Cambridge University Press, Cambridge.

Ward, C. V. 1991. Functional Anatomy of the Lower Back and Pelvis of the Miocene Hominoid *Proconsul nyanzae* from Mfangano Island, Kenya. Ph.D. Dissertation, Johns Hopkins University, Baltimore.

Ward, C. V., Walker, A., and Teaford, M. F. 1991. *Proconsul* did not have a tail. *J. Hum. Evol.* **21:**215–220.

Wood, B. A. 1976. The nature and basis of sexual dimorphism in the primate skeleton. *J. Zool. Lond.* **180:**15–34.

Yablokov, A. V. 1974. *Variability of Mammals,* Amerind Publishing Co., New Delhi.

Species Recognition in Middle Miocene Hominoids

16

LAWRENCE B. MARTIN AND PETER ANDREWS

Introduction

This chapter addresses the issue of species recognition in fossil samples. The particular example that is examined is the extensive collection of dental and gnathic remains from the middle Miocene locality at Paşalar, Turkey. This site is unusual in that available sedimentological and taphonomic evidence indicate that it was accumulated extremely rapidly and from a very localized area (Alpagut *et al.*, 1990a; Andrews and Alpagut, 1990; Andrews and Ersoy, 1990; Bestland, 1990), so that the mammalian fauna is comparable to a museum collection of modern animals made from one locality. In fact, the fauna may be sampled from within 3–5 km of the site, which would include only a few hundred meters of altitude at most. The sediments may have been deposited in hours or days, so that the temporal range of the fauna is determined by the period of predepositional skeletal preservation. Therefore, it might be anticipated that a species at Paşalar might be less variable than modern comparative samples collected from a variety of geographical localities. The Paşalar sample

LAWRENCE B. MARTIN • Departments of Anthropology and Anatomical Sciences, State University of New York at Stony Brook, Stony Brook, New York 11794. PETER ANDREWS • Department of Palaeontology, The Natural History Museum, London SW7 5BD, England.

Species, Species Concepts, and Primate Evolution, edited by William H. Kimbel and Lawrence B. Martin. Plenum Press, New York, 1993.

thus provides an unusual opportunity to examine analytical methods that may be used to determine species numbers in such an assemblage.

The question that we seek to answer in this chapter is, How many species exist within a large paleontological sample that can be shown to be temporally and geographically restricted at a level comparable to a collection of recent apes (in terms of time) from a single population (in terms of geography)? Modern hominoids, and particularly great apes, are taken as suitable models for the degrees of variation to be expected in a sexually dimorphic extinct ape. We consider the effect of geographic sampling by comparing levels of variation in samples of orangutans from Borneo and Sumatra and the effect of temporal sampling by comparing subfossil orangutans with living ones (Hooijer, 1948; Groves *et al.*, 1992). This analysis demonstrates that one or both of these factors, time and geography, can result in a dramatic increase in overall variation, and since the evidence is that the Paşalar fauna is more restricted both temporally and geographically than the combined sample of recent and subfossil orangutans, the expected level of variation should be lower.

In recent years it has become almost axiomatic that paleontologists will tend to underestimate the number of species in the fossil record. This premise is based on the fact that numerous examples of sympatric, morphologically similar, congeneric species are known among primates. The same is also true for suids, rodents, some carnivores, and some artiodactyls, although many other mammalian taxa show much less overlap in dental morphology, even with their sister species. A review of the literature on species diversity in modern faunas from tundra desert through tropical forests (Andrews, 1990) reveals a range of species numbers per locality from 22 to 76. In fact, only subtropical forests (52 species, number of localities sampled = 5), African tropical bush and woodland (75 species, n = 7), and Asian (60 species, n = 5) and African (76 species, n = 11) forests have species numbers exceeding 50. Some 57–60 mammalian species have already been recognized at Paşalar (Alpagut *et al.*, 1990a; Andrews, 1990), which suggests that species number cannot have been greatly underestimated at this site unless its species diversity exceeded that of any modern mammalian fauna, which seems unlikely. This fact suggests to us that current species recognition practices may not, on average, be producing results that are significantly in error. However, this does not help us in deciding whether we have one or five species of a particular taxon, as these figures would be able to be accommodated in almost all cases, but it does indicate that the axiom may be flawed.

Species Concepts

A variety of species concepts have been proposed and most receive support by various contributors to this volume. Four, in particular, receive attention: the Dobzhansky-Mayr biological species concept (BSC; see Szalay, this volume; Jolly, this volume; Costello *et al.*, this volume; Rose and Bown, this volume), the Paterson recognition concept (RC; see Masters, this volume), Simpson's evolutionary species concept (ESC; see Krishtalka, this volume, and Cracraft's phylogenetic species concept (PSC; see Kimbel and Rak, this volume). These various concepts potentially have different implications for the way that modern apes are divided into species, which could affect our analysis of modern species variation that we use as a reference for studies of variation in the fossil record.

Living Apes

The BSC does not work well for allopatric populations, and thus is of limited value in addressing the specific status of Bornean vs. Sumatran orangutans, for example, or of pygmy vs. common chimpanzees, in any meaningful way. The RC has not, as yet, been applied in any detail to extant hominoids, though the studies of Marshall and Sugardjito (1986) indicate that species-specific calls support their rather finely divided classification of gibbons (with nine species recognized). The ESC and the PSC might be employed to recognize some of the current hominoid subspecies, especially of orangs and of gorillas, as full species, and perhaps to collapse those of chimpanzees, excluding bonobos, into a single species without divisions, but no detailed studies have yet been done. Nonetheless, it is becoming increasingly clear that comparative samples need to be assembled with much greater care than has often been the case. A procedure such as the selection of the 20 most complete males and females to be found at one's local museum may not result in the ideal, or even really an appropriate, comparative sample. However, the available metrical data have not been gathered with such factors in mind, and this fact may serve to increase levels of variation, where samples are geographically widespread, or to decrease them, where samples are geographically localized. Perhaps the most important thing is to be aware of the potential for mismatches between the composition of the fossil and of the comparative samples to influence one's analysis of the fossil record. For the present study, the conventional identification of species groups of extant hominoids is employed (see Table 1).

Paleontology

Fossil samples pose three major problems with regard to species recognition: time, space, and levels of variation. These are briefly considered here with regard to topics that are more fully addressed elsewhere in this volume. The issue of species within lineages is not one with which those of us working with the Miocene ape record have had to contend, as few sites provide continuous stratigraphic records and those that do have a relatively low frequency of apes. However, the related problem of time-averaging has been raised with regard to the early Miocene collection from Rusinga Island, Kenya and the late Miocene assemblage from Lufeng, China (Kelley, 1986; Pickford, 1986; Kelley and Etler, 1989; Wood and Xu, 1991).

Lineages. It is easy to draw species boundaries within hypothetical lineages when gaps occur in the fossil record, as is usually the case. The problems that occur when such gaps are filled have been addressed most fully by Gingerich (1985), Krishtalka (this volume), and Rose and Bown (1986, this volume). This is an extreme case of a problem common to many fossil samples, that is, the extent to which time has operated to increase variation in a single evolving lineage (or species). Within the Willwood Formation, several cases are known where the start and end points of the lineage were historically recognized as distinct taxa, but with the filling in of gaps in the record, questions have been raised concerning the "biological reality" of the species boundaries that were originally drawn. Clearly some kind of subdivision is essential if we are to make meaningful interpretations of morphological, ecological, and adaptational shifts over the course of time in a lineage, and perhaps most obviously for the purposes of biostratigraphy, for which we would wish to use the finest divisions possible. There is not much debate as to whether such divisions should be made (e.g., Rose and Bown, this volume; Krishtalka, this volume); but rather, debate centers on whether these subgroups should be termed *species* or not. The answer to that question depends on one's concept of species and is thus as much philosophical as practical (see, e.g., Masters, Eldredge, Szalay, Jolly, Rose and Bown, Harrison, Kimbel and Rak, all in this volume). The fundamental disagreement is between those who see lineages as evolving, reproductively cohesive units (speciation does not occur no matter how much change accrues) and those who believe either that anagenetic (phyletic) speciation is a real process or that evolutionary species must be divided up into biological species-sized parcels in order to make the world manageable. There are strengths to both arguments, but we propose to sidestep further discussion of the issue, as none of the fossil samples that we consider here demonstrates continuous change over time.

Temporal Variation. What we encounter most frequently in the Miocene fossil ape record is a geographically constrained fossil site that has accumulated over a substantial, though usually unknown, period of time. It is difficult to establish how much time needs to be involved for significant change to occur, though studies to date indicate that these periods may be quite short (Gingerich, 1987). The argument usually made is that populations of animals exhibit relatively predictable levels of variation for any given parameter. As the mean characteristics of populations can be expected to change over time, it is argued that many fossil assemblages display a time-averaged quality in which variation exceeds modern standards as a result of the temporal duration of sampling of the species (Kelley and Etler, 1989; Kelley and Xu, 1991; Wood and Xu, 1991). Where the fossil record is sufficiently complete and sedimentation rates can be estimated, the degree to which these effects occur can be studied (Gingerich, 1985, 1987; Rose and Bown, 1986; Kelley and Gingerich, 1991), although results to date have not been conclusive. When applied to Miocene ape samples, however, such arguments have generally been made in the absence of any real idea as to the duration of sediment accumulation. It is not solely a question of the precision with which absolute dates for the sites have been calculated, nor of the magnitude of the error factors associated with these (Teaford *et al.*, this volume). What is needed is information concerning the rate and duration of sedimentation and also the taphonomic processes at work.

In many cases it might be supposed that change accrues with time, but this needs to be demonstrated rather than assumed, and in the absence of any independent demonstration that such is the case, it remains no more than a supposition. We contend that the time-averaged nature of fossil samples has been overstated in some cases, enabling researchers to maintain a single species interpretation of a highly variable sample. We also feel that it is misleading to term extant species as being *time slices* of lineages for studying variation without taking account of evolutionary history. This may be the case when a genetically defined population is being examined, but it is more likely that the levels of variation seen in a species at one moment in time reflect, to some degree, the time since all organisms of that species last shared a common ancestor. Thus, species variation may also be seen as a product not only of modern geographic variation but also of the species' evolutionary history.

For time-averaging to be a viable argument two criteria need to be satisfied: (1) The duration of sediment accumulation must be shown to have been prolonged through studies of sediment accumulation rates, as has been done, for example, in the Siwaliks (Kappelman, 1986; Flynn *et al.*, 1990; Kappelman *et al.*, 1991), or by absolute chronological studies; (2) the sample must show change over time: a shift in mean dimensions related to stratigraphic position rather than the maintenance of high variation as a constant pattern. In the absence of such evidence, especially of change up and down a stratigraphic section, the claim of time-averaging should not be taken as the null hypothesis.

Geographic Variation. Geographic variation is a very real problem in species recognition (see, e.g., Albrecht and Miller, Groves, Jolly, Tattersall, all in this volume). In the fossil record it may manifest itself in two main ways: (1) in making comparisons among samples from different sites and (2) where taphonomic and sedimentary evidence indicate that fossils may have been drawn from a large catchment area. In the former situation we may also encounter a problem of comparing sites of different ages and/or ecologies (see discussion of Çandır and Neudorf, below). The second problem probably rarely occurs at levels that are significantly different from extant taxa sampled at a single location, but where samples from a variety of localities are pooled (e.g., the Siwalik *Sivapithecus*) this could be a problem. The catchment area problem must be addressed prior to attempting an analysis of variation in a fossil assemblage. This requires that the taphonomic history of the fossils in relation to the sediments be understood as fully as possible. A wide catchment area might lead to incorporation of closely related, parapatric forms in a fossil assemblage, or it might sample several habitat types, which might combine populations of a single species that have geographically patterned characteristics of morphology (Martin, 1983; Plavcan, 1990; Albrecht and Miller, this volume; Groves, this volume; Jolly, this volume; Tattersall, this volume). The problem of comparing samples from different sites increases the likelihood of introducing errors into species recognition. Some allowance can be made by selecting comparative samples from geographically defined populations, either from the same regions as the fossils or from regions of equivalent geographic range. Detailed analyses such as these could provide considerable guidance in selecting appropriate comparative samples with which to compare variation between fossil sites.

In the happy circumstance that we can discount, or control for, the effects of geography and time on the level of variation in our fossil assemblage, as we believe is the case for the Paşalar sample (Andrews and Alpagut, 1990; Andrews and Ersoy, 1990; Bestland, 1990), we are left with the problem that is the focus of this chapter. This is the analysis of the level of variation in a morphologically homogeneous fossil sample with respect to the determination of how many species may be present.

Materials and Methods

Materials

Modern Samples

The modern hominoid samples that have been used for most of this study are shown in Table 1. Where additional published data have been used, this fact is noted in footnotes to the relevant tables.

Fossil Samples

The published collection of hominoids from Paşalar now totals over 700 described specimens, comprising jaws, dental remains, and some postcrania (Andrews and Tobien, 1977; Alpagut *et al.*, 1990b). The material comprises three maxillary specimens, two mandibles, 603 isolated permanent teeth, 42 deciduous teeth, and 11 postcranial elements, all of which were recovered during our excavations from 1983 to 1989 (Alpagut *et al.*, 1990b). There are, in addition, 86 isolated teeth from earlier collections (Andrews and Tobien, 1977), and these are included in the calculation of sample statistics. Sample size is based on complete, measurable specimens. The sample size for each dental element used in this study is shown in Table 2.

Most teeth in the Paşalar sample are easy to identify. The difficulties that are encountered are with the molars, which are similar to each other in crown morphology. The lower molars are among the best represented teeth in the Paşalar collection and M_3s are easily recognized, but there are sometimes problems in distinguishing the first from the second molars. Obviously it helps to have associated material, where there is no ambiguity, but because of the close similarity between M_1 and M_2 there will always be a problem with isolated teeth. The smallest elements in this pool must be M_1s, and the largest must be M_2s, but there are no natural breaks in the distribution that can be used reliably to make determinations based on size. However, many of the M_1s retain signs of a double mesial interproximal facet where they contacted first with the deciduous premolar and then with P_4, and the identity of these teeth is clear. In the absence of this it becomes more difficult to distinguish a large M_1 from a small M_2. For the upper molars, positive identifications of the third molar are based on the extreme distal abbreviation of the crown and absence of distal contact facets on heavily worn teeth. Beyond this, identifications have been made on the basis of minor morphological features, for we found no evidence for double mesial contact facets on upper molars, presumably because there is less difference in distal crown height between dp^4 and P^4 than is the case for dp_4 and P_4. Based on the numbers of M^3s, which are readily identifiable, it appears that we have too few M^1s and too many M^2s (Table 2). We feel, therefore, that we may have included some M^1s in the M^2 sample, and we are undertaking further, more detailed analyses of micromorphology to remedy this problem. The more nearly equal numbers of M_1s, M_2s, and M_3s in the collection than is the case for the

Table 1. Samples Sizes of Extant Hominoids Used for the Analyses in this Study, by Gender and Tooth Measurement

	P. troglodytes			*P. paniscus*			*G. gorilla*			*P. pygmaeus*			*H. syndactylus*			*H. lar*		
	m	f	m + f	m	f	m + f	m	f	m + f	m	f	m + f	m	f	m + f	m	f	m + f
I^1 M-D	10	10	20	15	20	41	12	12	24	12	9	21	6	5	11	14	7	21
I^1 B-L	10	10	20	15	21	42	12	12	24	12	9	21	6	5	11	14	7	21
I^2 M-D	10	10	20	13	20	38	11	12	23	11	10	21	6	5	11	14	7	21
I^2 B-L	10	10	20	13	20	38	11	12	23	12	10	22	6	5	11	14	7	21
C^1 Max. L.	14	12	26	15	18	33	20	20	40	12	9	21	7	5	12	13	7	20
C^1 Perp. B.	14	11	25	15	18	33	20	20	40	12	9	21	7	5	12	13	7	20
P^3 M-D	14	12	26	17	26	50	20	20	40	12	10	22	7	5	12	14	9	23
P^3 B-L	13	11	24	17	27	51	20	20	40	12	10	22	7	5	12	14	9	23
P^4 M-D	14	12	26	16	23	45	20	20	40	12	10	22	6	5	11	14	9	23
P^4 B-L	13	11	24	16	24	46	10	19	39	12	10	22	6	5	11	14	9	23
M^1 M-D	14	12	26	32	41	93	20	20	40	12	10	22	7	5	12	14	9	23
M^1 B-L	12	11	23	32	41	93	20	20	40	12	10	22	7	5	12	14	9	23
M^2 M-D	14	12	26	18	27	52	20	20	40	12	10	22	7	5	12	14	9	23
M^2 B-L	13	11	24	18	27	52	20	20	40	12	10	22	7	5	12	14	9	23
M^3 M-D	11	10	21	11	11	24	20	20	40	12	10	22	7	5	12	13	9	22
M^3 B-L	11	9	20	11	11	24	20	20	40	12	10	22	7	5	12	13	9	22
I_1 M-D	0	0	0	16	22	43	0	0	0	11	9	20	7	3	10	14	9	23
I_1 B-L	0	0	0	15	22	43	0	0	0	11	9	20	7	3	10	14	9	23
I_2 M-D	0	0	0	17	24	47	0	0	0	11	9	20	7	3	10	13	9	22
I_2 B-L	0	0	0	17	24	47	0	0	0	11	9	20	7	3	10	13	9	22
C_1 Max. L.	13	11	24	16	20	37	20	18	38	12	9	21	7	4	11	14	8	22
C_1 Perp. B.	14	11	25	16	20	37	20	18	38	12	7	19	7	4	11	14	8	22
P_3 Max. L	14	12	26	17	24	47	20	20	40	10	5	15	7	5	12	14	9	23
P_3 Perp. B.	14	12	26	17	24	47	20	20	40	10	5	15	7	5	12	14	9	23
P_4 M-D	14	12	26	18	24	48	20	20	40	11	8	19	7	4	11	14	9	23
P_4 B-L	14	12	26	18	24	48	20	20	40	11	8	19	7	4	11	14	9	23
M_1 M-D	14	12	26	33	40	92	20	20	40	12	10	22	7	5	12	14	9	23
M_1 B-L	14	12	26	34	39	91	20	20	40	12	10	22	7	5	12	14	9	23
M_2 M-D	14	12	26	17	25	49	20	20	40	12	10	22	7	5	12	13	9	22
M_2 B-L	14	12	26	17	26	50	20	20	40	12	10	22	7	5	12	13	9	22
M_3 M-D	13	12	25	12	13	26	20	20	40	12	10	22	7	5	12	13	9	22
M_3 B-L	13	12	25	12	14	28	20	20	40	12	10	22	7	5	12	13	9	22

m = males; f = females; m + f = mixed sex sample.
P. troglodytes and *G. gorilla* data from Pilbeam (1969), except incisor measurements, which are unpublished data of the authors. *P. paniscus* data are from Johanson (1974).
Hylobates data were kindly provided by T. Harrison. *Pongo* data were taken by the authors on B.M.N.H. Zoology Department specimens.

upper molars (Table 2) gives us greater confidence in our attributions of lower teeth.

Uniformitarianism and Falsifiability

The Null Hypothesis

Metrical studies by Martin (1983) on gibbons, Cope (1989; this volume) on sympatric guenons, and Plavcan (1990; this volume) on various catarrhine primates have shown that pooled-species samples of modern primates can exhibit

Table 2. Samples Sizes for Paşalar Permanent Teeth (to August 1989)

Tooth type	Upper	Lower
I1	40	19
I2	42	27
C1	26	36
P3	30	41
P4	49	47
M1	37	56
M2	78	62
M3	61	59
Total	363	347

The sample includes both isolated and associated teeth, some of which may be antimeres (Andrews and Tobien, 1977; Alpagut *et al.*, 1990b).
Total permanent teeth = 710.

levels of variation indistinguishable from those of single-species samples. If such morphologically homogeneous samples were encountered in the fossil record, they would appear to represent a single species by available criteria, the implication being that a single-species interpretation can never be demonstrated conclusively in a fossil sample, it can only be falsified. The single-species interpretation is therefore the null hypothesis. As Martin (1983, p. 49) pointed out

> CV (coefficient of variation) can only be used to say that there is probably more than one species present. It cannot be used to confirm that a single species is present or to determine the number of species present.

In some of our previous work (Martin and Andrews, 1984; Andrews and Martin, 1987), we did not apply this criterion, but rather took the position that failure to falsify the single-species hypothesis, in the cases of *Graecopithecus freybergi* and *Heliopithecus leakeyi,* demonstrated that a single species existed. We stand corrected!

The problem then is how to make species determinations in the fossil record on an objective basis. The null hypothesis that a single species is present in a morphologically homogeneous sample can be falsified by studies of variation in comparison with extant species (e.g., Kay, 1982a; Martin and Andrews, 1984; Kelley and Etler, 1989; Wood and Xu, 1991). There are three possible outcomes of such analyses: (1) levels of variation in the fossil sample exceed reference values, (2) levels of variation in the fossil sample are contained within reference values, and (3) the analysis is inconclusive. On the basis of recent studies, the third outcome might be considered to be the most common result! This is a consequence of what Cope (1989) and Cope and Lacy (1992) have discussed under the heading of "power."

The purpose of an analytical approach is to produce correct results. A correct result in this case means the recognition of a mixed taxon sample as such (Cope, 1989, this volume; Cope and Lacy, 1992). This may sound obvious, but it has profound implications for the evaluation of analytical methods.

Types of Error. Cope (1989) has recognized that two types of error may occur in tests of the single-species hypothesis: type 1 errors, whereby a sample drawn from a single species is incorrectly interpreted as containing multiple species; type 2 errors, whereby a sample that is in fact drawn from multiple taxa is not falsified as a single species by the analysis. Obviously, the nature of errors that occur can only be known for modern taxa where the taxonomic composition of the sample is known. Cope's (1989; this volume) work has shown that both type 1 and type 2 errors can easily occur.

Analytical Efficiency. The accepted definition of power is: power = 1 − type 2 error rate (Cope and Lacy, 1992). This means that an increase in the reference value used for species tests will decrease its power because it will increase the number of type 2 errors, although it will also reduce the number of type 1 errors. By contrast, reducing the reference value increases power as it decreases the number of type 2 errors, although it also increases the number of type 1 errors. Paleoprimatologists have primarily concerned themselves with trying to avoid type 1 errors and have thus used techniques with very low power (e.g., Kay, 1982a; Martin, 1983; Martin and Andrews, 1984; Kelley and Etler, 1989; Wood and Xu, 1991).

We suggest that efforts should be made to evaluate methods for single-species falsification in terms of analytical efficiency rather than power. By this term we mean a measure that considers both type 1 and type 2 errors, since both are undesirable. Our reasons for this are simple. If one examines the simulation plots of Cope and Lacy (1992) it can be seen that maximum power would be achieved using threshold values of CV <1 (at $n = 5$), CV ≈ 3 ($n = 10$), CV ≈ 4 ($n = 15$ and $n = 20$), and CV ≈ 5 ($n = 50$). All of these threshold values result in analyses with high power, since they have been chosen to minimize type 2 errors. However, this criterion alone is completely useless because these thresholds result in unacceptably high rates of type 1 error ($\approx 100\%$ at $n = 5$, $\approx 90\%$ at $n = 10$, $>60\%$ at $n = 15$, $>50\%$ at $n = 20$, $\approx 5\%$ at $n = 50$). Thus, if we concern ourselves solely with power, we might apply standards that have unacceptable type 1 error rates at sample sizes that we typically encounter in the fossil record. This is not what Cope and Lacy (1992) advocate, since they set their type 1 error rate at 5% and then measure power at that threshold level. This illustration indicates the desirability of inventing a way to measure analytical efficiency.

Falsification of the Single Species Hypothesis and a Second-Order Null Hypothesis. When the null hypothesis is falsified by the analysis there are several possible explanations that need to be considered: (1) the comparative data used are inappropriate, (2) the sample comprises a single species sampled over time so that a shifting mean has appeared to increase the overall level of variation when the sample is treated as a statistical population, (3) the sample comprises more than one species, and (4) the sample comprises a single species whose variation at one time exceeds known comparative species. How do we determine which of these factors is at work? The selection of appropriate comparative data is considered in more detail below, and for the CV has been the subject of detailed study by Cope and Lacy (1992). The bottom line is that mismatches in sample composition, that is, sample size, gender composition, etc., between the fossil sample and the comparative sample should be evaluated as possible causes of type 1 and type 2 errors.

We have discussed the criteria for establishing time-averaging above and these can address explanation (2). In the event that the third explanation is accepted, Cope and Lacy (1992) have suggested that a second-order null hypothesis should be established, that is, that the sample comprises two species. This seems eminently reasonable, although we would add that third-, fourth-, and fifth-order hypotheses should also be considered (three, four, and five species, etc.) as each tier is tested and falsified. For the present, no one, to our knowledge, has developed techniques with which to evaluate this second, or higher, order null hypothesis. It is clearly time that this be considered. It seems to us that we can select among explanations (1), (2), and (3) based on objective criteria. Explanation (4) is more difficult and is the subject of the section on uniformitarianism and falsifiability below.

Failure to Falsify the Null Hypothesis. In the case that the single species hypothesis is not falsified by a metrical analysis, the investigator should turn to other lines of evidence to see whether these can throw additional light on the matter, for example, evidence of adaptations, growth, etc. If everything known about the sample is compatible with its interpretation as a single species, then it is possible to treat it as such for studies of, e.g., sexual dimorphism, adaptation, etc., but only with the proviso that in some cases this taxonomic interpretation will be incorrect as the result of type 2 errors, which can be shown to be undetectable in many cases for extant primate genera with sympatric congeners.

A frequent problem is that the results of analyses of variation are considered to be inconclusive (e.g., Kay, 1982a; cf. Martin and Andrews, 1984; Kelley and Etler, 1989; Wood and Xu, 1991) because falsification is not absolute. This is a problem of power and analytical efficiency that was considered in more detail in the preceding section of this paper.

Uniformitarianism and Falsifiability

Another way in which the falsification of single-species hypotheses has been argued to be incorrect (i.e., interpreted as a type 1 error), in the absence of geographical or temporal factors, is seen in Kelley and Xu's (1991; Kelley, this volume) work. These authors have argued that the approach advocated here, and in fact employed by Kelley and Etler (1989), is flawed and represents a misapplication of the concept of uniformitarianism, one that limits past life forms to modern standards. This argument has some merit because it is clear from the fossil and subfossil record that animals such as *Gigantopithecus* and *Megaladapis* existed in the past that lie outside modern size limits for primates and extend considerably the total size range of primates. However, these authors suggest that this fact demonstrates the possibility that a species existed in the past that exhibited more variation than any modern taxon. This is a confusion of uniformitarianism as applied to arguments about interspecific variation with those about intraspecific variation. It seems to us that the fact that CV values for M1 dimensions, for example, are constrained not just in all hominoids, but in all primates and all land mammals that have been studied, suggests that what we are seeing is something fundamental in mammalian biology. It may be argued, as Kelley (this volume) does, that this is a modern pattern and that we should not assume that it existed in the past. In the absence of any mechanism to explain the remarkable convergence among mammalian lineages in levels of variation, we conclude that what is widely observed in the present was also typical of the past.

A related issue concerns the selection of appropriate modern species as models of variation. For example, Cope's work (1989; personal communication) shows that the pattern of variation seen at Lufeng, where size classes of canines correspond to sexes (Kelley and Xu, 1991), could be mimicked in mixed-species samples of *Cercopithecus,* the implication being that a study treating this sample as a single species *a priori* may be flawed. The question that paleoprimatologists must confront is whether the guenon situation is relevant to the Lufeng sample. Kelley (this volume) feels that hominoids alone provide a suitable model for the patterning and distribution of variation, and that guenons are inappropriate models for hominoid variation. The problem with the view that only hominoids are relevant is that it predetermines the outcome of all analyses; hominoids are low in species diversity compared to other groups of primates, and there are few sympatric taxa that can be used to model what might happen in a fossil assemblage.

At the crux of this debate is the question of why hominoids show this pattern of diversity. Kelley (this volume) believes that the fact that hominoids of similar morphology and similar size do not occur sympatrically today reflects some fundamental aspect of ape biology that would also prohibit such an arrangement in the past. However, we know that sympatry of morphologically similar species does occur in gibbons, guenons, macaques, langurs, and lemurs, so the argument that hominoid distribution is constrained by the biological characteristics of apes requires further analysis before it is accepted. There is also the question of when this supposed aspect of ape biology first appeared in the evolutionary record, which would determine the fossil groups to which it was relevant. This is particularly the case because the fossil and subfossil record shows that hominoid primates were once much more widely distributed than today, so that the distribution of modern hominoids is that of relict populations. Since species diversity in relict faunas is impoverished (Andrews *et al.,* 1979), an alternative view of recent hominoid variation and distribution to that advocated by Kelley is that extant apes cannot provide a suitable model with which to compare extinct hominoids that were both more diverse and more widely distributed. How should paleoprimatologists proceed in order to resolve this conundrum?

One way would be to look at the fossil record of, for example, Miocene apes, for evidence of multiple taxa at a single site. The problem is that where such examples occur (e.g., *Dryopithecus* in the Miocene of Spain; fossil apes from the middle Miocene of Paşalar, Turkey; *Sivapithecus* from Indo-Pakistan; and *Proconsul* from the early Miocene of Kenya), more than one interpretation of diversity has been offered (Andrews and Tobien, 1977; Andrews, 1978; Kay, 1982a; Martin, 1983; Kelley, 1986; Kelley and Pilbeam, 1986; Ruff *et al.,* 1989; Alpagut *et al.,* 1990b; Teaford *et al.,* this volume). Many workers would see these examples as falsifying Kelley's proposal that morphologically and metrically similar hominoid species cannot live sympatrically (Andrews, 1981; Martin, 1981; Kay, 1982a,b; Kay and Simons, 1983; Ruff *et al.,* 1989; Teaford *et al.,* this volume), and thus confirming that the present distribution of hominoids reflects their relict nature rather than some fundamental aspect of ape biology. Kelley, however, would argue that all of these cases represent single, highly variable species, confirming that life in the Miocene was very different than life today. One's analytical models are thus determined by one's perception of hominoid ecology past and present.

Test of Species Numbers

Selection of Statistics

In metric studies of species variation two classes of statistic have been used: the coefficient of variation and range-based measurements. The coefficient of variation for a dimension may be used as a size-independent expression of variance (Simpson *et al.*, 1960; Gingerich, 1974; Gingerich and Schoeninger, 1979). If one wished to use raw figures for range, it would be necessary to find an extant analog of similar mean value to the fossil sample under analysis. Due to the fact that relatively small numbers of hominoid species are alive today, it may not be possible to find such a comparison. Even if this were possible, there is no particular reason to believe that a species selected on the basis of its size alone provides the best model for variation. Instead, range may be converted to a size-independent measurement by expressing it as a percentage of the mean (R%; Martin, 1983; Martin and Andrews, 1984). Ranges of variation may also be compared numerically by the index of maximum and minimum values (MI) (Martin, 1983; Martin and Andrews, 1984).

The statistic of choice for many workers has been the CV, following the work of Simpson (1941) and Gingerich (1974). Simpson argued that range-based statistics always tend to underestimate variation in a taxon, especially at small sample sizes, and that these would be of little use in attempts to demonstrate that a sample constitutes a single species. However, Martin and Andrews (1984) argued that range-based statistics had some advantages over CV, particularly at smaller sample sizes. For example, the fact that range-based measurements potentially can falsify a single species interpretation for a sample as small as two permits quantitative evaluation of species numbers in cases where CV cannot be applied. It should be noted that Simpson's (1941) arguments relate to power and type 2 errors but do not address the question of the accuracy of a result when range-based measures falsify the null hypothesis.

Gingerich's (1974) study of variation in mammal dentitions established the CV as the statistic of choice for the evaluation of levels of metrical variation in fossil hominoid samples (Kay, 1981a,b; Kay and Simons, 1983; Kelley and Etler, 1989; Wood and Xu, 1991). In studies of extant catarrhine variation that were particularly aimed at the paleospecies question, Cope (1989) and Plavcan (1990) provided vigorous support for CV as the preferred statistic. Their basis for arguing that CV was the more robust statistic was the fact that it was more successful than range-based statistics in detecting multiple species in pooled samples (Cope, 1989), i.e., it had greater power, since it resulted in fewer type 2 errors, which is essentially the argument made by Simpson (1941). We do not think that this is the only relevant criterion.

Fossil samples at a locality or site are often very small for any tooth position. Martin (1983) argued that the effects of small sample size on CV can be very different from the effects on range-based statistics. The problem with both kinds of statistic is that they provide inaccurate estimates of variation at small sample sizes. On average, CV will be more accurate in predicting population variation than a range-based statistic. However, CV has the potential to overestimate population variation at small sample sizes (Martin, 1983; Martin and Andrews, 1984), which could lead to type 1 errors (the single-species hypothesis being incorrectly rejected), as well as to underestimate population variation, which will lead to type 2 errors (failing to reject the single-species hypothesis for a sample comprising more than one taxon; Cope and Lacy, 1992). R% may also overestimate population variation at small sample sizes because it depends on the value determined for the mean, which may shift up or down at small sample sizes. By contrast, range and MI can only underestimate the population or species range of variation, which makes type 2 errors (failure to reject the single-species hypothesis for a multiple taxon sample) common, but type 1 errors (incorrect rejection of the single-species hypothesis) uncommon or nonexistent.

In a recent computer simulation study based on Cope's (1989) *Cercopithecus* data, Cope and Lacy (1992) examined the effect of sample size on CV. Their results show that, on average, CV values for small samples underestimate true population variation, although they cover a range of values both above and below the actual value. Thus, at small sample sizes the reference maximum CV will occasionally falsify a single species interpretation incorrectly (a type 1 error) for one or two variables, though this is unlikely for the whole dentition if the comparative sample values are based on larger samples than are present in the fossil sample. It should be noted that while occasional type 1 errors will occur in any statistical procedure, they are quite rare compared to the total of variables tested. In principle, CVs for small samples could range from zero to an unspecified upper value. By contrast, range-based statistics can only vary between zero and the "true" value for the taxon. Larger samples will always give bigger estimates of range, given that no comparative sample is likely to contain the entire range of variation that exists. By their very nature, range-based statistics can only underestimate the true variation of the population; they can never provide a false overestimate. Thus, while CV analyses will result in both type 1 and type 2 errors, range-based statistics result in type 1 errors, only infrequently, providing that the comparative samples are as large or larger than the fossil sample and that the effects of temporal variation have been taken into account. This critical quality makes their use invaluable in studies of species numbers in the fossil record.

Cope (1989) argued that the CV was the statistic of choice over range-based statistics on the basis of its effectiveness in detecting multiple taxa, rather than on the confidence with which one could interpret the significance of a falsification of the single species hypothesis; that is, he was concerned primarily with the power of the technique and felt that range-based statistics lacked the power to make them of wide use. However, Cope (1989) recognized the accuracy of the results of range-based analyses when they did falsify the single species hypothesis. To quote from this study (Cope, 1989, p. 109):

> Thus, a failure to exceed single species maximum values cannot be used to support an argument that only a single species is present. However, if an R% or MI estimate in a small sample does exceed the single species maximum, *it strongly suggests that more than one species is present,* since both methods are highly biased against such an outcome. (emphasis added)

One way that these methods could be made more reliable would be to use modern samples that are constituted in exactly the same way as the fossil sample. Ideally this should extend to gender composition as well as sample size, although gender is rarely known for fossils. An alternative is to calculate confidence limits for the small sample, but Cope and Lacy (1992) have shown that this reduces the power of the technique to such an extent as to make it of little value. Perhaps the best method would be to use a simulation approach to calculate an appropriate critical value for the sample size of interest, as has been done by Cope and Lacy (1992) for *Cercopithecus*. Generally, it is the case that very large samples of modern apes are compared with much smaller samples of fossils (e.g., Kay, 1982a; Martin and Andrews, 1984; Kelley and Etler, 1989; Wood and Xu, 1991). Large sample sizes produce very accurate values of CV for the modern animals, but the figures for the smaller fossil samples are less accurate, which may increase type 1 and type 2 errors. If CVs for the modern material were computed at the same sample size as for the fossils, the reference values would decrease on average. This means that some fossil samples currently viewed as excessively variable by CV analysis might be found to fall within modern limits if a correction was made for the influence of sample size. Range-based measures provide a counter to this tendency, as they rarely produce type 1 errors, though they are prone to type 2 errors and thus have low power.

The recent study by Cope and Lacy (1992), which came to our attention after the present work was all but completed, has demonstrated important inadequacies in previous approaches using reference maxima (e.g., Kay, 1982a; Martin and Andrews, 1984; Kelley and Etler, 1989; Wood and Xu, 1991) and has cast doubt on the value of using confidence limits in the way that has been done in the past in tests of species numbers. Cope and Lacy's (1992) simulation study used multiple iterations at each sample size from statistical universes of known taxonomic composition and with known levels of variation. This showed that while the use of reference maxima was a very conservative approach in terms of minimizing the occurrence of type 1 errors, it had very limited power as a result of the fact that it greatly increased the frequency of type 2 errors. Cope and Lacy (1992) found that the median value for CV for simulation studies, using the largest *n* sample at the sample size that matches the fossil sample under investigation, provided the reference standard with the greatest power, i.e., with the least number of type 2 errors. Clearly there is a need for a similar study for hominoids, although published data are not at present available for specimens of known geographical locale. However, if we assume that the statistical results reported in the Cope and Lacy (1992) study of *Cercopithecus* also apply to the hominoids, which seems reasonable in terms of overall patterns, then the finding that the use of reference maxima is overly conservative is very important. Likewise, the suggestion that the use of confidence limits and correction factors for small sample sizes reduces the power of the analysis by increasing type 2 errors without increasing its accuracy requires attention. The implication of these findings is that a study that compares reference maxima for large samples of extant forms with either corrected or uncorrected values for smaller samples of fossils is strongly biased against falsification of the single species hypothesis (e.g., Kay, 1982a; Martin and Andrews, 1984; Kelley and Xu, 1991; Wood and Xu, 1991); that is, such studies result in a large number of type 2 errors and are constructed so that type 1 errors are virtually impossible. Thus, a high degree of confidence can be attached to any values in the fossil sample that exceed the reference maxima. This has implications for the work reported here, and also for the published studies of Kelley and Etler (1989) and Wood and Xu (1991).

Measures of Within-Species Variation in Hominoids

Our aim in providing comparative data on within-species variation in modern hominoids is to make available standards with which to compare the Paşalar variation. Given the restricted temporal and geographic range of the Paşalar apes, it would be most appropriate to use measures for geographic populations of each species rather than species' values across their entire range. These data are not at present available for most hominoids, although Groves *et al.* (1992) have given some idea of their importance in an analysis of orangutan M2 variation: Bornean orangs have greater M2 dimensions than Sumatran, but orangs from northwest Borneo differ more from Sumatran orangs than do those from southwest Borneo, so that it is likely that there are two size morphs within Borneo. Moreover, the subfossil orang teeth from Sumatra (Hooijer, 1948) are closer in size to the extant Bornean samples than to Sumatran ones. Although we have argued elsewhere (Alpagut *et al.,* 1990b) that the Paşalar hominoids are early representatives of the *Pongo* clade, there is no particularly compelling reason to view levels of variation in one species of extant ape as more relevant to the paleospecies problem at Paşalar than any other.

In the absence of locality data for our extant hominoid sample, we continue to take the most conservative approach by selecting the maximum value of each statistic for each parameter (Table 3) to create a hypothetical maximally variable hominoid (Table 4). We have also adjusted the values that we have calculated (Table 3) to reflect published data based on different and/or larger samples that produce results that exceed our own values (Kay 1982a; Wood and Xu, 1991). It is clear from Cope and Lacy's (1992) study that this approach is not really desirable. By definition, all species of modern apes are contained within, and in fact fall below, these values for the total set of parameters (since the maximum values reported come from several species) and every species falls below for some; that is, the maximum-variation hominoid is a construct that does not exist in real populations. This approach greatly reduces the power of the method because it will result in a large number of type 2 errors, but it has the benefit that it is almost inconceivable, in the absence of temporal or geographical biases, that type 1 errors could be made.

Coefficient of Variation. In the present study we have calculated CVs for males and females of each species separately as well as for mixed-sex (but not sex-balanced) samples of modern hominoids (Table 3). Kay (1982a,b), as well as Wood and Xu (1991), assumed that CV will be at a maximum in mixed-sex samples, but we did not find this to be the case for numerous measurements (Table 3). As the single-sex samples are about half the size of the mixed-sex samples, this may be showing the sensitivity of CV to sample size, since the same data points are included in both calculations. Clearly, studies of sexual dimorphism based on CV need to recognize this fact. For our work, we have used the maximum value encountered regardless of whether this derived from a

Table 3. Coefficients of Variation in Extant Hominoids (Table 1) Showing the Values for Each Sex and for a Mixed Sex Sample (not available for *P. paniscus*)

	P. troglodytes			*P. paniscus*		*G. gorilla*		
	m	f	m + f	m	f	m	f	m + f
I^1 M-D	5.44	***8.49***	7.97	8.74	6.73	7.56	6.94	8.24
I^1 B-L	5.31	6.78	7.69	7.59	5.16	9.45	7.38	11.27
I^2 M-D	7.66	7.63	8.32	8.86	8.86	12.11	***17.19***	15.52
I^2 B-L	4.18	5.73	7.01	8.22	5.63	12.69	15.71	17.65
C^1 Max. L.	8.81	4.70	13.15	8.11	5.56	6.46	5.66	20.00
C^1 Perp. B.	12.02	3.72	14.26	9.09	6.80	6.70	7.07	19.13
P^3 M-D	***6.89***	5.41	6.43	8.11	5.56	6.61	5.67	6.99
P^3 B-L	5.06	***5.60***	5.19	6.45	4.35	6.67	6.41	6.98
P^4 M-D	***4.55***	4.01	4.22	7.94	8.20	4.96	***6.43***	6.11
P^4 B-L	3.44	2.66	3.52	6.67	4.55	4.85	***6.51***	6.28
M^1 M-D	5.35	***5.39***	5.26	5.56	5.56	5.74	5.26	6.05
M^1 B-L	4.22	***5.26***	4.64	5.00	5.00	5.89	5.27	6.20
M^2 M-D	***6.40***	5.92	6.11	6.74	6.67	6.25	6.21	7.27
M^2 B-L	4.36	***5.68***	5.04	6.86	4.95	5.41	5.76	6.38
M^3 M-D	6.29	***6.55***	6.34	6.10	10.00	6.46	7.38	8.27
M^3 B-L	***9.11***	5.43	7.44	5.15	6.32	6.36	***7.84***	7.66
I_1 M-D	—	—	—	9.46	9.72	—	—	—
I_1 B-L	—	—	—	7.14	4.41	—	—	—
I_2 M-D	—	—	—	8.00	10.96	—	—	—
I_2 B-L	—	—	—	5.63	4.35	—	—	—
C_1 Max. L.	9.68	4.70	11.87	7.00	7.95	7.57	6.77	18.15
C_1 Perp. B.	8.29	4.41	11.55	5.26	10.77	6.45	6.39	16.65
P_3 Max. L	***5.42***	2.68	4.60	6.17	4.88	5.71	5.72	9.98
P_3 Perp. B.	***8.56***	5.76	8.41	16.22	12.86	9.29	9.04	11.41
P_4 M-D	***8.34***	3.60	6.86	5.63	11.43	5.84	5.02	6.71
P_4 B-L	5.12	***5.32***	5.20	8.97	7.89	4.79	***8.47***	7.49
M_1 M-D	***5.68***	5.03	5.37	4.08	6.12	3.89	4.23	4.79
M_1 B-L	***6.01***	4.78	5.48	5.62	6.82	3.56	***5.86***	5.48
M_2 M-D	***6.12***	4.50	5.41	6.12	5.88	6.00	4.99	6.52
M_2 B-L	***6.34***	5.39	6.01	6.52	6.59	3.93	***8.19***	7.01
M_3 M-D	***7.40***	3.47	5.99	4.44	6.59	5.68	7.00	8.00
M_3 B-L	***6.34***	3.51	5.51	4.76	6.98	5.35	***7.61***	7.50

Where the maximum value encountered in a species is from a single sex sample, this figure is set in bold italics.

single-sex or a mixed-sex sample. Values for these maxima are given in Table 4, together with the maximum values reported by Kay (1982a) and Wood and Xu (1991). These three data sets were combined to produce CV values that define the maximum variance found in living hominoids.

The least variable dental dimensions in hominoids are M1 length (CV = 7.1 for M^1 and 7.2 for M_1). When a fossil sample has a CV that exceeds these values, the sample should be further examined to determine whether its high CV results from small sample size, sex composition factors, or the inclusion in the sample of more than one species. A similar approach may be taken for each of the dental variables in turn (Tables 3 and 4). Since no published values for hominoid CV's

Table 3. (*Continued*)

P. pygmaeus			*H. syndactylus*			*H. lar*		
m	f	m + f	m	f	m + f	m	f	m + f
7.24	6.47	8.89	***9.92***	9.41	9.30	***8.57***	5.21	7.64
7.55	6.08	10.34	4.66	***7.71***	5.89	***12.83***	7.66	11.56
7.26	5.75	8.14	5.75	***6.12***	5.86	***8.62***	2.60	8.42
11.36	7.77	12.11	3.78	***5.20***	4.27	***10.16***	6.36	9.05
5.69	10.51	16.20	8.93	***14.21***	11.86	6.03	***7.86***	7.26
10.87	10.79	17.55	8.59	5.45	10.40	***9.36***	3.61	7.81
5.62	7.06	7.43	5.05	***9.06***	6.66	4.33	***6.73***	5.42
5.63	5.52	7.53	4.66	***5.81***	5.05	5.48	***5.82***	5.59
9.94	6.53	10.71	6.69	***8.02***	7.30	6.83	***7.90***	7.09
8.26	4.42	8.22	3.81	***7.27***	5.34	4.81	***5.75***	5.11
4.27	3.09	6.83	6.50	***6.89***	6.46	3.83	***6.01***	4.83
3.26	2.66	6.34	***3.75***	3.56	3.50	***4.41***	3.29	3.95
5.72	4.54	9.22	4.86	***5.38***	5.05	3.09	***7.42***	5.15
4.19	3.33	7.31	3.41	***5.06***	3.96	4.14	***4.46***	4.17
8.06	9.22	13.19	10.21	***11.02***	10.51	9.17	***12.73***	10.48
4.22	5.76	7.42	7.03	***7.53***	7.00	6.80	***7.26***	6.90
8.39	8.48	10.36	1.75	***6.57***	5.14	4.67	***6.16***	5.19
7.03	7.73	8.80	***6.34***	3.85	5.62	6.97	***9.45***	8.05
9.82	8.87	10.11	***13.09***	9.49	11.78	8.61	***8.73***	8.49
10.72	6.31	10.16	***9.64***	2.53	7.99	***10.28***	5.93	8.88
9.97	10.51	13.86	4.13	***5.37***	4.68	***6.75***	4.76	6.35
10.58	5.35	19.47	***10.48***	5.55	10.16	5.87	***6.95***	6.42
7.11	***16.27***	12.67	5.46	***9.03***	7.62	***5.32***	4.08	4.78
5.02	7.74	9.19	5.18	***7.95***	6.19	6.49	6.74	6.83
4.63	9.99	10.65	4.40	***14.05***	8.54	6.34	***8.94***	7.27
5.78	6.44	10.90	5.94	***6.45***	5.82	***7.40***	6.85	7.14
3.61	4.35	5.92	7.49	***8.85***	7.69	4.68	5.05	5.05
4.27	3.50	6.56	6.11	***8.20***	6.69	5.19	***5.78***	5.44
3.95	4.59	8.22	4.96	***8.07***	6.13	2.78	***6.04***	4.70
4.61	4.77	7.95	4.64	***6.59***	5.39	4.59	***6.17***	5.16
4.45	8.78	9.98	***11.95***	9.84	10.70	***7.48***	6.25	6.87
3.59	7.02	8.74	***8.70***	5.51	7.64	4.75	***5.39***	4.96

exceed this hypothetical, maximally variable hominoid (Table 4), a fossil sample whose CV lies beyond these values for even a single measurement should be treated as suspect as containing only a species.

Range and Range Expressed as a Percentage of Mean. Range statistics for modern hominoids are given in Table 5.

For the great ape species a regression of R% (Table 5) on CV (Table 3) for 24 variables (three taxa each) gave a correlation coefficient of 0.93, which confirms that R% is consistent with CV when moderately large samples (Table 1) are considered.

Maximum Divided by Minimum Index (MI) (Table 5). This has the advantage that even a pair of specimens can be compared to falsify a single species in-

Table 4. The Maximum Values of Coefficient of Variation Encountered in Hominoidea (Table 3)

	This study		Kay (1982a)	Wood and Xu, (1991)	Overall
I^1 M-D	*P. pygmaeus*	8.9	11.7	12.5	12.5
I^1 B-L	*H. lar*	12.8	13.9	11.2	13.9
I^2 M-D	*G. gorilla*	17.2	—	12.9	17.2
I^2 B-L	*G. gorilla*	17.7	—	12.9	17.7
C^1 Max. L.	*G. gorilla*	20.0	20.5	20.9	20.9
C^1 Perp. B.	*G. gorilla*	19.1	19.9	21.3	21.3
P^3 M-D	*P. paniscus*	8.1	9.1	11.0	11.0
P^3 B-L	*P. pygmaeus*	7.5	9.0	8.9	9.0
P^4 M-D	*P. pygmaeus*	10.7	8.3	10.7	10.7
P^4 B-L	*P. pygmaeus*	8.3	7.3	9.0	9.0
M^1 M-D	*P. pygmaeus*	6.8	6.5	7.1	7.1
M^1 B-L	*P. pygmaeus*	6.3	6.4	7.2	7.2
M^2 M-D	*P. pygmaeus*	9.2	8.1	10.2	10.2
M^2 B-L	*P. pygmaeus*	7.3	9.1	8.3	9.1
M^3 M-D	*P. pygmaeus*	13.2	9.2	12.2	13.2
M^3 B-L	*P. troglodytes*	9.1	10.7	9.8	10.7
I_1 M-D	*P. pygmaeus*	10.4	—	15.9	15.9
I_1 B-L	*H. lar*	9.5	—	12.8	12.8
I_2 M-D	*P. paniscus*	11.0	—	12.9	12.9
I_2 B-L	*P. pygmaeus*	10.7	—	13.3	13.3
C_1 Max. L.	*G. gorilla*	18.2	17.9	20.2	20.2
C_1 Perp. B.	*G. gorilla*	19.5	19.1	17.9	19.5
P_3 Max. L	*P. pygmaeus*	16.3	10.4	12.1	16.3
P_3 Perp. B.	*P. paniscus*	16.2	10.9	15.3	16.2
P_4 M-D	*P. paniscus*	11.4	7.3	9.3	11.4
P_4 B-L	*P. pygmaeus*	10.9	8.4	9.0	10.9
M_1 M-D	*P. paniscus*	6.1	6.1	7.2	7.2
M_1 B-L	*P. paniscus*	6.8	6.4	7.8	7.8
M_2 M-D	*P. pygmaeus*	8.2	6.7	7.9	8.2
M_2 B-L	*G. gorilla*	8.2	7.3	8.6	8.2
M_3 M-D	*P. pygmaeus*	10.0	8.7	9.7	10.0
M_3 B-L	*P. pygmaeus*	8.7	9.5	10.5	10.5

Notes: Column 1 shows the maximum CV for each tooth dimension for the samples in Table 3 (*H. syndactylus* was excluded as the sample is small). Columns 2 and 3 give the maximum values reported by Kay (1982a) for great apes and by Wood and Xu (1991). Column 4 is the maximum CV reported for any hominoid for each measurement.

terpretation. There is no prior division of the sample into male and female categories. The largest specimen in a sample is, however, most likely to be a male, and the smallest is most likely to be a female (this has been confirmed empirically, see Martin, 1983), and so by computing maximum/minimum × 100 (MI), an index that parallels the sexual dimorphism index is obtained. Like R%, this index will underestimate the variation at small sample sizes, giving it a tendency to yield type 2 errors. As with other range-based statistics, it is practically impossible for this statistic to produce type 1 errors, provided that the comparative sample is larger than the fossil sample.

Gender Determination and Sexual Dimorphism. Gender determination is critical for the quantification of sexual dimorphism and is also important for species recognition. It is particularly the case with Miocene hominoids that species identification problems have become intertwined with questions of overall levels of variation and also of sexual dimorphism. If two adult size categories of one sex can be demonstrated to exist in a fossil sample, then most biologists would accept that this sample could not be interpreted as a single species. Unfortunately gender determination for most teeth is difficult, except when relatively complete jaws are preserved. Relatively complete jaws may be sexed using canine to molar ratios (Martin, 1983: Table 2.9). This may provide a small sample of teeth the sex of which is known, but there are no Miocene ape samples in which this approach could be used to ameliorate the problem of species recognition.

The best approach would be one that enables sexing of isolated teeth. Kelley and Xu (1991) and Kelley (this volume) have recently reported a method that enables them to determine the sex of mandibular canine teeth. In this approach, indices of canine dimensions are plotted against one another to produce a bivariate plot in which there is no overlap between sexes. The indices used are maximum buccal crown height divided by maximum mesiodistal crown length on the Y-axis and mesial ridge length (from crown apex to the mesial corner of the lingual cingulum) divided by maximum buccal height on the X-axis. The method should allow sex determination of all unworn mandibular canines as well as any associated teeth. Unfortunately, canines are the most variable teeth in hominoids so that analyses of variation are complicated. However, if more than one size category of a sex were encountered, this would falsify the single species hypothesis. Unfortunately the converse is not necessarily true, as Cope's (1989) work has shown that mixed samples of sympatric *Cercopithecus* species fall into two size groups for canines, one all males and the other all females.

The overall level of sexual dimorphism is best measured by comparing mean values for males and females (Garn *et al.*, 1967; Martin, 1981). In their recent analysis of sexual dimorphism in the Lufeng hominoids, Kelley and Xu (1991) reported higher levels of molar sexual dimorphism than have been encountered in any extant anthropoid. These authors, and Wood and Xu (1991), however, have also argued that overall levels of variation at Lufeng exceed those of modern hominoids due to time averaging.

This raises the question of whether time averaging would affect measures of sexual dimorphism. It could be argued that mixing species, or combining sequential populations with differing means, would increase overall levels of variation but would not affect the ratio of male mean to female mean. If more than one species or population were included, the effect, if anything, would be to reduce the value of the dimorphism index to the average dimorphism value for both species/populations. Thus, if the sexes have been correctly determined then the dimorphism values must be at least as great in one of the component species or populations as for the combined sample, even if more than one taxon is included. Likewise, for a sample in which mean values are shifting over time, the ratio will be unaffected. The only real problem would be if distributions were very skewed, for example, if a sample comprised mainly males of a larger species and mainly females of a smaller species. In an evolving lineage this could happen if small specimens were more common at one end of the stratigraphic section and large specimens at the other. Such taphonomic and stochastic biases could occur but should not be assumed. It should be noted, however, that it has yet to be demonstrated that a high level of dimorphism results in a high level of species variation.

Bimodality. Bimodality has been used as a criterion for falsification of a sin-

Table 5. The Mean, Range, and Range-Based Statistics for the Hominoid Samples in Table 1[a]

	P. troglodytes					*P. paniscus*					*G. gorilla*				
	$\overline{X}$	Min	Max	R%	MI	$\overline{X}$	Min	Max	R%	MI	$\overline{X}$	Min	Max	R%	MI
I^1 M-D	11.5	9.6	12.8	28	1.33	10.4	8.9	11.9	19	1.32	13.2	11.2	15.7	***34***	***1.40***
I^1 B-L	9.1	7.7	10.3	29	1.34	7.7	6.8	9.2	31	1.35	10.4	8.5	12.6	39	1.48
I^2 M-D	8.7	7.2	10.0	32	1.39	7.9	6.9	10.1	41	1.30	9.2	6.0	11.6	***61***	***1.93***
I^2 B-L	8.4	7.2	9.3	25	1.29	7.2	6.3	8.5	31	1.33	9.5	6.2	12.7	***69***	***2.05***
C^1 Max. L.	12.7	10.5	15.6	40	1.49	10.0	8.2	13.3	51	1.62	17.8	12.5	23.2	***60***	***1.86***
C^1 Perp. B.	10.0	8.2	13.2	50	1.61	7.8	6.3	10.7	56	1.70	13.6	10.1	17.8	57	1.76
P^3 M-D	7.2	6.3	8.0	24	1.27	7.3	6.2	8.4	***30***	***1.35***	10.5	9.2	12.1	29	1.33
P^3 B-L	10.5	9.4	11.8	23	1.15	9.2	8.3	10.3	22	1.24	15.2	13.0	17.1	***27***	***1.32***
P^4 M-D	7.0	6.6	7.6	14	1.15	6.2	5.0	7.6	42	1.52	10.8	9.5	12.2	25	1.28
P^4 B-L	10.2	9.5	10.8	13	1.13	8.9	7.7	10.3	29	***1.34***	14.9	12.6	16.8	28	1.33
M^1 M-D	9.9	9.0	10.6	16	1.14	9.0	7.9	10.3	27	1.25	14.4	12.7	16.7	***28***	***1.31***
M^1 B-L	11.2	9.9	11.7	16	1.18	10.0	8.7	11.3	26	1.30	15.2	13.3	17.4	***27***	***1.31***
M^2 M-D	10.1	9.0	11.3	23	1.26	9.0	7.8	10.5	30	1.35	15.4	13.4	17.9	29	1.34
M^2 B-L	11.6	10.2	12.8	22	1.25	10.1	9.2	12.1	***29***	***1.32***	16.1	14.0	18.3	27	1.31
M^3 M-D	9.1	8.2	10.3	23	1.23	8.1	7.1	9.6	31	1.24	14.4	12.0	17.0	35	1.42
M^3 B-L	11.0	8.8	12.2	22	1.22	9.6	8.8	11.3	26	1.25	15.3	13.0	17.2	27	1.32
I_1 M-D	—	—	—	—	—	7.3	5.6	8.7	42	***1.55***	—	—	—	—	—
I_1 B-L	—	—	—	—	—	6.9	6.2	8.1	28	1.31	—	—	—	—	—
I_2 M-D	—	—	—	—	—	7.4	5.2	9.0	***51***	***1.60***	—	—	—	—	—
I_2 B-L	—	—	—	—	—	7.0	6.3	8.4	30	1.31	—	—	—	—	—
C_1 Max. L.	12.2	10.3	15.3	41	1.49	9.3	7.5	11.4	42	1.52	15.7	11.3	20.2	***57***	***1.79***
C_1 Perp. B.	9.8	8.0	12.2	43	1.53	7.0	5.8	8.9	44	1.47	12.3	9.1	15.5	52	1.70
P_3 Max. L	11.1	10.2	12.1	17	1.14	8.2	7.1	9.3	27	1.19	16.2	13.7	18.6	30	1.36
P_3 Perp. B.	7.8	7.0	9.8	36	1.40	7.2	5.2	9.7	***63***	***1.87***	11.2	8.2	13.8	50	1.68
P_4 M-D	7.7	6.8	9.0	29	1.29	7.0	5.4	9.1	***53***	***1.43***	11.2	9.8	13.0	29	1.33
P_4 B-L	8.9	8.0	9.8	20	1.23	7.7	5.6	9.2	***47***	***1.64***	12.9	10.5	14.4	30	1.37
M_1 M-D	10.6	9.8	12.2	23	1.24	9.8	8.5	11.9	***35***	1.22	15.3	13.6	16.8	21	1.24
M_1 B-L	9.7	8.8	11.0	23	1.25	8.8	7.4	9.8	***27***	***1.31***	13.2	11.6	14.4	21	1.24
M_2 M-D	11.0	10.2	12.6	22	1.24	10.0	8.3	11.4	***29***	1.18	16.7	14.7	19.2	27	1.31
M_2 B-L	10.6	9.5	12.3	17	1.29	9.1	7.7	10.5	***31***	***1.36***	15.0	12.8	16.8	27	1.31
M_3 M-D	10.3	9.2	11.6	23	1.22	9.1	8.3	10.6	25	1.14	16.9	14.2	19.2	30	1.35
M_3 B-L	10.1	9.4	11.4	20	1.19	8.5	7.4	9.4	24	1.27	14.8	12.4	17.5	***34***	***1.41***

[a]Min = the minimum value encountered; Max = the maximum value encountered. The maximum value for each measurement in all hominoids is set in bold italics.

gle species hypothesis (e.g., Oxnard, 1987). However, the data analyzed by Cope (1989) for *Cercopithecus* show that bimodal distributions of anterior dental dimensions reflect the sexual rather than the taxonomic composition of samples. For sympatric species, Cope (1989) found that bimodally distributed lengths of posterior teeth usually indicate multiple taxa in a sample. However, he also showed that spurious bimodal distributions can result from sampling error in single-species samples that match those actually containing more than one species (see also Martin and Andrews, 1984) and that some multiple-species samples show unimodal and highly symmetrical distributions. Overall, these results indicate that bimodality in postcanine tooth lengths may be taken to indicate the presence of multiple taxa in larger samples. In *Cercopithecus,* however, it is more common that pooled-species samples show overlapping distributions than two

Table 5. (*Continued*)

P. pygmaeus					*H. syndactylus*					*H. lar*				
$\bar{X}$	Min	Max	R%	MI	$\bar{X}$	Min	Max	R%	MI	$\bar{X}$	Min	Max	R%	MI
13.6	11.7	15.7	29	1.34	5.2	4.7	6.0	25	1.28	4.8	4.2	5.5	27	1.28
12.0	9.6	14.3	39	1.49	4.4	4.0	4.8	18	1.20	3.8	3.3	5.4	***55***	***1.64***
8.6	7.3	9.7	28	1.33	4.3	3.9	4.8	21	1.23	4.0	3.6	4.8	30	1.33
8.6	6.9	11.2	49	1.62	4.9	4.6	5.3	14	1.13	4.0	3.3	5.1	45	1.46
15.0	10.3	18.0	51	1.75	8.6	6.8	10.5	43	1.54	7.5	6.7	8.7	27	1.30
11.9	8.3	15.7	***62***	***1.89***	5.9	5.0	7.1	36	1.42	5.3	4.5	6.0	28	1.20
9.6	8.1	10.9	29	***1.35***	5.6	4.8	6.1	23	1.27	4.7	4.2	5.2	21	1.24
12.5	10.8	14.2	***27***	1.31	6.1	5.5	6.7	20	1.10	5.0	4.6	5.5	18	1.20
9.5	8.0	12.8	***51***	***1.60***	5.5	4.7	6.0	24	1.28	4.3	3.8	5.0	28	1.26
12.7	10.0	14.1	***32***	1.28	6.6	5.9	7.1	18	1.17	5.4	4.8	6.0	22	1.18
11.9	10.7	13.4	23	1.25	7.5	6.7	8.1	19	1.17	5.7	5.3	6.2	16	1.17
13.1	11.8	14.8	23	1.25	7.3	6.9	7.7	25	1.10	6.2	5.9	7.0	18	1.19
12.1	10.4	14.4	***33***	***1.38***	8.0	7.4	8.7	16	1.18	6.1	5.6	6.7	18	1.16
13.7	12.2	15.3	23	1.25	8.1	7.6	8.6	12	1.12	6.5	6.1	7.0	14	1.15
11.2	8.5	13.8	***47***	***1.62***	6.9	5.5	8.4	42	1.23	5.3	4.2	6.2	38	1.48
13.3	10.9	15.0	***31***	***1.38***	7.8	7.1	8.6	19	1.19	6.1	5.2	6.7	25	1.29
9.1	7.4	10.5	34	1.42	3.3	3.0	3.5	15	1.17	3.2	2.9	3.5	***44***	1.21
9.7	8.0	11.3	***34***	***1.41***	4.0	3.7	4.5	20	1.18	3.5	2.9	3.9	29	1.34
8.8	7.0	10.1	35	1.38	3.9	3.4	5.0	41	1.47	3.4	2.9	4.0	32	1.38
9.9	8.6	12.3	***37***	***1.41***	4.6	4.0	5.2	26	1.16	3.9	3.5	4.5	26	1.29
14.1	10.6	17.6	50	1.66	7.8	7.2	8.3	14	1.15	6.8	6.1	7.7	24	1.26
10.6	7.6	14.2	***62***	***1.87***	5.9	5.3	7.5	37	1.42	4.9	4.3	5.5	24	1.28
13.6	9.6	15.6	***44***	***1.63***	8.4	7.1	9.5	29	1.34	6.7	6.2	7.3	16	1.16
10.0	8.2	11.3	31	1.38	4.9	4.3	5.3	20	1.21	4.0	3.6	4.5	23	1.25
10.4	8.4	11.7	32	1.39	6.5	5.0	7.1	32	1.42	5.2	4.6	5.9	25	1.28
11.4	9.3	13.3	35	1.43	4.9	4.4	5.2	16	1.16	4.2	3.8	5.1	31	1.34
12.8	11.1	14.1	23	***1.27***	7.7	6.9	8.7	23	1.23	6.1	5.3	6.6	21	1.25
11.9	10.6	13.3	23	1.25	6.1	5.4	6.8	23	1.20	5.0	4.5	5.7	24	1.27
13.5	11.7	15.4	27	***1.32***	8.5	7.5	9.1	19	1.19	6.2	5.4	6.6	19	1.20
12.6	11.1	14.3	25	1.29	6.7	6.1	7.4	19	1.09	5.4	4.9	6.0	20	1.22
13.3	11.0	15.6	35	***1.42***	8.2	6.5	9.5	***37***	1.22	6.1	5.4	7.0	26	1.30
11.9	9.8	13.5	31	1.38	6.5	5.2	7.0	28	1.10	5.3	4.8	5.9	21	1.23

nonoverlapping clusters, one large and one small. We will now examine the Paşalar sample in light of this discussion.

The Paşalar Example

Introduction to the Site

The fossil site at Paşalar is situated on the edge of the Gönen basin in Western Anatolia about 75 kilometers west-southwest of Bursa (Andrews and Alpagut, 1990). The sediments are exposed in the side of a road cut, which led to their discovery by the Turkish–German lignite survey in 1968. Fossils collected there during the 1960s were described by Sickenberg *et al.* (1975), and the primates by Andrews and Tobien (1977), who assigned the hominoids to *Ramapithecus wickeri* and *Sivapithecus darwini* for the small and large specimens, respectively. Subsequently, *Ramapithecus* has been synonymized with *Sivapithecus* (e.g., Greenfield, 1979, Andrews and Cronin, 1982), with the specimens from East Africa being distinguished from *Sivapithecus/Ramapithecus* and returned to their original genus *Kenyapithecus* by Martin (1983). As a result of these changes, the specimens formerly attributed to *Ramapithecus wickeri* from Paşalar and Çandır were referred by Martin (1983) to the species *Sivapithecus alpani* (Tekkaya, 1974) described from Çandır. Similarly, the attribution of the large Paşalar hominoids to *Sivapithecus darwini* has been questioned (Kay, 1982a; Kay and Simons, 1983). In light of what we have learned about the morphology of the Paşalar hominoids as a result of the greatly enlarged samples, we concur with these workers' view that the Paşalar hominoids differ from both *K. wickeri* and from *S. darwini*. In addition, it is now clear that the pattern of subnasal morphology in the Paşalar sample requires that this material be removed from the genus *Sivapithecus* (Andrews *et al.*, in preparation).

In 1983, fieldwork was recommenced at Paşalar and has proceeded annually since that time. This has led to the recovery of many fossil mammals and to a more complete understanding of the geology of the site (Andrews and Alpagut, 1990), the nature of the sedimentary environment (Bestland, 1990), and the taphonomic processes that have influenced the composition of the Paşalar fauna (Andrews and Ersoy, 1990). For the purposes of the present paper, two major conclusions have emerged. First, the fossils were concentrated in a single unit of sands and gravels representing relatively high-energy conditions (Andrews and Alpagut, 1990; Andrews and Ersoy, 1990). The fossiliferous greenish-gray sand unit ranges from 1.1 to 1.6 m in thickness. On the basis of its sedimentary structure, Andrews and Alpagut (1990) and Bestland (1990) have argued that the sand unit was deposited extremely rapidly as the result of a single depositional event. Second, the evidence from regional geology (Andrews and Alpagut, 1990) and from the sediments themselves (Bestland, 1990) indicates that the fossils were derived from a restricted area no more than a few kilometers distant from the present site. In light of these findings, we feel that the animals represented in the Paşalar fauna sample an extremely limited geographical area ($\approx$ 25 km^2) that spanned only a limited range of altitude. As such, they offer a close approximation to the animals that utilize a small area over a period of, say, 100 years.

These conclusions are significant for the taxonomic interpretation of the hominoids, as they indicate that the fossils represent a nearly synchronous group of animals, so that the effects of time-averaging on sample variation is minimal or nonexistent. The restricted geographical, and probably ecological, extent of the catchment area also minimizes the effects of geographic variation. Both of these facts mean that the Paşalar fossils may be interpreted as both ecologically and geologically contemporaneous, i.e., they represent a close approximation to a living community of animals.

Hominoid Dental Morphology

The hominoid sample is morphologically homogeneous, particularly with regard to the molar teeth. However, two morphologies can be recognized for a number of teeth, particularly the upper central and lateral incisors, possibly the lower third premolar, and the M_3. Starting from the null hypothesis that a single species is represented in the Paşalar collection, the upper central incisors provide the most striking morphological evidence for the existence of at least two taxa at Paşalar.

Upper Central Incisors

There are two described types of upper central incisor (Alpagut *et al.*, 1990b). In one the tooth is moderately high crowned and spatulate (termed *morphology I*) (Fig. 1). It has a diffuse lingual swelling at the base of the crown that appears to be an outgrowth of the lingual cingulum in that traces of its origin from a pillarlike structure may still be seen. The lingual swelling is V-shaped in outline, broader towards the apex of the crown (i.e., incisally), but becoming less distinct as it broadens. Cervically, the swelling narrows to a point that is either continuous with the narrow lingual cingulum, which is also raised at this point, or is in close contact with it. The swelling is most sharply defined at this point, and the narrowness of the swelling and the adjoining cingulum give the crown the appearance of being relatively narrow when viewed from the lingual side. This impression is not justified by the overall length and breadth measurements of the crown, and because of the lack of repeatable landmarks on this part of the crown, it has not been possible to demonstrate this feature with more precise measurements. We have tried measuring height and breadth of the swelling for comparison with similar measurements of the lingual pillar present in morphology II, but this has proved ineffective. The swelling becomes shallower as it broadens out towards the apex of the crown, so that it rarely has any trace of wear on it. In the teeth with this morphology, wear is mainly restricted to the incisal edge, extending on to the lingual surface as far as, but not usually on to, the lingual swelling.

The second type represents more than 90% of the teeth in the Paşalar sample. The crowns appear relatively more robust due to a prominent lingual pillar (termed *morphology II*) (Fig. 1). The length/breadth dimensions of the crown show it to be labiolingually deep compared with mesiodistal length, but it is not certain whether this is because of a reduction in mesiodistal length or an expansion of the buccolingual dimension (Tables 6 and 7). Comparison of crown height shows that the crown is higher relative to the mesiodistal diameter than is the case for morphology I, and it may be that both this and the apparent greater robusticity can be attributed to the same phenomenon, namely, reduction in mesiodistal diameter. The lingual pillar is well defined throughout its length, and it runs from the base of the crown, where it is continuous with the very narrow lingual cingulum, to close to the incisal edge, gradually becoming narrower in the process. This is opposite to the shape of the lingual swelling in morphology I, which broadens towards the incisal edge of the crown. It is also different in that the pillar becomes more prominent incisally as well, and as a result it nearly always shows signs of wear as it encroaches from the incisal edge on to the lingual surface of the crown. The degree of wear is frequently so strong as to completely obliterate the whole of the pillar, leaving a pillar-shaped concave dentine pit in its place.

In comparison with other hominoids, morphology II is seen also in *Proconsul*

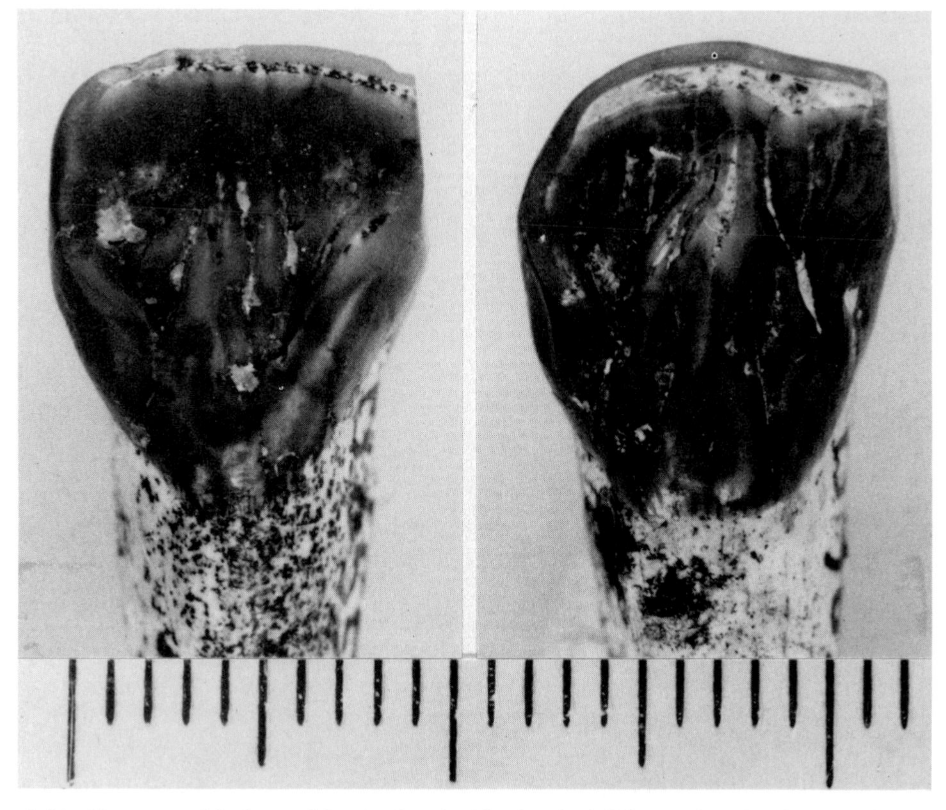

Fig. 1. Maxillary central incisors of Paşalar hominoids. On the left is an example of morphology I without a lingual pillar (D655, Rt I^1) and on the right is morphology II with a lingual pillar (D965, Rt I^1). The scale is in millimeters.

(Andrews, 1978), and *Dryopithecus* from Rudabánya, Hungary; Can Ponsic, Spain; and La Grive, France (an undescribed specimen currently on loan to the BMNH); *Lufengpithecus;* and *Paranthropus,* all of which have a prominent lingual pillar. It seems quite possible that this morphology represents the retention in the Paşalar material of an ancestral hominoid morphology. It may also be that the other morphology seen in the Paşalar sample, with a lingual swelling rather than a pillar, could be derived from this same ancestral morphology. It has not been observed in any other hominoid, fossil or modern, so that it appears to be unique to this part of the Paşalar sample. This morphological variation is not within the normal limits present in single primate species. The two morphologies are quite distinct, with no intermediates found so far, and since the sample size is better than is available for most samples of fossil hominoid, the lack of intermediates must be considered significant.

Upper Lateral Incisors

The lateral incisors display comparable differences to those seen in the centrals. The great majority of teeth have a narrow but distinct ridge running

Table 6. Samples Statistics for the Combined Paşalar Hominoid Sample for All Permanent Teeth

	n	$\bar{x}$	SD	CV	Min	Max	R%	MI
I^1 MD	43	9.43	0.87	9.23	8.2	11.5	***35.0***	1.40
I^1 BL	41	7.98	0.72	9.02	6.8	9.0	27.6	1.32
I^2 MD	43	5.60	0.49	8.75	4.5	6.3	32.1	1.40
I^2 BL	42	6.65	0.80	11.97	5.2	8.2	45.1	1.58
C^1 Max. L.	27	12.20	2.95	***24.18***	7.5	17.2	***79.5***	***2.29***
C^1 Perp. B.	26	9.51	2.17	***22.85***	7.0	12.7	59.9	1.81
P^3 MD buc.	33	7.70	0.58	7.55	6.5	8.7	28.6	1.34
P^3 BL	30	11.82	0.96	8.13	9.7	13.7	***33.8***	***1.41***
P^4 MD	50	7.33	0.70	9.52	6.2	8.6	32.7	1.39
P^4 BL	49	11.75	0.99	8.40	9.9	13.7	32.3	***1.38***
M^1 MD	37	10.06	0.90	***8.87***	8.7	12.0	***32.8***	***1.38***
M^1 BL	37	11.70	0.86	***7.34***	10.2	13.2	25.6	1.29
M^2 MD	78	11.43	1.17	10.24	9.1	14.0	***42.9***	***1.54***
M^2 BL	78	13.21	1.11	8.40	10.6	15.2	***34.8***	***1.43***
M^3 MD	61	10.99	1.25	11.39	9.0	13.0	36.4	1.44
M^3 BL	61	12.52	1.28	10.24	10.4	15.0	***36.7***	***1.44***
I_1 MD	19	5.15	0.43	8.37	4.4	5.8	27.2	1.32
I_1 BL	19	6.55	0.77	11.71	5.6	9.0	***51.9***	***1.61***
I_2 MD	32	5.47	0.49	8.87	4.0	6.2	40.2	1.55
I_2 BL	27	7.53	1.11	***14.70***	6.0	9.2	***67.7***	***1.53***
C_1 Max. L.	36	11.91	2.18	18.26	7.1	15.7	***72.1***	***2.21***
C_1 Perp. B.	36	8.78	2.02	***23.01***	5.4	12.0	***75.2***	***2.22***
P_3 Max. L	41	12.00	1.28	10.64	9.4	14.3	40.8	1.52
P_3 Perp. B.	43	7.56	0.67	8.85	6.3	8.5	29.1	1.35
P_4 MD	47	8.82	0.69	7.79	7.1	10.5	38.5	***1.48***
P_4 BL	47	9.76	0.71	7.30	8.1	11.1	30.7	1.37
M_1 MD	56	11.07	0.88	***7.93***	9.3	13.0	33.4	***1.40***
M_1 BL	56	10.03	0.79	***7.88***	8.4	11.5	***30.1***	***1.37***
M_2 MD	62	13.05	1.13	***8.69***	10.1	14.9	***36.8***	***1.48***
M_2 BL	62	11.77	1.22	***10.41***	9.4	14.0	***39.1***	***1.49***
M_3 MD	59	13.97	1.34	9.63	11.2	16.2	***35.8***	***1.45***
M_3 BL	60	11.46	1.26	***10.96***	9.2	14.0	***41.9***	***1.52***

Values of CV, R%, and MI for which the Paşalar sample exceeds the reference values for extant hominoids (Tables 4 and 5) are set in bold italics, those values of MI that also exceed variation in subfossil orangutans [data from Hooijer (1948), calculated by Alpagut *et al.* (1990b)] are underlined in addition.
n = Sample size; $\bar{x}$ = mean; SD = standard deviation; CV = coefficient of variation; min = minimum observed value; max = maximum observed value; R% = (range/mean × 100); MI = maximum observed value/minimum observed value.

down the lingual face of the crown, and this is the only feature in an otherwise gently rounded lingual surface. The second morphology differs in having the lingual face of the lateral incisors with multiple ridges of varying complexity. No dental size distinction has been found between samples with these two morphologies.

Molars

The lower molars can be divided into two groups on the basis of shape, and this is seen most clearly for the third molars. We have yet to complete the

Table 7. Sample Parameters of Upper Central Incisors for the Combined Sample (from Table 6) and for Separate Samples for Crowns with (Morphology II) and without (Morphology I) Lingual Pillars

	N	$\overline{X}$	SD	CV	Min	Max	R%	MI
I^1 MD								
Whole sample	43	9.4	0.87	9.23	8.2	11.5	35.0	1.40
Without pillar (I)	6	10.8	0.48	4.42	9.6	11.5	17.6	1.20
With pillar (II)	37	9.2	0.84	9.13	8.2	10.7	27.2	1.31
I^1 BL								
Whole sample	41	8.0	0.72	9.02	6.8	9.0	27.6	1.32
Without pillar (I)	6	8.5	0.26	3.04	7.1	8.9	21.2	1.25
With pillar (II)	35	7.9	0.72	9.02	6.8	9.0	27.6	1.32

analyses of these teeth, but preliminary indications are that the common morphology for third molars is having triangular crowns, tapering distally to a large and centrally placed hypoconulid. The second morphology, only observed on two teeth so far, has crowns with rectangular shape, with little or no distal taper, and hypoconulids small and buccally placed, as on the anterior molars. These crowns are also elongated and narrow (Fig. 2).

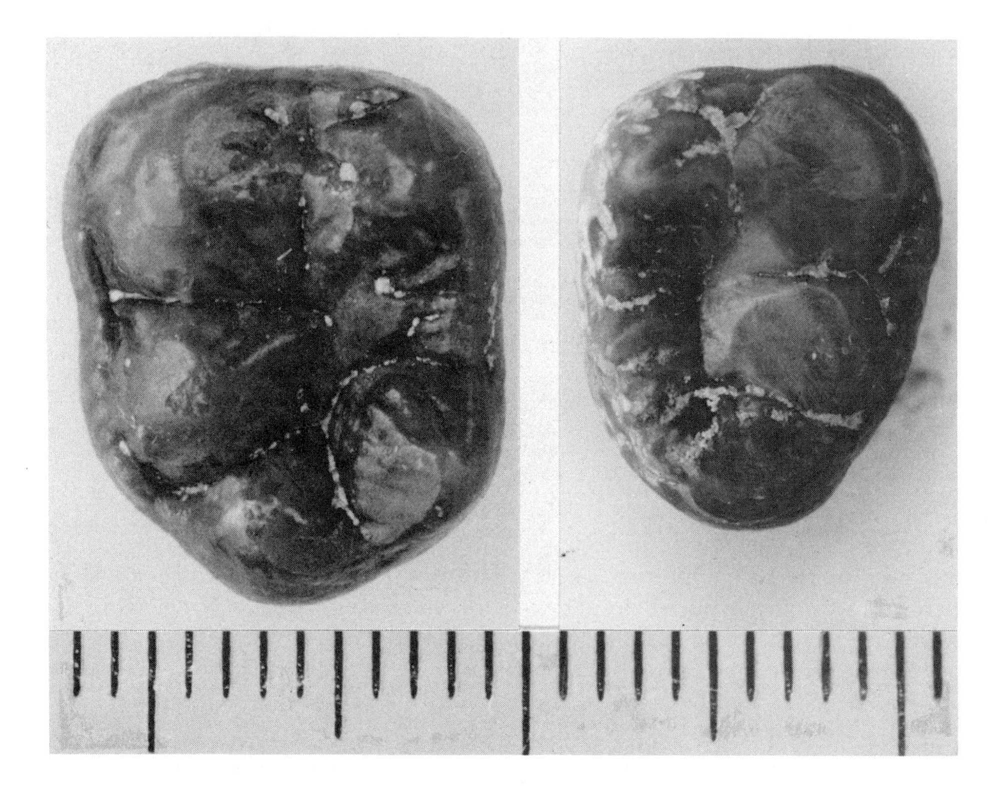

Fig. 2. Mandibular third molars of Paşalar hominoids. On the left is C281 (Lt M_3), which is rectangular with a small, buccally placed hypoconulid. On the right is C218 (Rt M_3), which is almost triangular in outline and has a large, distally placed hypoconulid. The scale is in millimeters.

As work progresses on the enormous sample from Paşalar, it is anticipated that further morphological differences will be documented. For the moment it is only possible to conclude that there are indications of more than one species at Paşalar on the basis of morphological heterogeneity. The problem is that we cannot identify differences on the majority of tooth types, so we are unable to form species hypodigms. Consequently, we are refraining, for the present, from recognizing this diversity taxonomically.

Metric Variation

Size variation in the Paşalar upper central incisors (Tables 6 and 7) is of the same order as the size variation of the same teeth in recent orangutans (Tables 3 and 5), and it exceeds the variation in subfossil orangutans (Table 6). This variation is comparable in overall extent to that of the whole of the Siwalik sample of central incisors, which on most interpretations contains at least two species (Greenfield, 1979; Kay, 1982a; Kelley and Pilbeam, 1986; Alpagut *et al.*, 1990b). Size variation does not correspond exactly with morphological differences. Morphology II is represented by both large and small teeth, with incisors lacking the lingual pillar (morphology I) clustered at the top end of the range of incisors having pillars, although newly recovered specimens include smaller examples (Alpagut, personal communication). The two samples are shown separately in Table 7. The CV values (Table 6) lie well within the range for modern hominoids. R% indicates the presence of more than one species based solely on size (Table 6), but when the I^1 sample is broken down into morphology I and II, both morphologies show levels of variation (as measured by R%) that are within the great ape range (Table 7). Thus, some of the size variation is correlated with the morphological groups, and when this is accounted for, size alone does not falsify single species hypotheses for either of the two groups. However, if one were to doubt the distinctiveness of the two morphological groups, then the single species hypothesis can be falsified purely on the basis of the overall degree of size variation (Table 6). The same kind of results apply for MI (Table 6), where the combined sample lies at the great ape maximum, which tends to support the presence of more than one species based on size alone. However, when the sample is separated into morphology I and II, both subsamples fall within the range of values for MI in extant great apes (Table 7). The great ape values exceed those encountered in the subfossil orangutan sample (Alpagut *et al.*, 1990b), so that it seems safe to conclude that a single species interpretation for Paşalar can be falsified based on the level of metric variation in I^1s. When I^1s are broken down into two morphologies, neither of these two subgroups show variation levels in excess of extant apes and cannot, therefore, be falsified as single species. Morphology II incisors show much greater variation in size than morphology I, but this may reflect the much greater sample size.

It has been argued above that the presence of more than one size group of canine teeth within a gender would falsify the single-species hypothesis. We have calculated the indices employed by Kelley and Xu (1991; Kelley, this volume) for all of the Paşalar lower canines that are complete and unworn or only moderately worn (Table 8). Of the lower canines assigned to females, the securely identified

Table 8. Gender Determinations for Paşalar Lower Canines

	MD length	Buc. ht	Mes. ridge	Ht/length	Mes. R./ht	Sex
BP 55	14.0	17.2	11.6	1.23	0.83	M
BP 56	13.5	17.1	12.6	1.27	0.74	?M
BP 57	13.2	17.0	11.8	1.29	0.69	?M
BP 59	14.0	20.0	13.6	1.42	0.68	M
C 173	12.1	19.1	13.3	1.58	0.70	M
C 203	13.7	16.4	12.9	1.20	0.79	M
D 115	9.3	11.1	5.5	1.19	0.50	F
D 179	11.7	16.9	13.8	1.44	0.82	M
D 770	13.5	19.7	15.2	1.45	0.77	M
E 310	9.0	10.0	4.9	1.11	0.49	F
E 428	13.1	16.4	11.7	1.25	0.71	?M
E 763	13.0	17.9	12.9	1.38	0.72	M
E 901	13.6	17.3	14.0	1.27	0.81	M
G 519	9.5	11.7	6.5	1.23	0.56	F
G 1313	12.3	16.7	12.7	1.36	0.76	M
G 1352	12.4	19.4	14.4	1.56	0.74	M
G 1793	15.3	17.8	13.8	1.16	0.78	?M
H 669	9.6	12.8	6.6	1.33	0.52	F
H 1646	12.8	16.0	11.4	1.25	0.71	?M

Gender determinations were made by plotting the values for the Paşalar specimens onto Kelley and Xu's (1991) Fig. 2. In general, the Paşalar values did not overlap the Lufeng values and only overlapped slightly modern mean values. Where values fell clearly with the male or female set, a sex was assigned. Where the specimen fell closer to one gender group than to another, it was determined as probable female/male (?F/?M).

teeth range in length from 9.0 to 9.6. The definite male teeth range between 11.7 and 14.0. Interestingly, 10 of the definitely sexed specimens are male and only four are female, and if the "probables" are included this ratio becomes 15:4. This fact may explain the low range for the females, and it also demonstrates a highly skewed sample, which might influence considerations of sexual dimorphism.

Size distributions for upper first molar dimensions show a division into two distinct size classes, a phenomenon not present in the samples of the other two molars. It would certainly be unusual for any single species sample based on an n of 37 to have a bimodal distribution for the length/breadth dimensions of the first molar, and so some weight must be attached to this distribution as indicating a pattern of variation that is not compatible with a single species interpretation.

All three lower molars show bimodality when numbers of specimens are plotted against size classes. The size distribution of M_2 is the most nearly continuous, and that of M_3 is most disjunct. Bimodality in the first molar, which is usually the least variable tooth in hominoid dentitions, is unusual for a single species, although the greater degree of bimodality in M_3 could be more easily explained because that tooth is usually more variable. Overall the discontinuous, bimodal distributions of the postcanine teeth are most compatible with a multiple species interpretation.

Coefficient of Variation

Values for CV in the Paşalar hominoid sample are compared with values reported for great apes in Table 6. Even though we have adopted reference maxima that make it extremely difficult to falsify a single species interpretation (see above), and thus make the analysis prone to type 2 errors, we find that 11 out of 32 CV values at Paşalar exceed the reference maxima. These include values for C^1, M^1, I_2, C_1, M_1, M_2, and M_3. The extant hominoid maximum values are derived from large comparative samples ($n = 64$; Wood and Xu, 1991) as well as those in Table 1. In addition, it seems likely that the Paşalar sample includes a number of antimeres, which would serve to reduce the variation at any sample size (Wood and Xu, 1991). These results effectively falsify the single-species interpretation on the basis of excessive metric variation.

Range/Mean Index

The values for R% in the Paşalar sample are compared with the range of values seen in great apes in Table 6. If we include I^1, then 16 of 32 values exceed the reference maxima. The teeth in question are I^1, C^1, P^3, M^1, M^2, M^3, I_1, I_2, C_1, M_1, M_2, and M_3, which simply represents an expansion of the CV data. However, the extant reference samples are smaller than those from Paşalar (not allowing for antimeres) for nine of these values, so that only seven are unequivocal. However, among these nine the samples of molars with values that exceed the reference samples' values also have CV values that falsify the single species hypothesis so that some of these may be significant in addition. This analysis thus falsifies the single species interpretation for the Paşalar sample because range-based statistics tend to have low power, but are also practically immune from type I errors when the fossil sample size is smaller than that of the comparative samples. In light of the morphological differences in I^1, separate values of R% were calculated for the two phena. This could be considered to represent a second-order null hypothesis (that two species are present) based on morphology. Interestingly, a single-species hypothesis cannot be falsified either for the morphology I or the morphology II sample (Table 7).

Maximum ÷ Minimum Index

The MI index values for the Paşalar sample are compared with the range of values in great apes in Table 6. Reference maxima are exceeded by the Paşalar sample in 18 out of 32 measurements (on C^1, P^3, P^4, M^1, M^2, M^3, I_1, I_2, C_1, P_4, M_1, M_2, M_3), which complements and further extends the results from CV and R%. It should be noted that of these 18, 11 are based on fossil sample sizes that are larger than those for the extant reference samples, which means that only seven can be taken to be unequivocal. However, it should also be noted that for five of these less certain values the single species interpretation was falsified by the CV analysis, and these may therefore be significant also. These results falsify the single species interpretation for the Paşalar hominoids.

If we ignore the geological and taphonomic evidence that the sample is temporally and geographically restricted, we can evaluate the variation at Paşalar in relation to the subfossil orangutan sample, which derives from a large geographic area and covers a great deal of time (Hooijer, 1948). The Paşalar sample shows more variation in MI than the maximum values in extant great apes and in subfossil orangs for 5 out of 32 measurements (Table 6). These include C^1, I_1,

C_1, M_1, and M_2, and in all cases the subfossil samples exceed the Paşalar sample size. We consider this an effective falsification of a single species interpretation, even if the sample at Paşalar were considered to be time-averaged, and/or to sample a range of habitants and geography.

Summary

Each approach to the analysis of metric variation in the Paşalar hominoid sample provides grounds to reject the single species hypothesis. The existence of two distinct I^1 morphs with no known intermediates is also evidence for the presence of more than one species. Thus, we reject the single species hypothesis for the Paşalar primates on two grounds: (1) excessive metrical variation, and (2) morphological heterogeneity in I^1. We should therefore examine the second-order null hypothesis that two species are represented.

In the absence of established metric criteria to evaluate this second-order hypothesis, and given our inability at present to sort most tooth types into discrete groups, we begin with the I^1s. When the I^1 sample is subdivided into morphology I and II, the metric variation of each subsample is within the range for a single species of great ape. Thus, the single species hypothesis cannot be falsified for either of the two subsamples. Morphology I is presently known only for medium to large specimens (MD = 9.6–11.5, BL = 7.1–8.9), and these lie at the upper end of the size distribution for the whole sample in mesiodistal diameter (8.2–11.5). Morphology II includes all of the smallest I^1s, but also a large sample of medium to large I^1s (MD = 8.2–10.7, BL = 6.8–9.0). Mesiodistal length separates the two samples best and shows that the morphology I species is medium to large, while the morphology II species is small to medium, at least in terms of I^1 MD length, with relatively little overlap in their size distributions. Thus, based on morphology and metrics, there is evidence for two species whose size ranges overlap. The larger species is rare; the smaller species, with morphology II I^1s, is much more common. If this pattern extends to canines and other parts of the dentition, then there may be no need to argue for the presence of more than two hominoid species at Paşalar, while recognizing that this is simply the second-order null hypothesis.

Comparison to Çandır

The Paşalar hominoids were originally compared with material assigned to *Ramapithecus wickeri* from Fort Ternan, Kenya and Çandır, Turkey (Andrews and Tobien, 1977). Martin (1983) argued that the Kenyan material should be returned to its original genus, *Kenyapithecus*, and that the Paşalar and Çandır species should be recognized as *Sivapithecus alpani*. The teeth in the single mandible from Çandır all fall at, or slightly below, the bottom of the range of variation of the Paşalar sample. Morphologically they are identical, and there is no evidence to reject the null hypothesis that the same species is represented at the two sites. The trivial name for this species is *alpani* Tekkaya 1974. It seems most likely that this is the common, smaller species at Paşalar.

Comparison to Neudorf

The Paşalar material was also originally compared with similar-sized specimens of *Sivapithecus darwini* from Neudorf, Czechoslovakia by Andrews and Tobien (1977). From the same site at Neudorf, however, is another hominoid called *Griphopithecus suessi,* based on a single tooth (numbered NHMW 15) that has been described variously as an M^1 or M^2, and most often as a deciduous dp^4 (Abel, 1902; Remane, 1921). The tooth is heavily worn but has little dentine exposure. Although much of the cervical enamel has been fractured away, the crown height is substantial where preserved, and we are confident that the tooth is a permanent molar. If the tooth were really a dp^4, then its size would preclude it from belonging to either of the species that is present at Paşalar. In any case, its morphology is completely different from that of the sample of dp^4s from Paşalar, although we feel that it differs in the direction of permanent molars. *Griphopithecus suessi* Abel 1902 has frequently been synonymized with *Sivapithecus darwini* on the grounds that its holotype is a deciduous tooth of the latter species. It has been considered that if the tooth is a permanent first molar then it would be too small to belong to the hypodigm of *S. darwini.* However, if the tooth is an M^1, it lies towards the low end of the range of variation at Paşalar for that tooth. The holotype of *S. darwini,* a lower third molar, also falls within the Paşalar range, somewhat below the mean value. If the relative size patterns from M^1 to M_3 in the Paşalar sample hold for the Neudorf hominoids, then it is impossible, at present, to reject the null hypothesis that all of the Neudorf material belongs to a single species. *G. suessi* has page priority over *S. darwini* but, following Remane's (1921) act as first reviser (Article 24 of the ICZN), *suessi* is the junior subjective synonym of *darwini,* which makes the genus *Griphopithecus* available through the type species *Griphopithecus darwini* Abel 1902.

The inferred thick enamel of the Neudorf teeth is one of the few characters shared with the Paşalar sample, but this has been shown to be general for middle to late Miocene hominoids (Kay, 1981, 1982a; Martin, 1983, 1985; Andrews and Martin, 1991). Similarly, the presence of lower molar cingula shared by the Paşalar sample and *G. darwini* from Neudorf is also present in other middle Miocene genera, e.g., *Kenyapithecus,* so that the two key characters used by Andrews and Tobien (1977) to assign the Paşalar sample to *S. darwini* have proved to be nearly ubiquitous for middle Miocene apes. The level of morphological similarity between the Neudorf and Paşalar hominoids, however, is sufficiently great to assign the Paşalar species to *Griphopithecus;* but we are unable to maintain the specific identification of the Paşalar hominoid proposed by Andrews and Tobien (1977), and we are following the proposal of Martin (1983) to refer the bulk of the Paşalar sample to the species *Griphopithecus alpani* Tekkaya 1974.

Taxonomic Conclusions

Thus, the Paşalar hominoid sample comprises at least two species. The smaller, and more common one of these is conspecific with the hominoid at Çandır and is provisionally assigned to *Griphopithecus alpani.* The larger and rarer species could be linked to a variety of middle Miocene taxa, but we cannot elaborate on its status further at the present time.

Conclusions

Studies of metric variation should be directed towards attempts to falsify the null hypothesis that a single species is represented in a morphologically homogeneous fossil sample. The CV has the greatest analytical power, but can potentially result in either type 1 or type 2 errors. Use of reference maxima and confidence limits reduces power by increasing type 2 errors. Simulation studies such as those of Cope and Lacy (1992) should be carried out to provide reference values at sample sizes appropriate to the fossil sample under analysis. Range-based statistics have low analytical power but are practically immune to type I errors if comparative sample sizes exceed those of the fossils. A combination of the two approaches can only improve the accuracy of determinations of species numbers. Time-averaging can be shown to increase levels of variation for whole samples that belong to a single lineage. Rather than being assumed as the standard effect, however, it should be demonstrated that variation overall exceeds variation within finer stratigraphic levels and that the mean for the sample shifts according to stratigraphic position. In the absence of these criteria, time-averaging should not be invoked to explain away unacceptably high levels of variation.

In the Paşalar sample, size variation falsifies the null, single species hypothesis based on three different statistics: CV, R%, and MI. In addition, two morphologies of I^1 are recognized. On these grounds, we reject the null, single species hypothesis and consider the second-order null hypothesis that at least two species of hominoid are represented at Paşalar. For the present, the morphology of I^1 provides the clearest indication of how the sample may be subdivided into two groups. When this is done the metric variation of each subgroup, morphology I and II, is contained within the variation levels of reference samples so that single species hypotheses cannot be rejected for either subgroup. If the pattern of size distribution for I^1 extends to the other teeth, it seems at least possible that division of other dental elements into subgroups might mean that the null hypothesis, that each of these subgroups represents a single species, could not be rejected based on levels of variation. Thus we believe that the secondary null hypothesis, that two species are represented may be robust. The most commonly represented species is provisionally recognized as *Griphopithecus alpani,* which is also known from Çandır. There is a second, rarer and larger species that is unnamed at the present, and whose generic affinities are presently unclear. It seems plausible that these two taxa are sister species, as we have so much difficulty in telling the hypodigms apart, but it is by no means clear to which hominoid genus either of them is most closely related.

ACKNOWLEDGMENTS

We are deeply grateful to Berna Alpagut for her indefatigable efforts to bring Turkish paleoanthropology back to life and for permission to work on the new hominoid fossils from Paşalar. We are also grateful to the many people who have helped with the field work at Paşalar, particularly to Libby Andrews and Wendy Martin, who have contributed so greatly to the collection of the primate specimens. We are grateful to Roshna Wunderlich for her photographic work and to Helen Giles for typing the tables. We thank Dana Cope, John Fleagle, Terry Harrison, Jukka Jernvall, Jay Kelley, Mike Lacy, and Mark Teaford for helpful comments on the manuscript and Bill Kimbel for thorough editorial work, in addition to valuable comments. The Paşalar excavation was funded by the Turkish Directorate of Antiquities (Ministry of Culture and Tourism) and the University of Ankara, Department of Palaeoanthropology. Our work on the Paşalar hominoids has been made possible through the generous support of the L.S.B. Leakey Foundation and the Boise Fund (Oxford University). The work reported here was supported in part by a grant from the National Science Foundation, BNS 89 18695.

References

Abel, O. 1902. Zwei neue Menscenaffen aus den Leithakalkbildungen des Wiener Beckens. *Sitzungsberichte der Akademie der Wissenschaften. Mathematisch-Naturwissenschaftliche Classe* **111:**1171–1207.

Alpagut, B., Andrews, P., and Martin, L. 1990a. Miocene Paleoecology of Paşalar, Turkey, in: E. H. Lindsay, V. Fahlbusch, and P. Mein (eds.), *European Neogene Mammal Chronology,* pp. 443–459. NATO ASI Series A: Life Sciences Vol. 180. Plenum Press, New York.

Alpagut, B., Andrews, P., and Martin, L. 1990b. New hominoid specimens from the Middle Miocene site at Paşalar, Turkey. *J. Hum. Evol.* **19:**397–422.

Andrews, P. J. 1978. A revision of the Miocene Hominoidea of East Africa. *Bull. Br. Mus. (Nat. Hist.) (Geology)* **30(2):**85–224.

Andrews, P. 1981. Species diversity and diet in monkeys and apes during the Miocene, in: C. B. Stringer (ed.), *Aspects of Human Evolution,* pp. 25–61. Taylor and Francis, London.

Andrews, P. 1990. Palaeoecology of the Miocene fauna from Paşalar, Turkey. J. Hum. Evol. **19:**569–582.

Andrews, P., and Alpagut, B. 1990. Description of the fossiliferous units at Paşalar, Turkey. *J. Hum. Evol.* **19:**343–361.

Andrews, P., and Cronin, J. E. 1982. The relationships of *Sivapithecus* and *Ramapithecus* and the evolution of the orang-utan. *Nature* **297:**541–546.

Andrews, P., and Ersoy, A. 1990. Taphonomy of the Miocene bone accumulations at Paşalar, Turkey. *J. Hum. Evol.* **19:**379–396.

Andrews, P. J., and Martin, L. 1987. The phyletic position of the Ad Dabtiyah hominoid. *Bull. Br. Mus. (Nat. Hist.) (Geology)* **41(4):**383–393.

Andrews, P., and Martin, L. 1991. Hominoid dietary evolution. *Philos. Trans. R. Soc. Lond.* **334:**199–209.

Andrews, P., and Tobien, H. 1977. New Miocene locality in Turkey with evidence on the origin of *Ramapithecus* and *Sivapithecus. Nature* **268:**699–701.

Andrews, P., Lord, J. M., and Nesbit Evans, E. M. 1979. Patterns of ecological diversity in fossil and modern mammalian faunas. *Biol. J. Linnean Soc.* **11:**177–205.

Bestland, E. A. 1990. Sedimentology and paleopedology of Miocene alluvial deposits at the Paşalar Hominoid site, Western Turkey. *J. Hum. Evol.* **19:**363–377.

Cope, D. A. 1989. Systematic Variation in *Cercopithecus* Dental Samples. Unpublished Ph.D. dissertation, The University of Texas at Austin.

Cope, D. A., and Lacy, M. G. (1992). Falsification of a single species hypothesis using the coefficient of variation: A simulation approach. *Am. J. Phys. Anthropol.* **89:**359–378.

Flynn, L. J., Pilbeam, D., Jacobs, L. L., Barry, J. C., Behrensmeyer, A. K., and Kappelman, J. W. 1990. The Siwaliks of Pakistan: Time and faunas in a Miocene terrestrial setting. *J. Geol.* **98:**589–604.

Garn, S. M., Lewis, A. B., Swindler, D. R., and Kerewsky, R. S. 1967. Genetic control of sexual dimorphism in tooth size. *J. Dent. Res.* **46(5, Suppl.)**:963–973.

Gingerich, P. D. 1974. Size variability of the teeth in living mammals and the diagnosis of closely related sympatric fossil species. *J. Paleontol.* **48**:895–903.

Gingerich, P. D. 1985. Species in the fossil record: concepts, trends, and transitions. *Paleobiology* **11**:27–41.

Gingerich, P. D. 1987. Evolution and the fossil record: patterns, rates, and processes. *Can. J. Zool.* **65**:1053–1060.

Gingerich, P. D., and Schoeninger, M. 1979. Patterns of tooth size variability in the dentitions of primates. *Am. J. Phys. Anthrop.* **51**:457–466.

Greenfield, L. O. 1979. On the adaptive pattern of "*Ramapithecus*." *Am J. Phys. Anthropol.* **50**:527–548.

Groves, C. P., Westwood, C., and Shea, B. T. 1992. Unfinished business: Mahalanobis and a clockwork orang. *J. Hum. Evol.* **22**:327–340.

Hooijer, D. A. 1948. Prehistoric teeth of man and of the orang-utan from central Sumatra, with notes on the fossil orang-utan from Java and southern China. *Zoologische Mededeelingen (Leiden)* **29**:175–301.

Johanson, D. C. 1974. Some metric aspects of the permanent and deciduous dentition of the pygmy chimpanzee (*Pan paniscus*). *Am. J. Phys. Anthropol.* **41**:39–48.

Kappelman, J. 1986. The Paleoecology and Chronology of the Middle Miocene Hominoids from the Chinji Formation of Pakistan. Unpublished Ph.D. dissertation, Harvard University.

Kappelman, J., Kelley, J., Pilbeam, D., Sheikh, K. A., Ward, S., Anwar, M., Barry, J. C., Brown, B., Hake, P., Johnson, N. M., Raza, S. M., and Shah, S. M. I. 1991. The earliest occurrence of *Sivapithecus* from the middle Miocene Chinji Formation of Pakistan. *J. Hum. Evol.* **21**:61–73.

Kay, R. F. 1981. The nut-crackers—A new theory of the adaptations of the Ramapithecinae. *Am. J. Phys. Anthropol.* **55**:141–151.

Kay, R. F. 1982a. *Sivapithecus simonsi,* a new species of Miocene Hominoid, with comments on the phylogenetic status of the Ramapithecinae. *Int. J. Primatol.* **3**:113–172.

Kay, R. K. 1982b. Sexual dimorphism in Ramapithecinae. *Proc. Natl. Acad. Sci. USA* **79**:209–212.

Kay, R. F., and Simons, E. L. 1983. A reassessment of the relationship between later Miocene and subsequent Hominoidea, in: R. L. Ciochon and R. S. Corruccini (eds.), *New Interpretations of Ape and Human Ancestry,* pp. 577–624. Plenum Press, New York.

Kelley, J. 1986. Species recognition and sexual dimorphism in *Proconsul* and *Rangwapithecus. J. Hum. Evol.* **15**:461–495.

Kelley, J., and Etler, D. 1989. Hominoid dental variability and species number at the Late Miocene site of Lufeng, China. *Am. J. Primatol.* **18**:15–34.

Kelley, J., and Gingerich, P. D. 1991. The effects of time-accumulation on metric variability in fossil samples (abstract). *Soc. Vertebr. Paleontol.* **11 (Suppl.)**: 39A.

Kelley, J., and Pilbeam, D. 1986. The Dryopithecines: Taxonomy, comparative anatomy, and phylogeny of Miocene large hominoids, in: D. R. Swindler and J. Erwin (eds.), *Comparative Primate Biology, Volume 1: Systematics, Evolution, and Anatomy,* pp. 361–411. Alan R. Liss, New York.

Kelley, J., and Xu, Q. 1991. Extreme sexual dimorphism in a Miocene hominoid. *Nature* **352**:151–153.

Marshall, J., and Sugardjito, J. 1986. Gibbon systematics, in: D. R. Swindler and J. Erwin (eds), *Comparative Primate Biology,* Vol. 1: *Systematics, Evolution, and Anatomy,* pp. 137–185. Alan R. Liss, New York.

Martin, L. 1981. New specimens of *Proconsul* from Koru, Kenya. *J. Hum. Evol.* **10**:139–150.

Martin, L. B. 1983. The Relationships of the Later Miocene Hominoidea. Unpublished Ph.D. dissertation, University College London.

Martin, L. 1985. Significance of enamel thickness in hominoid evolution. *Nature* **314**:260–263.

Martin, L., and Andrews, P. 1984. The phyletic position of *Graecopithecus freybergi* Koenigswald. *Courier Forschungsinstitut Senckenberg* **69**:25–40.

Oxnard, C. E. 1987. *Fossils, Teeth and Sex. New Perspectives on Human Evolution.* University of Washington Press, Seattle.

Pickford, M. 1986. Sexual dimorphism in *Proconsul, in:* M. Pickford and B. Chiarelli (eds.), *Sexual Dimorphism in Living and Fossil Primates,* pp. 133–170. Il Sedicesimo, Firenze.

Pilbeam, D. R. 1969. Tertiary Pongidae of East Africa: Evolutionary relationships and taxonomy. *Bull. Peabody Mus. Nat. Hist.* **31:**1–185.

Plavcan, J. M. 1990. Sexual Dimorphism in the Dentition of Extant Anthropoid Primates. Ph.D. Dissertation, Duke University.

Remane, A. 1921. Zur Beurteilung der fossilen anthropoiden. *Zbl. Min. Geol. Paleontol.* **11:**335–339.

Rose, K. D., and Bown, T. M. 1986. Gradual evolution and species discrimination in the fossil record. *Contrib. Geol., Univ. Wyom., Special Paper* **3:**119–130.

Ruff, C. B., Walker, A., and Teaford, M. F. 1989. Body mass, sexual dimorphism and femoral proportions of *Proconsul* from Rusinga and Mfangano Islands, Kenya. *J. Hum. Evol.* **18:**515–536.

Sickenberg, O., Becker-Platen, J. D., Benda, L., Berg, D., Engesser, B., Gaziry, W., Heissig, K., Hünermann, K. A., Sondaar, P. Y., Schmidt-Kittler, N., Staesche, K., Staesche, U., Steffens, P., and Tobien, H. 1975. Die Gliederung des höheren Jungtertiärs und Altquartärs in der Türkei nach Vertebraten und ihre Bedeutung für die internationale Neogen-Stratigraphie. *Geologisches Jahrbuch* **B15:**1–167.

Simpson, G. G. 1941. Range as a zoological character. *Am. J. Sci.* **239:**785–804.

Simpson, G. G., Roe, A., and Lewontin, R. C. 1960. *Quantitative Zoology*. Harcourt and Brace, New York.

Tekkaya, I. 1974. A new species of anthropoid (Primates, Mammalia) from Anatolia. *Bull. Min. Res. Explor. Inst. Turkey (Ankara)* **83:**148–165.

Wood, B. A., and Xu, Q. 1991. Variation in the Lufeng dental remains. *J. Hum. Evol.* **20:**291–311.

Taxonomic Implications of Sexual Dimorphism in *Lufengpithecus*

17

JAY KELLEY

Introduction

Sexual dimorphism frequently complicates the determination of species numbers in fossil samples. This has been particularly true for anthropoid primates, among whose living members are some of the most highly size dimorphic extant terrestrial mammals (Jungers and Susman, 1984; Markham and Groves, 1990). A number of Miocene hominoid fossil assemblages have been interpreted as sampling a single large-bodied species displaying substantial body-size dimorphism, usually inferred from apparently very high levels of dental dimorphism, since most hominoid samples preserve few or no postcranial remains. These include sites from the early (Songhor, Koru, Rusinga), middle (Maboko, Ft. Ternan, Nachola, Paşalar), and late (Siwaliks, Lufeng, Rudabanya, Ravin de la Pluie) Miocene (de Bonis, 1983, 1984; Kelley and Pilbeam, 1986; Kelley, 1986a,b; Pickford, 1985, 1986a,b; Wu and Wang, 1987; Kelley and Etler, 1989). In the case of at least one species, from the early Miocene site of Rusinga in Kenya, it has been suggested that dental and body-size dimorphism were perhaps greater than in any living anthropoid (Kelley and Pilbeam, 1986; Kelley, 1986a; Pickford, 1986a).

For many of these sites, the single-species alternative has been rejected because the necessary levels of dental variation and dimorphism were consid-

JAY KELLEY • Department of Oral Biology, College of Dentistry, The University of Illinois at Chicago, Chicago, Illinois 60612.

Species, Species Concepts, and Primate Evolution, edited by William H. Kimbel and Lawrence B. Martin. Plenum Press, New York, 1993.

ered to be too great by extant species standards (Kay, 1982a,b; Wu and Oxnard, 1983a,b). In the case of the Rusinga sample, the same argument has been made regarding the large postcranial sample (Teaford *et al.*, 1988, this volume; Ruff *et al.*, 1989). Proposed alternatives to single-species hypotheses have invariably involved a morphologically similar large and small species pair, each being either minimally dimorphic or showing essentially no dimorphism at all (Kay, 1982a,b; Wu and Oxnard, 1983a,b; Ruff *et al.*, 1989). Many of these samples were in fact originally described as containing two species, one large and one small, or more than two largely size-defined species. Partly as a result of these taxonomic disagreements, the case for very great sexual dimorphism in many Miocene hominoid species is still in doubt.

However, not all taxonomic situations are equally ambiguous, and not all levels of implied sexual dimorphism appear unacceptable by modern species standards. There seems to be general agreement, especially among those who work directly with the material and based on the most recent and thorough taxonomic analyses, that sites such as Maboko, Rudabanya, Ravin de la Pluie, and Lufeng contain a single large-bodied hominoid species (de Bonis, 1983; Martin and Andrews, 1984; Pickford, 1985, 1986b; Wu *et al.*, 1986; Wu, 1987; Kordos, 1987, 1988; Kelley and Etler, 1989; Schwartz, 1990; Wood and Xu, 1991). Other sites, such as Paşalar, may contain more than one large hominoid species, but with most of the material belonging to one highly dimorphic large-bodied species (Alpagut *et al.*, 1990; Martin and Andrews, this volume).

In spite of the near consensus concerning the taxonomy of these samples, and indications that sexual dimorphism may have been marked, there are reasons why estimates of sexual dimorphism that are both precise and reliable cannot be calculated and why, therefore, these species do not serve as convincing demonstrations of extreme sexual dimorphism. First, sample sizes are often small, particularly of more complete material such as associated dentitions. Second and more importantly, whether or not sample sizes are adequate, it has not been possible, for the most part, to reliably determine the sex of individual specimens (Kay 1982a,b; Oxnard, 1987). The reason for this is that in most skeletal and dental features, males and females overlap in size (Oxnard, 1987), and using only size extremes that can be sexed with some confidence obviously produces exaggerated measures of dimorphism. As an alternative to sexing individuals, several workers have attempted to estimate the degree of sexual dimorphism indirectly by using measures of overall species variation (e.g., Fleagle *et al.*, 1980; Kay, 1982a,b; Leutenegger and Shell, 1987) or a ratio based on the largest and smallest specimens (e.g., Martin and Andrews, 1984; Alpagut *et al.*, 1990). Such indirect measures are broadly useful for qualitative comparisons and approximate ranking by degree of dimorphism, but they lack the necessary precision to be useful and reliable quantitative measures.

To obtain reliable estimates of sexual dimorphism that are sufficiently precise to evaluate the claims of extreme or even unprecedented sexual dimorphism in fossil hominoids, we need a fossil sample that satisfies the following conditions:

1. The taxonomy must be secure and based on a number of different criteria; this requires large samples and relatively complete material from a temporally and geographically restricted setting.

2. There must be a means to sex reliably individual specimens.
3. Sample sizes must be sufficiently large and specimens sufficiently complete to provide accurate measures of sexual dimorphism that are amenable to statistical analysis.

Until recently no such fossil hominoid sample was available. Now, however, there is a sample, from the late Miocene locality of Lufeng, Yunnan Province, China that can be demonstrated to meet these conditions. The Lufeng site has produced a large hominoid sample comprising more than 750 specimens from a single outcrop. Included are several complete skulls and a large number of complete and partial jaws with dentitions, or isolated but associated teeth representing single individuals (Wu *et al.*, 1983, 1984, 1985; Wood and Xu, 1991). While the temporal span represented by the approximately 6 meters of hominoid-bearing sediment is uncertain (Badgley *et al.*, 1988), this is a relatively restricted stratigraphic interval and there are no indications of size changes over time (see below).

Species Number at Lufeng

One Species or Two?

The Lufeng large hominoid sample was originally described as comprising two species, one large and one small (Xu *et al.*, 1978; Xu and Lu, 1979). This followed prevailing notions about species composition in Miocene hominoid samples, particularly regarding the presence of *Sivapithecus* and "*Ramapithecus*" (Simons, 1976). Oxnard and his colleagues (Wu and Oxnard, 1983a,b; Oxnard *et al.*, 1985; Oxnard, 1987) indirectly affirmed the presence of two species by claiming that a single-species alternative for the sample would require an unacceptable degree of sexual dimorphism. However, a two-species alternative was never evaluated on its merits, other than to claim that samples of large and small dental specimens each showed evidence of sexual dimorphism. Most importantly, no one carried out a comparative analysis of metric variation in the large dental samples with respect to the one- and two-species alternatives.

Recently, such an analysis was carried out by Kelley and Etler (1989) using the dental metric data from Lufeng available in the literature. This study demonstrated that in a large and small two-species alternative, each of the species would have numerous measures of metric variation substantially below minimum values and lower 95% confidence intervals of extant apes. This was most evident in the absolute ranges of measurements for individual tooth positions (Table 1), based on samples of between 18 and 46 teeth at each position, with most of the samples having more than 25 teeth (Wu and Oxnard, 1983a). In contrast, a single-species hypothesis could not be rejected by these same measures of variation (Kelley and Etler, 1989). Wood and Xu (1991) carried out a complementary study of detailed occlusal morphology, cusp proportions, and dental metrics in the Lufeng teeth. They also failed to find any compelling reason for recognizing the presence of more than one species. Additionally, both studies noted features in the canine and premolar samples that conform to the

Table 1. Range/Mean: Lufeng Tooth Breadths (Two-Species Alternative)[a]

	Extant hominoid Lower 95% limit	Small species	Large species
I^1	0.22	**0.17**	**0.20**
I^2	0.28	**0.25**	—
C^1	0.34	**0.13**	**0.24**
P^4	0.15	0.22	—
C_1	0.27	**0.20**	**0.21**
P_3	0.21	**0.13**	**0.18**
P_4	0.20	—	**0.13**
M_3	0.18	0.21	0.20

[a]Modified from Kelley and Etler (1989). Dashed lines indicate tooth positions for which Lufeng data were unavailable. Lufeng values lower than the 95% confidence limit for extant hominoid species are shown in boldface. Extant hominoid range/mean confidence limits calculated from data in Martin (1983). Values for I^1 and I^2 incorporate range/mean data from one additional, unpublished sample of gorillas. Lufeng values calculated from data in Wu and Oxnard (1983a,b) and Oxnard (1987). Lufeng ranges determined from upper and lower limits, respectively, of highest and lowest histogram intervals for each tooth position; actual ranges, and therefore range/mean values, may in some cases even be slightly less.

patterns of within-species sexual dimorphism in extant apes. Regarding Wu and Oxnard's (1983a,b) claim of evidence for sexual dimorphism in the two dental size groups, Xu and Wood (1989) demonstrated that the bimodalities in tooth dimension histograms upon which this was based are likely distributional artifacts resulting from the choice of particular measurement intervals.

Furthermore, support for a single-species alternative does not rest entirely on dental criteria. The pattern of morphological differences that distinguishes the large and small Lufeng crania is precisely the same as that which distinguishes males from females among extant, sexually dimorphic great apes (Wu *et al.*, 1983, 1986). This in itself does not preclude the presence of more than one species, but it does demonstrate that in the crania, as in the dentition, the morphological variation that is present is of a kind to suggest gender rather than species differences.

As noted by both Kelley and Xu (1991) and Martin (1991), there are two-species alternatives for Lufeng other than that in which there is one large and one small species, alternatives in which the two species broadly overlap in size. This appears unlikely at Lufeng. Metric distributions for the postcanine teeth are distinctly bimodal and, in the case of M_3, even slightly disjunct (Fig. 1). Unlike the bimodal postcanine distributions demonstrated for the partitioned samples by Wu and Oxnard (1983a,b), those for the entire sample can be shown not to depend on the chosen histogram interval. While extant primates generally do not show bimodal postcanine distributions, they can occur in very highly dimorphic species (Martin and Andrews, 1984). If the sample comprised two species that differed somewhat in average size but broadly overlapped, one

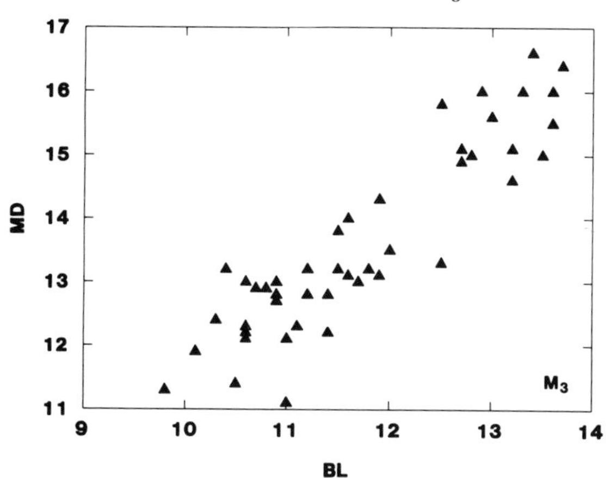

Fig. 1. Bivariate plot of Lufeng M_3 specimens.

would expect unimodal or platykurtic rather than bimodal distributions (see Plavcan, this volume; Cope, this volume). Still, bimodal distributions can be produced with two-species samples, particularly with unequal sampling of the two species and of males and females within one or both.

Equally important is the distribution of known males and females within the total sample of each postcanine tooth (see Kelley and Xu, 1991, and below for the method of sexing individuals). In the Lufeng sample, known males and females do not overlap and they are separated at the interval between the two modes (Fig. 2). This pattern is repeated at each postcanine tooth position, with sample sizes for P_4-M_2 of approximately ten each for both males and females. One can attempt to model this pattern using both single-species and two-species samples, and for the latter by varying the representation of the two species and of the sexes. This was attempted using a single-species sample in which the degree of sexual dimorphism was artificially heightened over that of the orangutan, a combined orangutan and gorilla sample, and an orangutan sample to which a number of male gorillas had been added to create a two-species sample with the greatest likelihood of duplicating the Lufeng pattern (larger species represented only by males). After repeated sampling of these data sets, only the heightened dimorphism, single-species sample produced patterns that approximated that of the Lufeng distribution of males and females within a bimodal distribution (Fig. 3). The results of this limited analysis suggest that the Lufeng distributions might not be easily reproduced in a two-species sample. Preliminary results from more comprehensive and analytically rigorous modeling experiments indicate that to obtain the Lufeng pattern from a two-species sample would require highly improbable sampling (Kelley and Plavcan, manuscript in preparation).

The reason that it is so difficult to produce the Lufeng pattern of dental metric distributions from a two-species sample is that the two salient characteristics of these distributions, bimodality and little or no overlap between males and females, increase in likelihood through opposing tendencies. The likelihood of a bimodal distribution increases as the size difference between the two species increases. However, as the two species diverge in size, the likelihood of male–female overlap also increases. Decreasing this likelihood requires that the two

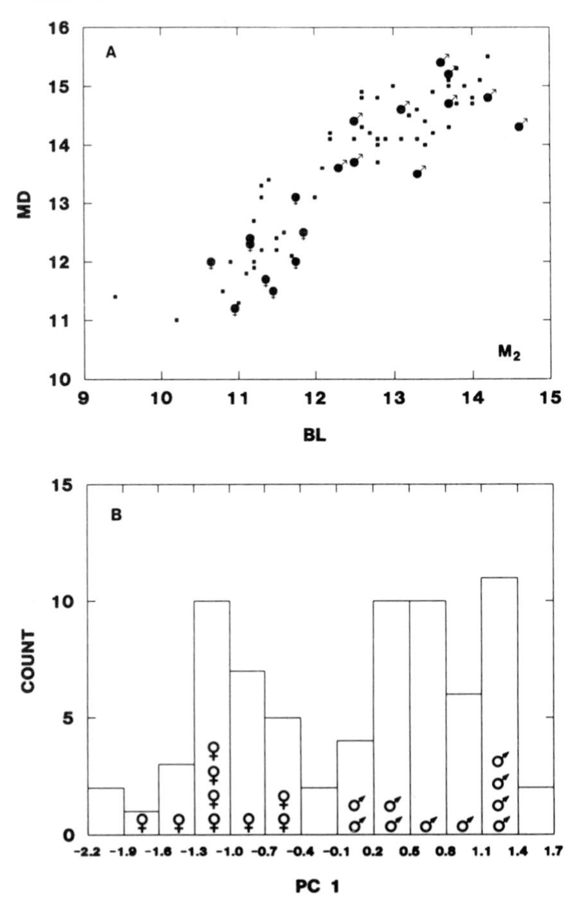

Fig. 2. (A) Bivariate plot of Lufeng M_2 specimens, with sexed specimens indicated by male and female symbols; sexing methods described in text. (B) Histogram of Lufeng M_2 specimens based on a principal components analysis of tooth length and breadth. PC1 equals the first principal component coefficient. The result is the equivalent of positioning each tooth along the major axis of the bivariate plot.

species be more nearly equal in size. The exercise is made even more difficult with the need to maintain the other sample limitations of the Lufeng sample, such as total range at each tooth position and amount of within-sex variation.

Martin (1991) claimed that the Lufeng coefficients of variation (CVs), some of which are very high (Kelley and Etler, 1989; Wood and Xu, 1991), indicate the presence of more than one species. The greater amount of information contained in the *patterning* of the variance illuminates sample characteristics that cannot be addressed through less revealing measures such as the CV or range–mean. The patterning of the variance reveals that a two-species taxonomy is actually unlikely. Since the elevated CVs were the only aspects of the sample presumed to require more than one species, there remains no compelling reason to question the remaining body of evidence indicating the presence of only one species at Lufeng (Kelley and Etler, 1989; Wood and Xu, 1991), *Lufengpithecus lufengensis* (Wu, 1987).

Extant Primate Models and Species Number at Fossil Hominoid Sites

Cope (this volume) and Plavcan (this volume) have both presented the results of modeling experiments using cases of sympatric, congeneric cer-

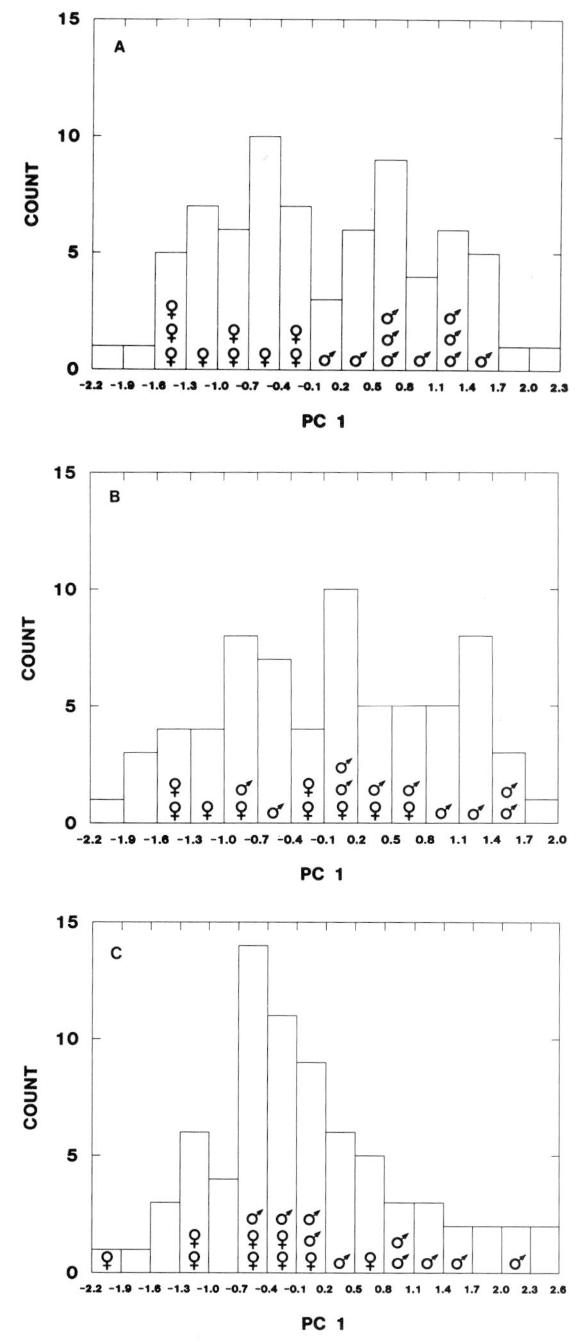

Fig. 3. Representative histograms of lower second molars from three simulated situations: (A) a single species that is more dimorphic than the orangutan; (B) two equally represented dimorphic species that overlap in dental size; and (C) two overlapping species with unequal representation. The smaller species comprises most of the individuals and is very dimorphic, while the larger species is represented only by male individuals; all sample sizes approximately equal to that for Lufeng M_2. For each trial in each of the three situations, ten males and nine females were randomly selected to equal the numbers of sexed specimens in the Lufeng sample. (A) was produced by removing eight of the largest females and eight of the smallest males in an orangutan sample and replacing them with an equal number of individuals that were, respectively, smaller than the mean female size and larger than the mean male size; (B) was produced by selecting equal numbers of male and female orangutans and gorillas; (C) was produced by randomly removing eight individuals from an orangutan sample and replacing them with eight male gorillas.

copithecoid monkeys as a means of testing the utility of measures of postcanine dental variation for determining species number in fossil samples. Each found that mixed-species samples can produce measures of variation (CV, range/mean) that lie within the ranges obtained from single-species samples. Thus, they conclude that, on the basis of dental metric criteria alone, it cannot be determined with confidence that fossil samples with measures of variation equivalent to those of modern single-species samples do in fact contain only one species, especially when those measures are near the upper limits of the single-species ranges. Further, Plavcan, in particular, concludes that species diversity in the fossil record, and presumably at single sites, is underestimated; that is, it is likely that there are numerous instances of what one might call disguised species (see also Plavcan, 1989).

These are well-conceived and meticulous analyses, but there are reasons to question the applicability of their results to the Miocene hominoid fossil record, particularly that of the middle and late Miocene. Species are biological entities, not mere statistical entities. The factors that govern species distributions, including patterns of sympatry and allopatry, are real biological attributes, not just abstract statistical probabilities. Thus, as in any area of biology, we must be extremely careful about applying indiscriminantly the conclusions derived from one group to other groups. Living apes and cercopithecoid monkeys have very different distribution patterns, and I would like to suggest several biological attributes that might in large part be responsible for these differences.

Life History

First is the suite of attributes encompassed under the heading of life history parameters. These include such things as length of gestation period, age at weaning, interbirth interval, age at sexual maturity, and longevity. Apes and monkeys have very different life history profiles (Harvey and Clutton-Brock, 1985; Harvey *et al.,* 1987). Apes have life history profiles characterized by a longer gestation period, advanced age at weaning with a subsequently longer interbirth interval, advanced age at sexual maturity and first birth, and a longer life span. This is particularly evident in a comparison of the life history profiles of gibbons with similarly sized or larger monkeys (Harvey *et al.,* 1987). The ape profile is one of prolonged maturation and slow population replacement.

If we assume that the very low population densities of apes (Clutton-Brock and Harvey, 1977; Waser, 1987) reflect saturation for the particular environment (Pianka, 1978) related in some way to their maturational profile combined with the distribution of food resources, then it may be that there is a low carrying capacity limit, not only for the numbers of individuals in any given ape species (population density), but also in the total number of species that can be supported in any given area (species density). This would place an effective limit on the number of ape species that could occur sympatrically unless population densities were reduced below their already very low levels. This may partly explain why when apes are found in sympatry today, they are fundamentally different ecologically (reflected in easily differentiated morphologies, including dental morphology), and, even with pronounced differences in ecology, there are comparatively few taxa, generally no more than two species as compared to six or more for monkeys (Gautier-Hion, 1978; Waser, 1987).

Admittedly, we do not know that Miocene apes possessed extant ape life history profiles. Nevertheless, the modern end points of the cercopithecoid/ hominoid divergence are more strikingly divergent in life history parameters than in many other attributes (Smuts *et al.,* 1987), and these had to become manifest at some point during the evolutionary history of the two groups. I suggest that the modern ape life history profile is characteristic of at least the great ape and human clade, and it is not unreasonable, given a similar profile in gibbons, to view it as among the defining elements, perhaps the paramount defining element, of the Hominoidea (Kelley, 1992a).

It may be, too, that we do not have to rely on conjecture in this matter. It has been documented that a life history profile associated with prolonged maturity and slow replacement, such as that of apes, is correlated with a relatively large brain (Sacher, 1959; Harvey *et al.*, 1987; Harvey and Bennett, 1983). Therefore, measures of relative brain size for fossil hominoids (e.g., Walker *et al.*, 1983) might allow some inferences to be drawn about life history parameters. However, this requires great precision and accuracy in estimates of both variables, and the uncertainties associated with estimates of body size (Ruff *et al.*, 1989; Jungers, 1990; Fortelius, 1990) and brain size (note two decades of controversies over hominid brain sizes) would probably render such measures unreliable. Recently, another method for inferring life history parameters from fossil data has been put forward, based on the time of eruption of the first molar (Smith, 1989). If the approximate age of eruption can be established using indicators of developmental increments on the teeth (Bromage and Dean, 1985; Beynon and Wood, 1987; Beynon and Dean, 1988), then it may be possible to infer whether various Miocene hominoids had life history profiles closer to those of living monkeys or apes. It may be possible to apply this method to certain taxa with presently available material.

Body Size

Another relevant biological attribute, and one that is closely correlated with life history parameters (Harvey *et al.*, 1987; Maiorana, 1990), is body size. There are a number of studies that document a decrease in numbers of species in sympatry with increasing body size (Hutchinson and MacArthur, 1959; Pianka, 1970; Stanley, 1973; Fleming, 1973; Van Valen, 1973; Peters, 1983; Maiorana, 1990). Interestingly, this trend may be more pronounced in tropical forests than in other types of habitats (Fleming, 1973; Maiorana, 1990).

It seems reasonable to presume that what holds for large animals as a whole holds for closely related species as well, but to my knowledge this has not been tested. Such a study was suggested some time ago by Hutchinson and MacArthur (1959), who proposed that "It might be possible to gain further information by an intensive examination of the incidence of related sympatric species of similar size in various groups of large and small animals in different kinds of environments." While this size effect appears to be most pronounced at very large body sizes (Fleming, 1973; Maiorana, 1990), most Miocene hominoids, particularly those of the later Miocene, were in the neighborhood of an order of magnitude or more larger than the monkeys used by Cope and Plavcan for their modeling (Fleagle, 1988). These differences in size must be taken into account when discussing the probability of multiple ape species at given fossil sites.

There are situations that run counter to the general trend regarding body size and species number, such as that of open country bovids in East Africa (Sinclair and Norton-Griffiths, 1979). Nearly 20 different bovid species occupy the Serengeti Plain over the course of the year, with body sizes ranging from under 10 kilograms to several hundred kilograms (Eisenberg, 1981; Janis, 1990). These fall into three general body size groups, with each group containing species pairs or multiple species with substantial body size overlap. Additionally, like cercopithecoid monkeys, bovids are notoriously difficult to diagnose from dental morphology (C. Janis, personal communication).

As fossil assemblages, determinations of species number within size groupings would be difficult based only on dental remains (although it would be fairly straightforward with the use of horn cores, which would also be included in the assemblages). However, this is a situation that would be unlikely to be duplicated among forest hominoids because of (1) the previously noted strengthening of the body size/sympatry trend within tropical forests, (2) the fact that many of the Serengeti bovids are transitory migrants, a lifestyle not found among forest species, and (3) the fact that many of the bovid species are specialist herbivores, which allows a greater degree of sympatry. Further, and probably due in large part to the latter two factors, savannah habitats are unusual in comparison to other kinds of habitats, particularly forests, in the numbers of large herbivorous species that they support (Maiorana, 1990).

Morphological Patterns

Apart from considerations of ape and monkey biology, there are other reasons to question the applicability of the cercopithecoid models of Cope and Plavcan to the Miocene hominoid record. Cercopithecoids are notable for the degree of morphological homogeneity in their dentitions, especially the molar dentition (Szalay and Delson, 1979). The dental homogeneity we see in sympatric cercopithecid species is a reflection of this overall homogeneity. Apes, by contrast, are quite heterogeneous, even if we restrict the comparison only to extant apes. Among fossil apes, there are easily recognizable dental differences between many sites, even those that are approximately contemporaneous (e.g., Haritalyangar/Siwalik U-level and Lufeng; Siwalik Chinji level and Rudabanya), and samples that are relatively homogeneous in the postcanine dentition often are strikingly different in the anterior teeth (Kelley and Pilbeam, 1986). Among apes, taxic diversity is accompanied by dental morphological diversity to a degree not seen in extant or fossil monkeys (Kelley and Pilbeam, 1986; Szalay and Delson, 1979).

If fossil hominoid sites have numerous cases of disguised species, then a very different pattern from that of monkeys would have to predominate, one in which a high degree of within-site dental homogeneity is accompanied by marked between-site heterogeneity. Indeed, what is most striking about many middle-late Miocene hominoid sites is that there is only one ape morphology represented. Further, large hominoids are relatively rare at most sites (although not at Lufeng), suggesting low population densities, even if only one species is represented. If these samples represent multiple species, then population densities would had to have been exceptionally low, even taking into account taphonomic factors (Kelley, 1992b).

In the early and middle Miocene of Africa, we do have sites (Rusinga, Koru,

Songhor, Maboko, Ft. Ternan) bearing multiple catarrhine taxa in numbers equal to those of modern sites with similarly sized sympatric anthropoid species (Harrison, 1989b; Waser, 1987; Oates *et al.*, 1990; Chapman and Chapman, 1990). At these fossil sites, species can be differentiated by their molar dentitions; in fact, they are diagnosed primarily by molar morphology (Andrews, 1978; Harrison, 1989a).* There is disagreement as to whether some, any, or all of these species are apes as opposed to primitive catarrhines (Harrison, 1987; Andrews and Martin, 1987), but the pattern is nonetheless different from that of cercopithecoid monkeys. Disguised species within these already diverse catarrhine faunas would necessarily entail remarkable degrees of species diversity at single locales.

Apes as a Relict Group

Within the same forest regions in which live the cercopithecoid species used by Cope and Plavcan for their modeling, there are apes. They have presumably been subject to the same history of biotic/abiotic pressures as the monkeys, and yet they have strikingly different patterns of species numbers and species distributions. Why choose the monkeys as models for the taxonomies and patterns of sympatry/allopatry of fossil hominoids when the (almost surely) biologically more similar apes are available? Why, in particular, use the genus *Cercopithecus,* which has the most unusual patterns of distribution of any anthropoid genus, including not only multiple-species sympatry but numerous mixed-species groups as well (Struhsaker, 1981; Gautier-Hion, 1988; Norconk, 1990)?

The reason usually offered is that the living apes are an impoverished group with a relict distribution and that they are therefore suspect as a comparative model in questions such as these (e.g., Martin, 1991). The implication is that apes have, for most of their history, been more speciose within the areas they occupy than they are today. The latter is a notion that seems to have had its genesis in the comparison of the modern ape and monkey faunas, and the sense from the fossil record that the community "roles" of apes in the Miocene were the same as those of monkeys in the present. It also seems to be tied into hypotheses of the competitive replacement of apes by monkeys (Andrews, 1981; Lovejoy, 1981). Not only is evidence to support these positions lacking, but there are good reasons for supposing them to be largely inaccurate (Kelley, 1992a,b).

It is true that apes were more geographically widespread throughout much of the middle to late Miocene than they are today, and during this period there were perhaps more species at any given point in time than at present. In this sense the extant apes do represent an impoverished group, particularly in Eurasia. But more species overall as a consequence of expanded range does not necessarily imply greater species density. It does not necessarily follow from this that patterns of species sympatry and allopatry were in any way different in the past than they are at present, and that apes were necessarily formerly more like extant cercopithecoid monkeys in this respect. In fact, the apparent cooccur-

*At each of these sites, by all proposed taxonomies, there is only one large anthropoid species, or one large and one moderately large species (Kelley, 1986a; Harrison, 1989b; Ruff *et al.*, 1989).

rence at many middle and late Miocene Eurasian sites of one large hominoid and one small catarrhine (*Pliopithecus, Dionysopithecus*) (Ginsburg, 1986; Mein, 1986; Wu and Pan, 1984; Rose, 1989) closely mimics what we find over much of present-day or late Pleistocene Southeast Asia with the sympatric occurrence of single species of *Pongo* and *Hylobates*.

It is perhaps more instructive, and more pertinent, to view this question from a continental rather than a global perspective. In Africa, excepting the genus *Papio* and to an extent *Cercopithecus aethiops*, apes and monkeys presently inhabit comparable geographic areas west of the rift. Are only the apes to be considered a relict fauna? If not, then their very much lower species densities cannot be attributed to this phenomenon. If the relative impoverishment of apes (in species numbers) *is* a result of relict status, why have monkeys not been similarly affected? It would seem that some explanation other than relict status for apes is necessary to explain differences in species density between the two groups (Kelley, 1992b).

In further consideration of the applicability of cercopithecoid models and the relict status of apes, New World cebids are more similar to cercopithecoid monkeys than to apes in species diversity and density (as expected, given their sizes and life history characteristics), but they are like apes in that instances of sympatric, congeneric species are uncommon (A. Rosenberger, personal communication). Excluding the callitrichines, these are apparently limited to the genus *Cebus* (Terborgh, 1983; Waser, 1987). Further, like apes, cebids as a whole display a great deal of dental morphological heterogeneity (Szalay and Delson, 1979), and sympatric species are, for the most part, easily distinguished by their dental morphology. Thus, even if extant apes are less speciose as a result of relict status, formerly greater species richness in itself would not constitute an argument for cercopithecoid-like patterns of sympatry and the likelihood of disguised species among fossil apes.

Thus, cercopithecoids are highly unusual among living anthropoids in their patterns of species distributions, and I would suggest that the models of Cope and Plavcan, while of the greatest importance for interpreting the cercopithecoid fossil record, are of questionable utility for the fossil record of hominoids. Species density and the patterns of sympatry/allopatry of the modern ape fauna, despite its impoverishment globally, might well be representative of ape distributions over much or most of their evolutionary history (Kelley, 1992b). Rather than speculating on the presumed impoverishment of the modern ape fauna and the possible reasons for this (Lovejoy, 1981), perhaps we should turn our attention to trying to better understand why extant apes have the distributions that they do, with relatively few sympatric species and pronounced ecological/morphological differences where they occur, and why they differ so greatly from cercopithecoid monkeys in this respect.

Sexual Dimorphism in Lufengpithecus

Determination of the degree of sexual dimorphism in *Lufengpithecus* has been reported elsewhere (Kelley and Xu, 1991) but will be repeated in some detail here. Indices of sexual dimorphism were calculated for the postcanine

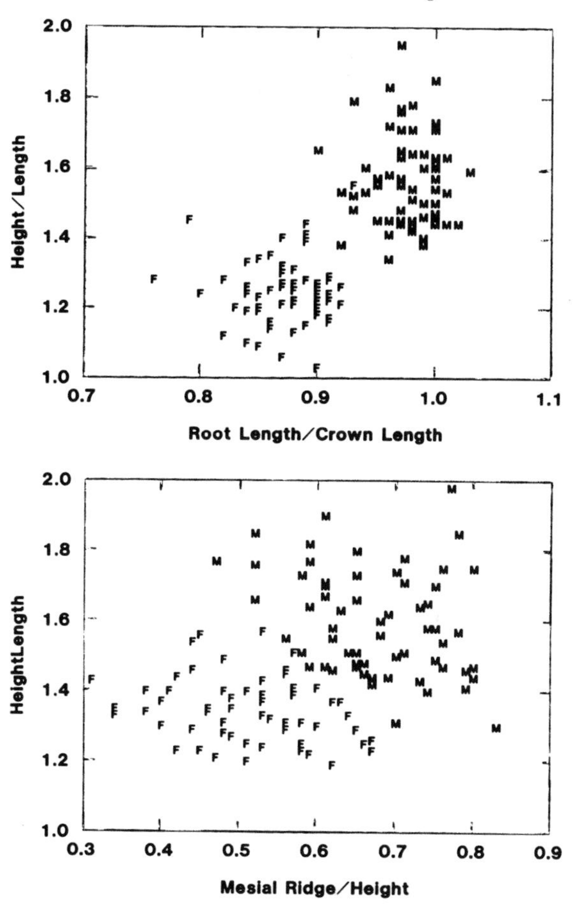

Fig. 4. Bivariate plots of upper (A) and lower (B) canine shape indices for extant great apes; M, male individuals; F, female individuals. Height/length: Maximum buccal crown height/ maximum mesiodistal length. Root length/crown length: Root length/crown length measured at the cervix. Mesial ridge/height: Mesial ridge length (measured from the crown apex to the mesial corner of the lingual cingulum)/maximum buccal height. Species composition with sample sizes in parentheses: (A) *Gorilla gorilla gorilla*, male (24), female (20); *Pan troglodytes troglodytes*, male (12), female (10); *Pongo pygmaeus pygmaeus*, male (28), female (32); (B) *G. g. gorilla*, male (24), female (16); *P. t. troglodytes*, male (15), female (12); *P. p. pygmaeus*, male (23), female (29).

dentition, ultimately to be used as an indicator of body-size dimorphism.* Since indirect measures of dental dimorphism do not produce estimates with sufficient precision, a direct measure was required. This, in turn, requires a means to sex individual specimens. As with most size criteria, postcanine tooth size itself is unreliable for determining the sex of specimens. However, certain features of the canine dentition can be used for this purpose.

I have elsewhere described features of upper and lower canine shape that, on average, differ between males and females of extant great apes and that can be used to distinguish size-sorted groups of fossil canines as either male or female (Kelley, 1986a; Kelley and Etler, 1989). These same criteria can be used to identify reliably the sex of individual canines (Kelley, manuscript in preparation) (Fig. 4). By these shape parameters, a subsample of all the complete, unworn, or minimally worn Lufeng lower canines, either isolated or associated with postcanine dentitions, fall into distinct male and female groups (Fig. 5). In terms of absolute size, the upper and lower canine samples are both characterized by

*A direct estimate of body size and body-size dimorphism in *L. lufengensis* is not yet possible as only four, mostly fragmentary, postcranial bones have been recovered (Wu *et al.*, 1986).

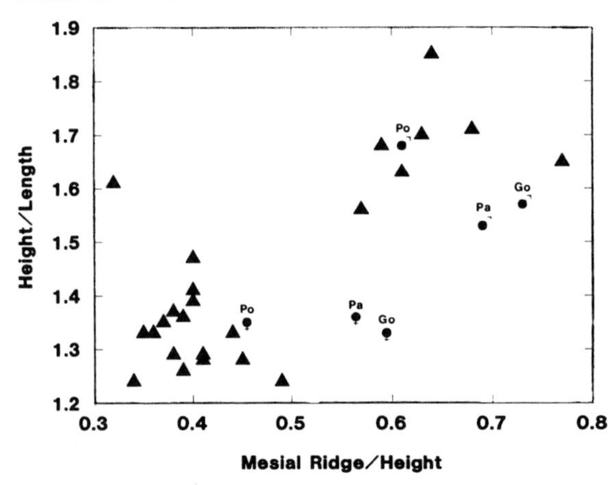

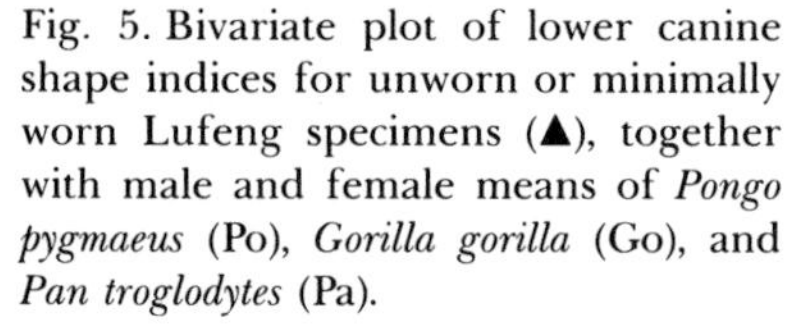
Fig. 5. Bivariate plot of lower canine shape indices for unworn or minimally worn Lufeng specimens (▲), together with male and female means of *Pongo pygmaeus* (Po), *Gorilla gorilla* (Go), and *Pan troglodytes* (Pa).

two nonoverlapping groups, with, in each case, the cluster of large specimens containing all those with a demonstrated male morphology and the cluster of small specimens containing all those with a female morphology (Fig. 6). Interpreted according to extant ape canine morphology, these two size groups in the fossil sample define male and female groups. Because in this species male and female canines do not overlap in size, even partial canines that cannot be characterized according to their shape can be confidently sexed, as can the teeth with which they are associated. In this way, it was possible to sex 20 partial or complete lower dentitions (11 female, 9 male).

The canines were then used to sex the lower third premolars. P_3s also fall into two nonoverlapping size groups (Fig. 7). All of those known to be male based on associated canines fall into the large-sized group, while all those known to be female fall into the small-sized group. Thus, lower third premolars and the teeth associated with them can also be sexed based only on size. This added eight further postcanine dentitions containing P_3s but not canines (5 female, 3 male) to the sample of sexed specimens.

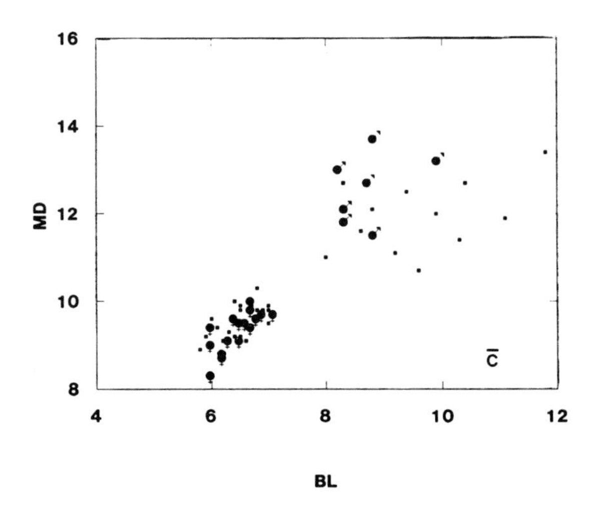

Fig. 6. Bivariate plot of mesiodistal length × buccolingual breadth for the entire Lufeng lower canine sample. Male–female symbols indicate specimens sexed according to shape indices and shown in Fig. 5. Apparent difference from Fig. 5 in number of sexed females results from two specimens having identical crown dimensions.

Table 2. Postcanine Dental Dimorphism in *L. lufengensis* and *P. pygmaeus*[a]

	P_4	M_1	M_2	M_3	M_{avg}	PC_{avg}
Lufengpithecus	1.35	1.40	1.40	1.39	1.40	1.39
N (M,F)	10,10	9,11	10,9	6,9		
Pongo	1.21	1.19	1.23	1.25	1.22	1.22
N (M,F)	38,35	38,35	38,36	32,32		

[a]The measure of postcanine tooth size used was tooth area (mean mesiodistal length × mean buccolingual breadth). The index of sexual dimorphism was calculated as male mean area/female mean area. Averages were calculated from the individual tooth positions for the whole molar row (M_{avg}) and for the entire masticatory postcanine (P4-M3) dentition (PC_{avg}). For *Lufengpithecus*, indices for the left and right sides were calculated separately and then averaged; the maximum difference between sides was 2.2%. In cases where one antimere was missing, the remaining tooth was used for both left and right side calculations. Sample sizes (N) for male (M) and female (F). Variability in sample sizes for both *Pongo* and *Lufengpithecus* results because not every specimen preserves the entire tooth row from P4-M3. Lufeng values based on measurements provided by Xu Qinghua. *Pongo* data from Mahler (1973).

Although in this species absolute size does differentiate males from females in the lower canine/premolar complex, this could not have been known with confidence without the use of size-independent measures to verify the sex-based partitioning, since canine size alone does not consistently separate males from females among the extant great apes (Oxnard *et al.*, 1985; Kelley, unpublished data).

Indices of sexual dimorphism were then calculated for the sexed postcanine dentitions of *L. lufengensis* and were compared to the same indices for the orang-utan, *Pongo pygmaeus* (Table 2). Averages were calculated from the individual tooth positions for the whole molar row and for the entire masticatory postcanine dentition (P_4-M_3). Because of the larger samples of lower dentitions in *L. lufengensis,* only these were used in the analysis. In the postcanine dentition, *Pongo* is the most dimorphic living hominoid (Martin, 1983; Oxnard *et al.*, 1985; Swindler, 1976), and it is among the most sexually dimorphic of all living anthropoids, as a species perhaps the most dimorphic (Plavcan, 1990). Only baboons show a roughly equivalent degree of dimorphism (Swindler, 1976; Phil-

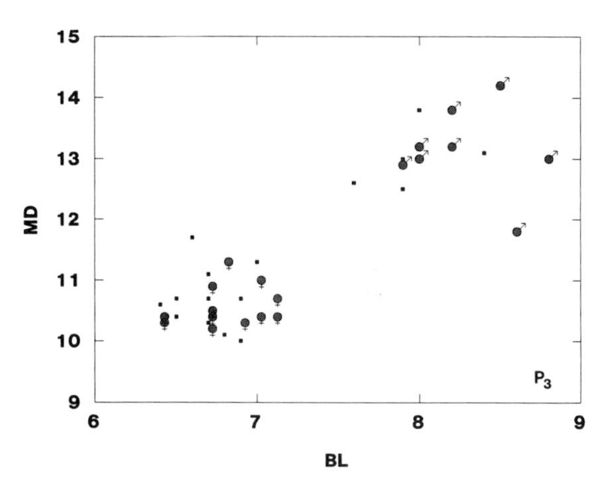

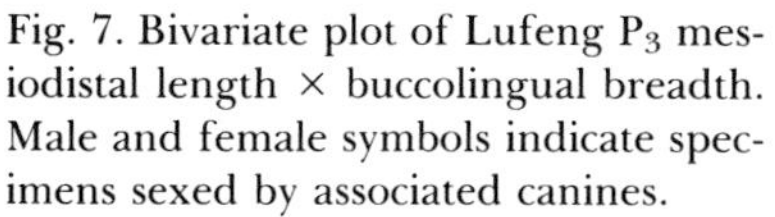
Fig. 7. Bivariate plot of Lufeng P_3 mesiodistal length × buccolingual breadth. Male and female symbols indicate specimens sexed by associated canines.

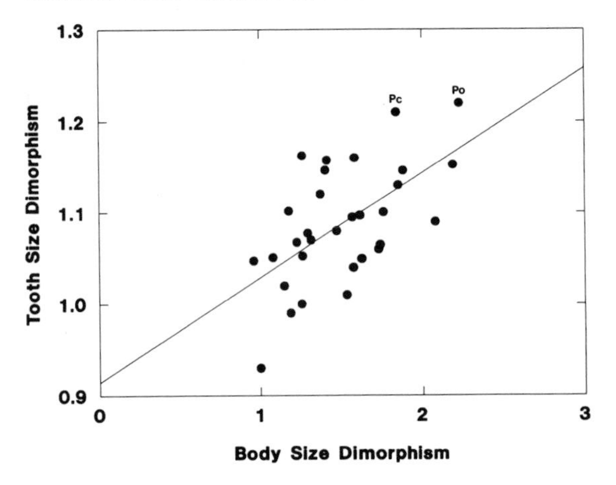

Fig. 8. Regression of mandibular postcanine tooth size dimorphism against body size dimorphism. Tooth size dimorphism equals PC_{avg} described in Table 2; body size dimorphism calculated as male weight/female weight. Tooth dimorphism = 0.914 + 0.115 (body size dimorphism); Kendall Tau-B = 0.411 ($p < 0.001$). Dental data from Swindler (1976), Mahler (1973), Leutenegger (1971), Lucas *et al.* (1986), Fleagle and Kitahara-Frisch (1984), and Pilbeam (1969). Body weight data from Leutenegger and Cheverud (1982), Jungers (1985), and Fleagle (1988). Po: *Pongo pygmaeus;* Pc: *Papio cynocephalus.*

lips-Conroy and Jolly, 1981) (Fig. 8), with individual populations occasionally having somewhat greater levels of dimorphism than the *Pongo* sample used here (Swindler *et al.,* 1967). Postcanine dental dimorphism is shown to be substantially greater in *Lufengpithecus* than in *Pongo.*

Since vagaries of sampling and differences in sample sizes can affect species comparisons, it was necessary to determine the probability that dental dimorphism in *Lufengpithecus* was in fact greater than in *Pongo.* The null hypothesis was that *L. lufengensis* was no more dimorphic than *P. pygmaeus.* For *P. pygmaeus,* a probability distribution of values for the index of dimorphism was created for each tooth position through repetitive sampling (with replacement) of the data base of 41 males and 39 females, selecting sample sizes approximately equal to those of the Lufeng samples of sexed specimens. The database included only Bornean orangutans measured by one individual (Mahler, 1973). Results are shown in Table 3. These confirm that sexual dimorphism in the postcanine

Table 3. Indices of Dental Dimorphism in *Lufengpithecus* and *Pongo*[a]

	P_4	M_1	M_2	M_3	M_{avg}	PC_{avg}
Lufengpithecus	1.35	1.40	1.40	1.39	1.40	1.39
N (M,F)	10,10	9,11	10,9	6,9		
Pongo sampling x̄	1.24	1.20	1.25	1.28	1.24	1.24
N (M & F ranges)	8–10	8–10	8–10	6–10		
S.D.					.066	.067
Sampling range					1.14–1.34	1.13–1.35

[a]Calculation of indices and sources of data as in Table 2. Individual tooth positions were not analyzed separately because they are not independent. Sampling x̄: Means of 20 random samples of 10 males and 10 females each generated using a subroutine of the Systat statistical package. N (M & F ranges): Ranges of sample sizes for each sex for the 20 samples; variability results because not every *Pongo* specimen preserves the entire tooth row from P4-M3. S.D.: Standard deviation. Sampling range: Range of values for the 20 samples.

dentition of *L. lufengensis* exceeds that of the orangutan. The two average values of dimorphism for the postcanine teeth of *Lufengpithecus* lie outside the ranges of values produced from the bootstrap procedure. They lie more than two standard deviations beyond the means of the *Pongo* samples.

Based on the results presented here, it is reasonable to conclude that *L. lufengensis* was more sexually dimorphic in the postcanine dentition than the most dimorphic living hominoid. Since the orangutan is one of the most dentally dimorphic primate species, *Lufengpithecus* likely had a level of dental dimorphism that is unprecedented among living anthropoids.

What are the implications of these findings to the question of body-size dimorphism in *L. lufengensis?* As noted, body size cannot be accurately estimated from the very limited postcranial remains from Lufeng. However, an examination of the relationship between postcanine tooth-size dimorphism and body-size dimorphism in 31 anthropoid species revealed that the two are highly significantly correlated (Fig. 8). Therefore, the extreme dental dimorphism in *L. lufengensis* quite likely is a reflection of, and perhaps a result of, an equally great and possibly unprecedented degree of body-size dimorphism.

While the correlation between tooth- and body-size dimorphism is highly significant, the residuals are large and only about 35% of the variance in tooth-size dimorphism is "explained" by body-size dimorphism. Thus, while a qualitative assessment of probable body-size dimorphism in *Lufengpithecus* is, I think, warranted, any attempt at a precise estimate from these data would have unacceptably large associated errors. One could, of course, attempt to estimate body sizes from the postcanine teeth themselves. Unfortunately, the various regression equations that have been generated for this purpose produce widely varying estimates for fossil taxa (e.g., Kay and Simons, 1980; Gingerich *et al.*, 1982; Conroy, 1987).

As a simple exercise, I would like to suggest a different approach using tooth size. Let us assume that tooth-size/body-size proportionality, and female relative tooth size (Cochard, 1987), are roughly the same in *Lufengpithecus* and *Pongo.* There is some justification for this in that (1) some Miocene hominoids, particularly thicker enameled hominoids, appear to have been somewhat mega-

Table 4. Estimates of Body Size and Body-Size Dimorphism in *Lufengpithecus*[a]

	M_1 area (mm)	M_2 area (mm)	Body weight (kg)
Pongo			
M	169.7	194.9	86
F	142.5	158.7	39
Lufengpithecus			
M	140.5	191.0	(71–84)
F	100.3	136.7	(27–34)

[a]*Pongo* tooth areas calculated from the database of 41 males and 39 females. *Lufengpithecus* tooth areas calculated from the sample of sexed specimens; sample sizes in Table 2. *Pongo* body weights from Markham and Groves (1990).

dont to the degree that *Pongo* is megadont relative to the African apes (Kelley, 1986b), and (2) *Lufengpithecus* may be on the orangutan lineage (Kelley and Etler, 1989; Schwartz, 1990), even though it is not likely to be a member of the *Sivapithecus-Pongo* clade (Brown and Ward, 1988; Kelly and Etler, 1989). Using a simple ratio, estimates of mean body weight can be generated for males and females of *L. lufengensis* (Table 4). Not unexpectedly, M_1 and M_2 produce somewhat different estimates, with an average male weight about 90% that of the orangutan male mean and an average female weight about 78% that of the female mean, but with similar male/female ratios of 2.6:1 and 2.5:1, respectively. I would not claim any greater degree of reliability with this method than with published regression equations, and I put these figures forward only as working estimates until adequate postcranial material is available.

Implications of Lufengpithecus Sexual Dimorphism for Miocene Hominoid Taxonomy

It follows from the results of the examination of *Lufengpithecus* that the greatest sexual dimorphism in tooth size, and probably body size, found among extant anthropoid primates does not represent some upper limit during primate evolutionary history. This has important implications for the taxonomic interpretation of other, less well-represented fossil hominoid samples. The presence of seemingly unacceptably high levels of sexual dimorphism in fossil samples that are morphologically homogeneous does not necessarily require the recognition of more than one species. It may be that patterns and degrees of sexual dimorphism not represented among extant primates characterized many fossil hominoid species (Kelley, 1986b). We have as yet only a limited understanding of how the developmental processes of bimaturism and differential growth rates (Shea, 1986); ecological pressures, such as feeding competition and female metabolic requirements (Demment, 1983; Pickford, 1986c); and behavioral demands, including mating (Jarman, 1983; Clutton-Brock, 1985; Kay *et al.*, 1988), predator deterence (Cheney and Wrangham, 1987), and intergroup aggression (Cheney, 1987), interact to produce particular patterns of sexual dimorphism. Different patterns from those known at present should not be unexpected.

Earlier suggestions of very great size dimorphism in other fossil hominoid samples (references cited in Introduction) may well be real and not the result of taxonomic error (Kay, 1982a,b; Wu and Oxnard, 1983a,b; Teaford *et al.*, this volume). The sample characteristics associated with Lufeng appear to typify the large hominoid samples at many of these less well-represented Miocene sites, including Rudabanya, Ravin de la Pluie, and different stratigraphic levels in the Siwaliks. These are morphologically homogeneous samples, always with measures of dental metric variation near or somewhat above those of the most variable extant catarrhines, and with a pattern of large canines that are morphologically male and small canines that are morphologically female (Kelley, 1986b; Kelley and Pilbeam, 1986). If, as Plavcan (1989, this volume) asserts, the probability is high that we are underestimating the numbers of fossil species and that some of these sites contain multiple species that cannot be differentiated, why do we not also find a single case of two large species that are as different in size and morphology as are, for example, the chimpanzee and gorilla? Why is dental variation always so tantalizingly near the limit between acceptance and rejection for a single species, without one clear case of metrically distinguishable species? Interpreted in the context of sexual dimorphism that is somewhat greater than the maximum observed among extant primates, these common sample attributes cease to be a problematic coincidence. Instead, the presence of only one large species at these and perhaps other sites not only conforms well with certain biological expectations developed earlier, but species' distribution patterns become more like those of the living apes as well. Rather than make the task of delimiting species more difficult, as one might expect, the likely presence of extreme sexual dimorphism in *Lufengpithecus* actually appears to clarify an otherwise troublesome taxonomic picture for Miocene hominoids.

Rusinga *Proconsul* and the Question of Body-Size Dimorphism

In light of the above, it might be instructive to reexamine the question of taxonomy and sexual dimorphism in one Miocene hominoid sample in particular, the *Proconsul* sample from Rusinga (plus Mfangano). This case is particularly interesting because of the abundant postcranial remains known from this site, including nearly complete remains of entire limbs (Walker and Pickford, 1983; Beard *et al.*, 1986) that allow more reliable size estimates to be made for several individuals. It is interesting too for the amount of controversy it has generated, being the only sample that was previously claimed to contain a species more dimorphic than any found today.

As noted in the Introduction, a single-species alternative for the Rusinga *Proconsul* sample has been questioned or rejected based on the presumed amount of body-size dimorphism that would be required (Teaford *et al.*, 1988, this volume; Ruff *et al.*, 1989). However, as this sample has expanded, the inferred degree of necessary body-size dimorphism has decreased. Walker and Pickford (1983), working with relatively complete remains of one large and one small individual, claimed a size ratio of 4:1. Kelley (1986a) and Pickford (1986a) pointed out that the body weight given by them for the extant reference species of the large individual was too high and that the ratio for these two individuals was actually close to 3:1. Teaford *et al.* (1988) later reaffirmed the 4:1 ratio, based not on simple size–analogue comparisons, but rather on an analysis of long-bone cross-sectional dimensions (Ruff, 1987), employing this time two large (~37 kg) and two small (~9.5 kg) specimens.

In both Walker and Pickford (1983) and Teaford *et al.* (1988), these individuals were implicitly assumed to represent the means of their respective groups, an assumption that was questioned by both Kelley (1986a) and Pickford (1986a). Subsequently, Ruff *et al.* (1989) published further data on cross-sectional dimensions and body weight in living primates and added three individuals to the sample of large Rusinga specimens. This resulted in a lowered mean body size for the larger animals (to ~31 kg) and reduced the body-size ratio between large and small individuals to 3.4:1.

However, even a 3.4:1 mean size ratio between the large and small *Proconsul*

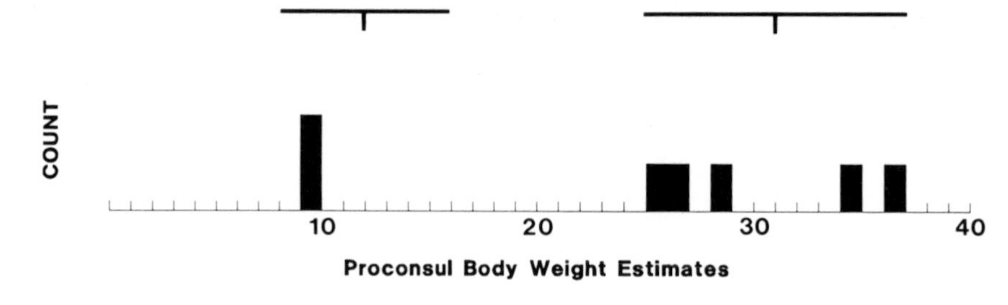

Fig. 9. Estimated body weights of *Proconsul* specimens from Rusinga and Mfangano, together with suggested body size ranges and means for the two groups of specimens (upper bars). *Proconsul* estimates from Ruff *et al.* (1989).

individuals is surely too great. What is immediately apparent from the plot of Rusinga body weights in Ruff *et al.* (1989) is that there is no body-size range associated with the two individuals that constitute the small-sized group (Fig. 9). Whether this group is a species or a single-sex group, there has to have been a range of body sizes. The question then is, are these two ~9-kg individuals nearer the bottom, middle, or top of this range? The fact that one of these individuals is not fully adult (albeit nearly so: Napier and Davis, 1959; Ruff *et al.*, 1989) argues for a position closer to the bottom of the actual adult range. That the other specimen is fully adult does not necessarily alter this inference (see table of age-classed body weights for female baboons in Smuts, 1985).

Further, Ruff *et al.* note that there are larger individuals than these two within the small-sized group (for them, they are likely to be males of the same species as the two estimated individuals that they and others regard as females), individuals for which body weight estimates are not yet available. They speculate that the upper end of the size range for this small-sized group might be as high as 18 kg, and from this they imply that the weight ratio between large and small Rusinga/Mfangano specimens might actually be closer to 3:1 than 3.4:1. Whether one regards all of the smaller Rusinga and Mfangano specimens as the females of the species represented by the larger individuals or as the males and females of a separate species, the two 9-kg specimens would lie near the bottom of the size range of this group as a whole.

Let us conservatively assume a body weight range of about 8–16 kg for the smaller Rusinga and Mfangano specimens, which gives an estimated mean body weight of 12 kg. With a mean weight of 31 kg for the large specimens, the ratio of mean body size between large and small individuals then becomes 2.6:1, identical to the rough estimate produced for *Lufengpithecus* based on dental size.

How does this compare with body-size dimorphism maxima among extant primate species? Using only reliable adult body weights for lowland gorillas (Jungers and Susman, 1984) and Bornean orangutans (Markham and Groves, 1990), calculated male: female weight ratios are 2.37:1 and 2.22:1, respectively. It should be noted that sample sizes for female gorillas and both male and female orangutans are quite small. Including two female orangutans from the collections of the Museum of Comparative Zoology that were omitted as questionable by Markham and Groves, but whose fully adult status I was able to confirm, elevates the figure for orangutans slightly to 2.26:1. If 2.6:1 is close to the actual body-size ratio between large and small *Proconsul* specimens, then the amount of body-size dimorphism necessitated by a single-species hypothesis would not be as unreasonable by extant primate standards as claimed by Teaford *et al.* (1988) and Ruff *et al.* (1989), CVs of small samples of tarsal measurements notwithstanding (Teaford *et al.*, this volume). Given the suggested degree of body-size dimorphism for *Lufengpithecus*, it might not be unreasonable at all.

To the argument that one would especially not expect such a degree of dimorphism in an animal no larger than a pygmy chimpanzee (Ruff *et al.*, 1989; Teaford *et al.*, 1988, this volume), one only has to look at other very highly size dimorphic anthropoid species. According to body-weight data presented for *Mandrillus sphinx* (Gautier-Hion, 1978; Jungers, 1985; Fleagle, 1988), which admittedly may not be as reliable as those for the great apes, the male/female weight ratio is 2.34:1, at a body weight only one-sixth that of the gorilla (and less, on average, than that of Rusinga *Proconsul*). Reliable weight data for *Nasalis larvatus* (Schultz, 1941) result in a weight ratio of 2.06:1 for an animal nearly an order of magnitude less massive than the gorilla.

To refer to monkeys here might be considered inconsistent, since I have argued that monkeys are inappropriate models for fossil apes in the matter of spatial distribution. However, references to monkeys are valid in this instance, I believe, because the underlying biological factors of sexual dimorphism are presumably the same in the two groups and presumably operate in similar ways to produce given patterns of dimorphism. In contrast, the underlying biology of spatial distribution seems to me to be quite different in the two groups, as argued earlier. By way of analogy, as in studies of allometry the appropriateness of the comparative sample depends entirely on the issue at hand.

While there is clearly a link between body size and body-size dimorphism for anthropoid primates as a whole (Leutenegger and Cheverud, 1982, 1985; *Mandrillus* and *Nasalis* are still among the largest nonhominoid anthropoids), the relationship does not lend itself to defining predictable limits of dimorphism within particular body-size categories. Indeed, species of *Pan*, to which Ruff *et al.* (1989) liken *Proconsul nyanzae* in body size and body-size dimorphism, are highly unusual among living primates in having so relatively little size dimorphism at such a large body size (Leutenegger, 1982).

These arguments are admittedly speculative; this is an exercise in possibilities, not probabilities. But given the presently missing body-size range within the Rusinga and Mfangano *Proconsul* data, I would suggest that they are at least as compelling as are arguments made for the presence of large and small species within this sample. If the only argument against a single-species hypothesis for Rusinga *Proconsul* is the amount of body-size dimorphism that would be required, then I suggest that this is still very much an open question. It should also be remembered that the large and small two-species hypothesis is not without problems. With regard to the large canine samples from Rusinga and Mfangano, it would either produce, according to Teaford *et al.* (1988), "a male-dominated *P. nyanzae* sample and a female-dominated *P. africanus* sample," or two species, both of which would be minimally or nondimorphic in the canine dentition, one having relatively large canines and the other having relatively small canines (Kelley, 1986a). Canine monomorphism is a rare condition among extant catarrhines (Leutenegger, 1982). Is it just coincidence that Teaford and co-worker's large-sized species is male dominated while the small-sized species is female dominated?

Uniformitarianism, Operational Limits, and Falsifiability

It has been implied or explicitly stated that to suggest greater degrees of sexual dimorphism in the past than at present somehow goes beyond using the present to interpret the past, that is, that it somehow violates or lies outside the principle of uniformitarianism (Kay, 1982a; Teaford *et al.*, 1988; Ruff *et al.*, 1989). There is nothing about the principle of uniformitarianism that would restrict us to the degrees or patterns of sexual dimorphism that we find today. Uniformitarianism is about processes, not outcomes or events (Shea, 1982). Therefore, to restrict ourselves only to the patterns displayed by present analogues (Teaford *et al.*, this volume) to describe past phenomena is a misapplication of the principle. There is seemingly a greater willingness to entertain patterns quite unlike any at present in other areas of paleobiology, such as postcranial skeletal interpretation and the reconstruction of positional behavior (e.g., Rose, 1983; Walker and Pickford, 1983; Beard *et al.*, 1986), perhaps because we perceive that we better understand the biomechanical rules governing the processes.

The limitations of a strict analogue approach can also be demonstrated with reference to more tractable questions relating to body size. To claim that there could not have been greater degrees of dimorphism in the past than at present is logically equivalent to claiming that there could not have been terrestrial animals larger than the largest animal living today. This is demonstrably not true. It is probably not even true within Mammalia; *Baluchitherium* was larger, perhaps substantially larger, than the modern elephant (Gingerich, 1990; C. Janis, personal communication). Similarly, the largest extant primate, the gorilla, is not the largest primate to have ever lived. While a cautionary note must again be sounded about body-size estimates based on craniodental remains, it is almost inconceivable that *Gigantopithecus blacki* would not have been larger than *Gorilla* (Fleagle, 1988).

It has been stated that hypotheses of unprecedented sexual dimorphism cannot be falsified (Plavcan, this volume; Martin, 1991). Strictly speaking, neither can much else that we do in paleobiology, despite frequent claims to the contrary. If this is to be the criterion of legitimacy for what we do, then we shall have a very narrow range of subjects that we can address. It would, for example, exclude all phylogenetic analysis, since there are no logically certain biological criteria for deciding questions of character-state homology and polarity (Cartmill, 1981).

Cope (this volume) and Martin (1991) claim that single-species hypotheses for fossil samples can be falsified, but this is true only within the framework of a certain premise, that there were no species in the past that were more variable or dimorphic than their extant comparative species. In nonhistorical sciences, premises can be shown to be untenable by experimental results or underlying principles or laws that are empirically derived and supported by theory. There are as yet no equivalents by which to evaluate premises about evolutionary outcomes. Suggested support for this premise is the general consistency of measures of variation in a variety of extant mammals (Cope, this volume). However, most mammals are not particularly size dimorphic. In the anthropoid primates, we are dealing with a group whose members are generally very highly size dimorphic, and, in the hominoids, a group with some of the most dimorphic extant mammals. In many cases, this is reflected in degrees of dental dimorphism (Fig. 8). Given the correspondence between dental dimorphism and variation (Kay, 1982a,b), I would therefore suggest that data on variation from many of the mammalian groups in the studies cited by Cope are not particularly relevant for assessing limits to variation in highly dimorphic groups.

Cope's work and that of Plavcan (this volume), are intended to facilitate the taxonomic process by providing some guiding operational limits for allowable amounts of variation and sexual dimorphism within species. However, their work, like that of Teaford *et al.* (this volume), is a methodology that seemingly precludes a demonstration of greater variation or sexual dimorphism in the past, whether or not such are considered possible or even likely. Fortelius (1990) addressed the same issue and stated the problem thus, "One may never know of interesting systematic differences between fossil and living species eliminated from the data by the assumption that they do not exist." Such approaches, while unquestionably bringing more rigour to the analysis of variation, also preclude certain kinds of evolutionary novelty. But I would argue that we should by now expect novelty from the fossil record.

If one allows the possibility of greater degrees of sexual dimorphism in the past, the question then becomes, how much *is* too much? Clearly, there are degrees of dimorphism that would be untenable, but the point at which we cross into that realm cannot be deduced from biological first principles because, as noted, we have only a limited understanding of the underlying biology of sexual dimorphism. The question can only be answered by the fossil species themselves, and this brings us back to questions of taxonomy. Degrees of sexual dimorphism derive from the adopted taxonomy and the determination of species hypodigms. Normally, there will be alternative taxonomic hypotheses, and ideally these should be discussed in a comparative fashion (e.g., Kelley, 1986a; Teaford *et al.*, 1988; Kelley and Etler, 1989). Alternatives will have particular strengths and weaknesses, and these can be used to rank them in a qualitatively probabilistic way according to explicit criteria. This, I would argue, constitutes most of what we do in paleobiology rather than truly falsifying hypotheses. There will not necessarily be agreement about the ranking, but at least the nature of disagreements will be explicit. These criteria, and the premises and analyses upon which the taxonomies are based, will always be based one way or another on the present, but hopefully on a present that is recognized as being a single evolutionary instant that does not encompass all known evolutionary phenomena.

We are limited in our use of the present to evaluate the past if we do not fully understand what governs the underlying biological processes, but I do not mean to imply that we should therefore abandon ties to the present. Without retreating from my earlier comments about uniformitarianism and analogues, in addition to debating the procedures leading to competing taxonomic hypotheses, I think that within limits we can usefully employ existing morphological patterns to assess, again qualitatively and comparatively, the likelihood of particular outcomes. After all, groups are defined by a certain biological commonality, which does place certain limits on the expression of morphology, behavior, etc. What is at issue is how to use this information from the modern world.

By way of illustration, I would like to return to the case of Rusinga *Proconsul*. While there has been no specific criticism of the taxonomic analysis concluding that there is a single *Proconsul* species at Rusinga (Kelley, 1986a), Teaford *et al.* (1988, this volume) and Ruff *et al.* (1989) have questioned the resulting outcome necessitating very high levels of sexual dimorphism. They assume that the greatest body-size dimorphism evident among extant primates is an actual or operational limit that makes a single-species alternative improbable. Given the degree of dental dimorphism displayed by *Lufengpithecus*, this assumption is now challengeable on more than just philosophical grounds.

I would approach the question of patterns of dimorphism in extant primates differently. If higher primates as a group, particularly catarrhines, were characterized typically by low levels of sexual dimorphism, then I too would question the likelihood of dimorphism that would be very great for any terrestrial mammal. This, however, is not the case. Higher primates, especially catarrhines, are generally very size dimorphic. Indeed, they are exceptional among mammals for the degrees of body-size dimorphism they exhibit at such relatively small body sizes. Again, two of the living great apes are among the most dimorphic living mammals (Jungers and Susman, 1984; Markham and Groves, 1990; Nowak and Paradiso, 1983), at body weights considerably less than most similarly or somewhat less dimorphic bovids, for example (Janis, 1990).

If an unprecedented degree of dimorphism is to be forthcoming from the fossil record, it might be expected to be found among the primates, especially the apes. Evaluated in this way, and given the indications of marked dimorphism in many other fossil anthropoids (Fleagle *et al.*, 1980; Kelley and Pilbeam, 1986; Harrison, 1989a; Pan *et al.*, 1989), I would contend that patterns of sexual dimorphism in living anthropoids do not suggest that the necessary level of dimorphism in a single-species alternative for Rusinga *Proconsul* is improbable. To the contrary, the suggestion by Ruff *et al.* (1989) that their large species, *P. nyanzae,* was about as dimorphic as similarly sized chimpanzees (incidentally, a maximum estimate, since all the smaller individuals were assumed to be female and the two larger individuals were assumed to be male) might be considered somewhat unlikely since, as noted, chimpanzees have unusually low levels of dimorphism for a primate this size. Similarly, the resulting pattern of essentially no canine dimorphism in either species might also reduce the likelihood of a large and small two-species alternative, since this is a rare condition among catarrhines and uncommon for anthropoids as a whole (Leutenegger, 1982).

With regard to ranking taxonomic alternatives, we can return to Lufeng. Kelley and Etler (1989) evaluated the relative likelihood of single-species and large and small two-species alternatives. We concluded that the two-species alternative was improbable and that there were insufficient reasons to question the single-species hypothesis. Arguments made earlier show this to be so for an overlapping two-species alternative as well. Subsequently, it became apparent that a single-species required exceptional levels of dental dimorphism. For the reasons outlined in this chapter and elsewhere (Kelley and Xu, 1991), it was decided that this observation did not alter the relative probabilities associated with the various taxonomic alternatives. There were still serious objections to the two-species alternatives and insufficient biological reasons to question the single-species alternative simply because it required such a high degree of dimorphism. Others can now question the explicit rationale leading to these conclusions, both for accepting a single-species alternative and rejecting a two-species alternative. For us, in the case of Lufeng, this still means accepting something that is unlike anything that we see today but, because the competing taxonomic alternatives are so much less likely, this becomes something to be explained rather than something to be explained away.

Acknowledgments

I would first like to thank my collaborator in the study of the Lufeng remains, Xu Qinghua from the Institute of Vertebrate Paleontology and Paleoanthropology, Beijing, China. I also thank Bill Kimbel and Lawrence Martin for the invitation to participate in this symposium. I wish to acknowledge the following individuals for comments or discussions on various aspects of this work: Gene Albrecht, Lisa Brooks, Terry Harrison, Paul Harvey, Christine Janis, Lawrence Martin, Joe Miller, David Pilbeam, Michael Plavcan, and Richard Smith. The following institutions kindly made available their collections of primate skeletal remains: The Museum of Comparative Zoology, Harvard University; The Smithsonian Institution; The British Museum (Natural History); Museum für Naturkunde der Humboldt-Universität zu Berlin; Anthropologie Staatssammlung München. Use of the computer facilities at the Erikson Biographical Institute, Providence, Rhode Island, is gratefully acknowledged. Work on the Lufeng fossils was supported by the National Academy of Sciences Foreign Scholar Exchange Program, the Chinese Academy of Sciences, and the L.S.B. Leakey Foundation.

References

Alpagut, B., Andrews, P., and Martin, L. 1990. New hominoid specimens from the Middle Miocene site at Paşalar, Turkey. *J. Hum. Evol.* **19:**397–422.

Andrews, P. 1978. A revision of the Miocene Hominoidea of East Africa. *Bull. Br. Mus. (Nat. Hist.) Geol. Ser.* **30:**85–224.

Andrews, P. 1981. Species diversity and diet in monkeys and apes during the Miocene, in: C. Stringer (ed.), *Aspects of Human Evolution,* pp. 25–61. Taylor and Francis, London.

Andrews, P., and Martin, L. B. 1987. Cladistic relationships of extant and fossil hominoids. *J. Hum. Evol.* **16:**101–118.

Badgley, C., Qi, G., Chen, W., and Han, D. 1988. Paleoecology of a Miocene, tropical, upland fauna. *Nat. Geogr. Res.* **4:**178–195.

Beard, K. C., Teaford, M. F., and Walker, A. 1986. New wrist bones of *Proconsul africanus* and *P. nyanzae* from Rusinga Island, Kenya. *Folia Primatol.* **47:**97–118.

Beynon, A. D., and Dean, M. C. 1988. Distinct dental development patterns in early fossil hominids. *Nature* **335:**509–514.

Beynon, A. D., and Wood, B. A. 1987. Patterns and rates of enamel growth in the molar teeth of early hominids. *Nature* **326:**493–496.

Bonis, L. de. 1983. Phyletic relationships of Miocene hominoids and higher primate classification, in: R. L. Ciochon and R. S. Corruccini (eds.), *New Interpretations of Ape and Human Ancestry,* pp. 625–649. Plenum, New York.

Bonis, L. de., and Melentis, J. 1984. La position phyletique d'*Ouranopithecus*. *Cour. Forsch. Inst. Senckenberg* **69:**13–23.

Bromage, T. G., and Dean, M. C. 1985. Re-evaluation of the age at death of immature fossil hominids. *Nature* **317:**525–527.

Brown, B., and Ward, S. C. 1988. Facial and basicranial relationships in *Sivapithecus* and *Pongo,* in: J. H. Schwartz (ed.), *Biology of the Orang-utan,* pp. 247–260. Oxford University Press, Oxford.

Cartmill, M. 1981. Hypothesis testing and phylogenetic reconstruction. *Z. Zool. Syst. Evol.* **19:**73–96.

Chapman, C. A., and Chapman, L. J. 1990. Dietary variability in primate populations. *Primates* **31:**121–128.

Cheney, D. L. 1987. Predation, in: B. B. Smuts, D. L. Cheney, R. M. Seyfarth, R. W. Wrangham, and T. T. Struhsaker (eds.), *Primates Societies,* pp. 227–239. University of Chicago Press, Chicago.

Cheney, D. L., and Wrangham, R. W. 1987. Interactions and relationships between groups, in: B. B. Smuts, D. L. Cheney, R. M. Seyfarth, R. W. Wrangham, and T. T. Struhsaker (eds.), *Primate Societies,* pp. 267–281. University of Chicago Press, Chicago.

Clutton-Brock, T. H. 1985. Size, sexual dimorphism, and polygyny in primates, in: W. L. Jungers (ed.), *Size and Scaling in Primate Biology,* pp. 51–60. Plenum, New York.

Clutton-Brock, T. H., and Harvey, P. H. 1977. Species differences in feeding and ranging behaviour in primates, in: T. H. Clutton-Brock (ed.), *Primate Ecology,* pp. 557–584. Academic Press, London.

Cochard, L. R. 1987. Postcanine tooth size in female primates. *Amer. J. Phys. Anthropol.* **74:**47–54.

Conroy, G. C. 1987. Problems of body-weight estimation in fossil primates. *Int. J. Primat.* **8:**115–137.

Demment, M. W. 1983. Feeding ecology and the evolution of body size of baboons. *Afr. J. Ecol.* **21:**219–233.

Eisenberg, J. F. 1981. *The Mammalian Radiations.* University of Chicago Press, Chicago.

Fleagle, J. G. 1988. *Primate Adaptation and Evolution.* Academic Press, New York.

Fleagle, J. G., and Kitahara-Frisch, J. 1984. Correlation and adaptation in the dentition of lar gibbons, in: H. Preuschoft, D. J. Chivers, W. Y. Brockelman, and N. Creel (eds.), *The Lesser Apes,* pp. 192–206. Edinburgh University Press, Edinburgh.

Fleagle, J. G., Kay, R. F., and Simons, E. L. 1980. Sexual dimorphism in early anthropoids. *Nature* **287:**328–330.

Fleming, T. H. 1973. Numbers of mammal species in north and central America forest communities. *Ecology* **54:**555–563.

Fortelius, M. 1990. Problems with using fossil teeth to estimate body sizes of extinct mammals, in: J. Damuth and B. J. MacFadden (eds.), *Body Size in Mammalian Paleobiology,* pp. 207–228. Cambridge University Press, Cambridge.

Gautier-Hion, A. 1978. Food niches and coexistence in sympatric primates in Gabon, in: D. J. Chivers and J. Herbert (eds.), *Recent Advances in Primatology, Vol. 1, Behaviour,* pp. 269–286. Academic Press, London.

Gautier-Hion, A. 1988. Polyspecific associations among forest guenons: ecological, behavioral and evolutionary aspects, in: A. Gautier-Hion, F. Bourliere, J. P. Gautier, and J. Kingdon (eds.), *Evolutionary Biology of the African Guenons,* pp. 452–475. Cambridge University Press, Cambridge.

Gingerich, P. D. 1990. Prediction of body mass in mammalian species from long bone lengths and diameters. *Contrib. Mus. Paleontol.* **28:**79–92.

Gingerich, P. D., Smith, B., and Rosenberg, K. 1982. Allometric scaling in the dentition of primates and prediction of body weight from tooth size in fossils. *Am. J. Phys. Anthropol.* **58:**81–100.

Ginsburg, L. 1986. Chronology of the European pliopithecids, in: J. G. Else and P. C. Lee (eds.), *Selected Proceedings of the Tenth Congress of the International Primatological Society, Vol. 1, Primate Evolution,* pp. 47–57. Cambridge University Press, Cambridge.

Harrison, T. 1987. The phylogenetic relationships of the early catarrhine primates: A review of the current evidence. *J. Hum. Evol.* **16:**41–80.

Harrison, T. 1989a. A taxonomic revision of the small catarrhine primates from the early Miocene of East Africa. *Folia Primatol.* **50:**59–108.

Harrison, T. 1989b. A new species of *Micropithecus* from the middle Miocene of Kenya. *J. Hum. Evol.* **18:**537–558.

Harvey, P. H., and Bennett, P. M. 1983. Brain size, energetics, ecology and life history patterns. *Nature* **306:**314–315.

Harvey, P. H., and Clutton-Brock, T. H. 1985. Life history variation in primates. *Evolution* **39:**559–581.

Harvey, P. H., Martin, R. D., and Clutton-Brock, T. H. 1987. Life histories in comparative perspective, in: B. B. Smuts, D. L. Cheney, R. M. Seyfarth, R. W. Wrangham, and T. T. Struhsaker (eds.), *Primate Societies,* pp. 181–196. University of Chicago Press, Chicago.

Hutchinson, G. E., and MacArthur, R. H. 1959. A theoretical ecological model of size distributions among species of animals. *Am. Nat.* **93:**117–125.

Janis, C. M. 1990. Correlation of cranial and dental variables with body size in ungulates and macropodoids, in: J. Damuth and B. J. MacFadden (eds.), *Body Size in Mammalian Paleobiology,* pp. 255–300. Cambridge University Press, Cambridge.

Jarman, P. 1983. Mating system and sexual dimorphism in large, terrestrial, mammalian herbivores. *Biol. Rev.* **58:**485–520.

Jungers, W. L. 1985. Body size and scaling of limb proportions in primates, in: W. L. Jungers (ed.), *Size and Scaling in Primate Biology,* pp. 345–381. Plenum, New York.

Jungers, W. L. 1990. Problems and methods in reconstructing body size in fossil primates, in: J. Damuth and B. J. MacFadden (eds.), *Body Size in Mammalian Paleobiology,* pp. 103–118. University of Cambridge Press, Cambridge.

Jungers, W. L., and Susman, R. L. 1984. Body size and skeletal allometry in African apes, in: R. L. Susman (ed.), *The Pygmy Chimpanzee,* pp. 131–178. Plenum, New York.

Kay, R. F. 1982a. *Sivapithecus simonsi,* a new species of Miocene hominoid with comments on the phylogenetic status of the Ramapithecinae. *Int. J. Primat.* **3:**113–173.

Kay, R. F. 1982b. Sexual dimorphism in Ramapithecinae. *Proc. Nat. Acad. Sci. USA* **79:**209–212.

Kay, R. F., and Simons, E. L. 1980. The ecology of African Oligocene Anthropoidea. *Int. J. Primat.* **1:**21–37.

Kay, R. F., Plavcan, J. M., Glander, K. E., and Wright, P. C. 1988. Sexual selection and canine dimorphism in New World monkeys. *Am. J. Phys. Anthropol.* **77:**385–397.

Kelley, J. 1986a. Species recognition and sexual dimorphism in *Proconsul* and *Rangwapithecus*. *J. Hum. Evol.* **15:**461–495.

Kelley, J. 1986b. Paleobiology of Miocene Hominoids. Ph.D. Dissertation, Yale University.

Kelley, J. 1992a. Evolution of the apes, in: R. D. Martin, D. Pilbeam, and J. S. Jones (eds.), *The Cambridge Encyclopedia of Human Evolution,* pp. 223–230. Cambridge University Press, Cambridge.

Kelley, J. 1992b. A biological hypothesis of ape species density. *Abstracts of the XIVth Congress of the International Primate Society,* pp. 72–73.

Kelley, J., and Etler, D. 1989. Hominoid dental variability and species number at the late Miocene site of Lufeng, China. *Am. J. Primat.* **18:**15–34.

Kelley, J., and Pilbeam, D. 1986. The dryopithecines: taxonomy, anatomy and phylogeny of Miocene large hominoids, in: D. R. Swindler and J. Erwin (eds.), *Comparative Primate Biology, Vol. 1, Systematics, Evolution and Anatomy,* pp. 361–411. Alan R. Liss, New York.

Kelley, J., and Xu, Q. 1991. Extreme sexual dimorphism in a Miocene hominoid. *Nature* **352:**151–153.

Kordos, L. 1987. Description and reconstruction of the skull of *Rudapithecus hungaricus* Kretzoi (Mammalia). *Ann. Hist.-Natur. Mus. Nat. Hung.* **79:**77–88.

Kordos, L. 1988. Comparison of early primate skulls from Rudabanya (Hungary) and Lufeng (China). *Anthropol. Hung.* **20:**9–22.

Leutenegger, W. 1971. Metric variability in the postcanine dentition in colobus monkeys. *Am. J. Phys. Anthropol.* **35:**91–100.

Leutenegger, W. 1982. Scaling of sexual dimorphism in body weight and canine size in primates. *Folia Primat.* **37:**163–176.

Leutenegger, W., and Cheverud, J. M. 1982. Correlates of sexual dimorphism in primates. *Int. J. Primat.* **3:**387–402

Leutenegger, W., and Cheverud, J. M. 1985. Sexual dimorphism in primates: The effects of size, in: W. L. Jungers (ed.), *Size and Scaling in Primate Biology,* pp. 33–50. Plenum, New York.

Leutenegger, W., and Shell, B. 1987. Variability and sexual dimorphism in canine size of *Australopithecus* and extant hominoids. *J. Hum. Evol.* **16:**359–367.

Lovejoy, C. O. 1981. The origin of man. *Science* **211:**341–350.

Lucas, P. W., Corlett, R. T., and Luke, D. A. 1986. Postcanine tooth size and diet in anthropoid primates. *Z. Morph. Anthropol.* **76:**253–276.

Mahler, P. E. 1973. Metric Variation in the Pongid Dentition. Ph.D. Dissertation, The University of Michigan.

Maiorana, V. C. 1990. Evolutionary strategies and body size in a guild of mammals, in: J. Damuth and B. J. MacFadden (eds.), *Body Size in Mammalian Paleobiology,* pp. 69–102. Cambridge University Press, Cambridge.

Markham, R., and Groves, C. P. 1990. Brief communication: Weights of wild orang utans. *Am. J. Phys. Anthropol.* **81:**1–3.

Martin, L. B. 1983. The Relationships of the Later Miocene Hominoidea. Ph.D. Dissertation, University of London.

Martin, L. 1991. Teeth, sex, and species. *Nature* **352:**111–112.

Martin, L., and Andrews, P. 1984. The phyletic position of *Graecopithecus freybergi* Koenigswald. *Cour. Forsch. Sencken.* **69:**25–40.

Mein, P. 1986. Chronological succession of hominoids in the European Neogene, in: J. G. Else and P. C. Lee (eds.), *Selected Proceedings of the Tenth Congress of the International Primatological Society, Vol. 1, Primate Evolution,* pp. 60–70. Cambridge University Press, Cambridge.

Napier, J. R., and Davis, P. R. 1959. The forelimb skeleton and associated remains of *Proconsul africanus. Fossil Mammals of Africa, Vol. 16.* British Museum (Natural History), London.

Norconk, M. A. 1990. Introductory remarks: Ecological and behavioral correlates of polyspecific primate troops. *Am. J. Primat.* **21:**81–85.

Nowak, R. M., and Paradiso, J. L. 1983. *Walker's Mammals of the World, 4th ed.* Johns Hopkins University Press, Baltimore.

Oates, J. F., Whitesides, G. H., Davies, A. G., Waterman, P. G., Green, S. M., Dasilva, G. L., and Mole, S. 1990. Determinants of variation in tropical forest primate biomass: New evidence from West Africa. *Ecology* **71:**328–343.

Oxnard, C. E. 1987. *Fossils, Teeth and Sex: New Perspectives on Human Evolution.* University of Washington Press, Seattle.

Oxnard, C. E., Lieberman, S. S., and Gelvin, B. R. 1985. Sexual dimorphisms in dental dimensions of higher primates. *Am. J. Primat.* **8:**127–152.

Pan, Y., Waddle, D. M., and Fleagle, J. G. 1989. Sexual dimorphism in *Laccopithecus robustus,* a late Miocene hominoid from China. *Am. J. Phys. Anthropol.* **79:**137–158.

Peters, R. H. 1983. *The Ecological Implications of Body Size.* Cambridge University Press, Cambridge.

Phillips-Conroy, J. E., and Jolly, C. J. 1981. Sexual dimorphism in two subspecies of Ethiopian baboons (Papio hamadryas) and their hybrids. *Am. J. Phys. Anthropol.* **56:**115–129.

Pianka, E. R. 1970. On r- and K-selection. *Am. Nat.* **104:**592–597.

Pianka, E. R. 1978. *Evolutionary Ecology.* Harper and Row, New York.

Pickford, M. 1985. A new look at *Kenyapithecus* based on recent discoveries in western Kenya. *J. Hum. Evol.* **14:**113–143.

Pickford, M. 1986a. Sexual dimorphism in *Proconsul. Hum. Evol.* **1:**111–148.

Pickford, M. 1986b. Hominoids from the Miocene of East Africa and the phyletic position of *Kenyapithecus. Z. Morph. Anthropol.* **76:**117–130.

Pickford, M. 1986c. On the origins of body size dimorphism in primates. *Hum. Evol.* **1:**77–90.

Pilbeam, D. 1969. Tertiary Pongidae of East Africa: Evolutionary relationships and taxonomy. *Peabody Mus. Bull.* **31:**1–185.

Plavcan, J. M. 1989. The coefficient of variation as an indicator of interspecific variability in fossil assemblages (abstr.). *Am. J. Phys. Anthropol.* **78:**285.

Plavcan, J. M. 1990. Sexual dimorphism in the dentition of extant anthropoid primates. Ph.D. Dissertation, Duke University.

Rose, M. D. 1983. Miocene hominoid postcranial morphology: Monkey-like, ape-like, neither, or both?, in: R. L. Ciochon and R. S. Corruccini (eds.), *New Interpretations of Ape and Human Ancestry,* pp. 405–417. Plenum, New York.

Rose, M. D. 1989. New Postcranial specimens of catarrhines from the middle Miocene Chinji Forma-

tion, Pakistan: Descriptions and a discussion of proximal humeral functional morphology in anthropoids. *J. Hum. Evol.* **18:**131–162.
Ruff, C. B. 1987. Structural allometry of the femur and tibia in Hominoidea and *Macaca. Folia Primat.* **48:**9-49.
Ruff, C. B., Walker, A., and Teaford, M. F. 1989. Body mass, sexual dimorphism and femoral proportions of *Proconsul* from Rusinga and Mfangano Islands, Kenya. *J. Hum. Evol.* **18:**515–536.
Sacher, G. A. 1959. Relation of lifespan to brain weight and body weight in mammals, in: G. E. Wolstenholme and M. O'Connors (eds.), *CIBA Colloquia on Ageing 5,* pp. 115–141. Churchill, London.
Schultz, A. H. 1941. The relative size of the cranial capacity in primates. *Am. J. Phys. Anthropol.* **28:**273–287.
Schwartz, J. H. 1990. *Lufengpithecus* and its potential relationship to an orang-utan clade. *J. Hum. Evol.* **19:**591–605.
Shea, B. T. 1986. Ontogenetic approaches to sexual dimorphism in anthropoids. *Hum. Evol.* **1:**97–110.
Shea, J. H. 1982. Twelve fallacies of uniformitarianism. *Geology* **10:**455–460.
Simons, E. L. 1976. The nature of the transition in the dental mechanism from pongids to hominids. *J. Hum. Evol.* **5:**511–528.
Sinclair, A. R. E., and Norton-Griffiths, M. (eds.), 1979. *Serengeti: Dynamics of an Ecosystem.* University of Chicago Press, Chicago.
Smith, B. H. 1989. Dental development as a measure of life history in primates. *Evolution* **43:**683–688.
Smuts, B. B. 1985. *Sex and Friendship in Baboons.* Aldine, New York.
Smuts, B. B., Cheney, D. L., Seyfarth, R. M., Wrangham, R. W., and Struhsaker, T. T. (eds.). 1987. *Primate Societies.* University of Chicago Press, Chicago.
Stanley, S. M. 1973. An explanation for Cope's rule. *Evolution* **27:**1–26.
Struhsaker, T. T. 1981. Polyspecific associations among tropical rain-forest primates. *Z. Tierpsychol.* **57:**268–304.
Swindler, D. R. 1976. *Dentition of Living Primates.* Academic Press, New York.
Swindler, D. R., McCoy, H. A., and Hornbeck, P. V. 1967. The dentition of the baboon (*Papio anubis*), in: H. Vagtbord (ed.), *The Baboon in Medical Research,* Vol. II, pp. 133–150. University of Texas Press, Austin.
Szalay, F. S., and Delson, E. 1979. *Evolutionary History of the Primates.* Academic Press, New York.
Teaford, M. F., Beard, K. C., Leakey, R. E., and Walker, A. 1988. New hominoid facial skeleton from the early Miocene of Rusinga Island, Kenya, and its bearing on the relationship between *Proconsul nyanzae* and *Proconsul africanus. J. Hum. Evol.* **17:**461–477.
Terborgh, J. 1983. *Five New World Primates.* Princeton University Press, Princeton, NJ.
Van Valen, L. 1973. Body size and numbers of plants and animals. *Evolution* **27:**27–35.
Walker, A. C., and Pickford, M. 1983. New postcranial fossils of *Proconsul africanus* and *Proconsul nyanzae,* in: R. L. Ciochon and R. S. Corruccini (eds.), *New Interpretations of Ape and Human Ancestry,* pp. 325–351. Plenum, New York.
Walker, A. C., Falk, D., Smith, R., Pickford, M. 1983. The skull of *Proconsul africanus:* Reconstruction and cranial capacity. *Nature* **305:**525–527.
Waser, P. M. 1987. Interactions among primate species, in: B. B. Smuts, D. L. Cheney, R. M. Seyfarth, R. W. Wrangham, and T. T. Struhsaker (eds.), *Primate Societies,* pp. 210–226. University of Chicago Press, Chicago.
Wood, B. A., and Xu, Q. 1991. Variation in the Lufeng dental remains. *J. Hum. Evol.* **20:**291–311.
Wu, R. 1987. A revision of the classification of the Lufeng great apes. *Acta Anthropol. Sinica* **6:**265–271.
Wu, R., and Oxnard, C. E. 1983a. Ramapithecines from China: Evidence from tooth dimensions. *Nature* **306:**258–260.
Wu, R., and Oxnard, C. E. 1983b. *Ramapithecus* and *Sivapithecus* from China: Some implications for higher primate evolution. *Am. J. Primat.* **5:**303–344.
Wu, R., and Pan, Y. 1984. A late Miocene gibbon-like primate from Lufeng, Yunnan Province. *Acta Anthropol. Sinica* **3:**193–200.
Wu, R., Xu, Q., and Lu, Q. 1983. Morphological features of *Ramapithecus* and *Sivapithecus* and their

phylogenetic relationships—morphology and comparison of the crania. *Acta Anthropol. Sinica* **2**:1–10.

Wu, R., Lu, Q., and Xu, Q. 1984. Morphological features of *Ramapithecus* and *Sivapithecus* and their phylogenetic relationships—morphology and comparison of the mandibles. *Acta Anthropol. Sinica* **3**:1–10.

Wu, R., Xu, Q., and Lu, Q. 1985. Morphological features of *Ramapithecus* and *Sivapithecus* and their phylogenetic relationships—morphology and comparison of the teeth. *Acta Anthrop. Sinica* **4**:197–204.

Wu, R., Xu, Q., and Lu, Q. 1986. Relationships between Lufeng *Sivapithecus* and *Ramapithecus* and their phylogenetic position. *Acta Anthropol. Sinica* **5**:1–30.

Xu, Q., and Lu, Q. 1979. The mandibles of *Ramapithecus* and *Sivapithecus* from Lufeng, Yunnan. *Vertebr. Palasiat.* **17**:1–13.

Xu, Q., and Wood, B. A. 1989. Dental variation in the Lufeng hominoid dental remains (abstr.). *Am. J. Phys. Anthropol.* **78**:286.

Xu, Q., Lu, Q., Pan, Y., Zhang, X., and Zheng, L. 1978. Fossil mandible of the Lufeng *Ramapithecus*. *Kexue Tongbao* **9**:544–556.

4

Species and Species Recognition in the Hominid Fossil Record

The Importance of Species Taxa in Paleoanthropology and an Argument for the Phylogenetic Concept of the Species Category 18

WILLIAM H. KIMBEL AND YOEL RAK

> Assuming for the moment that history could be analyzed completely into a single set of atomistic elements, there are infinitely many ways these elements can be organized into historical sequences. The role of the central subject is to form the main strand around which the historical narrative is woven. . . The important feature of central subjects is that from the point of view of the historical narrative associated with them, they are *individuals*. The identity and continuity of such individuals can be and must be determined independently of the events that make up the narrative.
>
> David Hull (1975, p. 255)

Introduction

Paleoanthropology attempts to describe the diversity of extinct primate forms, to interpret this diversity in a phylogenetic framework based on the distribution of shared evolutionary novelties, and to explain the emergence and transformation

WILLIAM H. KIMBEL • Institute of Human Origins, Berkeley, California 94709.
YOEL RAK • Department of Anatomy, Sackler Faculty of Medicine, Tel Aviv University, Tel Aviv, Israel, and Institute of Human Origins, Berkeley, California 94709.
Species, Species Concepts, and Primate Evolution, edited by William H. Kimbel and Lawrence B. Martin. Plenum Press, New York, 1993.

of novelties in terms of a positive causal relationship between changes in structure/function and enhanced organismal fitness (i.e., adaptation). These components may be viewed as a sequence of steps toward a "complete" explanation of evolutionary change, each step logically contingent on those preceding it. Thus, hypotheses seeking to explain the adaptive basis of evolutionary morphological change necessarily depend on the prior acceptance of a hypothesis of vectored phylogenetic change. In turn, a phylogenetic hypothesis must be grounded in some theoretical concept of the units of diversity, among which the pattern of phylogenetic relationships is sought.

As paleontologists used to looking at evolution from the "bottom up," we concede that our theoretical historical world view may appear different from that of scientists preoccupied with contemporary organisms and the evolutionary dynamics of their natural groupings. However, the realms of "today" and "yesterday" are not separate biological realities, and as the history of evolutionary theory has amply illustrated, there is opportunity for conceptual synthesis between disciplines concerned with different scopes and scales of evolutionary time. The bases for such synthesis are the sequence of events and the processes governing those events, which connect the present to the past. It is thus our firm conviction that evolutionary history must inform our judgments about the biology of contemporary organisms. But what stands in the way of happy agreement on the "process" and "pattern" of evolution is the fact that the processes operating today are often as imperfectly understood as the historical events that, through our uniformitarian assumptions, were shaped by the very same processes.

Nevertheless, we believe that evolutionary theory as presently understood constrains how we conceptualize the entities that are postulated to play a role in it. Inherent in the Darwinian theory of descent with modification by natural selection is the existence of organisms, among whom differential reproduction accounts for the biased perpetuation of heritable phenotypic variation through time and space. From a scientific point of view, the ontology and epistemology of organisms has been relatively unproblematic. This is not the case with species, however, which are minimally (and neutrally, we believe) conceived as closed communities of organismal ancestry and descent. Although genetic and morphological discontinuity among these communities strikes us as an empirical fact, it has proven difficult to achieve consensus about the status of species in evolutionary theory. Judgments range from species being direct participants in evolutionary change, and even targets of selection, to species being merely the historical effects of processes occurring at or below the level of organisms. Such uncertainty has not deterred a number of systematists (e.g., Nelson and Platnick, 1981) from claiming that evolutionary theory (however broadly or vaguely construed) is irrelevant to the ordering of the empirical facts of biological diversity. That this ordering turns out to be hierarchical is also an empirical fact, but one that we believe cannot be divorced from some overarching process theory about why life should be structured this way. It is descent with modification, coupled with speciation (minimally conceived as the production of new closed communities of organismal ancestry and descent), that provides the theoretical context for the hierarchy observed in nature. Thus, the species is the conceptual pivot about which turn all discussions of the generation and organization of biological diversity, both past and present.

Elsewhere, one of us (Kimbel, 1991) has critically reviewed competing species concepts in the context of paleoanthropology. In this chapter we further discuss the benefits of the phylogenetic concept of the species category developed by Cracraft (1983, 1987, 1989a,b). We preface this discussion with a brief review of the ontological status of species in evolutionary theory (see Eldredge, this volume, and Szalay, this volume, for fuller, albeit conflicting, treatments of this topic). We then discuss the extension of the phylogenetic species concept (PSC) to paleontology and finally illustrate its utility by applying it to the fossil record of hominid evolution.

The Nature of Species Taxa

Critical to clear thinking about species concepts and the role of species in evolutionary theory is the distinction between the species *taxon* and the species *category*. Mayr (1969, p. 5) puts it this way: "A category [of the Linnaean hierarchy] . . . is an abstract term, a class name, while the taxa placed in these categories are concrete zoological objects." Species concepts are definitions of the species category, not of particular species taxa. What does it mean to say that a species taxon is a "concrete zoological object?" Philosophers have generally recognized two metaphysical categories to partition states of existence: classes and individuals. In common parlance a class is a "kind of thing," and a "kind of thing" is spatiotemporally dimensionless and unchanging because the characteristics that define a "kind" are not subject to change. On the other hand, an individual is a particular, a spatiotemporally integrated and finite object.* An instance of a "kind of thing" cannot change without becoming a different "kind of thing," whereas an individual can change and still be the same individual. "Male" is an example of a class. Bill Kimbel and Yoel Rak are paradigm individuals who are members of this class (by most accounts) by virtue of certain defining properties, but we are not ourselves classes.

What of species? Michael Ghiselin's (1969, 1974, 1987) "radical solution" to the problem of the ontological status of species taxa was to drive home the point that species have the characteristics of individuals in the metaphysical sense (see also Hull, 1976, 1978, 1987). Species are individuals because they are "spatiotemporally restricted wholes bound together . . . by relations of ancestry and descent, with determinable beginnings, extensions and extinctions in space and time" (Splitter, 1988, p. 329). Species have these characteristics by virtue of the spatiotemporally continuous network of reproduction among their constituent organisms; that is, they are unique, historical lineages. Thus, species taxa are individuals composed of less inclusive individuals (organisms).

*It is important to understand that in this context the term *individual* is not equated with *organism*, which is but one level in the hierarchy of individuals in nature (e.g., Eldredge, 1985). The perception of individuals in this hierarchy is scale-dependent.

The ontology of species taxa as individuals does not attach to any particular set of taxonomic principles (*contra* Szalay and Bock, 1991). Rather, it is a direct corollary of evolutionary theory. As stated by Hull (1978, p. 342–343),

> Even if entire species are not sufficiently well integrated to function as units of selection, they are the entities which evolve as a result of selection at lower levels. The requirements of selection at these lower levels place constraints on the manner in which species can be conceptualized. Species as the results of selection are necessarily lineages, not sets of similar organisms.

In order for selection to operate as an evolutionary force, there must be descent (replication and reproduction). Descent requires spatiotemporal continuity, implying that the organisms connected by descent form a lineage. A lineage, itself, is thus spatiotemporally localized and is also capable of change due to processes operating on the heritable properties of organisms. In principle, then, similarity is a "red herring" because what really matters in deciding whether two organisms belong to the same species is knowledge that they are part of the same genealogical network, rather than what they look like. The central notion in this distinction is one of *cause* (Sober, 1980). Morphological similarity does not cause an organism to be a member of a particular species; being part of a particular genealogy causes an organism to be part of the species and to be phenotypically similar to other organisms within it. [An apt analogy from Simpson [1961] is that we consider two individuals twins because they derive from the same ovum (the cause), not because of phenotypic similarity. We infer common descent from phenotypic similarity.] Morphological similarity wedded to the concept of descent distinguishes homologous characters (retained ancestral as well as newly derived) from nonhomologous characters. If species taxa were conceptualized as classes of morphologically similar organisms, the distinction between the two would be irrelevant.

Paleontologists must (and neontologists usually do), however, rely on morphological characters to individuate species taxa. Characters, whether phenotypic or genotypic, are properties of organisms that are diagnostic of their part-whole relationship of organisms to species. The distinction between *diagnosis* and *definition* is biologically fundamental (Mayr, 1969, p. 6; de Queiroz and Gauthier, 1990; Ghiselin, 1984; Rowe, 1987; Wiley, 1989).* Ghiselin (1984, p. 106) notes that a diagnosis in systematics functions much the same way as it does in medicine: "When a physician diagnoses an illness, he infers what the illness is, but he does not claim that the patient has it by definition. A patient could have all of the diagnostic traits (symptoms) and not have the disease, or none of the diagnostic traits yet still have the disease." Thus, the diagnosis of a species taxon is, in effect, the "symptoms" of genealogical cohesion; diagnostic characters are those that are held to distinguish one genealogically cohesive system of parental ancestry and descent (lineage) from other, similar and/or closely related systems.

*The use of the term *definition* here implies definition by *intension:* a list of the properties necessary and sufficient for the name to apply to a class of objects. An individual can only be defined *ostensively,* that is, by "pointing" to or "showing" it. The names attached to objects defined ostensively are proper names.

Defining the Species Category: The Phylogenetic Concept

As noted above, the species category of the Linnaean hierarchy is the class whose members are taxa ranked at the species level. Beyond this uncontroversial statement, there isn't much agreement over which concept of the species category best describes the nature of species taxa, although in our view a theoretically acceptable species concept must account for the characteristics of species taxa as individuals. However, at the same time, all species concepts are constructed to recognize *discontinuities* between species taxa, whether by virtue of reproductive isolation (Mayr, 1970), different fertilization systems (Paterson, 1985), evolutionary roles and historical fates (Simpson, 1961), ecological niches (Van Valen, 1976), or patterns of character variation (Nelson and Platnick, 1981). Thus, the species category is first and foremost a catalog of regular discontinuities that permits us to formulate generalizations across species taxa. It is these discontinuities that also enable us to internest species in monophyletic groups, and in this restricted sense, to paraphrase E. O. Wiley (1981), monophyletic higher taxa are the products of history, while species produce it.

As paleontology is the only evolutionary discipline necessarily concerned with time on the geological scale, its practitioners have generally sought species concepts that explicitly recognize the notion of species as lineages, such as G. G. Simpson's evolutionary species concept (ESC): "[A species is] a lineage (an ancestral-descendant sequence of populations) evolving separately from others and with its own unitary evolutionary role and tendencies (Simpson, 1961, p. 151; see also Wiley, 1981, for discussion). Such lineages qualify as species when, by virtue of the morphological cohesiveness of organismal clusters, they are inferred to have had "their own unitary evolutionary role and tendencies" and thus are further inferred to have been (or to be) reproductively isolated from one another. Simpson argued that the "nondimensional" biological species of Mayr (e.g., 1970, p. 12) is a special case of the evolutionary species: Biological species are synchronic and sympatric instances of the larger class of separately evolving lineages. An important distinction between the BSC of Mayr (1970, but *not* 1982) and the ESC of Simpson is that the latter implicitly invokes the notion of the adaptive ("economic" *sensu* Eldredge, 1985) response of organisms linked by descent ("evolutionary role") as a ranking criterion.

Although both the biological species concept (BSC) and the ESC are commonly viewed as consistent with the ontology of species taxa as individuals, many advocates of these concepts assert that species lose their discreteness when observed vertically (Sylvester-Bradley, 1956; Cain, 1960; Simpson, 1961; Mayr, 1969, 1982; Bock, 1979; Gingerich, 1985; Rose and Brown, 1986, this volume; Häuser, 1987; Szalay, 1989, this volume; Szalay and Bock, 1991). Thus, according to Bock (1979, p. 29):

> A cross-section of a phyletic lineage at any point in time is a species. However, cross-sections of the same phyletic lineage at different points in time are not different species nor are they the same species. These are simply different cross-sections of the same phyletic lineage at different times. . . [I]t is not possible to speak of the age of a species, or the origin of a species, or of the life and death of a species.

Mayr (1982, p. 294) criticizes the ESC by claiming "the definition is that of a phyletic lineage, but not of a species." And Gingerich (1985, p. 29) states "Species are in many cases objective evolutionary units on a time plane and at the same time arbitrary units crossing time planes."

Apparently, what these authors have in mind is two ontologically distinct, noncomparable entities: phyletic lineages and species. A phyletic lineage is a species only in the taxonomic (character-based) sense and is categorically distinct from a biological species (an interbreeding population that is reproductively isolated from other such populations). It is important to note that such a division is not due to practical difficulties in distinguishing species taxa over evolutionary time. Rather, it stems from the view that *theoretically* species do not have a time dimension. Thus, although it is frequently said that paleontologists have trouble "applying" the Biological Species Concept to the fossil record, we contend that paleontologists are expected to employ the concept to make inferences about a class of things (lineages) that the concept is explicitly designed *not* to embrace.

For paleontologists, this conclusion implies, *inter alia,* that attempts to derive rules for species identification in the fossil record from the regularities of morphological variation within extant species are undermined by the BSC. The comparison of incomparable entities allows claims that extinct species lineages, due to their time depth, are liable to harbor a degree or quality of variation not observed in contemporary biological species. This has led to the concept of "successional" species (also chronospecies)—time-successive segments of a single evolving lineage that are thought to differ phenotypically to the same degree as closely related biological species and therefore are given different names (Simpson, 1961, p. 165–166). Wiley (1981, p. 38–41) has noted that although the chronospecies concept began as a convenient taxonomic device for the paleontologist, it has attained the status of a real, if thus far undescribed, biological process (phyletic speciation). The key and internally contradictory point of this notion is that while phyletic speciation allegedly produces discrete units (species), a continuous lineage can in theory only be divided arbitrarily, since there is no point at which reproductive continuity is broken. Dividing up extinct species lineages into morphological packages equivalent to contemporary "biological" sister species would be expedient if our primary interest is to produce phenotypic uniformity between the past and the present, but we cannot see how such expediency can claim theoretical justification from evolutionary biology.

The notion that phyletic lineages and species constitute different kinds of entities is a necessary consequence of the failure of the BSC to explain the temporal persistence criterion of individuality. The BSC cannot be reconciled with the fact that species have spatio*temporal* continuity, that they are, in fact lineages. Recognizing species taxa as individuals with history may remedy the ontological problem of the reality and persistence of species in time and space, but it is no argument for the BSC. Under the BSC, the "species" of the paleontologist will differ conceptually from that of the neontologist, precluding any meaningful generalizations about species from the past to the present. We do not believe that this is a defensible solution to the "species problem." We do believe that a univocal species concept is possible, at least for sexually reproducing organisms. Such a concept will acknowledge the central role that species play in phylogeny.

Vrba (1985, p. ix) phrased it well when she wrote, "The concept that species are discrete 'limbs' in the sexual parts of life's phylogeny makes sense . . . It is appropriate that the boundaries of species in time and space should have a one-to-one relationship with discrete parts of the tree of life." Put another way, it is the sequential production of new reproductively cohesive lineages that yields a hierarchical pattern of phylogenetic relationships. If one adopts the view (as we do) that recovering the historical pattern of relationships should logically and operationally be prior to the exploration of specific hypotheses about adaptation, then a species concept that acknowledges species taxa as fundamental historical units might be preferred. Such a phylogenetic species concept has been provided by Cracraft (1983, 1987, 1989a,b): "A species is an irreducible (basal) cluster of organisms, diagnosably distinct from other such clusters, and within which there is a parental pattern of ancestry and descent" (Cracraft, 1989a, p. 34–35). According to this concept, species "function" in phylogeny as the smallest complete lineages of ancestry and descent, among which there is a hierarchical pattern of relationships (see also Eldredge and Cracraft, 1980, p. 92; McKitrick and Zink, 1988; Nixon and Wheeler, 1990).

The phylogenetic species concept (PSC) is theoretically rooted in the thesis that species are reproductively cohesive lineages. Although it is also operational, the PSC is neither typological nor essentialistic, in that it does not treat taxa as classes defined by characters that each and every one of their members must possess. The PSC accommodates sex, age, and other morphs derived from the same system of parental ancestry and descent because it emphasizes precisely that set of characters from which are inferred the existence and structure of the underlying (causal) reproductive continuity among a population of otherwise heteromorphic organisms. Thus, diagnostic characters are signs of reproductive cohesion that enable the grouping of organisms by virtue of uniquely shared ancestry and descent.

Given the species-as-individuals ontology described in the previous section, the species category must be used to recognize taxonomic diversity resulting from the nonarbitrary origin of new reproductively cohesive lineages. We agree with arguments that such lineages can arise nonarbitrarily only through cladogenesis. So, to the question, "Which biological entities qualify for species rank?," one might answer: "Reproductively cohesive lineages." But there is more than one level of reproductive cohesion in nature; local breeding units (demes) as well as "biological" species form, respectively, smaller and larger reproductively cohesive lineages. Under the BSC, the largest reproductively cohesive units are called *species* because the criterion of species rank is reproductive isolation; multiple demes may be reproductively compatible, gene flow bonding them together, and the outer limit of gene flow constituting the species' boundary. Thus, "biological" species are reproductively cohesive entities (usually) composed of less inclusive reproductively cohesive entities.

It is here that we confront a conceptual clash between the phenomenon of intraspecific geographic variation and the taxonomic category of subspecies. Under the BSC distinct allopatric populations (nominal subspecies) are sometimes viewed as "incipient" species (i.e., a stage in the speciation process) and thus as units of evolution. However, labeling a deme a "subspecies" is a strictly taxonomic operation that does not impart special evolutionary significance to the unit thus named. There is no known biological process that converts a deme into a "subspecies," and there are no objectively determinable criteria of subspecies status. All demes of a species differ to one extent or another, but it would clearly be inappropriate to recognize every "different" deme taxonomically. This is why the subspecies category is considered a more or less arbitrary tool of convenience for the taxonomist (e.g., Wilson and Brown, 1953; Simpson, 1961, p. 176; Mayr, 1963, pp. 347–348, 1969, p. 42, 1982, p. 289; Wiley, 1981, p. 28; Cracraft, 1983; McKitrick and Zink, 1988; see Jolly, this volume, and Shea *et al.*, this volume, for contrary arguments). The subspecies category denotes a rank in the Linnaean hierarchy; demes are local genealogical units of which species are comprised. *Every* deme of a species is a candidate future species (although the vast majority are ephemeral over the long term), and this, of course, is true whether or not *any* deme of a species is considered worthy of subspecific designation by a taxonomist. Thus, "subspecies" is a taxonomic unit and "deme" is a genealogical unit. We stress that we are not denying the fact that species are geographically variable, or even polytypic. We are instead addressing an issue of ontology, and to make this clear, we pose the question "Does evolutionary theory presuppose the existence of units called 'subspecies?'" We contend that the answer is "no." Substituting either of the terms *demes* or *species* in the question, however, the answer is "yes."

Species occupy the lowest level of biological organization at which phylogenetic relationships can meaningfully be sought. This is because, as recognized by Hennig (1966, pp. 18–20), ancestor-descendant relationships within a community of interbreeding organisms are inherently reticulate rather than hierarchical (see also de Queiroz, 1988; Kluge, 1990; Nixon and Wheeler, 1990). From the point of view of cladistic analysis, phylogenetic relationships among organisms or demes comprising a reproductively cohesive unit will be unresolvable due to discordant character distributions and the widespread prevalence of homoplasy (as in *Homo sapiens,* for example).* It is only when the reproductive network is severed that local demes can achieve phylogenetic independence. Although geographic isolation does not guarantee genetic divergence, the genetic changes underlying a shift in specific mate-recognition systems evolve in allopatry (e.g., Paterson, 1985; Masters, this volume), thus "signaling" a new reproductively cohesive lineage. Therefore, under the BSC's ranking criterion of reproductive isolation, a single "biological" species composed of allopatric demes may (but does not necessarily) contain more than one evolutionary unit.

The PSC does not imply that all primate "subspecies" would automatically be elevated to species status. While all primate species exhibit polymorphisms in a number of behavioral, morphological, or allelic characters, in many cases it is still unclear how these variants correspond to the actual geographic distributions of populations. To the extent that particular subspecies diagnosed on the basis of museum specimens turn out to refer to arbitrary segments of inadequately sampled ecogeographic clines or to be the product of geographically discordant character variation, the need for subspecies names will fall away (e.g., Thorpe,

*Which is not to say that single character lineages cannot be represented hierarchically. But within an interbreeding group such lineages will cut across actual lines of descent at the organismal level. An example is maternal mtDNA lineages (Nixon and Wheeler, 1990).

1987). However, to the extent that discrete allopatric or parapatric populations reveal historical independence, it would be more meaningful, in our view, to consider them as species.

Let us be clear that while the PSC invokes organismal characters as a grouping criterion, it does not provide an infallible method of attributing individual organisms to species. No theoretically coherent species concept can make this claim because all species concepts are about populations (or lineages) of variable organisms. However, the grouping of organisms into phena is a logically and operationally necessary first step in all systematic analyses, and we do not see a clear distinction between how this is accomplished under the PSC vs. other species concepts. The more important question relates to ranking phena taxonomically. How "diagnosable" must a population be in order to qualify for species rank under the PSC? This question is problematic because, whether in neontology or paleontology, we must (under certain statistical assumptions) infer population limits from the characteristics of samples. This, too, is a limitation common to all species concepts (even under the BSC, we do not "observe" reproductive isolation in entire populations). Perhaps a more meaningful way to ask the question is, "How diagnosable must a sample be in order to be considered representative of a historically independent population?," since historical independence focuses more clearly on the central idea of the phylogenetic species. Mayr (1969, p. 190) advocated a necessarily arbitrary "75% rule" in the recognition of subspecies: "a population is recognized as a valid subspecies if 75 percent of the individuals differ from 'all' (= 97 percent) of the individuals of a previously recognized subspecies." A lesser degree of difference did not warrant taxonomic recognition of the populations, while greater difference signified specific distinction. In modern practice, however, multivariate techniques distinguish populations at a much finer level of resolution, with degrees of difference much less than that called for by the "75% rule" commonly viewed as justifying subspecific distinction (e.g., Shea *et al.*, this volume, on the "subspecies" of *Pan troglodytes*). The OTU (operational taxonomic unit) of numerical phenetic taxonomy may or may not be coextensive with a phylogenetic species (OTUs may initially consist of separate sexes or age classes, for example). The decision will be based on the distribution of characters within and among the relevant samples of organisms, as we address in the next section.

Discovering Phylogenetic Species in Paleontology

The ontology of species taxa as spatiotemporally discrete, reproductively cohesive entities means that extinct species are potentially discoverable, but our methodology must be such that we are prepared to make such discoveries. Nevertheless, the fossil record conspires to make this task difficult at best. Obviously every fossil organism was once part of a deme of a species, but in no case are we actually able to perceive entire demes, let alone whole species, in the fossil record. This is why the epistemology of the paleontologist is always a matter of inference; we cannot observe an extinct species, but we can infer its existence and determine its spatiotemporal locale. The proper use of characters is the key to this enterprise.

The (usually) implicit invocation of the biological species concept or the Simpsonian evolutionary species concept (and the associated taxonomic chronospecies notion) has led to the spurious claim that species and phyletic lineages are not comparable entities: Species taxa, having no reality in time, are construed in paleontology in much the same way as subspecies in neontology, as mere taxonomic artifices. This, in turn, has resulted in a tendency to view *characters,* rather than taxa, as the focus of explanation: If species lack temporal extension, then at least morphology can be tracked through time (in a phyletic lineage) and explanations for perceived change be presented in terms of the evolutionary (usually adaptive) response of organisms over time. The consequence of this approach is that the temporal ordering of a morphocline is presumed to be isomorphic with phylogeny, whereas in fact phylogenetic history is deterministic with respect to a morphocline and its temporal pattern.

In this section we explore ways in which the phylogenetic concept of species can be brought to bear on questions of species recognition in the hominid fossil record. We emphasize the proper (and logically primary) role of characters in diagnosing species taxa that stems from considering these taxa as spatiotemporally extended wholes. Our examples are by no means exhaustive, although some of them sample more extensive work by us in progress.

There is potentially an infinite number of characters on which a taxonomic analysis could draw (i.e., the purely phenetic "total morphological pattern"). The uniform or continuous distribution of characters is commonly interpreted as positive evidence for the species-level unity of a sample of organisms. However, as demonstrated by Cope (this volume) and Plavcan (this volume), this criterion can in practice (using osteological attributes, at least) fail to detect multiple congeneric "biological" (sympatric) species in samples artificially mixed to simulate paleontological cases. In contrast, Tattersall (1986) advocated employing the degree of morphological difference between extant species—an implicit invocation of phylogenetic criteria—as a guide to species recognition in paleontology. We must, of course, pay heed to both within-species and among-species character distributions in the evaluation of species composition in order to avoid confusing gender- and age-related characters with interspecific differences. Accordingly, we identify three kinds of characters that will be useful in making decisions about the species composition of fossil samples:

1. Apomorphies (or combinations thereof), which will support a hypothesis of a single system of phylogenetic ancestry and descent (a monophyletic clade). These are the diagnostic characters referred to in the discussion above. Their utility stems from the empirical generalization that in the process of differentiation, new lineages become "fixed" for novel characters (genotypic or phenotypic). This does not imply that every organism in the sample must bear all of the characters, only that the distribution of the characters will delimit relatively homogeneous (nonoverlapping) groups within the sample.

2. Sexually dimorphic and ontogenetically labile characters, which will be consistent with a hypothesis of a single system of parental ancestry and descent. Based on our knowledge of the biology of a wide range of extant taxa, we should

be able to identify with confidence those attributes likely to vary according to sex and age within species. Subsequently, we may be able to identify taxon-specific patterns of sexual dimorphism, for example, and to use these as additional diagnostic characters.

3. Polymorphic characters, whose variation cannot reasonably be attributed to either sexual dimorphism or ontogenetic change, which *might* be consistent with more than one system of parental ancestry and descent. The conditional nature of this set of characters is stressed because of the possibility that within-species geographic or temporal factors are responsible for the observed variation. Although this should never be assumed *a priori,* the finding of a large number of such polymorphic characters will require a deeper investigation into the probable cause(s).

With respect to variation between contemporaneous, geographically disjunct samples, the pattern of character variation will guide decision making. For example, a pattern of congruence among states of several independent character systems in each sample is more likely to reflect phylogenetic differentiation than ecological differences due to local adaptation (in which case there would be no reason to expect a pattern of congruence between independent systems; Thorpe, 1987). What is needed here are criteria for character interdependence, which implies the use of sound biological judgement in the definition of characters. There are at least two organism-level paths leading to the correlation among character states, functional and ontogenetic. The functional integration of a particular set of characters, which may be rooted historically in an environmentally mediated selective regime, may reasonably be assumed to be ontogenetically integrated as well. However, the converse is not necessarily true: The ontogenetic integration of characters does not imply a unitary functional cause, since genetic phenomena such as pleiotropy and linkage disequilibrium can produce a developmental bond between morphological characters that have no common functional background. In either case, character analysis must reveal a pattern of correlation among character states (otherwise the question of interdependence is moot). However, establishing a correlation between characters is a necessary but not sufficient condition for a hypothesis of functional or ontogenetic interdependence. A combination of experimental and ontogenetic evidence (which, in paleontology, must derive from neontological analogs) must be brought to bear on the causal background of the observed correlation.

As an example, we draw the reader's attention to our analysis of the asterionic region of the hominoid calvaria (Kimbel and Rak, 1985). Through the use of ontogenetic and functional criteria, it was shown that certain taxon-specific patterns in the sutural arrangement at the asterion are intimately connected to compound temporal/nuchal crest formation on the temporal bone, both hypothesized to reflect the differential posterior migration of the temporalis muscle fibers during growth. Although it would not appear critical to have knowledge of this sort in the initial sorting of specimens into phena, it would be an important part of the broader process of evaluating character independence and judging whether geographically disjunct samples represent populations that had differentiated phylogenetically.

A more intractable problem is the temporal influence on character variation. Given two samples from the same geographic region, but from different stratigraphic horizons, how can we tell whether the observed variation is chronoclinal (anagenetic) or not? This is obviously of some importance because, under the ontology of species adopted here, speciation cannot be said to have occurred in the absence of cladogenesis. We see no nonarbitrary resolution to this dilemma short of appealing to the "gaps in the record" argument, which claims that it is the patchiness of the fossil record that enables us to demarcate ancestral from descendant species. In light of this, we recognize temporally disjunct clusters of specimens that are separately diagnosable by apomorphies as distinct species. We think this is better than to operate under the assumption that missing evidence would extinguish the distinctiveness of the clusters through a morphologically graded series of stratigraphic intermediates (which is tantamount to assuming an ancestor–descendant relationship). In this case, one also has recourse to the findings of cladistic analysis. Although *a posteriori,* the demonstration that two putative species are not sister taxa weighs against the hypothesis that they sample the same system of parental ancestry and descent.

On the other hand, the continuously graded stratigraphic series of fossils documented, for example, by Gingerich (e.g., 1985), Rose and Bown (e.g., 1986, this volume), and Krishtalka (this volume) are among the most challenging problems for systematics in paleontology. The hypothesis of a single anagenetically evolving lineage can be refuted by the fossil record itself, i.e., when cladogenesis is demonstrated to separate hypothesized ancestral and descendant species [for example, see Kimbel (1991) on *Homo habilis* and *H. erectus*], but the failure to refute the hypothesis does not mean that anagenesis has thereby been demonstrated. To support the anagenesis hypothesis one must always appeal to missing evidence of cladogenesis. This does not mean that anagenetic evolution does not occur, only that it is difficult to substantiate unequivocally. Part of the problem no doubt lies in the fact that the continuously distributed quantitative characters usually used to document anagenetically "evolving" lineages are difficult to employ in phylogenetic analysis at the species level. It would clearly be preferable to examine the evidence for trends in such quantitative characters within the context of an independently derived phylogenetic hypothesis, but in the absence of the characters on which to build such a hypothesis, one is more or less obliged to accept stratophenetic interpretations of phylogeny at face value. Once this is accepted, temporally contiguous clusters of specimens that are not separately diagnosable, but rather evince continuous character change over time, represent a single, unbroken lineage (species) in our view.

Sexual Dimorphism in *Australopithecus boisei*

The null hypothesis that one species is present in a given sample of fossils should always be preferred because it is rarely, if ever, possible to falsify the alternative, multiple-species, hypothesis based on the statistical distribution of organismal characters (Cope, this volume; Plavcan, this volume). The observation of a shared set of apomorphic states in a sample of fossils may lead to either of two conclusions: The states may be autapomorphies of a single species or they may be synapomorphies of two or more sister species (and thus be primitive at the level of the individual species comprising a monophyletic higher taxon). However, the identification of discrete subsamples on the basis of characters inferred to be sexually dimorphic would leave the null hypothesis unchallenged, supporting the conclusion that the observed novelties are autapomorphies of a single species. Kimbel (1991) discussed how this process corroborates the conspecific status of the putative sex morphs of *Australopithecus boisei.* The key differences between specimens KNM-ER 406, ER 13750, and OH 5, on the one hand, and ER 407 and ER 732, on the other hand, are concordant with a hypothesis of sexual dimorphism within a single system of parental ancestry and descent: The two groups differ in facial size (but not calvarial size), the presence or absence of ectocranial crests, the degree of pneumatization, etc. However, a subset of the apomorphies shared by the two groups are uniquely derived, indicating a proximally shared phylogenetic heritage: the details of facial shape and topography (Rak, 1983), as well as aspects of the glenoid region of the cranial base (Kimbel, 1986). Taken together, the two aspects of character variation make the null hypothesis of a single sexually dimorphic species a more likely explanation of the observed intergroup differences than the two-species hypothesis (cf. Groves, 1989).

The Two Species of Homo habilis

When the components of character variation fail to support one another, the single-species hypothesis becomes suspect. This is the case with regard to the different morphs conventionally subsumed in the species *Homo habilis.* Here, a relatively heterogeneous group of fossils has been treated by most specialists until very recently as a single species under the assumption that observed variation may be explained by sexual dimorphism, intraspecific geographic, and temporal variation, or some unspecified combination of these factors (but see Walker and Leakey, 1978). However, two discrete subsets can be formed from this sample on the basis of the following cranial differences that cannot readily be accounted for by a hypothesis of sexual dimorphism: supraorbital structure, facial shape and topography, calvarial size, preglenoid plane size and orientation, and mandibular premolar morphology (Wood, 1985, this volume; Wood and Uytterschaut, 1987; Lieberman *et al.,* 1988; Chamberlain, 1989; Kimbel and Rak, in preparation; see Table 1). Moreover, there are few, if any, apomorphies common to the two groups that are not also shared with other species in the genus *Homo:* The morphology of the cranial base, the prominence of the nasal bridge, the anterolaterally directed frontal process of the zygomatic, and reduced subnasal prognathism are all apomorphic for the *Homo* clade.

It is unlikely that geographic and temporal factors are wholly responsible for the intergroup variation, since both morphs are present in the same basin and are at least in part contemporary within the limits of radioisotopic dating errors (Feibel *et al.,* 1989). Moreover, appeals to intraspecific geographic variation are tenuous in view of the fact that the South African skull Stw. 53, which in our judgment is clearly distinct from *A. africanus* (Kimbel and Rak, in prepara-

Table 1. Cranial Characters of *Homo rudolfensis* and *Homo habilis*

Homo rudolfensis	*Homo habilis*
No supraorbital torus or supratoral sulcus (thickened superior orbital rim only)	Superiorly protruding supraorbital torus, bilaterally arched and thickest above midorbit, with supratoral sulcus
Frontal squama flat in coronal plane of postorbital constriction	Frontal squama convex in coronal plane of postorbital constriction
Superior facial breadth (Martin no. 43) less than midfacial breadth (Martin no. 46)	Superior facial breadth greater than midfacial breadth
Infraorbital plate inflated to coronal plane anterior to nasal bridge	Infraorbital plate flat and vertical, in coronal plane at or posterior to nasal bridge.
Anterior edge of maxillary zygomatic process over premolars and anterior to coronal plane of inferior orbital margin	Anterior edge of zygomatic process over P4/M1 or M1, in coronal plane of inferior orbital margin
Preglenoid plane extensive and horizontal	Preglenoid plane restricted and steep
Molarized mandibular P4[a]	Mandibular P4 not molarized

[a]See Wood and Uytterschaut (1987) and Wood (this volume).

tion), is difficult to distinguish from East African specimens OH 24 and OH 62 in relevant morphology (Johanson *et al.*, 1987). While we do not dismiss the possibility that the present samples of *H. habilis* demonstrate polytypism (Miller, 1991), this argument loses credibility in the face of a consistent pattern of morphological differentiation between samples that overlap in time as well as space. The conclusion that such a pattern reflects geographic variation within a single species would force us into the position of denying that speciation was an important factor in generating morphological diversity within major hominid clades unless it could be demonstrated that two closely related populations were genuinely (ecologically) sympatric, and it is doubtful whether the Pliocene hominid record affords the resolution necessary to justify such a claim. The most likely explanation for the observed variation is that two species are represented in the sample: *H. habilis* (the Olduvai specimens plus ER 1813, 3734, 3735, 3891, L894-1 and Stw. 53) and *H. rudolfensis* Alexeev 1986 (ER 1470, 1590, 1802, 3732).* The corroboration of the suggestion by Leakey *et al.* (1989) that different postcranial morphologies distinguish these craniodentally diagnosed taxa would constitute independent evidence of phylogenetic differentiation.

As it presently diagnosed, *H. habilis* occurs in both eastern and southern Africa and spans at least a 200 Kyr interval between ≥1.9 (e.g., KNM-ER 1813, OH 24) and ≤1.7 (e.g., OH 13) Myr (Feibel *et al.*, 1989; Walter *et al.*, 1991). *Homo rudolfensis* is thus far known only from the Koobi Fora Fm. in the slim 100 Kyr interval between ≤1.9 and ca. 1.8 Myr (Feibel *et al.*, 1989). Dental remains, possibly of *Homo* from the Shungura Fm., predate the Koobi Fora and Olduvai samples, but it is not clear whether these specimens are attributable to either of the species under consideration here. It would thus appear that these two early

*We differ with Wood (this volume) on the attribution of KNM-ER 3891. Other investigators (e.g., Walker and Leakey, 1978; Stringer, 1986; Chamberlain, 1989) have proposed different taxonomic divisions of the conventional *H. habilis* hypodigm than that presented here. The principal difference concerns the composition of the Olduvai sample.

Homo species are approximately synchronic, and at least one of them, *H. habilis*, overlaps the known temporal range of *H. erectus* (see Kimbel, 1991). Although the phylogenetic arrangement of these taxa may be debated, the African late Pliocene record clearly does not present a classic picture of gradual, unilinear evolution.

The Problem of Variation in *Australopithecus africanus*

The search for patterns of diversity using the criteria discussed above is sometimes frustrated by conflicting evidence within a paleontological sample. It may be that character states are not distributed concordantly among the specimens such that different characters yield different divisions of the sample. In such cases of polymorphism, the null hypothesis of a single species may successfully resist refutation because no clear evidence of independent systems of ancestry and descent can be provided. Indeed, this result is very much what would be expected from a nonhierarchical intraspecific genealogy (see above). A detailed character analysis will reveal a pattern that "gestalt" or quantitative phenetic analyses will not. *Australopithecus africanus* will serve as an illustration of how these difficulties may be addressed.

Several workers have lately commented on the large degree of morphological variation in the skull of *A. africanus sensu stricto* (Clarke, 1988; Kimbel and White, 1988). Most of this variation, which is best manifested in the face and cranial base, is observed within the large Sterkfontein sample and is unusually patterned when compared either to that of extant hominoids or other early hominid species represented by adequate samples. We recently restudied the *A. africanus* hypodigm; one goal was to sort out this variation and to attempt to identify diagnostic characters of the species.

Facial Skeleton

We found the following derived character states to be highly stable in distribution and thus, in combination, to be diagnostic of the species:

1. Moderately to strongly developed anterior pillars
2. Maxillary furrow
3. Zygomatic prominence
4. Strong superoinferior angulation between anteriorly directed frontal process and malar surface of zygomatic
5. High, straight zygomaticoalveolar crest
6. Transversely flat to concave infraorbital plate
7. Thin supraorbital rim (not a supraorbital torus)
8. Superior facial breadth less than midfacial breadth
9. Deep, flexed anterior palate

This diagnosis essentially confirms Rak's (1983) previous findings. These character states are derived with respect to the primitive condition of outgroup taxa, but many of them are shared with *A. robustus:* Only the strong development of anterior pillars, the maxillary furrow, and the deep, flexed anterior palate are not encountered in the South African "robust" species. Of course, the facial skeleton of *A. robustus* is otherwise distinguished by derived morphology of the frontal squama, peripheral face, nasal cavity, and the relative position of the palate, in which areas *A. africanus* retains the primitive state. It is thus a unique combination of derived and primitive character states that permits the diagnosis of the species *A. africanus.*

Relevant to the question of diversity *within* the *A. africanus* hypodigm is the observation that not all specimens possess all of the derived character states in the list given above. For example, Sts. 52a is generalized in lacking anterior pillars and maxillary furrows (#1 and 2), but its possession of #3–6 is a sufficient basis for attributing it to the species. We might hesitate to draw this conclusion if it was true that anterior pillars and maxillary furrows (or lack thereof) were always correlated, but this is not the case. Thus, Sts. 71 has anterior pillars but no maxillary furrows, whereas Stw. 391 lacks a pillar but has a furrow. Furthermore, within the sample of facial remains, midfacial and subnasal prognathism, as well as the morphology of the inferior nasal margin, vary considerably, but randomly, with respect to the set of diagnostic character states enumerated above. It is clear that the *A. africanus* facial skeleton is polymorphic, but the distribution of the polymorphisms within the available sample does not permit the recognition of diagnosable subsets within the hypodigm.

Cranial Base

Dean and Wood's (1982) metrical analysis concluded that the cranium of *A. africanus* is primitive (relative to the extant great apes) in its narrow base, sagittally oriented petrous pyramids, and coronally oriented tympanic axes, but derived in its short base, anteriorly positioned foramen magnum, and mediolaterally short tympanic plates. Picq (1990, pp. 155–158, 162) studied the glenoid region of the hominid cranial base from a phylogenetic point of view and reported a number of hominid synapomorphies in *A. africanus.*

In our study of the cranial base, we focused on the glenoid region as a possible source of diagnostic characters of *A. africanus.* We identified 16 characters and 38 character states that assist in the attribution of specimens to taxa. Compared to extant catarrhine outgroup taxa as well as other early hominid species (*A. afarensis, A. robustus, A. boisei*) the glenoid region of *A. africanus* exhibits an extremely high degree of polymorphism, with 11 of the 16 characters showing two or more states and three of the five monomorphic characters manifesting the primitive state. No other hominid species is polymorphic for more than three characters. The extant hominoid glenoid region presents a very conservative morphological pattern, differing little from the generalized catarrhine condition, and exhibits very little polymorphism within species; it is certainly not an area of the cranium manifesting sexual dimorphism other than in overall size in the most sexually differentiated taxa (*Gorilla, Pongo*). In contrast with the situation in the facial skeleton, seven of the nine glenoid region characters are rendered polymorphic by the inclusion in the *A. africanus* hypodigm of a

Table 2. Glenoid Region Characters Relevant to Specific Status of Sts. 19

Outgroups[a]	*A. africanus*	Sts. 19	*H. habilis*
Tympanic horizontal	Tympanic horizontal or inclined	Tympanic vertical	Tympanic vertical
Tympanic posterior to PGP[b]	Tympanic posterior to PGP	Tympanic and PGP merge superiorly	Tympanic and PGP merge superiorly
PGP large	PGP large	PGP large	PGP reduced
Petrous crest absent	Petrous crest weak or moderate	Petrous crest strong	Petrous crest moderate or strong
Vaginal process absent	Vaginal process absent	Vaginal process present	Vaginal process absent or present
Mastoid fissure open laterally[c]	Mastoid fissure open laterally	Mastoid fissure closed laterally	Mastoid fissure closed laterally
Petrous sagittally oriented	Petrous sagittally oriented	Petrous more coronal	Petrous more coronal
Preglenoid plane horizontal/extensive	Preglenoid plane horizontal/extensive	Preglenoid plane horizontal/extensive	Preglenoid plane steep/restricted
F. ovale indents lat. pterygoid plate	F. ovale indents lat. pterygoid plate	F. ovale posterior to pterygoid	F. ovale indents or posterior to pterygoid
Lat. pterygoid plate triangular in lat. view (long base superiorly)[d]	Lat. pterygoid plate triangular in lat. view (long base superiorly)	Lat. pterygoid plate rectangular in lat. view (abbreviated base superiorly)	Lat. pterygoid plate rectangular in lat. view (abbreviated base superiorly)
Eustachian process absent	Eustachian process present	Eustachian process absent	Eustachian process absent
Medially, tympanic faces inferior	Medially, tympanic faces anteroinferior	Medially, tympanic faces inferior	Medially, tympanic faces inferior

[a]Outgroup taxa consist of *Pan troglodytes* and *Gorilla gorilla* from Hamann-Todd Osteological Collection, Cleveland Museum of Natural History.
[b]PGP = postglenoid process.
[c]The mastoid fissure (also tympanomastoid groove) is the gap between the petrous crest of the tympanic and the anterior face of the mastoid process. Medially, this space contains the stylomastoid foramen; it either continues as a fissure to the lateral margin of the cranial base or terminates if the petrous crest merges with the anterior face of the mastoid process.
[d]This observation on Sts. 19 is originally Clarke's (1977).

single cranium. This specimen, Sts. 19, exhibits highly derived morphology relative to the other specimens in the sample (Taung, TM 1511, Sts. 5, Sts. 25, Sts. 71, Stw. 266, MLD 37/38),* as noted previously by Broom and Robinson (1950, pp. 27–33), Clarke (1977, pp. 225, 237, 243, 259–260, 263), as well as Dean and Wood (1982). Table 2 lists the diagnostic characters of *A. africanus* and provides comparative data for the hominoid outgroup taxa, *H. habilis* and Sts. 19. The Sts. 19 glenoid region exhibits several character states shared with *H. habilis (sensu stricto),* but in other respects retains plesiomorphic glenoid region morphology in comparison to this taxon. Although we acknowledge the potential hazards of drawing taxonomic conclusions on the basis of a single specimen, we nevertheless believe that the null hypothesis is refuted in this case: The conventional *A. africanus* hypodigm from the Sterkfontein Type Site (Member 4)

*The Taung juvenile retains diagnostic osseous glenoid region morphology on the ventral aspect of the natural endocast, which will be discussed elsewhere (Kimbel and Rak, in preparation).

contains at least one specimen attributable to a second species, apparently of the genus *Homo*.*

Clades, Grades, and Homo erectus

Although African and non-African hominids conventionally attributed to *Homo erectus* share a broadly similar phenetic pattern, questions have been raised as to whether it is a paraphyletic (grade) species recognized on the basis of primitive characters (Andrews, 1984; Stringer, 1984; Wood, 1984; Hublin, 1986; Tattersall, 1986). It is alleged that some or all of the African specimens of the taxon (e.g., OH 9; OH 12; KNM-ER 730, 3733, 3883; SK 847) lack the diagnostic cranial autapomorphies of the Asian sample, such as thickened calvarial bone and pronounced ectocranial buttressing (*torus angularis,* sagittal keeling, etc.). Andrews (1984) and Tattersall (1986) restrict *H. erectus* to Asia, whereas Wood (1984) includes the Bed II and Bed IV Olduvai fossils in this species but contends that the Koobi Fora specimens lack several components of a *H. erectus* "combination definition" (i.e., a unique combination of primitive and derived character states; see below) based on the Asian material. However, the extent to which the African fossils are devoid of Asian character states is debatable.

On the basis of our own (admittedly brief) survey of the evidence, we share doubts about the actual extent of inter-regional clustering of many of these diagnostic characters (Rightmire, 1984, 1990; Turner and Chamberlain, 1989; Bräuer, 1990; Kennedy, 1991; Harrison, this volume). In our judgement, OH 9 and 12 can easily be accommodated in *H. erectus* based on vault thickness, incipient ectocranial buttressing, supraorbital form, and occipital morphology. With regard to the Koobi Fora material, we are particularly impressed by the occipital morphology (including the occipital torus configuration), supraorbital form and bone thickness of ER 730, the glenoid region (especially tympanic) anatomy and bone thickness of ER 3883, and the supraorbital form, glenoid morphology, and incipient frontal keeling of ER 3733, all of which closely parallel conditions documented in the Asian *H. erectus* sample (Table 3).

The African specimens attributed to *H. erectus,* however, are not identical in

*The Sts. 19 cranial base was recovered from a lime dump "which was probably the remains of quarrying operations of 30 or 40 years ago" (Broom and Robinson, 1950, p. 27), rather than from *in situ* breccia of the Type Site excavations of the 1940s. This led Clarke (1977, p. 260) to wonder whether the specimen might have derived from Member 5 sediments, which are known to contain *Homo* (e.g., Stw. 53). Subsequently, Clarke (1990) fit the Stw. 73/Sts. 22 maxillary fragment from the same lime dump to Sts. 19 based on the claimed match of M2/M3 interproximal wear facets. We regard this fit as tenuous due to the fact that the Sts. 19 M3 lacks most of its mesial face, retaining only a sliver of contact facet buccally; we imagine that a number of Sterkfontein maxillary M2s would "fit" the Sts. 19 M3. Moreover, the Sts. 19 M3 appears to bear significantly more occlusal wear than the M2 of Stw. 73/Sts. 22 and to approach more closely the wear of the latter specimen's M1. On the basis of facial characters, we regard Stw. 73/Sts. 22 as *A. africanus,* the phylogenetic relationships of which we are presently studying (Kimbel and Rak, in preparation). As Stw. 73/Sts. 22 is one of the "small toothed" Sterkfontein palates, we suspect that Clarke would include it in that part of the Sterkfontein sample he identifies as *A. africanus,* which is considered by him (1988) to have been ancestral to *Homo*.

Table 3. Cranial Characters of *Homo erectus* and *Homo habilis*

Homo erectus	*Homo habilis* (*sensu stricto*)[a]
Thick cranial vault bones	Thin cranial vault bones
Tympanic plate thick and vertically deep; petrous crest strong	Tympanic plate thinner, shallow; petrous crest relatively weak
Nuchal torus continuous between asteria, thickened superiorly and inferiorly, with midline bulge[b]	Prominent superior nuchal line or nuchal crest (not torus), weak laterally
Incipient to well-developed metopic (frontal) and sagittal (parietal) keeling; *torus angularis*[c]	Absent
Preglenoid plane extensive and horizontal	Preglenoid plane restricted and sleep
Supraorbital torus continuous, with continuous, gutter-like supratoral sulcus[d]	Supraorbital torus bilaterally arched; supratoral sulcus divided bilaterally by locally thickened torus above midorbit
Index of occipital scales ≥100[e]	Index of occipital scales <100
Nasal bridge strongly projecting	Nasal bridge projecting, but less prominent

[a]KNM-ER 1813, ER 3891, OH 13, OH 16, OH 24, OH 62, Stw. 53.

[b]We follow Robinson's (1958, p. 398) distinction between nuchal torus and nuchal crest: "A torus is a bony thickening which is part of the skull architecture and is usually rounded, not sharply crested . . . [A torus is] not necessarily concerned with musculature except indirectly. Crests, on the other hand, are directly related to muscles . . ." A well-developed nuchal torus is characteristic of Asian *H. erectus*, and it is also present, albeit less conspicuously, on African specimens ER 730, ER 3733, OH 9, and OH 12. Although the occipital of *H. habilis* cranium ER 1813 possesses a rounded midline prominence, only a nuchal crest is developed laterally.

[c]A *torus angularis* is not present on ER 3733 or ER 3883, but it is found on OH 9. As described by Rightmire (1990), this feature also varies within the Asian sample.

[d]Supraorbital and supratoral form is essentially equivalent in ER 3883, Sangiran 17, and Sambungmachan. Other than the gracility of the torus, the supraorbital region of ER 3733 is equivalent in morphology to that of Asian specimens and is different from that of *H. habilis* (as described in this table). Both East African crania show incipient metopic keels. In superior view, the *H. erectus* frontal squama is transversely convex behind the supratoral sulcus, and where the metopic keel is developed it shows an anterior midline "peak." The convex frontal with a variably expressed "peak" is expressed on Asian as well as African crania here considered to represent *H. erectus*. However, a convex frontal is also present in *H. habilis* (ER 1813, Stw. 53, OH 16) and the majority of grest apes; in *H. rudolfensis* and *Australopithecus* spp., on the other hand, the squama is transversely flat or even slightly concave.

[e]The index is computed as inion-opisthion chord/lambda-inion chord × 100. Data on early hominids indicate that variation in this index is in part due to both gender and ontogenetic age (Kimbel *et al.*, 1984, pp. 367–369). Thus if ER 3883 is a male and ER 3733 a female, a relatively low index in the latter (Rightmire, 1990) should not be a surprise. Kimbel *et al.* also discussed the likelihood that a high index is produced in different taxa by dissimilar occipital growth patterns. Thus, an index value of greater than 100 in *H. erectus* does not automatically qualify as a primitive character (*contra* Andrews, 1984).

every way to those from Asia. For example, the *torus angularis* is poorly expressed, if not absent, on ER 3733 and 3883, and these specimens (but not ER 730) have a more rounded sagittal occipital contour (due to the relatively weak nuchal torus) than is usually encountered in the Asian sample. The hypothesis that the African late Pliocene-early Pleistocene record contains a third distinct species of *Homo* (the Turkana specimens, SK 847), contemporary with the more generalized *H. habilis* but preceding the first appearance of *H. erectus* (OH 9, 12, etc.), remains a viable one. However, based on presently available evidence of character distributions within and among the relevant samples, we do not see the case for two phylogenetically independent taxa as clear cut.

The example of *H. erectus* brings to the forefront a more fundamental issue, that of whether a species must have autapomorphies in order to be a clade as opposed to a grade. If all extinct hominid species must be autapomorphic for at least one character, then from a strict cladistic point of view none of them could be considered ancestral to subsequent species, since ancestral species by definition lack autapomorphic states. We take a different view, both of species taxon integrity and the ancestral status of extinct species. Among the supraspecific categories, a clade is a monophyletic (= holophyletic) set of species united by the common possession of apomorphies. These apomorphies are primitive within the clade, but they are autapomorphic (unique) for the clade as a whole. At the species level a clade is something quite different: It is a basal group of organisms within which there is a parental pattern of ancestry and descent.

The question of whether species are monophyletic in the sense of the higher categories is not relevant to their existence as discrete lineages, nor to their individuation as such. The phylogenetic species concept makes this clear: It does not require that the characters of organismal clusters in a species be autapomorphic, only that the organismal clusters themselves be diagnosable as representing a distinct lineage. As Wood (1984) and Harrison (this volume) emphasize, a species may be diagnosable by any unique set of character states that subsequently is shown to be a combination of symplesiomorphies and synapomorphies relative to its more derived descendant and primitive ancestral taxa, respectively. Hublin (1986, p. 184) suggests that "from a purely cladistic perspective there is probably no *H. erectus*" because it is completely plesiomorphic relative to its putative descendant, *H. sapiens.* If this description is true (and we do not insist that it is), the phylogenetic relationships of *H. erectus* might remain mysterious, but we maintain that prior knowledge of the cladistic relationships of species is not relevant to their identification as the basal taxa of phylogenetic analysis. All that is required is that a species be diagnosable in space and time, and Rightmire's (1990, pp. 188–190) diagnosis of *H. erectus* seems to accomplish this successfully (see also Table 3). Thus, the failure to find autapomorphies in *H. erectus* fossils does not make this taxon a grade, nor does it logically entail the conclusion that this taxon is the phyletic ancestor of *H. sapiens* in a unilinear evolutionary scheme, unless we impose the prior assumption that a "perfect" fossil record would fail to reveal a speciation event with the persistence of the ancestral species.*

Conclusions

Tattersall (1986, 1991) has argued persuasively that the fossil record of hominid evolution contains more species than paleoanthropologists have been willing to admit. His conclusion stems chiefly from the empirical observation of a low degree of osteological differentiation between living primate sister species in comparison to the number of distinct hominid morphs in the paleontological

*A pressing objective in this context is the resolution of the taxonomic and phylogenetic status of the Neandertals (see Rak, this volume) and the so-called archaic *Homo sapiens* group (i.e., Bodo, Kabwe, Petralona, Arago, Dali, Saldanha, etc.), which would shed light on proposed phylogenetic relationships between *H. erectus* and *H. sapiens.*

record. We believe that the greater part of this reluctance can be traced to the widespread perception in paleoanthropology that taxa in the fossil record are transitory manifestations of character evolution.

This suggestion stems from an approach to the fossil record that places primary emphasis on the tracking of sets of characters over time, rather than on the identification and subsequent examination of the relationships among taxa—a tendency encoded in the frequent use of process-laden terms, such as *transition* and *shift,* which convey the notion of selection operating to mold phenotypes *within* reproductively continuous lineages. We can find no theoretical justification for the (implicit or explicit) suggestion by some paleoanthropologists (e.g., Clark, 1988, p. 365; Clark and Lindly, 1989, pp. 630–631; Smith and Paquette, 1989, p. 182; Trinkaus, 1990, p. 6) that the elucidation of "shifts" or "transitions" in functional–adaptive character complexes is, *a priori,* key to the resolution of systematic issues at the species level, because speciation is not reducible to the effects of selection acting on "economic" phenotypic attributes within populations.

We submit that such an approach prejudges the most fundamental evolutionary issue in paleoanthropology: the taxonomy of, and the phylogenetic relationships among, species. It exemplifies the use of characters to reify a phyletic lineage when none was presupposed in the first place. Questions of phylogenetic history cannot be answered by a nontaxic approach because the inevitable outcome of treating characters as fragments of a temporally ordered morphocline is a unilinear character phylogeny. The possibility that temporally successive populations may have a relationship to one another other than that of phyletic ancestor and descendant cannot even be entertained under this approach because the populations themselves are viewed as epiphenomena of character phylogenesis.

We have argued that in adopting the stance that the phyletic lineage and the "biological" species are categorically distinct entities, the student of the fossil record is forced to treat species as operationally necessary taxonomic constructs rather than ontologically meaningful units and is then left with building phylogenies on the basis of characters disembodied from any theoretically significant notion of evolutionary diversity. We endorse Delson's (1993) contention that what paleontologists need in a species concept is an integration of theoretical and operational viewpoints. We believe that the phylogenetic concept of species fills this need. The PSC is completely consistent with the theoretical notion of species as spatiotemporally extended, reproductively cohesive lineages of parental ancestry and descent, and it emphasizes the prior role of characters in delineating these basic units of phylogenetic analysis. Perhaps the most important implication of the PSC is that species in the present are conceptually equivalent to species in the geological past.

Acknowledgments

We thank Eric Delson, Lawrence Martin, and Ian Tattersall for insightful discussion of some of the ideas presented here, although agreement with our conclusions, either general or specific, is not implied. The manuscript benefitted from thorough and thoughtful critiques by Steve Ward, Lawrence Martin, Bill Jungers, Jon Marks, and David Pilbeam. We acknowledge an intellectual debt to Ian Tattersall, whose important 1986 paper, "Species Recognition in Human Paleontology," sparked our thinking about species.

A National Science Foundation grant (BNS 8820113) to WHK supported our research on *A. africanus*. We thank Professor Phillip Tobias and the late Mr. Alun Hughes, University of the Witwatersrand, Johannesburg, and Dr. C. K. Brain and Mr. David Panagos, Transvaal Museum, Pretoria, for permission to study fossils in their care and many kindnesses during our 1989 visit. Dr. Bruce Latimer kindly made the great ape samples in the Cleveland Museum of Natural History's Hamann-Todd Osteological Collection available for study.

References

Andrews, P. 1984. An alternative interpretation of the characters used to define *Homo erectus*. *Cour. Forsch.-Inst. Senckenberg* **69:**167–175.

Bock, W. J. 1979. The synthetic explanation of macroevolutionary change: a reductionist approach. *Bull. Carnegie Mus. Nat. Hist.* **13:**20–69.

Bräuer, G. 1990. The occurrence of some controversial *Homo erectus* cranial features in the Zhoukoudian and East African hominids. *Acta Anthropol. Sinica* **9:**350–358.

Broom, R., and Robinson, J. T. 1950. Further evidence of the structure of the Sterkfontein ape-man *Plesianthropus*. *Transvaal Mus. Mem.* **4:**11–84.

Cain, A. J. 1960. *Animal Species and Their Evolution.* Harper, New York.

Chamberlain, A. T. 1989. Variations within *Homo habilis*, in: Hominidae: Proceedings of the 2nd International Congress of Human Paleontology, pp. 175–181. Jaca Book, Milan.

Clark, G. A. 1988. Some thoughts on the Black Skull: An archeologist's assessment of WT-17000 (*A. boisei*) and systematics in human paleontology. *Am. Anthropol.* **90:**357–371.

Clark, G. A., and Lindly, J. M. 1989. The case for continuity: observations on the biocultural transition in Europe and western Asia, in: P. Mellars, and C. Stringer (eds.), *The Human Revolution: Behavioural and Biological Perspectives in the Origins of Modern Humans*, pp. 626–676. Princeton University Press, Princeton, NJ.

Clarke, R. J. 1977. The Cranium of the Swartkrans Hominid SK 847 and its Relevance to Human Origins. Ph.D. thesis, University of the Witwatersrand.

Clarke, R. J. 1988. A new *Australopithecus* cranium from Sterkfontein and its bearing on the ancestry of *Paranthropus*, in: F. E. Grine (ed.), *Evolutionary Biology of the "Robust" Australopithecines*, pp. 285–292. Aldine de Gruyter, New York.

Clarke, R. J. 1990. Observations on some restored hominid specimens in the Transvaal Museum, Pretoria, in: G. H. Sperber (ed.), *From Apes to Angels: Essays in Anthropology in Honor of Phillip V. Tobias*, pp. 135–151. Wiley-Liss, New York.

Cracraft, J. 1983. Species concepts and speciation analysis. *Curr. Ornithol.* **1:**159–187.

Cracraft, J. 1987. Species concepts and the ontology of evolution. *Biol. Philos.* **2:**329–346.

Cracraft, J. 1989a. Speciation and its ontology: the empirical consequences of alternative species concepts for understanding patterns and processes of differentiation, in: D. Otte and J. A. Endler (eds.), *Speciation and its Consequences*, pp. 28–59. Sinauer Associates, Sunderland, MA.

Cracraft, J. 1989b. Species as entities in biological theory, in: M. Ruse (ed.), *What the Philosophy of Biology Is*, pp. 31–52. Kluwer Academic Publishers, Dordrecht.

de Queiroz, K. 1988. Systematics and the Darwinian revolution. *Philos. Sci.* **55:**238–259.

de Queiroz, K., and Gauthier, J. 1990. Phylogeny as a central principle in taxonomy: phylogenetic definitions of taxon names. *Syst. Zool.* **39:**307–322.

Dean, M. C., and Wood, B. A. 1982. Basicranial anatomy of Plio-Pleistocene hominids from East and South Africa. *Am. J. Phys. Anthropol.* **59:**157–174.

Delson, E. 1993. How many species of *Homo* in Europe? NATO Scientific Affairs Division Advanced Research Workshop: "Les Premiers Peuplements Humains de l'Europe" (in press).
Eldredge, N. 1985. *Unfinished Synthesis: Biological Hierarchies and Modern Evolutionary Thought.* Oxford University Press, New York.
Eldredge, N., and Cracraft, J. 1980. *Phylogenetic Patterns and the Evolutionary Process.* Columbia University Press, New York.
Feibel, C., Brown, F. H., and McDougall, I. 1989. Stratigraphic context of fossil hominids from the Omo Group deposits: northern Turkana basin, Kenya and Ethiopia. *Am. J. Phys. Anthropol.* **78:**595–622.
Ghiselin, M. T. 1969. *The Triumph of the Darwinian Method.* University of California Press, Berkeley.
Ghiselin, M. T. 1974. A radical solution to the species problem. *Syst. Zool.* **23:**536–544.
Ghiselin, M. T. 1984. "Definition," "character," and other equivocal terms. *Syst. Zool.* **33:**104–110.
Ghiselin, M. T. 1987. Species concepts, individuality and objectivity. *Biol. Philos.* **2:**127–143.
Häuser, C. L. 1987. The debate about the biological species concept: a review. *Z. Zool. Syst. Evolut.-Forsch.* **25:**241–257.
Hennig, W. 1966. *Phylogenetic Systematics.* University of Illinois Press, Chicago.
Hublin, J. J. 1986. Some comments on the diagnostic features of *Homo erectus. Anthropos (Brno)* **23:**175–187.
Hull, D. L. 1975. Central subjects and historical narratives. *Hist. Theory* **14:**253–274.
Hull, D. L. 1976. Are species really individuals? *Syst. Zool.* **25:**174–191.
Hull, D. L. 1978. A matter of individuality. *Philos. Sci.* **45:**335–360.
Hull, D. L. 1987. Genealogical actors in ecological roles. *Biol. Philos.* **2:**168–184.
Johanson, D. C., Masao, F. T., Eck, G. G., White, T. D., Walter, R. C., Kimbel, W. H., Asfaw, B., Manega, P., Ndessokia, P., and Suwa, G. 1987. A new partial skeleton of *Homo habilis* from Olduvai Gorge, Tanzania. *Nature* **327:**205–209.
Kennedy, G. E. 1991. On the autapomorphic traits of *Homo erectus. J. Hum. Evol.* **20:**375–412.
Kimbel, W. H. 1986. Calvarial Remains of *Australopithecus afarensis:* A Comparative Phylogenetic Study. Ph.D. thesis, Kent State University.
Kimbel, W. H. 1991. Species, species concepts and hominid evolution. *J. Hum. Evol.* **20:**355–371.
Kimbel, W. H., and Rak, Y. 1985. Functional morphology of the asterionic region in extant hominoids and fossil hominids. *Am. J. Phys. Anthropol.* **66:**31–54.
Kimbel, W. H., and White, T. D. 1988. Variation, sexual dimorphism and the taxonomy of *Australopithecus,* in: F. E. Grine (ed.), *Evolutionary Biology of the "Robust" Australopithecines,* pp. 175–192. Aldine de Gruyter, New York.
Kimbel, W. K., White, T. D., and Johanson, D. C. 1984. Cranial morphology of *Australopithecus afarensis:* a comparative study based on a composite reconstruction of the adult skull. *Am. J. Phys. Anthropol.* **64:**337–388.
Kluge, A. G. 1990. Species as historical individuals. *Biol. Philos.* **5:**417–431.
Leakey, R. E., Walker, A., Ward, C. V., and Grausz, H. M. 1989. A partial skeleton of a gracile hominid from the upper Burgi Member of the Koobi Fora Formation, East Lake Turkana, Kenya, 167–173. Hominidae. In: *Proceedings of the 2nd International Congress of Human Paleontology.* Jaca Book, Milan.
Lieberman, D. E., Pilbeam, D. R., and Wood, B. A. 1988. A probabilistic approach to the problem of sexual dimorphism in *Homo habilis:* a comparison of KNM-ER 1470 and KNM-ER 1813. *J. Human Evol.* **17:**503–511.
Mayr, E. 1963. *Animal Species and Evolution.* Harvard University Press, Cambridge, MA.
Mayr, E. 1969. *The Principles of Systematic Zoology.* McGraw-Hill, New York.
Mayr, E. 1970. *Populations, Species and Evolution.* Harvard University Press, Cambridge, MA.
Mayr, E. 1982. *The Growth of Biological Thought.* Harvard University Press, Cambridge, MA.
McKitrick, M. C., and Zink, R. M. 1988. Species concepts in ornithology. *Condor* **90:**1–14.
Miller, J. A. 1991. Does brain size variability provide evidence of multiple species in *Homo habilis? Am. J. Phys. Anthropol.* **84:**385–398.
Nelson, G., and Platnick, N. 1981. *Systematics and Biogeography.* Columbia University Press, New York.
Nixon, K. C., and Wheeler, Q. D. 1990. An amplification of the phylogenetic species concept. *Cladistics* **6:**211–223.
Paterson, H. E. 1985. The recognition concept of species, in: E. S. Vrba (ed.), *Species and Speciation,* pp. 21–29. Transvaal Museum, Pretoria.

Picq, P. G. 1990. *L'Articulation Temporo-mandibulaire des Hominidés.* CNRS, Paris.

Rak, Y. 1983. *The Australopithecine Face.* Academic Press, New York.

Rightmire, G. P. 1984. Comparison of *Homo erectus* from Africa and southeast Asia. *Cour. Forsch-Inst. Senckenberg* **69:**83–98.

Rightmire, G. P. 1990. *The Evolution of* Homo erectus. Cambridge University Press, Cambridge.

Robinson, J. T. 1958. Cranial cresting patterns and their significance in the Hominoidea. *Am. J. Phys. Anthropol.* **16:**397–428.

Rose, K. D., and Brown, T. M. 1986. Gradual evolution and species discrimination in the fossil record, in: K. M. Flanagan and J. A. Lillegraven (ed.), *Vertebrates, Phylogeny and Philosophy: Contributions to Geology, University of Wyoming, Special Paper 3,* pp. 119–130. University of Wyoming, Laramie, WY.

Rowe, T. 1987. Definition and diagnosis in the phylogenetic system. *Syst. Zool.* **36:**208–211.

Simpson, G. G. 1961. *The Principles of Animal Taxonomy.* Columbia University Press, New York.

Smith, F. H., and Paquette, S. P. 1989. The adaptive basis of Neandertal facial form, with some thoughts on the nature of modern human origins, in: E. Trinkaus (ed.), *The Emergence of Modern Humans: Biocultural Adaptations in the Later Pleistocene,* pp. 181–210. Cambridge University Press, Cambridge.

Sober, E. 1980. Evolution, population thinking, and essentialism. *Philos. Sci.* **47:**350–383.

Splitter, L. J. 1988. Species and identity. *Philos. Sci.* **55:**323–348.

Stringer, C. B. 1984. The definition of *Homo erectus* and the existence of the species in Africa and Europe. *Cour. Forsch-Inst. Senckenberg* **69:**131–143.

Stringer, C. B. 1986. The credibility of *Homo habilis,* in: B. A. Wood, L. B. Martin, and P. Andrews (eds.), *Major Topics in Primate and Human Evolution,* pp. 266–294. Cambridge University Press, Cambridge.

Sylvester-Bradley, P. C. 1956. The new paleontology, in: P. C. Sylvester-Bradley (ed.), *The Species Concept in Palaeontology,* pp. 1–8. The Systematics Association, London.

Szalay, F. S. 1989. Comment on "Human fossil history and evolutionary paradigms" by D. R. Pilbeam, in: M. K. Hecht (ed.), *Evolutionary Biology at the Crossroads,* pp. 145–148. Queens College Press, Flushing, NY.

Szalay, F. S., and Bock, W. J. 1991. Evolutionary theory and systematics: relationships between process and patterns. *Z. Zool. Evolut-Forsch.* **29:**1–39.

Tattersall, I. 1986. Species recognition in human paleontology. *J. Hum. Evol.* **15:**165–176.

Tattersall, I. 1991. What was the Human Revolution? *J. Hum. Evol.* **20:**77–83.

Thorpe, R. S. 1987. Geographic variation: a synthesis of cause, data, pattern, and congruence in relation to subspecies, multivariate analysis and phylogenesis. *Boll. Zool.* **54:**3–11.

Trinkaus, E. 1990. Cladistics and the hominid fossil record. *Am. J. Phys. Anthropol.* **83:**1–11.

Turner, A., and Chamberlain, A. 1989. Speciation, morphological change and the status of African *Homo erectus. J. Hum. Evol.* **18:**115–130.

Van Valen, L. M. 1976. Ecological species, multispecies, and oaks. *Taxon* **25:**233–239.

Vrba, E. S. 1985. Introductory comments on species and speciation, in: E. S. Vrba (ed.), *Species and Speciation,* pp. ix–xviii. Transvaal Museum, Pretoria.

Walker, A. C., and Leakey, R. E. 1978. The hominids of East Turkana. *Sci. Am.* **239:**44–56.

Walter, R. C., Manega, P. C., Hay, R. L., Drake, R. E., and Curtis, G. H. 1991. Laser-fusion $^{40}Ar/^{39}Ar$ dating of Bed I, Olduvai Gorge, Tanzania. *Nature* **354:**145–149.

Wiley, E. O. 1981. *Phylogenetics.* Wiley, New York.

Wiley, E. O. 1989. Kinds, individuals and theories, in: M. Ruse (ed.), *What the Philosophy of Biology Is,* pp. 289–300. Kluwer Academic Publishers, Dordrecht.

Willman, R. 1989. Evolutionary or biological species? *Abh. Naturwiss. Ver. Hamburg* **28:**95–110.

Wilson, E. O., and Brown, W. L. 1953. The subspecies concept and its taxonomic application. *Syst. Zool.* **2:**97–111.

Wood, B. A. 1984. The origin of *Homo erectus. Cour. Forsch-Inst. Senckenberg* **69:**99–111.

Wood, B. A. 1985. Early *Homo* in Kenya, and its systematic relationships, in: E. Delson (ed.), *Ancestors: The Hard Evidence,* pp. 206–214. Alan R. Liss, New York.

Wood, B. A., and Uytterschaut, H. 1987. Analysis of the dental morphology of Plio-Pleistocene hominids. III. Mandibular premolar crowns. *J. Anat.* **154:**121–156.

Early *Homo*
How Many Species?

19

BERNARD WOOD

Introduction

The problem of deciding when phenotypic variation exceeds that which can be tolerated within a single species is a familiar one to both paleontologists in general (Mayr *et al.*, 1953; Sylvester-Bradley, 1956; Simpson, 1961) and to paleoanthropologists in particular (e.g., Weidenreich, 1946; Campbell, 1962; Zwell and Pilbeam, 1972; Wolpoff, 1978; Pilbeam, 1978). Some authors have regarded these taxonomic difficulties as intrinsic to the practice of equating fossil with neontological species and have proposed that any resolution lies in the direction of redefining the fossil species concept. For example, Cain (1954) and George (1956) used the terms *paleospecies* and *chronospecies,* respectively, to refer to the whole (George, 1956), or part (Pilbeam, 1972), of an evolutionary lineage. These devices thus incorporate the element of time within the definition of a paleontological species.

A more radical suggestion, that chronological information should supplant morphology as the principal factor determining species has, perhaps understandably, attracted little support (Campbell, 1972). A third approach has suggested the use of Bayesian probability theory as a formal test of whether samples are derived from one or more taxa (Pilbeam and Vaisnys, 1975; Pilbeam, 1978; Kennedy, 1990); this latter strategy has not received the interest it deserves. In the present study paleontological species are modeled on neontological species with the proviso that fossil species incorporate any additional variation that is related to the temporal dimension of the hypodigm (see below).

BERNARD WOOD • Hominid Palaeontology Research Group, Department of Human Anatomy and Cell Biology, University of Liverpool, Liverpool L69 3BX, England.

Species, Species Concepts, and Primate Evolution, edited by William H. Kimbel and Lawrence B. Martin. Plenum Press, New York, 1993.

This chapter reports the results of an analysis of the taxonomic significance of the variation displayed by early *Homo* remains from East Africa. Early *Homo* is interpreted as including material that has been allocated to, likened to, or compared with *Homo habilis.* These remains are derived with respect to "gracile" australopithecine species, yet they lack the specializations of either the "robust" australopithecines or later *Homo* taxa. The study concentrates on cranial remains and compares both the degree and the pattern of the observed variation in the fossil hypodigm with that in relevant living analogues.

Early Homo: The Problem

The original hypodigm of *Homo habilis* was recovered from Beds I and II of Olduvai Gorge, Tanzania (Leakey *et al.*, 1964). The holotype, OH7, was found at site FLKNN I in 1960 and included parts of both parietals and most of the alveolar process and associated tooth crowns, but little else, of a juvenile mandible. The paratypes included both cranial, OH4 (MNKI), OH6 (FLKI), and OH13 (MNKII), and postcranial, OH8 (FLKNNI), remains. Robinson (1965) expressed misgivings about the allocation of OH13 to *H. habilis,* suggesting instead that its affinities lay with *H. erectus.* Walker and Leakey have subsequently also doubted its links with the holotype and, instead, have chosen to refer it to *Australopithecus* and have classified it (along with OH24, see below) as a late-surviving, small-bodied, member of that genus.

In addition to the type series, other hominids from Beds I and II at Olduvai have been attributed to, or compared with, *H. habilis* (Table 1). The fragmented cranium and associated dentition OH16 was, along with OH14, at first only

Table 1. List of Cranial, Mandibular, and Dental Remains of Early *Homo* from East Africa that have been Allocated or Likened to *Homo habilis*

	Koobi Fora (KNM-)	Olduvai (OH)	Omo
Skulls and crania	1470, 1478, 1590, 1805, 1313, 3732, 3735, 3891	7,13, 16, 24, 52, 62	L894-1
Mandibles	819, 1482, 1483, 1501, 1502, 1506, 1801, 1802, 3734	37	Omo 75-14, Omo 222-2744
Associated and isolated teeth	807, 808, 809, 1462, 1480, 1508, 1814	4, 6, 15, 21, 27, 39, 40, 41, 42, 44, 45	L26-1g, L28-30, L28-31, L398-573, L398-1699, Omo 29-43, Omo 33-740, Omo 33-3282, Omo 5496, Omo 47-47, Omo 74-18, Omo 75s-15, Omo 75s-16, Omo 123-5495, Omo 166-781, Omo 177-4525, Omo 195-1630, Omo K7-19, Omo SH1-17, P933-1

"provisionally referred" to *H. habilis* (Leakey *et al.*, 1964, p. 9). Subsequently doubts were expressed that OH16 may not belong with *H. habilis* (Leakey, 1966; Tobias and von Koenigswald, 1964), but more recently it has been consistently attributed to that taxon (Tobias, 1980a, 1985, 1991). The initial allocation of OH24 to *H. habilis* was tentative (Leakey *et al.*, 1971), and subsequent assessments have reached differing conclusions. While Tobias (1972, 1980a, 1985) has been convinced of its resemblance to *H. habilis,* others (Leakey, 1974; Howell, 1978; Olson, 1978; Leakey and Walker, 1980) have stressed its australopithecine affinities. The partial skeleton, OH62, comprises a fragmented skull and numerous postcranial elements (Johanson *et al.*, 1987; Johanson, 1989). This specimen has been allocated to *H. habilis* because of "similarities of the OH62 face, palate, and dentition to *H. habilis* (especially Stw 53)" by Johanson *et al.* (1987, p. 208), but Stw 53 has yet to be formally allocated to *H. habilis* (see below).

The largest and best-preserved component of the early *Homo* hypodigm comes from Koobi Fora (Table 1), but, as with the Olduvai remains, there is no unanimity about the wisdom of allocating some of the specimens to *H. habilis,* or even to *Homo.* Walker (1976) and Walker and Leakey (1978) have drawn attention to the australopithecine affinities of some of these remains (e.g., KNM-ER 1470 and 1813), but most commentators have not been deterred from including many, if not all, of this material in *H. habilis* (e.g., Howell, 1978; Johanson, 1989).

Remains from the Shungura Formation have also either been likened to *H. habilis* or allocated to it either implicitly (e.g., Coppens, 1980) or explicitly (e.g., Howell, 1978; Tobias, 1989). A fragmentary cranium, L894-1, from Member G of the Shungura Formation, was judged to represent "a member of an early species of the genus *Homo* (*Homo habilis* or *Homo modjokertensis*)" on the basis of "similarities to Olduvai hominids 24 and 13" (Boaz and Howell, 1977, p. 93). Other hominid remains, mostly isolated teeth listed in Coppens (1980), have been assessed as *Homo* cf. *habilis;* these are included in Table 1. Martyn and Tobias (1967) did not refer the Chemeron temporal fragment to *H. habilis,* but they did suggest that some of its traits indicate a "closer affinity with an evolved *H. habilis* (Old. H 13)" (Martyn and Tobias, 1967, p. 480).

The taxon *H. habilis* has been invoked as a possible solution for SK 847 from Swartkrans (Clarke and Howell, 1972; Clarke 1977) and has been cited as the probable taxonomic allocation of Stw 53 from member 5 at Sterkfontein (Hughes and Tobias, 1977). Subsequent assessments of Stw 53 have reinforced the association with *H. habilis,* and after initial efforts at reconstruction it was judged to be "almost identical to the cranium OH24" (Clarke, 1985, p. 175). Clarke (1977) has also counseled caution over the allocation of other cranial material from Swartkrans to *Paranthropus.* These remains, along with SK 847, must be regarded as candidates for allocation to *H. habilis;* so, too, must material from members 1 and 2, which has been recovered during the latest phase in the recovery of hominid remains from the Swartkrans deposits (Brain *et al.*, 1988).

Fossils from the Near East and Asia, found at Ubeidiyah and in Indonesia, have been canvassed as belonging to *H. habilis* (Leakey *et al.*, 1964; Tobias and von Koenigswald, 1964), but support for both proposals has since been withdrawn (Tobias, 1966, 1991).

In addition to those who have expressed doubts about the allocation of individual specimens to *H. habilis,* several authors have made more general propositions about the material that has been widely accepted as belonging to *H. habilis*. Some observers have speculated that the *H. habilis* hypodigm should be restricted to a "bigger-brained" subset of this material (Wood, 1985; Stringer, 1986, 1987), whereas others have supported alternative proposals, for example, that *H. habilis* should be defined either temporally (Stringer, 1986) or geographically (Groves and Mazak, 1975; Chamberlain, 1989).

This brief review has demonstrated that there is now widespread acceptance that the hypodigm of *H. habilis* includes material that can justifiably be included in *Homo* (Tobias, 1989). However, doubts do persist about the hypodigm as it stands. First, can the material that has been referred to *H. habilis* be sensibly referred to a single hominid species, or does its range of variability suggest that more than one taxon should be invoked to accommodate it? Second, if it is concluded that more than one species has been subsumed within the *H. habilis* hypodigm, do both taxa belong to *Homo*, or have *Australopithecus* remains been inadvertently referred to *H. habilis?* This study has concentrated on *H. habilis* from East African sites for two reasons. First, its taxonomic allocations are generally less controversial than those from sites elsewhere, and second, because by doing so it reduces the potentially confounding effects of geographical variation. The analysis reported here has concentrated on cranial evidence for it is both more plentiful than postcranial evidence and because relevant comparative data are presently available to test the significance of the observed variation in the hypodigm. Nonetheless, postcranial evidence is not ignored and is referred to in the subsequent discussion.

Taxonomic Hypotheses

The particular taxonomic problem posed by the East African material allocated to early *Homo* can be usefully summarized and exemplified in the form of five hypotheses that are set out in Fig. 1. In these schemes, early *Homo* cranial material from three sites—Koobi Fora, Olduvai, and the Shungura Formation—has been allocated to two "regional" samples. The Shungura and Koobi Fora Formation hominid samples have been grouped together to form a regional subset from the Omo Group deposits (OM; de Heinzelin, 1983; Feibel *et al.*, 1989).

A prerequisite of this investigation of variation in early *Homo* is the ability to subdivide the sample into two, or more, time bands. When choosing the temporal subdivisions a balance must be struck between the need to maintain a sensible sample size and the desirability of having enough time bands to be able to detect temporally related trends in variation. The smallest time-band width that still gave workable sample sizes (i.e., three or more) is 200 Kyr. Two 200 Kyr subdivisions were used, one beginning with a time band between 2.59 and 2.4 Kyr (Table 2), the other starting with a time band some 100 Kyr younger, i.e., 2.49–2.3 Myr (Table 3); all dates are extracted from Feibel *et al.* (1989).

The first of the five hypotheses, A, suggests that the early *Homo* fossil evidence samples a single species, namely, *H. habilis* (Fig. 1A). The implications of hypothesis A (Table 4) are that variation along the temporal axis, T (Fig. 2), or

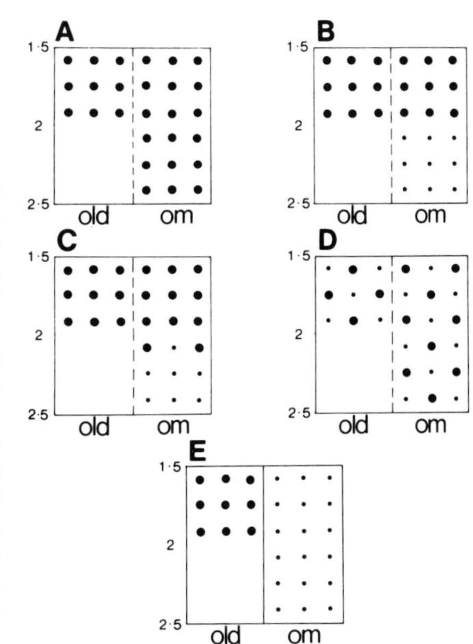

Fig. 1. Diagram to illustrate five hypotheses, A–E, to explain the variation within East African early *Homo* cranial remains. Each box is divided into the two regional subsets of the fossil evidence: OLD = Olduvai, OM = Omo Group deposits (i.e., Shungura and Koobi Fora Formations). The vertical axis is time in Myr. The lower left-hand corner of each box is blank because hominids of that age are not known from Olduvai Gorge. The larger solid circles denote *H. habilis;* smaller solid circles denote a second early *Homo* taxon. Hypothesis A suggests a single-taxon solution; Hypotheses B–E all imply more than one early *Homo* taxon. Hypotheses B and C specify that the two taxa are time successive, whereas hypotheses D and E see them as synchronic. Hypotheses D and E differ in that hypothesis E suggests that the two taxa are geographical variants, i.e., synchronic but allopatric.

Table 2. Early *Homo* Cranial, Mandibular, and Dental Remains from East Africa Set Out in 200 kyr Time-Band Divisions, Registered at 2.59–2.4 Myr

Time band	Time span	Specimen numbers	
1.	2.4–2.59 my	L26-1g; Omo 166-781	
2.	2.2–2.39 my	L28-30; L28-31; L398-573; L398-1699; Omo 33-740; Omo 33-3282; Omo 33-5496; Omo 123-5495; Omo 177-4525; P933-1	Omo 75s-15; Omo 75s-16
3.	2.0–2.19 my	Omo 29-43; Omo 47-47; Omo 75-14; Omo 195-1630; Omo 222-2744; Omo SH1-17	
4.	1.8–1.99 my	L894-1	
		KNM-ER 1462; 1470; 1478; 1480; 1482; 1483; 1501; 1502; 1590; 1801; 1802; 1805; 1814; 3732; 3734; 3735	Omo 74-18
		OH 4; 6; 7; 24; 44; 45; 52; 62	KNM-ER 1508; 1814; 3891
5.	1.6–1.79 my	KNM-ER 807; 819; 1506	
		OH 16; 21; 27; 37; 39; 40; 41; 42	
6.	1.4–1.59 my	KNM-ER 808; 809	
		OH 13; 15	
7.	1.0–1.19 my	Omo K7-19 (<1.05 my)	

Temporal data from Feibel *et al.* (1989).

Table 3. Early *Homo* Cranial, Mandibular, and Dental Remains from East Africa Set Out in 200 kyr Time-Band Divisions, Registered at 2.49–2.3 Myr

Time band	Time span	Specimen numbers
1.	2.3–2.49 my	L26-1g; L28-30; L28-31; L398-573; L398-1699; Omo 33-740; Omo 33-3282; Omo 33-5496; Omo 123-5495; Omo 166-781; Omo 177-4525; P933-1
2.	2.1–2.29 my	Omo 29-43; Omo 47-47; Omo 195-1630; Omo 222-2744; Omo SH1-17
		Omo 75s-15; Omo 75s-16
3.	1.9–2.09 my	Omo 75-14
4.	1.7–1.89 my	L894-1; Omo 74-18
		KNM-ER 1462; 1470; 1478; 1480; 1482; 1483; 1501; 1502; 1506; 1508; 1590; 1801; 1802; 1805; 1813; 1814; 3732; 3734; 3735; 3891
		OH 4; 6; 7; 24; 27; 39; 42; 44; 45; 52; 62
5.	1.5–1.69 my	KNM-ER 807; 808; 809; 819
		OH 13; 15; 16; 21; 37; 40; 41
6.	0.9–1.09 my	Omo K7-19 (<1.05 my)

Temporal data from Feibel *et al.* (1989).

along the whole, Z, or the regional parts, X and Y, of the time-band axes does not exceed what would be expected in a single hominid species. The lack of any temporal trends in the expression of character states in this model suggests that hypothesis A is compatible with stasis in early *Homo* and is thus more consistent with a punctuational (Gould and Eldredge, 1977) than with a gradualistic (Cronin *et al.*, 1981) interpretation of hominid evolution.

All the remaining hypotheses (B–E) suggest that variation in early *Homo* is such that more than one species should be recognized. Hypotheses B and C are, respectively, punctuational and gradualistic variants of a more general hypothesis that suggests early *Homo* in East Africa consists of two, time-successive, species, the more recent of which would be *H. habilis.* The implications of hypotheses B and C are identical, except that the overlap implicit in the gradualist interpretation is responsible for the differences in categories (3, 4, and 7) (Table 4). Apart from these differences, hypotheses B and C postulate that, whereas variation in axis T would exceed that in a single species, variation within time bands (axes X, Y, and Z) would not. The two models also suggest that, in cladistic terms, there should be consistent differences in the expression of nonmetric traits, or character states, between the taxa.

Hypothesis D suggests that early *Homo* is represented by two synchronic and sympatric species, one of which is *H. habilis.* Because both species extend through the one million year span of early *Homo,* and because they are both present in the two regions post-1.9 Myr, variation in all four axes, T–Z, would exceed that expected in a single taxon. There should, however, be no temporal differences in variation nor in the expression of nonmetric traits, for both taxa are present throughout the time range; the model is an explicitly punctuational

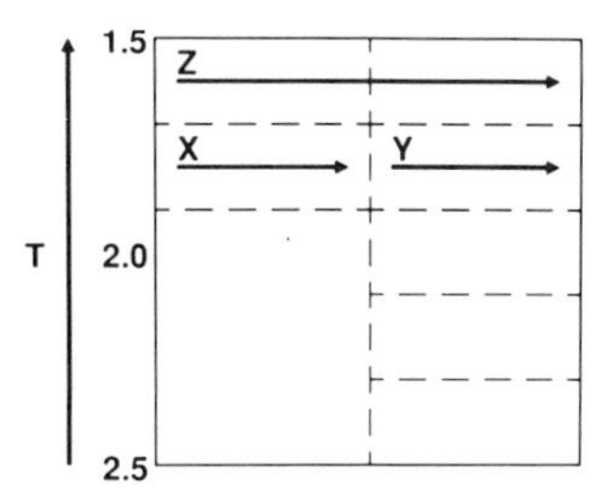

Fig. 2. Diagram to illustrate how each box in Figure 1 has been divided up horizontally into 200 Kyr time bands. The letters refer to temporal and geographical subgroups of the East African early *Homo* sample. T refers to the whole temporal sample, which may also be divided into regional subsets. Z refers to a subsample spanning one of the 200 Kyr time bands, and X and Y to, respectively, the Olduvai and Omo regional subsets of any single time band. The vertical axis is time in Myr.

one. The last of the five hypothetical examples, E, suggests that early *Homo* consists of two synchronic, but allopatric, species, each confined to one of the two main regions, *H. habilis* at Olduvai and a separate species being sampled in the Omo basin. This hypothesis suggests that a speciation event took place prior to 2.5 Myr or 2.3 Myr (Tobias, 1989) and that neither of the resulting, regionally based, early *Homo* taxa bears evidence of excessive variation in the temporal, T, axis. Whereas variation across both regions (axis Z) in any time band post-1.9 Myr would exceed that expected within a single species, within each region (axes X and Y) variation would correspond to that in a single taxon.

The five hypotheses, A–E, by no means exhaust the possible arrangements for early *Homo,* but it is suggested that they do include and summarize the components of any more complex hypothesis, including those that suggest there are more than two species represented in early *Homo* or that different numbers of species may be represented in the fossil record of the two regions.

Table 4. Implications of Taxonomic Hypotheses A–E

Implications	Hypotheses				
	A	B	C	D	E
1. Overall variation (axis T) in the combined hypodigm is greater than that in suitable analogues.	N	Y	Y	Y	N[b]
2. Overall variation (axis T) in both of the regional hypodigms is greater than that in suitable analogues.	N	N	N	Y	N
3. Variation across a time band (axis Z) is greater in degree than that in suitable analogues.	N	N	N[a]	Y	Y
4. Variation across a time band (axis Z) is different in character from that in suitable analogues.	N	N	N[a]	Y	Y
5. Variation across a time band within each geographical region (axes Y and Z) is greater than that in suitable analogues.	N	N	N	Y	N
6. Temporal trend(s) in the expression of character states, including nonmetric traits.	N	Y	Y	N	N
7. Successive supraregional time bands show approximately equal variability.	Y	Y	N	Y	Y
8. Stasis (S)/gradual (G) change.	S	S	G	S	S

[a]Except in the area of species overlap.
[b]Within each region the variation should not exceed that of the analogue.

Hypothesis Testing

Continuous morphometric data as well as nonmetrical expressions of preserved morphology can be used to discriminate between the five hypotheses. Variability is traditionally compared using the coefficient of variation (CV). The CV is a readily comprehended measure of variation (Gingerich and Schoeninger, 1979), but it takes no account of the distribution of the data (Pilbeam, 1969; Wolpoff, 1969), is sample-size dependent (Martin and Andrews, 1984), and, as Sokal and Braumann (1980) showed, *F*-tests based on the CV are especially sensitive to departures from normality. Schultz (1985) compared the efficacy of a range of tests of variability and concluded that Levene's test (Van Valen, 1978), employing the median rather than the mean as an estimate of central location, was the most robust comparator. Two forms were recommended, the "median ratio" and the "median log," and each had advantages depending on the ways the data depart from normality. In the event, the author advised that "it will probably often be easier to simply perform both tests and choose the most significant result (Gans, 1984)" (Schultz, 1985, p. 456), and this is the strategy adopted in this study. The 95% confidence limits of the CV values are also provided for each of the fossil samples and subsamples.

Several authors have pointed out that no neontological species represents an entirely suitable analogue for assessing the significance of variation within a fossil species for the reason that time, even a few tens of thousands of years, may generate an additional source of variation in fossil collections (Pilbeam, 1978; Lieberman *et al.,* 1988; Wood and Xu, 1991). Within the higher primates, the degree of intraspecific variation is lowest in *H. sapiens* and *Pan* (Wood, 1976). However, although these two taxa are probably the forms most closely related to early hominids (Sibley and Ahlquist, 1984, 1987; Miyamoto *et al.,* 1987, 1988; Caccone and Powell, 1989), it is sensible to be conservative and to use a more dimorphic ape, in this case, *Gorilla,* as an additional comparator for assessing the extent of intraspecific variation (Lieberman *et al.,* 1988; Wood *et al.,* 1991). Details of the *Gorilla* comparative sample used in this study are set out in Table 5 and also are given in Wood *et al.* (1991). A second strategy has been used to simulate any time-related component of variation, which is to use the variation observed within other hominid taxa, for this would also include the temporal element, which is lacking in the extant analogues. Two taxa have been used as analogues: *Paranthropus boisei,* because it is approximately sympatric and synchronic with early *Homo,* and *H. erectus,* because it is widely canvassed as the sister taxon of *H. habilis.* Details of the *P. boisei* hypodigm used for comparison are given in Table 5 and in Wood (1991a), and for *H. erectus* the collections from Java and Zhoukoudian have been pooled to calculate CV values for its Asian hypodigm. Leaving aside the continuing debate about what African and European fossils should be included in *H. erectus,* it is plainly sensible to use the Asian hypodigm alone in order to avoid the possibly confounding effects of geographical variation.

Variability has been compared using a maximum of 78 measurements. These include length and breadth measurements of the maxillary (n = 16) and mandibular (n = 16) dentitions, computed crown areas of the canine and

Table 5. Details of Comparative Samples Used for Assessing the Significance of the Degree of Variation within the Early *Homo* Hypodigm

Gorilla	A sample of *Gorilla gorilla gorilla* from the Powell-Cotton Museum, Birchington, Kent (Wood, 1975) was supplemented by specimens from Natural History Museum, London; Rijksmuseum von Natuurlyke Historie, Leiden; Liverpool Museum, Liverpool; Department of Zoology, Edinburgh University; The Royal Museum of Scotland, Edinburgh. Total 34 males/30 females
Paranthropus boisei	Koobi Fora
	KNM-ER 403, 404, 405, 406, 407, 725, 726, 727, 728, 729, 732, 733, 801, 802, 805, 810, 812, 814, 818, 1170, 1171, 1467, 1468, 1469, 1477, 1479, 1509, 1803, 1804, 1806, 1816, 1819, 1820, 3229, 3230, 3729, 3737, 3885, 3886, 3890, 3954, 5429, 5679, 5887, 6082, 13750, 15930, 15940, 15950, 16841, 17760
	West Turkana
	KWM-WT 17396, 17400
	Peninj
	Chesowanja
	KNM-CH 1, 302, 304
	Olduvai
	OH 5, 26, 30
	Omo
	F203-1, L64-2, L74A-21, L338y-6, L427-7, L628-1, L628-1, L628-2, L628-5, L704-2, L726-11, Omo 47-46, Omo 76-37, Omo 141-2, Omo 323-896

postcanine teeth (n = 12), the lengths of the premolar and molar tooth rows (n = 4), and the summed areas of the premolars and molars (n = 4). The mandibular (n = 7) and cranial (n = 19) measurements are listed in the caption to Table 8. Detailed definitions of all the measurements are given in Wood (1991b). Data are recorded for the subsets of fossil data only if three or more specimens are available.

Suitable extant analogues for assessing the significance of differences in the pattern of variation must also be sought. Oxnard (1985, 1987, 1988) has noted significant species differences in the pattern of dental sexual dimorphism in higher primates, but the results of a recent study of cranial, mandibular, and dental variation (Wood *et al.*, 1991) suggest that while interspecific variations in the pattern of sexual dimorphism do exist, they may not be as marked as Oxnard suggested. Nonetheless, the results of Oxnard and Wood *et al.* are consistent to the extent that both studies concur that there are two basic patterns of dental sexual dimorphism in higher primates, one for *H. sapiens* and a second for *Pan, Gorilla,* and *Pongo.* Mandibular and cranial measurements also point to a similar basic pattern, with, in addition, the suggestion that *Pan* and *Homo* may share a pattern of sexual dimorphism that is different from that seen in *Gorilla* and *Pongo* (Wood *et al.*, 1991). Thus, when considering models for the pattern of sexual dimorphism among extant taxa, *H. sapiens* and *Pan* are the forms closest (see above) to the predicted pattern of intraspecific variation for early hominids.

Multivariate analysis has also been used to compare the pattern of cranial variation (Oxnard, 1985; O'Higgins, 1989; Wood *et al.*, 1991). Details of the comparative data used in the canonical analyses performed as part of this study are given in Wood *et al.* (1991). The cranial variables used in each analysis are those that can be recorded for each of the pairs of fossil crania, i.e., KNM-ER 1470/OH24 and 1470/1813. These are listed in the caption accompanying each figure.

Degree of Variation

The crucial comparison is that between the combined regional hypodigms of early *Homo* (Fig. 2, Axis T) and the three comparators *Gorilla, P. boisei,* and *H. erectus sensu stricto.* If variation within the whole sample of early *Homo* does not exceed that in the comparative material, either in degree, pattern, or in respect of well-defined morphological characters, then Hypothesis A, supporting a single-taxon solution, cannot be refuted.

Early *Homo* CV values have been computed for maxillary dental (Table 6), mandibular dental (Table 7), and skull variables (Table 8). Many of the mean CV values of the dental variables of early *Homo* (Tables 6 and 7) exceed equivalent values for the four comparative samples, but in only a small number of variables is the variation in the fossil sample significantly greater than that in the comparative samples (Tables 9 and 10). Comparisons of the variation within 19 cranial and seven mandibular variables measured on the combined regional samples of early *Homo* are presented in Table 8. Excessive variation relative to the extant and fossil comparative samples is concentrated on breadth measurements of the face and on measurements that reflect the size and shape of the horizontal ramus of the mandible (Tables 8 and 11).

The cranial capacity of early *Homo* is more variable (CV = 13.1) than one would expect for a single hominid species; the comparable CV value for *P. boisei* is just 10.5 (Table 11).

Pattern of Variation

The observed degree of variation within the overall early *Homo* hypodigm suggests that few of the measurements considered are so variable that a single taxon solution is out of the question. What, then, of the pattern of variation within the hypodigm? Does it resemble, or not, patterns of intraspecific variation within the comparative populations? The study of Wood *et al.* (1991) suggests that there are consistent differences in the way that variables differ within and between taxa. Within the dentition, for example, canine variables and the crown breadths of the noncanine teeth are generally good sex-discriminating variables, whereas crown mesiodistal diameters are better taxonomic discriminators. Consideration of Tables 6 and 7 shows that while canine CVs are high in early *Homo,* they are no higher than values for the postcanine teeth. The breadths of tooth crowns are not consistently less variable than crown lengths. Thus, the pattern of

Table 6. Details of the Variability within the Early *Homo* Sample and Its Subsets: Maxillary Dentition

		Combined sample					"Old" Region					"OM" region					Time band "4"					Time band "5"				
		N	Mean	SD	CV	95% CL	N	Mean	SD	CV	95% CL	N	Mean	SD	CV	95% CL	N	Mean	SD	CV	95% CL	N	Mean	SD	CV	95% CL
I1	LL	3	7.9	0.7	9.9	−7.7–27.5	—	—	—	—	—	—	—	—	—	—	—	—	—	—	—	—	—	—	—	—
	MD	3	11.2	1.1	10.7	−8.3–29.7	—	—	—	—	—	—	—	—	—	—	—	—	—	—	—	—	—	—	—	—
I2	LL	5	6.9	1.0	15.7	1.6–29.8	3	6.8	1.0	15.7	−12.6–44.0	—	—	—	—	—	3	6.2	0.3	4.6	−3.5–12.7	—	—	—	—	—
	MD	5	7.5	0.7	9.3	1.1–17.5	3	7.6	0.7	9.7	− 7.5–26.9	—	—	—	—	—	3	7.2	0.8	11.8	−9.2–32.8	—	—	—	—	—
C	LL	9	10.2	1.6	15.6	6.9–24.3	—	—	—	—	—	7	10.5	1.7	16.4	5.4–27.4	8	10.1	1.6	16.6	6.5–26.7	—	—	—	—	—
	MD	9	9.3	1.3	14.4	6.4–22.4	—	—	—	—	—	7	9.3	1.5	16.7	5.5–27.9	8	9.4	1.4	14.9[a]	5.9–23.9	—	—	—	—	—
	CA	9	96.0	27.6	29.6	12.2–47.0	—	—	—	—	—	7	98.7	31.3	32.8	9.2–56.5	8	96.4	29.5	31.6[a]	11.1–52.1	—	—	—	—	—
P3	BL	13	12.0	0.9	7.2	4.1–10.3	5	11.5	0.6	5.0	0.6– 9.4	8	12.4	0.9	7.2	2.9–11.5	9	12.1	1.0	8.2	3.7–12.7	4	11.9	0.6	5.7	−0.7–12.1
	MD	13	9.1	0.8	9.4	5.4–13.5	5	9.2	0.4	4.2	0.5– 7.9	8	9.0	1.0	12.0[a]	4.8–19.2	9	9.0	1.0	11.1[a]	5.0–17.2	4	9.2	0.5	5.6	−0.7–11.9
	CA	13	110.0	16.8	15.5	8.7–22.3	5	116.5	8.9	8.7							9	110.2	19.7	18.4	8.1–28.7	4	109.5	0.9	9.0	−1.2–19.2
P4	BL	12	12.1	0.9	7.9	4.3–11.5	6	11.7	0.7	6.2	1.6–10.8	6	12.5	1.0	8.7	2.2–15.2	9	12.2	1.1	9.1	4.1–14.1	3	12.0	0.5	4.1	−3.1–11.3
	MD	12	9.5	0.7	7.8	4.3–11.3	6	9.4	0.5	5.9	1.5–10.3	6	9.6	0.9	10.0[a]	2.5–17.5	9	9.5	0.8	8.4	3.8–13.0	3	9.5	0.7	7.6	−5.8–21.0
	CA	12	116.0	17.4	15.3	8.3–22.3	6	11.0	11.2	10.5	2.6–18.4	6	121.5	21.9	18.8	4.4–33.2	9	117.0	19.4	17.1	7.5–26.7	3	114.0	12.3	11.7	−9.1–32.5
M1	BL	19	13.3	0.8	5.7	3.8– 7.7	9	12.9	0.6	4.9	2.2– 7.6	10	13.6	0.7	5.1	2.5– 7.7	11	13.3	0.9	6.9	3.6–10.2	6	13.1	0.5	4.1	1.1– 7.2
	MD	19	12.9	0.8	6.3	4.1– 8.5	9	12.6	0.7	5.6	2.6– 8.7	10	13.2	0.8	6.2	3.1– 9.4	11	12.9	0.8	6.2	3.2– 9.2	6	12.7	0.7	6.0	1.5–10.5
	CA	19	172.0	19.7	11.6	7.6–15.6	9	162.0	15.8	10.0	4.5–15.5	10	181.0	19.2	10.9	5.3–16.5	11	171.5	21.5	12.8	6.2–19.4	6	167.5	15.8	9.8	2.5–17.1
M2	BL	11	14.9	1.4	9.5	5.0–14.1	6	14.6	1.3	9.1	2.3–15.9	5	15.4	1.7	11.3	1.3–21.4	7	14.4	1.5	10.8	3.7–17.9	4	15.1	1.3	8.8	−1.2–18.8
	MD	12	12.9	0.8	6.4	3.5– 9.3	6	12.6	0.6	4.6	1.2– 8.0	6	13.6	0.9	7.1	1.8–12.4	8	12.9	1.0	7.9	3.2–12.6	4	12.9	0.3	2.6	−0.3– 5.5
	CA	11	193.0	29.3	15.5	8.0–23.0	6	183.5	23.4	13.3	3.3–23.3	5	210.5	36.6	18.3	1.7–34.9	7	186.5	34.5	19.1	6.2–32.0	4	195.0	20.5	11.2	−1.6–24.0
M3	BL	15	14.7	1.4	9.6	5.8–13.4	7	15.3	1.6	10.9	3.7–18.1	8	14.1	0.9	6.8	2.8–10.8	7	14.1	0.6	4.2	1.5– 7.0	7	15.0	1.9	12.8	4.3–21.3
	MD	15	12.7	1.1	8.4	5.1–11.7	7	12.8	1.0	8.4	2.9–13.9	8	12.6	1.1	9.2	3.7–14.7	7	12.6	1.0	8.0[a]	2.7–13.3	7	12.7	1.2	9.5	3.2–15.8
	CA	15	188.0	31.1	16.8	10.0–23.6	7	197.5	34.9	18.3	5.9–30.7	8	179.5	26.6	15.3	6.0–24.6	7	178.0	18.5	10.7	3.6–17.8	7	192.5	39.8	21.4	6.8–36.0
	PRL	7	18.1	1.2	6.9	2.4–11.4	5	18.6	0.9	5.0	0.6– 9.4	—	—	—	—	—	8	18.7	1.8	9.9[a]	4.0–15.8	3	18.7	1.2	6.9	−5.3–19.1
	MRL	7	38.1	2.1	5.7	2.0– 9.4	3	37.9	2.4	6.8	− 5.2–18.8	4	38.3	2.3	6.4	−0.8–13.6	4	38.3	2.3	6.4	−0.8–13.6	3	37.9	2.4	6.8	−5.2–18.8
	PCA	7	210.0	19.4	9.6	3.3–15.9	5	215.5	20.3	9.9	1.1–18.7	—	—	—	—	—	8	228.5	40.6	18.3	7.1–29.5	3	223.0	23.6	11.4	−8.9–31.7
	MCA	6	531.0	61.8	12.1	3.0–21.2	3	536.0	81.2	16.3	−13.1–45.7	3	528.0	55.2	11.3	−8.8–31.4	3	526.5	53.5	11.0	−8.6–30.6	3	536.5	81.2	16.3	−13.1–45.7

BL = maximum buccolingual diameter; MD = corrected mesiodistal diameter; CA = computed crown area; PRL = length of premolar row; MRL = length of molar row; PCA = combined computed area of premolar crowns; MCA = combined computed area of molar crowns.

[a]Variables in which the level of variation in the fossil sample exceeds variation in one, or more, of the comparative samples, i.e., *Gorilla*, *P. boisei*, and *H. erectus sensu stricto*, identified using the criteria suggested by Schultz (1985).

Table 7. Details of the Variability with the Early *Homo* Sample and Its Subsets: Mandibular Dentition

		Combined sample					"Old" Region					"OM" region					Time band "4"					Time band "5"				
		N	Mean	SD	CV	95% CL	N	Mean	SD	CV	95% CL	N	Mean	SD	CV	95% CL	N	Mean	SD	CV	95% CL	N	Mean	SD	CV	95% CL
I_1	LL	3	6.8	0.2	3.3	−2.5– 9.1	3	6.8	0.2	3.3	−2.5– 9.1	—	—	—	—	—	—	—	—	—	—	—	—	—	—	—
	MD	3	6.4	0.1	1.0	−0.8– 2.8	3	6.4	0.1	1.0	−0.8– 2.8	—	—	—	—	—	—	—	—	—	—	—	—	—	—	—
I_2	LL	4	7.6	0.1	0.8	−0.1– 1.7	4	7.6	0.1	0.8	−0.1– 1.7	—	—	—	—	—	—	—	—	—	—	—	—	—	—	—
	MD	4	7.4	0.3	3.9	−0.5– 8.3	4	7.4	0.3	3.9	−0.5– 8.3	—	—	—	—	—	—	—	—	—	—	—	—	—	—	—
C	LL	—	—	—	—	—	—	—	—	—	—	—	—	—	—	—	—	—	—	—	—	—	—	—	—	—
	MD	5	8.6	0.8	9.3[a]	1.1–17.5	4	8.8	0.8	10.1	−1.4–21.6	—	—	—	—	—	3	8.7	0.4	4.6[a]	−3.5–12.7	—	—	—	—	—
	CA	—	—	—	—	—	—	—	—	—	—	—	—	—	—	—	—	—	—	—	—	—	—	—	—	—
P_3	BL	12	10.1	1.4	14.0	7.6–20.4	6	9.5	1.0	10.7	2.7–18.7	6	10.7	1.5	14.8	3.6–26.0	6	9.8	1.2	13.0	3.2–22.8	3	9.4	1.5	17.1	−13.8–48.0
	MD	12	9.9	0.9	9.1	5.0–13.2	6	9.5	0.6	6.8	1.7–11.9	6	10.3	1.0	9.9	25.–17.3	6	9.5	0.7	7.1	1.8–12.4	3	9.5	0.9	10.5	− 8.2–29.2
	CA	12	101.0	21.9	22.2	11.8–32.7	6	91.0	15.2	17.4	4.1–30.7	6	110.0	24.3	22.8	5.0–40.6	6	94.3	17.0	18.7	4.3–33.1	3	91.0	23.7	28.3	−25.3–81.9
P_4	BL	14	11.0	1.3	11.7	6.9–16.5	6	10.5	0.5	5.1	1.3– 8.9	8	11.3	1.6	14.2	5.6–22.8	9	11;.0	1.4	12.9	5.8–20.0	4	10.4	0.6	6.5	− 0.8–13.8
	MD	14	10.2	1.0	9.5	5.6–13.4	6	9.9	0.7	7.4	1.9–12.9	8	10.4	1.1	10.9	4.4–17.4	9	10.3	1.0	9.6	4.3–14.9	4	9.6	0.7	7.4	− 1.0–15.8
	CA	13	111.5	21.4	19.6	10.9–28.3	6	104.0	11.9	11.9	2.9–20.9	7	117.5	26.4	23.2	7.2–39.2	8	113.0	21.8	19.9	7.7–32.1	4	99.8	13.0	13.8	− 2.0–29.6
M_1	BL	16	12.5	1.0	7.8	4.8–10.8	4	12.4	0.5	4.6	−0.6– 9.8	12	12.5	1.1	8.9	4.9–12.9	10	12.5	0.9	7.2	3.5–10.9	3	12.3	0.6	5.6	− 4.3–15.5
	MD	15	14.0	0.7	5.1	3.1– 7.1	4	14.0	0.17	5.3	−0.7–11.3	11	14.0	0.7	5.4	2.8– 8.0	10	14.0	0.6	4.0	2.0– 6.0	—	—	—	—	—
	CA	15	175.0	20.7	12.0	7.2–16.8	4	173.5	15.7	9.6	−1.3–20.5	11	175.5	22.9	13.3	6.9–19.7	10	174.0	15.8	9.3	4.6–14.0	—	—	—	—	—
M_2	BL	16	13.6	1.9	8.1	5.0–11.2	4	13.5	1.3	10.0	−1.4–21.4	12	13.6	1.1	8.1	4.4–11.8	8	13.4	0.9	6.6	2.7–10.5	3	13.5	1.6	12.5	− 9.8–34.8
	MD	16	15.5	1.1	7.5	4.7–10.3	4	15.0	0.7	4.8	−0.6–10.2	12	15.7	1.2	8.1	4.4–11.8	8	15.5	1.3	8.5	3.4–13.6	3	14.8	0.6	4.4	− 3.3–12.1
	CA	15	211.1	30.6	14.8	8.9–20.7	4	203.5	27.0	14.1	−2.1–30.3	11	214.0	32.6	15.6	8.0–23.2	7	206.0	27.7	13.9	4.6–23.2	3	199.5	31.2	17.2	−13.9–48.3
M_3	BL	14	13.2	1.0	7.8	4.6–11.0	6	13.3	0.9	7.1	1.8–12.4	8	13.2	1.1	8.9	3.6–14.2	7	13.4	0.9	7.3	2.5–12.1	4	13.4	1.6	9.2	− 1.2–19.6
	MD	13	15.4	0.9	5.6	3.2– 8.0	6	15.4	0.5	3.3	0.9– 5.8	7	15.4	1.1	7.5	2.6–12.4	6	15.4	0.9	5.8	1.5–10.1	4	15.4	0.6	4.4	− 0.6– 9.4
	CA	13	202.0	23.6	11.9[a]	6.7–17.1	6	204.5	20.3	10.3[a]	2.6–18.0	7	200.5	27.6	14.2	4.7–23.7	6	204.0	24.4	12.5	3.1–21.9	4	205.5	26.0	13.4	− 1.9–28.7
	PRL	9	19.8	1.7	8.9	4.0–13.8	5	19.4	1.3	7.0	0.8–13.2	4	20.4	2.2	11.4[a]	−1.6–24.4	5	19.9	1.5	7.9	0.9–14.9	3	18.9	1.6	9.2	− 7.1–25.5
	MRL	—	—	—	—	—	—	—	—	—	—	—	—	—	—	—	—	—	—	—	—	—	—	—	—	—
	PCA	9	206.5	43.0	21.4	9.3–33.6	5	194.2	28.0	15.1	1.6–28.7	4	229.4	57.6	26.6	−5.4–58.6	5	205.0	40.8	20.9	1.8–40.0	3	186.5	36.7	21.3	−17.8–60.4
	MCA	—	—	—	—	—	—	—	—	—	—	—	—	—	—	—	—	—	—	—	—	—	—	—	—	—

For details of the abbreviations and keys to the superscripts, see caption to Table 5.

variation in early *Homo* dental variables provides equivocal evidence about whether the differences are likely to be interspecific or intraspecific.

Turning to the cranium, Wood *et al.* (1991) proposed that facial width measurements are generally good sex discriminators, whereas cranial breadths, e.g., minimum frontal and maximum parietal, are better taxonomic discriminators. In Table 8 it is apparent that the pattern of difference in early *Homo* is a confusing one, with high levels of variation in some variables suggesting an intraspecific pattern (e.g., interorbital breadth), whereas others (e.g., minimum frontal breadth) point to an interspecific basis for the observed variation. Although the degree of cranial capacity variation in early *Homo* does not exceed that in the extant comparative sample, the pattern of distribution of the values is different. The extent of bimodality in the smaller fossil sample is not matched, even in the extreme sexual dimorphism of *Gorilla* (Fig. 3).

Evidence from facial shape is particularly crucial, not only because the face is a region that varies in different ways within and between taxa (Wood *et al.*, 1991), but also because the relatively well-preserved facial skeletons of specimens such as OH 24, KNM-ER 1470, and KNM-ER 1813 provide some of the best evidence to test the null hypothesis that these crania belong to a single species. When variables are considered in the univariate sense, there are evident taxonomic differences in the way facial shape varies within a species. In modern *H. sapiens* males generally show less basal and alveolar prognathism than females (Luboga, 1986). Early hominid faces are all relatively prognathic, and OH 24, KNM-ER 1470, and KNM-ER 1813 are no exception. However by some criteria, e.g., alveolar projection, KNM-ER 1470 is less prognathic than KNM-ER 1813 and OH 24, while in others, e.g., midfacial projection, KNM-ER 1470 is more prognathic than the latter crania (Bilsborough and Wood, 1988). One of the obvious differences between the proportions of the face of KNM-ER 1470 and the faces of OH 24 and KNM-ER 1813 is the relationship between the widths of the upper and midface. In the latter crania the upper face exceeds the midface in width, whereas in KNM-ER 1470 the reverse is the case. Within the comparative nonhuman primate groups, males tend to have a relatively broader midface (Wood, 1975), but across four modern *H. sapiens* populations there are no significant sex differences in regional facial widths (Luboga, 1986). Likewise, the difference in orientation of the malar region between the fossil crania does not correspond to patterns of intraspecific variation within the most relevant extant samples.

Perhaps the clearest indication of the significance of the differences in facial form between KNM-ER 1470 and the smaller early *Homo* crania can be found when the multivariate relationships between the fossil crania are compared with those between separate sex samples of the extant comparative taxa. Canonical analysis provides this facility and the relationships shown in diagram form in Fig. 4 are based on an analysis using 14 cranial variables. It is apparent that the slope of the line joining KNM-ER 1470 and OH 24 is approximately orthogonal to the axes uniting the group means of the male and females of *H. sapiens, Pan,* and *Gorilla,* respectively. The angles subtended to the X, Y, and Z axes by the slopes for the comparative samples and the two fossil crania, together with the Mahalanobis distance between the group means and the fossil crania, are given in Table 12. Pairwise comparisons between comparative individual males and

Table 8. Details of Variability with the Early *Homo* Sample and Its Subsets: Cranial and Mandibular

		Combined sample					Time band "4"					Time band "5"				
		N	Mean	SD	CV	95% CL	N	Mean	SD	CV	95% CL	N	Mean	SD	CV	95% CL
GL-OP	(1)	3	153	11.2	7.9	−6.1–21.9	3	153	11.2	7.9	−6.1–21.9					
PCL	(3)	4	61	5.4	9.4	−1.3–20.1	4	61	5.4	9.4	−1.3–20.1					
MFB	(8)	5	80	8.0	10.5	1.2–19.8	5	80	8.0	10.5	1.2–19.8					
MPB	(9)	4	110	9.6	9.3	−1.3–19.9	4	110	9.6	9.3	−1.3–19.9					
BB	(11)	4	110	12.0	11.6	−1.6–24.8	4	110	12.0	11.6	−1.6–24.8					
BW	(13)	3	120	7.5	6.8	−5.2–18.8	3	120	7.5	6.8	−5.2–18.8					
BOB	(50)	4	94	6.7	7.6	−1.0–16.2	4	94	6.7	7.6	−1.0–16.2					
BJB	(51)	3	117	13.1	12.1	−9.5–33.7	3	117	13.1	12.1	−9.5–33.7					
IOB	(55)	5	26	7.0	28.3	1.5–55.1	5	26	7.0	28.3	1.5–55.1					
OAB	(54)	3	67	2.5	4.0	−3.0–11.0	3	67	2.5	4.0	−3.0–11.0					
OB	(56)	4	36	3.3	9.7	−1.3–20.7	4	36	3.3	9.7	−1.3–20.7					
OH	(57)	4	32	2.2	7.3	−1.0–15.6	4	32	2.2	7.3	−1.0–15.6					
MNW	(68)	4	26	2.4	9.8	−1.3–20.9	4	26	2.4	9.8	−1.3–20.9					
NH	(69)	4	48	7.5	16.6	−2.6–35.8	4	48	7.5	16.6	−2.6–35.8					
PL	(90)	4	43	4.3	10.6	−1.5–22.7	4	43	4.3	10.6	−1.5–22.7					
ICD	(98)	5	31	3.7	12.5	1.4–23.6	4	31	4.2	14.5	−2.2–31.2					
PH	(103)	5	14	4.9	36.8[a]	0.4–73.2	4	13	4.9	41.9[a]	−12.9–96.7					
IAB	(93)	4	35	4.5	13.7	−2.0–29.4	3	36	5.2	15.6	−12.5–43.7					
CC	(—)	9	657	86.1	13.1	6.0–21.0	7	660	99.1	15.6	5.2–26.1					
SH	(141)	6	33	5.4	17.0	4.0–30.0	4	36	3.3	9.7	−1.3–20.7	—	—	—	—	—
ST	(142)	5	22	3.2	15.3	1.6–29.0	3	24	1.6	7.2	−5.5–19.9	—	—	—	—	—
HM_1	(150)	12	30	10.3	35.0[a]	17.5–52.6	9	30	11.7	40.1	15.0–65.2	3	32	5.8	19.6	−16.1– 55.3
TM_1	(151)	12	20	7.0	35.7[a]	17.7–53.7	9	19	7.7	41.7	15.4–68.0	3	22	4.8	23.6	−20.1– 67.3
AM_1	(152)	11	541	162.2	30.7	14.8–46.6	8	535	152.4	29.4	10.6–48.2	3	557	223.2	43.4	−46.1–132.9
RIM_1	(—)	11	64	4.5	7.2	3.8–10.6	8	63	4.2	6.9	2.8–11.0	3	67	5.1	8.2	−6.3– 22.7
ICB	(166)	4	18	1.7	10.0	−1.4–21.4	—	—	—	—	—	—	—	—	—	—

Table 8. *(Continued)*

		"Om" region					"Old" region				
		N	Mean	SD	CV	95% CL	N	Mean	SD	CV	95% CL
GL-OP	(1)	—	—	—	—	—					
PCL	(3)	3	62	5.1	8.9	−6.9– 24.7					
MFB	(8)	4	82	8.5	11.0	−1.5– 23.5					
MPB	(9)	4	110	9.6	9.3	−1.3– 19.9					
BB	(11)	3	111	14.2	13.8	−10.9– 38.5					
BW	(13)	—	—	—	—	—					
BOB	(50)	3	95	7.9	9.0	−6.9– 24.9					
BJB	(51)	3	117	13.1	12.1	−9.5– 33.7					
IOB	(55)	4	27	7.9	31.6	−7.3– 70.5					
OAB	(54)	—	—	—	—	—					
OB	(56)	3	36	3.3	9.9	−7.7– 27.5					
OH	(57)	3	32	2.6	8.7	−6.7– 24.1					
MNW	(68)	3	26	2.6	10.8	−8.4– 30.0					
NH	(69)	3	50	7.6	16.4	−13.2– 46.0					
PL	(90)	3	43	5.3	13.3	−10.5– 37.1					
ICD	(98)	3	32	4.0	13.5	−10.6– 37.6					
PH	(103)	3	12	6.0	54.0[a]	−65.4–173.4					
IAB	(93)	—	—	—	—	—					
CC	(—)	5	669	115.0	18.1	1.7– 34.5	4	641	40.0	6.6	−0.9–14.1
SH	(141)	4	36	3.3	9.6	−1.3– 20.5					
ST	(142)	3	24	1.6	7.3	−5.6– 20.2					
HM_1	(150)	10	31	11.3	37.9	16.2– 59.7					
TM_1	(151)	9	19	8.0	42.3[a]	15.5– 69.1	3	21	3.7	19.0	−15.6–53.6
AM_1	(152)	9	565	168.8	30.7	12.5– 48.9					
RIM_1	(—)	9	64	4.7	7.6	3.5– 11.8					
ICB	(166)	—	—	—	—	—					

For keys to superscript see caption to Table 5.

Keys to abbreviations: Cranial measures—GL-OP = glabella-opisthocranion; PCL = posterior cranial length; MFB = minimum frontal breadth; MPB = maximum parietal breadth; BB = biporionic breadth; BW = basal breadth; BOB = biorbital breadth; BJB = bijugal breadth; IOB = interorbital breadth; OAB = outer alveolar breadth; OB = orbital breadth; OH = orbital height; MNW = maximum nasal width; NH = nasal height; PL = palate length; ICD = intercanine distance; PH = palatal height; IAB = internal alveolar bradth; CC = cranial capacity.

Mandibular measures—SH = symphyseal height; ST = symphyseal thickness; HM_1 = corpus height; TM_1 = corpus thickness; AM_1 = computed corpus area; RIM_1 = corpus robusticity index; ICB = intercanine breadth.

All measurements are in millimeters.

Numbers in parentheses after the measurement abbreviations refer to their number in Wood (1991b), to which reference should be made for details of measurement definitions and measuring technique.

Table 9. Details of the Variability within the One Extant Hominoid and Two Fossil Hominid Comparative Samples: Maxillary Dentition

		Gorilla			*Paranthropus boisei* (*sensu stricto*)			*Homo erectus* (*sensu stricto*)		
		N	CV	95% CL	N	CV	95% CL	N	CV	95% CL
I1	LL	64	8.8	7.2–10.3	—	—	—	6	3.9	1.0– 6.8
	MD	64	12.5	10.2–14.7	—	—	—	5	5.0	0.6– 9.4
I2	LL	64	11.5	9.4–13.5	5	13.1	1.4–24.8	3	1.3	−1.0– 3.6
	MD	64	12.9	10.6–15.2	5	6.5	0.8–12.2	3	18.7	−15.3–52.7
C	LL	64	20.4	16.6–24.1	6	10.5	2.6–18.4	9	9.0	4.1–13.9
	MD	64	20.9	17.1–24.8	7	10.6	3.6–17.6	9	5.5	2.5– 8.5
	CA	64	39.0	31.1–46.9	6	13.6	3.3–23.9	9	12.3	5.5–19.1
P3	BL	64	8.3	6.9– 9.8	8	8.0	3.2–12.8	10	9.7	4.8–14.7
	MD	64	9.1	7.4–10.7	8	5.7	2.3– 9.1	10	8.8	4.3–13.3
	CA	64	15.5	12.7–18.3	8	9.6	3.9–15.3	10	17.7	8.5–26.9
P4	BL	64	7.6	6.3– 9.0	7	8.3	2.8–13.8	14	6.2	3.7– 8.7
	MD	64	8.4	6.9– 9.8	7	4.5	1.6– 7.5	14	7.7	4.5–10.9
	CA	64	13.9	11.4–16.3	7	12.0	4.0–20.0	14	12.4	7.3–17.5
M1	BL	64	7.2	6.0– 8.5	8	6.9	2.8–11.0	9	5.7	2.6– 8.8
	MD	64	7.1	5.9– 8.4	8	5.6	2.3– 8.9	10	8.1	4.0–12.2
	CA	64	13.3	10.9–15.6	8	11.0	4.4–17.6	9	11.8	5.3–18.3
M2	BL	64	7.6	6.2– 8.9	8	10.2	4.1–16.3	10	7.1	3.5–10.7
	MD	64	8.3	6.8– 9.8	8	7.0	2.8–11.2	10	9.9	4.8–15.0
	CA	64	14.3	11.7–16.9	8	16.1	6.3–25.9	10	16.3	7.8–24.8
M3	BL	64	9.8	8.1–11.6	4	14.0	−2.1–30.1	12	10.7	5.8–15.6
	MD	64	10.2	8.4–12.1	4	7.4	−1.0–15.8	12	7.2	4.0–10.5
	CA	64	19.4	15.9–23.0	4	17.7	−2.8–38.2	12	17.5	9.4–25.6
	PRL	64	7.5	6.2– 8.9	6	2.8	0.7– 4.9	5	6.0	0.7–11.3
	MRL	64	7.7	6.3– 9.1	4	6.1	−0.8–13.0	5	7.2	0.9–13.6
	PCA	64	13.7	11.2–16.1	6	8.9	2.3–16.3	5	12.7	1.4–24.0
	MCA	64	14.8	12.1–17.4	4	16.6	−2.8–38.0	5	14.0	1.5–26.5

females results in a large "family" of axes with which the axis between the fossils can sensibly be compared. The differences between the fossil axis and these comparative axes suggest that it is highly unlikely that the shape differences between the fossil crania resemble the pattern of difference between male and female crania in the comparative sample. These findings provide strong evidence that sexual variation may be an insufficient explanation for the differences in facial shape between KNM-ER 1470 and crania such as OH 24.

Temporal Trends in Character States

This section examines the extent to which the expression of character states, including nonmetric traits, varies within the hypodigm of early *Homo*. Given the nature of the hypodigm, with its emphasis on dental and mandibular remains, the characters considered are dental and concentrate on the mandibular dentition.

Postcanine Crown Morphology

Previous investigations have suggested that there are taxonomically significant differences in the details of the mandibular premolar and molar tooth-crown morphology of early hominid taxa (Wood and Abbott, 1983; Wood *et al.*, 1983; Wood and Uytterschaut, 1987), although Hartman (1988) has recently suggested that the same morphological features may act to obscure the phylogenetic relationships between extant hominoid genera. Posterior probability values, generated from canonical analysis data, provide a guide to the relative reliability of these features as taxonomic discriminators (Wood *et al.*, 1983; Wood and Uytterschaut, 1987). Using this criterion, the form of the P_4 crown, specifically the relative areas of the main cusp elements, is the most reliable of this type of taxonomic indicator. The multivariate relationships between specimens are best displayed using principal components analysis (PCA), and a plot of the first two principal components generated from an analysis of 32 early hominid tooth crowns is illustrated in Wood and Uytterschaut (1987). This plot suggests that while some early *Homo* specimens from Koobi Fora (e.g., KNM-ER 1802) have

Table 10. Details of the Variability within the One Extant Hominoid and Two Fossil Hominid Comparative Samples: Mandibular Dentition

		Gorilla			*Paranthropus boisei* (*sensu stricto*)			*Homo erectus* (*sensu stricto*)		
		N	CV	95% CL	N	CV	95% CL	N	CV	95% CL
I_1	LL	64	11.6	9.6–13.7	7	12.3	4.1–20.5	8	4.8	2.0– 7.6
	MD	64	14.4	11.8–17.0	7	9.7	3.3–16.1	8	7.9	3.2–12.6
I_2	LL	64	13.3	10.9–15.7	4	12.3	−1.8–26.4	8	4.4	1.8– 7.0
	MD	64	11.6	9.5–13.7	5	3.3	0.4– 6.2	7	6.1	2.1–10.1
C	LL	64	17.9	14.6–21.1	9	9.4	4.3–14.6	7	8.9	3.0–14.8
	MD	64	18.9	15.4–22.3	8	9.1	3.7–14.5	7	8.6	2.9–14.3
	CA	64	35.0	28.1–41.9	8	14.2	5.6–22.8	6	17.5	4.1–30.9
P_3	BL	64	15.3	12.6–18.1	7	7.3	2.5–12.1	16	10.2	6.3–14.1
	MD	64	12.1	9.9–14.2	7	7.6	2.6–12.6	17	8.2	5.2–11.2
	CA	64	22.9	18.7–27.2	7	12.8	4.3–21.3	16	16.4	10.1–22.7
P_4	BL	64	8.8	7.3–10.4	14	7.7	4.5–10.9	10	13.8	6.7–20.9
	MD	64	8.7	7.1–10.2	14	7.3	4.3–10.3	9	4.5	2.1– 7.0
	CA	64	15.4	12.6–18.2	14	13.1	7.7–18.5	9	13.2	5.9–20.5
M_1	BL	64	7.8	6.4– 9.2	11	6.9	3.6–10.2	19	7.6	5.0–10.2
	MD	64	7.2	5.9– 8.4	11	5.1	2.7– 7.5	19	8.8	5.8–11.8
	CA	64	13.6	11.1–16.0	11	10.3	5.4–15.2	18	15.6	10.0–21.2
M_2	BL	64	8.1	6.6– 9.5	14	7.8	4.6–11.0	14	7.4	4.4–10.4
	MD	64	7.1	5.8– 8.3	14	8.8	5.2–12.4	14	6.4	3.8– 9.0
	CA	64	13.9	11.4–16.4	14	16.4	9.5–23.3	14	12.2	7.2–17.3
M_3	BL	64	10.5	8.6–12.4	18	8.5	5.5–11.5	14	8.9	5.2–12.6
	MD	64	8.3	6.8– 9.8	18	8.8	5.7–11.9	14	13.2	7.7–18.7
	CA	64	16.5	13.5–19.5	18	17.2	11.0–23.4	14	11.9	9.4–29.5
	PRL	64	9.8	8.1–11.6	4	6.7	−0.9–14.3	6	3.4	0.9– 5.9
	MRL	64	6.4	5.3– 7.6	5	8.3	1.0–15.6	4	8.2	−1.1–17.5
	PCA	64	22.8	18.6–27.1	3	9.0	−6.9–24.9	6	13.9	3.4–24.4
	MCA	64	13.8	11.3–16.3	5	17.0	1.7–32.4	3	17.9	−14.5–50.3

Table 11. Details of the Variability within the One Extant Hominoid and Two Fossil Hominid Comparative Samples: Cranial and Mandibular

		Gorilla			*Paranthropus boisei* (*sensu stricto*)			*Homo erectus* (*sensu stricto*)		
		N	CV	95% CL	N	CV	95% CL	N	CV	95% CL
GL-OP	(1)	64	11.8	9.7–13.9	4	12.7	−1.8–27.2	8	4.8	2.0– 7.6
PCL	(3)	64	17.9	14.6–21.2	3	2.3	−1.7– 6.3	5	18.0	1.7–34.3
MFB	(8)	64	6.2	5.1– 7.3	6	8.2	2.1–14.3	10	7.3	3.6–11.0
MPB	(9)	64	4.7	3.9– 5.6	6	8.0	2.0–14.0	10	3.9	1.9– 5.9
BB	(11)	64	10.5	8.6–12.4	5	10.9	1.2–20.6	8	4.2	1.7– 6.7
BW	(13)	64	11.1	9.1–13.1	4	14.3	−2.1–30.7	4	6.9	−0.9–14.7
BOB	(50)	64	8.7	7.2–10.3	3	11.0	−8.6–30.6	—	—	—
BJB	(51)	64	10.5	8.17–12.4	3	13.0	−10.2–36.2	—	—	—
IOB	(55)	64	21.7	17.7–25.7	4	8.8	−1.2–18.8	—	—	—
OAB	(54)	64	7.3	6.0– 8.6	3	14.8	−11.8–41.4	—	—	—
OB	(56)	64	7.1	5.9– 8.4	3	12.3	−9.6–34.2	—	—	—
OH	(57)	64	6.9	5.7– 8.2	3	10.2	−7.9–28.3	—	—	—
MNW	(68)	64	10.2	8.4–12.1	4	10.3	−1.4–22.0	—	—	—
NH	(69)	64	15.9	13.0–18.8	—	—	—	—	—	—
PL	(90)	64	12.7	10.4–15.0	3	16.8	−13.5–47.1	—	—	—
ICD	(98)	64	10.8	8.9–12.8	3	5.6	−4.3–15.5	—	—	—
PH	(103)	64	13.9	11.4–16.4	3	3.2	−2.4– 8.8	—	—	—
IAB	(93)	64	9.2	7.6–10.9	3	5.1	−3.9–14.1	—	—	—
CC	(—)	64	14.1	11.6–16.7	7	10.5	3.6–17.4	11	11.7	6.1–17.3
SH	(141)	64	13.4	11.0–15.8	11	15.0	7.7–22.3	7	19.3	6.2–32.4
ST	(142)	64	13.1	10.7–15.4	11	9.5	5.0–14.1	4	20.1	−3.4–43.6
HM_1	(150)	64	11.8	9.7–13.9	25	9.9	7.0–12.8	7	16.1	5.3–26.9
TM_1	(151)	64	9.8	8.0–11.5	25	11.8	8.3–15.3	7	17.4	5.7–29.1
AM_1	(152)	64	18.7	15.3–22.2	25	21.3	14.8–27.8	7	29.4	8.6–50.2
RIM_1	(—)	64	10.0	8.2–11.8	25	7.5	5.3– 9.7	7	12.1	4.1–20.1
ICB	(166)	64	14.3	11.7–16.9	5	5.0	0.6– 9.4	—	—	—

more "molarized" P_4 crowns than any of the teeth from Olduvai, so that they fall within, or adjacent to, the "envelope" enclosing the combined samples of East (EAFROB) and South (SAFROB) African "robust" australopithecines, the overall spread of the early *Homo* data points is no greater than for teeth attributed to other single species, for example, *P. boisei* or *P. robustus.* Thus, this evidence does not especially support a multiple-taxon solution.

Postcanine Root Morphology

The potential of using variations in root form for taxonomic discrimination has been explored by Wood *et al.* (1988). The study of Abbott (1984) provides the necessary comparative framework.

Higher primates are apparently consistent in their P_4 root morphology, nonhuman catarrhine primates always being two-rooted, with mesial and distal elements, whereas modern human P_4s are single rooted. Anterior premolar roots show more intraspecific variation. *Gorilla* and *Pongo* P_3s are typically two-rooted, with mesiobuccal and distal roots. The majority of *Pan* P_3s also takes this form, but approximately one third are single-rooted, like the P_3s of modern *H. sapiens*. The incidence of Tomes' root form, in which the apex of the root is bifurcate, in modern human P_3s varies from 3% to 6% (Abbott, 1984). Against this background, the mandibular premolar root form of early *Homo* is remarkably variable, with its expression in both P_3s and P_4s varying from a single root, via Tomes' root form, to teeth with separate mesial and distal roots (Table 13).

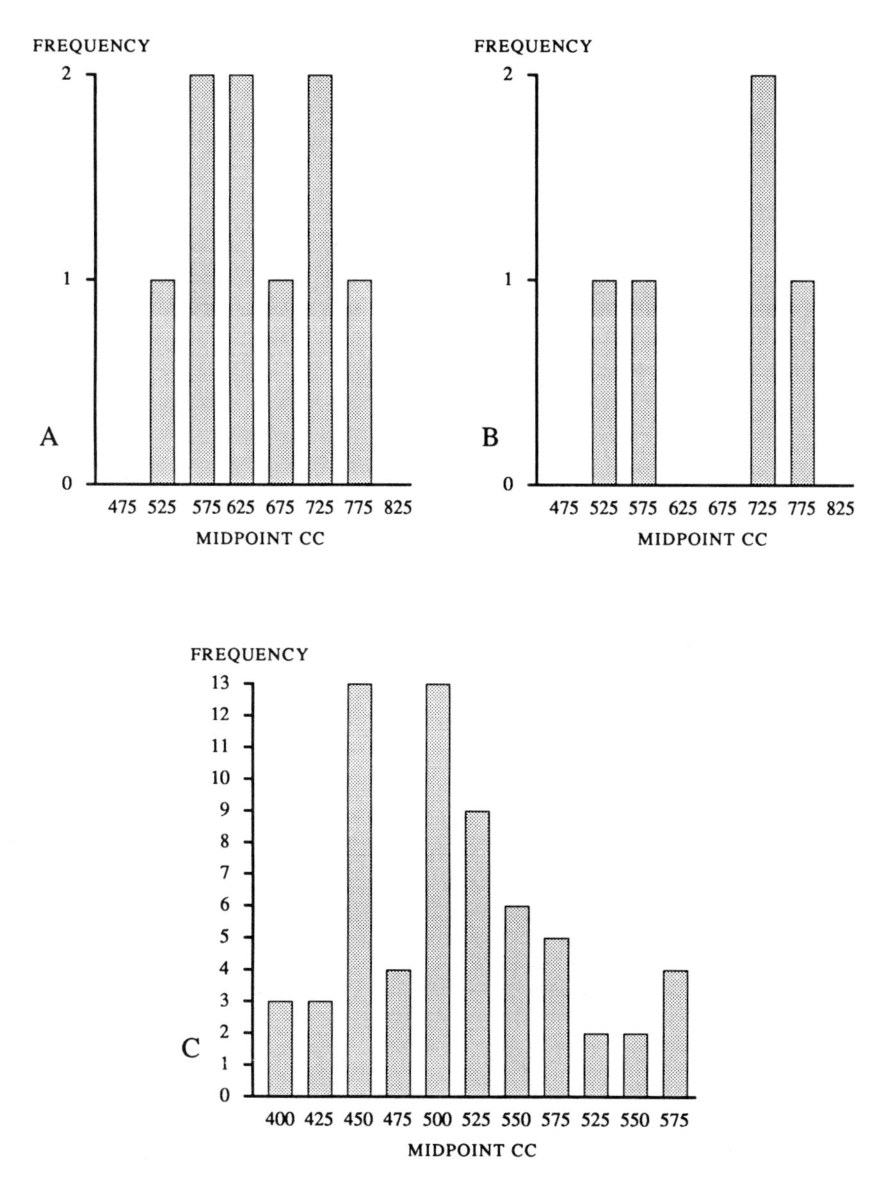

Fig. 3. Histograms of cranial capacity values within early *Homo* and an extant *Gorilla* sample. All volumes are expressed in cubic centimeters. (A) Combined regional early *Homo* hypodigm; (B) Koobi Fora early *Homo* hypodigm; (C) *Gorilla gorilla* (M = 34; F = 30).

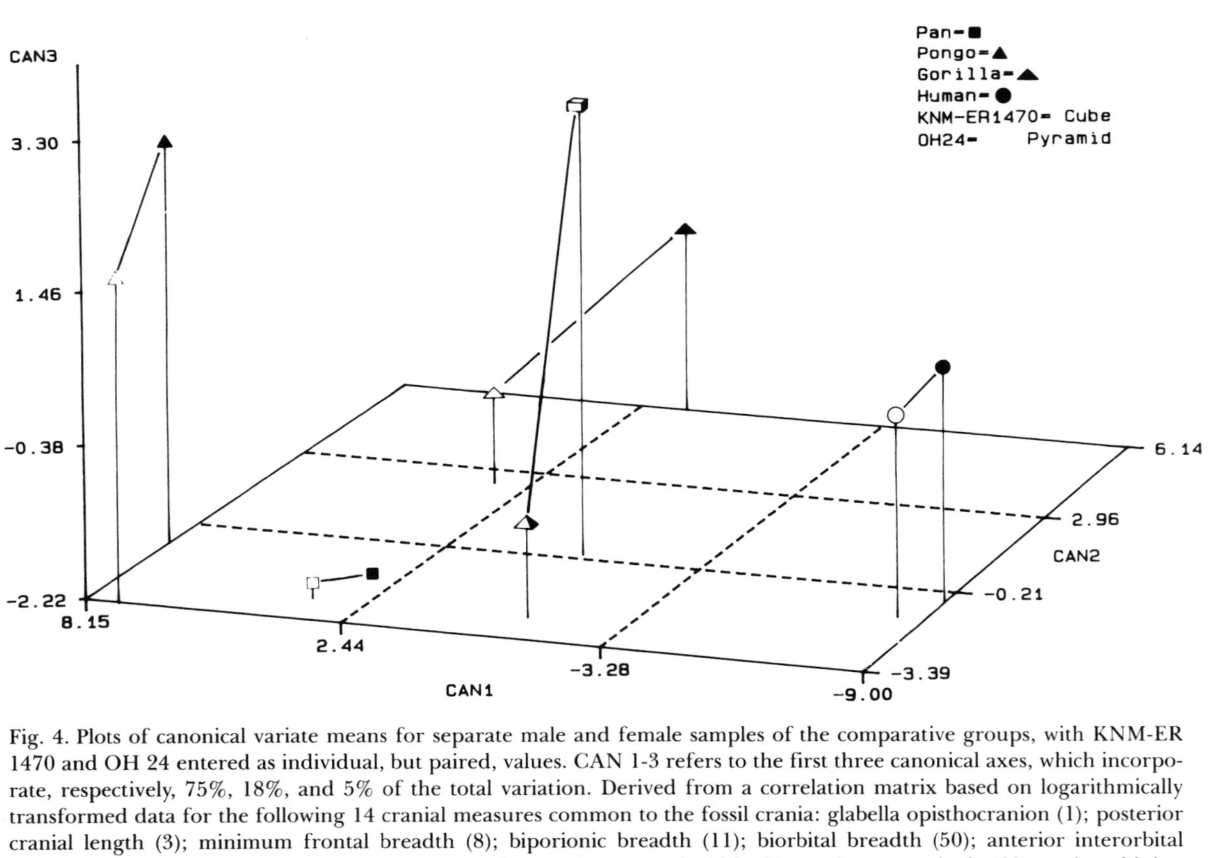

Fig. 4. Plots of canonical variate means for separate male and female samples of the comparative groups, with KNM-ER 1470 and OH 24 entered as individual, but paired, values. CAN 1-3 refers to the first three canonical axes, which incorporate, respectively, 75%, 18%, and 5% of the total variation. Derived from a correlation matrix based on logarithmically transformed data for the following 14 cranial measures common to the fossil crania: glabella opisthocranion (1); posterior cranial length (3); minimum frontal breadth (8); biporionic breadth (11); biorbital breadth (50); anterior interorbital breadth (55); orbital breadth (56); orbital height (57); maximum nasal width (68); nasion-nasospinale (69); nasion-rhinion (C) (71); inner palatal breadth (91); upper intercanine distance (98); palate depth (103). The numbers in parentheses refer to the measurement definitions that are given in Wood (1991b). Details of the comparative groups are given in Wood *et al.* (1991).

Table 12. Angles Subtended to the X, Y, and Z Axes by the Slopes Joining Male and Female Group Means and the Slopes Joining the Fosssil Early *Homo* Cranium KNM-ER 1470 to (A) OH24 and (B) KNM-ER 1813, Together with the Relevant Mahalanobis' Distances

	A				B			
	X	Y	Z	D	X	Y	Z	D
1.	23.9	−64.3	8.9	12	15.9	73.8	− 3.1	13
2.	−16.3	−60.0	−24.5	27	−16.7	−64.4	−18.9	27
3.	25.7	−58.1	−17.6	44	23.6	−62.0	−14.2	46
4.	−35.0	−42.1	−28.1	10	40.6	−42.8	−19.9	10
5.	− 6.3	−29.8	−59.4	51	−19.4	−64.0	−16.7	29

1. *Pan troglodytes verus* (males = 18; females = 33).
2. *Pongo pygmaeus pygmaeus* (males = 21; females = 22).
3. *Gorilla gorilla gorilla* (males = 34; females = 30).
4. *Homo sapiens* (males = 40; females = 35).
5. (A) KNM-ER 1470/OH 24; (B) KNM-ER 1470/1813.
D = Mahalanobis D^2 distance.

Two details are noteworthy. First, complex root forms are apparently confined to specimens from Koobi Fora, and second, within that group one specimen, KNM-ER 1802, has a premolar root size sequence that closely resembles that of the "robust" australopithecines (Wood *et al.,* 1988: Fig. 7). Root form shows little intraspecific variation in the higher primates, but the P_3s of *Pan* do provide some evidence of sexual dimorphism. While the incidence of single-rooted P_3s is 20% (n = 20) in males, nearly half, 46% (n = 11), of *Pan* females are single-rooted.

Table 13. Mandibular Premolar Root Form of Specimens of Early *Homo* from Koobi Fora and Olduvai

	P_3	P_4
Koobi Fora		
KNM-ER 819	2T	2R: M + D
KNM-ER 1482	2T	2R: M + D
KNM-ER 1483	1R	2T
KNM-ER 1501	1R	1R
KNM-ER 1801	2T	2R: M + D
KNM-ER 1802	?2R: M + D	?2R: M + D
KNM-ER 1805	?2R: MB + D(DL)	?
KNM-ER 3734	2T	2R: M + D
Olduvai Gorge		
OH 7	2T	1R
OH 13	1R	1R
OH 16	1R	?1R
OH 37	1R	?1R

T = Tome's root form; R = root number; M = mesial root; D = distal root; MB = mesiobuccal root; DL = distolingual root.

Table 14. Linear Measurements (Raw and Corrected) of Enamel Exposed on Naturally Fractured Early *Homo* Teeth

	LT			CT			OT		
	Raw	Cor1	Cor2	Raw	Cor1	Cor2	Raw	Cor1	Cor2
Molars									
KNM-ER 807 (A)	1.9	1.7	2.2						
KNM-ER 808 (G)	1.5	1.3	1.8				1.1	1.0	1.3
KNM-ER 809 (A)	1.2	1.1	1.4				0.8	0.7	0.9
KNM-ER 1482 (A)	1.6	1.3[a]	1.8[a]				1.7	1.4[a]	1.9[a]
(D)	1.8	1.6[a]	2.0				1.5	1.3	1.6[a]
KNM-ER 1483 (E)	1.7	1.5	2.0				1.7	1.5	2.0
KNM-ER 1802 (B)	1.7	1.4	1.8				2.0	1.6	2.2
KNM-ER 1805 (E)	2.1	1.9	2.5	1.7	1.5	2.0			
OH 6	1.6	1.4	1.9	1.9	1.7	2.2	1.6	1.4	1.9
OH 16									
(M_1)	1.9	1.7	2.2						
(M_2)	1.8	1.5	1.9						
N	11	11	11				7	7	7
Mean	1.7	1.5	2.0				1.5	1.3	1.7
SD	0.2	0.2	0.3				0.4	0.3	0.5
CV (C)	12.0	13.6	15.3				27.8	24.0	30.6
95% CL	6.2–17.8	7.0–20.2	7.9–22.7				8.3–47.3	7.4–40.6	8.8–52.4
Premolars									
KNM-ER 1482 (A)									
P_3	1.5	1.3[a]	1.6[a]				1.2	1.1[a]	1.3[a]
P_4	1.5	1.3[a]	1.6[a]	1.6	1.4[a]	1.8[a]	1.5	1.3[a]	1.6[a]
KNM-ER 1802 (D)	1.5	1.3	1.8	1.9	1.7	2.3			
KNM-ER 1805 (E)	2.1	1.9	2.5	1.7	1.5	2.1			
N	4	4	4	3	3	3			
Mean	1.7	1.5	1.9	1.7	1.5	2.1			
SD	0.3	0.3	0.4	0.2	0.2	0.3			
CV (C)	18.7	21.2	22.4	12.7	14.4	15.4			
95% CL	−3.1–40.5	−3.7–46.1	−4.0–48.8	−10.0–35.4	−11.4–40.2	−12.3–43.1			
Deciduous molars									
KNM-ER 808 (H)	0.5								

[a]Corrected values for EAFHOM (see Beynon and Wood, 1986).
For details of the correction factors, see Beynon and Wood (1986).

However, in the early *Homo* hypodigm it is the P_4 roots and not those of P_3 that display the excessive variability.

Enamel Morphology

Enamel thickness measurements and patterns of enamel growth have proved to be useful hominid taxonomic discriminators (Beynon and Wood, 1986, 1987; Grine and Martin, 1988; Conroy, 1991). Martin (1983, Table 4.13, p. 299) has provided "average enamel thickness" data from sectioned teeth. These data, calculated as the area between the enamel surface and the enamel–dentine junction, while they differ in detail from observations made on naturally fractured teeth, apparently provide similar measures of enamel thickness (Grine and Martin, 1988). The CV values of the area measure are generally high, ranging from 13 in *Pongo* to 24 in a sample of 13 modern *H. sapiens* teeth.

The enamel thicknesses of naturally fractured early *Homo* teeth are shown in Table 14. The thickness of the occlusal enamel, which best discriminates *Paranthropus* teeth, is particularly variable, with mean CV values greater than 28. However, Martin's (1983) data suggest that equivalent levels of variability are also seen in his small modern human comparative sample. Thus, while these enamel thickness values are suggestive of "thicker" and "thinner" enamel varieties of early *Homo,* they do not prove strong evidence for multiple early *Homo* species.

One, or More, Early Homo Taxa?

The case for taxonomic heterogeneity within early *Homo* is finely balanced. The only other published studies that have considered this range of material in as much detail are those of Tobias (1980b, 1985, 1988, 1991) and Chamberlain (1987). Tobias considered the distribution of a great many morphological features and concluded that the "total morphological pattern" of *H. habilis* did not justify any subdivision. Chamberlain (1987) adopted an explicitly objective, quantitative, approach and used size-corrected "shape" data from a maximum of 90 linear cranial and mandibular variables. A relatively simple distance measure, the mean character difference (Cain and Harrison, 1958), was used to establish parameters for intraspecific variation across four higher primates and three other extant catarrhine taxa. From this test it was proposed that there were empirical grounds for setting a general threshold for shape variation within a species at an MCD value of 0.7.

Chamberlain (1987) found that this value was regularly either matched, or exceeded, within the early *Homo* sample. Several other authors have used a much narrower range of morphological evidence to assess the taxonomic significance of variation within the same sample. For example, both Wood (1985) and Stringer (1986) reviewed endocranial volume measurements in early *Homo* and commented on how much more variable they were than those in extant samples, but see Miller (1991) for a contrary interpretation.

There is no clear-cut solution to the conundrum posed by the variation displayed by the early *Homo* hypodigm. However, the extent of the differences between the patterns of dental, gnathic, and facial variation within early *Homo* and those observed in the relevant comparators do suggest that the variability is more likely to be interspecific rather than intraspecific. Thus, on these grounds, hypothesis A (Fig. 1) is rejected.

Multiple Taxa—Geographical and Temporal Distributions

The most efficient way to pursue the implications of the conclusion that taxonomic heterogeneity does exist within early *Homo* is to examine the regional hypodigms to test if either, or both, show any evidence of degree, pattern, and character variation that is in excess of that expected in a single species.

Dental variation in the regional hypodigms is summarized in Tables 6 and 7. Whereas just two of the dental variables from the Olduvai sample (OLD) are more variable in degree than the comparative samples, with the exception of the P_3, the premolar crowns of the Omo basin hypodigm (OM) are excessively variable. Only the Omo basin hypodigm is large enough to justify calculating CV values for cranial and mandibular variables (Table 8), and several variables display more variation than was observed in the total sample. The degree of difference in facial form between KNM-ER 1470 and 1813 can be gauged in the multivariate sense using the Mahalanobis D^2 distances between the specimens, based on a set of 24 cranial variables (Bilsborough and Wood, 1988). The D^2 distance, 11.6, is comparable to that between early *H. erectus* and modern *H. sapiens,* or between the former and KNM-ER 732 (Bilsborough and Wood, 1988). However, D^2 distance values must be interpreted with some care, for in the same study KNM-ER 406 and 732, which are judged to be most probably intraspecific variants, are separated by no less than 14.4 D^2 units.

While the mandibular symphysis of the Omo basin mandibles shows relatively little variation (Table 8), the horizontal corpus at M_1 is excessively variable in both height and width. That this is due to greater variation in size rather than shape is evidence from the CV values of computed cross-sectional area and robusticity. The CV values of the former exceed that in the *P. boisei* hypodigm (Wood, 1991a), whereas the robusticity index values show much less variation.

Turning to the observed pattern of variation in the early *Homo* subset from Koobi Fora and the Shungura Formation, this is also indicative of the presence of more than one taxon. Taxonomic diversity is suggested because it is the mesiodistal diameters of the premolars that are excessively variable. Likewise, the suite of cranial variables showing high levels of variation includes orbital height and intercanine distance, both of which are apparently indicators of taxonomic and not intraspecific heterogeneity (Wood *et al.,* 1991). In addition, the cranial capacities of early *Homo* crania from Koobi Fora lie at the extremes of the range for the combined regional hypodigm (Fig. 3B).

The nature of the variables differentiating the facial morphology of KNM-ER 1470 and OH 24 has already been referred to in detail, and much the same interpretations can be placed upon the observed differences between KNM-ER

1470 and 1813. The two fossil crania are well-enough preserved to allow their relationship to be compared with the pattern of sex differences observed between closely related extant species (Wood *et al.*, 1991) (Fig. 5). The result is much like the comparison between KNM-ER 1470 and OH 24, for the slope of the axis of variation joining the KNM-ER 1470 and 1813 is orientated differently to the between-sex slopes for the extant *Homo, Pan,* and *Gorilla* samples (Table 12). Canonical analysis using logarithmically transformed values of the variables in common between the mandibles KNM-ER 1802 and 3734 (Table 15) results in additional evidence for taxonomic heterogeneity within the Koobi Fora subsample. The degree of morphological difference between the two fossil mandibles is more than twice that between the means of sexually dimorphic higher primates.

Turning to dental measures, the early *Homo* subsample from the Omo group deposits displays as great a range of crown shape, root form, or enamel thickness as does the combined regional sample. Mandibular posterior premolar crown size varies by a factor of two (KNM-ER 1802 and 3734), and while some Koobi Fora early *Homo* mandibles have a mandibular premolar root form that is as highly derived as that in modern *H. sapiens* (KNM-ER 1501), other specimens (KNM-ER 1802) apparently have a root morphology that resembles the derived condition otherwise only seen in *P. boiseï* (Wood, 1988). Size-corrected occlusal enamel thickness measured on the available, naturally fractured, surfaces of KNM-ER 809 and 1802 also varies by a factor of two.

Table 15. Angles Subtended to the X, Y, and Z Axes by the Slopes Joining Male and Female Group Means and the Slopes Joining the Early *Homo* Mandible KNM-ER 1802 to (A) 3734, and (B) OH13, and the Mandible OH 7 to (C) KNM-ER 3734 and (D) OH 13, Together with the Relevant Mahalanobis Distances

	A				B			
	X	Y	Z	D	X	Y	Z	D
1.	15.5	−74.2	− 3.1	10	− 2.3	−26.5	−63.3	8
2.	−45.1	−33.8	−25.8	18	−47.6	−19.8	−35.7	23
3.	−45.3	− 0.4	−44.7	16	−50.2	3.6	−39.6	23
4.	−49.6	−37.2	−13.5	9	−48.4	−32.2	−23.2	9
5.	−48.5	26.3	−29.5	43	−58.4	− 3.0	−31.4	40
	C				**D**			
	X	**Y**	**Z**	**D**	**X**	**Y**	**Z**	**D**
1.	− 1.8	30.7	−59.2	26	− 1.9	31.0	−59.0	27
2.	−29.1	27.9	−47.6	37	−29.1	28.6	−47.0	37
3.	−29.4	42.2	−33.7	35	−29.4	42.4	−33.5	35
4.	−50.1	−12.8	−37.0	9	−50.0	−12.1	−37.5	9
5.	−24.6	−65.4	0.5	25	−29.4	−38.6	−37.5	31

1. *Pan troglodytes verus* (males = 18; females = 33).
2. *Pongo pygmaeus pygmaeus* (males = 21; females = 22).
3. *Gorilla gorilla gorilla* (males = 34; females = 30).
4. *Homo sapiens* (males = 40; females = 35).
5. (A) KNM-ER 1802/3734; (B) KNM-ER 1802/OH13; (C) OH7/KNM-ER 3734; (D) OH7/OH 13.
D = Mahalanobis D^2 distance.

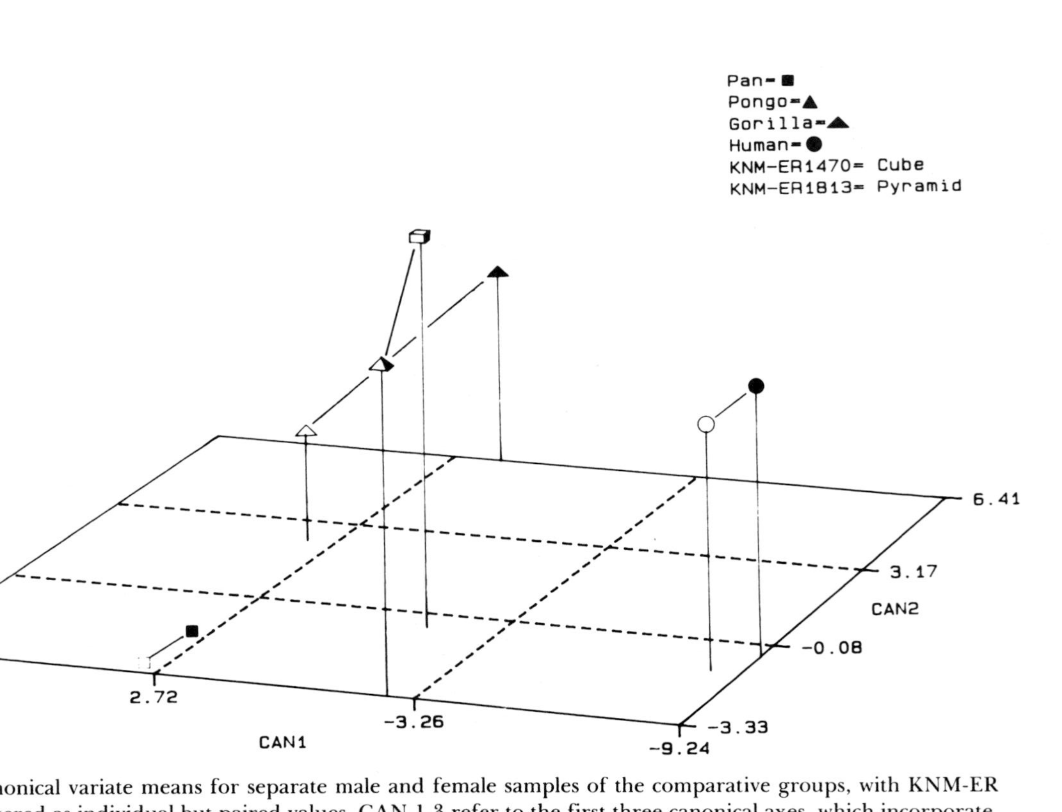

Fig. 5. Plots of canonical variate means for separate male and female samples of the comparative groups, with KNM-ER 1470 and 1813 entered as individual but paired values. CAN 1-3 refer to the first three canonical axes, which incorporate, respectively, 77%, 17%, and 5% of the total variation. Derived from a correlation matrix based on logarithmically transformed data for 16 cranial measures common to the fossil crania: glabella-opisthocranion (1); posterior cranial length (3); minimum frontal breadth (8); maximum parietal breadth (9); biporionic breadth (11); bisupramastoid breadth (12); opisthion-inion (C) (37); biorbital breadth (50); anterior interorbital breadth (55); orbital breadth (56); orbital height (57); maximum nasal width (68); nasion-nasospinale (69); inner palatal breadth (91); upper intercanine distance (98); palate depth (103). Details of the comparative groups are given in Wood *et al.* (1991).

With respect to the question of whether there is taxonomic heterogeneity in the Olduvai early *Homo* subsample, while there is some positive evidence for homogeneity in that subsample (e.g., cranial capacity and mandibular premolar root form are relatively invariant), regrettably, with the numbers of specimens from Olduvai generally being approximately half those from the Omo group of deposits, it is more often a case of absence of evidence.

The next step in the analysis is to establish which hypothesis—B, C, or D—applies to the Omo basin hypodigm. Are the multiple early *Homo* taxa it apparently samples synchronic or allochronic? Hypotheses B and C would be supported if there were temporal trends in the expression of a significant number of character states. Likewise, if there were enough specimens to test an early and a late time-band subsample, then neither would show evidence of taxonomic heterogeneity if hypotheses B and C are to be supported. The only two time bands for which data about variability are available, numbers 4 and 5, are sequential, so that hypothesis C cannot easily be distinguished from hypothesis D.

Turning first to the examination of the time-band data for evidence of any excessive degree of variation, while the maxillary dentition does provide some positive evidence for heterogeneity in time band 4, no such evidence is forthcoming from time band 5. So modest are the total numbers of cranial and mandibular remains that when they are split into the two time bands, only the latter material furnishes data sets of three or more specimens. Once again it is the size of the mandibular corpus that is excessively variable.

In terms of the pattern of variation, time band 4, no matter which way the intervals are registered (Tables 2 and 3), contains both KNM-ER 1470 and 1813, and it is the approximate synchronicity of these two crania that provides the strongest evidence for two taxa being represented in that time band. Likewise, the same time band includes mandibular premolars with simple (KNM-ER 1501) and complex (KNM-ER 1802) root systems (Table 13). These observations clearly betray in advance the likely result of an inquiry into whether there are temporal trends in the expression of character states. For, with the possible exception of reduced enamel thickness (KNM-ER 807 and 808), which is confined to time band 4 in one of the sample-division systems, the gamut of expressions of the other characters can be seen within time band 5.

Thus, the conclusions of this analysis, while agreeing in part with assessments made by others, notably about the taxonomic homogeneity of the early *Homo* hypodigm from Olduvai (Chamberlain, 1987; Tobias, 1989; Johanson, 1989), also suggest, *contra* the preceding authors, that the hypodigm from Koobi Fora most probably samples more than one, and probably two, species of early *Homo,* not counting *H. ergaster* (see below). The same data suggest that these two taxa are marginally more likely to be synchronic than time successive. The three most likely hypotheses to explain variation in early *Homo,* in reverse alphabetical order of probability, are summarized in Fig. 6.

Taxonomic Implications

The lack of compelling evidence for subdividing the Olduvai early *Homo* hypodigm simplifies the taxonomy, for the Olduvai hypodigm must, therefore,

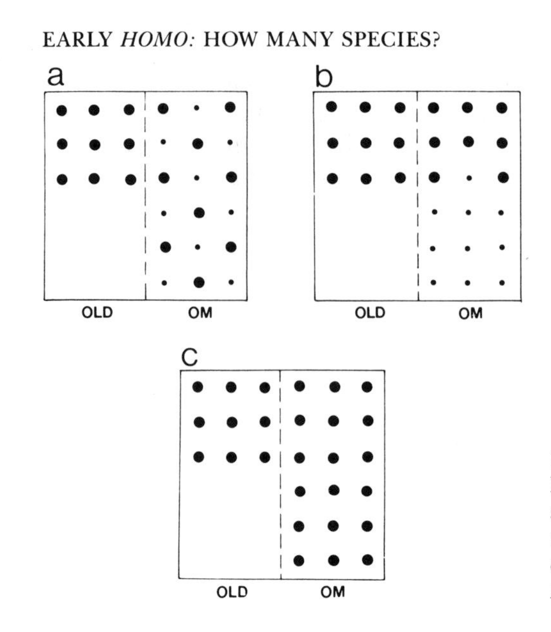

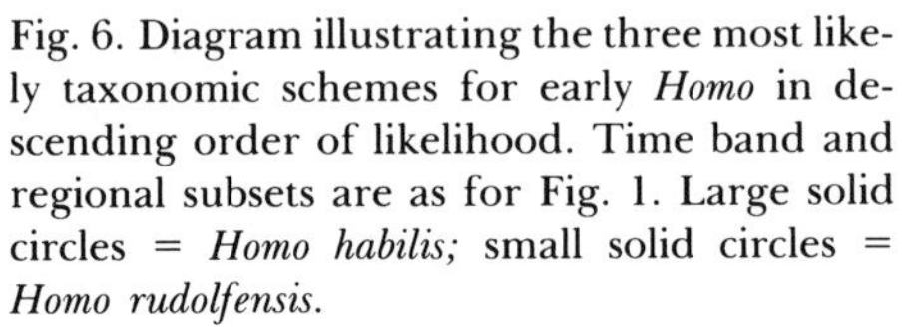

Fig. 6. Diagram illustrating the three most likely taxonomic schemes for early *Homo* in descending order of likelihood. Time band and regional subsets are as for Fig. 1. Large solid circles = *Homo habilis;* small solid circles = *Homo rudolfensis.*

belong to *H. habilis* (Leakey *et al.,* 1964). That being accepted, the next task is to examine the balance of the Omo region subsample, but particularly the remains from Koobi Fora, to see if they can be assigned to one, or the other, of the two taxa, one presumably represented by KNM-ER 1470 and the other by KNM-ER 1813 (see above).

The associated skull, KNM-ER 1805, is frustratingly enigmatic in some taxonomically important aspects of morphology, for example, the form of the mandibular premolar tooth roots. However, the shape of the reconstructed face suggests that the affinities of KNM-ER 1805 lie with KNM-ER 1813 rather than with KNM-ER 1470. The Mahalanobis D^2 data also suggest this, for the distance between KNM-ER 1805 and 1813, 6.5, is half that between KNM-ER 1805 and 1470 (Bilsborough and Wood, 1988).

The fragmented cranium KNM-ER 1478 provides a little diagnostic morphology, but the probable size of the M^2 crown, the lengths of the molar and premolar tooth rows, and the form of the temporal bone all point to links with a KNM-ER 1813 type of cranium.

The vault and dental remains of the juvenile KNM-ER 1590 provide several clues to its taxonomic affinities. The tooth crowns are all large and correspond more closely to the size expectations of what is presumed to be the more megadont of the two morphs, i.e., KNM-ER 1470 and its ilk, than to remains such as KNM-ER 1813. For a relatively immature individual (the range of estimated chronological age is 4.0–8.2 years), the parietals are impressively large, and this too would suggest that KNM-ER 1590 is best allocated to the early *Homo* species represented by KNM-ER 1470. The calotte and the partial face of KNM-ER 3732 allow useful comparisons to be made between it, KNM-ER 1470, and 1813. The size of the calotte, the width of the frontal bone, and the anterior inclination of the malar region all point to affinities with KNM-ER 1470.

Despite its fragmentary nature, KNM-ER 3735 has sufficiently preserved morphology to assess its relationships with KNM-ER 1470 and 1813. The overall size and form of the temporal fragment (D) are not diagnostic, but the major sphenoid contribution to the entoglenoid process is closer to the morphology observed in KNM-ER 1813. Likewise, the vertically inclined malar process (G) and the form of the frontal fragment (N) are also suggestive of affinities with the smaller of the two reference crania.

The reconstructed, but weathered, palate of KNM-ER 3891 provides some diagnostic information. The evidence of a stout canine jugum and the anteriorly situated (P^4) "take-off" of the zygomatic process point to affinities with KNM-ER 1470. The form of the temporal fragment (A), for example, the substantial root of the zygomatic process and the large entoglenoid, is also compatible with an allocation to a KNM-ER 1470-like taxon, but the major sphenoid contribution to that process is a point of contrast with KNM-ER 1470. Nonetheless, the overall impression is that KNM-ER 3891 is more likely to be linked with the latter cranium.

Turning to the mandibular evidence, it is now almost traditional to regard KNM-ER 1470 and 1802 as conspecific. Evidence for that assumption is mainly based on functional inference, which interprets the long, orthognathic face of KNM-ER 1470 as being adapted to withstand heavy mastication stress. Likewise, the large cross-sectional area of the corpus of KNM-ER 1802 at M_1, the degree of talonid formation on the premolar crowns, and the complexity of the underlying roots are also interpreted as adaptations for heavy postcanine tooth use. Tenuous as this evidence is, and in the absence of associated mandibles, the grounds for linking KNM-ER 1470 and 1802 in the same species are as sound, and probably sounder, than those supporting any of the other combinations of crania and mandibles that may be postulated.

Extensive erosion of the horizontal ramus of KNM-ER 819 has reduced its overall size and most probably also has affected its shape, so that its allocation to early *Homo* was, in any case, somewhat tentative. If it does, however, as seems likely, belong in this group then its affinities are clearly with mandibles like KNM-ER 1802. It is similar in size and shape, sharing the latter's everted base, and the form of the premolar roots (P_3 = 2T; P_4 = 2R: M + D; Table 13) is also strongly in favor of an association with mandibles lacking the derived condition of mandibular premolar root reduction seen in later *Homo* taxa. Much the same arguments and conclusions apply to the better-preserved mandible, KNM-ER 1482. It too shows some superficial phenotypic resemblances to the mandibles of *P. boisei,* but the body, P_4 crowns, and premolar roots lack the extreme specializations associated with that taxon. These same features align the specimen firmly with the group epitomized by KNM-ER 1802.

The large size of the corpus, the robusticity of the horizontal ramus, and the form of the mandibular symphysis all suggest that KNM-ER 1483 is part of the hypodigm being gathered around KNM-ER 1802. However, unlike the other members of that group of specimens, the premolar roots of KNM-ER 1483 are relatively simple (P_3 = 1R; P_4 = 2T). Its inclusion in that group should, therefore, be made in the knowledge that it would widen the group's range of expression of premolar root form.

The affinities of the two adult mandible fragments, KNM-ER 1501 and the 1502, can sensibly be considered together. The small canine crown, the simplicity of its premolar root system, and the size of the corpus set KNM-ER 1501 apart from KNM-ER 1802. Grounds for the distinction of KNM-ER 1502 from the latter mandible include size, but also relate to the form of its M_1 crown. The affinities of KNM-ER 1501 and 1502 are with a hominid with a reduced premolar root system and smaller molar crowns than KNM-ER 1802. The best size match is with mandibles such as KNM-ER 1805.

The specimen KNM-ER 1506 comprises a fragment of the right side of a mandibular body with two molar tooth crowns, together with two isolated upper premolars. The preserved bone of the horizontal ramus is bulbous and robust, but the form of the tooth crowns, both upper and lower, is strongly suggestive of early *Homo*. Its most likely association is with the KNM-ER 1802-like group of mandibles, but the dearth of available evidence suggests that *Homo* sp. indet. may be a more prudent taxonomic verdict on this specimen.

The next specimen to be considered, KNM-ER 1801, is the left side of an adult mandibular corpus. The size of the corpus and the M_3 crown, the buttressing of the symphysis, and the form of the premolar roots (P_3 = 2T; P_4 = 2R: M + D) all suggest that KNM-ER 1801 bears much stronger resemblances to KNM-ER 1802 than to smaller mandibles with less complex root systems.

The mandible KNM-ER 3734, though more gracile than KNM-ER 1802, shares several features with it. Their symphyseal profiles are similar, as are the eversion of the base and the form of the premolar roots (P_3 = 2T; P_4 = 2R: M + D). There are, however, some features of KNM-ER 3734 that set it apart from KNM-ER 1802; these include the relatively small size of the premolar and molar tooth crowns, and the buccolingual narrowing of the molars and premolars. The ambiguity about the affinities of KNM-ER 3734 is reflected in the results of multivariate analysis comparing it with KNM-ER 1802 and OH7 (Table 15).

The isolated teeth KNM-ER 807, 808, and 809 are clearly *Homo* by virtue of their size, fissure pattern, and cusp morphology, but this evidence alone is insufficient for them to be resolved into one, or other, of the morphs that have been defined mainly on cranial and mandibular evidence. Their taxonomic attribution should be *Homo* sp. indet. The balance of the isolated teeth in the Koobi Fora early *Homo* sample and KNM-ER 1462, 1480, 1508, and 1814 offer insufficient information to allocate them at the species level; they too should be treated as *Homo* sp. indet.

Identification of Early Homo Taxa from Koobi Fora

The conclusion of this review of early *Homo* is that while there is evidence from the face, mandible, and mandibular postcanine dentition for taxonomic heterogeneity within the Omo region hypodigm, it is not apparent among the Olduvai subset. Inferences about functional correlations between the face and mandible have resulted in the proposal that the two species within the Koobi Fora early *Homo* hypodigm are exemplified by specimens KNM-ER 1470 and 1802, and KNM-ER 1813 and 1501, respectively. In the previous section, the balance of the Koobi Fora early *Homo* collection was reviewed in the light of this proposal, specifically to see if these specimens can also be sorted into the two proposed morphs. Eight of the 24 specimens comprising isolated teeth and mandible fragments, lack sufficient evidence to be assigned to one or the other morph, and it is proposed that their taxonomic attributions be confirmed as *Homo* sp. indet. Fifteen of the cranial and mandibular remains can be linked to one or the other morph, and just one specimen, the mandible KNM-ER 3734, appears to possess a combination of features, some of which are shared with KNM-ER 1802 and others of which link it with KNM-ER 1501. The characters that define and discriminate between the two morphs, such as facial shape, sphenoid expansion, mandible size, and robusticity, and the form of the mandibular postcanine dentition are summarized in Table 16.

The next task is to determine whether either of the two early *Homo* taxa represented at Koobi Fora can be identified as *H. habilis*. Phenetic and cladistic evidence both suggest that KNM-ER 1813 more closely resembles *H. habilis* than does KNM-ER 1470. The face of KNM-ER 1813 and *H. habilis* are only 6.5 Mahalanobis D^2 units apart (Bilsborough and Wood, 1988), but, in contrast, Stringer (1986, p. 287) has stressed the phenetic differences between them. The

Table 16. Morphological Features of the Cranium, Mandible, and Dentition of *Homo habilis* and *Homo rudolfensis*

	Homo habilis	*Homo rudolfensis*
Absolute brain size	X = 610 cc	X = 751 cc
Relative brain size	EQ approx 4	EQ approx 4
Overall cranial vault morphology	Enlarged occipital sagittal contribution	Primitive
Endocranial morphology	Primitive sulcal pattern (Falk, 1983, but see Holloway, 1985)	Frontal lobe asymmetry (Falk, 1983; Tobias, 1987)
Suture pattern	Complex	Simple
Frontal	Incipient supraorbital torus	Torus absent
Parietal	Coronal > sagittal chord	Primitive
Face—overall	Upper face > midface breadth	Midface > upper face breadth; markedly orthognathic
Nose	Margins sharp and everted; evident nasal sill	Less marginal eversion; no nasal sill
Malar surface	Vertical or near vertical	Anteriorly inclined
Palate	Foreshortened	Large
Upper teeth	Probably two-rooted premolars	Premolars three-rooted; absolutely and relatively large anterior teeth
Mandibular fossa	Relatively deep	Shallow
Foramen magnum	Orientation variable	Anteriorly inclined
Mandibular corpus	Moderate relief on external surface	Marked relief on external surface
	Rounded base	Everted base
Lower teeth	Buccolingually narrowed postcanine crowns	Broad postcanine crowns
	M_3 reduction	No M_3 reduction
	Reduced talonid on P_4	Relatively large P_4 talonid
	Mostly single rooted	Twin, plate-like, P_4 roots, and 2T, or even twin, plate-like P_3 roots

original description of *H. habilis* made no mention of cranial characters unique to that taxon, but the relative narrowness of the midface is a detailed feature shared by *H. habilis* and KNM-ER 1813. Leakey *et al.* (1989) have previously suggested that KNM-ER 3735 resembles *H. habilis,* and this study has stressed the affinities of the former specimen with KNM-ER 1813.

The mandibular hypodigm of *H. habilis* (e.g., OH 7, 13, 37, and 62) suggests a range of corpus size and shape that overlaps that of KNM-ER 1501 and 1802, but the size of 1802-like mandibles extends well above the range of *H. habilis.* With regard to evidence from the tooth crowns, while it is true that the mandibular premolars and molars of *H. habilis* are buccolingually narrow relative to australopithecines, the ranges of P_3, P_4, and M_1 crown shape indices of *H. habilis* embrace the index values of KNM-ER 1502 and 1802 (Wood, 1991b). The root form of the premolars and molars are more useful taxonomically, and they suggest strong links between the derived root reduction seen in *H. habilis* and the root form in equivalent mandibles such as KNM-ER 1501 and 1805. In summary, most evidence suggest that the smaller of the cranial and mandibular morphs at Koobi Fora corresponds with *H. habilis.* Thus, it is suggested that six of the Koobi Fora early *Homo* specimens (Table 17) should be included within *H. habilis.*

The identity of the second early *Homo* taxon at Koobi Fora is a more complex problem. It is apparently not, as some have hinted, *H. ergaster.* That species name is indissolubly linked with KNM-ER 992, and if *H. ergaster* is a good taxon at all, proposals put forward elsewhere would suggest that it should either be the species name for *H. erectus*-like specimens, such as KNM-ER 3733, 3883 etc., or sunk into *H. erectus;* further details are given in Wood (1991b). The cranium KNM-ER 1470 has already been proposed as the type specimen of a new species, *H. rudolfensis,* by Alexeev (1986); the suggestion has since been supported by Groves (1989). If *H. rudolfensis* is a "good" species, and it seems to be in that it was promulgated correctly with respect to the ICZN, then the second early *Homo* taxon at Koobi Fora is *H. rudolfensis* Alexeev, 1986.

Postcranial Evidence

The postcranial remains of *H. habilis* have been surrounded by as much, if not more, controversy than the cranial evidence. The first accounts (Leakey *et*

Table 17. List of Early *Homo* Cranial Specimens from Olduvai and Koobi Fora Allocated to *Homo habilis, Homo rudolfensis,* and *Homo* sp. indet.

	Homo habilis	*Homo rudolfensis*	*Homo* sp. indet.
Skulls and crania	KNM-ER 1478, 1805, 1813, 3735 OH7, 13, 16, 24, 52, 62	KNM-ER 1470, 1590, 3732, 3891	—
Mandibles	KNM-ER 1501, 1502 OH 37	KNM-ER 819, 1482, 1483, 1801, 1802	KNM-ER 1506, 3734
Isolated teeth	OH 4, 6, 15, 21, 27, 39, 40, 41, 42, 44, 45	—	KNM-ER 807, 808, 809, 1462, 1480, 1508, 1814

al., 1964; Day and Napier, 1964) tended to emphasize the resemblance between these remains and modern *H. sapiens*, but subsequent studies have generally stressed the mosaic nature of this material. The pedal remains, in particular, have been reported to show a mixture of primitive and derived features (Wood, 1974; Lewis, 1989). The school of interpretation that stressed the modernity of the postcranial skeleton of *H. habilis* could always claim that the more primitive looking of the Olduvai postcranial fossils belonged not to *H. habilis* but to *P. boisei*. However, the recent discovery of a partial skeleton, OH 62, from Bed I at Olduvai, has provided very firm evidence that the cranium of a "gracile" hominid is associated with postcranial bones that are apparently primitive in both detailed morphology and overall proportions (Johanson *et al.*, 1987; Johanson, 1989).

Turning to Koobi Fora, the evidently *Homo*-like postcranial remains, such as the femora KNM-ER 1472 and 1481, which are apparently contemporary with crania such as KNM-ER 1470, differ in both size and shape from specimens such as OH 62 (Wood, 1987). The differences in the anatomy of the femur head and neck are such that OH 62 resembles specimens attributed to *Paranthropus* from Swartkrans, whereas KNM-ER 1472 and 1481 are sufficiently derived in the direction of *Homo* that some workers have attributed them to *H. erectus* (Kennedy, 1983). It could be argued that the shape differences in proximal femoral anatomy are a consequence of allometry, and thus likely to be of reduced taxonomic significance, but such allometric trends as have been observed within collections of modern *H. sapiens* femora do not support this explanation (Wood and Wilson, 1986).

Thus, the range of femoral and other postcranial morphology in early *Homo* is also suggestive of multiple taxa. There is little evidence to suggest that *H. habilis* postcranial remains from Olduvai are excessively variable, but there is evidence that the postcranial remains of the two posited early *Homo* taxa sampled at Koobi Fora may be significantly different. Leakey *et al.* (1989) have also suggested that there may be taxonomic heterogeneity within the early *Homo* postcranial remains from Koobi Fora, and they propose that one of the two taxa, the one represented by KNM-ER 3735, is conspecific with OH 62. This association would link the more primitive of the two postcranial morphs with *H. habilis*. Thus, no current interpretation of the relevant postcranial remains runs counter to the taxonomic proposals made on the basis of the cranial evidence.

Summary

East African remains that have been allocated to or compared with *H. habilis* have been investigated to test whether the hypodigm shows excessive variation. While there is only weak evidence that the degree of variation within the sample is greater than that predicted from extant analogues, evidence from several anatomical complexes suggest that the pattern of variation within early *Homo* is more like that observed between species than within them.

Examination of two regional subsamples, one from Olduvai and the other made up of the combined samples from the Omo and Koobi Fora, suggests that while there is no evidence that the Olduvai subsample is excessively variable, the collection from the Omo basin does demonstrate evidence of interspecific variation. Early *Homo* remains from Koobi Fora can be resolved into two taxa, *H. habilis* and *H. rudolfensis* Alexeev, 1986. The former includes the crania KNM-ER 1805 and 1813, and the mandible KNM-ER 1501; the latter embraces the cranium, KNM-ER 1470 and the mandible KNM-ER 1802. In addition to differences in overall size, the two early *Homo* taxa differ in features of the cranial vault, face, mandible, and mandibular dentition.

Conclusions based on the cranial evidence are mirrored by taxonomic inferences drawn from postcranial remains. These suggest that the limb skeleton of *H. habilis* is relatively primitive, resembling that of *Australopithecus* and *Paranthropus,* whereas the postcranial remains believed to be representative of *H. rudolfensis* are apparently derived in the manner of *H. erectus.*

ACKNOWLEDGMENTS

The author is grateful to Richard Leakey and the Trustees of the National Museums of Kenya for the opportunity to undertake a detailed study of the Koobi Fora cranial remains, and to the curators and custodians of the fossil hominid remains that were examined during the course of this research. Craig Engleman, Alan Turner, Andrew Chamberlain, Yu Li, and Paula Guest all contributed to either the research or the preparation of this chapter. The long-term support of The Natural Environment Research Council and The Leverhulme Trust is gratefully acknowledged.

References

Abbott, S. A. 1984. A Comparative Study of Tooth Root Morphology in the Great Apes, Modern Man and Early Hominids. Ph.D. Thesis, University of London.

Alexeev, V. P. 1986. *The Origin of the Human Race.* Progress Publishers, Moscow.

Beynon, A. D., and Wood, B. A. 1986. Variations in enamel thickness and structure in East African hominids. *Am. J. Phys. Anthropol.* **70:**177–193.

Beynon, A. D., and Wood, B. A. 1987. Patterns and rates of enamel growth in the molar teeth of early hominids. *Nature* **326:**493–496.

Bilsborough, A. and Wood, B. A. 1988. Cranial morphometry of early hominids I. Facial region. *Am. J. Phys. Anthropol.* **76:**61–86.

Boaz, N. T. and Howell, F. C. 1977. A gracile hominid cranium from Upper Member G of the Shungura Formation, Ethiopia. *Am. J. Phys. Anthropol.* **46:**93–108.

Brain, C. R., Churcher, C. S., Clark, J. D., Grine, F. E., Shipman, P., Susman, R. L., Turner, A., and Watson, V. 1988. New evidence of early hominids, their culture and environment from the Swartkrans Cave, South Africa. *S. Afr. J. Sci.* **84:**828–835.

Caccone, A., and Powell, J. R. 1989. DNA divergence among hominoids. *Evolution* **43:**925–942.

Cain, A. J. 1954. *Animal Species and Their Evolution.* London, Hutchinson.

Cain, A. J., and Harrison, G. A. 1958. An analysis of the taxonomists' judgement of affinity. *Proc. Zool. Soc. Lond.* **131:**85–98.

Campbell, B. 1962. The systematics of man. *Nature* **194:**225–232.

Campbell, B. G. 1972. Conceptual progress in physical anthropology: fossil man. *Ann. Rev. Anthropol.* **1:**27–54.

Chamberlain, A. T. 1987. A Taxonomic Review and Phylogenetic Analysis of *Homo habilis.* Ph.D. Thesis, The University of Liverpool.

Chamberlain, A. T. 1989. Variations within *Homo habilis,* in: G. Giacobini (ed.), *Hominidae,* pp. 175–181. Jaca Books, Milan.

Clarke, R. J. 1977. A juvenile cranium and some adult teeth of early *Homo* from Swartkrans, Transvaal. *S. Afr. J. Sci.* **73:**46–49.

Clarke, R. J. 1985. *Australopithecus* and early *Homo* in Southern Africa, in: E. Delson (ed.), *Ancestors: The Hard Evidence,* pp. 171–177. Alan R. Liss, New York.

Clarke, R. J., and Howell, F. C. 1972. Affinities of the Swartkrans 847 hominid cranium. *Am. J. Phys. Anthropol.* **37:**319–336.

Conroy, G. C. (1991). Enamel thickness in South African australopithecines: non-invasive determination by computed tomography. *Pal. Africana* **28:**53–59.

Coppens, Y. 1980. The differences between *Australopithecus* and *Homo;* preliminary conclusions from the Omo Research Expedition's studies, in: L.-K. Konigsson (ed.), *Current Argument on Early Man,* pp. 207–225. Pergamon, Oxford.

Cronin, J. E., Boaz, N. T., Stringer, C. B., and Rak, Y. 1981. Tempo and mode in hominid evolution. *Nature* **292:**113–122.

de Heinzelin, J. (ed.) 1983. The Omo Group. Stratigraphic and related earth sciences studies in the Lower Omo Basin, Southern Ethiopia. Musee Royal de L'Afrique Central, Tervuren, Belgique. Annales, **8** *Sciences Geologues, No. 85,* pp. 1–365.

Day, M. H. and Napier, J. R. 1964. Fossil foot bones. *Nature* **201:**969–970.

Falk, D. 1983. Cerebral cortices of East African early hominids. *Science* **221:**1072–1074.

Feibel, C. S., Brown, F. H., and McDougall, I. 1989. Stratigraphic context of fossil hominids from the Omo Group deposits: Northern Turkana Basin, Kenya and Ethiopia. *Am. J. Phys. Anthropol.* **78:**595–622.

Gans, D. J. 1984. The search for significance: different tests on the same data. *J. Stat. Comput. Simulation* **19:**1–21.

George, T. N. 1956. Biospecies, chronospecies and morphospecies, in: P. C. Sylvester-Bradley (ed.), *The Species Concept in Paleontology,* pp. 123–137. The Systematics Association, London.

Gingerich, P. D., and Schoeninger, M. J. 1979. Patterns of tooth size variability in the dentition of primates. *Am. J. Phys. Anthropol.* **51:**457–465.

Gould, S. J., and Eldredge, N. 1977. Punctuated equilibria: the tempo and mode of evolution considered. *Paleobiology* **3:**115–151.

Grine, F. E., and Martin, L. B. 1988. Enamel thickness and development in *Australopithecus* and *Paranthropus,* in: F. E. Grine (ed.), *Evolutionary History of the "Robust" Australopithecines,* pp. 3–42. Aldine de Gruyter, New York.

Groves, C. P. 1989. *A Theory of Human and Primate Evolution.* Clarendon Press, Oxford.

Groves, C. P., and Mazak, V. 1975. An approach to the taxonomy of the Hominidae: Gracile Villafranchian hominids of Africa. *Cas. Miner. Geol.* **20:**225–247.

Hartman, S. E. 1988. A cladistic analysis of hominoid molars. *J. Hum. Evol.* **17:**489–502.

Holloway, R. L. 1985. The past, present, and future significance of the lunate sulcus in early hominid evolution, in: P. V. Tobias (ed.), *Hominid Evolution,* pp. 47–62. Alan R. Liss, New York.

Howell, F. C. 1978. Hominidae, in: V. J. Maglio and H. B. S. Cooke (eds.), *Evolution of African Mammals,* pp. 154–248. Harvard University Press, Cambridge.

Hughes, A. R., and Tobias, P. V. 1977. A fossil skull probably of the genus *Homo* from Sterkfontein, Transvaal. *Nature* **265:**310–312.

Johanson, D. C. 1989. A partial *Homo habilis* skeleton from Olduvai Gorge, Tanzania: a summary of preliminary results, in: G. Giacobini (ed.), *Hominidae: Proceedings of the Second International Congress of Human Paleontology,* pp. 155–166. Jaca Book, Milan.

Johanson, D. C., Masao, F. T., Eck, G. G., White, T. D., Walker, R. C., Kimbel, W. H., Asfaw, B., Manega, P., Ndessokia, P., and Suwa, G. 1987. New partial skeleton of *Homo habilis* from Olduvai Gorge, Tanzania. *Nature* **327:**205–209.

Kennedy, G. E. 1983. A morphometric and taxonomic assessment of a hominine femur from the Lower Member, Koobi Fora, Lake Turkana. *Am. J. Phys. Anthropol.* **61:**429–434.

Kennedy, K. A. R. 1990. Narmada Man fossil skull from India: dating, morphology, taxonomy. *Am. J. Phys. Anthropol.* **81:**248–249.

Leakey, L. S. B. 1966. *Homo habilis, Homo erectus* and the australopithecines. *Nature* **209:**1279–1281.

Leakey, L. S. B., Tobias, P. V., and Napier, J. R. 1964. A new species of the genus *Homo* from Olduvai Gorge. *Nature* **202:**7–9.

Leakey, M. D., Clarke, R. J., and Leakey, L. S. B. 1971. New hominid skull from Bed I, Olduvai Gorge, Tanzania. *Nature* **232:**308–312.

Leakey, R. E. F. 1974. Further evidence of Lower Pleistocene hominids from East Rudolf, North Kenya, 1973. *Nature* **248:**653–656.

Leakey, R. E. F., and Walker, A. 1980. On the status of *Australopithecus afarensis. Science* **207:**1103.

Leakey, R. E., Walker, A., Ward, C. V., and Grausz, H. M. 1989. A partial skeleton of a gracile hominid from the Upper Burgi Member of the Koobi Fora Formation, East Lake Turkana, Kenya. G. Giacobini (ed.), *Proc. 2nd. Int. Cong. Human Paleontology,* pp. 167–173. Jaca Book, Milan.

Lewis, O. J. 1989. *Functional Morphology of the Evolving Hand and Foot.* Clarendon Press, Oxford.

Lieberman, D. E., Pilbeam, D. R., and Wood, B. A. 1988. A probabilistic approach to the problem of sexual dimorphism in *Homo habilis:* a comparison of KNM-ER 1470 and KNM-ER 1813. *J. Hum. Evol* **17:**503–511.

Luboga, S. A. 1986. Morphometric Variation in the Cranial Base and Facial Skeleton of Higher Primates with Special Reference to Modern Humans. Ph.D. thesis, Makerere University.

Martin, L. B. 1983. The Relationships of the Later Miocene Hominoidea. Ph.D. thesis, University of London.

Martyn, J., and Tobias, P. V. 1967. Pleistocene deposits and new fossil localities in Kenya. *Nature* **215:**479–480.

Mayr, E., Linsley, E. G., and Usinger, R. L. 1953. *Methods and Principles of Systematic Zoology.* McGraw-Hill, New York.

Miller, J. A. 1991. Does brain size variability provide evidence of multiple species in *Homo habilis? Am. J. Phys. Anthropol.* **84:**385–398.

Miyamoto, M. M., Slightom, J. L., and Goodman, M. 1987. Phylogenetic relations of humans and African apes from DNA sequences in the un-globin gene. *Science* **238:**369–373.

Miyamoto, M. M., Koop, B. F., Slightom, J. L., Goodman, M., and Tennant, M. R. 1988. Molecular systematics of higher primates: geneological relationships and classification. *Proc. Natl. Acad. Sci. USA* **85:**7627–7631.

O'Higgins, P. 1989. A Morphometric Study of Cranial Shape in the Hominidea. Ph.D. Thesis, University of London.

O'Higgins, P., Moore, W. J., Johnson, D. R., McAndrew, T. J., and Flinn, R. M. 1989. Patterns of cranial sexual dimorphism in certain groups of extant hominoids. *J. Zool. Lond.* **222:**353–362.

Olson, T. R. 1978. Hominid phylogenetics and the existence of *Homo* in Member 1 of the Swartkrans Formation, South Africa. *J. Hum. Evol.* **7:**159–178.

Oxnard, C. 1985. Hominids and hominoids, lineages and radiations, in: P. V. Tobias (ed.), *Hominid Evolution: Past, Present and Future,* pp. 271–278. Alan R. Liss, New York.

Oxnard, C. E. 1987. *Fossils, Teeth and Sex. New Perspectives on Human Evolution.* University of Washington Press, Seattle.

Oxnard, C. 1988. Fossils, teeth and sex: new perspectives in human evolution. *Proc. Austr. Soc. Hum. Biol.* **1:**23–73.

Pilbeam, D. R. 1969. Tertiary Pongidae of East Africa: evolutionary relationships and taxonomy. *Bull. Peabody Mus. Nat. Hist.* **31:**1–185.

Pilbeam, D. 1972. *The Ascent of Man. An Introduction to Human Evolution.* Collier Macmillan, London.

Pilbeam, D. 1978. Recognizing specific diversity in heterogeneous fossil samples, in: C. Jolly (ed.), *Early Hominids of Africa,* pp. 505–515. Duckworth, London.

Pilbeam, D. and Vaisnys, J. R. 1975. Hypothesis testing in paleoanthropology, in R. H. Tuttle (ed.), *Paleoanthropology, Morphology and Paleoecology,* pp. 3–13. Mouton, The Hague.

Robinson, J. T. 1965. *Homo 'habilis'* and the australopithecines. *Nature* **205:**121–124.

Schultz, B. B. 1985. Levene's test for relative variation. *Syst. Zool.* **34:**449–456.

Sibley, C. G., and Ahlquist, J. E. 1984. The phylogeny of the hominoid primates, as indicated by DNA-DNA hybridization. *J. Mol. Evol.* **20:**2–15.

Sibley, C. G., and Ahlquist, J. E. 1987. DNA hybridization evidence of hominoid phylogeny: Results from an expanded data set. *J. Mol. Evol.* **26:**99–121.

Simpson, G. G. 1961. *Principles of Animal Taxonomy.* Columbia University Press, New York.

Sokal, R. R., and Braumann, C. A. 1980. Significance tests for coefficients of variation and variability profiles. *Syst. Zool.* **29:**50–66.

Stringer, C. B. 1986. The credibility of *Homo habilis,* in: B. Wood, L. Martin, and P. Andrews (eds.), *Major Topics in Primate and Human Evolution,* pp. 266–294. Cambridge University Press, Cambridge.

Stringer, C. B. 1987. A numerical cladistic analysis for the genus *Homo. J. Hum. Evol.* **16:**135–146.

Sylvester- Bradley, P. C. (ed.) 1956. *The Species Concept in Palaeontology.* The Systematics Association, London.

Tobias, P. V. 1966. Fossil hominid remains from Ubeidiya, Israel. *Nature* **211:**130–133.

Tobias, P. V. 1972. "Dished faces," brain size and early hominids. *Nature* **239:**468–469.

Tobias, P. V. 1980a. The natural history of the helicoidal occlusal plane and its evolution in early *Homo. Am. J. Phys. Anthropol.* **53:**173–187.

Tobias, P. V. 1980b. A survey and synthesis of the African hominids of the late Tertiary and early Quaternary periods, in: L.-K. Konigsson (ed.), *Current Argument on Early Man,* pp. 86–113, Pergamon, Oxford.

Tobias, P. V. 1985. Single characters and the total morphological pattern redefined: the sorting effected by a selection of morphological features of the early hominids, in: E. Delson (ed.), *Ancestors: The Hard Evidence,* pp. 94–101. Alan R. Liss, New York.

Tobias, P. V. 1987. The brain of *Homo habilis:* a new level of organization in cerebral evolution. *J. Hum. Evol.* **16:**741–761.

Tobias, P. V. 1988. Numerous apparently synapomorphic features in *Australopithecus robustus, Australopithecus boisei* and *Homo habilis:* support for the Skelton-McHenry-Drawhorn hypothesis, in: F. E. Grine (ed.), *Evolutionary History of the "Robust" Australopithecines,* pp. 293–308. Aldine de Gruyter, New York.

Tobias, P. V. 1989. The status of *Homo habilis* in 1987 and some outstanding problems, in: G. Giacobini (ed.), *Proc. 2nd Int. Congress of Human Paleontology,* pp. 141–149. Jaca Book, Milan.

Tobias, P. V. 1991. *Olduvai Gorge, Volume IV. The Skulls, Endocasts and Teeth of Homo Habilis.* Cambridge University Press, Cambridge.

Tobias, P. V., and von Koenigswald, G. H. R. 1964. A comparison between the Olduvai hominids and those of Java and some implications for hominid phylogeny. *Nature,* **204:**515–518.

Van Valen, L. 1978. The statistics of variation. *Evol. Theor.* **4:**33–43.

Walker, A. 1976. Remains attributable to *Australopithecus* in the East Rudolf succession, in: Y. Coppens, F. Clark Howell, G. L. Isaac and R. E. Leakey (eds.), *Earliest Man and Environments in the Lake Rudolf Basin,* pp. 484–489. Chicago University Press, Chicago.

Walker, A., and Leakey, R. E. F. 1978. The hominids of East Turkana. *Sci. Am.* **239(2):**44–56.

Weidenrich, F. 1946. Generic, specific and subspecific characters in human evolution. *Am. J. Phys. Anthropol.* **4:**413–431.

Wolpoff, M. H. 1969. Cranial capacity and taxonomy of Olduvai hominid 7. *Nature* **223:**182–183.

Wolpoff, M. H. 1978. Analogies and interpretation in palaeoanthropology, in: C. J. Jolly (ed.), *Early Hominids of Africa,* pp. 461–503. Duckworth, London.

Wood, B. A. 1974. Olduvai Bed I post-cranial fossils: a reassessment. *J. Hum. Evol.* **3:**373–378.

Wood, B. A. 1975. An Analysis of Sexual Dimorphism in Primates. Ph.D. thesis, University of London.

Wood, B. A. 1976. The nature and basis of sexual dimorphism in the primate skeleton. *J. Zool. Lond.* **180:**15–34.

Wood, B. A. 1985. Early *Homo* in Kenya, and its systematic relationships, in: E. Delson (ed.), *Ancestors: The Hard Evidence,* pp. 206–214. Alan R. Liss, New York.

Wood, B. A. 1987. Who is the 'real' *Homo habilis? Nature* **327:**187–188.

Wood, B. A. 1988. Are "robust" australopithecines a monophyletic group?, in: F. Grine (ed.), *Evolutionary History of the "Robust" Australopithecines,* pp. 269–284. Aldine de Gruyter, New York.

Wood, B. A. 1991a. A palaeontological model for determining the limits of early hominid taxonomic variability. *Palaeont. Afr.* **28:**69–75.

Wood, B. A. 1991b. *Koobi Fora Research Project Volume 4: Hominid Cranial Remains.* Clarendon Press, Oxford.

Wood, B. A., Abbott, S. A. 1983. Analysis of the dental morphology of Plio-Pleistocene hominids. I: Mandibular molars: crown area measurements and morphological traits. *J. Anat.* **136:**197–219.

Wood, B. A., and Uytterschaut, H. 1987. Analysis of the dental morphology of Plio-Pleistocene hominids. III. Mandibular premolar crowns. *J. Anat.* **154:**121–156.

Wood, B. A., and Xu, Q. 1991. Variation in the Lufeng dental remains. *J. Hum. Evol.* **20:**291–311.

Wood, B. A., and Abbott, S. A., and Graham, S. H. 1983. Analysis of the dental morphology of Plio-Pleistocene hominids. II: Mandibular molars—study of cusp areas, fissure pattern and cross-sectional shape of the crown. *J. Anat.* **137:**287–314.

Wood, B. A., and Wilson, G. B. 1986. Patterns of allometry in modern human femora, in: R. Singer and J. K. Lundy (eds.), *Variation, Culture and Evolution in African Populations,* pp. 101–108. Witwatersrand University Press, Johannesburg.

Wood, B. A., Abbott, S. A., and Uytterschaut, H. 1988. Analysis of the dental morphology of Plio-Pleistocene hominids. IV: Mandibular postcanine tooth morphology. *J. Anat.* **156:**107–139.

Wood, B. A., Yu, L., and Willoughby, C. 1991. Intraspecific variation and sexual dimorphism in cranial and dental variables among higher primates and their bearing on the hominids fossil record. *J. Anat.* **174:**185–205.

Zwell, M., and Pilbeam, D. R. 1972. The single species hypothesis, sexual dimorphism, and variability in early hominids. *Yb. Phys. Anthropol.* **16:**69–79.

Morphological Variation in *Homo neanderthalensis* and *Homo sapiens* in the Levant

20

A Biogeographic Model

YOEL RAK

Introduction

It was only the geographic proximity of the Mount Carmel specimens that spared them from being assigned to many different taxa as "separate form[s] of humanity." Were it not for these circumstances, McCown and Keith (1939) would not have conceived of accommodating such a great range of variation in one taxon. However, a clear morphological dichotomy between the hominids from Skhul and those from Tabun (C-1, the female skeleton, and C-2, the isolated mandible) emerges from their monograph. The existence of two kinds of hominids in a relatively small geographic area of the Middle East has since been confirmed through discoveries at several additional sites. Other hominids have been found, including specimens from Amud (Suzuki and Takai, 1970) and Shanidar (Trinkaus, 1983), which can be grouped comfortably with the Tabun specimens, whereas specimens uncovered at Qafzeh (Vandermeersch, 1981) can be added to the Skhul group. Until quite recently, McCown and Keith's basic contention—that the Neanderthal-looking Tabun group represented the earlier, primitive anatomy, and the modern-looking Skhul group represented the later, derived anatomy—was generally accepted.

YOEL RAK • Department of Anatomy, Sackler Faculty of Medicine, Tel Aviv University, Tel Aviv, Israel, and Institute of Human Origins, Berkeley, California 94709.

Species, Species Concepts, and Primate Evolution, edited by William H. Kimbel and Lawrence B. Martin. Plenum Press, New York, 1993.

The purpose of this chapter is threefold. The first is to demonstrate that the Neanderthals differ, on the specific level, from *Homo sapiens;* in other words, to show that from the perspective of the traditional morphological standards used in primate taxonomy, the magnitude of difference in the face and in the pelvis, justifies such a taxonomic decision. The second goal is to demonstrate that the phylogenetic relationship between *H. sapiens* and Neanderthals, which is based on their morphology and the emerging timetable, cannot be anagenetic. Even when Neanderthals are recognized as a separate species, the morphological and temporal factors do not allow us to consider the Neanderthal an ancestral species to *H. sapiens*, as did some investigators in the past who endorsed the Neanderthal's species status. The third goal of the chapter is to propose an explanation for the morphological variation characterizing the hominids of the Mousterian period in the Middle East.

Morphological and Temporal Considerations

Although in the past the Qafzeh specimens were claimed by some to be contemporary with or even older than the Neanderthals (much older than commonly thought at the time; Haas, 1972; Tchernov, 1984), there was a general reluctance to accept this possibility. At the root of this reluctance were the facts that the chronology was based on biostratigraphic considerations rather than absolute dates and, perhaps even more significantly, the difficulty that many investigators have in accepting a cladogenetic scenario for such a late segment of human evolution. Indeed, this combination of factors led McCown and Keith (1939) to argue with Garrod about the need to view the Tabun skeleton as chronologically earlier than the specimens from Skhul; and, many years later, it also led Trinkaus (1984, 258) to write as follows:

> All of the techniques that have been used to date the Qafzeh remains have limitations, and each is insufficient to confirm a date. However, the younger date would alleviate the need to account for how two distinct groups of *Homo sapiens* could have coexisted for several millennia in the small area of northern Israel, using the same cultural adaptive complex and yet remaining biologically distinct. For the purposes of this discussion, I will employ Occam's Razor, use the date for the Qafzeh remains that creates the *fewest problems*, and consider them to date to around 40 ky BP, about the same age as the Skhul layer B remains. [my emphasis]

In recent years, excavations were reopened at the Kebara Cave in Israel (Bar-Yosef *et al.*, 1986). As a result, the Neanderthal presence on Mount Carmel was reconfirmed through diagnostic elements of the skeleton, and the chronology and the stratigraphy of the Middle Paleolithic in Israel were reassessed with the aid of modern techniques of absolute dating (Valladas et al., 1987, 1988). In its general appearance and anatomical detail, the mandible of the skeleton discovered in Kebara in 1983 can easily be identified as that of a "classical" Neanderthal. Tillier *et al.* (1989) reached this conclusion and thus lumped this specimen with the other Neanderthal specimens of the Middle East (those from Amud, Shanidar, and Tabun). On the basis of the mandible's typical appearance, I am confident that the face of this individual, whose cranium is missing, was that of a "classical" Neanderthal.

Similarly, the specimen's pelvis exhibits the very long and slender superior pubic ramus found in other Neanderthals, in contrast to the short, thick ramus seen in modern humans and the Mousterian hominids from Qafzeh and Skhul (Rak, 1990; see also Trinkaus, 1976). The pelvis's superb state of preservation provides new information on Neanderthal pelvic anatomy and permits us to evaluate the functional significance of these differences. I believe that the anatomical differences between the Neanderthal pelvis and the modern pelvis indicate profound biomechanical differences.

The length of the superior pubis, about 90 mm, is greater by far than what we would expect, given the size of the inlet. Specimens of modern *H. sapiens* having an inlet size similar to that of the Kebara Neanderthal exhibit an average ramus length of only 70 mm! The ratio of ramus length to inlet width in the Neanderthal, whose index is 60.3, is much greater than that in modern humans, where the mean index for males is 55.3 (n = 43; SD = 3.1). The pelvic inlet of the Neanderthal is only 13% wider (transversely) than the average inlet of modern males, whereas the superior pubic ramus of the former is 31% longer than that of the latter. Clearly, it is not the size of the inlet, of which the ramus constitutes a major part, that dictates the length of the ramus. In an attempt to understand how to accommodate such a long ramus with a normal-sized inlet, I came to the conclusion that it is not the ramus length in itself that is significant but, rather, the distance from the acetabulum to the *symphysis pubis*. In other words, the long ramus of the Neanderthal stems from the fact that the acetabulum is located farther posterolaterally from the symphysis (along the circumference of the inlet rim) than in the modern male. Indeed, most of the anatomical differences that distinguish the Neanderthal pubis from that of modern humans stem from that shift in the position of the acetabulum.

The position of the acetabulum can be evaluated metrically through the use of an index that relates it to the anteroposterior dimension of the inlet. With the pelvis held such that the inlet rim is in a horizontal orientation, perpendiculars are dropped from the anterior edge of the acetabulum and from the anterior and posterior ends of the inlet at the midline. An index with a small value denotes a relatively large distance between the coronal plane of the acetabulum and that of the *symphysis pubis*, and a relatively small distance between the coronal plane of the acetabulum and the sacrum. The average index in modern males is 81.0 (n = 57; SD = 3.5), and in the Kebara Neanderthal, it is 61.5. In other words, the acetabulum in the modern male is located closer to the coronal plane of the symphysis and more anterior to the sacrum than in the Neanderthal. It is my contention (Rak, manuscript in preparation) that the forward migration of the acetabula in modern humans has made the pelvis into a specialized cushioning device that absorbs the weight of the cyclic dropping of the trunk during walking (Fig. 1). The discrepancy between the coronal planes of the acetabula and the sacrum prevents the falling center of mass from landing directly on and pounding against the femur heads. The forward position of the acetabula and the concomitantly short superior pubic ramus result in a short lever arm. The result is that the ramus is subjected to great bending moments. The thickened modern ramus can be viewed, therefore, as a necessary reinforcement of this

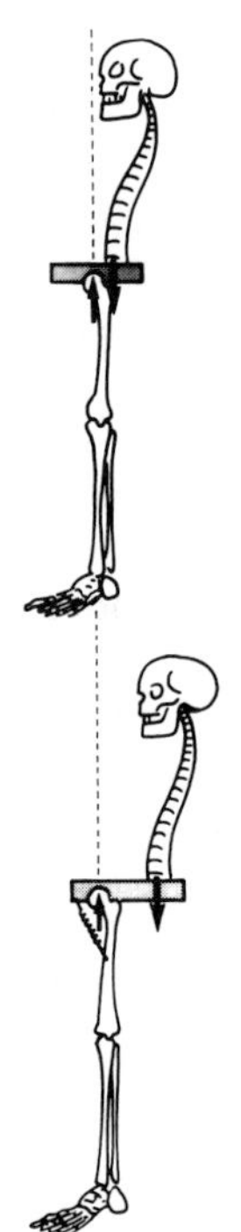

Fig. 1. A schematic view of two pelvic configurations depicted in an exaggerated manner. Upper: The primitive configuration as in the Neanderthal. Lower: The derived state, as in the modern human, where a relatively large distance is formed between the acetabula and the vertebral column.

short lever arm (Rak and Rosenberg, manuscript in preparation). As the significance of the length and thickness of the superior pubic ramus becomes clear, this element ceases to be merely an anatomical feature of vague functional value. Even its smallest fragments are of great importance because of their biomechanical implications.

As we cannot carry out the traditional practice of comparing these pelvises with specimens from an outgroup,* we are relying on biomechanic assessments to help us determine the polarity of the two bipedal pelvic configurations—the primitive configuration as opposed to the derived—and are thus, in essence, applying Ridley's line of thought exposed in his chapter entitled "The Functional Criterion" (1986). Ridley portrays the usefulness of functional criteria not in categorically determining polarity, but rather in increasing the level of confidence in a given phylogenetic decision. He (1986:136) writes:

> The functional criterion will not usually give certainty. It is virtually certain in the case of vestigal organs, but they are an exception. Usually, the method will not allow us to conclude anything more than that one direction of evolution is more probable that the other. No doubt the further study of any particular case might often make the conclusion more certain, but at any one moment we may have to live with uncertainty.

*There are no Middle Pleistocene pelvises complete enough for comparison, and although in some aspects the pelvic anatomy of *Australopithecus* is reminiscent of that found in *H. neanderthalensis*, I am reluctant to go so far as to view the Pliocene hominid condition as the ancestral morphotype for the Kebara and other Neanderthal pelvises.

On the basis of the morphological comparison and evaluation of the functional advantage of one form over the other, I believe that the Neanderthal pelvis still represents the primitive configuration, and the modern pelvis represents the derived character state. Indeed, in long-distance walking, the modern human pelvis appears to have the edge over the Neanderthal pelvis.

What I view as the primitive morphology, that of the Neanderthal pelvis, could easily have evolved through a relatively simple modification—the forward migration of the acetabula—into the derived form exhibited by the modern pelvis. However, the morphology of the Qafzeh pelvis, which is identical in every way to that of the modern pelvis (Rak, 1990), combined with the fossil specimen's geologic age (Valladas *et al.*, 1987), render unlikely the scenario of the modern pelvis having evolved from the pelvis of the known Neanderthals. In the evaluation of the relationship between the Qafzeh and the Kebara specimens, chronological and anatomical considerations impose a cladogenetic geometry.

A cladogenetic picture also emerges when the face of the Neanderthal is examined and compared with that of modern humans. In this case, however, the Neanderthal face represents the derived morphology. The Neanderthal face emerges as unique, deviating both morphologically and architecturally from the pattern seen in other hominid taxa (and in the primates in general; Rak, 1986). Howells (1975) concludes his metric analysis with the comment that the faces of *Homo erectus* and *H. sapiens* have more in common than either of them has with the Neanderthal face. Even a cursory examination of hominids, such as KNM-ER 3733 and SK 847, reveals a morphology more akin to *H. sapiens* than to the Neanderthal. From the point of view of phylogenetic analysis, it is much more economical to link *H. sapiens* directly to the hominid form represented by KNM-ER 3733 and SK 847 than to interpose the unique facial and cranial anatomy of the Neanderthal between them. If we accept this simpler phylogenetic hypothesis, then the discovery of hominids (Zuttiyeh, Skhul, and Qafzeh) that are contemporary with the Neanderthals and yet have a primitive that is, modern-looking, face should not come as a surprise and should even be expected.

This idea, of course, is not new and has come up in almost every discussion of the phylogenetic position of the Neanderthals since their discovery (see reviews by Bowler, 1986; Mann and Trinkaus, 1974; Spencer and Smith, 1981). Most notably, it was promoted by Boule (1911–1913), who influenced the views of many. Among the more modern investigators, Howells was the one who in 1942 came to the conclusion that the Neanderthals play no part in the evolutionary history of modern humans and actually constitute a separate species from *H. sapiens*. Weidenreich (1943, p. 39), arguing against Howells' hypothesis, said that it "proves that the view of the independency of Neanderthal Man is *still* advocated, in spite of all that has been written against it in recent years." With his famous diagram (reproduced here as Fig. 2), Weidenreich attempted to demonstrate the Neanderthal's intermediate position between *H. erectus* and *H. sapiens*. So eager was he to force the Neanderthal into a unilinear evolutionary scheme that he failed to distinguish between a morphocline with a defined polarity and the actual phylogeny itself. The advocates of the two slightly different theories—the "presapiens" and the "preneanderthal"—maintained, to varying degrees, that the Neanderthals, specifically the European Neanderthals, constituted a

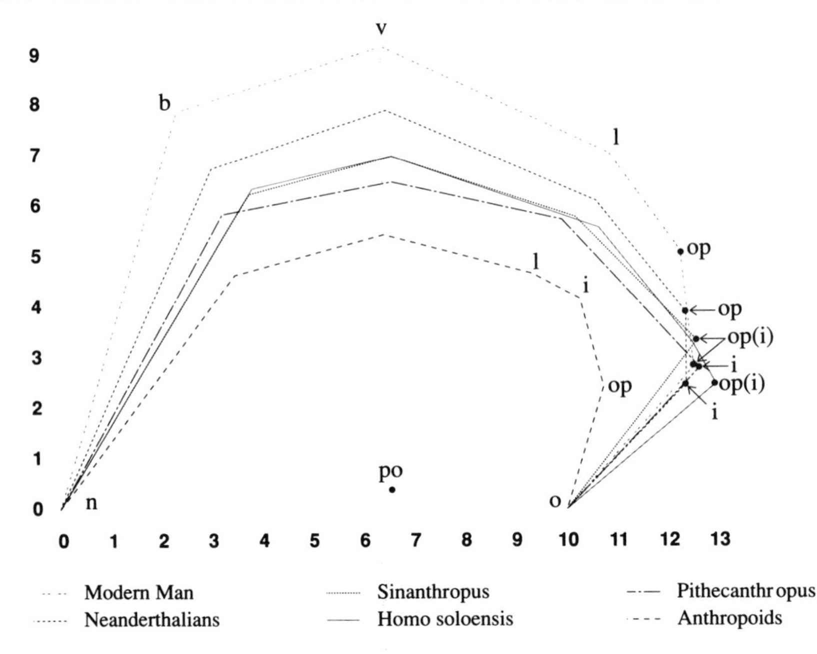

Fig. 2. Weidenreich's famous diagram showing "the gradual expansion of the braincase in the course of human evolution" (1943, p. 46). Weidenreich offered this diagram as support of his idea of unilinear evolution, in which Neanderthals precede *H. sapiens* phylogenetically. Apparently, he failed to distinguish between a morphocline and the actual phylogeny.

side branch of the main stem *leading* to modern humans (Howell, 1952; Vallois, 1954; Boule and Vallois, 1957).

The truth of the matter is that even among the more enthusiastic advocates of a unilinear model of human evolution (Weidenreich, 1947, 1949; Weinert, 1953), certain cladogenetic aspects are to be found that undoubtedly stem from the vague anatomical definition of the Neanderthal taxon itself. One can readily understand how Hrdlicka's definition of the Neanderthal—a definition still accepted by some scholars—might be an endless source of taxonomic and phylogenetic confusion. He writes that "the only workable definition of Neanderthal man and period seems to be, for the time being, the man and period of the Mousterian culture" (1927, p. 251). In the 1970s, a period in which the taxonomic pendulum seemed to rest on the side of the unilinear model, F. C. Howell (1973, p. 123) still asked, in the heading of a chapter, "Just who was the Neanderthal?" In this choice of a title, he signaled his discontent with, and the problematics of, viewing the Neanderthal's unique morphology as a direct predecessor of modern human morphology.

Once again, the reluctance to remove the Neanderthal from our ancestry and to accept the cladogenetic model seems to be what leads Trinkaus (1983, 1984, 1987) and investigators who endorse his view (Smith and Paquette, 1989) to see the Neanderthal facial configuration as resulting primarily from two distinct trends: the posterior withdrawal of the zygomatic bones to their position in modern humans and the persistence of the prognathic configuration of *H. erectus* in the rest of the face. The Neanderthal, according to this approach, can *still* serve as an intermediate stage between *H. erectus* and modern humans. Without entering into the details of their argument, I would like to stress that the retreat of the zygomatic bones has never been convincingly demonstrated. For example, the position of the zygomatic process of the maxilla relative to the dental arcade probably does not indicate a retreat of the zygomatic bone, as claimed by Trinkaus (1987), but rather stems from the more sagittal orientation of both the zygomatic bone and the zygomatic process of the maxilla, and the attachment of these elements, at that orientation, to a wide maxillary body (see also Rak, 1986, 1991). Furthermore, it is not at all clear how this supposed retreat could have resulted in the particular morphology manifested in certain Neanderthal facial elements that can only be viewed as derived because they differ from corresponding elements that *H. sapiens* and *H. erectus* have in common. For example, let us note the immense width of the pyriform aperture, the flaring nasal apophysis, and the horizontally oriented nasal bones; all of these features combine to produce the dramatic prominence of the nasal bridge and the typical midfacial prognathism, on the one hand, and the distinct subdivision of the infraorbital region, on the other (Rak, 1986).

Although many attempts have been made to interpret the uniqueness of the Neanderthal face (see discussion in Rak, 1986), the question of the functional significance is of little relevance here. A comparison with the faces of other hominids and, indeed, other primates, is what reveals this uniqueness.

It is the combination of a derived face with a primitive pelvis that enables us to diagnose Neanderthal as a separate species from *H. sapiens,* which is characterized by the reverse—a primitive face and a derived pelvis (Fig. 3). These opposite combinations and the emerging chronological picture point to the contemporaneity of two distinct species and compel us to recognize that two lineages were evolving separately. It is still the traditional taxonomic considerations—the magnitude of morphological differences between the Neanderthal and *H. sapiens*—that constitute the basis for proposing species status for the Neanderthal. Both the morphology and the basic architectural plan of the face, as well as fundamental differences in the pelvis, have led me to consider the Neanderthals as a distinct species—*Homo neanderthalensis* King. In recent years, other investigators have reached the same conclusion (see, for example, Stringer and Grün, 1991; Tattersall, 1986; Howell, 1991). On the basis of anatomical considerations and paleontological criteria, along with the revised geologic age, we cannot but falsify the hypothesis that *H. neanderthalensis* represents a stage in an evolutionary sequence leading to *H. sapiens.*

Some of the Middle Eastern Neanderthal specimens (most notably Tabun C-1, Shanidar 5, and Kebara) are "classical" in appearance and resemble those of Western Europe. However, other specimens (Tabun C-2, Shanidar 2 and 4, and Amud I) do not adhere to this classical image. Although a broad consensus maintains that the latter are Neanderthals, their face and mandible do exhibit a more modern-looking morphology, and hence, they are considered by some investigators as closer to modern *H. sapiens* than the former group is. However, I believe that these specimens (Tabun C-2, Shanidar 2 and 4, and Amud I) constitute the more *primitive* segment of the spectrum characterizing the species; and since the face of modern *H. sapiens* (like that of *H. erectus*) shows, in essence, the primitive configuration, the more generalized morphology of these spec-

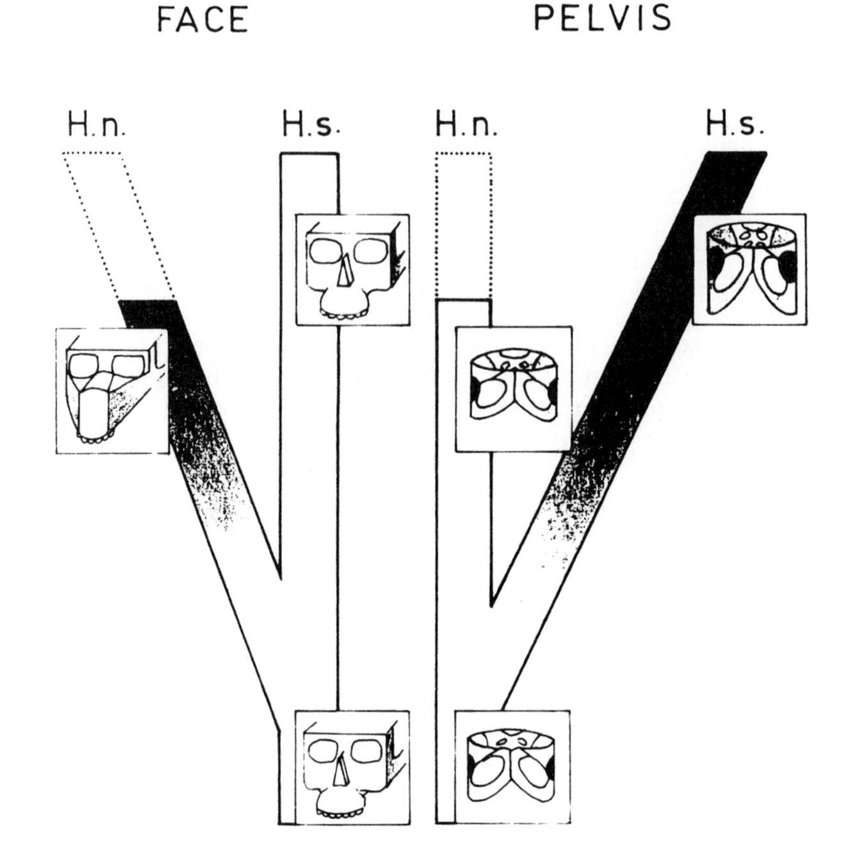

Fig. 3. The proposed phylogenetic relationship between *H. sapiens* and Neanderthal. The diagram on the left is based on the face and shows the Neanderthal as the derived form. The diagram on the right is based on the pelvis and shows the Neanderthal as the primitive form and *H. sapiens* as the derived. Only the true pelvis is depicted.

imens is phylogenetically uninformative. They do not provide evidence of interbreeding, and they certainly do not represent a late or even terminal stage in the *H. neanderthalensis* evolutionary sequence—an intermediate link between the "classical" Neanderthal and modern *H. sapiens*. Their somewhat modern-looking face does not suggest that the anatomy *appeared* late, despite the fact that it resembles the anatomy of the only surviving species, the latest in the phylogeny. The face of these individuals is simply more primitive, with no chronological implications necessarily attached. Similarly, and for the same reasons, we can reasonably expect to find that in the parallel lineage of *H. sapiens*, part of the morphological spectrum of derived pelvises will include specimens with a somewhat more primitive appearance, closer to that of *H. neanderthalensis*. The presence of a *sapiens*-like morphology in the spectrum of Neanderthal faces (like the predicted, more generalized pelvis in part of the *H. sapiens* clade) is not evidence of a chronologically late stage; on the contrary, this morphology is simply a reflection of phylogenetic heritage.

A Biogeographic Model

The model that I propose offers an explanation for the great variation characterizing the Mousterian hominids in the Middle East: how the caves of the period could have been inhabited by hominids exhibiting such a wide morphological range, or, in other words, how such a relatively restricted geographic region could have accommodated *H. sapiens, H. neanderthalensis,* and those populations that purportedly fill in the morphological gap between the two.

A generalized appearance can stem not only from an early geologic age. Sexual dimorphism and young ontogenetic age are well-known agents that introduce generalized morphology and thus, also, variation into the spectrum. Another cause of variation is geographic distance, which has been recognized as "a powerful mechanism for achieving genetic divergence" (Mayr, 1963, p. 517; Wright, 1943). The gradual change of a given trait in contemporaneous populations produces a morphocline that begins at a generalized source and exhibits differentiation across a broad geographic area, as the distance introduces new ecological niches. A specific case of such distribution is a long chain of clines that curves back on itself. The chain runs along a restricted geographic route (such as the shoreline of an island or the margins of a valley), and the ends overlap in sympatric populations, producing the well-known "chain of races," or *rassenkreis.**

What Mayr (1963) calls "speciation by distance" has been demonstrated frequently. It is expressed in an increase in sterility over geographical distance. The far ends of the sequence do not interbreed, and they are intersterile. Nevertheless, Goldschmidt (1940), for example, insisted that because of genetic continuity, the ends should not be considered distinct species. Dobzhansky (1964, p. 270), on the other hand, recognized the problem of the systematist who "is confronted with the impossibility of drawing the 'line' anywhere between the two [species]" but yet claims that since the far ends are genetically isolated, they should be considered separate species.

One model that may be able to account for hominid diversity in the Middle East is based, therefore, on the phenomenon of geographic differentiation along a cline. If we consider the face, this model places the primitive end of the morphocline in Africa in the form of an early *H. sapiens* and the derived end deep within Europe in the form of *H. neanderthalensis.* The model predicts that the face of specimens from the Middle East, which is situated between these two regions, will be more generalized, that is, less differentiated, than that of the more western Neanderthals. However, even such a restricted site as the northern part of Israel presents a more complicated picture of morphological variation.

This proposed zoogeographic model is horizontal and thus lacks the depth necessary to reflect the time span of the Mousterian period, during which the territorial range of *H. neanderthalensis* habitation might well have fluctuated. The caves of Israel, situated on the periphery of the Neanderthal range, thus alternatively "sampled" whichever of the populations prevailed at a given time:

*Mayr and Ashlock (1991, p. 427) draw attention to the misuse of the German word *rassenkreis.* They maintain that the term translates as "polytypic species" rather than "circle of races."

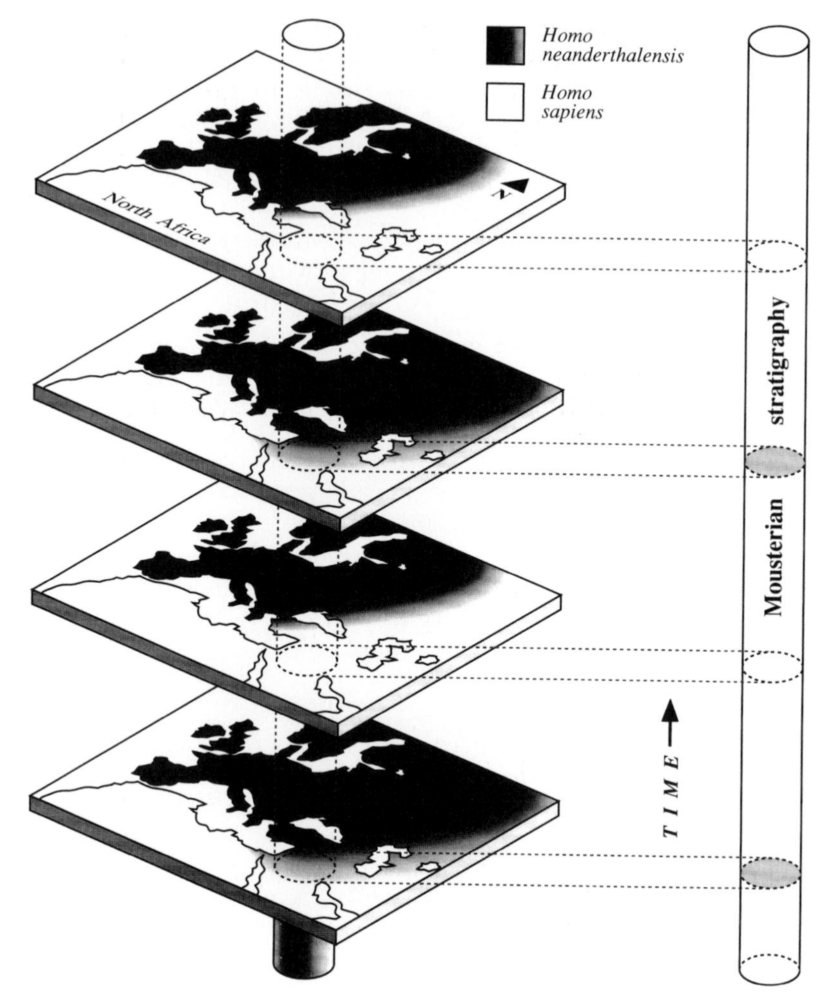

Fig. 4. Fluctuations in the territory of the Neanderthal (black) and of early *H. sapiens* (white) shown on a map of circa Mediterranea. The Middle East Mousterian stratigraphy is indicated by the cylinder.

Expansion of the Neanderthal territory brought a *H. neanderthalensis* presence into the caves of the region, and shrinkage of the territory, with the corresponding expansion of the *H. sapiens* territory, led to the presence of *H. sapiens* in the caves. The model actually predicts that the varying size of the Neanderthal territory will be manifested in a corresponding degree of anatomical differentiation in a given cave. In other words, a particulary extensive expansion of the Neanderthal territory would have brought a much more classical-looking Neanderthal into the Galilee caves than a smaller expansion would have. Hence, oscillation in the size of the territory can be expected to have introduced different anatomical forms that interdigitate stratigraphically, even within a single cave (Fig. 4).

In essence, the effect of the dimension of time (on a greater scale than the periods described above) on a vertical evolutionary lineage does not differ from the effect of geographic distance on a horizontal morphocline. According to Mayr (1942, 1949), populations are thus expected to show increasing differentiation along an evolving lineage. As in the case of a geographic morphocline, researchers such as Mayr (1942), Simpson (1943), and Campbell (1963) consider the opposite ends of a segment in the lineage, or chronocline, as two distinct species [termed by Simpson (1961) as *chronospecies* or *successional species*, and by Mayr (1942) as *allochronic species*], whereas others refuse to recognize the two ends as distinct species, no matter how long the sequence is nor how many differences accumulate from one end of it to the other (very much analogous to the way Goldschmidt views the geographic *rassenkreis*). According to the latter view, speciation can only result from a cladogenetic event (Eldredge and Cracraft, 1980; Hull, 1978; Szalay and Bock, 1991; Wiley, 1981).

Conclusions

No matter which view (Goldschmidt's vs. Dobzhansky's) one accepts in reaching a taxonomic decision, both the geographic and the temporal effects must be taken into account when evaluating the fossil record. The presence in the Middle East of a highly specialized Neanderthal such as Shanidar 5, for example, can result from its late geologic age or from an extensive southward geographic expansion of the *H. neanderthalensis* territory. The biogeographic model suggests that the combination of a late specimen and great territorial expansion can theoretically result in a Neanderthal with an even more classical, i.e., derived, appearance in a given cave. On the other hand, the presence in the Amud cave of a more generalized (modern-looking) *H. neanderthalensis* specimen, such as Amud I (Suzuki, 1970), can stem from the specimen's great chronological age or from the cave's location on the periphery of the modestly expanded Neanderthal territory. The combination of these two factors may, again, introduce even more generalized specimens. Similarly, morphological differences are expected to occur in the *H. sapiens* lineage. The difference between the hominids from Zuttiyeh and Qafzeh, two caves that are geographically close, can be attributed to the time difference between them.

It is, therefore, the specific geographic location of the Middle East that introduces such morphological complexity into the caves. The model suggests that at a given point in time, the picture of facial morphology is much less complex at the extremities of the geographic range of the populations; and, as we reach the ends of this spectrum, the situation is, indeed, simpler. In western Europe, only Neanderthals were present almost to the time of their disappearance, whereas in North Africa (for example, Djebel Irhoud and Hauafteah), early *H. sapiens* was the sole inhabitant during the same period. It was apparently due to the fact that the hominids were unable to cross the Straits of Gibraltar that the complete ring species was not achieved.

If it is true that no particular degree of morphological change is associated with the genetic event of speciation (Lambert and Paterson, 1982), it follows that most morphological changes succeed the actual speciation event itself. This is apparently the basis for the paleontological phenomenon of underestimating the number of species on the basis of osteological characters: There may be more speciation events than the morphological changes that record them (Tattersall, 1986). From this point of view of the fossil record, this implies that a particular specimen need only reveal its species identity incipiently in order to be considered part of the taxon. Hence, even incipient Neanderthal morphology is sufficient evidence to merit a specimen's inclusion in *H. neanderthalensis.** It follows, therefore, that it is less excusable to attribute specimens with such morphology to *H. sapiens* than to exclude more generalized specimens—those with few apparent *H. neanderthalensis* characters—from the hypodigm of *H. neanderthalensis,* although phylogenetically the latter actually belong to this species.

In regard to the combination of features mentioned above (a derived pelvis and primitive face in *H. sapiens,* and the opposite in *H. neanderthalensis;* see Fig. 3) the cladogenetic phylogeny predicts that those specimens on the *H. neanderthalensis* clade that show a more primitive face will still have a typically *H. neanderthalensis,* i.e., primitive, pelvis. On the other hand, an anagenetic model, in which such specimens are viewed as late, that is, "advanced" Neanderthals, on the verge of becoming *H. sapiens,* predicts that they will exhibit a more derived pelvis. It is difficult to reach solid conclusions based on the very fragmentary evidence available. The Amud I specimen, which is often used as an example of an "advanced" Neanderthal on the basis of cranial morphology, has little left of its pelvis (Endo and Kimura, 1970). Nevertheless, this fragment, as well as the Shanidar 4 specimen (Trinkaus, 1983), show hallmarks of the diagnostic *H. neanderthalensis,* or primitive, pelvis.

The sense of eternal stability conveyed by a cave and its fossil-bearing deposits should not mislead us into thinking that the physical and biotic realms were likewise invariant across this time span. Although the Mousterian period was short from an evolutionary perspective, it was long enough to be characterized by periodic fluctuations in niche size and shape. It is very likely, given the geographic position of the Middle East, that such a process was responsible for the distributional pattern of *H. neanderthalensis* and *H. sapiens* during the upper Pleistocene of this region.

Acknowledgments

I would like to thank Dr. William Kimbel, Dr. Eric Delson, and two anonymous reviewers for their meticulous comments, which were of great help. Figs. 2 and 4 were skillfully drawn by Douglas Beckner. Finally, I thank the L. S. B. Leakey Foundation for supporting fieldwork in Israel, especially the Kebara and Amud projects.

*Referring again to the biogeographic model, we can suppose, therefore, that a taxonomist who is attempting to draw the geographic boundary between *H. neanderthalensis* and *H. sapiens* at a given point in time will tend to place it within the geographic range of *H. neanderthalensis* rather than at the true (i.e., genetic) boundary between the two species.

References

Bar-Yosef, O., Vandermeersch, B., Arensburg, B., Goldberg, P., Laville, H., Meignen, L., Rak, Y., Tchernov, E., and Tillier, A.-M. 1986. New data on the origin of modern man in the Levant. *Curr. Anthropol.* **27:**63–64.

Bock, W. J. 1979. The synthetic explanation of macroevolutionary change: A reductionist approach. *Bull. Carnegie Mus. Nat. Hist.* **13:**20–69.

Boule, M. 1911–1913. L'homme fossile de la Chapelle-aux-Saints. *Ann. Paléontol.* **6:**111–172; **7:**21–192; **8:**1–70.

Boule, M., and Vallois, H. 1957. *Fossil Man.* Dryden, New York.

Bowler, J. P. 1986. *Theories of Human Evolution: A Century of Debate, 1844–1944.* The Johns Hopkins University Press, Baltimore.

Campbell, B. 1963. Quantitative taxonomy and human evolution, in: S. L. Washburn (ed.), *Classification and Human Evolution,* pp. 50–74. Aldine, Chicago.

Dobzhansky, T. 1964. *Genetics and The Origin of Species.* Columbia University Press, New York.

Eldredge, N., and Cracraft, J. 1980. *Phylogenetic Patterns and the Evolutionary Process.* Columbia University Press, New York.

Endo, B., and Kimura, T. 1970. Postcranial skeleton of the Amud man, in: H. Suzuki and F. Takai (ed.), *The Amud Man and His Cave Site,* pp. 231–406. Academic Press, Tokyo.

Goldschmidt, R. 1940. *The Material Basis of Evolution.* Yale University Press, New Haven.

Haas, G. 1972. The microfauna of Jebel Qafzeh. *Palaeovertebrata* **5:**261–270.

Howell, F. 1952. Pleistocene glacial ecology and the evolution of "Classic Neandertal" man. *Southwest. J. Anthropol.* **8:**377–410.

Howell, F. 1973. *Early Man.* Time-Life Books, New York.

Howell, F. 1991. The integration of archeology with paleontology. Paper delivered at Spring Systematics Symposium, Field Museum of Natural History, Chicago.

Howells, W. 1942. Fossil man and the origin of races. *Am. Anthropol.* **44:**182–193.

Howells, W. 1975. Neanderthal man: Facts and figures, in: R. H. Tuttle (ed.), *Paleoanthropology: Morphology and Paleoecology,* pp. 389–407. Mouton, Paris.

Hrdlicka, A. 1927. The Neanderthal phase of man. *J. R. Anthropol. Inst.* **57:**249–274.

Hull, D. L. 1978. A matter of individuality. *Philos. Sci.* **45:**335–360.

Lambert, D. M., and Paterson, H. E. 1982. Morphological resemblance and its relationship to genetic distance measures. *Evol. Theory* **5:**291–300.

Mann, A. and Trinkaus, E. 1974. Neanderthal and Neanderthal-like fossils from the upper Pleistocene. *Yrbk. Phys. Anthropol.* **17:**169–193.

Mayr, E. 1942. *Systematics and the Origin of Species.* Columbia University Press, New York.

Mayr, E. 1949. Speciation and evolution, in: G. L. Jepsen, E. Mayr and G. Simpson (ed.), *Genetics, Paleontology, and Evolution,* pp. 281–298. Princeton University Press, Princeton, NJ.

Mayr, E. 1963. *Animal Species and Evolution.* Harvard University Press, London.

Mayr, E., and Ashlock, P. D. 1991. *Principles of Systematic Zoology,* 2nd ed. McGraw-Hill, New York.

McCown, T. D., and Keith, A. 1939. *The Stone Age of Mount Carmel.* Clarendon Press, Oxford.

Rak, Y. 1986. The Neandertal: A new look at an old face. *J. Hum. Evol.* **15:**151–164.

Rak, Y. 1990. On the differences of two pelvises of Mousterian context from the Qafzeh and Kebara caves, Israel. *Am. J. Phys. Anthropol.* **81:**323–332.

Rak, Y. 1991. Sergio Sergi's method and its bearing on the question of zygomatic bone position in the Neandertal face, in: M. Piperno and G. Scichilone (eds.), *The Circeo 1 Neandertal Skull Studies and Documentation,* pp. 301–310. Rome, Instituto Poligrafico e Zecca Dello Stato.

Ridley, M. 1986. *Evolution and Classification: The Reformation of Cladism.* Longman, London.

Simpson, G. G. 1943. Criteria for genera, species and subspecies in zoology and palaeozoology. *Am. N.Y. Acad. Sci.* **44:**145–178.

Simpson, G. G. 1961. *Principles of Animal Taxonomy.* Columbia University Press, New York.

Smith, F. H., and Paquette, S. P. 1989. The adaptive basis of Neandertal facial form, with some thoughts on the nature of modern human origins, in: E. Trinkaus (ed.), *The Emergence of Modern Humans,* pp. 181–210. Cambridge University Press, Cambridge.

Spencer, F., and Smith, F. 1981. The significance of Ales Hrdlicka's "Neanderthal Phase of Man": a historical and current assessment. *Am. J. Phys. Anthropol.* **56:**435–459.

Stringer, C., and Grün, R. 1991. Time for the last Neanderthals. *Nature* **351:**701–702.
Suzuki, H. 1970. The skull of Amud man, in: H. Suzuki and F. Takai (ed.), *The Amud Man and His Cave Site,* pp. 123–206. University of Tokyo Press, Tokyo.
Suzuki, H., and Takai, F. 1970. *The Amud Man and His Cave Site.* University of Tokyo Press, Tokyo.
Szalay, F. S., and Bock, W. J. 1991. Evolutionary theory and systematics: relationships between process and patterns. *Z. Zool. Syst. Evolut.-Forsch.* **29:**1–39.
Tattersall, I. 1986. Species recognition in human paleontology. *J. Hum. Evol.* **15:**165–176.
Tchernov, E. 1984. Faunal turnover and extinction rate in the Levant, in: P. S. Martin and R. Klein (eds.), *Quaternary Extinctions,* pp. 528–552. University of Arizona Press, Tucson.
Tillier, A.-M., Arensburg, B., and Duday, H. 1989. La Mandibule et les dents du Neanderthalien de Kebara (Homo 2). *Paleorient* **15:**39–58.
Trinkaus, E. 1976. The morphology of European and Southwest Asian Neandertal pubic bones. *Am. J. Phys. Anthropol.* **44:**95–103.
Trinkaus, E. 1983. *The Shanidar Neanderthals.* Academic Press, New York.
Trinkaus, E. 1984. Western Asia, in: F. H. Smith and F. Spencer (eds.), *The Origin of Modern Humans,* pp. 251–293. Alan R. Liss, New York.
Trinkaus, E. 1987. The Neandertal face: evolutionary and functional perspective on a recent hominid face. *J. Hum. Evol.* **16:**429–443.
Valladas, H., Joron, J., Valladas, G., Arensburg, B., Bar-Yosef, O., Belfer-Cohen, A., Goldberg, P., Laville, H., Meignen, L., Rak, Y., Tchernov, E., Tillier, A.-M., and Vandermeersch, B. 1987. Thermoluminescence dates for the Neandertal burial site at Kebara (Mount Carmel), Israel. *Nature* **330:**159–160.
Valladas, H., Reyss, J., Valladas, G., Bar-Yosef, O. and Vandermeersch, B. 1988. Thermoluminescence dating of the Mousterian Proto-Cro-Magnon remains of Qafzeh Cave (Israel). *Nature* **331:**614–616.
Vallois, H. 1954. Neanderthals and presapiens. *J. R. Anthropol. Inst.* **84:**111–130.
Vandermeersch, B. 1981. *Les Hommes Fossiles de Qafzeh (Israel).* Editions du Centre de la Recherche Scientifique, Paris.
Weidenreich, F. 1943. The "Neanderthal Man" and the ancestors of "*Homo sapiens.*" *Am. Anthropol.* **42:**375–383.
Weidenreich, F. 1947. Facts and speculations concerning the origin of *Homo sapiens. Am. Anthropol.* **49:**187–203.
Weidenreich, F. 1949. Interpretations of the fossil material. *Am. Anthropol. Assoc. Stud. Phys. Anthropol.* **1:**47–59.
Weinert, H. 1953. Der fossile Mensch., in: A. Kroeber (ed.), *Anthropology Today,* pp. 101–119. University of Chicago Press, Chicago.
Wiley, E. O. 1981. *Phylogenetics.* John Wiley and Sons, New York.
Wright, S. 1943. Isolation by distance. *Genetics* 28:114–138.

Summary

5

Species and Speciation 21

Conceptual Issues and Their Relevance for Primate Evolutionary Biology

WILLIAM H. KIMBEL AND LAWRENCE B. MARTIN

Introduction

When confronted with the bewildering array of life forms, whether living or extinct, the evolutionary scientist must attempt to create order out of apparent chaos before addressing any other of the myriad questions that stem from a dedication to understanding the organic world. As "the lowest level of genuine discontinuity above the level of the individual [organism]" (Mayr, 1982, p. 251), the species is the basic unit of evolutionary diversity. One would think, therefore, that there is a large degree of unanimity regarding the nature of the species and the role it plays in evolutionary theory. But this is not the case. As amply illustrated by the contributions to this book, there remains a great deal of disagreement on the so-called "species problem." Questions such as (1) what *are* species in the context of evolutionary theory, (2) what processes are responsible for species' origins (and extinctions), and (3) what role do species play in evolution over geological time have occupied the center of debate among evolutionary biologists for decades, and the discussion shows no signs of abatement as we close in on the millennium.

WILLIAM H. KIMBEL • Institute of Human Origins, Berkeley, California 94709.
LAWRENCE B. MARTIN • Departments of Anthropology and Anatomical Sciences, State University of New York at Stony Brook, Stony Brook, New York 11794.

Species, Species Concepts, and Primate Evolution, edited by William H. Kimbel and Lawrence B. Martin. Plenum Press, New York, 1993.

The investigation of these questions should be far more than a exercise in cerebration for theoreticians and philosophers, since it carries important implications for the data of the "real world." *How should nature be carved up into entities called species*—arguably the most fundamental operational problem in evolutionary biology—hinges on the results of this investigation. This is why practitioners of species biology must be explicit about the concepts they use in their work.

Historically, paleontologists have tended to focus on the transformation of morphology through geologic time and on the origin of major morphological complexes that delineate higher taxa. Neontologists, on the other hand, have concentrated on the analysis of organic diversity (genetic, organismic, and taxonomic) and have attempted to understand the biological processes and interactions responsible for its genesis. Although adaptation by natural selection within populations has been hailed by most neo-Darwinists as the common thread linking these enterprises conceptually, speciation has occupied a zone of uncertainty between them. This uncertainty would appear to have both ontological and epistemological sources. On the one hand, speciation is thought not to be caused merely by natural selection acting on heritable phenotypic variation within populations. Such concepts as genetic drift, gene flow, allopatry, reproductive isolating mechanisms, and mate recognition systems, among others, are usually invoked in discussions of how speciation occurs. On the other hand, precise knowledge of the genetic mechanisms responsible for reproductive and other character changes accompanying speciation does not presently exist. Moreover, speciation is a process usually sufficiently extended in time to mask its observation during a human lifetime, yet it is virtually instantaneous from the perspective of the paleontological record. To appreciate speciation as a historical process, we must work backward, as it were, from the products of speciation to gain insight into the process itself. However, the existence of both the product and the process arises out of purely theoretical considerations [much as in subatomic physics, as Eldredge (this volume) notes], and although we are rightly concerned with practical problems of species recognition in our daily work, this task cannot be tackled without attempting to understand the nature of the basic entities posited to play a role in evolutionary theory.

It is at this ontological level that a new division in the conceptual consensus has appeared. Rieppel (1986, p. 313) characterized the rift between two mutually exclusive but complementary "ways of seeing" evolution: "While looking at nature, it is possible to interpret it either as 'being' (in terms of pattern), or as 'becoming' (in terms of process)." In this regard, it would be difficult to overestimate the impact of Hennig's (1966) *Phylogenetic Systematics* and Eldredge and Gould's (1972) "Punctuated Equilibria: An Alternative to Phyletic Gradualism," contributions that emphasized new "ways of seeing" evolution for paleontologists and neontologists alike. Hennig's cladistic approach to phylogeny reconstruction logically entailed an isomorphic relationship between the relative temporal ordering of speciation events and the hierarchical structure of the Linnaean system. Although critics have charged that this juxtaposition attempts to map a dynamic and continuous evolutionary process on a static and discrete systematization of taxa (e.g., Szalay, this volume; Rose and Bown, this volume), Eldredge and Gould's punctuated equilibria model, based on the marriage of paleontological data and allopatric speciation theory, suggested that most major organismal phenotypic change occurs during the geologically instantaneous interval of speciation and is succeeded by a relatively long, quiescent period of phenotypic stasis.

As a consequence—and notwithstanding the ongoing debate over the generality, if not the validity, of punctuated equilibria (e.g., Somit and Peterson, 1992)—many paleontologists have come to feel more comfortable with importing a horizontal perspective to the interpretation of their data. According to this perspective, knowledge of the origins and extinctions of species (historical events) and of the phylogenetic patterns of species diversity (the cumulative results) is logically prior to the formulation of specific hypotheses regarding process (adaptation, competition, dispersal, and the transformation of morphology over time). The validity of the ontology that divorces pattern from process is a relatively new topic of discussion and debate in evolutionary biology; a commitment to one side or the other of the issue would appear to underlie many of the arguments about species presented in this book. At the center of this debate lies that mystery wrapped in an enigma: the species.

Conceptual Foundations of the Species Notion

In attempting to understand how *the* species functions in evolutionary theory, it would seem important to inquire about the attributes of the units in nature (taxa) that are called *species*. The species *taxon* has again become a favorite topic for philosophical discussion in recent years, and to the extent that the growing literature has been digested by practicing evolutionists, it has had a significant impact on debates within the science itself, as documented by many of the contributions to this book.

In this volume, Eldredge and Kimbel and Rak explicitly ground their discussions in the *individuality thesis* propounded initially by Ghiselin (1974) and extended by Hull (e.g., 1976, 1978), among others, which holds that a species taxon is an "individual" in the logical sense of being a spatiotemporally discrete entity, capable of (theoretically) infinite change in character (via descent with modification). What gives a species these attributes of individuality is the integration afforded by the unbroken network of reproductive ties among organisms throughout its existence. The Ghiselin–Hull ontology has several consequences for theoretical ideas of species:

1. Species are internally cohesive by virtue of the shared plexus of reproduction over time and space.
2. Species are placed firmly within the realm of ancestry and descent, and are thus seen to constitute lineages.
3. Species have theoretically definable origins (at speciation) and terminations (at extinction), i.e., they are discrete.

There are authors who deny that species have the attributes of individuality. For example, Bock (1986), whose general views on systematics are endorsed by Szalay (this volume), declares that species taxa are in fact classes. A full exploration of the issue is beyond the scope of this book, but we believe that a commitment to the idea that species are classes underlies Bock's and Szalay's rejection of the second and third attributes of species in the list above. However, we join Bock and others in believing that the individual versus class debate with respect to species amounts to more than philosophical esoterica; resting on its outcome is the ontological legitimacy of recent ideas about macroevolution and the hierarchy of evolutionary entities and processes (e.g., Eldredge, 1985; Salthe, 1985; Grene, 1987).

Reproductive Cohesion or Reproductive Isolation?

The internal cohesion attribute of species raises the issue of *reproductive isolation,* the ranking criterion of the biological species concept (BSC) (Szalay, this volume). There has always been disagreement on the extent to which reproductive isolation should play a role in the theoretical definition of the species category, and this has consequences for species recognition in paleontology as well as neontology, as discussed by Eldredge (this volume). Szalay argues the view that reproductive isolation (the failure or unwillingness of organisms to interbreed) between synchronic and sympatric populations is the only known test by which we can recognize units unequivocally conforming to the theoretical structure of the BSC. Inherent in this view is the notion that "species" is a *relational* concept—like "brother" (Mayr, 1957)—that is, the "fit" of species taxa to the theoretical species concept can be assessed only when populations are evaluated in relation to other populations from which they are reproductively isolated in sympatry. Selection for mechanisms that reinforce reproductive isolation between sympatric populations (i.e., selection against organisms that "mismate") is seen to be the chief causal agent in the final stage of the speciation process. Therefore, distinct populations dispersed in time and/or space are beyond the reach of direct test and according to Szalay fall under the appellation "taxonomic species": character-based, multidimensional approximations to the theoretical "biological species."

In contrast, following Paterson's (e.g., 1985) recognition concept (RC) of species, both Masters (this volume) and Eldredge (this volume) maintain that a species is *self-defining* by virtue of the unique fertilization system (manifested phenotypically in a set of signaling mechanisms between the sexes, the specific mate-recognition system, or SMRS) shared by all its component organisms, and that the limits of the species (that is, its reproductive isolation from other species) are purely incidental to this. Species, in this view, are an *effect* of strong positive selection for niche-specific attributes that promote mating and ensure syngamy. Speciation is held to be completed in allopatry at the time when a new fertilization system becomes stabilized in the daughter population; selection against mismating with organisms of the parent population in neosympatry (i.e., selection for postzygotic isolating mechanisms) is not accorded a causal role in species formation. Were we able to define components of the SMRS, we would be able to test conclusions as to the rank of allopatric and/or allochronic populations, although the applicability of the RC in paleontology is limited to fossil groups in which the SMRS is in part skeletal (e.g., Turner and Chamberlain, 1989).* According to Masters, the SMRS (or at least the crucial aspects of it) will be fixed throughout the geographic range of a recognition species, even if the species is divided into many otherwise differentiated geographic demes. This is rather different to the view of biological species, according to which the distribution of mechanisms promoting postzygotic isolation should be restricted to populations occurring sympatrically with nonconspecifics, as pointed out by Moore (1957).

Szalay (this volume) sees the RC as "redundant" with respect to the BSC, a charge brought by others who argue that the SMRS is equivalent to the premating isolating mechanisms invoked in explanations of allopatric speciation under the BSC (Raubenheimer and Crowe, 1987). But strong contrasts between the BSC and the RC remain, particularly with respect to the divergent views on the role in speciation of postzygotic isolating mechanisms. The RC's emphasis on the reproductive interactions among organisms *within populations,* rather than on reproductive isolation *between populations,* underscores the idea that the RC represents an ontological departure from the BSC, a shift that invites further critical scrutiny. Of course, the reproductive coherence of interbreeding populations is fundamental to the BSC (Eldredge, this volume), but "reproductive isolation" is always invoked in response to the questions, "What separates species from other, smaller, interbreeding units (i.e., infraspecific taxa) in nature?" The contention that the RC provides the epistemology to resolve the status of distinct, closely related, allopatrically distributed populations, long viewed as a problematic area for the BSC (hence the requirement for the "taxonomic species" notion), underscores the difference between the two concepts.

Masters (this volume) contends that the SMRS embraces "those communicatory signals responsible for both the establishment and maintenance of the social structure within which sexual recognition and mating take place" (p. 51, this volume), which is a somewhat broader conception than that proposed by Paterson (1985), and may be more akin to the "demonstrative communication paradigm" described by Brockelman and Gittins (1984; see Groves, this volume). This leads Jolly (this volume) to conclude that the criteria of the RC are sufficiently liberal to justify using almost any fixed character to identify species under the concept. Thus, for example, Masters upholds the specific distinctiveness of hamadryas and anubis baboons, even though in areas of contact along the Awash River in Ethiopia they produce fertile hybrids. For Masters, the facts that the parental populations maintain distinct, ecologically mediated social organizations, which presumably evolved in allopatry, and that crossmating of the parental populations and backcrosses involving F_1 hybrids are not random with respect to parental phenotypes implies that separate SMRSs are involved. Jolly, who employs the "isolation" criterion of the BSC, concludes that despite a poorly understood history of population divergence within the papionins, fertile hy-

*We are intrigued by the possibility that bipedalism itself may have functioned as part of a SMRS during the speciation event that gave rise to the hominid clade, although the extent to which this behavior–anatomical complex would have been the target of selection *for* mate signaling and recognition is another question. If so, then bipedalism *qua* SMRS would be primitively retained in all subsequent hominids, notwithstanding other shifts in the fertilization system that would have characterized these descendant taxa.

bridization in zones of parapatry implies that speciation has not occurred, or at least is incomplete.

We must be cognizant of factors potentially behind such differences of opinion. One is that "reproductive isolation" (absence of interbreeding, a behavioral phenomenon) and "genetic isolation" (lack of introgression, a genetic phenomenon) are not the same thing. Richard Harrison (1990, p. 97) characterizes the common situation of "hybrid zones [in which] effective premating barriers are apparently absent and individuals of mixed ancestry are common. Yet introgression is often very limited, i.e., outside of a narrow zone the hybridizing taxa remain distinct." In such cases, decisions as to the specific status of the parental populations will largely depend on which criterion is employed. Both the biological species concept and the recognition concept ostensibly concern "genetic species," but this distinction is not always observed in practice. Another factor is that while naturally occurring hybrids may backcross with the parental populations, it is important to know the extent to which the hybrids are disadvantaged with respect to reproductive success over the long-term (Godfrey and Marks, 1991). However, this is unknown for most cases of hybridizing primate populations.

Among the monophyletic *lar*-group gibbons, ecological requirements are uniform and there is little or no adaptive "economic" diversification, but there is substantial differentiation in pelage and/or vocalization patterns that have evolved in allopatry as the result of geographic isolation of populations by river systems. In a discussion of hybridization between groups in secondary contact, Groves (this volume) notes that hybrid units are relatively rare and restricted geographically in two cases where the parental groups are characterized by marked differences in pelage and/or vocalization (*lar* × *pileatus; lar* × *agilis*), but are much more common and geographically widespread where such differentiation is minimal (*albibarbis* × *muelleri*). Furthermore, pelage and vocalization appear to be geographically uniform within *lar, pileatus,* and *agilis,* including areas of overlap and hybridization. Brockelman and Gittins (1984) interpret this as evidence that there has been no "reinforcement" of isolating mechanisms acting to accentuate reproductive behavioral differences in the hybrid zone.

In their survey of *Saimiri* systematics in this volume, Costello and coauthors do not find "reproductive behaviors that may function to isolate groups" of mainly parapatric, occasionally hybridizing, populations with distinct social organizations, vocalizations, pelage coloration, and ecological preferences. This conclusion will likely draw fire from RC advocates. However, Costello and colleagues sound an important cautionary note when they point out that the unique vocalizations of different *Saimiri* populations are not mating calls, which squirrel monkeys do not produce. It is one thing to invoke a phenotypic character in the definition of an SMRS; it is another to demonstrate that the character evolved (or was co-opted) in response to selection for mate recognition in a specified environment.

One aspect of Costello and coauthors' work underlines the conceptual differences between the BSC and the RC. In their discussion of reproductive behavioral distinctions among *Saimiri* populations, they usually frame their characterizations in terms of "function," such as "*function to isolate groups,*" "*function to preclude hybridization,*" and so forth. The idea that species-specific reproductive behavior functions to isolate populations from one another is central to the BSC paradigm.* It is, however, squarely opposed to the function of the SMRS posited by the recognition concept. The point here is that two distinct conceptual matrices potentially channel the investigation of primate population biology along divergent pathways. Different questions may be asked of the same phenomena, resulting in disparate interpretations of those phenomena. Under the RC, the search for *barriers* to reproduction between populations is misguided in the first place, whereas investigation into the existence of separate SMRSs in hybridizing populations may not be considered under the BSC because the fact of hybridization may imply that no differences in SMRS ought to exist.

It is somewhat ironic that despite their obvious differences, the RC and the phylogenetic species concept (PSC) (Kimbel and Rak, this volume; Cracraft, 1983) will often identify the same units as species. The PSC is concerned with a level of differentiation potentially finer than that based on SMRSs, with the result that reproductive compatibility between distinct phylogenetic species is interpreted as a symplesiomorphy (see Eldredge, this volume). However, both concepts jointly differ from the BSC to the extent that they adopt the ontological stance that species are units of reproductive cohesion as opposed to isolation. The point is driven home by the fact that some researchers working within the BSC paradigm seem positively disposed toward the PSC, yet confine it to the subspecies level, preferring to implement the isolation criterion of the BSC when it comes to the species level (e.g., Jolly, this volume; Avise and Ball, 1990).

Species and Lineage: The Temporal Dimension

According to the biological species concept, in the vertical dimension a population theoretically belongs to no species. Characters allow us merely to divide a lineage (the time path) into taxonomic chunks that have no relationship to "biological" species (Szalay, this volume). Thus, although biological species can arise only through cladogenesis, the BSC permits the existence of "species" as segments of undivided lineages. It is this paradox that forms the basis for the difficult arguments over species and speciation in the fossil record.

The extensive stratigraphic sequences of North American Paleogene mammals have provided grist for the mills of paleobiologists on both sides of the species debate. This volume's chapters on Eocene primates by Rose and Bown and Krishtalka provide further arguments for the view that nontrivial, directional morphological change can accrue in undivided lineages over long periods of geological time, a conclusion that presents the classic species recognition problem in paleontology. Although Rose and Bown recognize cladogenesis as the source of biological diversity at any one moment in time, they argue that ana-

*Costello and coauthors' use of the term "function" with respect to reproductive behavior would appear to correspond to what Bock and von Wahlert (1965) call the "biological role." This does not necessarily imply that the feature is an adaptation, however, although certain interpretations of the BSC do see "isolating mechanisms" as adaptive in the strict sense (see Masters, this volume; Paterson, 1985).

genesis is both a mechanism of character change within-species *and* a process of speciation (phyletic speciation). They argue that without dividing an unbranched lineage into species based on the amount of change observed, we inevitably confront Simpson's (1961) dilemma of having to include a protist in the evolutionary species *Homo sapiens* due to the unbroken chain of ancestry and descent that links them. Thus, Rose and Bown segment the anagenetic lineages of Willwood Formation primates according to Simpson's rule of thumb that the morphological difference between stratigraphically successive "chronospecies" is equivalent to that separating contemporaneous biological species. However, recent advocates of the evolutionary species concept (ESC), including Krishtalka (this volume), recognize a fallacy in Simpson's dilemma: the "lineage" linking *H. sapiens* and the protist is an abstraction; from the phylogenetic point of view, it does not exist except as a tiny part of the history of innumerable cladogenetic events of which the monophyletic taxon Eukaryota is comprised (*all* those lineages, living and extinct, that ultimately trace back to the same protist ancestor) (Wiley, 1978). Krishtalka thus refuses to divide single lineages into species, opting instead for numbered stages or semiformal "lineage segments."

From the vertical perspective, species boundaries are of course drawn between parental and daughter populations. However, from the point of view of the theoretical species concept—whether biological, evolutionary, cladistic, or phylogenetic—species boundaries are never crossed purely as a function of time and organismal change within a population; two populations only stand in the relation of species when they are phylogenetic sisters or daughter and surviving parent.* The crucial issue for "phyletic speciation" is that in an undivided lineage at no point are any two populations ever in either of these relationships. Thus, a disruption of reproductive continuity between descendant and ancestral populations of a single lineage can only be established by nomenclatorial convention.

The sole process by which speciation is known to occur in nature is cladogenesis, usually promoted by geographic isolation. It is the (genetic) divergence that occurs *subsequent* to geographic isolation that launches a population on a separate, irreversible, historical trajectory, which is germane to Tattersall's (this volume) observation that freshly minted species will not initially increase morphological diversity beyond that already evolved in a geographically differentiated parent species. The details of how these changes occur remain one of the outstanding unsolved problems of evolutionary biology (Coyne, 1992). But inasmuch as this evolution occurs *within* populations, anagenesis is essential to speciation even under cladistic or phylogenetic concepts of species. So, the central issue is not whether or not anagenesis occurs. It does, and it is an important component of the speciation process. But anagenesis cannot be synonymous with speciation. Under the Ghiselin–Hull ontology, the contrast between species and lineage does not exist, and therefore attempts to carve out separate ontological niches for biological species, taxonomic species, phyletic lineages, and the like are misdirected.

*This is not the same as the reproductive isolation criterion of the BSC, which requires ecological competition and reproductive character displacement between two populations in neosympatry.

Origins and Extinctions: Recognizing Species in the Fossil Record

If species are expected to have definable origins and extinctions, recognizing them as such is another story. With respect to the paleontological record, the nub of the problem is expressed by Szalay (this volume): "there are no known empirical means to ascertain that branching was responsible for the evolution of sample differences unless the samples are assuredly sympatric and synchronic." In other words, given two morphologically distinct but spatiotemporally disjunct samples that might be called species, how can the possibility be excluded that the differences were due to phyletic evolution (anagenesis), with the samples therefore actually representing populations of a single evolving lineage? This is less of a problem for BSC advocates because the "taxonomizing" of a lineage by the application of taxon names is not thought of as recognizing biological species. But it does amount to a criticism of the ESC and PSC (Kimbel and Rak, this volume), which attempt more or less explicitly to incorporate the multidimensional attributes of the species taxon into the category definition. These species concepts lack an epistemology to resolve the conundrum expressed above, Szalay argues.

Several of this volume's contributions on the paleontological record illustrate the issues. In examining species diversity among Plio-Pleistocene hominids, both Wood and Kimbel and Rak promote a taxic conception of species, which emphasizes the prior recognition of differentiated (= phylogenetic) units in the fossil record on the basis of (1) discrete character (apomorphy) distributions and (2) comparison of degrees of morphological variation between the fossil samples and closely related extant primate species (see below). This "horizontal" approach yields species that are represented by the termini of hierarchically arrayed branches of a cladogram, with extant and extinct species treated as equivalent entities ontologically and epistemologically. Absent, however, is an explicit consideration of "vertical" comparisons, which express temporally ordered sequences of morphological evolution in hypotheses of character *transformation* from an ancestral condition to one or more apomorphic conditions. Whereas species are revealed by stable, hierarchically deployed morphological patterns under the horizontal approach, transformation hypotheses do not in and of themselves reveal taxa. Without stratigraphic, geographic, and ecologic information indicating that the phena at hand satisfy the BSC's strict tests of sympatry and synchrony, speciation (cladogenesis) cannot be said to have been responsible for the observed differences. This is where, in paleontology, the relational nature of the BSC is revealed operationally: Whereas the taxic perspective views apomorphy as a signpost of speciation, the BSC calls upon the physical relations between populations in time and space to settle the issue of speciation decisively.

Although the BSC permits us to "taxonomize" morphological differences among spatiotemporally disjunct samples in the fossil record, it must be remembered that in so doing we do not discover "species" in any conceptually meaningful sense of the term. According to the taxic view, we do indeed discover conceptually meaningful units in the fossil record, irrespective of spatiotemporal relationships between samples, but the conceptual context is markedly different. The phylogenetic species concept turns on the distinction between phylogeny and *tokogeny,* Hennig's (1966) term for the process of reticulating (nonhierarchic) ancestry and descent within biparental species (see Eldredge, this volume; Harrison, this volume; Kimbel and Rak, this volume). Within sexually reproducing species, the nonhierarchic distribution of individual character lineages results from recombination and segregation during meiosis. The PSC, which defines the species as the minimal diagnosable cluster within which there is a parental pattern of ancestry and descent (Cracraft, 1983), draws a sharp line between the within-species tokogenetic pattern and the among-species, hierarchic, phylogenetic pattern of descent. Phylogenetic species are those populations that cannot be further subdivided cladistically, owing to a tokogenetic pattern of descent; they are basal phylogenetic entities. It follows, then, that the tokogenetic pattern itself is not an appropriate subject of cladistic analysis, as stressed here by Harrison (this volume). Given the phylogenetic species concept, the operational problem is where to draw the line between the tokogenetic and phylogenetic signals (Eldredge, this volume). That this is no less of a problem in studies of extant primates than in paleontology is well shown by Tattersall's (this volume) analysis of lemur species, which, as a group, exhibit a high degree of homoplasy and thus an incompletely resolved cladistic structure. When the level of homoplasy is high, the tokogenetic and phylogenetic signals will be difficult to disentangle.

Nevertheless, the PSC will tend to recognize species irrespective of their spatiotemporal relationships to one another. "Diagnosably distinct" allopatric and/or allochronic populations are species under the PSC, whereas they are theoretically indeterminate under the BSC (as suggested by Kimbel and Rak, this volume, the use of "subspecies" concept is not motivated by theoretical considerations under the BSC). For example, under the PSC, *Homo rudolfensis* and *Homo habilis,* differentiated allochronic taxa (Wood, this volume), are accorded the same meaning as *Pan troglodytes* and *Homo sapiens,* differentiated synchronic–sympatric taxa. Only the latter pair of "reproductively isolated" taxa are meaningful evolutionary units according to the theoretical structure of the BSC. The principal point is that under the BSC, the reproductive isolation of populations implies cladogenesis, while under the PSC the differentiation of populations implies cladogenesis. Consequently, at the level of the cladogram the BSC either must incorporate various kinds of "taxa" or exclude certain kinds that it treats as problematic, such as differentiated allopatric and allochronic taxa, while the PSC includes only one kind and is thus inclusive.

It is ironic that one of human paleontology's better examples of phenetic, temporal, and geographic evidence jointly pointing to cladogenesis in the fossil record—and here we have in mind the west Asian Upper Pleistocene hominids (Rak, this volume)—elicits so much controversy on the inference that two "biological species" have been sampled.

Methodological Issues

In the earlier years of this century, considerable research was devoted to describing and quantifying the normal morphological variation (intrasexual, intersexual, and ontogenetic) of primate species in the wild. Although much of this work (epitomized by the prodigious efforts of Adolph Schultz, W. C. Osman Hill, and others) was not placed in an explicitly theoretical evolutionary context, it merged smoothly with the growing emphasis on population biology that figured centrally in the neo-Darwinian synthesis of the 1930s. Perhaps the most obvious consequence of this disciplinary convergence for primate studies was the final abandonment of the typological approach to the fossil record and the simplification of the taxonomy of extinct forms. As shown by a number of the contributions to this volume, the population approach is still very much with us, albeit enhanced by powerful quantitative techniques for grouping and discrimination (see especially Albrecht and Miller, this volume; Shea *et al.*, this volume).

Two contrasting empirical claims about the geographic variation within and among primate species are presented in this volume. Albrecht and Miller (this volume, p. 125) promote a "hierarchy" of variation for primates that depicts a "general correspondence between morphological and taxonomic differentiation among demes, subspecies, and species," within a given higher taxon. On the other hand, Tattersall (this volume) states that the morphological differences between primate sister species is often no greater than those that separate differentiated geographical populations within a species. On balance, we suspect that the inventory of living primates can be marshalled in support of both positions, which may merely mean that it is difficult to generalize about primates as a whole. Again, however, we must be more cognizant than we have been of the different paradigmatic influences (not to mention handed-down classificatory conventions) that may shape conclusions in particular cases. This is especially true since both representations of reality are said to hold lessons for the interpretation of the fossil record.

The central theme of chapters in this volume by Albrecht and Miller and Shea and coauthors is that the importance of geographically based morphological variation within primate species is often underappreciated, especially by paleoanthropologists, whose judgments of taxonomic diversity in the fossil record are frequently suspected of being quasi-typological by students of extant primate variability. As a result of underestimating intraspecific geographic variation among living primates, so the argument goes, paleoanthropologists are prone to overstate species diversity among extinct primates. Albrecht and Miller urge the adoption by paleoprimatologists of strategies for the appropriate selection of modern comparative samples based on the questions being asked of a given fossil assemblage (e.g., for investigations of sexual dimorphism, the comparative samples would be drawn from single demes; for analyses of species diversity, the entire geographic ranges of the comparative taxa would be sampled). This strategy naturally would require a prior understanding of the systematics of the relevant comparative taxa, something the paleontologist may not be able to take for granted. However, as Albrecht and Miller's morphometric rubicons of taxonomic differentiation do not translate between higher taxa, we wonder whether theirs is a powerful strategy for paleoprimatology.

Contrasting with the Albrecht and Miller position is Tattersall's (1986) belief that species diversity in the hominid fossil record, at least, has been underestimated because of the tendency by paleoanthropologists to reflexively ascribe phylogenetic differentiation to intraspecific, microevolutionary causes. This notion stems from the following arguments: (1) speciation—a genetic event—does not demand a predictable or prescribed amount of morphological divergence [i.e., the "zygostructure" and the "phenostructure" of a population are not tightly correlated (Jolly, this volume)]; (2) the level of intraspecific morphological variation is an unhelpful guide to the past species diversity because there is no theoretical expectation that a certain amount of within-species variation is associated with the attainment of species rank; (3) however, empirically, reproductively isolated primate sister species often show only subtle osteological differentiation; (4) therefore, Tattersall concludes, when we encounter discrete morphs in the hominid fossil record, we should expect them to signify differentiation at the species level. Tattersall's position would seem to receive limited support from the studies on primate dental metric variation by Cope (this volume) and Plavcan (this volume), which demonstrate that for some extant primate groups, mixed dental samples assembled from sympatrically occurring species pairs frequently hide their heterospecific origins when the usual standards of relative metric variation are employed (e.g., coefficient of variation).

The statistical techniques employed in the recognition of species diversity in morphologically homogeneous fossil samples have recently come under increased scrutiny. Two approaches to the question of detecting species numbers are espoused in this volume. The conventional approach is an increasingly refined application of standards of morphological variation documented for extant species to the analysis of variation in fossil samples (e.g., chapters in this volume by Albrecht and Miller; Cope; Martin and Andrews; Plavcan; Shea *et al.*; Teaford *et al.*). A second approach, promoted here by Kelley, takes the view that within-species variation in the past should not necessarily be constrained by the range of within-species variation observed in the extant fauna. Kelley argues that too strict a "uniformitarian" approach removes the ability of paleontology to expand our knowledge of the past products of evolution on their own terms. Although Kelley is not completely explicit about how his approach to past species diversity is to be operationalized, his recognition of Miocene hominoid species depends on a prior attribution of dental elements (particularly canines, on morphological grounds) to sex and the empirical observation that within-sex distributions of canine size are unimodal for extant primate species. Cope (this volume), however, demonstrates that bimodal distributions of canine size—what would commonly be interpreted as separate male and female modes of a single dimorphic species distribution—can be generated from data for pooled dental samples of multiple, sympatric *Cercopithecus* species.

Notwithstanding Kelley's dismissal of the *Cercopithecus* case as atypical, a requirement for a fuller exploration of how we might resolve similar cases in the fossil record would seem to be indicated. The view taken by Cope, Martin and Andrews, and Teaford *et al.* is that because one can never prove statistically that a given sample of fossils is comprised of only a single species, neither can we demonstrate that a species in the past showed greater synchronic variation than any living species. Traditionally, then, the null hypothesis is that a given assemblage of fossils is comprised of a single species. The widely accepted *ceteris paribus* rule is that the failure to falsify the null hypothesis is tantamount to the corroboration of the view that a single species interpretation best explains the observed variation. The empirical basis for this dictum requires testing. To date, nearly all metric analyses of species numbers in the primate fossil record have tested the null hypothesis using reference values that yield low type I error probabilities; i.e., values resulting in few incorrect rejections of a true null hypothesis. This has been accomplished by using the maximum permissible reference levels (based on living primate species) or by setting type I error probability at 5%. However, the simulation studies of Cope and Lacy (1992) demonstrate empirically that such values result in very high type II error rates; i.e., a large number of failures to reject a false null hypothesis that a sample is comprised of but one species. In other words, our justifiable concern to avoid oversplitting a paleontological assemblage has in many cases rendered the failure to refute the null hypothesis virtually meaningless. Accepting a higher type I error rate would increase the probability that a sample for which the null hypothesis cannot be rejected can reasonably be accepted as containing a single species. It would thus appear that a large number of published metric tests of species numbers that accept the one-species composition of a paleontological sample based on the failure to reject the null hypothesis will have to be reevaluated in light of this work.

While future work resulting from this conclusion may eventually support the position that the primate fossil record contains more species than have been recognized so far, our ability to recognize species in the fossil record is only as adequate as are our conceptual foundations and inferential tools. We do not think that problems with taxonomic analysis in paleoanthropology (for example) will evaporate once our hard disks are filled to overflowing with morphometric data on primate geographic variation, for such data merely beg the issue. In the absence of a clearly formed concept of how nature ought to be validly carved up, no amount of data can be expected to lead to the resolution of paleoanthropological, or any other, taxonomic issues of interest.

Conclusions

In this final chapter of *Species, Species Concepts, and Primate Evolution,* we have not tried to carve a consensus out of the diverse views on species and speciation expressed by the contributors to the volume. We have instead concentrated mainly on the diversity because we believe that many of the contemporary debates about primate systematics and evolution are strongly influenced by different evolutionary and systematic philosophies. This is not a novel finding, of course. The trouble is that within the arena of primate evolutionary studies, and paleoanthropology in particular, seldom are these philosophical differences acknowledged, let alone articulated in context. It has been our goal to try to bring to the surface some of these underlying biases in context. In the process, some of our own biases have naturally surfaced, although we have tried to be fair to those with whom we disagree.

In spite of the profound conceptual differences that drive debates about species in these pages and elsewhere, we believe that there is a consensus on the following general conclusions:

1. It is critical to the success of scientific dialogue that all investigators of primate species biology (extinct and extant) explicitly articulate their conceptual view of species and speciation and explore the epistemological ramifications of this view for their research designs.

2. Research on primate species biology must integrate the investigations of population genetics, socioeconomic and sociosexual structure, hybridization, vocalization patterns, morphology, phylogeny, and historical biogeography. However, inasmuch as speciation is the author of evolutionary history, the delineation of historical factors behind the current patterns of diversity through phylogenetic and historical biogeographic research must play a larger role than it has to date if our goal is a more complete understanding of speciation among the primates.

3. We require a more secure anchor in biogeographic reality for the soft- and hard-tissue variation documented in museum collections around the world. The taxonomic classification of morphological discontinuity at and especially below the species level based on museum samples must correspond to actual distributions of discrete populations in the wild. The extent to which this is currently the case no doubt varies from taxon to taxon and collection to collection, but we should assiduously evaluate and verify where possible, rather than unquestioningly perpetuate, the classifications of the past.

4. Theoretical and practical systematics ought to be a dynamic and well-funded component of primate species biology. Lest there be doubts raised about the importance of such research, we refer readers to recent discussions of taxonomy and endangered species policy (e.g., O'Brien and Mayr, 1991; Geist, 1992).

5. Although the past record of life can be interpreted only through the phenomenological filter of the present, this does not imply that actual past objects and events in life history are constrained to lie within the known range of contemporary phenomena. The unique historical events of the evolutionary past are recoverable and, through the proper use of inference, are comprehensible. This extends to the discovery of extinct species. It is only by virtue of taphonomy that extinct species communicate a restricted range of information on which to base our inferences, yet the process of knowledge acquisition about extinct species is precisely the same as that employed in the acquisition of knowledge about living species. While we cannot observe the exchange of genes among dead organisms, this does not deprive extinct species of either empirical or conceptual significance; after all, the observation of gene exchange among the living is often no less equivocal when it comes to making taxonomic judgments. The crucial realization is that both enterprises are equally dependent on an accepted ontology of the species. A view of species as historical individuals is one that sees neontology and paleontology as equal though distinctive partners in the pursuit of knowledge about evolutionary history.

References

Avise, J., and Ball, R. 1990. Principle of genealogical concordance in species concepts and biological taxonomy, in: D. Futuyma and J. Antonovics (eds.), *Oxford Surveys in Evolutionary Biology*, Vol. 7, pp. 45–67. Oxford University Press, Oxford.

Bock, W. 1986. Species concepts, speciation, and macroevolution, in: K. Iwartsuki, P. Raven, and W. Bock (eds.), *Modern Aspects of Species*, pp. 31–57. University of Tokyo Press, Tokyo.

Bock, W., and von Wahlert, G. 1965. Adaptation and the form–function complex. *Evolution* **19:**269–299.

Brockelman, W., and Gittins, S. 1984. Natural hybridization in the *Hylobates lar* species group: Implications for speciation in gibbons, in: H. Preuschoft, D. Chivers, W. Brockelman, and N. Creel (eds.), *The Lesser Apes: Evolutionary and Behavioral Biology*, pp. 498–532. Edinburgh University Press, Edinburgh.

Cope, D., and Lacy, M. 1992. Falsification of a single species hypothesis using the coefficient of variation: A simulation approach. *Am. J. Phys. Anthrop.* **89:**359–378.

Coyne, J. 1992. Genetics and speciation. *Nature* **355:**511–515.

Cracraft, J. 1983. Species concepts and speciation analysis. *Curr. Ornithol.* **1:**159–187.

Eldredge, N. 1985. *Unfinished Synthesis: Biological Hierarchies and Modern Evolutionary Thought.* Oxford University Press, New York.

Eldredge, N., and Gould, S. 1972. Punctuated equilibria: An alternative to phyletic gradualism, in: T. Schopf (ed.), *Models in Paleobiology*, pp. 305–322. Freeman: San Francisco.

Geist, V. 1992. Endangered species and the law. *Nature* **357:**274–276.

Ghiselin, M. 1974. A radical solution to the species problem. *Syst. Zool.* **23:**536–544.

Godfrey, L., and Marks, J. 1991. The nature and origin of primate species. *Yrbk. Phys. Anthrop.* **34:**39–68.

Grene, M. 1987. Hierarchies in biology. *Am. Sci.* **75:**504–510.

Harrison, R. 1990. Hybrid zones: Windows on evolutionary process, in: D. Futuyma and J. Antonovics (eds.), *Oxford Surveys in Evolutionary Biology*, Vol. 7, pp. 69–128. Oxford University Press, Oxford.

Hennig, W. 1966. *Phylogenetic Systematics.* University of Illinois Press, Chicago.

Hull, D. 1976. Are species really individuals? *Syst. Zool.* **25:**174–191.

Hull, D. 1978. A matter of individuality. *Philos. Sci.* **45:**335–360.

Mayr, E. 1957. Species concepts and definitions, in: E. Mayr (ed.), *The Species Problem*, pp. 1–22. AAAS Publication No. 50. American Association for the Advancement of Science, Washington, D.C.

Mayr, E. 1982. *The Growth of Biological Thought.* Harvard University Press, Cambridge, MA.

Moore, J. 1957. An embryologist's view of the species problem, in: E. Mayr (ed.), *The Species Problem*, pp. 325–338. AAAS Publication No. 50. American Association for the Advancement of Science, Washington, D.C.

O'Brien, S., and Mayr, E. 1991. Bureaucratic mischief: Recognizing endangered species and subspecies. *Science* **251:**1187–1188.

Paterson, H. 1985. The recognition concept of species, in: E. S. Vrba (ed.), *Species and Speciation*, pp. 21–29. Transvaal Museum, Pretoria.

Raubenheimer, D., and Crowe, T. 1987. The recognition species concept: Is it really an alternative? *S. Afr. J. Sci.* **83:**550–534.

Rieppel, O. 1986. Species are individuals: A review and critique of the argument, in: M. Hecht, B. Wallace, and G. Prance (eds.), *Evolutionary Biology*, Vol. 20, pp. 283–317. Plenum Press, New York.

Salthe, S. 1985. *Evolving Hierarchical Systems.* Columbia University Press, New York.

Simpson, G. 1961. *The Principles of Animal Taxonomy.* Columbia University Press, New York.

Somit, A., and Peterson, S. (eds.) 1992. *The Dynamics of Evolution: The Punctuated Equilibrium Debate in the Natural and Social Sciences.* Cornell University Press, Ithaca, New York.

Tattersall, I. 1986. Species recognition in human paleontology. *J. Hum. Evol.* **15:**165–176.

Turner, A., and Chamberlain, A. 1989. Speciation, morphological change and the status of African *Homo erectus*. *J. Hum. Evol.* **18:**115–130.

Wiley, E. 1978. The evolutionary species concept reconsidered. *Syst. Zool.* **27:**17–26.

Index

Absarokius, 309, 311, 317, 334, 338
Adaptive diversity, 7, 348–350, 353–354
Afropithecus, 351, 353
Albrecht, G. H., 280, 284, 289–290, 549
Allometry, 271–272, 275–279, 281, 517
Allouatta, 135–136
Amud, 523–524, 533
Anagenesis, 22, 235, 259, 300–301, 309–312, 321, 324–325, 332–339, 342, 472, 545–547
Analytical efficiency, 401–402
Anemorhysis, 314–315, 317, 319–320, 326
Archaeolemur, 154, 156
Assortative mating, 115–116, 205
Ateles, 134–135
Atlanthropus mauritanicus, 361
Australopithecus, 308, 355–356, 486, 488, 518, 526
  afarensis, 212–213, 239, 267–269, 279, 287, 355, 476
  africanus, 239, 287, 308, 355, 473, 475–478
  boisei, 472–473, 476
  robustus, 476
  See also Paranthropus
Avatars, 17
Awash hybrid zone, 56–57, 79–82, 84, 98, 543–544

Baboons: *see Papio*
Bayesian probability theory, 485
Bighorn Basin, 312, 338
Bilsborough, A., 288
Biogeography, 151–152, 531–533
Biological species concept, 4–18, 22–29, 33–34, 36–39, 46–47, 68–69, 83–85, 87, 92–93, 102, 165, 177–178, 266, 300, 302–307, 309, 311, 324, 341, 390, 394–395, 465–470, 542–548
Biospecies, 290, 301, 338–339, 341
Bock, W. J., 26, 28, 178, 305, 465, 541–542
Boule, M., 527
Bown, T. M., 334–338, 340, 545–546
Brace, C. L., 14
BSC: *see* Biological species concept

Callithrix jacchus, 131–132
Çandır, 423, 425
Cantius, 311–314, 334
Carpolestes, 312
Category, 4–5, 9; *see also* Species category
Centrifugal speciation pattern, 109, 112
Cercocebus, 49, 53, 80, 92, 97
  dental variation in, 242–244
Cercopithecus, 48–49, 53, 99, 101, 141, 213–215, 350, 357, 403, 439
  dental variation in, 213–232, 242–245, 247, 250–251
Chamberlain, A. T., 49, 60, 363, 507
Cheetham, A. H., 284, 289–290
Cheilostome bryozoans, 289–290
Chemeron, 487
Chimpanzees: *see Pan*
Chromosomes, 46, 111–112, 185–186, 191
Chronocline, 310, 322, 472, 533
Chronospecies, 13–18, 23, 37, 127, 300, 304, 466, 470, 485, 533, 546; *see also* Paleospecies
Clade, 479–480
Cladistic species concept, 24, 301–302, 304, 306–308, 315, 322–324, 339, 546
Cladistics, 14–16, 26–27, 30–38, 345, 468
  and classification, 302, 322–323, 345–360, 365–366, 540
  pattern, 44–45, 57, 462
  and species recognition, 360–365
Cladogenesis, 11, 18, 22, 30, 92, 300–301, 305, 321, 325, 332, 335–336, 339, 467, 472, 533, 545–548
Classes, 5, 9–10, 13–14, 36, 463
Classification, 68–70, 301–302, 325–326, 339–342
  of *Australopithecus africanus*, 475–478
  of *Australopithecus boisei*, 472–473
  and cladistics, 302, 322–323, 345–360, 365–366, 540
  of early *Homo*, 473–475, 507–509, 511–518
  of early Miocene catarrhines, 351–360
  of galagos, 51–52
  of *Homo erectus*, 360–365, 478–480
  of *Homo habilis*, 473–475, 507–508, 511–512, 515–518
  of *Lemur*, 164–167, 173–175
  of Neanderthals, 527–530
  of *Pan*, 270–280
  of *Papio*, 83–103
  of Paşalar hominoids, 415, 423–425
  of *Saimiri*, 179–182, 205–206, 544
Coefficient of variation, 211–213, 215–216, 218–225, 228–236, 240–252, 254–256, 268, 379–380, 400–402, 404–411, 421–422, 425, 434–436, 492, 494, 550

Cohesion species concept, 303
Colobus, 53, 136–137, 148–149
  dental variation in, 242, 244–245, 251, 253, 256
Coolidge, H. J., 274, 279
Coon, C. S., 14
Cope, D. A., 240, 379, 400–407, 411, 414, 425, 434, 438, 440, 450–451, 470, 550–551
Costello, R. K., 544–545
Cracraft, J., 91, 307, 463, 467
Cranial variation
  in *Australopithecus africanus*, 475–478
  in early *Homo*, 474, 494, 497–500, 503–505, 507–511
  in *Gorilla*, 497, 502–503
  in *Homo erectus*, 478–480, 502
  in lemurs, 167, 169, 171, 173
  in *Pan*, 269–280
  in *Paranthropus boisei*, 502
  in *Presbytis entellus*, 153–154
  in *Saimiri*, 184
Crompton, R. H., 51, 55
Crovello, T. J., 4, 58, 70–71
CV: *see* Coefficient of variation

Darwin, C. R., 6, 9, 68, 285
De Queiroz, K., 17, 27, 308
Definition, of species, 27, 68–71, 100, 103–104, 123–124, 302, 463–464
Delson, E., 266–267, 288, 481
Demes, 16–17, 81, 88–89, 124, 467–468
Demonstrative communication paradigm, 117–118, 543
Dendropithecus, 351–352, 354, 357, 359, 384
Dental variation, 211–213, 232–236, 239–241, 253–260
  in *Cercocebus*, 242, 244
  in *Cercopithecus*, 213–232, 242–245, 247, 250–251
  in *Colobus*, 242, 244, 251, 253, 256
  in early *Homo*, 494–497, 501–503, 505–509, 511
  in Eocene primates, 312–318
  in *Gorilla*, 242, 247, 374, 408–410, 412–413, 500–501
  in *Homo erectus*, 500–501
  in *Hylobates*, 242, 244, 408–410, 412–413
  in lemurs, 167, 169–171, 173
  in *Lufengpithecus*, 431–434
  in *Macaca*, 242, 244–245, 247, 250–251
  in *Pan*, 242, 244–245, 247, 251, 374, 408–410, 412–413, 441–442, 505
  in *Paranthropus boisei*, 500–501
  in Paşalar hominoids, 418–423
  in *Pongo*, 242, 247, 251–252, 257, 408–410, 412–413
  in *Presbytis*, 242, 244–245, 249, 251–252, 254–255
  in *Proconsul*, 374–377, 383, 388
  in *Saimiri*, 184, 192–194, 198–202
Diacodexis, 287, 314, 322, 332–335, 340, 342
Diagnosis, of species, 70, 100, 363, 464, 469, 480
Dobzhansky, T., 6–8, 13, 24, 46–47, 96, 303, 531, 533
Donoghue, M. J., 17, 308
Donrussellia, 315
Dryopithecus, 403, 417

Early *Homo*, 102–103, 473–475, 486–488
  classification of, 473–475, 507–509, 511–518
  cranial variation in, 474, 494, 497–500, 503–505, 507–511
  dental variation in, 494–497, 501–503, 505–509, 511
  postcranial variation in, 516–517
  taxonomic hypotheses concerning, 488–494
Eckhardt, R. B., 269
Ecogeographic variation, 151–152, 156, 468
Ecological niches, 5, 28
Ecological species concept, 25, 27–28, 308
ECSC: *see* Ecological species concept
Eldredge, N., 33, 43, 91, 96, 337, 540–542
Epistemology and species, 3–4, 23, 25–26, 29, 32, 43, 68–69, 469–470
ESC: *see* Evolutionary species concept
Evolutionary species concept, 10, 24–25, 28–29, 33, 300–307, 310–311, 319, 323–324, 333, 341, 394–395, 465–466, 546–547
EVSC: *see* Evolutionary species concept
Extinction, 10–11, 18, 25, 37, 90–91, 541; *see also* Pseudoextinction

Fertilization system, 8, 11, 15, 48–49, 97, 542
Founder effect, 339
Futuyma, D. J., 44, 90
Fuzzy sets, 36

Galago, 49–52, 130, 281
Geladas: *see Theropithecus*
Gene flow, 151
Genetic distances, 45–46
  within *Saimiri*, 202–204
Genetic species, 44–47, 96–97
Geographic races, 45
Geographic variation, 123–128, 397, 549
  in *Alouatta*, 135–136
  in *Archaeolemur*, 156
  in *Ateles*, 134–135
  in *Callithrix*, 131–132
  in *Cercopithecus*, 141
  in *Colobus*, 136–137, 148–149
  and fossil primates, 152–157
  in *Gorilla*, 145–146
  in *Homo sapiens*, 146–147
  in *Hylobates*, 110–118, 142–144
  in *Lepilemur*, 129–130
  in *Macaca*, 136–141, 150
  in *Otolemur*, 130
  in *Pan*, 144–146
  in *Papio*, 75–80, 141–142
  in *Pongo*, 143–144
  in *Presbytis*, 153–154
  in primates, 147–153, 549–551
  in *Saguinus*, 131–132
  in *Saimiri*, 133–134, 181–184
  and speciation, 164
Ghiselin, M. T., 9, 337, 463–464, 541
Gibbons: *see Hylobates*
Gigantopithecus, 450
Gingerich, P. D., 12–13, 239–240, 312–314, 334–338, 404, 466

Godfrey, L. R., 124, 126
Goldschmidt, R., 531, 533
Gorilla, 118–119, 145–146, 395, 492–493
  cranial variation in, 497, 502–503
  dental variation in, 242, 244, 247, 399, 408–410, 412–413, 441–442, 500–501
  postcranial variation in, 380–381, 383–388
Gould, S. J., 540
Gradualism, 34, 60, 300–301, 307, 310–312, 335–337, 339, 490
Graecopithecus freybergi, 400
Griphopithecus, 424–425
Groves, C. P., 67, 91, 126, 146, 269–271, 361, 544
Guenons: *see Cercopithecus*

Hapalemur, 167, 169–170, 172
Heliopithecus leakeyi, 400
Hennig, W., 25, 29–30, 34, 305, 346, 362, 468, 540
Hershkovitz, P., 178–179, 181–183
Holophyly, 25, 30, 32, 35, 37, 302, 308, 346, 480
Homo, 14, 53, 60, 355, 363, 486
  capensis, 361
  erectus, 14, 60, 324, 475, 486, 492, 500–502, 517–518, 527–529
    classification of, 360–365, 478–480
  ergaster, 361, 511, 516
  habilis, 53, 239, 268–269, 288, 308, 324, 363, 477, 479, 486–492, 515–516, 548
    classification of, 473–475, 507–508, 511–512, 515–518
    See also Early *Homo*
  leakeyi, 361
  neanderthalensis, 102, 529–534
    classification of, 524, 527–530
    See also Neanderthals
  rudolfensis, 474–475, 479, 512, 515–518, 548; *see also* Early *Homo*
  sapiens, 14, 54, 98, 146–147, 267, 324, 360–361, 363–364, 480, 492–493, 497, 524, 527–534, 548
Homoplasy, 171–175, 289, 360, 468, 548
Howell, F. C., 528
Howells, W. W., 146, 288–289, 527
Howler monkeys, 135–136
Hrdlička, A., 528
Hull, D. L., 9, 337, 341, 464, 541
Huxley, J. S., 24, 89–90
Hybrid zones, 72, 544
  in *Hylobates*, 113–118, 544
  in *Papio*, 56–57, 74, 79, 81–84, 98, 543–544
Hybridization, 17, 55–56, 94, 99, 543–545
  in *Lemur*, 165–167
  in *Macaca*, 55, 125–126
  in *Saimiri*, 185–186, 196–198, 206, 544–545
Hylobates, 49, 53, 395, 544
  dental variation in, 242, 244, 408–410, 412–413
  geographic variation in, 110–118, 142–144
  pelage of, 115–116
Hypothesis testing
  power in, 232–235, 282, 400–401, 404–407, 425
  for single species hypotheses, 178, 212–213, 228–236, 257–258, 373–375, 379, 400–402, 405–

Hypothesis testing (*Cont.*)
  for single species hypothesis (*Cont.*)
      407, 414, 420–423, 425, 429–430, 434–436, 447–452, 472–473, 550–551
  types I and II errors in, 230–233, 235, 281–282, 401–402, 404–407, 411, 422, 425, 551
Hypothetico-deductive method, 44–45

$I_{max/min}$: *see* Maximum/minimum index
IC: *see* Isolation concept
Intra- and interspecific variation, 58, 123–128, 147–151, 211–213, 241, 259–260, 266–270, 280–291, 362, 402, 494, 497, 549–550
Isolating mechanisms, 7–8, 47–49, 96
  postmating, 58, 165, 186, 542–543
  premating, 50, 54, 150, 165, 543
Isolation concept, 28, 48–50, 52, 55–58, 69, 99, 303; *see also* Biological species concept

Jackson, J. B. C., 284, 289–290
Johanson, D. C., 267–268
Jolly, C. J., 56–57, 125–126, 283, 285–286, 543–544

Kalepithecus, 351–352
Kay, R. F., 240, 257
Kebara, 524–525, 529
Keith, A., 523–524
Kelley, J., 212, 235, 256–259, 374, 376, 379, 387, 389, 402–403, 411, 414, 550
Kenyapithecus, 423–424
Kimbel, W. H., 56–57, 93, 102, 154, 212, 280, 283, 285–288
Koobi Fora, 474, 487–488
Krishtalka, L., 287, 319, 322, 545–546

Language, 53–54
Langurs: *see Presbytis*
Lemur, 53, 101, 164–175, 357
Lepilemur, 53, 129–130, 167, 169–170, 172
Lieberman, D. E., 269
Limnopithecus, 351–352, 357, 359
Lineages, 10–14, 23–26, 28–32, 35, 37–39, 300–307, 310–312, 324–326, 333, 338–342, 395–396, 463–467, 472, 481, 541, 545–546
Linnaean system, 4, 26, 321–323, 325–326, 341, 540
Linnaeus, C., 5
Lufeng, 429–431
Lufengpithecus, 212, 256, 417, 434
  dental variation in, 403, 431–434
  sexual dimorphism in, 258, 414, 440–447, 452–453
Lyell, C., 43

Macaca, 53, 55, 91, 97, 99, 125–126, 136–141, 150
  dental variation in, 242, 244–245, 247, 250–251
Mandrillus, 67, 80, 91–92, 449
Mangabeys: *see Cercocebus*
Marmosets, 131–132
Martin, L. B., 212, 450
Masters, J. C., 96–99, 542–543
Mate recognition, 49–51, 54, 97–99, 115, 543–544
Maximum/minimum index, 212–213, 215, 229–230, 379, 404–405, 411, 420, 422–423
Mayr, E., 7–8, 12–13, 16, 24, 26–27, 36, 43–44, 46–

Mayr (*Cont.*)
47, 68–69, 90, 177, 285–286, 302–306, 311, 463, 465–466, 469, 533
McCown, T. D., 523–524
Mean character difference, 507
Metaspecies, 308, 320
Mfangano, 373–377, 447
MI: *see* Maximum/minimum index
Microchoerus, 312
Micropithecus, 351–352, 357, 359
Miller, J. M. A., 268–269, 280, 284, 289–290, 549
Modern synthesis, 22, 24, 33, 39, 69
Monophyletic species concept, 308, 315, 320, 323–324, 339
Monophyly, 15, 23, 25, 30–33, 57, 300, 302, 308, 346–347, 349, 354, 470, 480
Morphocline, 333, 470, 481, 531, 533
Morphocryptic species, 45, 49, 53, 60, 96, 289
Morphospecies, 289–290
Mount Carmel, 523–524
Mousterian, 528
Multivariate statistical methods, 128–129, 271–272, 493–494

Nasalis, 449
Natural selection, 7, 48
Neanderthals, 102, 267, 288, 363, 480, 523–534; *see also Homo neanderthalensis*
Necrolemur, 312
Neodarwinian synthesis, 22, 24, 33, 39, 69
Neudorf, 424
Nixon, K. C., 57, 91
Nomenclature, 321–326, 335, 339–342
Nyanzapithecus, 351, 353, 355, 384

Olduvai Gorge, 486
Olson, T. R., 269
Omomyids, 309, 314–321, 334
Ontology and species, 3–6, 9, 11–18, 23, 25–26, 29, 32–34, 43–44, 68–69, 462–464, 468–469, 540–542, 552
Operational taxonomic unit, 469
Orangutans: *see Pongo*
Oreopithecus, 355
Otolemur, 130

Paleospecies, 100–101, 153, 259, 300–301, 304, 309–312, 341, 365, 375, 485; *see also* Chronospecies
Pan, 118–119, 144–146, 268, 283, 290, 395, 492–493, 548
cranial variation in, 269–280
dental variation in, 242, 244–245, 247, 251, 374, 408–410, 412–413, 441–442, 505
postcranial variation in, 380–381, 383–388
Papio, 67, 141–142
Anubis form of, 55–57, 73, 75–76, 80–85, 87–88, 94, 98, 100
Chacma form of, 73, 78–81, 83, 88
classification of, 83–103
clines in, 78–79, 87, 91
Guinea form of, 73, 75, 79, 83, 85, 100
habitats of, 79–80
Hamadryas form of, 56–57, 73, 75–77, 79–87, 92, 94, 98, 100, 104

Papio (*Cont.*)
hybrid zones in, 56–57, 74, 79, 81–84, 98, 543–544
pelage of, 73–79
phenostructure of, 73–80, 103–104
postcranial variation in, 380–387
Yellow form of, 73, 75, 77–80, 82–83, 87, 94, 100–101, 104
zygostructure of, 74, 80–83, 103–104
Paranthropus, 269, 355, 417, 487, 518
boisei, 492–493, 500–502, 517
See also Australopithecus
Paraphyly, 15, 34, 302, 308, 339, 354
Paşalar, 393–394, 397–398, 407, 414–415, 430
fossil hominoids from
classification of, 415, 423–425
dental morphology of, 415–420
dental variation in, 418–423
Paterson, H. E. H., 5, 7–8, 28, 48, 55, 97, 542–543
Patterson, C., 45
Pelage
of *Hylobates*, 115–116
of *Papio*, 73–79
of *Saimiri*, 181–184, 196–198
Pelycodus, 312
Phenacolemur, 312
Phenetic species concept, 57
Phenostructure, 68, 71–72, 125, 550
of *Papio*, 73–80, 103–104
Phillips-Conroy, J. E., 56–57, 81
Phyletic lineage, 304–305, 325, 466, 470, 481, 546; *see also* Lineages
Phyletic speciation, 300–301, 304, 312, 325, 332, 334–336, 339–342, 396, 466, 545–546
Phylogenetic species concept, 24–25, 29–34, 44, 57–60, 69, 91–96, 98, 102, 291, 307, 315, 319, 323, 394–395, 463, 467–472, 480–481, 545–548
Phylogenetic systematics: *see* Cladistics
Pickford, M., 256–257, 374–375, 379
Plavcan, J. M., 212, 228, 379, 404, 434, 436, 438, 440, 446, 451, 470, 550
Polyphyly, 354
Polytypic species, 91–93, 95, 103, 129, 149, 300, 531
Pongo, 118–119, 143–144, 395, 407, 493
dental variation in, 242, 244, 247, 251–252, 257, 408–410, 412–413, 441–446
Postcranial variation, 375–377, 380
in early *Homo*, 516–517
in *Gorilla*, 380–381, 383–388
in *Pan*, 380–381, 383–388
in *Papio*, 380–387
in *Proconsul*, 376–377, 380–388, 390
Power: *see* Hypothesis testing
Praeanthropus africanus, 355–356
Presbytis, 101, 153–154
dental variation in, 242, 244–245, 249, 251–252, 254–255
Proconsul, 212, 239, 243, 246, 256–257, 351–354, 357, 359, 373–377, 390, 403, 416, 447–449, 452
dental variation in, 374–377, 383, 388
postcranial variation in, 376–378, 380–388, 390, 447–449
Protein electrophoresis, 93, 110, 186–187, 194–196, 202–204, 289

PSC: *see* Phylogenetic species concept
Pseudoextinction, 11, 25, 30–31, 33, 37, 336, 339–340
Pseudospeciation, 336, 339–340
Pseudotetonius, 315–322, 324, 334, 338, 340–341
Punctuated equilibrium, 11, 13, 25, 28, 32, 37, 60, 286, 300, 302, 307, 310, 335–336, 490, 540–541

Qafzeh, 523–524, 527, 533

R%: *see* Range as a percentage of the mean
Ramapithecus, 282, 291
Range as a percentage of the mean, 212–213, 215, 229, 379, 404–405, 411, 420, 422
Range-based statistics, 212, 216, 229–230, 232–235, 379–380, 384, 404–406, 425
Rangwapithecus, 243, 351, 353, 357, 359
Rassenkreis, 83, 85, 94, 103, 531, 533; *see also* Ring species
Ravin de la Pluie, 212, 239, 243, 430, 446
RC: *see* Recognition concept
Recognition concept, 5, 8–10, 25, 27–28, 48–60, 69, 96–99, 102, 165, 205, 266, 303, 394–395, 542–545
Reproductive criteria of species, 5–8
Reproductive isolation, 7, 9, 13, 46–48, 125–126, 311, 339, 542–546, 548
Ridley, M., 306, 311, 526
Rightmire, G. P., 363–364, 480
Ring species, 129, 134, 167, 534; *see also* Rassenkreis
Rose, K. D., 334–338, 340, 545–546
Rudabanya, 430, 438, 446
Rusinga, 373–377, 429–430, 439, 447, 452

Saguinus fuscicollis, 131–132
Saimiri, 133–134, 177–178, 544
behavior of, 187–192
biochemical diversity of, 186–187, 194–196, 202–204
chromosomes of, 185–186, 191
classification of, 179–182, 205–206, 544
dental variation in, 184, 192–194, 198–202
geographic variation in, 133–134, 178–184
pelage of, 181–184, 196–198
Schwartz, J. H., 269
Semispecies, 45, 83, 90, 95–96, 103–104, 113, 120
Sexual dichromatism, 115, 165, 182
Sexual dimorphism, 125, 155, 222–223, 240–241, 247, 251–252, 254–258, 373–375, 379–380, 388, 407, 410–411, 414, 429–433, 440–453, 470–473, 493–494, 497, 500, 505, 531
Shanidar, 523–524, 529, 534
Shea, B. T., 549
Sibling species, 7, 45–46, 150, 284, 300, 309
Simiolus, 351–352
Simpson, G. G., 5, 9–11, 24, 29, 302, 304–306, 310–311, 315, 322–324, 333, 337, 341, 347, 349, 379, 404, 464–465, 533, 546
Single species hypotheses: *see* Hypothesis testing
Sivapithecus, 239, 257, 291, 397, 403, 424, 446
Siwaliks, 397, 438, 446
Skhul, 523–525, 527
SMRS: *see* Specific mate recognition system
SMRSC: *see* Specific mate recognition system
Sokal, R. R., 4, 58, 70–71
Speciation, 7–8, 11, 15, 23–24, 31–33, 36–37, 45–48, 55, 57–60, 96, 103, 109, 119–120, 123–127, 185, 300–301, 336, 339, 396, 462, 481, 533, 540–542, 546, 552
and morphological differences, 25, 101, 150, 163–164, 259–260, 266, 283–284, 309–310, 534, 550
Species
as classes, 9–11, 13–14, 541–542
as individuals, 9–12, 27, 33, 36, 69, 337, 341, 463–467, 541–542, 546, 552
Species category, 4–5, 9, 24, 26, 68, 70, 92, 463, 465, 467, 542
Species concepts, 4–6, 22–25, 29, 38–39, 44–46, 67–68, 103–104, 123–126, 302, 308–309, 323, 332, 341, 396, 462, 465, 485, 539–542; *see also* Biological species concept, Cladistic species concept, Cohesion species concept, Ecological species concept, Evolutionary species concept, Monophyletic species concept, Phenetic species concept, Phylogenetic species concept, Recognition concept
Species discreteness, 5–7
Species numbers
and body size, 437–438
and life history parameters, 436–437
Species recognition, 4, 37–38, 266–267, 281–283, 287–290, 300, 309, 393–397, 411, 470–472, 542, 547–551
and cladistics, 360–365
Species taxon, 4–5, 9, 27, 69–70, 463–465, 541
Species-lineage, 14, 333–334, 340–342
Specific mate recognition system, 8, 11, 15–17, 28, 48–49, 53–54, 59–60, 95–97, 150, 165, 468, 542–545
in galagos, 51–52
in *Hylobates*, 53, 112, 116–118, 120
in *Papio*, 55–57, 82, 97–98, 543
Spider monkeys, 134–135
Sportive lemurs: *see Lepilemur*
Squirrel monkeys, 133–134, 177–206, 544; *see also Saimiri*
Steinius, 315, 320
Sterkfontein, 475, 478, 487
Stratigraphic gaps, 13, 310, 338, 472
Stratophenetic studies, 312, 314, 472
Strigorhysis, 334, 338
Stringer, C. B., 361
Subspecies, 17–18, 45, 103–104, 120, 124–125, 127–128, 154–155, 241, 255, 280–287, 307, 310, 322, 340–341, 467–470, 545, 548
in *Papio*, 67, 83–96, 104
Swartkrans, 487, 517
Szalay, F. S., 541–543, 547

Tabun, 523–524, 529
Tamarins, 131–132
Tatmanius, 315, 317–320, 324
Tattersall, I., 52, 60, 90, 100–102, 155, 240, 280, 283–289, 311, 363–364, 470, 480, 546, 549–550
Taxogram, 27, 30
Taxon, 4–5, 9, 28–29; *see also* Species taxon

Taxonomic species, 23, 25, 27, 29, 32, 37–38, 44, 542–543, 546
Taxonomy, 22–30, 35, 38–39
Tchadanthropus uxoris, 361
Teilhardina, 314–317, 319–320, 326, 334, 342
Telanthropus capensis, 361
Tetonius, 309, 314–322, 324, 334, 338, 340–342
Theropithecus, 55, 67, 83, 87–88, 97, 104
Thorington, R. W., 178–179, 181–184, 205
Tobias, P. V., 287, 340–341, 361, 507
Tokogeny, 6–7, 10, 17–18, 59, 71, 548
Total morphological pattern, 470, 507
Trinkaus, E., 524, 528–529
Turkanapithecus, 351, 353
Turner, A., 49, 53–54, 60, 363
Types I and II errors: *see* Hypothesis testing
Typology, 69, 338, 340, 467

Uniformitarianism, 389–390, 402–404, 450–453, 550–551

Van Valen, L. M., 35–36, 308
Van Vark, G. N., 288
Varecia, 173, 175
Vocalizations, 49–55, 97, 116–117, 188–190
Vrba, E. S., 34, 467

Weidenreich, F., 527
Wheeler, Q. D., 57, 91
Whewell, W., 5–6
White, T. D., 268, 287
Wiley, E. O., 10–11, 305–306, 315, 337, 341, 465
Willman, R., 33–34
Willwood Formation, 302, 312, 321, 334, 395, 546
Wilson, A. C., 46
Wind River Basin, 338
Wolpoff, M. H., 14
Wright, S., 16–17

Zimmermann, E., 52
Zuttiyeh, 527, 533
Zygostructure, 68, 71–72, 125, 550
  of *Papio*, 74, 80–83, 103–104